CAMBRIDGE LIBRARY COLLECTION

Books of enduring scholarly value

Mathematical Sciences

From its pre-historic roots in simple counting to the algorithms powering modern
desktop computers, from the genius of Archimedes to the genius of Einstein, advances
in mathematical understanding and numerical techniques have been directly responsible
for creating the modern world as we know it. This series will provide a library of the mo
influential publications and writers on mathematics in its broadest sense. As such, it wi'
not only the deep roots from which modern science and technology have grown, but a
astonishing breadth of application of mathematical techniques in the humanities and
sciences, and in everyday life.

Scientific Papers

This final volume of papers by Lord Rayleigh covers the period from 1911 to his d
1919. The first of the Solvay Conferences in 1911 played a key role in the foundai
quantum theory. Although invited, Rayleigh did not attend. His principal achiev
in development and consolidation across classical physics, in which he continu
research. In a 1917 paper, he used electromagnetic theory to derive a formula f
the reflection properties from a regularly stratified medium. In 1919, he inves
for the natural phenomenon of the stunning iridescent colours of birds and i
continued his longstanding participation in the Society for Psychical Resear
been founded in 1882 for the study of 'debatable phenomena'. One of his la'
also from 1919, was his presidential address to that society; this considers
unorthodox views and practices. He concludes by asserting the importan
maintaining open minds in the pursuit of truth.

Scientific Papers

VOLUME 6: 1911–1919

BARON JOHN WILLIAM STRUTT RAYLEIGH

CAMBRIDGE
UNIVERSITY PRESS

CAMBRIDGE UNIVERSITY PRESS

Cambridge New York Melbourne Madrid Cape Town Singapore São Paolo Delhi

Published in the United States of America by Cambridge University Press, New York

www.cambridge.org
Information on this title: www.cambridge.org/9781108005470

© in this compilation Cambridge University Press 2009

This edition first published 1920
This digitally printed version 2009

ISBN 978-1-108-00547-0

SCIENTIFIC PAPERS

CAMBRIDGE UNIVERSITY PRESS

C. F. CLAY, Manager

LONDON : FETTER LANE, E.C. 4

NEW YORK : THE MACMILLAN CO.
BOMBAY
CALCUTTA } MACMILLAN AND CO., Ltd.
MADRAS
TORONTO : THE MACMILLAN CO. OF
CANADA, Ltd.
TOKYO : MARUZEN-KABUSHIKI-KAISHA

SCIENTIFIC PAPERS

BY

JOHN WILLIAM STRUTT,

BARON RAYLEIGH,

O.M., D.Sc., F.R.S.,

CHANCELLOR OF THE UNIVERSITY OF CAMBRIDGE,
HONORARY PROFESSOR OF NATURAL PHILOSOPHY IN THE ROYAL INSTITUTION.

VOL. VI.

1911—1919

CAMBRIDGE
AT THE UNIVERSITY PRESS
1920

PREFACE

THIS volume completes the collection of my Father's published papers. The two last papers (Nos. 445 and 446) were left ready for the press, but were not sent to any channel of publication until after the Author's death.

Mr W. F. Sedgwick, late Scholar of Trinity College, Cambridge, who had done valuable service in sending corrections of my Father's writings during his lifetime, kindly consented to examine the proofs of the later papers of this volume [No. 399 onwards] which had not been printed off at the time of the Author's death. He has done this very thoroughly, checking the numerical calculations other than those embodied in tables, and supplying footnotes to elucidate doubtful or obscure points in the text. These notes are enclosed in square brackets [ ] and signed W. F. S. It has not been thought necessary to notice minor corrections.

RAYLEIGH.

Sept. 1920.

CONTENTS

* [1914. It would have been in better accordance with usage to have said " of Relative Index differing little from Unity."]

* [1917. It would be more correct to say $P_n(\cos \theta)$, where $\cos \theta$ lies between ± 1.]

PAGE

ERRATA

(INCLUDING THE ERRATA NOTED IN VOLUME V. PAGE XIII.)

VOLUME I.

Page viii, line 4.} *For* end lies *read* ends lie.
 ,, 64, line 8.}

,, 86, last line. *For* 1882 *read* 1881.

,, 89, line 10. *Insert* comma *after* maximum.

,, 144, line 6 from bottom. *For* D *read* D,.

,, 324, equation (8). *Insert* negative sign *before the single*
  integral.

,, ,, line 2 from bottom. *For* (1) *read* (5).

,, 325, line 10. *For* $-nVH$ *read* $+n\rho VH$.

          And *Theory of Sound*, Vol. I.
          (1894), p. 477, equation (8) and
          last line, and p. 478, line 12.

,, 442, line 9. *After* $\dfrac{\rho'-\rho}{\rho'}$ *insert* y.

,, 443, line 9. *For* (7) *read* (8).

,, 443, line 10. *For* η *read* ξ.

,, 446, line 10. *For* ϕ *read* ϕ'.

,, 448, line 5. *For* v *read* c.

,, 459, line 17. *For* 256, 257 *read* 456, 457.

,, 492, line 7 from bottom. *For* $r\sqrt{2n}$ *read* $r/\sqrt{2n}$.

,, 494, lines 10 and 12. *For* $-\dfrac{2mr^2}{n^2-4m^2}\cos 2\theta$ *read* $+\dfrac{2mr^2}{n^2-4m^2}\cos 2\theta$.

,, 523, line 9. *For* n/λ *read* n/k.

,, 524. In the second term of equations (32) and following *for* ΔK^{-1} *read* $\Delta\mu^{-1}$.

,, 525, line 11. *For* f *read* f_1.

,, 526, line 13. *For* $f:g$ *read* $f_1:g_1$.

,, 528, line 3 from bottom. *For* e^{int} *read* $e^{i(nt-kr_0)}$.

,, 538, line 11 from bottom. This passage is incorrect (*see* Vol. VI. Art. 355, p. 41).

,, 556. In line 8 *after* (15) *add* with $s\phi - \frac{1}{2}\pi$ for $s\phi$; in line 9 *for* δA_s *read* $\delta A_s'$; and *for* line
  10 *substitute* $+\delta A_s'$ *as* $\{\cos \frac{1}{2}s\pi + \cos(\frac{1}{2}s\pi + s\pi)\}$ F.
  Throughout lines 12—25 *for* A_s, A_1, A_2, ... A_6, δA_s *read* A_s', A_1', A_2', ... A_6', $\delta A_s'$;
  for sin $\frac{1}{2}s\pi$ *read* $-\cos\frac{1}{2}s\pi$; and *reverse* the signs of the expressions for A_2', A_4', A_6'.
  Similarly, in *Theory of Sound*, Vol. I. (1894), p. 427, *substitute* $s\phi + \frac{1}{2}\pi$ for $s\phi$ in (32)
  (*see* p. 424), and in lines 11—26 *for* A_s', A_s, δA_s *read* A_s, A_s', $\delta A_s'$, and *for* sin *read*
  $+\cos$. Also in (43) and (47) *for* s^2-s *read* s^3-s. [In both cases the work done corresponding to δA_s vanishes whether s be odd or even.]

VOLUME II.

,, 197, line 19. *For* nature *read* value.

,, 240, line 22. *For* dp/dx *read* dp/dy.

,, 241, line 2. *For* du/dx *read* du/dy.

,, 244, line 4. *For* k/n *read* n/k.

,, 323, lines 7 and 16 from bottom. *For* Thomson *read* C. Thompson.

,, 345, line 8 from bottom. *For* as pressures *read* at pressures.

,, 386, lines 12, 15, and 19. *For* cos *CBD* *read* cos *CBB'*.

,, 389, line 6. *For* minor *read* mirror.

,, 414, line 5. *For* favourable *read* favourably.

,, 551, first footnote. *For* 1866 *read* 1886.

VOLUME III.

Page 11, footnote. *For* has *read* have.

,, 92, line 4. *For* Vol. I. *read* Vol. II.

,, 129, equation (12). *For* $e^{u\,(i-x)}dx$ *read* $e^{u\,(i-x)}du$.

,, 162, line 19, and p. 224, second footnote. *For* Jellet *read* Jellett.

,, 179, line 15. *For* Provostaye *read* De la Provostaye.

,, 224, equation (20). *For* 2χ *read* χ. $\}$ And *Theory of Sound*, Vol. I. (1894),

,, ,, second footnote. *For* p. 179 *read* p. 343. $\}$ p. 412, equation (12), and p. 423 (footnote).

,, 231, line 5 of first footnote. *For* 171 *read* 172.

,, 273, lines 15 and 20. *For* $\{\phi\,(x)\}^2$ *read* $\int_{-\infty}^{+\infty}\{\phi\,(x)\}^2\,dx$.

,, 314, line 1. *For* (38) *read* (39).

,, 326. In the lower part of the Table, under Ampton *for* $c\flat+4$ *read* $e\flat+4$, and under Terling
(3) *for* $b\flat+6$ *read* $b+6$ (and in *Theory of Sound*, Vol. I. (1894), p. 393).

,, 522, equation (31). *Insert as factor of last term* $1/R$.

,, 548, second footnote. *For* 1863 *read* 1868.

,, 569, second footnote. *For* alcohol *read* water.

,, 580, line 3. Prof. Orr remarks that a is a function of r.

VOLUME IV.

,, 14, lines 6 and 8. *For* 38 *read* 42.

,, 267, lines 6, 10, and 20, and p. 269, line 1. *For* van t' Hoff *read* van 't Hoff. Also in
Index, p. 604 (the entry should be under Hoff).

,, 277, equation (12). *For* dz *read* dx.

,, 299, first footnote. *For* 1887 *read* 1877.

,, 369, footnote. *For* 1890 *read* 1896.

,, 400, equation (14). A formula equivalent to this was given by Lorenz in 1890.

,, 418. In table opposite 6 *for* ·354 *read* ·324.

,, 453, line 8 from bottom. *For* $\dfrac{2}{n-1}$ *read* $-\dfrac{2}{n-1}$.

,, 556, line 8 from bottom. *For* reflected *read* rotated.

,, 570, line 7 (Section III). *For* 176 *read* 179.

,, 576, line 7 from bottom.$\}$ *For* end lies *read* ends lie.
,, 586, line 20.

,, 582, last line. *For* 557 *read* 555.

,, 603. *Transfer* the entry under Provostaye *to* De la Provostaye.

,, 604. *Transfer the entry* II 553 *from* W. Weber *to* H. F. Weber.

VOLUME V.

,, 43, line 19. *For* (5) *read* (2).

,, 137, line 14. μ is here used in two senses, which must be distinguished.

,, 149, line 3. *For* P_0 *read* P_1.

,, 209, footnote. *For* XLX. *read* XIX.

,, 241, line 10 from bottom. *For* position *read* supposition.

,, 255, first footnote. *For* Matthews *read* Mathews.

,, 256, line 6. *For* 1889 *read* 1899.

,, 265, line 16 from bottom. *For* § 351 *read* § 251.

,, ,, ,, 15 ,, ,, *For* solution *read* relation.

,, 266, lines 5 and 6, and *Theory of Sound*, § 251. An equivalent result had at an earlier date
been obtained by De Morgan (*see* Volume VI. p. 233).

,, 286, line 7. *For* a *read* x.

VOLUME V—*continued.*

Page 364, title, and p. ix, Art. 320. *After* Acoustical Notes *add* vii.

,,   409, first line of P.S. *For* anwer *read* answer.

,,   444, line 2 of footnote. *For* p. 441, line 9 *read* p. 442, line 9.

,,   496, equation (4). *Substitute* equation (19) on p. 253 of Volume vi. (*see* pp. 251—253),

    reading $\left(\dfrac{l'}{l} - \dfrac{l}{l'}\right)$ *for* $\left(\dfrac{l'}{l} + \dfrac{l}{l'}\right)$.

,,   549, equation (48). *For* e^{-ikr} *read* e^{-ikr_0}.

,,   619, line 3. *Omit* the second expression for $J_n(n)$.

,,   ,,   lines 11, 12, 19. *For* 2·1123 *read* 1·3447.   ⎫ *See* the first footnote on p. 211 of

,,   ,,   line 12. *For* 1·1814 *read* 1·8558.   ⎬ Volume vi.

,,   ,,   line 19. *For* ·51342 *read* ·8065.   ⎭

VOLUME VI.

,,   4, first footnote. *After* equation (8) *add*:—*Scientific Papers*, Vol. v. p. 619. *See also* Errata last noted above.

,,   5, line 3. *For* $(2n+1)z^2 = 4n(n+1)(n+2)$ *read* $z^2 = 2n(n+2)$, so that z^2 is an integer.

,,   11, last footnote. *For* § 230 *read* § 250 (fourth edition).

,,   13, equation (17). *For* $\frac{1}{8}k^4a^4$ *read* $\frac{3}{4}k^4a^4$.

,,   14, footnote. *For* § 247 *read* § 251 (fourth edition).

,,   78, footnote. *Add*:—*Scientific Papers*, Vol. v. p. 400.

,,   87, footnote. *Add*:—Thomson and Tait's *Natural Philosophy*, Vol. i. p. 497.

,,   89, second footnote. *For* 328 *read* 329.

,,   90, second footnote. *Add*:—*Math. and Phys. Papers*, Vol. iv. p. 77.

,,   138, footnote. *For* 1868 *read* 1865, *and for* Vol. ii. p. 128, *read* Vol. i. p. 526.

,,   148, footnote. *Add*:—*Scientific Papers*, Vol. iv. p. 407, and this Volume, p. 47.

,,   155, footnote. *For* Vol. iv. *read* Vol. iii.

,,   222, second footnote. *For* Vol. ii. *read* Vol. i. And in *Theory of Sound*, Vol. i. (1894), last line of § 207, *for* 4·4747 *read* 4·4774

,,   223, line 5 from bottom. *For* 0·5772156 *read* 0·5772157.

,,   225, line 1. *For* much greater *read* not much greater.

,,   ,,   line 6 from bottom. *For* 13·094 *read* 3·3274.

,,   253, equation (19). *For* $\left(\dfrac{l'}{l} + \dfrac{l}{l'}\right)$ *read* $\left(\dfrac{l'}{l} - \dfrac{l}{l'}\right)$.

,,   259, line 5. *For* $-\dfrac{2}{a}\dfrac{dy}{dz}$ *read* $\mp\dfrac{2}{a}\dfrac{dy}{dz}$.

,,   263, equation (24). *For* $\dfrac{\omega^2 a}{2T}$ *read* $\dfrac{\omega^2 a^3}{2T}$.

,,   ,,   ,,   (25). *For* $\left(1 - \dfrac{3r^2}{a^2}\right)$ *read* $\left(1 + \dfrac{3}{4}\dfrac{r^2}{a^2}\right)$.

,,   282, footnote. *For* p. 77 *read* p. 71.

,,   303, line 17. *For* $\surd(bvc/\kappa)$ *read* $\surd(bvc\kappa)$.

,,   307, line 8. *For* $\dfrac{d\phi}{dy}$ *read* $-\dfrac{d\phi}{dy}$.

,,   315, line 2. *Delete* 195.

,,   341, second footnote. *Add*:—[This Volume, p. 275].

,,   351, line 13 from bottom. *For* $Tg\rho$ *read* $T/g\rho$.

350.

NOTE ON BESSEL'S FUNCTIONS AS APPLIED TO THE VIBRATIONS OF A CIRCULAR MEMBRANE.

[Philosophical Magazine, Vol. XXI. pp. 53—58, 1911.]

IT often happens that physical considerations point to analytical conclusions not yet formulated. The pure mathematician will admit that arguments of this kind are suggestive, while the physicist may regard them as conclusive.

The first question here to be touched upon relates to the dependence of the roots of the function $J_n(z)$ upon the order n, regarded as susceptible of continuous variation. It will be shown that each root increases continually with n.

Let us contemplate the transverse vibrations of a membrane fixed along the radii $\theta = 0$ and $\theta = \beta$ and also along the circular arc $r = 1$. A typical simple vibration is expressed by *

$$w = J_n(z_n^{(s)} r) . \sin n\theta . \cos (z_n^{(s)} t), \quad \dots\dots\dots\dots(1)$$

where $z_n^{(s)}$ is a finite root of $J_n(z) = 0$, and $n = \pi/\beta$. Of these finite roots the lowest $z_n^{(1)}$ gives the principal vibration, *i.e.* the one without internal circular nodes. For the vibration corresponding to $z_n^{(s)}$ the number of internal nodal circles is $s - 1$.

As prescribed, the vibration (1) has no internal nodal diameter. It might be generalized by taking $n = \nu\pi/\beta$, where ν is an integer; but for our purpose nothing would be gained, since β is at disposal, and a suitable reduction of β comes to the same as the introduction of ν.

In tracing the effect of a diminishing β it may suffice to commence at $\beta = \pi$, or $n = 1$. The frequencies of vibration are then proportional to the roots of the function J_1. The reduction of β is supposed to be effected by

---

* *Theory of Sound,* §§ 205, 207.

increasing without limit the potential energy of the displacement (w) at every point of the small sector to be cut off. We may imagine suitable springs to be introduced whose stiffness is gradually increased, and that without limit. During this process every frequency originally finite must increase*, finally by an amount proportional to $d\beta$; and, as we know, no zero root can become finite. Thus before and after the change the finite roots correspond each to each, and every member of the latter series exceeds the corresponding member of the former.

As β continues to diminish this process goes on until when β reaches $\frac{1}{2}\pi$, n again becomes integral and equal to 2. We infer that every finite root of J_2 exceeds the corresponding finite root of J_1. In like manner every finite root of J_3 exceeds the corresponding root of J_2, and so on†.

I was led to consider this question by a remark of Gray and Mathews‡— "It seems probable that between every pair of successive real roots of J_n there is exactly one real root of J_{n+1}. It does not appear that this has been strictly proved; there must in any case be an odd number of roots in the interval." The property just established seems to allow the proof to be completed.

As regards the latter part of the statement, it may be considered to be a consequence of the well-known relation

$$J_{n+1}(z) = \frac{n}{z} J_n(z) - J_n{}'(z). \qquad \ldots\ldots\ldots\ldots\ldots\ldots\ldots(2)$$

When J_n vanishes, J_{n+1} has the opposite sign to $J_n{}'$, both these quantities being finite§. But at consecutive roots of J_n, $J_n{}'$ must assume opposite signs, and so therefore must J_{n+1}. Accordingly the number of roots of J_{n+1} in the interval must be *odd*.

The theorem required then follows readily. For the first root of J_{n+1} must lie between the first and second roots of J_n. We have proved that it exceeds the first root. If it also exceeded the second root, the interval would be destitute of roots, contrary to what we have just seen. In like manner the second root of J_{n+1} lies between the second and third roots of J_n, and so on. The roots of J_{n+1} *separate* those of J_n ‖.

---

* *Loc. cit.* §§ 88, 92 *a*.

† [1915. Similar arguments may be applied to tesseral spherical harmonics, proportional to cos $s\phi$, where ϕ denotes longitude, of fixed order n and continuously variable s.]

‡ *Bessel's Functions*, 1895, p. 50.

§ If J_n, J_{n+1} could vanish together, the sequence formula, (8) below, would require that every succeeding order vanish also. This of course is impossible, if only because when n is great the lowest root of J_n is of order of magnitude n.

‖ I have since found in Whittaker's *Modern Analysis*, § 152, another proof of this proposition, attributed to Gegenbauer (1897).

The physical argument may easily be extended to show in like manner that all the finite roots of $J_n'(z)$ increase continually with n. For this purpose it is only necessary to alter the boundary condition at $r = 1$ so as to make $dw/dr = 0$ instead of $w = 0$. The only difference in (1) is that $z_n^{(s)}$ now denotes a root of $J_n'(z) = 0$. Mechanically the membrane is fixed as before along $\theta = 0$, $\theta = \beta$, but all points on the circular boundary are free to slide transversely. The required conclusion follows by the same argument as was applied to J_n.

It is also true that there must be at least one root of J'_{n+1} between any two consecutive roots of J_n', but this is not so easily proved as for the original functions. If we differentiate (2) with respect to z and then eliminate J_n between the equation so obtained and the general differential equation, viz.

$$J_n'' + \frac{1}{z} J_n' + \left(1 - \frac{n^2}{z^2}\right) J_n = 0, \quad \dots\dots\dots\dots\dots(3)$$

we find

$$\left(1 - \frac{n^2}{z^2}\right) J'_{n+1} + \frac{n}{z^3}(n^2 - 1 - z^2) J_n' + \left(1 - \frac{n^2 + n}{z^2}\right) J_n'' = 0. \quad \dots(4)$$

In (4) we suppose that z is a root of J_n', so that $J_n' = 0$. The argument then proceeds as before if we can assume that $z^2 - n^2$ and $z^2 - n(n+1)$ are both positive. Passing over this question for the moment, we notice that J_n'' and J'_{n+1} have opposite signs, and that both functions are finite. In fact if J_n'' and J_n' could vanish together, so also by (3) would J_n, and again by (2) J_{n+1}; and this we have already seen to be impossible.

At consecutive roots of J_n', J_n'' must have opposite signs, and therefore also J'_{n+1}. Accordingly there must be at least one root of J'_{n+1} between consecutive roots of J_n'. It follows as before that the roots of J'_{n+1} separate those of J_n'.

It remains to prove that z^2 necessarily exceeds $n(n+1)$. That z^2 exceeds n^2 is well known[*], but this does not suffice. We can obtain what we require from a formula given in *Theory of Sound*, 2nd ed. § 339. If the finite roots taken in order be $z_1, z_2, \dots z_s \dots$, we may write

$$\log J_n'(z) = \text{const.} + (n-1) \log z + \Sigma \log (1 - z^2/z_s^2),$$

the summation including all finite values of z_s; or on differentiation with respect to z

$$\frac{J_n''(z)}{J_n'(z)} = \frac{n-1}{z} - \Sigma \frac{2z}{z_s^2 - z^2}.$$

This holds for all values of z. If we put $z = n$, we get

$$\Sigma \frac{2n}{z_s^2 - n^2} = 1, \quad \dots\dots\dots\dots\dots\dots\dots(5)$$

---

[*] Riemann's *Partielle Differentialgleichungen*; *Theory of Sound*, § 210.

since by (3)

$$J_n''(n) \div J_n'(n) = -n^{-1}.$$

In (5) all the denominators are positive.   We deduce

$$\frac{z_1{}^2 - n^2}{2n} = 1 + \frac{z_1{}^2 - n^2}{z_2{}^2 - n^2} + \frac{z_1{}^2 - n^2}{z_3{}^2 - n^2} + \dots > 1 ; \quad \dots\dots\dots\dots(6)$$

and therefore

$$z_1{}^2 > n^2 + 2n > n(n+1).$$

Our theorems are therefore proved.

If a closer approximation to $z_1{}^2$ is desired, it may be obtained by sub-stituting on the right of (6) $2n$ for $z_1{}^2 - n^2$ in the numerators and neglecting n^2 in the denominators.   Thus

$$\frac{z_1{}^2 - n^2}{2n} > 1 + 2n \left(z_2{}^{-2} + z_3{}^{-2} + \dots \right)$$

$$> 1 + 2n \left\{ z_1{}^{-2} + z_2{}^{-2} + z_3{}^{-3} + \dots - \frac{1}{n(n+2)} \right\}.$$

Now, as is easily proved from the ascending series for J_n',

$$z_1{}^{-2} + z_2{}^{-2} + z_3{}^{-2} + \dots = \frac{n+2}{4n(n+1)};$$

so that finally

$$z_1{}^2 > n^2 + 2n + \frac{n^3}{(n+1)(n+2)} . \quad \dots\dots\dots\dots(7)$$

When n is very great, it will follow from (7) that $z_1{}^2 > n^2 + 3n$.   However the approximation is not close, for the ultimate form is[*]

$$z_1{}^2 = n^2 + [1\cdot6130] \, n^{4/3}.$$

As has been mentioned, the sequence formula

$$\frac{2n}{z} J_n(z) = J_{n-1}(z) + J_{n+1}(z) \quad \dots\dots\dots\dots\dots(8)$$

prohibits the simultaneous evanescence of J_{n-1} and J_n, or of J_{n-1} and J_{n+1}. The question arises—can Bessel's functions whose orders (supposed integral) differ by more than 2 vanish simultaneously?   If we change n into $n+1$ in (8) and then eliminate J_n, we get

$$\left\{ \frac{4n(n+1)}{z^2} - 1 \right\} J_{n+1} = J_{n-1} + \frac{2n}{z} J_{n+2}, \quad \dots\dots\dots(9)$$

from which it appears that if J_{n-1} and J_{n+2} vanish simultaneously, then either $J_{n+1} = 0$, which is impossible, or $z^2 = 4n(n+1)$.   Any common root of J_{n-1} and J_{n+2} must therefore be such that its square is an integer.

---

[*] *Phil. Mag.* Vol. xx. p. 1003, 1910, equation (8). [1913. A correction is here introduced. See Nicholson, *Phil. Mag.* Vol. xxv. p. 200, 1913.]

Pursuing the process, we find that if J_{n-1}, J_{n+3} have a common root z, then

$$(2n + 1) z^2 = 4n (n + 1) (n + 2),$$

so that z^2 is rational. And however far we go, we find that the simultaneous evanescence of two Bessel's functions requires that the common root be such that z^2 satisfies an algebraic equation whose coefficients are integers, the degree of the equation rising with the difference in order of the functions. If, as seems probable, a root of a Bessel's function cannot satisfy an integral algebraic equation, it would follow that no two Bessel's functions have a common root. The question seems worthy of the attention of mathematicians.

351.

HYDRODYNAMICAL NOTES.

[*Philosophical Magazine*, Vol. XXI. pp. 177—195, 1911.]

Potential and Kinetic Energies of Wave Motion.—Waves moving into Shallower Water.—Concentrated Initial Disturbance with inclusion of Capillarity.—Periodic Waves in Deep Water advancing without change of Type.—Tide Races.—Rotational Fluid Motion in a Corner.—Steady Motion in a Corner of Viscous Fluid.

IN the problems here considered the fluid is regarded as incompressible, and the motion is supposed to take place in two dimensions.

Potential and Kinetic Energies of Wave Motion.

When there is no dispersion, the energy of a progressive wave of any form is half potential and half kinetic. Thus in the case of a long wave in shallow water, "if we suppose that initially the surface is displaced, but that the particles have no velocity, we shall evidently obtain (as in the case of sound) two equal waves travelling in opposite directions, whose total energies are equal, and together make up the potential energy of the original dis‑ placement. Now the elevation of the derived waves must be half of that of the original displacement, and accordingly the potential energies less in the ratio of 4 : 1. Since therefore the potential energy of each derived wave is one quarter, and the total energy one half that of the original displacement, it follows that in the derived wave the potential and kinetic energies are equal " *.

The assumption that the displacement in each derived wave, when separated, is similar to the original displacement fails when the medium is dispersive. The equality of the two kinds of energy in an infinite pro‑ gressive train of simple waves may, however, be established as follows.

* "On Waves," *Phil. Mag.* Vol. I. p. 257 (1876); *Scientific Papers*, Vol. I. p. 254.

Consider first an infinite series of simple stationary waves, of which the energy is at one moment wholly potential and [a quarter of] a period later wholly kinetic. If t denote the time and E the total energy, we may write

$$\text{K.E.} = E \sin^2 nt, \qquad \text{P.E.} = E \cos^2 nt.$$

Upon this superpose a similar system, displaced through a quarter wave-length in space and through a quarter period in time. For this, taken by itself, we should have

$$\text{K.E} = E \cos^2 nt, \qquad \text{P.E.} = E \sin^2 nt.$$

And, the vibrations being *conjugate*, the potential and kinetic energies of the combined motion may be found by simple addition of the components, and are accordingly independent of the time, and each equal to E. Now the resultant motion is a simple progressive train, of which the potential and kinetic energies are thus seen to be equal.

A similar argument is applicable to prove the equality of energies in the motion of a simple conical pendulum.

It is to be observed that the conclusion is in general limited to vibrations which are infinitely small.

Waves moving into Shallower Water.

The problem proposed is the passage of an infinite train of simple infinitesimal waves from deep water into water which shallows gradually in such a manner that there is no loss of energy by reflexion or otherwise. At any stage the whole energy, being the double of the potential energy, is proportional per unit length to the square of the height; and for motion in two dimensions the only remaining question for our purpose is what are to be regarded as corresponding lengths along the direction of propagation.

In the case of long waves, where the wave-length (λ) is long in comparison with the depth (l) of the water, corresponding parts are as the velocities of propagation (V), or since the periodic time (τ) is constant, as λ. Conservation of energy then requires that

$$(\text{height})^2 \times V = \text{constant}; \quad \ldots\ldots\ldots\ldots\ldots\ldots\ldots(1)$$

or since V varies as $l^{\frac{1}{2}}$, height varies as $l^{-\frac{1}{4}}$*.

But for a dispersive medium corresponding parts are not proportional to V, and the argument requires modification. A uniform regime being established, what we are to equate at two separated places where the waves are of different character is the *rate of propagation of energy* through these places. It is a general proposition that in any kind of waves the ratio of the energy propagated past a fixed point in unit time to that resident in unit

---

* *Loc. cit.* p. 255.

length is U, where U is the *group-velocity*, equal to $d\sigma/dk$, where $\sigma = 2\pi/\tau$, $k = 2\pi/\lambda$*. Hence in our problem we must take

$$\text{height varies as } U^{-\frac{1}{2}}, \quad\dots\dots\dots\dots\dots\dots\dots(2)$$

which includes the former result, since in a non-dispersive medium $U = V$.

For waves in water of depth l,

$$\sigma^2 = gk \tanh kl, \quad\dots\dots\dots\dots\dots\dots\dots(3)$$

whence

$$2\sigma U/g = \tanh kl + kl\,(1 - \tanh^2 kl). \quad\dots\dots\dots\dots(4)$$

As the wave progresses, σ remains constant, (3) determines k in terms of l, and U follows from (4). If we write

$$\sigma^2 l/g = l', \quad\dots\dots\dots\dots\dots\dots\dots\dots(5)$$

(3) becomes

$$kl \,.\, \tanh kl = l', \quad\dots\dots\dots\dots\dots\dots\dots(6)$$

and (4) may be written

$$2\sigma U/g = kl + (l' - l'^2)/kl. \quad\dots\dots\dots\dots\dots(7)$$

By (6), (7) U is determined as a function of l' or by (5) of l.

If kl, and therefore l', is very great, $kl = l'$, and then by (7) if U_0 be the corresponding value of U,

$$2\sigma U_0/g = 1, \quad\dots\dots\dots\dots\dots\dots\dots\dots(8)$$

and in general

$$U/U_0 = kl + (l' - l'^2)/kl. \quad\dots\dots\dots\dots\dots(9)$$

Equations (2), (5), (6), (9) may be regarded as giving the solution of the problem in terms of a known σ. It is perhaps more practical to replace σ in (5) by λ_0, the corresponding wave-length in a great depth. The relation between σ and λ_0 being $\sigma^2 = 2\pi g/\lambda_0$, we find in place of (5)

$$l' = 2\pi l/\lambda_0 = k_0 l. \quad\dots\dots\dots\dots\dots\dots\dots(10)$$

Starting in (10) from λ_0 and l we may obtain l', whence (6) gives kl, and (9) gives U/U_0. But in calculating results by means of tables of the hyperbolic functions it is more convenient to start from kl. We find

kl	l'	U/U_0	kl	l'	U/U_0
∞	kl	1·000	·6	·322	·964
10	kl	1·000	·5	·231	·855
5	4·999	1·001	·4	·152	·722
2	1·928	1·105	·3	·087	·566
1·5	1·358	1·176	·2	·039	·390
1·0	·762	1·182	·1	·010	·200
·8	·531	1·110	kl	$(kl)^2$	$2kl$
·7	·423	1·048	—	—	—

* *Proc. Lond. Math. Soc.* Vol. ix. 1877 ; *Scientific Papers*, Vol. i. p. 326.

It appears that U/U_0 does not differ much from unity between $l' = \cdot 23$ and $l' = \infty$, so that the shallowing of the water does not at first produce much effect upon the height of the waves. It must be remembered, however, that the wave-length is diminishing, so that waves, even though they do no more than maintain their height, grow *steeper*.

Concentrated Initial Disturbance with inclusion of Capillarity.

A simple approximate treatment of the general problem of initial linear disturbance is due to Kelvin*. We have for the elevation η at any point x and at any time t

$$\eta = \frac{1}{\pi} \int_0^\infty \cos kx \cos \sigma t\, dk$$

$$= \frac{1}{2\pi} \int_0^\infty \cos(kx - \sigma t)\, dk + \frac{1}{2\pi} \int_0^\infty \cos(kx + \sigma t)\, dk, \quad \ldots\ldots(1)$$

in which σ is a function of k, determined by the character of the dispersive medium—expressing that the initial elevation $(t = 0)$ is concentrated at the origin of x. When t is great, the angles whose cosines are to be integrated will in general vary rapidly with k, and the corresponding parts of the integral contribute little to the total result. The most important part of the range of integration is the neighbourhood of places where $kx \pm \sigma t$ is stationary with respect to k, *i.e.* where

$$x \pm t\frac{d\sigma}{dk} = 0. \quad \ldots\ldots\ldots\ldots\ldots\ldots\ldots\ldots\ldots\ldots\ldots(2)$$

In the vast majority of practical applications $d\sigma/dk$ is positive, so that if x and t are also positive the second integral in (1) makes no sensible contribution. The result then depends upon the first integral, and only upon such parts of that as lie in the neighbourhood of the value, or values, of k which satisfy (2) taken with the lower sign. If k_1 be such a value, Kelvin shows that the corresponding term in η has an expression equivalent to

$$\eta = \frac{\cos(\sigma_1 t - k_1 x - \frac{1}{4}\pi)}{\sqrt{\{-2\pi t\, d^2\sigma/dk_1^2\}}}, \quad \ldots\ldots\ldots\ldots\ldots\ldots(3)$$

σ_1 being the value of σ corresponding to k_1.

In the case of deep-water waves where $\sigma = \sqrt{(gk)}$, there is only one predominant value of k for given values of x and t, and (2) gives

$$k_1 = gt^2/4x^2, \qquad \sigma_1 = gt/2x, \quad \ldots\ldots\ldots\ldots\ldots(4)$$

making

$$\sigma_1 t - k_1 x - \tfrac{1}{4}\pi = gt^2/4x - \tfrac{1}{4}\pi, \ldots\ldots\ldots\ldots\ldots(5)$$

and finally

$$\eta = \frac{g^{\frac{1}{2}}t}{2\pi^{\frac{1}{2}}x^{\frac{3}{2}}}\cos\left\{\frac{gt^2}{4x} - \frac{\pi}{4}\right\}, \quad \ldots\ldots\ldots\ldots\ldots(6)$$

the well-known formula of Cauchy and Poisson.

* *Proc. Roy. Soc.* Vol. XLII. p. 80 (1887) ; *Math. and Phys. Papers*, Vol. IV. p. 303.

In the numerator of (3) σ_1 and k_1 are functions of x and t. If we inquire what change (Λ) in x with t constant alters the angle by 2π, we find

$$\Lambda \left\{ k_1 + \left(x - t \frac{d\sigma}{dk_1} \right) \frac{dk_1}{dx} \right\} = 2\pi,$$

so that by (2) $\Lambda = 2\pi/k_1$, i.e. the effective wave-length Λ coincides with that of the predominant component in the original integral (1), and a like result holds for the periodic time*. Again, it follows from (2) that $k_1 x - \sigma_1 t$ in (3) may be replaced by $\int k_1 dx$, as is exemplified in (4) and (6).

When the waves move under the influence of a capillary tension T in addition to gravity,

$$\sigma^2 = gk + Tk^3/\rho, \quad \dots\dots\dots\dots\dots\dots\dots(7)$$

ρ being the density, and for the wave-velocity (V)

$$V^2 = \sigma^2/k^2 = g/k + Tk/\rho, \quad \dots\dots\dots\dots\dots(8)$$

as first found by Kelvin. Under these circumstances V has a minimum value when

$$k^2 = g\rho/T. \quad \dots\dots\dots\dots\dots\dots\dots(9)$$

The group-velocity U is equal to $d\sigma/dk$, or to $d(kV)/dk$; so that when V has a minimum value, U and V coincide. Referring to this, Kelvin towards the close of his paper remarks " The working out of our present problem for this case, or any case in which there are either minimums or maximums, or both maximums and minimums, of wave-velocity, is particularly interesting, but time does not permit of its being included in the present communication."

A glance at the simplified form (3) shows, however, that the special case arises, not when V is a minimum (or maximum), but when U is so, since then $d^2\sigma/dk_1^2$ vanishes. As given by (3), η would become infinite—an indication that the approximation must be pursued. If $k = k_1 + \xi$, we have in general in the neighbourhood of k_1,

$$kx - \sigma t = k_1 x - \sigma_1 t + \left(x - t \frac{d\sigma}{dk_1} \right)\xi - \frac{t}{1.2}\frac{d^2\sigma}{dk_1^2}\xi^2 - \frac{t}{1.2.3}\frac{d^3\sigma}{dk_1^3}\xi^3. \quad \dots(10)$$

In the present case where the term in ξ^2 disappears, as well as that in ξ, we get in place of (3) when t is great

$$\eta = \frac{\cos(k_1 x - \sigma_1 t)}{2\pi \{\frac{1}{6} t\, d^3\sigma/dk_1^3\}^{\frac{1}{3}}} \int_{-\infty}^{+\infty} \cos \alpha^3 . d\alpha, \quad \dots\dots\dots(11)$$

varying as $t^{-\frac{1}{3}}$ instead of as $t^{-\frac{1}{2}}$.

The definite integral is included in the general form

$$\int_{-\infty}^{+\infty} \cos \alpha^m . d\alpha = \frac{2}{m}\Gamma\left(\frac{1}{m}\right)\cos\frac{\pi}{2m}, \quad \dots\dots\dots\dots(12)$$

* Cf. Green, Proc. Roy. Soc. Ed. Vol. xxix. p. 445 (1909).

giving

$$\int_{-\infty}^{+\infty} \cos \alpha^2 . d\alpha = \sqrt{\left(\frac{\pi}{2}\right)}; \qquad \int_{-\infty}^{+\infty} \cos \alpha^3 . d\alpha = \frac{1}{\sqrt{3}} \Gamma(\tfrac{1}{3}). \quad \dots\dots(13)$$

The former is employed in the derivation of (3).

The occurrence of stationary values of U is determined from (7) by means of a quadratic. There is but one such value (U_0), easily seen to be a minimum, and it occurs when

$$k^2 = \{\sqrt{\tfrac{4}{3}} - 1\}\frac{g\rho}{T} = \cdot 1547 \frac{g\rho}{T}. \quad \dots\dots\dots\dots\dots(14)$$

On the other hand, the minimum of V occurs when $k^2 = g\rho/T$ simply.

When t is great, there is no important effect so long as x (positive) is less than $U_0 t$. For this value of x the Kelvin formula requires the modification expressed by (11). When x is decidedly greater than $U_0 t$, there arise two terms of the Kelvin form, indicating that there are now two systems of waves of different wave-lengths, effective at the same place.

It will be seen that the introduction of capillarity greatly alters the character of the solution. The quiescent region inside the annular waves is easily recognized a few seconds after a very small stone is dropped into smooth water*, but I have not observed the duplicity of the annular waves themselves. Probably the capillary waves of short wave-length are rapidly damped, especially when the water-surface is not quite clean. It would be interesting to experiment upon truly linear waves, such as might be generated by the sudden electrical charge or discharge of a wire stretched just above the surface. But the full development of the peculiar features to be expected on the inside of the wave-system seems to require a space larger than is conveniently available in a laboratory.

Periodic Waves in Deep Water advancing without change of Type.

The solution of this problem when the height of the waves is infinitesimal has been familiar for more than a century, and the pursuance of the approximation to cover the case of moderate height is to be found in a well-known paper by Stokes†. In a supplement published in 1880‡ the same author treated the problem by another method in which the space coordinates x, y are regarded as functions of ϕ, ψ the velocity and stream functions, and carried the approximation a stage further.

In an early publication§ I showed that some of the results of Stokes' first memoir could be very simply derived from the expression for the

---

* A checkered background, e.g. the sky seen through foliage, shows the waves best.

† Camb. Phil. Soc. Trans. Vol. VIII. p. 441 (1847); Math. and Phys. Papers, Vol. I. p. 197.

‡ Loc. cit. Vol. I. p. 314.

§ Phil. Mag. Vol. I. p. 257 (1876); Scientific Papers, Vol. I. p. 262. See also Lamb's Hydrodynamics, § 230.

stream-function in terms of x and y, and lately I have found that this method may be extended to give, as readily if perhaps less elegantly, all the results of Stokes' Supplement.

Supposing for brevity that the wave-length is 2π and the velocity of propagation unity, we take as the expression for the stream-function of the waves, reduced to rest,

$$\psi = y - \alpha e^{-y} \cos x - \beta e^{-2y} \cos 2x - \gamma e^{-3y} \cos 3x, \quad \dots\dots\dots(1)$$

in which x is measured horizontally and y vertically downwards. This expression evidently satisfies the differential equation to which ψ is subject, whatever may be the values of the constants α, β, γ. From (1) we find

$$U^2 - 2gy = (d\psi/dx)^2 + (d\psi/dy)^2 - 2gy$$
$$= 1 - 2\psi + 2(1 - g)y + 2\beta e^{-2y} \cos 2x + 4\gamma e^{-3y} \cos 3x$$
$$+ \alpha^2 e^{-2y} + 4\beta^2 e^{-4y} + 9\gamma^2 e^{-6y} + 4\alpha\beta e^{-3y} \cos x$$
$$+ 6\alpha\gamma e^{-4y} \cos 2x + 12\beta\gamma e^{-5y} \cos x. \quad \dots\dots\dots\dots\dots\dots\dots\dots(2)$$

The condition to be satisfied at a free surface is the constancy of (2).

The solution to a moderate degree of approximation (as already referred to) may be obtained with omission of β and γ in (1), (2). Thus from (1) we get, determining ψ so that the mean value of y is zero,

$$y = \alpha(1 + \tfrac{3}{8}\alpha^2) \cos x - \tfrac{1}{2}\alpha^2 \cos 2x + \tfrac{3}{8}\alpha^3 \cos 3x, \quad \dots\dots\dots(3)$$

which is correct as far as α^3 inclusive.

If we call the coefficient of $\cos x$ in (3) a, we may write with the same approximation

$$y = a \cos x - \tfrac{1}{2}a^2 \cos 2x + \tfrac{3}{8}a^3 \cos 3x. \quad \dots\dots\dots\dots(4)$$

Again from (2) with omission of β, γ,

$$U^2 - 2gy = \text{const.} + 2(1 - g - \alpha^2 - \alpha^4)y + \alpha^4 \cos 2x - \tfrac{4}{3}\alpha^5 \cos 3x. \quad \dots\dots(5)$$

It appears from (5) that the surface condition may be satisfied with α only, provided that α^4 is neglected and that

$$1 - g - \alpha^2 = 0. \quad \dots\dots\dots\dots\dots\dots\dots\dots\dots(6)$$

In (6) α may be replaced by a, and the equation determines the velocity of propagation. To exhibit this we must restore generality by introduction of $k (= 2\pi/\lambda)$ and c the velocity of propagation, hitherto treated as unity. Consideration of " dimensions " shows that (6) becomes

$$kc^2 - g - a^2c^2k^3 = 0, \quad \dots\dots\dots\dots\dots\dots\dots\dots(7)$$

or
$$c^2 = g/k \cdot (1 + k^2a^2). \quad \dots\dots\dots\dots\dots\dots\dots\dots(8)$$

Formulæ (4) and (8) are those given by Stokes in his first memoir.

By means of β and γ the surface condition (2) can be satisfied with inclusion of α^4 and α^5, and from (5) we see that β is of the order α^4 and γ of the order α^5. The terms to be retained in (2), in addition to those given in (5), are

$$2\beta (1 - 2y) \cos 2x + 4\gamma \cos 3x + 4\alpha\beta \cos x$$
$$= 2\beta \cos 2x - 2\alpha\beta (\cos x + \cos 3x) + 4\gamma \cos 3x + 4\alpha\beta \cos x.$$

Expressing the terms in $\cos x$ by means of y, we get finally

$$U^2 - 2gy = \text{const.} + 2y (1 - g - \alpha^2 - \alpha^4 + \beta)$$
$$+ (\alpha^4 + 2\beta) \cos 2x + (4\gamma - \tfrac{4}{3}\alpha^5 - 2\alpha\beta) \cos 3x. \quad ......(9)$$

In order to satisfy the surface condition of constant pressure, we must take

$$\beta = -\tfrac{1}{2}\alpha^4, \qquad \gamma = \tfrac{1}{12}\alpha^5, \quad ....(10)$$

and in addition

$$1 - g - \alpha^2 - \tfrac{3}{2}\alpha^4 = 0, \quad ....(11)$$

correct to α^5 inclusive. The expression (1) for ψ thus assumes the form

$$\psi = y - \alpha e^{-y} \cos x + \tfrac{1}{2}\alpha^4 e^{-2y} \cos 2x - \tfrac{1}{12}\alpha^5 e^{-3y} \cos 3x, \quad ....(12)$$

from which y may be calculated in terms of x as far as α^5 inclusive.

By successive approximation, determining ψ so as to make the mean value of y equal to zero, we find as far as α^4

$$y = (\alpha + \tfrac{5}{8}\alpha^3) \cos x - (\tfrac{1}{2}\alpha^2 + \tfrac{4}{3}\alpha^4) \cos 2x + \tfrac{3}{8}\alpha^3 \cos 3x - \tfrac{1}{3}\alpha^4 \cos 4x, \quad ...(13)$$

or, if we write as before a for the coefficient of $\cos x$,

$$y = a \cos x - (\tfrac{1}{2}a^2 + \tfrac{17}{24}a^4) \cos 2x + \tfrac{3}{8}a^3 \cos 3x - \tfrac{1}{3}a^4 \cos 4x, \quad ...(14)$$

in agreement with equation (20) of Stokes' Supplement.

Expressed in terms of a, (11) becomes

$$g = 1 - a^2 - \tfrac{1}{4}a^4, \quad ....(15)$$

or on restoration of k, c,

$$g = kc^2 - k^3 a^2 c^2 - \tfrac{1}{4}k^5 a^4 c^2. \quad ....(16)$$

Thus the extension of (8) is

$$c^2 = g/k . (1 + k^2 a^2 + \tfrac{4}{5}k^4 a^4), \quad ....(17)$$

which also agrees with Stokes' Supplement.

If we pursue the approximation one stage further, we find from (12) terms in α^5, additional to those expressed in (13). These are

$$y = \alpha^5 \left\{ \frac{373}{6 . 32} \cos x + \frac{243}{128} \cos 3x + \frac{125}{12 . 32} \cos 5x \right\}. \quad ......(18)*$$

It is of interest to compare the potential and kinetic energies of waves

* [1916. Burnside (*Proc. Lond. Math. Soc.* Vol. xv. p. 26, 1916) throws doubts upon the utility of Stokes' series.]

that are not infinitely small. For the stream-function of the waves regarded as progressive, we have, as in (1),

$$\psi = -\alpha e^{-y} \cos(x - ct) + \text{terms in } \alpha^4,$$

so that

$$(d\psi/dx)^2 + (d\psi/dy)^2 = \alpha^2 e^{-2y} + \text{terms in } \alpha^5.$$

Thus the mean kinetic energy per length x measured in the direction of propagation is

$$\frac{\alpha^2}{2} \int dx \int_y^\infty e^{-2y} dy = \frac{\alpha^2}{4} \int dx \, e^{-2y} = \frac{\alpha^2}{4} \int dx \, (1 - 2y + 2y^2) = \frac{\alpha^2}{4} \left\{ x + 2 \int y^2 dx \right\},$$

where y is the ordinate of the surface. And by (3)

$$\int y^2 dx = \{\tfrac{1}{2}(\alpha^2 + \tfrac{5}{4}\alpha^4) + \tfrac{1}{8}\alpha^4\} x.$$

Hence correct to α^4,

$$\text{K.E.} = \tfrac{1}{4}\alpha^2(1 + \alpha^2) x. \quad \dots\dots\dots\dots\dots\dots(19)$$

Again, for the potential energy

$$\text{P.E.} = \tfrac{1}{2}g \int y^2 dx = \tfrac{1}{2}gx(\tfrac{1}{2}\alpha^2 + \tfrac{3}{4}\alpha^4);$$

or since $g = 1 - \alpha^2$,

$$\text{P.E.} = \tfrac{1}{4}\alpha^2(1 + \tfrac{1}{2}\alpha^2) x. \quad \dots\dots\dots\dots\dots(20)$$

The kinetic energy thus exceeds the potential energy, when α^4 is retained.

Tide Races.

It is, I believe, generally recognized that seas are apt to be exceptionally heavy when the tide runs against the wind. An obvious explanation may be founded upon the fact that the relative motion of air and water is then greater than if the latter were not running, but it seems doubtful whether this explanation is adequate.

It has occurred to me that the cause may be rather in the motion of the stream relatively to itself, e.g. in the more rapid movement of the upper strata. Stokes' theory of the highest possible wave shows that in non-rotating water the angle at the crest is 120° and the height only moderate. In such waves the surface strata have a mean motion forwards. On the other hand, in Gerstner and Rankine's waves the fluid particles retain a mean position, but here there is *rotation* of such a character that (in the absence of waves) the surface strata have a relative motion backwards, i.e. against the direction of propagation*. It seems possible that waves moving against the tide may approximate more or less to the Gerstner type and thus be capable of acquiring a greater height and a sharper angle than would otherwise be expected. Needless to say, it is the steepness of waves, rather than their

---

* Lamb's *Hydrodynamics*, § 247.

mere height, which is a source of inconvenience and even danger to small craft.

The above is nothing more than a suggestion. I do not know of any detailed account of the special character of these waves, on which perhaps a better opinion might be founded.

Rotational Fluid Motion in a Corner.

The motion of incompressible inviscid fluid is here supposed to take place in two dimensions and to be bounded by two fixed planes meeting at an angle α. If there is no rotation, the stream-function ψ, satisfying $\nabla^2\psi = 0$, may be expressed by a series of terms

$$r^{\pi/a} \sin \pi\theta/\alpha, \qquad r^{2\pi/a} \sin 2\pi\theta/\alpha, \ldots r^{n\pi/a} \sin n\pi\theta/\alpha,$$

where n is an integer, making $\psi = 0$ when $\theta = 0$ or $\theta = \alpha$. In the immediate vicinity of the origin the first term predominates. For example, if the angle be a right angle,

$$\psi = r^2 \sin 2\theta = 2xy, \quad \ldots\ldots\ldots\ldots\ldots\ldots\ldots(1)$$

if we introduce rectangular coordinates.

The possibility of irrotational motion depends upon the fixed boundary not being closed. If $\alpha < \pi$, the motion near the origin is finite; but if $\alpha > \pi$, the velocities deduced from ψ become infinite.

If there be rotation, motion may take place even though the boundary be closed. For example, the circuit may be completed by the arc of the circle $r = 1$. In the case which it is proposed to consider the rotation ω is *uniform*, and the motion may be regarded as steady. The stream-function then satisfies the general equation

$$\nabla^2\psi = d^2\psi/dx^2 + d^2\psi/dy^2 = 2\omega, \quad \ldots\ldots\ldots\ldots\ldots(2)$$

or in polar coordinates

$$\frac{d^2\psi}{dr^2} + \frac{1}{r}\frac{d\psi}{dr} + \frac{1}{r^2}\frac{d^2\psi}{d\theta^2} = 2\omega. \quad \ldots\ldots\ldots\ldots\ldots(3)$$

When the angle is a right angle, it might perhaps be expected that there should be a simple expression for ψ in powers of x and y, analogous to (1) and applicable to the immediate vicinity of the origin; but we may easily satisfy ourselves that no such expression exists*. In order to express the motion we must find solutions of (3) subject to the conditions that $\psi = 0$ when $\theta = 0$ and when $\theta = \alpha$.

For this purpose we assume, as we may do, that

$$\psi = \Sigma R_n \sin n\pi\theta/\alpha, \quad \ldots\ldots\ldots\ldots\ldots\ldots\ldots(4)$$

* In strictness the satisfaction of (2) *at* the origin is inconsistent with the evanescence of ψ on the rectangular axes.

where n is integral and R_n a function of r only; and in deducing $\nabla^2\psi$ we may perform the differentiations with respect to θ (as well as with respect to r) under the sign of summation, since $\psi = 0$ at the limits. Thus

$$\nabla^2\psi = \Sigma \left(\frac{d^2 R_n}{dr^2} + \frac{1}{r}\frac{dR_n}{dr} - \frac{n^2\pi^2}{a^2 r^2} R_n \right) \sin\frac{n\pi\theta}{a}. \quad\dots\dots\dots\dots(5)$$

The right-hand member of (3) may also be expressed in a series of sines of the form

$$2\omega = \delta\omega/\pi \,.\, \Sigma n^{-1} \sin n\pi\theta/a, \quad\dots\dots\dots\dots\dots\dots(6)$$

where n is an *odd* integer; and thus for all values of n we have

$$r^2 \frac{d^2 R_n}{dr^2} + r\frac{dR_n}{dr} - \frac{n^2\pi^2 R_n}{a^2} = \frac{4\omega}{n\pi}\{1-(-1)^n\}. \quad\dots\dots\dots\dots(7)$$

The general solution of (7) is

$$R_n = A_n r^{n\pi/a} + B_n r^{-n\pi/a} + \frac{4\omega a^2 r^2 \{1-(-1)^n\}}{n\pi\,(4a^2 - n^2\pi^2)}, \quad\dots\dots\dots\dots(8)$$

the introduction of which into (4) gives ψ.

In (8) A_n and B_n are arbitrary constants to be determined by the other conditions of the problem. For example, we might make R_n, and therefore ψ, vanish when $r = r_1$ and when $r = r_2$, so that the fixed boundary enclosing the fluid would consist of two radii vectores and two circular arcs. If the fluid extend to the origin, we must make $B_n = 0$; and if the boundary be completed by the circular arc $r = 1$, we have $A_n = 0$ when n is even, and when n is odd

$$A_n = \frac{8\omega a^2}{n\pi\,(4a^2 - n^2\pi^2)} = 0. \quad\dots\dots\dots\dots\dots(9)$$

Thus for the fluid enclosed in a circular sector of angle a and radius unity

$$\psi = 8\omega a^2 \,\Sigma\, \frac{r^{n\pi/a} - r^2}{n\pi\,(n^2\pi^2 - 4a^2)} \sin\frac{n\pi\theta}{a}, \quad\dots\dots\dots\dots(10)$$

the summation extending to all odd integral values of n.

The above formula (10) relates to the motion of uniformly *rotating* fluid bounded by *stationary* radii vectores at $\theta = 0$, $\theta = a$. We may suppose the containing vessel to have been rotating for a long time and that the fluid (under the influence of a very small viscosity) has acquired this rotation so that the whole revolves like a solid body. The motion expressed by (10) is that which would ensue if the rotation of the vessel were suddenly stopped. A related problem was solved a long time since by Stokes[*], who considered the *irrotational* motion of fluid in a *revolving* sector. The solution of Stokes' problem is derivable from (10) by mere addition to the latter of $\psi_0 = -\frac{1}{2}\omega r^2$, for then $\psi + \psi_0$ satisfies $\nabla^2(\psi + \psi_0) = 0$; and this is perhaps the simplest

---

[*] *Camb. Phil. Trans.* Vol. VIII. p. 533 (1847); *Math. and Phys. Papers*, Vol. I. p. 305.

method of obtaining it. The results are in harmony; but the fact is not immediately apparent, inasmuch as Stokes expresses the motion by means of the velocity-potential, whereas here we have employed the stream-function.

That the subtraction of $\frac{1}{2}\omega r^2$ makes (10) an harmonic function shows that the series multiplying r^2 can be summed. In fact

$$8\alpha^2 \Sigma \frac{\sin(n\pi\theta/\alpha)}{n\pi(n^2\pi^2 - 4\alpha^2)} = \frac{\cos(2\theta - \alpha)}{2\cos\alpha} - \frac{1}{2},$$

so that $\quad \psi/\omega = \frac{1}{2}r^2 - \dfrac{r^2\cos(2\theta - \alpha)}{2\cos\alpha} + 8\alpha^2 \Sigma \dfrac{r^{n\pi/\alpha}\sin n\pi\theta/\alpha}{n\pi(n^2\pi^2 - 4\alpha^2)}. \quad \ldots\ldots(11)$

In considering the character of the motion defined by (11) in the immediate vicinity of the origin we see that if $\alpha < \frac{1}{2}\pi$, the term in r^2 preponderates even when $n = 1$. When $\alpha = \frac{1}{2}\pi$ exactly, the second term in (11) and the first term under Σ corresponding to $n = 1$ become infinite, and the expression demands transformation. We find in this case

$$\psi/\omega = \frac{1}{2}r^2 + \frac{2r^2}{\pi}(\theta - \frac{1}{4}\pi)\cos 2\theta + r^2\sin 2\theta\left(\frac{2}{\pi}\log r - \frac{3}{2\pi}\right) + \frac{2}{\pi}\Sigma \frac{r^{2n}\sin 2n\theta}{n(n^2 - 1)},$$
$$\ldots\ldots\ldots(12)$$

the summation commencing at $n = 3$. On the middle line $\theta = \frac{1}{4}\pi$, we have

$$\psi/\omega = \frac{1}{2}r^2 - \frac{3r^2}{2\pi} + \frac{2}{\pi}\left\{r^2\log r - \frac{r^6}{3.8} + \frac{r^{10}}{5.24} - \ldots\right\}. \quad \ldots\ldots(13)$$

The following are derived from (13):

r	$-\frac{1}{2}\pi\psi$	r	$-\frac{1}{2}\pi\psi$	r	$-\frac{1}{2}\pi\psi$
0·0	·00000	0·4	·14112	0·8	·13030
0·1	·02267	0·5	·16507	0·9	·07641
0·2	·06296	0·6	·17306	1·0	·00000
0·3	·10521	0·7	·16210		

The maximum value occurs when $r = ·592$. At the point $r = ·592$, $\theta = \frac{1}{4}\pi$, the fluid is stationary.

A similar transformation is required when $\alpha = 3\pi/2$.

When $\alpha = \pi$, the boundary becomes a semicircle, and the leading term $(n = 1)$ is

$$\psi/\omega = -\frac{8}{3\pi}r\sin\theta = -\frac{3}{8\pi}y \ldots, \quad \ldots\ldots\ldots\ldots(14)$$

which of itself represents an irrotational motion.

When $\alpha = 2\pi$, the two bounding radii vectores coincide and the containing vessel becomes a circle with a single partition wall at $\theta = 0$. In this case again the leading term is irrotational, being

$$\psi/\omega = -\frac{32}{15\pi} r^{\frac{1}{2}} \sin \tfrac{1}{2}\theta. \quad\dotfill(15)$$

Steady Motion in a Corner of a Viscous Fluid.

Here again we suppose the fluid to be incompressible and to move in two dimensions free from external forces, or at any rate from such as cannot be derived from a potential. If in the same notation as before ψ represents the stream-function, the general equation to be satisfied by ψ is

$$\nabla^4 \psi = 0 ; \quad\dotfill(1)$$

with the conditions that when $\theta = 0$ and $\theta = \alpha$,

$$\psi = 0, \qquad d\psi/d\theta = 0. \quad\dotfill(2)$$

It is worthy of remark that the problem is analytically the same as that of a plane elastic plate clamped at $\theta = 0$ and $\theta = \alpha$, upon which (in the region considered) no external forces act.

The general problem thus represented is one of great difficulty, and all that will be attempted here is the consideration of one or two particular cases. We inquire what solutions are possible such that ψ, as a function of r (the radius vector), is proportional to r^m. Introducing this supposition into (1), we get

$$\left\{ m^2 + \frac{d^2}{d\theta^2} \right\} \left\{ (m-2)^2 + \frac{d^2}{d\theta^2} \right\} \psi = 0, \quad\dotfill(3)$$

as the equation determining the dependence on θ. The most general value of ψ consistent with our suppositions is thus

$$\psi = r^m \{ A \cos m\theta + B \sin m\theta + C \cos (m-2)\theta + D \sin (m-2)\theta \}, \quad\dots(4)$$

where A, B, C, D are constants.

Equation (4) may be adapted to our purpose by taking

$$m = n\pi/\alpha, \quad\dotfill(5)$$

where n is an integer. Conditions (2) then give

$$A + C = 0, \qquad A + C \cos 2\alpha - D \sin 2\alpha = 0,$$

$$\frac{n\pi}{\alpha} B + \left(\frac{n\pi}{\alpha} - 2 \right) D = 0,$$

$$\frac{n\pi}{\alpha} B + \left(\frac{n\pi}{\alpha} - 2 \right) C \sin 2\alpha + \left(\frac{n\pi}{\alpha} - 2 \right) D \cos 2\alpha = 0.$$

When we substitute in the second and fourth of these equations the values of A and B, derived from the first and third, there results

$$C(1 - \cos 2\alpha) + D \sin 2\alpha = 0,$$
$$C \sin 2\alpha - D (1 - \cos 2\alpha) = 0 ;$$

and these can only be harmonized when $\cos 2\alpha = 1$, or $\alpha = s\pi$, where s is an integer. In physical problems, α is thus limited to the values π and 2π. To these cases (4) is applicable with C and D arbitrary, provided that we make

$$A + C = 0, \qquad B + \left(1 - \frac{2s}{n}\right) D = 0. \quad \ldots\ldots\ldots\ldots(5 \text{ bis})$$

Thus
$$\psi = Cr^{n/s} \left\{\cos\left(\frac{n\theta}{s} - 2\theta\right) - \cos\frac{n\theta}{s}\right\}$$
$$+ Dr^{n/s} \left\{\sin\left(\frac{n\theta}{s} - 2\theta\right) - \left(1 - \frac{2s}{n}\right) \sin\frac{n\theta}{s}\right\}, \quad \ldots\ldots(6)$$

making
$$\nabla^2\psi = 4 \left(\frac{n}{s} - 1\right) r^{-2+n/s} \left\{C \cos\left(\frac{n\theta}{s} - 2\theta\right) + D \sin\left(\frac{n\theta}{s} - 2\theta\right)\right\}. \quad \ldots(7)$$

When $s = 1$, $\alpha = \pi$, the corner disappears and we have simply a straight boundary (fig. 1). In this case $n = 1$ gives a nugatory result. When $n = 2$, we have

$$\psi = Cr^2 (1 - \cos 2\theta) = 2Cy^2, \quad \ldots\ldots\ldots\ldots\ldots(8)$$

Fig. 1.   Fig. 2.

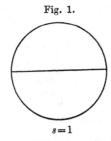

$s = 1$ $s = 2$

and $\nabla^2\psi = 4C$. When $n = 3$,

$$\psi = Cr^3 (\cos\theta - \cos 3\theta) + Dr^3 (\sin\theta - \tfrac{1}{3} \sin 3\theta), \quad \ldots\ldots(9)$$
$$\nabla^2\psi = 8r (C \cos\theta + D \sin\theta) = 8 (Cx + Dy). \quad \ldots\ldots\ldots\ldots(10)$$

In rectangular coordinates
$$\psi = 4Cxy^2 + \tfrac{4}{3} Dy^3, \quad \ldots\ldots\ldots\ldots\ldots\ldots(11)$$

solutions which obviously satisfy the required conditions.

When $s = 2$, $\alpha = 2\pi$, the boundary consists of a straight wall extending from the origin in one direction (fig. 2). In this case (6) and (7) give

$$\psi = Cr^{\frac{1}{2}n} \{\cos(\tfrac{1}{2}n\theta - 2\theta) - \cos\tfrac{1}{2}n\theta\}$$
$$+ Dr^{\frac{1}{2}n} \left\{\sin(\tfrac{1}{2}n\theta - 2\theta) - \left(1 - \frac{4}{n}\right) \sin\tfrac{1}{2}n\theta\right\}, \quad \ldots\ldots(12)$$

$$\nabla^2\psi = (2n - 4) r^{\frac{1}{2}n-2} \{C \cos(\tfrac{1}{2}n\theta - 2\theta) + D \sin(\tfrac{1}{2}n\theta - 2\theta)\}. \quad \ldots(13)$$

2—2

Solutions of interest are afforded in the case $n = 1$. The C-solution is ($C = \frac{1}{4}$)

$$\psi = \tfrac{1}{4} r^{\frac{1}{2}} (\cos \tfrac{3}{2} \theta - \cos \tfrac{1}{2} \theta) = - r^{\frac{1}{2}} \cos \tfrac{1}{2} \theta \sin^2 \tfrac{1}{2} \theta, \quad \ldots\ldots\ldots(14)$$

vanishing when $\theta = \pi$, as well as when $\theta = 0$, $\theta = 2\pi$, and for no other admissible value of θ. The values of ψ are reversed when we write $2\pi - \theta$ for θ. As expressed, this value is negative from 0 to π and positive from π to 2π. The minimum occurs when $\theta = 109° 28'$. Every stream-line which enters the circle ($r = 1$) on the left of this radius leaves it on the right.

The velocities, represented by $d\psi/dr$ and $r^{-1} d\psi/d\theta$, are infinite at the origin.

For the D-solution we may take

$$\psi = r^{\frac{1}{2}} \sin^3 \tfrac{1}{2} \theta. \quad \ldots\ldots\ldots\ldots\ldots\ldots\ldots\ldots(15)$$

Here ψ retains its value unaltered when $2\pi - \theta$ is substituted for θ. When r is given, ψ increases continuously from $\theta = 0$ to $\theta = \pi$. On the line $\theta = \pi$ the motion is entirely transverse to it. This is an interesting example of the flow of viscous fluid round a sharp corner. In the application to an elastic plate ψ represents the displacement at any point of the plate, supposed to be clamped along $\theta = 0$, and otherwise free from force within the region considered. The following table exhibits corresponding values of r and θ such as to make $\psi = 1$ in (15):

θ	r	θ	r
180°	1·00	60°	64·0
150°	1·23	20°	$10^4 \times 3·65$
120°	2·37	10°	$10^6 \times 2·28$
90°	8·00	0°	∞

When $n = 2$, (12) appears to have no significance.

When $n = 3$, the dependence on θ is the same as when $n = 1$. Thus (14) and (15) may be generalized:

$$\psi = (Ar^{\frac{1}{2}} + Br^{\frac{3}{2}}) \cos \tfrac{1}{2} \theta \sin^2 \tfrac{1}{2} \theta, \quad \ldots\ldots\ldots\ldots(16)$$

$$\psi = (A'r^{\frac{1}{2}} + B'r^{\frac{3}{2}}) \sin^3 \tfrac{1}{2} \theta. \quad \ldots\ldots \ldots\ldots\ldots\ldots(17)$$

For example, we could satisfy either of the conditions $\psi = 0$, or $d\psi/dr = 0$, on the circle $r = 1$.

For $n = 4$ the D-solution becomes nugatory; but for the C-solution we have

$$\psi = Cr^2 (1 - \cos 2\theta) = 2Cr^2 \sin^2 \theta = 2Cy^2. \quad \ldots\ldots\ldots\ldots(18)$$

The wall (or in the elastic plate problem the clamping) along $\theta = 0$ is now without effect.

It will be seen that along these lines nothing can be done in the apparently simple problem of a horizontal plate clamped along the rectangular axes of x and y, if it be supposed free from force*. Ritz† has shown that the solution is not developable in powers of x and y, and it may be worth while to extend the proposition to the more general case when the axes, still regarded as lines of clamping, are inclined at any angle α. In terms of the now oblique coordinates x, y the general equation takes the form

$$(d^2/dx^2 + d^2/dy^2 - 2\cos\alpha\, d^2/dx\,dy)^2\, w = 0, \quad \ldots\ldots\ldots\ldots(19)$$

which may be differentiated any number of times with respect to x and y, with the conditions

$$w = 0, \qquad dw/dy = 0, \qquad \text{when } y = 0, \quad\ldots\ldots\ldots\ldots(20)$$

$$w = 0, \qquad dw/dx = 0, \qquad \text{when } x = 0. \quad\ldots\ldots\ldots\ldots(21)$$

We may differentiate, as often as we please, (20) with respect to x and (21) with respect to y.

From these data it may be shown that at the origin *all* differential coefficients of w with respect to x and y vanish. The evanescence of those of zero and first order is expressed in (20), (21). As regards those of the second order we have from (20) $d^2w/dx^2 = 0$, $d^2w/dx\,dy = 0$, and from (21) $d^2w/dy^2 = 0$. Similarly for the third order from (20)

$$d^3w/dx^3 = 0, \qquad d^3w/dx^2dy = 0,$$

and from (21)

$$d^3w/dy^3 = 0, \qquad d^3w/dx\,dy^2 = 0.$$

For the fourth order (20) gives

$$d^4w/dx^4 = 0, \qquad d^4w/dx^3\,dy = 0,$$

and (21) gives

$$d^4w/dy^4 = 0, \qquad d^4w/dx\,dy^3 = 0.$$

So far d^4w/dx^2dy^2 might be finite, but (19) requires that it also vanish. This process may be continued. For the $m+1$ coefficients of the mth order we obtain four equations from (20), (21) and $m-3$ by differentiations of (19), so that all the differential coefficients of the mth order vanish. It follows that every differential coefficient of w with respect to x and y vanishes at the origin. I apprehend that the conclusion is valid for all angles α less than 2π. That the displacement at a distance r from the corner should diminish rapidly with r is easily intelligible, but that it should diminish more rapidly than any power of r, however high, would, I think, not have been expected without analytical proof.

* If indeed gravity act, $w = x^2y^2$ is a very simple solution.
† *Ann. d. Phys.* Bd. xxviii. p. 760, 1909.

352.

ON A PHYSICAL INTERPRETATION OF SCHLÖMILCH'S THEOREM IN BESSEL'S FUNCTIONS.

[*Philosophical Magazine*, Vol. XXI. pp. 567—571, 1911.]

THIS theorem teaches that any function $f(r)$ which is finite and continuous for real values of r between the limits $r = 0$ and $r = \pi$, both inclusive, may be expanded in the form

$$f(r) = a_0 + a_1 J_0(r) + a_2 J_0(2r) + a_3 J_0(3r) + \ldots, \quad \ldots\ldots\ldots(1)$$

J_0 being the Bessel's function usually so denoted; and Schlömilch's demonstration has been reproduced with slight variations in several text-books[*]. So far as I have observed, it has been treated as a purely analytical development. From this point of view it presents rather an accidental appearance; and I have thought that a physical interpretation, which is not without interest in itself, may help to elucidate its origin and meaning.

The application that I have in mind is to the theory of aerial vibrations. Let us consider the most general vibrations in one dimension ξ which are periodic in time 2π and are also symmetrical with respect to the origins of ξ and t. The condensation s, for example, may be expressed

$$s = b_0 + b_1 \cos \xi \cos t + b_2 \cos 2\xi \cos 2t + \ldots, \quad \ldots\ldots\ldots\ldots(2)$$

where the coefficients b_0, b_1, &c. are arbitrary. (For simplicity it is supposed that the velocity of propagation is unity.) When $t = 0$, (2) becomes a function of ξ only, and we write

$$F(\xi) = b_0 + b_1 \cos \xi + b_2 \cos 2\xi + \ldots, \quad \ldots\ldots\ldots\ldots(3)$$

in which $F(\xi)$ may be considered to be an arbitrary function of ξ from 0 to π. Outside these limits F is determined by the equations

$$F(-\xi) = F(\xi + 2\pi) = F(\xi). \quad \ldots\ldots\ldots\ldots\ldots(4)$$

---

[*] See, for example, Gray and Mathews' *Bessel's Functions*, p. 30; Whittaker's *Modern Analysis*, § 165.

We now superpose an infinite number of components, analogous to (2) with the same origins of space and time, and differing from one another only in the direction of ξ, these directions being limited to the plane xy, and in this plane distributed uniformly. The resultant is a function of t and r only, where $r = \sqrt{(x^2 + y^2)}$, independent of the third coordinate z, and therefore (as is known) takes the form

$$s = a_0 + a_1 J_0(r) \cos t + a_2 J_0(2r) \cos 2t + a_3 J_0(3r) \cos 3t + \ldots, \quad \ldots(5)$$

reducing to (1) when $t = 0$*. The expansion of a function in the series (1) is thus definitely suggested as probable in all cases and certainly possible in an immense variety. And it will be observed that no value of ξ greater than π contributes anything to the resultant, so long as $r < \pi$.

The relation here implied between F and f is of course identical with that used in the purely analytical investigation. If ϕ be the angle between ξ and any radius vector r to a point where the value of f is required, $\xi = r \cos \phi$, and the *mean* of all the components $F(\xi)$ is expressed by

$$f(r) = \frac{2}{\pi} \int_0^{\frac{1}{2}\pi} F(r \cos \phi) \, d\phi. \ldots\ldots\ldots\ldots\ldots\ldots(6)$$

The solution of the problem of expressing F by means of f is obtained analytically with the aid of Abel's theorem. And here again a physical, or rather geometrical, interpretation throws light upon the process.

Equation (6) is the result of averaging $F(\xi)$ over all directions indifferently in the xy plane. Let us abandon this restriction and take the average when ξ is indifferently distributed in all directions whatever. The result now becomes a function only of R, the radius vector in space. If θ be the angle between R and one direction of ξ, $\xi = R \cos \theta$, and we obtain as the mean

$$\int_0^{\frac{1}{2}\pi} F(R \cos \theta) \sin \theta \, d\theta = \frac{1}{R} \{F_1(R) - F_1(0)\}, \quad \ldots\ldots\ldots\ldots(7)$$

where $F_1' = F$.

This result is obtained by a direct integration of $F(\xi)$ over all directions in space. It may also be arrived at indirectly from (6). In the latter $f(r)$ represents the averaging of $F(\xi)$ for all directions in a certain plane, the result being independent of the coordinate perpendicular to the plane. If we take the average again for all possible positions of this plane, we must recover (7). Now if θ be the angle between the normal to this plane and the radius vector R, $r = R \sin \theta$, and the mean is

$$\int_0^{\frac{1}{2}\pi} f(R \sin \theta) \sin \theta \, d\theta. \ldots\ldots\ldots\ldots\ldots\ldots(8)$$

---

* It will appear later that the a's and b's are equal.

We conclude that

$$R \int_0^{\frac{1}{2}\pi} f(R \sin \theta) \sin \theta \, d\theta = F_1(R) - F_1(0), \quad \dots\dots\dots(9)$$

which may be considered as expressing F in terms of f.

If in (6), (9) we take $F(R) = \cos R$, we find*

$$\int_0^{\frac{1}{2}\pi} J_0(R \sin \theta) \sin \theta \, d\theta = R^{-1} \sin R.$$

Differentiating (9), we get

$$F(R) = \int_0^{\frac{1}{2}\pi} f(R \sin \theta) \sin \theta \, d\theta + R \int_0^{\frac{1}{2}\pi} f'(R \sin \theta)(1 - \cos^2 \theta) \, d\theta. \quad \dots(10)$$

Now

$$R \int_0^{\frac{1}{2}\pi} \cos^2 \theta . f'(R \sin \theta) \, d\theta = \int \cos \theta . df(R \sin \theta)$$

$$= -f(0) + \int_0^{\frac{1}{2}\pi} f(R \sin \theta) \sin \theta \, d\theta.$$

Accordingly $\qquad F(R) = f(0) + R \int_0^{\frac{1}{2}\pi} f'(R \sin \theta) \, d\theta. \quad \dots\dots\dots(11)$

That $f(r)$ in (1) may be arbitrary from 0 to π is now evident. By (3) and (6)

$$f(r) = \frac{2}{\pi} \int_0^{\frac{1}{2}\pi} d\phi \, \{b_0 + b_1 \cos(r \cos \phi) + b_2 \cos(2r \cos \phi) + \dots\}$$

$$= b_0 + b_1 J_0(r) + b_2 J_0(2r) + \dots, \quad \dots\dots\dots\dots\dots\dots(12)$$

where $\qquad b_0 = \frac{1}{\pi} \int_0^\pi F(\xi) \, d\xi, \quad b_n = \frac{2}{\pi} \int_0^\pi \cos n\xi \, F(\xi) \, d\xi. \quad \dots\dots(13)$

Further, with use of (11)

$$b_0 = f(0) + \frac{1}{\pi} \int_0^\pi d\xi . \xi . \int_0^{\frac{1}{2}\pi} f'(\xi \sin \theta) \, d\theta, \quad \dots\dots\dots(14)$$

$$b_n = \frac{2}{\pi} \int_0^\pi d\xi . \xi \cos n\xi . \int_0^{\frac{1}{2}\pi} f'(\xi \sin \theta) \, d\theta, \quad \dots\dots\dots(15)$$

by which the coefficients in (12) are completely expressed when f is given between 0 and π.

The physical interpretation of Schlömilch's theorem in respect of two-dimensional aerial vibrations is as follows:—Within the cylinder $r = \pi$ it is possible by suitable movements at the boundary to maintain a symmetrical motion which shall be strictly periodic in period 2π, and which at times $t = 0$, $t = 2\pi$, &c. (when there is no velocity), shall give a condensation which

---

* *Enc. Brit.* Art. "Wave Theory," 1888; *Scientific Papers*, Vol. III. p. 98.

is arbitrary over the whole of the radius. And this motion will maintain itself without external aid if outside $r = \pi$ the initial condition is chosen in accordance with (6), $F(\xi)$ for values of ξ greater than π being determined by (4). A similar statement applies of course to the vibrations of a stretched membrane, the transverse displacement w replacing s in (5).

Reference may be made to a simple example quoted by Whittaker. Initially let $f(r) = r$, so that from 0 to π the form of the membrane is conical. Then from (12), (14), (15)

$$b_0 = \frac{\pi^2}{4}, \quad b_n = 0 \,(n \text{ even}), \quad b_n = -\frac{2}{n^2} \,(n \text{ odd});$$

and thus

$$f(r) = \frac{\pi^2}{4} - 2\left\{ J_0(r) + \frac{1}{9} J_0(3r) + \frac{1}{25} J_0(5r) + \dots \right\}, \quad \dots\dots(16)$$

the right-hand member being equal to r from $r = 0$ to $r = \pi$.

The corresponding vibration is of course expressed by (16) if we multiply each function $J_0(nr)$ by the time-factor $\cos nt$.

If this periodic vibration is to be maintained without external force, the initial condition must be such that it is represented by (16) for all values of r, and not merely for those less than π. By (11) from 0 to π, $F(\xi) = \frac{1}{2}\pi\xi$, from which again by (4) the value of F for higher values of ξ follows. Thus from π to 2π, $F(\xi) = \frac{1}{2}\pi(2\pi - \xi)$; from 2π to 3π, $F(\xi) = \frac{1}{2}\pi(\xi - 2\pi)$; and so on. From these f is to be found by means of (6). For example, from π to 2π,

$$f(r) = r \int_0^{\sin\theta = \pi/r} \sin\theta \, d\theta + \int_{\sin\theta = \pi/r}^{\sin\theta = 1} (2\pi - r\sin\theta) \, d\theta$$

$$= r - 2\sqrt{(r^2 - \pi^2)} + 2\pi \cos^{-1}(\pi/r), \quad \dots\dots\dots\dots\dots\dots(17)$$

where $\cos^{-1}(\pi/r)$ is to be taken in the first quadrant.

It is hardly necessary to add that a theorem similar to that proved above holds for aerial vibrations which are symmetrical in all directions about a centre. Thus within the sphere of radius π it is possible to have a motion which shall be strictly periodic and is such that the condensation is initially arbitrary at all points along the radius.

353.

BREATH FIGURES.

[*Nature*, Vol. LXXXVI. pp. 416, 417, 1911.]

THE manner in which aqueous vapour condenses upon ordinarily clean surfaces of glass or metal is familiar to all. Examination with a magnifier shows that the condensed water is in the form of small lenses, often in pretty close juxtaposition. The number and thickness of these lenses depend upon the cleanness of the glass and the amount of water deposited. In the days of wet collodion every photographer judged of the success of the cleaning process by the uniformity of the dew deposited from the breath.

Information as to the character of the deposit is obtained by looking through it at a candle or small gas flame. The diameter of the halo measures the angle at which the drops meet the glass, an angle which diminishes as the dew evaporates. That the flame is seen at all in good definition is a proof that some of the glass is uncovered. Even when both sides of a plate are dewed the flame is still seen distinctly though with much diminished intensity.

The process of formation may be followed to some extent under the microscope, the breath being led through a tube. The first deposit occurs very suddenly. As the condensation progresses, the drops grow, and many of the smaller ones coalesce. During evaporation there are two sorts of behaviour. Sometimes the boundaries of the drops contract, leaving the glass bare. In other cases the boundary of a drop remains fixed, while the thickness of the lens diminishes until all that remains is a thin lamina. Several successive formations of dew will often take place in what seems to be precisely the same pattern, showing that the local conditions which determine the situation of the drops have a certain degree of permanence.

An interesting and easy experiment has been described by Aitken (*Proc. Ed. Soc.* p. 94, 1893). Clean a glass plate in the usual way until the breath deposits equally.

"If we now pass over this clean surface the point of a blow-pipe flame, using a very small jet, and passing it over the glass with sufficient quickness to prevent the sudden heating breaking it; and if we now breathe on the glass after it is cold, we shall find the track of the flame clearly marked. While most of the surface looks white by the light reflected from the deposited moisture, the track of the flame is quite black; not a ray of light is scattered by it. It looks as if there were no moisture condensed on that part of the plate, as it seems unchanged; but if it be closely examined by a lens, it will be seen to be quite wet. But the water is so evenly distributed, that it forms a thin film, in which, with proper lighting and the aid of a lens, a display of interference colours may be seen as the film dries and thins away."

"Another way of studying the change produced on the surface of the glass by the action of the flame is to take the [plate], as above described, after a line has been drawn over it with the blow-pipe jet, and when cold let a drop of water fall on any part of it where it showed white when breathed on. Now tilt the plate to make the drop flow, and note the resistance to its flow, and how it draws itself up in the rear, leaving the plate dry. When, however, the moving drop comes to the part acted on by the flame, all resistance to flow ceases, and the drop rapidly spreads itself over the whole track, and shows a decided disinclination to leave it."

The impression thus produced lasts for some days or weeks, with diminishing distinctness. A permanent record may be obtained by the deposit of a very thin coat of silver by the usual chemical method. The silver attaches itself by preference to the track of the flame, and especially to the *edges* of the track, where presumably the combustion is most intense. It may be protected with celluloid, or other, varnish.

The view, expressed by Mr Aitken, which would attribute the effect to very fine dust deposited on the glass from the flame, does not commend itself to me. And yet mere heat is not very effective. I was unable to obtain a good result by strongly heating the *back* of a thin glass in a Bunsen flame. For this purpose a long flame on Ramsay's plan is suitable, especially if it be long enough to include the entire width of the plate.

It seems to me that we must appeal to varying degrees of cleanliness for the explanation, cleanliness meaning mainly freedom from grease. And one of the first things is to disabuse our minds of the idea that anything wiped with an ordinary cloth can possibly be clean. This subject was ably treated many years ago by Quincke (*Wied. Ann.* II. p. 145, 1877), who, however, seems to have remained in doubt whether a film of air might not give rise to the same effects as a film of grease. Quincke investigated the maximum edge-angle possible when a drop of liquid stands upon the surface of a solid. In general, the cleaner the surface, the smaller the maximum edge-angle. With alcohol and petroleum there was no difficulty in reducing the maximum angle to zero. With water on glass the angle could be made small, but increased as time elapsed after cleaning.

As a detergent Quincke employed hot sulphuric acid. A few drops may be poured upon a thin glass plate, which is then strongly heated over a Bunsen burner. When somewhat cooled, the plate may be washed under the tap, rinsed with distilled water, and dried over the Bunsen without any kind of wiping. The parts wetted by the acid then behave much as the track of the blow-pipe flame in Aitken's experiment.

An even better treatment is with hydrofluoric acid, which actually renews the surface of the glass. A few drops of the commercial acid, diluted, say, ten times, may be employed, much as the sulphuric acid, only without heat. The parts so treated condense the breath in large laminæ, contrasting strongly with the ordinary deposit.

It must be admitted that some difficulties remain in attributing the behaviour of an ordinary plate to a superficial film of grease. One of these is the comparative permanence of breath figures, which often survive wiping with a cloth. The thought has sometimes occurred to me that the film of grease is not entirely superficial, but penetrates in some degree into the substance of the glass. In that case its removal and renewal would not be so easy. We know but little of the properties of matter in thin films, which may differ entirely from those of the same substance in mass. It may be recalled that a film of oil, one or two millionths of a millimetre thick, suffices to stop the movements of camphor on the surface of water, and that much smaller quantities may be rendered evident by optical and other methods.

354.

ON THE MOTION OF SOLID BODIES THROUGH VISCOUS LIQUID.

[*Philosophical Magazine*, Vol. XXI. pp. 697—711, 1911.]

§ 1. THE problem of the uniform and infinitely slow motion of a sphere, or cylinder, through an unlimited mass of incompressible viscous liquid otherwise at rest was fully treated by Stokes in his celebrated memoir on Pendulums*. The two cases mentioned stand in sharp contrast. In the first a relative steady motion of the fluid is easily determined, satisfying all the conditions both at the surface of the sphere and at infinity; and the force required to propel the sphere is found to be finite, being given by the formula (126)

$$- F = 6\pi\mu a V, \quad \dots\dots\dots\dots\dots\dots\dots\dots\dots(1)$$

where μ is the viscosity, a the radius, and V the velocity of the sphere. On the other hand in the case of the cylinder, moving transversely, no such steady motion is possible. If we suppose the cylinder originally at rest to be started and afterwards maintained in uniform motion, finite effects are propagated to ever greater and greater distances, and the motion of the fluid approaches no limit. Stokes shows that more and more of the fluid tends to accompany the travelling cylinder, which thus experiences a continually decreasing resistance.

§ 2. In attempting to go further, one of the first questions to suggest itself is whether similar conclusions are applicable to bodies of other forms. The consideration of this subject is often facilitated by use of the well-known analogy between the motion of a viscous fluid, when the square of the motion is neglected, and the displacements of an elastic solid. Suppose that in the latter case the solid is bounded by two closed surfaces, one of which completely envelopes the other. Whatever displacements (α, β, γ) be imposed at these two surfaces, there must be a corresponding configuration

* *Camb. Phil. Trans.* Vol. IX. 1850; *Math. and Phys. Papers*, Vol. III. p. 1

of equilibrium, satisfying certain differential equations. If the solid be incompressible, the otherwise arbitrary boundary displacements must be chosen subject to this condition. The same conclusion applies in two dimensions, where the bounding surfaces reduce to cylinders with parallel generating lines. For our present purpose we may suppose that at the outer surface the displacements are zero.

The contrast between the three-dimensional and two-dimensional cases arises when the outer surface is made to pass off to infinity. In the former case, where the inner surface is supposed to be limited in all directions, the displacements there imposed diminish, on receding from it, in such a manner that when the outer surface is removed to a sufficient distance no further sensible change occurs. In the two-dimensional case the inner surface extends to infinity, and the displacement affects sensibly points however distant, provided the outer surface be still further and sufficiently removed.

The nature of the distinction may be illustrated by a simple example relating to the conduction of heat through a uniform medium. If the temperature v be unity on the surface of the sphere $r = a$, and vanish when $r = b$, the steady state is expressed by

$$v = \frac{a}{b-a}\left(\frac{b}{r} - 1\right). \quad\dots\dots\dots\dots\dots\dots\dots(2)$$

When b is made infinite, v assumes the limiting form a/r. In the corresponding problem for coaxal cylinders of radii a and b we have

$$v = \frac{\log b - \log r}{\log b - \log a}. \quad\dots\dots\dots\dots\dots\dots\dots(3)$$

But here there is no limiting form when b is made infinite. However great r may be, v is small when b exceeds r by only a little; but when b is great enough v may acquire any value up to unity. And since the distinction depends upon what occurs at infinity, it may evidently be extended on the one side to oval surfaces of any shape, and on the other to cylinders with any form of cross-section.

In the analogy already referred to there is correspondence between the displacements (α, β, γ) in the first case and the velocities (u, v, w) which express the motion of the viscous liquid in the second. There is also another analogy which is sometimes useful when the motion of the viscous liquid takes place in two dimensions. The *stream-function* (ψ) for this motion satisfies the same differential equation as does the transverse displacement (w') of a plane elastic plate. And a surface on which the fluid remains at rest ($\psi = 0$, $d\psi/dn = 0$) corresponds to a curve along which the elastic plate is clamped.

In the light of these analogies we may conclude that, provided the square of the motion is neglected absolutely, there exists always a unique steady motion of liquid past a solid obstacle of any form limited in all directions, which satisfies the necessary conditions both at the surface of the obstacle and at infinity, and further that the force required to hold the solid is finite. But if the obstacle be an infinite cylinder of any cross-section, no such steady motion is possible, and the force required to hold the cylinder in position continually diminishes as the motion continues.

§ 3. For further developments the simplest case is that of a material plane, coinciding with the coordinate plane $x = 0$ and moving parallel to y in a fluid originally at rest. The component velocities u, w are then zero; and the third velocity v satisfies (even though its square be not neglected) the general equation

$$\frac{dv}{dt} = \nu \frac{d^2 v}{dx^2}, \quad\dots\dots\dots\dots\dots\dots\dots\dots(4)$$

in which ν, equal to μ/ρ, represents the kinematic viscosity. In § 7 of his memoir Stokes considers periodic oscillations of the plane. Thus in (4) if v be proportional to e^{int}, we have on the positive side

$$v = A e^{int} e^{-x\sqrt{(in/\nu)}}. \quad\dots\dots\dots\dots\dots\dots\dots\dots(5)$$

When $x = 0$, (5) must coincide with the velocity (V) of the plane. If this be $V_n e^{int}$, we have $A = V_n$; so that in real quantities

$$v = V_n e^{-x\sqrt{(n/2\nu)}} \cos \{nt - x\sqrt{(n/2\nu)}\} \quad\dots\dots\dots\dots(6)$$

corresponds with
$$V = V_n \cos nt \quad\dots\dots\dots\dots\dots\dots(7)$$

for the plane itself.

In order to find the tangential force $(- T_3)$ exercised upon the plane, we have from (5) when $x = 0$

$$\left(\frac{dv}{dx}\right)_0 = - V_n e^{int} \sqrt{(in/\nu)}, \quad\dots\dots\dots\dots\dots\dots(8)$$

and $\quad T_3 = - \mu (dv/dx)_0 = \rho V_n e^{int} \sqrt{(in\nu)}$

$$= \rho \sqrt{(\tfrac{1}{2}n\nu)} . (1 + i) V_n e^{int} = \rho \sqrt{(\tfrac{1}{2}n\nu)} . \left(V + \frac{1}{n}\frac{dV}{dt}\right), \quad\dots\dots(9)$$

giving the force per unit area due to the reaction of the fluid upon one side. "The force expressed by the first of these terms tends to diminish the amplitude of the oscillations of the plane. The force expressed by the second has the same effect as increasing the inertia of the plane." It will be observed that if V_n be given, the force diminishes without limit with n.

In note B Stokes resumes the problem of § 7: instead of the motion of the plane being periodic, he supposes that the plane and fluid are initially at rest, and that the plane is then $(t = 0)$ moved with a constant velocity V.

This problem depends upon one of Fourier's solutions which is easily verified*. We have

$$\frac{dv}{dx} = -\frac{V}{\sqrt{(\pi \nu t)}} e^{-x^2/4\nu t}, \quad \ldots\ldots\ldots\ldots\ldots\ldots(10)$$

$$v = V - \frac{2V}{\sqrt{\pi}} \int_0^{x/2\sqrt{(\nu t)}} e^{-z^2} dz. \quad \ldots\ldots\ldots\ldots\ldots(11)$$

For the reaction on the plane we require only the value of dv/dx when $x = 0$. And

$$\left(\frac{dv}{dx}\right)_0 = -\frac{V}{\sqrt{(\pi \nu t)}}. \quad \ldots\ldots\ldots\ldots\ldots\ldots(12)$$

Stokes continues† "now suppose the plane to be moved in any manner, so that its velocity at the end of the time t is $V(t)$. We may evidently obtain the result in this case by writing $V'(\tau) d\tau$ for V, and $t - \tau$ for t in [12], and integrating with respect to τ. We thus get

$$\left(\frac{dv}{dx}\right)_0 = -\frac{1}{\sqrt{(\pi \nu)}} \int_{-\infty}^t \frac{V'(\tau) d\tau}{\sqrt{(t - \tau)}} = -\frac{1}{\sqrt{(\pi \nu)}} \int_0^\infty V'(t - t_1) \frac{dt_1}{\sqrt{t_1}}; \quad \ldots(13)\text{"}$$

and since $T_3 = -\mu \, dv/dx_0$, these formulæ solve the problem of finding the reaction in the general case.

There is another method by which the present problem may be treated, and a comparison leads to a transformation which we shall find useful further on. Starting from the periodic solution (8), we may generalize it by Fourier's theorem. Thus

$$\left(\frac{dv}{dx}\right)_0 = -\int_0^\infty V_n e^{int} \sqrt{(in/\nu)} \, dn \quad \ldots\ldots\ldots\ldots\ldots(14)$$

corresponds to

$$V(t) = \int_0^\infty V_n e^{int} dn, \quad \ldots\ldots\ldots\ldots\ldots\ldots(15)$$

where V_n is an arbitrary function of n.

Comparing (13) and (14), we see that

$$\int_0^\infty V_n e^{int} n^{\frac{1}{2}} \, dn = \frac{1}{\sqrt{(i\pi)}} \int_{-\infty}^t \frac{V'(\tau) d\tau}{\sqrt{(t - \tau)}}. \quad \ldots\ldots\ldots\ldots(16)$$

It is easy to verify (16). If we substitute on the right for $V'(\tau)$ from (15), we get

$$\frac{1}{\sqrt{(i\pi)}} \int_{-\infty}^t \frac{d\tau}{\sqrt{(t - \tau)}} \int_0^\infty in V_n e^{in\tau} \, dn;$$

and taking first the integration with respect to τ,

$$\int_{-\infty}^t \frac{e^{in\tau} d\tau}{\sqrt{(t - \tau)}} = \int_0^\infty \frac{e^{in(t-t_1)}}{\sqrt{t_1}} dt_1 = \sqrt{\left(\frac{\pi}{in}\right)} \cdot e^{int},$$

when (16) follows at once.

---

* Compare Kelvin, *Ed. Trans.* 1862; Thomson and Tait, Appendix D.

† I have made some small changes of notation.

As a particular case of (13), let us suppose that the fluid is at rest and that the plane starts at $t = 0$ with a velocity which is uniformly accelerated for a time τ_1 and afterwards remains constant. Thus from $-\infty$ to 0, $V(\tau) = 0$; from 0 to τ_1, $V(\tau) = h\tau$; from τ_1 to t, where $t > \tau_1$, $V(\tau) = h\tau_1$. Thus $(0 < t < \tau_1)$

$$\left(\frac{dv}{dx}\right)_0 = -\frac{1}{\sqrt{(\pi\nu)}} \int_0^t \frac{h\, d\tau}{\sqrt{(t-\tau)}} = -\frac{2h\sqrt{t}}{\sqrt{(\pi\nu)}}; \quad \dots\dots\dots\dots(17)$$

and $(t > \tau_1)$

$$\left(\frac{dv}{dx}\right)_0 = -\frac{1}{\sqrt{(\pi\nu)}} \int_0^{\tau_1} \frac{h\, d\tau}{\sqrt{(t-\tau)}} = -\frac{2h}{\sqrt{(\pi\nu)}} \{\sqrt{t} - \sqrt{(t-\tau_1)}\}. \quad \dots(18)$$

Expressions (17), (18), taken negatively and multiplied by μ, give the force per unit area required to propel the plane against the fluid forces acting upon *one* side. The force increases until $t = \tau_1$, that is so long as the acceleration continues. Afterwards it gradually diminishes to zero. For the differential coefficient of $\sqrt{t} - \sqrt{(t-\tau_1)}$ is negative when $t > \tau_1$; and when t is great,

$$\sqrt{t} - \sqrt{(t-\tau_1)} = \tfrac{1}{2}\tau_1 t^{-\frac{1}{2}} \quad \text{ultimately.}$$

§ 4. In like manner we may treat any problem in which the motion of the material plane is prescribed. A more difficult question arises when it is the *forces* propelling the plane that are given. Suppose, for example, that an infinitely thin vertical lamina of superficial density σ begins to fall from rest under the action of gravity when $t = 0$, the fluid being also initially at rest. By (13) the equation of motion may be written

$$\frac{dV}{dt} + \frac{2\rho\nu^{\frac{1}{2}}}{\sigma\pi^{\frac{1}{2}}} \int_0^t \frac{V'(\tau)\, d\tau}{\sqrt{(t-\tau)}} = g, \quad \dots\dots\dots\dots\dots(19)$$

the fluid being now supposed to act on *both* sides of the lamina.

By an ingenious application of Abel's theorem Boggio has succeeded in integrating equations which include (19)*. The theorem is as follows:— If $\psi(t)$ be defined by

$$\psi(t) = \int_0^t \frac{\phi'(\tau)\, d\tau}{(t-\tau)^{\frac{1}{2}}}, \quad \dots\dots\dots\dots\dots\dots(20)$$

then

$$\int_0^t \frac{\psi(\tau)\, d\tau}{(t-\tau)^{\frac{1}{2}}} = \pi \{\phi(t) - \phi(0)\}. \quad \dots\dots\dots\dots\dots(21)$$

For by (20), if $(t-\tau)^{\frac{1}{2}} = y$,

$$\psi(t) = 2\int_0^{\sqrt{t}} \phi'(t-y^2)\, dy;$$

---

* Boggio, *Rend. d. Accad. d. Lincei*, Vol. XVI. pp. 613, 730 (1907); also Basset, *Quart. Journ. of Mathematics*, No. 164, 1910, from which I first became acquainted with Boggio's work.

so that

$$\int_0^t \frac{\psi(\tau)\, d\tau}{(t-\tau)^{\frac{1}{2}}} = 2\int_0^{\sqrt{t}} \psi(t-x^2)\, dx = 4\int_0^{\sqrt{t}} dx \int_0^{\sqrt{(t-x^2)}} \phi'(t-x^2-y^2)\, dy$$

$$= 2\pi \int_0^{\sqrt{t}} \phi'(t-r^2)\, r\, dr = \pi \{\phi(t) - \phi(0)\},$$

where $r^2 = x^2 + y^2$.

Now, if t' be any time between 0 and t, we have, as in (19),

$$V'(t') + \frac{2\rho \nu^{\frac{1}{2}}}{\sigma \pi^{\frac{1}{2}}} \int_0^{t'} \frac{V'(\tau)\, d\tau}{\sqrt{(t'-\tau)}} = g.$$

Multiplying this by $(t-t')^{-\frac{1}{2}}\, dt'$ and integrating between 0 and t, we get

$$\int_0^t \frac{V'(t')\, dt'}{(t-t')^{\frac{1}{2}}} + \frac{2\rho \nu^{\frac{1}{2}}}{\sigma \pi^{\frac{1}{2}}} \int_0^t \frac{dt'}{(t-t')^{\frac{1}{2}}} \int_0^{t'} \frac{V'(\tau)\, d\tau}{(t'-\tau)^{\frac{1}{2}}} = g \int_0^t \frac{dt'}{(t-t')^{\frac{1}{2}}}. \quad \ldots(22)$$

In (22) the first integral is the same as the integral in (19). By Abel's theorem the double integral in (22) is equal to $\pi V(t)$, since $V(0) = 0$. Thus

$$\int_0^t \frac{V'(\tau)\, d\tau}{\sqrt{(t-\tau)}} + \frac{2\rho \nu^{\frac{1}{2}} \pi^{\frac{1}{2}}}{\sigma} V(t) = 2g\sqrt{t}. \quad \ldots\ldots\ldots\ldots(23)$$

If we now eliminate the integral between (19) and (23), we obtain simply

$$\frac{dV}{dt} - \frac{4\rho^2 \nu}{\sigma^2} V = g - \frac{4\rho \nu^{\frac{1}{2}}}{\sigma \pi^{\frac{1}{2}}} g\sqrt{t} \quad \ldots\ldots\ldots\ldots(24)$$

as the differential equation governing the motion of the lamina.

This is a linear equation of the first order. Since V vanishes with t, the integral may be written

$$\frac{4\rho^2 \nu V}{g\sigma^2} = e^{t'} \int_0^{t'} e^{-t} \left(1 - \frac{2\sqrt{t}}{\sqrt{\pi}}\right) dt$$

$$= \frac{2\sqrt{t'}}{\sqrt{\pi}} - 1 + \frac{2}{\sqrt{\pi}} e^{t'} \int_{\sqrt{t'}}^{\infty} e^{-x^2}\, dx, \quad \ldots\ldots\ldots\ldots(25)$$

in which $t' = t \cdot 4\rho^2 \nu / \sigma^2$. When t, or t', is great,

$$\int_{\sqrt{t'}}^{\infty} e^{-x^2}\, dx = \frac{e^{-t'}}{2\sqrt{t'}} \left(1 - \frac{1}{2t'} + \ldots\right); \quad \ldots\ldots\ldots\ldots(26)$$

so that

$$\frac{4\rho^2 \nu V}{g\sigma^2} = \frac{2\sqrt{t'}}{\sqrt{\pi}} - 1 + \frac{1}{\sqrt{(\pi t')}} \left(1 - \frac{1}{2t'} + \ldots\right). \quad \ldots\ldots\ldots(27)$$

Ultimately, when t is very great,

$$V = \frac{g\sigma}{\rho} \sqrt{\left(\frac{t}{\pi \nu}\right)}. \quad \ldots\ldots\ldots\ldots\ldots\ldots(28)$$

§ 5. The problem of the sphere moving with arbitrary velocity through a viscous fluid is of course more difficult than the corresponding problem of the plane lamina, but it has been satisfactorily solved by Boussinesq[*] and by Basset [†]. The easiest road to the result is by the application of Fourier's theorem to the periodic solution investigated by Stokes. If the velocity of the sphere at time t be $V = V_n e^{int}$, a the radius, M' the mass of the liquid displaced by the sphere, and $s = \sqrt{(n/2\nu)}$, ν being as before the kinematic viscosity, Stokes finds as the total force at time t

$$F = - M' V_n n \left\{ \left(\frac{1}{2} + \frac{9}{4sa} \right) i + \frac{9}{4sa} \left(1 + \frac{1}{sa} \right) \right\} e^{int} . \quad \dots\dots\dots(29)$$

Thus, if
$$V = \int_0^\infty V_n e^{int} \, dn, \quad \dots\dots\dots\dots\dots(30)$$

$$F = - M' \int_0^\infty V_n n e^{int} \left\{ \left(\frac{1}{2} + \frac{9}{4sa} \right) i + \frac{9}{4sa} \left(1 + \frac{1}{sa} \right) \right\} dn. \quad \dots\dots(31)$$

Of the four integrals in (31),

$$\text{the first} \quad = \tfrac{1}{2} \int_0^\infty in \, V_n e^{int} \, dn = \tfrac{1}{2} V' ;$$

$$\text{the fourth} = \frac{9\nu}{2a^2} \int_0^\infty V_n e^{int} \, dn \; = \; \frac{9\nu}{2a^2} V.$$

Also the second and third together give

$$\frac{9 (1 + i) \sqrt{(2\nu)}}{4a} \int_0^\infty V_n n^{\frac{1}{2}} e^{int} \, dn,$$

and this is the only part which could present any difficulty. We have, however, already considered this integral in connexion with the motion of a plane and its value is expressed by (16). Thus

$$F = - M' \left\{ \frac{1}{2} \frac{dV}{dt} + \frac{9\nu}{2a^2} V + \frac{9\nu^{\frac{1}{2}}}{2a\pi^{\frac{1}{2}}} \int_{-\infty}^t \frac{V'(\tau) \, d\tau}{(t - \tau)^{\frac{1}{2}}} \right\} \quad \dots\dots\dots(32)$$

The first term depends upon the inertia of the fluid, and is the same as would be obtained by ordinary hydrodynamics when $\nu = 0$. If there is no acceleration at the moment, this term vanishes. If, further, there has been no acceleration for a long time, the third term also vanishes, and we obtain the result appropriate to a uniform motion

$$F = - \frac{9\nu M' V}{2a^2} = - 6\pi a \rho \nu V = - 6\pi \mu a V,$$

as in (1). The general result (32) is that of Boussinesq and Basset.

* C. R. t. c. p. 935 (1885); Théorie Analytique de la Chaleur, t. II. Paris, 1903.
† Phil. Trans. 1888; Hydrodynamics, Vol. II. chap. XXII. 1888.

As an example of (32), we may suppose (as formerly for the plane) that $V(t) = 0$ from $-\infty$ to 0; $V(t) = ht$ from 0 to τ_1; $V(t) = h\tau_1$, when $t > \tau_1$. Then if $t < \tau_1$,

$$F = -hM' \left[\frac{1}{2} + \frac{9\nu t}{2a^2} + \frac{9\nu^{\frac{1}{2}} t^{\frac{1}{2}}}{a\pi^{\frac{1}{2}}} \right] ; \qquad \qquad (33)$$

and when $t > \tau_1$,

$$F = -hM' \left[\frac{9\nu\tau_1}{2a^2} + \frac{9\nu^{\frac{1}{2}}}{a\pi^{\frac{1}{2}}} \{ \sqrt{t} - \sqrt{(t-\tau_1)} \} \right] . \qquad \ldots \ldots (34)$$

When t is very great (34) reduces to its first term.

The more difficult problem of a sphere falling under the influence of gravity has been solved by Boggio (*loc. cit.*). In the case where the liquid and sphere are initially at rest, the solution is comparatively simple; but the analytical form of the functions is found to depend upon the ratio of densities of the sphere and liquid. This may be rather unexpected; but I am unable to follow Mr Basset in regarding it as an objection to the usual approximate equations of viscous motion.

§ 6. We will now endeavour to apply a similar method to Stokes' solution for a *cylinder* oscillating transversely in a viscous fluid. If the radius be a and the velocity V be expressed by $V = V_n e^{int}$, Stokes finds for the force

$$F = - M' in V_n e^{int} (k - ik'). \qquad \ldots \ldots \ldots (35)$$

In (35) M' is the mass of the fluid displaced; k and k' are certain functions of m, where $m = \frac{1}{2} a \sqrt{(n/\nu)}$, which are tabulated in his § 37. The cylinder is much less amenable to mathematical treatment than the sphere, and we shall limit ourselves to the case where, all being initially at rest, the cylinder is started with unit velocity which is afterwards steadily maintained.

The velocity V of the cylinder, which is to be zero when t is negative and unity when t is positive, may be expressed by

$$V = \frac{1}{2} + \frac{1}{\pi} \int_0^\infty \frac{\sin nt}{n} \, dn, \qquad \ldots \ldots \ldots (36)$$

in which the second term may be regarded as the real part of

$$\frac{1}{\pi i} \int_0^\infty \frac{e^{int}}{n} \, dn. \qquad \ldots \ldots \ldots (37)$$

We shall see further below, and may anticipate from Stokes' result relating to uniform motion of the cylinder, that the first term of (36) contributes nothing to F; so that we may take

$$F = - \frac{M'}{\pi} \int_0^\infty e^{int} (k - ik') \, dn,$$

corresponding to (37). Discarding the imaginary part, we get, corresponding to (36),

$$F = -\frac{M'}{\pi} \int_0^\infty (k \cos nt + k' \sin nt)\, dn. \qquad \text{...............(38)}$$

Since k, k' are known functions of m, or (a and ν being given) of n, (38) may be calculated by quadratures for any prescribed value of t.

It appears from the tables that k, k' are positive throughout. When $m = 0$, k and k' are infinite and continually diminish as m increases, until when $m = \infty$, $k = 1$, $k' = 0$. For small values of m the limiting forms for k, k' are

$$k = 1 + \frac{\frac{1}{4}\pi}{m^2 (\log m)^2}, \qquad k' = -\frac{1}{m^2 \log m}; \qquad \text{............(39)}$$

from which it appears that if we make n vanish in (35), while V_n is given, F comes to zero.

We now seek the limiting form when t is very great. The integrand in (38) is then rapidly oscillatory, and ultimately the integral comes to depend sensibly upon that part of the range where n is very small. And for this part we may use the approximate forms (39).

Consider, for example, the first integral in (38), from which we may omit the constant part of k. We have

$$\int_0^\infty k \cos nt\, dn = \frac{\pi}{4} \int_0^\infty \frac{\cos nt\, dn}{m^2 (\log m)^2} = \frac{4\pi\nu}{a^2} \int_0^\infty \frac{\cos(4\nu a^{-2} t \cdot x)\, dx}{x (\log x)^2} \quad \text{....(40)}$$

Writing $4\nu t/a^2 = t'$, we have to consider

$$\int_0^\infty \frac{\cos t' x \cdot dx}{x (\log x)^2}. \qquad \text{.............................(41)}$$

In this integral the integrand is positive from $x = 0$ to $x = \pi/2t'$, negative from $\pi/2t'$ to $3\pi/2t'$, and so on. For the first part of the range, if we omit the cosine,

$$\int_0^{\pi/2t'} \frac{dx}{x (\log x)^2} = \int \frac{d \log x}{(\log x)^2} = \frac{1}{\log(2t'/\pi)}; \qquad \text{..............(42)}$$

and since the cosine is less than unity, this is an over estimate. When t' is very great, $\log(2t'/\pi)$ may be identified with $\log t'$, and to this order of approximation it appears that (41) may be represented by (42). Thus if quadratures be applied to (41), dividing the first quadrant into three parts, we have

$$\frac{\cos \pi/12}{\log 6t'/\pi} + \cos \frac{3\pi}{12}\left[\frac{1}{\log 3t'/\pi} - \frac{1}{\log 6t'/\pi}\right] + \cos \frac{5\pi}{12}\left[\frac{1}{\log 2t'/\pi} - \frac{1}{\log 3t'/\pi}\right],$$

of which the second and third terms may ultimately be neglected in comparison with the first. For example, the coefficient of $\cos(3\pi/12)$ is equal to

$$\log 2 \div \log \frac{3t'}{\pi} \cdot \log \frac{6t'}{\pi}.$$

Proceeding in this way we see that the cosine factor may properly be identified with unity, and that the value of the integral for the first quadrant may be equated to $1/\log t'$. And for a similar reason the quadrants after the first contribute nothing of this order of magnitude. Accordingly we may take

$$\int_0^\infty k \cos nt \, dn = \frac{4\pi\nu}{a^2 \log t'}. \quad\ldots\ldots\ldots\ldots\ldots\ldots(43)$$

For the other part of (38), we get in like manner

$$\int_0^\infty k' \sin nt \, dn = -\frac{8\nu}{a^2}\int_0^\infty \frac{\sin t'x \,.\, dx}{x \log x} = \frac{8\nu}{a^2}\int_0^\infty \frac{\sin x' \, dx'}{x' \log (t'/x')}. \quad\ldots\ldots(44)$$

In the denominator of (44) it appears that ultimately we may replace $\log (t'/x')$ by $\log t'$ simply. Thus

$$\int_0^\infty k' \sin nt \, dn = \frac{4\pi\nu}{a^2 \log t'}, \quad\ldots\ldots\ldots\ldots\ldots\ldots(45)$$

so that the two integrals (43), (45) are equal. We conclude that when t is great enough,

$$F = -\frac{8\nu M'}{a^2 \log t'} = -\frac{8\nu M'}{a^2 \log (4\nu t/a^2)}. \quad\ldots\ldots\ldots\ldots(46)$$

But a better discussion of these integrals is certainly a desideratum.

§ 7.   Whatever interest the solution of the approximate equations may possess, we must never forget that the conditions under which they are applicable are very restricted, and as far as possible from being observed in many practical problems. Dynamical similarity in viscous motion requires that Va/ν be unchanged, a being the linear dimension. Thus the general form for the resistance to the uniform motion of a sphere will be

$$F = \rho\nu Va \,.\, f(Va/\nu), \quad\ldots\ldots\ldots\ldots\ldots\ldots(47)$$

where f is an unknown function. In Stokes' solution (1) f is constant, and its validity requires that Va/ν be small[*]. When V is rather large, experiment shows that F is nearly proportional to V^2. In this case ν disappears. "The second power of the velocity and independence of viscosity are thus inseparably connected"[†].

The general investigation for the sphere moving in any manner (in a straight line) shows that the departure from Stokes' law when the velocity is not very small must be due to the operation of the neglected terms involving the squares of the velocities; but the manner in which these act has not yet been traced. Observation shows that an essential feature in rapid fluid motion past an obstacle is the formation of a *wake* in the rear of the obstacle; but of this the solutions of the approximate equations give no hint.

[*] *Phil. Mag.* Vol. xxxvi. p. 354 (1893); *Scientific Papers*, Vol. iv. p. 87.

[†] *Phil. Mag.* Vol. xxxiv. p. 59 (1892); *Scientific Papers*, Vol. iii. p. 576.

Hydrodynamical solutions involving surfaces of discontinuity of the kind investigated by Helmholtz and Kirchhoff provide indeed for a wake, but here again there are difficulties. Behind a blade immersed transversely in a stream a region of "dead water" is indicated. The conditions of steady motion are thus satisfied; but, as Helmholtz himself pointed out, the motion thus defined is unstable. Practically the dead and live water are continually mixing; and if there be viscosity, the layer of transition rapidly assumes a finite width independently of the instability. One important consequence is the development of a suction on the hind surface of the lamina which contributes in no insignificant degree to the total resistance. The amount of the suction does not appear to depend much on the degree of viscosity. When the latter is small, the dragging action of the live upon the dead water extends to a greater distance behind.

§ 8. If the blade, supposed infinitely thin, be moved edgeways through the fluid, the case becomes one of "skin-friction." Towards determining the law of resistance Mr Lanchester has put forward an argument * which, even if not rigorous, at any rate throws an interesting light upon the question. Applied to the case of two dimensions in order to find the resistance F per unit length of blade, it is somewhat as follows. Considering two systems for which the velocity V of the blade is different, let n be the proportional width of corresponding strata of velocity. The momentum communicated to the wake per unit length of travel is as nV, and therefore on the whole as nV^2 per unit of time. Thus F varies as nV^2. Again, having regard to the law of viscosity and considering the strata contiguous to the blade, we see that F varies as V/n. Hence, nV^2 varies as V/n, or V varies as n^{-2}, from which it follows that F varies as $V^{3/2}$. If this be admitted, the general law of dynamical similarity requires that for the whole resistance

$$F = c\rho\nu^{\frac{1}{2}} l b^{\frac{1}{2}} V^{\frac{3}{2}}, \quad\dots\dots\dots\dots\dots\dots\dots\dots(48)$$

where l is the length, b the width of the blade, and c a constant. Mr Lanchester gives this in the form

$$F/\rho = c\nu^{\frac{1}{2}} A^{\frac{3}{4}} V^{\frac{3}{2}}, \quad\dots\dots\dots\dots\dots\dots\dots\dots(49)$$

where A is the area of the lamina, agreeing with (48) if l and b maintain a constant ratio.

The difficulty in the way of accepting the above argument as rigorous is that complete similarity cannot be secured so long as b is constant as has been supposed. If, as is necessary to this end, we take b proportional to n, it is bV/n, or V (and not V/n), which varies as nV^2, or bV^2. The conclusion is then simply that bV must be constant (ν being given). This is merely the usual condition of dynamical similarity, and no conclusion as to the law of velocity follows.

* *Aerodynamics*, London, 1907, § 35.

But a closer consideration will show, I think, that there is a substantial foundation for the idea at the basis of Lanchester's argument. If we suppose that the viscosity is so small that the layer of fluid affected by the passage of the blade is very small compared with the width (b) of the latter, it will appear that the communication of motion at any stage takes place much as if the blade formed part of an infinite plane moving as a whole. We know that if such a plane starts from rest with a velocity V afterwards uniformly maintained, the force acting upon it at time t is per unit of area, see (12),

$$\rho V \sqrt{(\nu/\pi t)}. \dots\dots\dots\dots\dots\dots\dots\dots\dots(50)$$

The supposition now to be made is that we may apply this formula to the element of width dy, taking t equal to y/V, where y is the distance of the element from the leading edge.   Thus

$$F = l\rho \, (\nu/\pi)^{\frac{1}{2}} \, V^{\frac{3}{2}} \int y^{-\frac{1}{2}} \, dy = 2l\rho \, (\nu/\pi)^{\frac{1}{2}} \, V^{\frac{3}{2}} b^{\frac{1}{2}}, \dots\dots\dots(51)$$

which agrees with (48) if we take in the latter $c = 2/\sqrt{\pi}$.

The formula (51) would seem to be justified when ν is small enough, as representing a possible state of things ; and, as will be seen, it affords an absolutely definite value for the resistance.   There is no difficulty in extending it under similar restrictions to a lamina of any shape.   If b, no longer constant, is the width of the lamina in the direction of motion at level z, we have

$$F = 2\rho \, (\nu/\pi)^{\frac{1}{2}} \, V^{\frac{3}{2}} \int b^{\frac{1}{2}} \, dz. \dots\dots\dots\dots\dots\dots(52)$$

It will be seen that the result is not expressible in terms of the *area* of the lamina.   In (49) c is not constant, unless the lamina remains always similar in shape.

The fundamental condition as to the smallness of ν would seem to be realized in numerous practical cases; but any one who has looked over the side of a steamer will know that the motion is not usually of the kind supposed in the theory.   It would appear that the theoretical motion is subject to instabilities which prevent the motion from maintaining its simply stratified character.   The resistance is then doubtless more nearly as the square of the velocity and independent of the value of ν.

When in the case of bodies moving through air or water we express V, a, and ν in a consistent system of units, we find that in all ordinary cases ν/Va is so very small a quantity that it is reasonable to identify $f(\nu/Va)$ with $f(0)$.   The influence of linear scale upon the character of the motion then disappears.   This seems to be the explanation of a difficulty raised by Mr Lanchester (*loc. cit.* § 56).

355.

ABERRATION IN A DISPERSIVE MEDIUM.

[*Philosophical Magazine*, Vol. XXII. pp. 130—134, 1911.]

THE application of the theory of group-velocity to the case of light was discussed in an early paper* in connexion with some experimental results announced by Young and Forbes†. It is now, I believe, generally agreed that, whether the method be that of the toothed wheel or of the revolving mirror, what is determined by the experiment is not V, the wave-velocity, but U, the group-velocity, where

$$U = d\,(kV)/dk,$$

k being inversely as the wave-length. In a dispersive medium V and U are different.

I proceeded:—"The evidence of the terrestrial methods relating exclusively to U, we turn to consider the astronomical methods. Of these there are two, depending respectively upon aberration and upon the eclipses of Jupiter's satellites. The latter evidently gives U. The former does not depend upon observing the propagation of a peculiarity impressed upon a train of waves, and therefore has no relation to U. If we accept the usual theory of aberration as satisfactory, the result of a comparison between the coefficient found by observation and the solar parallax is V—the wave-velocity."

The above assertion that stellar aberration gives V rather than U has recently been called in question by Ehrenfest‡, and with good reason. He shows that the circumstances do not differ materially from those of the toothed wheel in Fizeau's method. The argument that he employs bears, indeed, close affinity with the method used by me in a later paper§. "The

---

* *Nature*, Vols. XXIV., XXV. 1881; *Scientific Papers*, Vol. I. p. 537.

† These observers concluded that blue light travels *in vacuo* 1·8 per cent. faster than red light.

‡ *Ann. d. Physik*, Bd. XXXIII. p. 1571 (1910).

§ *Nature*, Vol. XLV. p. 499 (1892); *Scientific Papers*, Vol. III. p. 542.

explanation of stellar aberration, as usually given, proceeds rather upon the basis of the corpuscular than of the wave-theory. In order to adapt it to the principles of the latter theory, Fresnel found it necessary to follow Young in assuming that the æther in any vacuous space connected with the earth (and therefore practically in the atmosphere) is undisturbed by the earth's motion of 19 miles per second. Consider, for simplicity, the case in which the direction of the star is at right angles to that of the earth's motion, and replace the telescope, which would be used in practice, by a pair of perforated screens, on which the light falls perpendicularly. We may further imagine the luminous disturbance to consist of a single plane pulse. When this reaches the anterior screen, so much of it as coincides with the momentary position of the aperture is transmitted, and the remainder is stopped. The part transmitted proceeds upon its course through the æther independently of the motion of the screens. In order, therefore, that the pulse may be transmitted by the aperture in the posterior screen, it is evident that the line joining the centres of the apertures must not be perpendicular to the screens and to the wave-front, as would be necessary in the case of rest. For, in consequence of the motion of the posterior screen in its own plane, the aperture will be carried forward during the time of passage of the light. By the amount of this motion the second aperture must be drawn backwards, in order that it may be in the place required when the light reaches it. If the velocity of light be V, and that of the earth be v, the line of apertures giving the apparent direction of the star must be directed forwards through an angle equal to v/V."

If the medium between the screens is dispersive, the question arises in what sense the velocity of light is to be taken. Evidently in the sense of the group-velocity; so that, in the previous notation, the aberration angle is v/U. But to make the argument completely satisfactory, it is necessary in this case to abandon the extreme supposition of a single pulse, replacing it by a group of waves of approximately given wave-length.

While there can remain no doubt but that Ehrenfest is justified in his criticism, it does not quite appear from the above how my original argument is met. There is indeed a peculiarity imposed upon the regular wave-motion constituting homogeneous light, but it would seem to be one imposed for the purposes of the argument rather than inherent in the nature of the case. The following analytical solution, though it does not relate directly to the case of a simply perforated screen, throws some light upon this question.

Let us suppose that homogeneous plane waves are incident upon a "screen" at $z = 0$, and that the effect of the screen is to introduce a reduction of the amplitude of vibration in a ratio which is slowly periodic both with respect to the time and to a coordinate x measured in the plane of the screen, represented by the factor $\cos m (vt - x)$. Thus, when $t = 0$, there is no effect when $x = 0$, or a multiple of 2π; but when x is an odd multiple of π, there is a reversal of sign, equivalent to a change of phase of half a period. And the places where these particular effects occur travel along the screen with a velocity v which is supposed to be small relatively to that of light. In the absence of the screen the luminous vibration is represented by

$$\phi = \cos(nt - kz), \quad \ldots\ldots\ldots\ldots\ldots\ldots\ldots\ldots(1)$$

or at the place of the screen, where $z = 0$, by

$$\phi = \cos nt \quad \text{simply.}$$

In accordance with the suppositions already made, the vibration just behind the screen will be

$$\phi = \cos m(vt - x) . \cos nt$$
$$= \tfrac{1}{2}\cos\{(n + mv)t - mx\} + \tfrac{1}{2}\cos\{(n - mv)t + mx\}; \quad \ldots\ldots(2)$$

and the question is to find what form ϕ will take at a finite distance z behind the screen.

It is not difficult to see that for this purpose we have only to introduce terms proportional to z into the arguments of the cosines. Thus, if we write

$$\phi = \tfrac{1}{2}\cos\{(n + mv)t - mx - \mu_1 z\} + \tfrac{1}{2}\cos\{(n - mv)t + mx - \mu_2 z\}, \quad \ldots(3)$$

we may determine μ_1, μ_2 so as to satisfy in each case the general differential equation of propagation, viz.

$$\frac{d^2\phi}{dt^2} = V^2\left(\frac{d^2\phi}{dx^2} + \frac{d^2\phi}{dz^2}\right). \quad \ldots\ldots\ldots\ldots\ldots\ldots\ldots(4)$$

In (4) V is constant when the medium is non-dispersive; but in the contrary case V must be given different values, say V_1 and V_2, when the coefficient of t is $n + mv$ or $n - mv$. Thus

$$(n + mv)^2 = V_1^2(m^2 + m_1^2), \qquad (n - mv)^2 = V_2^2(m^2 + m_2^2). \quad \ldots\ldots(5)$$

The coefficients μ_1, μ_2 being determined in accordance with (5), the value of ϕ in (3) satisfies all the requirements of the problem. It may also be written

$$\phi = \cos\{mvt - mx - \tfrac{1}{2}(\mu_1 - \mu_2)z\} . \cos\{nt - \tfrac{1}{2}(\mu_1 + \mu_2)z\}, \quad \ldots\ldots(6)$$

of which the first factor, varying slowly with t, may be regarded as the amplitude of the luminous vibration.

The condition of constant amplitude at a given time is that $mx + \tfrac{1}{2}(\mu_1 - \mu_2)z$ shall remain unchanged. Thus the amplitude which is to be found at $x = 0$ on the screen prevails also behind the screen along the line

$$-x/z = \tfrac{1}{2}(\mu_1 - \mu_2)/m, \quad \ldots\ldots\ldots\ldots\ldots\ldots\ldots(7)$$

so that (7) may be regarded as the angle of aberration due to v. It remains to express this angle by means of (5) in terms of the fundamental data.

When m is zero, the value of μ is n/V; and this is true approximately when m is small. Thus, from (5),

$$\mu_1 - \mu_2 = \frac{\mu_1^2 - \mu_2^2}{2n/V} = \frac{2mv}{V} + \frac{nV}{2}\left(\frac{1}{V_1^2} - \frac{1}{V_2^2}\right)$$

and

$$\frac{\mu_1 - \mu_2}{2m} = \frac{v}{V}\left\{1 + \frac{n}{2mv}\frac{V_2 - V_1}{V}\right\} \quad\dots\dots\dots\dots\dots\dots(8)$$

with sufficient approximation.

Now in (8) the difference $V_2 - V_1$ corresponds to a change in the coefficient of t from $n + mv$ to $n - mv$. Hence, denoting the general coefficient of t by σ, of which V is a function, we have

$$V_1 - V_2 = 2mv \cdot dV/d\sigma;$$

and (8) may be written

$$\frac{\mu_1 - \mu_2}{2m} = \frac{v}{V}\left\{1 - \frac{\sigma}{V}\frac{dV}{d\sigma}\right\} \cdot \quad\dots\dots\dots\dots\dots\dots(9)$$

Again,

$$V = \sigma/k, \qquad U = d\sigma/dk,$$

and thus

$$\frac{\sigma}{V}\frac{dV}{d\sigma} = k\frac{dV}{d\sigma} = 1 - \frac{\sigma}{k}\frac{dk}{d\sigma},$$

and

$$1 - \frac{\sigma}{V}\frac{dV}{d\sigma} = \frac{\sigma}{k}\frac{dk}{d\sigma} = \frac{V}{U},$$

where U is the group-velocity.

Accordingly,

$$- x/z = v/U \quad\dots\dots\dots\dots\dots\dots\dots(10)$$

expresses the aberration angle, as was to be expected. In the present problem the peculiarity impressed is not uniform over the wave-front, as may be supposed in discussing the effect of the toothed wheel; but it exists nevertheless, and it involves for its expression the introduction of more than one frequency, from which circumstance the group-velocity takes its origin.

A development of the present method would probably permit the solution of the problem of a series of equidistant moving apertures, or a single moving aperture. Doubtless in all cases the aberration angle would assume the value v/U.

356.

LETTER TO PROFESSOR NERNST.

[Conseil scientifique sous les auspices de M. Ernest Solvay, Oct. 1911.]

DEAR PROF. NERNST,

Having been honoured with an invitation to attend the Conference at Brussels, I feel that the least that I can do is to communicate my views, though I am afraid I can add but little to what has been already said upon the subject.

I wish to emphasize the difficulty mentioned in my paper of 1900* with respect to the use of generalized coordinates. The possibility of representing the state of a body by a finite number of such (short at any rate of the whole number of molecules) depends upon the assumption that a body may be treated as rigid, or incompressible, or in some other way simplified. The justification, and in many cases the sufficient justification, is that a departure from the simplified condition would involve such large amounts of potential energy as could not occur under the operation of the forces concerned. But the law of equi-partition lays it down that every mode is to have its share of kinetic energy. If we begin by supposing an elastic body to be rather stiff, the vibrations have their full share and this share cannot be diminished by increasing the stiffness. *For this purpose* the simplification fails, which is as much as to say that the method of generalized coordinates cannot be applied. The argument becomes, in fact, self-contradictory.

Perhaps this failure might be invoked in support of the views of Planck and his school that the laws of dynamics (as hitherto understood) cannot be applied to the smallest parts of bodies. But I must confess that I do not like this solution of the puzzle. Of course I have nothing to say against following out the consequences of the [quantum] theory of energy—a procedure which has already in the hands of able men led to some interesting

* *Phil. Mag.* Vol. XLIX. p. 118; *Scientific Papers*, Vol. IV. p. 451.

conclusions. But I have a difficulty in accepting it as a picture of what actually takes place.

We do well, I think, to concentrate attention upon the diatomic gaseous molecule. Under the influence of collisions the molecule freely and rapidly acquires rotation. Why does it not also acquire vibration along the line joining the two atoms? If I rightly understand, the answer of Planck is that in consideration of the stiffness of the union the amount of energy that should be acquired at each collision falls below the minimum possible and that therefore none at all is acquired—an argument which certainly sounds paradoxical. On the other hand Boltzmann and Jeans contend that it is all a question of time and that the vibrations necessary for full statistical equilibrium may be obtained only after thousands of years. The calculations of Jeans appear to show that there is nothing forced in such a view. I should like to inquire is there any definite experimental evidence against it? So far as I know, ordinary laboratory experience affords nothing decisive.

I am yours truly,

RAYLEIGH.

357.

ON THE CALCULATION OF CHLADNI'S FIGURES FOR A SQUARE PLATE.

[*Philosophical Magazine*, Vol. XXII. pp. 225—229, 1911.]

In my book on the *Theory of Sound*, ch. x. (1st ed. 1877, 2nd ed. 1894) I had to speak of the problem of the vibrations of a rectangular plate, whose edges are free, as being one of great difficulty, which had for the most part resisted attack. An exception could be made of the case in which μ (the ratio of lateral contraction to longitudinal elongation) might be regarded as evanescent. It was shown that a rectangular plate could then vibrate after the same law as obtains for a simple bar, and by superposition some of the simpler Chladni's figures for a square plate were deduced. For glass and metal the value of μ is about $\frac{1}{4}$, so that for such plates as are usually experimented on the results could be considered only as rather rough approximations.

I wish to call attention to a remarkable memoir by W. Ritz[*] in which, somewhat on the above lines, is developed with great skill what may be regarded as a practically complete solution of the problem of Chladni's figures on square plates. It is shown that to within a few per cent. all the proper tones of the plate may be expressed by the formulæ

$$w_{mn} = u_m(x)\, u_n(y) + u_m(y)\, u_n(x),$$

$$w'_{mn} = u_m(x)\, u_n(y) - u_m(y)\, u_n(x),$$

the functions u being those proper to a free bar vibrating transversely. The coordinate axes are drawn through the centre parallel to the sides of the square. The first function of the series $u_0(x)$ is constant; the second $u_1(x) = x$. const.; $u_2(x)$ is thus the fundamental vibration in the usual sense, with two nodes, and so on. Ritz rather implies that I had overlooked the

---

[*] "Theorie der Transversalschwingungen einer quadratischen Platte mit freien Rändern," *Annalen der Physik*, Bd. xxviii. S. 737 (1909). The early death of the talented author must be accounted a severe loss to Mathematical Physics.

necessity of the first two terms in the expression of an arbitrary function. It would have been better to have mentioned them explicitly; but I do not think any reader of my book could have been misled. In § 168 the inclusion of *all** particular solutions is postulated, and in § 175 a reference is made to zero values of the frequency.

For the gravest tone of a square plate the coordinate axes are nodal, and Ritz finds as the result of successive approximations

$$w = u_1 v_1 + \cdot 0394 \, (u_1 v_3 + v_1 u_3)$$
$$- \cdot 0040 \, u_3 v_3 - \cdot 0034 \, (u_1 v_5 + u_5 v_1)$$
$$+ \cdot 0011 \, (u_3 v_5 + u_5 v_3) - \cdot 0019 \, u_5 v_5 \, ;$$

in which u stands for $u(x)$ and v for $u(y)$. The leading term $u_1 v_1$, or xy, is the same as that which I had used (§ 228) as a rough approximation on which to found a calculation of pitch.

As has been said, the general method of approximation is very skilfully applied, but I am surprised that Ritz should have regarded the method itself as new. An integral involving an unknown arbitrary function is to be made a minimum. The unknown function can be represented by a series of known functions with arbitrary coefficients—accurately if the series be continued to infinity, and approximately by a few terms. When the number of coefficients, also called generalized coordinates, is finite, they are of course to be determined by ordinary methods so as to make the integral a minimum. It was in this way that I found the correction for the open end of an organ-pipe[†], using a series with two terms to express the velocity at the mouth. The calculation was further elaborated in *Theory of Sound*, Vol. II. Appendix A. I had supposed that this treatise abounded in applications of the method in question, see §§ 88, 89, 90, 91, 182, 209, 210, 265; but perhaps the most explicit formulation of it is in a more recent paper[‡], where it takes almost exactly the shape employed by Ritz. From the title it will be seen that I hardly expected the method to be so successful as Ritz made it in the case of higher modes of vibration.

Being upon the subject I will take the opportunity of showing how the gravest mode of a square plate may be treated precisely upon the lines of the paper referred to. The potential energy of bending per unit area has the expression

$$V = \frac{qh^3}{3(1-\mu^2)} \left[(\nabla^2 w)^2 + 2(1-\mu) \left\{ \left(\frac{d^2 w}{dx\,dy}\right)^2 - \frac{d^2 w}{dx^2}\frac{d^2 w}{dy^2} \right\} \right], \quad \ldots\ldots(1)$$

* Italics in original.

† *Phil. Trans.* Vol. CLXI. (1870); *Scientific Papers*, Vol. I. p. 57.

‡ "On the Calculation of the Frequency of Vibration of a System in its Gravest Mode, with an Example from Hydrodynamics," *Phil. Mag.* Vol. XLVII. p. 556 (1899); *Scientific Papers*, Vol. IV. p. 407.

in which q is Young's modulus, and $2h$ the thickness of the plate (§ 214). Also for the kinetic energy per unit area we have

$$T = \rho h \dot{w}^2, \quad \dots\dots\dots\dots\dots\dots\dots\dots\dots\dots(2)$$

ρ being the volume-density. From the symmetries of the case w must be an odd function of x and an odd function of y, and it must also be symmetrical between x and y. Thus we may take

$$w = q_1 xy + q_2 xy (x^2 + y^2) + q_3 xy (x^4 + y^4) + q_4 x^3 y^3 + \dots. \quad \dots\dots(3)$$

In the actual calculation only the two first terms will be employed.

Expressions (1) and (2) are to be integrated over the square; but it will suffice to include only the first quadrant, so that if we take the side of the square as equal to 2, the limits for x and y are 0 and 1. We find

$$\iint (\nabla^2 w)^2 \, dx \, dy = 16 q_2^2, \quad \dots\dots\dots\dots\dots\dots\dots(4)$$

$$\iint \left\{ \left(\frac{d^2 w}{dx \, dy} \right)^2 - \frac{d^2 w}{dx^2} \frac{d^2 w}{dy^2} \right\} dx \, dy = q_1^2 + 4 q_1 q_2 + \tfrac{8}{5} q_2^2. \quad \dots\dots\dots(5)$$

Thus, if we set

$$V = \frac{4qh^3}{3(1+\mu)} V', \quad \dots\dots\dots\dots\dots\dots(6)$$

we have

$$V' = \tfrac{1}{2} q_1^2 + 2 q_1 q_2 + \tfrac{4}{5} q_2^2 + \frac{4 q_2^2}{1-\mu}. \quad \dots\dots\dots\dots(7)$$

In like manner, if

$$T = \frac{2\rho h}{9} T', \quad \dots\dots\dots\dots\dots\dots\dots(8)$$

$$T' = \tfrac{1}{2} \dot{q}_1^2 + \tfrac{6}{5} \dot{q}_1 \dot{q}_2 + \dot{q}_2^2 \left(\tfrac{3}{7} + \tfrac{9}{25} \right). \quad \dots\dots\dots\dots\dots(9)$$

When we neglect q_2 and suppose that q_1 varies as $\cos pt$, these expressions give

$$p^2 = \frac{6qh^2}{\rho(1+\mu)} = \frac{96qh^2}{\rho(1+\mu) a^4}, \quad \dots\dots\dots\dots\dots(10)$$

if we introduce a as the length of the side of the square. This is the value found in *Theory of Sound*, § 228, equivalent to Ritz's first approximation.

In proceeding to a second approximation we may omit the factors already accounted for in (10). Expressions (7), (9) are of the standard form if we take

$$A = 1, \qquad B = 2, \qquad C = \frac{8}{5} + \frac{8}{1-\mu},$$

$$L = 1, \qquad M = \frac{6}{5}, \qquad N = \frac{6}{7} + \frac{18}{25};$$

and Lagrange's equations are

$$(A - p^2L) q_1 + (B - p^2M) q_2 = 0, \\ (B - p^2M) q_1 + (C - p^2N) q_2 = 0, \Big\} \quad \dots\dots\dots\dots(11)$$

while the equation for p^2 is the quadratic

$$p^4(LN - M^2) + p^2(2MB - LC - NA) + AC - B^2 = 0. \quad \dots\dots(12)$$

For the numerical calculations we will suppose, following Ritz, that $\mu = \cdot225$, making $C = 11\cdot9226$. Thus

$$LN - M^2 = \cdot13714, \qquad AC - B^2 = 7\cdot9226,$$
$$2MB - LC - NA = -2 \times 4\cdot3498.$$

The smaller root of the quadratic as calculated by the usual formula is $\cdot9239$, in place of the 1 of the first approximation; but the process is not arithmetically advantageous. If we substitute this value in the first term of the quadratic, and determine p^2 from the resulting simple equation, we get the confirmed and corrected value $p^2 = \cdot9241$. Restoring the omitted factors, we have finally as the result of the second approximation

$$p^2 = \frac{96qh^2 \times \cdot9241}{\rho(1+\mu) a^4}, \quad \dots\dots\dots\dots\dots\dots(13)$$

in which $\mu = \cdot225$.

The value thus obtained is not so low, and therefore not so good, as that derived by Ritz from the series of u-functions. One of the advantages of the latter is that, being *normal* functions for the simple bar, they allow T to be expressed as a sum of squares of the generalized coordinates q_1, &c. As a consequence, p^2 appears only in the diagonal terms of the system of equations analogous to (11).

From (11) we find further

$$q_2/q_1 = -\cdot0852,$$

so that for the approximate form of w corresponding to the gravest pitch we may take

$$w = xy - \cdot0852\, xy\, (x^2 + y^2), \quad \dots\dots\dots\dots\dots(14)$$

in which the side of the square is supposed equal to 2.

358.

PROBLEMS IN THE CONDUCTION OF HEAT.

[*Philosophical Magazine*, Vol. XXII. pp. 381—396, 1911.]

THE general equation for the conduction of heat in a uniform medium may be written

$$\frac{dv}{dt} = \frac{d^2v}{dx^2} + \frac{d^2v}{dy^2} + \frac{d^2v}{dz^2} = \nabla^2 v, \quad \dots\dots\dots\dots\dots\dots(1)$$

v representing temperature. The coefficient (ν) denoting diffusibility is omitted for brevity on the right-hand of (1). It can always be restored by consideration of "dimensions."

Kelvin[*] has shown how to build up a variety of special solutions, applicable to an infinite medium, on the basis of Fourier's solution for a point-source. A few examples are quoted almost in Kelvin's words:

I. Instantaneous simple point-source; a quantity Q of heat suddenly generated at the point (0, 0, 0) at time $t = 0$, and left to diffuse through an infinite homogeneous solid:

$$v = \frac{Q e^{-r^2/4t}}{8\pi^{3/2}\, t^{3/2}}, \quad \dots\dots\dots\dots\dots\dots\dots\dots(2)$$

where $r^2 = x^2 + y^2 + z^2$. [The thermal capacity is supposed to be unity.] Verify that

$$\iiint\limits_{-\infty}^{+\infty} v\, dx\, dy\, dz = 4\pi \int_0^\infty v r^2\, dr = Q\,;$$

and that $v = 0$ when $t = 0$; unless also $x = 0$, $y = 0$, $z = 0$. *Every other solution is obtainable from this by summation.*

II. Constant simple point-source, rate q:

$$v \left[= q \int_0^\infty dt\, \frac{e^{-r^2/4t}}{8\pi^{3/2}\, t^{3/2}} \right] = \frac{q}{4\pi r}. \quad \dots\dots\dots\dots(3)$$

The formula within the brackets shows how this obvious solution is derivable from (2).

* "Compendium of Fourier Mathematics, &c.," *Enc. Brit.* 1880; *Collected Papers*, Vol. II. p. 44.

III.   Continued point-source; rate per unit of time at time t, an arbitrary function, $f(t)$:

$$v = \int_0^\infty d\chi f(t-\chi)\,\frac{e^{-r^2/4\chi}}{8\pi^{3/2}\,\chi^{3/2}}. \quad\ldots\ldots\ldots\ldots\ldots\ldots(4)$$

IV.   Time-periodic simple point-source, rate per unit of time at time t, $q \sin 2nt$:

$$v = \frac{q}{4\pi r}\,e^{-\sqrt{n}\,.\,r}\sin\left[2nt - n^{\frac12}.r\right]. \quad\ldots\ldots\ldots\ldots\ldots(5)$$

Verify that v satisfies (1); also that $-4\pi r^2 dv/dr = q \sin 2nt$, where $r = 0$.

V.   Instantaneous spherical surface-source; a quantity Q suddenly generated over a spherical surface of radius a, and left to diffuse outwards and inwards:

$$v = Q\,\frac{e^{-(r-a)^2/4t} - e^{-(r+a)^2/4t}}{8\pi^{3/2}\,ar t^{1/2}}. \quad\ldots\ldots\ldots\ldots\ldots(6)$$

To prove this most easily, verify that it satisfies (1); and further verify that

$$4\pi \int_0^\infty v r^2\,dr = Q\,;$$

and that $v = 0$ when $t = 0$, unless also $r = a$.   Remark that (6) becomes identical with (2) when $a = 0$; remark further that (6) is obtainable from (2) by integration over the spherical surface.

VI.   Constant spherical surface-source; rate per unit of time for the whole surface, q:

$$v\left[= q\int_0^\infty dt\,\frac{e^{-(r-a)^2/4t} - e^{-(r+a)^2/4t}}{8\pi^{3/2}\,ar t^{1/2}}\right]$$

$$= q/4\pi r \quad (r > a) \quad = q/4\pi a \quad (r < a).$$

The formula within the brackets shows how this obvious solution is derivable from (6).

VII.   Fourier's "Linear Motion of Heat"; instantaneous plane-source; quantity per unit surface, σ:

$$v = \frac{\sigma e^{-x^2/4t}}{2\pi^{1/2}\,t^{1/2}}. \quad\ldots\ldots\ldots\ldots\ldots\ldots\ldots(7)$$

Verify that this satisfies (1) for the case of v independent of y and z, and that

$$\int_{-\infty}^{+\infty} v\,dx = \sigma.$$

Remark that (7) is obtainable from (6) by putting $Q/4\pi a^2 = \sigma$, and $a = \infty$; or directly from (2) by integration over the plane.

In Kelvin's summary linear sources are passed over. If an instantaneous source be uniformly distributed along the axis of z, so that the rate per unit length is q, we obtain at once by integration from (2)

$$v = \int_{-\infty}^{+\infty} \frac{q\,dz\,e^{-(z^2+x^2+y^2)/4t}}{8\pi^{3/2}\,t^{3/2}} = \frac{q\,e^{-(x^2+y^2)/4t}}{4\pi t}. \qquad \dots\dots\dots(8)$$

From this we may deduce the effect of an instantaneous source uniformly distributed over a circular *cylinder* whose axis is parallel to z, the superficial density being σ. Considering the cross-section through Q—the point where v is to be estimated, let O be the centre and a the radius of the circle. Then if P be a point on the circle, $OP = a$, $OQ = r$, $PQ = \rho$, $\angle POQ = \theta$; and

$$\rho^2 = a^2 + r^2 - 2ar\cos\theta,$$

so that

$$v = \int_0^{2\pi} \frac{\sigma a\,d\theta\,e^{-\rho^2/4t}}{4\pi t} = \frac{\sigma a}{2t}\,e^{-\frac{r^2+a^2}{4t}}\,I_0\!\left(\frac{ra}{2t}\right), \qquad \dots\dots\dots(9)$$

$I_0(x)$, equal to $J_0(ix)$, being the function usually so denoted. From (9) we fall back on (8) if we put $a = 0$, $2\pi a\sigma = q$. It holds good whether r be greater or less than a.

When x is very great and positive,

$$I_n(x) = \frac{e^x}{\sqrt{(2\pi x)}}, \qquad \dots\dots\dots\dots\dots(10)$$

so that for very small values of t (9) assumes the form

$$v = \frac{\sigma a}{2\sqrt{(\pi rat)}}\,e^{-\frac{(r-a)^2}{4t}},$$

vanishing when $t = 0$, unless $r = a$.

Again, suppose that the instantaneous source is uniformly distributed over the *circle* $\zeta = 0$, $\xi = a\cos\phi$, $\eta = a\sin\phi$, the rate per unit of arc being q, and that v is required at the point x, 0, z. There is evidently no loss of generality in supposing $y = 0$. We obtain at once from (2)

$$v = \int_0^{2\pi} \frac{qa\,d\phi\,e^{-r^2/4t}}{8\pi^{3/2}\,t^{3/2}}, \qquad \dots\dots\dots\dots(11)$$

where

$$r^2 = (\xi - x)^2 + \eta^2 + z^2 = a^2 + x^2 + z^2 - 2ax\cos\phi.$$

Thus

$$v = \frac{qa}{4\pi^{1/2}\,t^{3/2}}\,e^{-\frac{a^2+x^2+z^2}{4t}}\,I_0\!\left(\frac{ax}{2t}\right), \qquad \dots\dots\dots(12)$$

from which if we write $q = \sigma\,dz$, and integrate with respect to z from $-\infty$ to $+\infty$, we may recover (9).

If in (12) we put $q = \sigma\, da$ and integrate with respect to a from 0 to ∞, we obtain a solution which must coincide with (7) when in the latter we substitute z for x.   Thus

$$\int_0^\infty a\, da\, e^{-a^2/4t}\, I_0\left(\frac{ax}{2t}\right) = 2t\, e^{x^2/4t}, \quad\ldots\ldots\ldots\ldots\ldots\ldots(13)$$

a particular case of one of Weber's integrals*.

It may be worth while to consider briefly the problem of initial instantaneous sources distributed over the plane $(\zeta = 0)$ in a more general manner.   In rectangular coordinates the typical distribution is such that the rate per unit of area is

$$\sigma \cos l\xi\,.\,\cos m\eta. \quad\ldots\ldots\ldots\ldots\ldots\ldots\ldots\ldots\ldots(14)$$

If we assume that at $x,\, y,\, z$ and time t, v is proportional to $\cos lx\,.\,\cos my$, the general differential equation (1) gives

$$\frac{dv}{dt} + (l^2 + m^2)\, v = \frac{d^2v}{dz^2},$$

or

$$\frac{d}{dt}\left\{e^{(l^2+m^2)\, t}\, v\right\} = \frac{d^2}{dz^2}\left\{e^{(l^2+m^2)\, t}\, v\right\};$$

so that, as for conduction in one dimension,

$$v = A \cos lx \cos my\, e^{-(l^2+m^2)\, t}\, \frac{e^{-z^2/4t}}{\sqrt{t}}, \quad\ldots\ldots\ldots\ldots\ldots(15)$$

and

$$\int_{-\infty}^{+\infty} v\, dz = 2\sqrt{\pi}\,.\, A \cos lx \cos my\, e^{-(l^2+m^2)\, t}.$$

Putting $t = 0$, and comparing with (14), we see that

$$A = \frac{\sigma}{2\sqrt{\pi}}\,. \quad\ldots\ldots\ldots\ldots\ldots\ldots\ldots\ldots\ldots(16)$$

By means of (2) the solution at time t may be built up from (14).   In this way, by aid of the well-known integral

$$\int_{-\infty}^{+\infty} e^{-a^2x^2} \cos 2cx\, dx = \frac{\sqrt{\pi}}{a}\, e^{-c^2/a^2}, \quad\ldots\ldots\ldots\ldots\ldots(17)$$

we may obtain (15) independently.

The process is of more interest in its application to polar coordinates. If we suppose that v is proportional to $\cos n\theta\,.\,J_n(kr)$,

$$\frac{d^2v}{dr^2} + \frac{1}{r}\frac{dv}{dr} + \frac{1}{r^2}\frac{d^2v}{d\theta^2} = -k^2v, \quad\ldots\ldots\ldots\ldots\ldots(18)$$

---

* Gray and Mathews' *Bessel's Functions*, p. 78, equation (160). Put $n=0$, $\lambda=0$. See also (31) below.

so that (1) gives

$$\frac{dv}{dt} + k^2 v = \frac{d^2 v}{dz^2}, \dots\dots\dots\dots\dots\dots\dots\dots\dots(19)$$

and

$$v = A \cos n\theta \, J_n(kr) \, e^{-k^2 t} \frac{e^{-z^2/4t}}{\sqrt{t}}. \dots\dots\dots\dots\dots(20)$$

From (20)

$$\int_{-\infty}^{+\infty} v\, dz = 2\sqrt{\pi} \cdot A \cos n\theta \, J_n(kr) \, e^{-k^2 t}. \dots\dots\dots\dots(21)$$

If the initial distribution on the plane $z = 0$ be per unit area

$$\sigma \cos n\theta \, J_n(kr), \dots\dots\dots\dots\dots\dots\dots(22)$$

it follows from (21) that as before

$$A = \frac{\sigma}{2\sqrt{\pi}}. \dots\dots\dots\dots\dots\dots\dots\dots\dots(23)$$

We next proceed to investigate the effect of an instantaneous source distributed over the circle for which

$$\zeta = 0, \quad \xi = a \cos \phi, \quad \eta = a \sin \phi,$$

the rate per unit length of arc being $q \cos n\phi$. From (2) at the point x, y, z

$$v = \int_0^{2\pi} \frac{q \cos n\phi \, e^{-r^2/4t} \, a\, d\phi}{8\pi^{3/2} t^{3/2}}, \dots\dots\dots\dots\dots\dots(24)$$

in which

$$r^2 = (\xi - x)^2 + (\eta - y)^2 + z^2 = a^2 + \rho^2 + z^2 - 2a\rho \cos(\phi - \theta),$$

if $x = \rho \cos\theta$, $y = \rho \sin\theta$. The integral that we have to consider may be written

$$\int_0^{2\pi} \cos n\phi \, e^{\rho' \cos(\phi - \theta)} \, d\phi = \int \cos n(\theta + \psi) \, e^{\rho' \cos\psi} \, d\psi$$

$$= \cos n\theta \int \cos n\psi \, e^{\rho' \cos\psi} \, d\psi - \sin n\theta \int \sin n\psi \, e^{\rho' \cos\psi} \, d\psi, \dots\dots(25)$$

where $\psi = \phi - \theta$, and $\rho' = a\rho/2t$. In view of the periodic character of the integrand, the limits may be taken as $-\pi$ and $+\pi$. Accordingly

$$\int_{-\pi}^{+\pi} \cos n\psi \, e^{\rho' \cos\psi} \, d\psi = 2\int_0^{\pi} \cos n\psi \, e^{\rho' \cos\psi} \, d\psi,$$

$$\int_{-\pi}^{+\pi} \sin n\psi \, e^{\rho' \cos\psi} \, d\psi = 0;$$

and

$$\int_0^{2\pi} \cos n\phi \, e^{\rho' \cos(\phi - \theta)} \, d\phi = 2\cos n\theta \int_0^{\pi} \cos n\psi \, e^{\rho' \cos\psi} \, d\psi. \dots\dots(26)$$

The integral on the right of (26) is equivalent to $\pi I_n(\rho')$, where

$$i^n I_n(\rho') = J_n(i\rho'), \dots\dots\dots\dots\dots\dots\dots(27)$$

J_n being, as usual, the symbol of Bessel's function of order n. For, if n be *even*,

$$\int_0^\pi \cos n\psi \, e^{\rho' \cos \psi} \, d\psi = \tfrac{1}{2} \int_0^\pi \cos n\psi \, (e^{\rho' \cos \psi} + e^{-\rho' \cos \psi}) \, d\psi$$

$$= \int_0^\pi \cos n\psi \cos (i\rho' \cos \psi) \, d\psi = \pi i^{-n} J_n (i\rho') = \pi I_n (\rho');$$

and, if n be *odd*,

$$\int_0^\pi \cos n\psi \, e^{\rho' \cos \psi} \, d\psi = -\tfrac{1}{2} \int_0^\pi \cos n\psi \, (e^{-\rho' \cos \psi} - e^{\rho' \cos \psi}) \, d\psi$$

$$= -i \int_0^\pi \cos n\psi \sin (i\rho' \cos \psi) \, d\psi = \pi I_n (\rho').$$

In either case

$$\int_0^\pi \cos n\psi \, e^{\rho' \cos \psi} \, d\psi = \pi I_n (\rho'). \quad \dots\dots\dots\dots\dots (28)$$

Thus

$$\int_0^{2\pi} \cos n\phi \, e^{\rho' \cos (\phi - \theta)} \, d\phi = 2\pi \cos n\theta \, I_n (\rho'), \quad \dots\dots\dots (29)$$

and (24) becomes

$$v = \frac{qa \cos n\theta}{4\pi^{1/2} \, t^{3/2}} I_n \left(\frac{a\rho}{2t}\right) e^{-\frac{a^2 + \rho^2 + z^2}{4t}}. \quad \dots\dots\dots\dots (30)$$

This gives the temperature at time t and place (ρ, z) due to an initial instantaneous source distributed over the circle a.

The solution (30) may now be used to find the effect of the initial source expressed by (22). For this purpose we replace q by $\sigma \, da$, and introduce the additional factor $J_n (ka)$, subsequently integrating with respect to a between the limits 0 and ∞. Comparing the result with that expressed in (20), (23), we see that

$$\frac{\sigma \cos n\theta \, e^{-z^2/4t}}{2 \sqrt{(\pi t)}}$$

is a common factor which divides out, and that there remains the identity

$$\frac{e^{-\rho^2/4t}}{2t} \int_0^\infty a \, da \, e^{-a^2/4t} J_n (ka) \, I_n \left(\frac{a\rho}{2t}\right) = J_n (k\rho) \, e^{-k^2 t}. \quad \dots\dots (31)$$

This agrees with the formula given by Weber, which thus receives an interesting interpretation.

Reverting to (30), we recognize that it must satisfy the fundamental equation (1), now taking the form

$$\frac{d^2 v}{dz^2} + \frac{d^2 v}{d\rho^2} + \frac{1}{\rho} \frac{dv}{d\rho} + \frac{1}{\rho^2} \frac{d^2 v}{d\theta^2} = \frac{dv}{dt}; \quad \dots\dots\dots\dots (32)$$

and that when $t = 0$ v must vanish, unless also $z = 0$, $\rho = a$.

If we integrate (30) with respect to z between $\pm \infty$, setting $q = \sigma dz$, so that $\sigma \cos n\theta$ represents the superficial density of the instantaneous source distributed over the *cylinder* of radius a, we obtain

$$v = \frac{\sigma a \cos n\theta}{2t} I_n \left(\frac{a\rho}{2t} \right) e^{-\frac{a^2 + \rho^2}{4t}}, \quad \dots\dots\dots\dots(33)$$

which may be regarded as a generalization of (9). And it appears that (33) satisfies (32), in which the term d^2v/dz^2 may now be omitted.

In V. Kelvin gives the temperature at a distance r from the centre and at time t due to an instantaneous source uniformly distributed over a spherical surface. In deriving the result by integration from (2) it is of course simplest to divide the spherical surface into elementary circles which are symmetrically situated with respect to the line OQ joining the centre of the sphere O to the point Q where the effect is required. But if the circles be drawn round another axis OA, a comparison of results will give a definite integral.

Adapting (12), we write $a = c \sin \theta$, c being the radius of the sphere, $x = OQ \sin \theta' = r \sin \theta'$, $z = r \cos \theta' - c \cos \theta$, so that

$$v = \frac{qc \sin \theta\, e^{-(c^2 + r^2)/4t}}{4\pi^{1/2}\, t^{3/2}} I_0 \left(\frac{cr \sin \theta \sin \theta'}{2t} \right) e^{\frac{rc \cos \theta \cos \theta'}{2t}} \quad \dots\dots\dots(34)$$

This has now to be integrated with respect to θ from 0 to π. Since the result must be independent of θ', we see by putting $\theta' = 0$ that

$$\int_0^\pi I_0 \left(\rho \sin \theta \sin \theta' \right) e^{\rho \cos \theta \cos \theta'} \sin \theta\, d\theta$$

$$= \int_0^\pi e^{\rho \cos \theta} \sin \theta\, d\theta = \frac{1}{\rho} (e^\rho - e^{-\rho}). \quad \dots\dots\dots(35)$$

Using the simplified form and putting $q = \sigma c\, d\theta$, where σ is the superficial density, we obtain for the complete sphere

$$v = \frac{\sigma c}{2\pi^{\frac{1}{2}} t^{\frac{1}{2}} r} \left(e^{-\frac{(c-r)^2}{4t}} - e^{-\frac{(c+r)^2}{4t}} \right), \quad \dots\dots\dots\dots(36)$$

agreeing with (6) when we remember that $Q = 4\pi c^2 \sigma$.

We will now consider the problem of an instantaneous source arbitrarily distributed over the surface of the sphere whose radius is c. It suffices, of course, to treat the case of a spherical harmonic distribution; and we suppose that per unit of area of the spherical surface the rate is S_n. Assuming that v is everywhere proportional to S_n, we know that v satisfies

$$\frac{1}{\sin \theta} \frac{d}{d\theta} \left(\sin \theta \frac{dv}{d\theta} \right) + \frac{1}{\sin^2 \theta} \frac{d^2v}{d\omega^2} + n(n+1)v = 0, \quad \dots\dots\dots(37)$$

θ, ω being the usual spherical polar coordinates. Hence from (1) v as a function of r and t satisfies

$$\frac{dv}{dt} = \frac{d^2 v}{dr^2} + \frac{2}{r}\frac{dv}{dr} - \frac{n(n+1)v}{r^2} = 0,$$

or

$$\frac{d(rv)}{dt} = \frac{d^2(rv)}{dr^2} - \frac{n(n+1)}{r^2}(rv) = 0 \dots\dots\dots\dots(38)$$

When $n = 0$, this reduces to the same form as applies in one dimension. For general values of n the required solution appears to be most easily found indirectly.

Let us suppose that S_n reduces to Legendre's function $P_n(\mu)$, where $\mu = \cos\theta$, and let us calculate directly from (2) the value of v at time t and at a point Q distant r from the centre of the sphere *along the axis of μ*. The exponential term is

$$e^{-\frac{r^2+c^2}{4t}} e^{\frac{rc\mu}{2t}} = e^{-\frac{r^2+c^2}{4t}} e^{\rho\mu}, \dots\dots\dots\dots\dots(39)$$

if $\rho = rc/2t$.   Now (*Theory of Sound*, § 334)

$$\int_{-1}^{+1} P_n(\mu)\, e^{i\rho\mu}\, d\mu = 2i^n \sqrt{\left(\frac{\pi}{2\rho}\right)} J_{n+\frac{1}{2}}(\rho), \dots\dots\dots\dots(40)$$

whence

$$\int_{-1}^{+1} P_n(\mu)\, e^{\rho\mu}\, d\mu = 2i^{n+\frac{1}{2}} \sqrt{\left(\frac{\pi}{2\rho}\right)} J_{n+\frac{1}{2}}(-i\rho), \dots\dots\dots\dots(41)$$

or, as it may also be written by (27),

$$= 2 \sqrt{\left(\frac{\pi}{2\rho}\right)} I_{n+\frac{1}{2}}(\rho). \dots\dots\dots\dots\dots(42)$$

Substituting in (2)

$$Q = 2\pi c^2\, P_n(\mu)\, d\mu, \dots\dots\dots\dots\dots(43)$$

we now get for the value of v at time t, and at the point for which $\rho = r$, $\mu = 1$,

$$rv = \frac{i^{n+\frac{1}{2}}\, c^{3/2}\, r^{\frac{1}{2}}\, e^{-(r^2+c^2)/4t}}{2t}\, J_{n+\frac{1}{2}}\left(-\frac{irc}{2t}\right). \dots\dots\dots\dots(44)$$

It may be verified by trial that (44) is a solution of (38). When μ is not restricted to the value unity, the only change required in (44) is the introduction of the factor $P_n(\mu)$.

When $n = 0$, $P_n(\mu) = 1$, and we fall back upon the case of *uniform* distribution. We have

$$J_{\frac{1}{2}}(x) = \sqrt{\left(\frac{2}{\pi x}\right)} \sin x, \dots\dots\dots\dots(45)$$

or

$$J_{\frac{1}{2}}(-ix) = i^{-\frac{1}{2}} \frac{e^x - e^{-x}}{\sqrt{(2\pi x)}}. \dots\dots\dots\dots(46)$$

Using this in (44), we obtain a result in accordance with (6), in which Q, representing the integrated magnitude of the source, is equal to $4\pi c^2$ in our present reckoning.

When $n = 1$, $P_1(\mu) = \mu$, and

$$J_{3/2}(x) = \sqrt{\left(\frac{2}{\pi x}\right)} \left\{\frac{\sin x}{x} - \cos x\right\}; \quad \dots\dots\dots\dots(47)$$

and whatever integral value n may assume $J_{n+\frac{1}{2}}$ is expressible in finite terms.

We have supposed that the rate of distribution is represented by a Legendre's function $P_n(\mu)$. In the more general case it is evident that we have merely to multiply the right-hand member of (44) by S_n, instead of P_n.

So far we have been considering instantaneous sources. As in II., the effect of *constant* sources may be deduced by integration, although the result is often more readily obtained otherwise. A comparison will, however, give the value of a definite integral. Let us apply this process to (33) representing the effect of a cylindrical source.

The required solution, being independent of t, is obtained at once from (1). We have inside the cylinder

$$v = A\rho^n \cos n\theta,$$

and outside

$$v = B\rho^{-n} \cos n\theta,$$

with $Aa^n = Ba^{-n}$. The intensity of the source is represented by the difference in the values of $dv/d\rho$ just inside and just outside the cylindrical surface. Thus

$$\sigma' \cos n\theta = n \cos n\theta \left(Ba^{-n-1} + Aa^{n-1}\right),$$

whence

$$Aa^n = Ba^{-n} = \sigma'a/2n,$$

$\sigma' \cos n\theta$ being the constant time rate. Accordingly, within the cylinder

$$v = \frac{\sigma'a}{2n} \left(\frac{\rho}{a}\right)^n \cos n\theta, \quad \dots\dots\dots\dots\dots(48)$$

and without the cylinder

$$v = \frac{\sigma'a}{2n} \left(\frac{\rho}{a}\right)^{-n} \cos n\theta. \quad \dots\dots\dots\dots\dots(49)$$

These values are applicable when n is any positive integer. When n is zero, there is no permanent distribution of temperature possible.

These solutions should coincide with the value obtained from (33) by putting $\sigma = \sigma' \, dt$ and integrating with respect to t from 0 to ∞. Or

$$\int_0^\infty \frac{dt}{t} I_n\left(\frac{a\rho}{2t}\right) e^{-\frac{a^2+\rho^2}{4t}} = \frac{1}{n} \left(\frac{\rho}{a}\right)^{\pm n}, \quad \dots\dots\dots\dots(50)$$

the $+$ sign in the ambiguity being taken when $\rho < a$, and the $-$ sign when $\rho > a$. I have not confirmed (50) independently.

In like manner we may treat a constant source distributed over a *sphere*. If the rate per unit time and per unit of area of surface be S_n, we find, as above, for inside the sphere (c)

$$v = \frac{c}{2n+1} \left(\frac{r}{c}\right)^n S_n, \quad\dots\dots\dots\dots\dots(51)$$

and outside the sphere

$$v = \frac{c}{2n+1} \left(\frac{c}{r}\right)^{n+1} S_n, \quad\dots\dots\dots\dots\dots(52)$$

and these forms are applicable to any integral n, *zero included*. Comparing with (44), we see that

$$i^{n+\frac12} \int_0^\infty \frac{dt}{t} e^{-\frac{r^2+c^2}{4t}} J_{n+\frac12}\left(-\frac{irc}{2t}\right) = \frac{2}{2n+1} \left(\frac{r}{c}\right)^{\pm(n+\frac12)}, \quad\dots\dots\dots(53)$$

which does not differ from (50), if in the latter we suppose $n = $ integer $+\frac12$.

The solution for a time-periodic simple point-source has already been quoted from Kelvin (IV.). Though derivable as a particular case from (4), it is more readily obtained from the differential equation (1) taking here the form—see (38) with $n = 0$—

$$\frac{d^2(rv)}{dt} = \frac{d^2(rv)}{dr^2},$$

or if v is assumed proportional to e^{ipt},

$$d^2(rv)/dr^2 - ip(rv) = 0, \quad\dots\dots\dots\dots\dots(54)$$

giving

$$rv = A e^{ipt} e^{-i^{\frac12}p^{\frac12}r}, \quad\dots\dots\dots\dots\dots(55)$$

as the symbolical solution applicable to a source situated at $r = 0$. Denoting by q the magnitude of the source, as in (5), we get to determine A,

$$\left[-4\pi r^2 \frac{dv}{dr}\right]_{r=0} = q e^{ipt} = 4\pi A,$$

so that

$$v = \frac{q}{4\pi r} e^{ipt} e^{-i^{\frac12}p^{\frac12}r}. \quad\dots\dots\dots\dots\dots(56)$$

If from (56) we discard the imaginary part, we have

$$v = \frac{q}{4\pi r} e^{-r\sqrt{(p/2)}} \cos\{pt - r\sqrt{(p/2)}\}, \quad\dots\dots\dots\dots(57)$$

corresponding to the source $q \cos pt$.

From (56) it is possible to build up by integration solutions relating to various distributions of periodic sources over lines or surfaces, but an independent treatment is usually simpler. We will, however, write down the integral corresponding to a uniform linear source coincident with the axis of z. If $\rho^2 = x^2 + y^2$, $r^2 = z^2 + \rho^2$, and (ρ being constant) $r\,dr = z\,dz$. Thus putting in (56) $q = q_1 dz$, we get

$$v = \frac{q_1 e^{ipt}}{2\pi} \int_\rho^\infty \frac{e^{-r\sqrt{(ip)}}\,dr}{\sqrt{(r^2-\rho^2)}}. \quad\dots\dots\dots\dots(58)$$

In considering the effect of periodic sources distributed over a plane xy, we may suppose

$$v \propto \cos lx \cdot \cos my, \quad\quad\quad\quad\quad\quad (59)$$

or again

$$v \propto J_n(kr) \cdot \cos n\theta, \quad\quad\quad\quad\quad (60)$$

where $r^2 = x^2 + y^2$. In either case if we write $l^2 + m^2 = k^2$, and assume v proportional to e^{ipt}, (1) gives

$$d^2v/dz^2 = (k^2 + ip)\, v. \quad\quad\quad\quad (61)$$

Thus, if

$$k^2 + ip = R\,(\cos\alpha + i\sin\alpha), \quad\quad\quad (62)$$

$$v = A e^{-\sqrt{R} \cdot (\cos\frac{1}{2}\alpha + i\sin\frac{1}{2}\alpha) z}\, e^{ipt}, \quad\quad\quad (63)$$

where A includes the factors (59) or (60). If the value of v be given on the plane $z = 0$, that of A follows at once. If the magnitude of the source be given, A is to be found from the value of dv/dz when $z = 0$.

The simplest case is of course that where $k = 0$. If $V e^{ipt}$ be the value of v when $z = 0$, we find

$$v = V e^{ipt}\, e^{-z\sqrt{(ip)}}\,; \quad\quad\quad\quad (64)$$

or when realized

$$v = V e^{-z\sqrt{(p/2)}} \cos\{pt - z\sqrt{(p/2)}\}, \quad\quad\quad (65)$$

corresponding to

$$v = V \cos pt \quad\quad \text{when } z = 0.$$

From (64)

$$-\left(\frac{dv}{dz}\right)_0 = \sqrt{(ip)} \cdot V e^{ipt} = \tfrac{1}{2}\sigma e^{ipt}, \quad\quad (66)$$

if σ be the source per unit of area of the plane regarded as operative in a medium indefinitely extended in both directions. Thus in terms of σ,

$$v = \frac{\sigma}{2\sqrt{p}}\, e^{i\,(pt - \frac{1}{4}\pi)}\, e^{-z\sqrt{(ip)}}, \quad\quad\quad\quad (67)$$

or in real form

$$v = \frac{\sigma}{2\sqrt{p}}\, e^{-z\sqrt{(p/2)}} \cos\{pt - \tfrac{1}{4}\pi - z\sqrt{(p/2)}\}, \quad\quad (68)$$

corresponding to the uniform source $\sigma\cos pt$.

In the above formulæ z is supposed to be positive. On the other side of the source, where z itself is negative, the signs must be changed so that the terms containing z may remain negative in character.

When periodic sources are distributed over the surface of a *sphere* (radius $= c$), we may suppose that v is proportional to the spherical surface harmonic S_n. As a function of r and t, v is then subject to (38); and when we introduce the further supposition that as dependent on t, v is proportional to e^{ipt}, we have

$$\frac{d^2(rv)}{dr^2} - \frac{n(n+1)}{r^2}\,(rv) - ip\,(rv) = 0. \quad\quad (69)$$

When $n = 0$, that is in the case of symmetry round the pole, this equation takes the same form as for one dimension; but we have to distinguish between the inside and the outside of the sphere.

On the inside the constants must be so chosen that v remains finite at the pole $(r = 0)$.   Hence

$$rv = A\, e^{ipt} (e^{r\sqrt{(ip)}} - e^{-r\sqrt{(ip)}}),\quad \dots\dots\dots\dots\dots\dots(70)$$

or in real form

$$rv = Ae^{r\sqrt{(p/2)}} \cos\{pt + r\sqrt{(p/2)}\} - Ae^{-r\sqrt{(p/2)}} \cos\{pt - r\sqrt{(p/2)}\}.\ \dots(71)$$

Outside the sphere the condition is that rv must vanish at infinity.   In this case

$$rv = B\, e^{ipt}\, e^{-r\sqrt{(ip)}},\quad \dots\dots\dots\dots\dots\dots\dots(72)$$

or in real form

$$rv = Be^{-r\sqrt{(p/2)}} \cos\{pt - r\sqrt{(p/2)}\}.\quad \dots\dots\dots\dots(73)$$

When n is not zero, the solution of (69) may be obtained as in Stokes' treatment of the corresponding acoustical problem (*Theory of Sound*, ch. XVII). Writing $r\sqrt{(ip)} = z$, and assuming

$$rv = A\, e^{z} + B\, e^{-z},\quad \dots\dots\dots\dots\dots\dots\dots(74)$$

where A and B are functions of z, we find for B

$$\frac{d^2B}{dz^2} - 2\frac{dB}{dz} - \frac{n(n+1)}{z^2}\, B = 0.\quad \dots\dots\dots\dots\dots(75)$$

The solution is

$$B = B_0 f_n(z),\quad \dots\dots\dots\dots\dots\dots\dots\dots(76)$$

where B_0 is independent of z and

$$f_n(z) = 1 + \frac{n(n+1)}{2\,.\,z} + \frac{(n-1)\,n\,(n+1)\,(n+2)}{2\,.\,4\,.\,z^2} + \dots,\quad \dots\dots(77)$$

as may be verified by substitution.   Since n is supposed integral, the series (77) terminates.   For example, if $n = 1$, it reduces to the first two terms.

The solution appropriate to the exterior is thus

$$rv = B_0 S_n e^{ipt}\, e^{-r\sqrt{(ip)}}\, f_n(i^{\frac{1}{2}} p^{\frac{1}{2}} r).\quad \dots\dots\dots\dots\dots(78)$$

For the interior we have

$$rv = A_0 S_n e^{ipt}\, \{e^{-r\sqrt{(ip)}}\, f_n(i^{\frac{1}{2}} p^{\frac{1}{2}} r) - e^{r\sqrt{(ip)}}\, f_n(-i^{\frac{1}{2}} p^{\frac{1}{2}} r)\},\quad \dots\dots(79)$$

which may also be expressed by a Bessel's function of order $n + \frac{1}{2}$.

In like manner we may treat the problem in two dimensions, where everything may be expressed by the polar coordinates r, θ.   It suffices to consider the terms in $\cos n\theta$, where n is an integer.   The differential equation analogous to (69) is now

$$\frac{d^2v}{dr^2} + \frac{1}{r}\frac{dv}{dr} - \frac{n^2}{r^2}\, v = ipv,\quad \dots\dots\dots\dots\dots\dots(80)$$

which, if we take $r \sqrt{(ip)} = z$, as before, may be written

$$\frac{d^2 (z^{\frac{1}{2}} v)}{dz^2} - \frac{(n - \frac{1}{2})(n + \frac{1}{2})}{z^2} (z^{\frac{1}{2}} v) = z^{\frac{1}{2}} v, \dots\dots\dots\dots\dots(81)$$

and is of the same form as (69) when in the latter $n - \frac{1}{2}$ is written for n.

As appears at once from (80), the solution for the interior of the cylinder may be expressed

$$v = A \cos n\theta\, e^{ipt} J_n (i^{3/2} p^{1/2} r), \dots\dots\dots\dots\dots(82)$$

J_n being as usual the Bessel's function of the nth order.

For the exterior we have from (81)

$$r^{\frac{1}{2}} v = B \cos n\theta\, e^{ipt} e^{-r \sqrt{(ip)}} f_{n - \frac{1}{2}} (i^{\frac{1}{2}} p^{\frac{1}{2}} r), \dots\dots\dots\dots(83)$$

where

$$f_{n - \frac{1}{2}} (z) = 1 + \frac{4n^2 - 1^2}{1 \cdot 8z} + \frac{(4n^2 - 1^2)(4n^2 - 3^2)}{1 \cdot 2 \cdot (8z)^2}$$

$$+ \frac{(4n^2 - 1^2)(4n^2 - 3^2)(4n^2 - 5^2)}{1 \cdot 2 \cdot 3 \cdot (8z)^3} + \dots \dots\dots\dots(84)$$

The series (84), unlike (77), does not terminate. It is ultimately divergent, but may be employed for computation when z is moderately great.

In these periodic solutions the sources distributed over the plane, sphere, or cylinder are supposed to have been in operation for so long a time that any antecedent distribution of temperature throughout the medium is without influence. By Fourier's theorem this procedure may be generalized. Whatever be the character of the sources with respect to time, it may be resolved into simple periodic terms; and if the character be known through the whole of past time, the solution so obtained is unambiguous. The same conclusion follows if, instead of the magnitude of the sources, the temperature at the surfaces in question be known through past time.

An important particular case is when the character of the function is such that the superficial value, having been constant (zero) for an infinite time, is suddenly raised to another value, say unity, and so maintained. The Fourier expression for such a function is

$$\frac{1}{2} + \frac{1}{\pi} \int_0^\infty \frac{\sin pt}{p}\, dp, \dots\dots\dots\dots\dots(85)$$

the definite integral being independent of the arithmetical value of t, but changing sign when t passes through 0; or, on the understanding that only the real part is to be retained,

$$\frac{1}{2} + \frac{1}{i\pi} \int_0^\infty \frac{e^{ipt}}{p}\, dp. \dots\dots\dots\dots\dots(86)$$

We may apply this at once to the case of the plane $z = 0$ which has been at 0 temperature from $t = -\infty$ to $t = 0$, and at temperature 1 from $t = 0$ to $t = \infty$. By (64)

$$v = \tfrac{1}{2} + \frac{1}{i\pi} \int_0^\infty \frac{e^{ipt - z\sqrt{(ip)}}}{p} \, dp. \quad \dots\dots\dots\dots\dots(87)$$

By the methods of complex integration this solution may be transformed into Fourier's, viz.

$$\frac{dv}{dz} = - \frac{1}{\sqrt{(\pi t)}} e^{-z^2/4t}, \dots\dots\dots\dots\dots\dots\dots\dots(88)$$

$$v = 1 - \frac{2}{\sqrt{\pi}} \int_0^{z/2\sqrt{t}} e^{-\alpha^2} d\alpha, \quad \dots\dots\dots\dots\dots(89)$$

which are, however, more readily obtained otherwise.

In the case of a cylinder $(r = c)$ whose surface has been at 0 up to $t = 0$ and afterwards at $v = 1$, we have from (83) with $n = 0$

$$v = \tfrac{1}{2} + \frac{c^{\frac{1}{2}}}{i\pi r^{\frac{1}{2}}} \int_0^\infty e^{ipt + (c-r)\sqrt{(ip)}} \frac{f_{-\frac{1}{2}}(i^{\frac{1}{2}} p^{\frac{1}{2}} r) \, dp}{f_{-\frac{1}{2}}(i^{\frac{1}{2}} p^{\frac{1}{2}} c) \, p}, \quad \dots\dots\dots(90)$$

of which only the real part is to be retained. This applies to the region outside the cylinder.

It may be observed that when t is negative (87) must vanish for positive z and (90) for $r > c$.

359.

ON THE GENERAL PROBLEM OF PHOTOGRAPHIC REPRODUCTION, WITH SUGGESTIONS FOR ENHANCING GRADATION ORIGINALLY INVISIBLE.

[*Philosophical Magazine*, Vol. XXII. pp. 734—740, 1911.]

In copying a subject by photography the procedure usually involves two distinct steps. The first yields a so-called *negative*, from which, by the same or another process, a second operation gives the desired *positive*. Since ordinary photography affords pictures in monochrome, the reproduction can be complete only when the original is of the same colour. We may suppose, for simplicity of statement, that the original is itself a transparency, *e.g.* a lantern-slide.

The character of the original is regarded as given by specifying the transparency (t) at every point, *i.e.* the ratio of light transmitted to light incident. But here an ambiguity should be noticed. It may be a question of the place at which the transmitted light is observed. When light penetrates a stained glass, or a layer of coloured liquid contained in a tank, the direction of propagation is unaltered. If the incident rays are normal, so also are the rays transmitted. The action of the photographic image, constituted by an imperfectly aggregated deposit, differs somewhat. Rays incident normally are more or less diffused after transmission. The effective transparency in the half-tones of a negative used for contact printing may thus be sensibly greater than when a camera and lens is employed. In the first case all the transmitted light is effective; in the second most of that diffused through a finite angle fails to reach the lens*. In defining t—the transparency at any place—account must in strictness be taken of the manner in which the picture is to be viewed. There is also another point to be considered. The transparency may not be the same for different kinds

---

* In the extreme case a negative seen against a dark background and lighted obliquely from behind may even appear as a positive.

of light. We must suppose either that one kind of light only is employed, or else that t is the same for all the kinds that need to be regarded. The actual values of t may be supposed to range from 0, representing complete opacity, to 1, representing complete transparency.

As the first step is the production of a negative, the question naturally suggests itself whether we can define the ideal character of such a negative. Attempts have not been wanting; but when we reflect that the negative is only a means to an end, we recognize that no answer can be given without reference to the process in which the negative is to be employed to produce the positive. In practice this process (of printing) is usually different from that by which the negative was itself made; but for simplicity we shall suppose that the same process is employed in both operations. This require- ment of identity of procedure in the two cases is to be construed strictly, extending, for example, to duration of development and degree of intensifica- tion, if any. Also we shall suppose for the present that the *exposure* is the same. In strictness this should be understood to require that both the intensity of the incident light and the time of its operation be maintained; but since between wide limits the effect is known to depend only upon the product of these quantities, we may be content to regard exposure as defined by a single quantity, viz. *intensity of light × time.*

Under these restrictions the transparency t' at any point of the negative is a definite function of the transparency t at the corresponding point of the original, so that we may write

$$t' = f(t), \quad\dotfill(1)$$

f depending upon the photographic procedure and being usually such that as t increases from 0 to 1, t' decreases continually. When the operation is repeated upon the negative, the transparency t'' at the corresponding part of the positive is given by

$$t'' = f(t'). \quad\dotfill(2)$$

Complete reproduction may be considered to demand that at every point $t'' = t$. Equation (2) then expresses that t must be the same function of t' that t' is of t. Or, if the relation between t and t' be written in the form

$$F(t, t') = 0, \quad\dotfill(3)$$

F must be a *symmetrical* function of the two variables. If we regard t, t' as the rectangular coordinates of a point, (3) expresses the relationship by a *curve* which is to be symmetrical with respect to the bisecting line $t' = t$.

So far no particular form of f, or F, is demanded; no particular kind of negative is indicated as ideal. But certain simple cases call for notice. Among these is

$$t + t' = 1, \quad\dotfill(4)$$

which obviously satisfies the condition of symmetry. The representative curve is a straight line, equally inclined to the axes. According to (4), when $t = 0$, $t' = 1$. This requirement is usually satisfied in photography, being known as freedom from fog—no photographic action where no light has fallen. But the complementary relation $t' = 0$ when $t = 1$ is only satisfied approximately. The relation between negative and positive expressed in (4) admits of simple illustration. If both be projected upon a screen from independent lanterns of equal luminous intensity, so that the images fit, the pictures obliterate one another, and there results a field of uniform intensity.

Another simple form, giving the same limiting values as (4), is

$$t^2 + t'^2 = 1 ; \quad \dots\dots\dots\dots\dots\dots\dots\dots\dots(5)$$

and of course any number of others may be suggested.

According to Fechner's law, which represents the facts fairly well, the visibility of the difference between t and $t + dt$ is proportional to dt/t. The gradation in the negative, constituted in agreement with (4), is thus quite different from that of the positive. When t is small, large differences in the positive may be invisible in the negative, and *vice versâ* when t approaches unity. And the want of correspondence in gradation is aggravated if we substitute (5) for (4). All this is of course consistent with complete final reproduction, the differences which are magnified in the first operation being correspondingly attenuated in the second.

If we impose the condition that the gradation in the negative shall agree with that in the positive, we have

$$dt/t = - dt'/t', \quad \dots\dots\dots\dots\dots\dots\dots\dots\dots(6)$$

whence

$$t \cdot t' = C, \quad \dots\dots\dots\dots\dots\dots\dots\dots\dots(7)$$

where C is a constant. This relation does not fully meet the other requirements of the case. Since t' cannot exceed unity, t cannot be less than C. However, by taking C small enough, a sufficient approximation may be attained. It will be remarked that according to (7) the negative and positive obliterate one another when superposed in such a manner that light passes through them in succession— a combination of course entirely different from that considered in connexion with (4). This equality of gradation (within certain limits) may perhaps be considered a claim for (7) to represent the ideal negative; on the other hand, the *word* accords better with definition (4).

It will be remembered that hitherto we have assumed the exposure to be the same in the two operations, viz. in producing the negative and in copying from it. The restriction is somewhat arbitrary, and it is natural to inquire whether it can be removed. One might suppose that the removal would allow a greater latitude in the relationship between t and t'; but a closer scrutiny seems to show that this is not the case.

5—2

The effect of varying the exposure (e) is the same as of an inverse alteration in the transparency; it is the product et with which we really have to do. This refers to the first operation; in the second, t'' is dependent in like manner upon $e't'$. For simplicity and without loss of generality we may suppose that $e = 1$; also that $e'/e = m$, where m is a numerical quantity greater or less than unity. The equations which replace (1) and (2) are now

$$t' = f(t), \qquad t = t'' = f(mt'); \qquad \dots\dots\dots\dots\dots\dots(8)$$

and we assume that f is such that it decreases continually as its argument increases. This excludes what is called in photography *solarization*.

We observe that if t, lying between 0 and 1, anywhere makes $t' = t$, then m must be taken to be unity. For in the case supposed

$$t = f(t) = f(mt);$$

and this in accordance with the assumed character of f cannot be true, unless $m = 1$. Indeed without analytical formulation it is evident that since the transparency is not altered in the negative, it will require the same exposure to obtain it in the second operation as that by which it was produced in the first. Hence, if anywhere $t' = t$, the exposures must be the same.

It remains to show that there is no escape from a local equality of t and t'. When $t = 0$, $t' = 1$, or (if there be fog) some smaller positive quantity. As t increases from 0 to 1, t' continually decreases, and must therefore pass t at some point of the range. We conclude that complete reproduction requires $m = 1$, *i.e.* that the two exposures be equal; but we must not forget that we have assumed the photographic procedure to be exactly the same, except as regards exposure.

Another reservation requires a moment's consideration. We have interpreted complete reproduction to demand equality of t'' and t. This seems to be in accord with usage; but it might be argued that *proportionality* of t'' and t' is all that is really required. For although the pictures considered in themselves differ, the effect upon the eye, or upon a photographic plate, may be made identical, all that is needed being a suitable variation in the intensity of the luminous background. But at this rate we should have to regard a white and a grey paper as equivalent.

If we abandon the restriction that the photographic process is to be the same in the two operations, simple conclusions of generality can hardly be looked for. But the problem is easily formulated. We may write

$$t' = f_1(et), \qquad t = t'' = f_2(e't'), \qquad \dots\dots\dots\dots\dots(9)$$

where e, e' are the exposures, not generally equal, and f_1, f_2 represent two functions, whose forms may vary further with details of development and intensification. But for some printing processes f_2 might be treated as a fixed function. It would seem that this is the end at which discussion should begin. When the printing process is laid down and the character of the results yielded thereby is determined, it becomes possible to say what is required in the negative; but it is not possible before.

In many photographs it would appear that gradation tends to be lost at the ends of the scale, that is in the high lights and deep shadows, and (as a necessary consequence, if the full range is preserved) to be exaggerated in the half-tones. For some purposes, where precise reproduction is not desired, this feature may be of advantage. Consider, for example, the experimental problem, discussed by Huggins, of photographing the solar corona without an eclipse. The corona is always present, but is overpowered by atmospheric glare. The problem is to render evident a very small relative difference of luminous intensity. If the difference is exaggerated in a suitably exposed and developed photograph, so much the better. A repetition of successive copyings might render conspicuous a difference originally invisible. At each operation we may suppose a factor a to be introduced, a being greater than unity. After n copyings dt/t becomes $a^n dt/t$. Unless the gain each time were very decided, this would be a slow process, and it would be liable to fail in practice owing to multiplication of slight irregular photographic markings. But a method proposed by Mach[*] and the present writer[†] should be of service here. By the aid of reflexion light at each stage is transmitted *twice* through the picture. By this means alone a is raised to equality with 2, and upon it any purely photographic exaggeration of gradation is superposed. Three successive copyings on this plan should ensure at least a ten-fold exaltation of contrast.

Another method, simpler in execution, consists in superposing a considerable number (n) of similar pictures. In this way the contrast is multiplied n times. Rays from a small, but powerful, source of light fall first upon a collimating lens, so as to traverse the pile of pictures as a parallel beam. Another condensing lens brings the rays to a focus, at which point the eye is placed. Some trials on this plan made a year ago gave promising results. Ten lantern-slides were prepared from a portrait negative. The exposure (to gas-light) was for about 3 seconds through the negative and for 30 seconds bare, *i.e.* with negative removed, and the development was rather light. On single plates the picture was but just visible. Some rough photometry indicated that each plate transmitted about one-third of the incident light. In carrying out the exposures suitable stops, cemented to the negative, must be provided to guide the lantern-plates into position, and thus to ensure their subsequent exact superposition by simple mechanical means.

When only a few plates are combined, the light of a Welsbach mantle suffices; but, as was to be expected, the utilization of the whole number (ten)

---

[*] Eder's *Jahrbuch f. Photographie.*

[†] *Phil. Mag.* Vol. XLIV. p. 282 (1897); *Scientific Papers*, Vol. IV. p. 333.

requires a more powerful source. Good results were obtained with a lime-light; the portrait, barely visible at all on the single plates, came out fairly well under this illumination. If it were proposed to push the experiment much further by the combination of a larger number of plates, it would probably be advantageous to immerse them in benzole contained in a tank, so as to obviate the numerous reflexions at the surfaces.

It has been mentioned that in the above experiment the development of the plates was rather light. The question may be raised whether further development, or intensification, might not make one plate as good as two or three superposed. I think that to a certain extent this is so. When in a recent experiment one of the plates above described was intensified with mercuric chloride followed by ferrous oxalate, the picture was certainly more apparent than before, when backed by a sufficiently strong light. And the process of intensification may be repeated. But there is another point to be considered. In the illustrative experiment it was convenient to copy all the plates from the same negative. But this procedure would not be the proper one in an attempt to render visible the solar corona. For this purpose a good many independent pictures should be combined, so as to eliminate slight photographic defects. As in many physical measurements, when it is desired to enhance the delicacy, the aim must be to separate feeble constant effects from chance disturbances.

It may be that, besides that of the corona, there are other astronomical problems to which one or other of the methods above described, or a combination of both, might be applied with a prospect of attaining a further advance.

360.

ON THE PROPAGATION OF WAVES THROUGH A STRATIFIED MEDIUM, WITH SPECIAL REFERENCE TO THE QUESTION OF REFLECTION.

[*Proceedings of the Royal Society*, A, Vol. LXXXVI. pp. 207—266, 1912.]

THE medium is supposed to be such that its properties are everywhere a function of but one coordinate x, being of one uniform quality where x is less than a certain value x_1, and of another uniform quality (in general, different from the first) where x exceeds a greater value x_{m-1}; and the principal problem is the investigation of the reflection which in general ensues when plane waves in the first medium are incident upon the stratifications. For the present we suppose the quality to be uniform through strata of finite thickness, the first transition occurring when $x = x_1$, the second at $x = x_2$, and the last at $x = x_{m-1}$.

The expressions for the waves in the various media in order may be taken to be

$$\left. \begin{aligned}
\phi_1 &= A_1 e^{i[ct+by-a_1(x-x_1)]} + B_1 e^{i[ct+by+a_1(x-x_1)]}, \\
\phi_2 &= A_2 e^{i[ct+by-a_2(x-x_1)]} + B_2 e^{i[ct+by+a_2(x-x_1)]}, \\
\phi_3 &= A_3 e^{i[ct+by-a_3(x-x_2)]} + B_3 e^{i[ct+by+a_3(x-x_2)]},
\end{aligned} \right\} \quad \ldots\ldots\ldots\ldots(1)$$

and so on, the A's and B's denoting arbitrary constants. The first terms represent the waves travelling in the positive direction, the second those travelling in the negative direction; and our principal aim is the determination of the ratio B_1/A_1 imposed by the conditions of the problem, including the requirement that in the final medium there shall be no negative wave.

As in the simple transition from one uniform medium to another (*Theory of Sound*, § 270), the symbols c and b are common to all the media, the first depending merely upon the periodicity, while the constancy of the second is required in order that the traces of the various waves on the surfaces of transition should move together—equivalent to the ordinary law of refraction. In the usual optical notation, if V be the velocity of propagation and θ the angle of incidence,

$$c = 2\pi V/\lambda, \qquad b = (2\pi/\lambda) \sin \theta, \qquad a = (2\pi/\lambda) \cos \theta, \quad \ldots \ldots (2)$$

where V/λ, $\lambda^{-1} \sin \theta$ are the same in all the strata. On the other hand a is variable and is connected with the direction of propagation within the stratum by the relation

$$a = b \cot \theta. \quad \ldots \ldots \ldots \ldots \ldots \ldots \ldots \ldots \ldots \ldots \ldots \ldots \ldots (3)$$

The a's are thus known in terms of the original angle of incidence and of the various refractive indices.

Since the factor $e^{i\,(ct+by)}$ runs through all our expressions, we may regard it as understood and write simply

$$\phi_1 = A_1 e^{-ia_1\,(x-x_1)} + B_1 e^{ia_1\,(x-x_1)}, \quad \ldots \ldots \ldots \ldots \ldots \ldots (4)$$

$$\phi_2 = A_2 e^{-ia_2\,(x-x_1)} + B_2 e^{ia_2\,(x-x_1)}, \quad \ldots \ldots \ldots \ldots \ldots \ldots (5)$$

$$\phi_3 = A_3 e^{-ia_3\,(x-x_2)} + B_3 e^{ia_3\,(x-x_2)}, \quad \ldots \ldots \ldots \ldots \ldots \ldots (6)$$

$$\ldots \ldots \ldots \ldots \ldots \ldots \ldots \ldots \ldots \ldots \ldots \ldots \ldots$$

$$\phi_m = A_m e^{-ia_m\,(x-x_{m-1})} + B_m e^{ia_m\,(x-x_{m-1})}. \quad \ldots \ldots \ldots \ldots \ldots (7)$$

In the problem of reflection we are to make $B_m = 0$, and (if we please) $A_m = 1$.

We have now to consider the boundary conditions which hold at the surfaces of transition. In the case of sound travelling through gas, where ϕ is taken to represent the velocity-potential, these conditions are the continuity of $d\phi/dx$ and of $\sigma\phi$, where σ is the density. Whether the multiplier attaches to the dependent variable itself or to its derivative is of no particular significance. For example, if we take a new dependent variable ψ, equal to $\sigma\phi$, the above conditions are equivalent to the continuity of ψ and of $\sigma^{-1}d\psi/dx$. Nor should we really gain generality by introducing a multiplier in both places. We may therefore for the present confine ourselves to the acoustical form, knowing that the results will admit of interpretation in numerous other cases.

At the first transition $x = x_1$ the boundary conditions give

$$a_1 (B_1 - A_1) = a_2 (B_2 - A_2), \qquad \sigma_1 (B_1 + A_1) = \sigma_2 (B_2 + A_2). \quad \ldots \ldots (8)$$

If we stop here, we have the simple case of the juxtaposition of two media both of infinite depth. Supposing $B_2 = 0$, we get

$$\frac{B_1}{A_1} = \frac{\sigma_2/\sigma_1 - a_2/a_1}{\sigma_2/\sigma_1 + a_2/a_1} = \frac{\sigma_2/\sigma_1 - \cot \theta_2/\cot \theta_1}{\sigma_2/\sigma_1 + \cot \theta_2/\cot \theta_1}. \quad \ldots \ldots \ldots \ldots (9)$$

For a further discussion of (9) reference may be made to *Theory of Sound* (*loc. cit.*). In the case of the simple gases the compressibilities are the same, and $\sigma_1 \sin^2 \theta_1 = \sigma_2 \sin^2 \theta_2$. The general formula (9) then identifies itself with Fresnel's expression

$$\frac{\tan (\theta_1 - \theta_2)}{\tan (\theta_1 + \theta_2)}. \quad \dots\dots\dots\dots\dots\dots\dots(10)$$

On the other hand, if $\sigma_2 = \sigma_1$, the change being one of compressibility only, we find

$$(9) = \frac{\sin (\theta_2 - \theta_1)}{\sin (\theta_2 + \theta_1)}, \quad \dots\dots\dots\dots\dots(11)$$

Fresnel's other expression.

In the above it is supposed that a_2 (and θ_2) are real. If the wave be incident in the more refractive medium and the angle of incidence be too great, a_2 becomes imaginary, say $-ia_2'$. In this case, of course, the reflection is total, the modulus of (9) becoming unity. The change of phase incurred is given by (9). In accordance with what has been said these results are at once available for the corresponding optical problems.

If there are more than two media, the boundary conditions at $x = x_2$ are

$$a_2 \left\{ B_2 e^{ia_2 (x_2 - x_1)} - A_2 e^{-ia_2 (x_2 - x_1)} \right\} = a_3 (B_3 - A_3), \quad \dots\dots\dots(12)$$

$$\sigma_2 \left\{ B_2 e^{ia_2 (x_2 - x_1)} + A_2 e^{-ia_2 (x_2 - x_1)} \right\} = \sigma_3 (B_3 + A_3), \quad \dots\dots\dots(13)$$

and so on. For extended calculations it is desirable to write these equations in an abbreviated shape. We set

$$B_2 - A_2 = H_2, \qquad B_2 + A_2 = K_2, \quad \text{etc.,} \quad \dots\dots\dots\dots(14)$$

$$\cos a_2 (x_2 - x_1) = c_1, \qquad i \sin a_2 (x_2 - x_1) = s_1, \quad \text{etc.} \dots\dots\dots(15)$$

$$a_3 / a_2 = \alpha_2, \qquad \sigma_3 / \sigma_2 = \beta_2, \quad \text{etc.;} \quad \dots\dots\dots\dots(16)$$

and the series of equations then takes the form

$$H_1 = \alpha_1 H_2, \qquad\qquad K_1 = \beta_1 K_2, \quad \dots\dots\dots\dots(17)$$

$$c_1 H_2 + s_1 K_2 = \alpha_2 H_3, \qquad s_1 H_2 + c_1 K_2 = \beta_2 K_3, \quad \dots\dots\dots\dots(18)$$

$$c_2 H_3 + s_2 K_3 = \alpha_3 H_4, \qquad s_2 H_3 + c_2 K_3 = \beta_3 K_4, \quad \dots\dots\dots\dots(19)$$

and so on. In the reflection problem the special condition is the numerical equality of H and K of highest suffix. We may make

$$H = -1, \qquad K = +1. \quad \dots\dots\dots\dots\dots\dots(20)$$

As we have to work backwards from the terms of highest suffix, it is convenient to solve algebraically each pair of simple equations. In this way, remembering that $c^2 - s^2 = 1$, we get

$$H_1 = \alpha_1 H_2, \qquad\qquad K_1 = \qquad\qquad \beta_1 K_2, \quad \dots\dots\dots(21)$$

$$H_2 = c_1 \alpha_2 H_3 - s_1 \beta_2 K_3, \qquad K_2 = -s_1 \alpha_2 H_3 + c_1 \beta_2 K_3, \quad \dots\dots\dots(22)$$

$$H_3 = c_2 \alpha_3 H_4 - s_2 \beta_3 K_4, \qquad K_3 = -s_2 \alpha_3 H_4 + c_2 \beta_3 K_4, \quad \dots\dots\dots(23)$$

and so on. In these equations the c's and the β's are real, and also the α's, unless there is "total reflection"; the s's are pure imaginaries, with the same reservation.

When there are three media, we are to suppose in the problem of reflection that $H_3 = -1$, $K_3 = 1$. Thus from (21), (22),

$$H_1 = -\alpha_1 (c_1\alpha_2 + s_1\beta_2), \qquad K_1 = \beta_1 (s_1\alpha_2 + c_1\beta_2);$$

so that
$$\frac{B_1}{A_1} = \frac{K_1 + H_1}{K_1 - H_1} = \frac{c_1 (\beta_1\beta_2 - \alpha_1\alpha_2) + s_1 (\alpha_2\beta_1 - \alpha_1\beta_2)}{c_1 (\beta_1\beta_2 + \alpha_1\alpha_2) + s_1 (\alpha_2\beta_1 + \alpha_1\beta_2)}. \quad\dots\dots(24)$$

If there be no "total reflection," the relative intensity of the reflected waves is

$$\frac{c_1{}^2 (\beta_1\beta_2 - \alpha_1\alpha_2)^2 - s_1{}^2 (\alpha_2\beta_1 - \alpha_1\beta_2)^2}{c_1{}^2 (\beta_1\beta_2 + \alpha_1\alpha_2)^2 - s_1{}^2 (\alpha_2\beta_1 + \alpha_1\beta_2)^2}, \quad\dots\dots\dots(25)$$

where
$$c_1{}^2 = \cos^2 a_2 (x_2 - x_1), \qquad -s_1{}^2 = \sin^2 a_2 (x_2 - x_1). \quad\dots\dots(26)$$

The reflection will vanish independently of the values of c_1 and s_1, i.e., whatever may be the thickness of the middle layer, provided

$$\beta_1\beta_2 - \alpha_1\alpha_2 = 0, \quad \alpha_2\beta_1 - \alpha_1\beta_2 = 0; \quad \text{or} \quad \beta_1 = \alpha_1, \quad \beta_2 = \alpha_2,$$

since these quantities are all positive. Reference to (9) shows that these are the conditions of vanishing reflection at the two surfaces of transition considered separately.

If these conditions be not satisfied, the evanescence of (25) requires that either c_1 or s_1 be zero. The latter case is realized if the intermediate layer be abolished, and the remaining condition is equivalent to $\sigma_3/\sigma_1 = a_3/a_1$, as was to be expected from (9). We learn now that, if there would be no reflection in the absence of an intermediate layer, its introduction will have no effect provided $a_2 (x_2 - x_1)$ be a multiple of π. An obvious example is when the first and third media are similar, as in the usual theory of "thin plates."

On the other hand, if c_1, or $\cos a_2 (x_2 - x_1)$, vanish, the remaining requirement for the evanescence of (25) is that $\beta_2/\alpha_2 = \beta_1/\alpha_1$.

In this case
$$\frac{\beta_1 - \alpha_1}{\beta_1 + \alpha_1} = \frac{\beta_2 - \alpha_2}{\beta_2 + \alpha_2};$$

so that by (9) the reflections at the two faces are equal in all respects.

In general, if the third and first media are similar, (25) reduces to

$$\frac{\{\beta_1/\alpha_1 - \alpha_1/\beta_1\}^2 \sin^2 a_2 (x_2 - x_1)}{4 \cos^2 a_2 (x_2 - x_1) + \{\beta_1/\alpha_1 + \alpha_1/\beta_1\}^2 \sin^2 a_2 (x_2 - x_1)}, \quad\dots\dots(27)$$

which may readily be identified with the expression usually given in terms of (9).

It remains to consider the cases of so-called total reflection. If this occurs only at the *second* surface of transition, a_1, a_2 are real, while a_3 is a pure imaginary. Thus α_1 is real, and α_2 is imaginary; c_1 is real always, and s_1 is imaginary as before; the β's are always real. Thus, if we separate the real and imaginary parts of the numerator and denominator of (24), we get

$$\frac{B_1}{A_1} = \frac{c_1\beta_1\beta_2 + s_1\alpha_2\beta_1 - c_1\alpha_1\alpha_2 - s_1\alpha_1\beta_2}{c_1\beta_1\beta_2 + s_1\alpha_2\beta_1 + c_1\alpha_1\alpha_2 + s_1\alpha_1\beta_2}, \quad \ldots\ldots\ldots\ldots(28)$$

of which the modulus is unity. In this case, accordingly, the reflection back in the first medium is literally total, whatever may be the thickness of the intermediate layer, as was to be expected.

The separation of real and imaginary parts follows the same rule when a_2 is imaginary, as well as a_3. For then α_1 is imaginary, while σ_2, s_1 are real. Thus $s_1\alpha_2\beta_1$ remains real, and $c_1\alpha_1\alpha_2$, $s_1\alpha_1\beta_2$ remain imaginary. The reflection back in the first medium is total in this case also.

The only other case requiring consideration occurs when a_2 is imaginary and a_3 real. The reflection is then total if the middle layer be thick enough, but if this thickness be reduced, the reflection cannot remain total, as is evident if we suppose the thickness to vanish. The ratios α_1, α_2 are now both imaginary, while s_1 is real. The separation of real and imaginary parts stands as in (24), and the intensity of reflection is still expressed by (25). If we take $a_2 = -ia_2'$, we may write in place of (25),

$$\frac{(\beta_1\beta_2 - \alpha_1\alpha_2)^2\cosh^2 a_2'(x_2 - x_1) - (\alpha_2\beta_1 - \alpha_1\beta_2)^2\sinh^2 a_2'(x_2 - x_1)}{(\beta_1\beta_2 + \alpha_1\alpha_2)^2\cosh^2 a_2'(x_2 - x_1) - (\alpha_2\beta_1 + \alpha_1\beta_2)^2\sinh^2 a_2'(x_2 - x_1)} . \ldots(29)$$

When $x_2 - x_1$ is extremely small, this reduces to

$$\frac{(\beta_1\beta_2 - \alpha_1\alpha_2)^2}{(\beta_1\beta_2 + \alpha_1\alpha_2)^2}, \quad \text{or} \quad \frac{(\sigma_3/\sigma_1 - a_3/a_1)^2}{(\sigma_3/\sigma_1 + a_3/a_1)^2},$$

in accordance with (9).

When on the other hand $x_2 - x_1$ exceeds a few wave-lengths, (29) approaches unity, as we see from a form, equivalent to (29), viz.,

$$\frac{(\beta_1^2 - \alpha_1^2)(\beta_2^2 - \alpha_2^2)\cosh^2 a_2'(x_2 - x_1) + (\alpha_2\beta_1 - \alpha_1\beta_2)^2}{(\beta_1^2 - \alpha_1^2)(\beta_2^2 - \alpha_2^2)\cosh^2 a_2'(x_2 - x_1) + (\alpha_2\beta_1 + \alpha_1\beta_2)^2} . \ldots\ldots(30)$$

It is to be remembered that in (30), α_1^2, α_2^2, $\alpha_1\alpha_2$ have negative values.

The form assumed when the third medium is similar to the first may be noted. In this case $\alpha_1\alpha_2 = 1$, $\beta_1\beta_2 = 1$, and we get from (29)

$$\frac{(\beta_1/\alpha_1 - \alpha_1/\beta_1)^2\sinh^2 a_2'(x_2 - x_1)}{(\beta_1/\alpha_1 - \alpha_1/\beta_1)^2\sinh^2 a_2'(x_2 - x_1) - 4} . \ldots\ldots\ldots\ldots(31)$$

In this case, of course, the reflection vanishes when $x_2 - x_1$ is sufficiently reduced.

Equations (21), etc., may be regarded as constituting the solution of the general problem. If there are m media, we suppose $H_m = -1$, $K_m = 1$, thence calculate in order from the pairs of simple equations H_{m-1}, K_{m-1}; H_{m-2}, K_{m-2}, etc., until H_1 and K_1 are reached; and then determine the ratio B_1/A_1. The procedure would entail no difficulty in any special case numerically given; but the algebraic expression of H_1 and K_1 in terms of H_m and K_m soon becomes complicated, unless further simplifying conditions are introduced. Such simplification may be of two kinds. In the first it is supposed that the total thickness between the initial and final media is small relatively to the wave-lengths, so that the phase-changes occurring within the layer are of subordinate importance. In the second kind of simplification the thicknesses are left arbitrary, but the changes in the character of the medium, which occur at each transition, are supposed small.

The problem of a thin transitional layer has been treated by several authors, L. Lorenz[*], Van Ryn[†], Drude[‡], Schott[§], and Maclaurin[||]. A full account will be found in *Theory of Light* by the last named. It will therefore not be necessary to treat the subject in detail here; but it may be worth while to indicate how the results may be derived from our equations. For this purpose it is convenient to revert to the original notation so far as to retain a and σ. Thus in place of (17), etc., we write

$$a_1 H_1 = a_2 H_2, \qquad\qquad \sigma_1 K_1 = \sigma_2 K_2, \qquad \dots\dots(32)$$

$$a_2(c_1 H_2 + s_1 K_2) = a_3 H_3, \qquad \sigma_2(s_1 H_2 + c_1 K_2) = \sigma_3 K_3, \text{ etc. } \dots(33)$$

In virtue of the supposition that all the layers are thin, the c's are nearly equal to unity and the s's are small. Thus, for a first approximation, we identify c with 1 and neglect s altogether, so obtaining

$$a_1 H_1 = a_2 H_2 = \dots = a_m H_m, \qquad \sigma_1 K_1 = \sigma_2 K_2 = \dots = \sigma_m K_m. \dots(34)$$

The relation of H_1, K_1 to H_m, K_m is the same as if the transition between the extreme values took place without intermediate layers, and the law of reflection is not disturbed by the presence of these layers, as was to be expected.

For the second approximation we may still identify the c's with unity, while the s's are retained as quantities of the first order. Adding together the column of equations constituting the first members of (32), (33), etc., we find

$$a_1 H_1 + a_2 s_1 K_2 + a_3 s_2 K_3 + \dots + a_{m-1} s_{m-2} K_{m-1} = a_m H_m; \dots\dots(35)$$

and in like manner, with substitution of σ for a and interchange of K and H,

$$\sigma_1 K_1 + \sigma_2 s_1 H_2 + \dots + \sigma_{m-1} s_{m-2} H_{m-1} = \sigma_m K_m. \dots\dots(36)$$

[*] *Pogg. Ann.* 1860, Vol. CXI. p. 460.

[†] *Wied. Ann.* 1883, Vol. XX. p. 22.

[‡] *Wied. Ann.* 1891, Vol. XLIII. p. 126.

[§] *Phil. Trans.* 1894, Vol. CLXXXV. p. 823.

[||] *Roy. Soc. Proc.* A, 1905, Vol. LXXVI. p. 49.

In the small terms containing s's we may substitute the approximate values of H and K from (34). For the problem of reflection we suppose $H_m + K_m = 0$. Hence

$$\frac{H_1}{K_1} = -\frac{\sigma_1 a_m}{\sigma_m a_1} \frac{1 + \frac{\sigma_m}{a_m} \Sigma \frac{a_2 s_1}{\sigma_2}}{1 + \frac{a_m}{\sigma_m} \Sigma \frac{\sigma_2 s_1}{a_2}}. \qquad (37)$$

In (37), $s_1 = i a_2 (x_2 - x_1)$, and so on, so that

$$\Sigma \frac{a_2 s_1}{\sigma_1} = i \int \frac{a^2 dx}{\sigma}, \qquad \Sigma \frac{\sigma_2 s_1}{a_2} = i \int \sigma \, dx, \qquad (38)$$

the integration extending over the layer of transition.

One conclusion may be drawn at once. To this degree of approximation the reflection is independent of the order of the strata. It will be noted that the sums in (37) are pure imaginaries. In what follows we shall suppose that a_m is real.

As the final result for the reflection, we find

$$\frac{B_1}{A_1} = \frac{K_1 + H_1}{K_1 - H_1} = R e^{i\delta}, \qquad (39)$$

where

$$R = \frac{\sigma_m/\sigma_1 - a_m/a_1}{\sigma_m/\sigma_1 + a_m/a_1}, \qquad (40)$$

$$\tan \delta = 2 \frac{\dfrac{a_m}{\sigma_m} \displaystyle\int \sigma \, dx - \dfrac{\sigma_m}{a_m} \displaystyle\int \dfrac{a^2 dx}{\sigma}}{\dfrac{a_1 \sigma_m}{a_m \sigma_1} - \dfrac{a_m \sigma_1}{a_1 \sigma_m}}. \qquad (41)$$

To this order of approximation the *intensity* of the reflection is unchanged by the presence of the intermediate layers, unless, indeed, the circumstances are such that (40) is itself small. If $\sigma_m/\sigma_1 = a_m/a_1$ absolutely, we have

$$R = \tfrac{1}{2} \left\{ \frac{a_m}{\sigma_m} \int \sigma \, dx - \frac{\sigma_m}{a_m} \int \frac{a^2 dx}{\sigma} \right\} \qquad (42)$$

and $\delta = \tfrac{1}{2}\pi$. This case is important in Optics, as representing the reflection at the polarising angle from a contaminated surface.

The two important optical cases: (i) where σ is constant, leading (when there is no transitional layer) to Fresnel's formula (11), and (ii) where $\sigma \sin^2 \theta$ is constant, leading to (10), are now easily treated as special examples. Introducing the refractive index μ, we find after reduction for case (i)

$$\delta' = -\frac{4\pi \cos \theta_1}{\lambda_1 (\mu_m^2 - \mu_1^2)} \int (\mu_m^2 - \mu^2) \, dx, \qquad (43)$$

where λ_1, μ_1 relate to the first medium, μ_m is the index for the last medium, and the integration is over the layer of transition. The application of (43) should be noticed when the layer is in effect abolished, either by supposing $\mu = \mu_m$, or, on the other hand, $\mu = \mu_1$.

In the second case (42), corresponding to the polarising angle, becomes

$$R = \frac{\pi}{\lambda_1 \mu_1 \sqrt{(\mu_m{}^2 + \mu_1{}^2)}} \int \frac{(\mu^2 - \mu_1{}^2)(\mu^2 - \mu_m{}^2)}{\mu^2}\, dx. \quad \dots\dots\dots(44)$$

In general for this case

$$\delta'' = -\frac{4\pi \cos\theta_1}{\lambda_1} \frac{\cos^2\theta_1 \int (\mu_m{}^2 - \mu^2)\, dx - \sin^2\theta_1 \int (\mu_m{}^2 - \mu^2)\left(\dfrac{\mu_1{}^2}{\mu^2} + \dfrac{\mu_1{}^2}{\mu_m{}^2} - 1\right) dx}{(\mu_m{}^2 - \mu_1{}^2)\left(\cos^2\theta_1 - \dfrac{\mu_1{}^2}{\mu_m{}^2}\sin^2\theta_1\right)}.$$

$$\dots\dots(45)$$

The second fraction in (45) is equal to the thickness of the layer of transition simply, when we suppose $\mu = \mu_1$.

Further, $\quad \delta'' - \delta' = -\dfrac{4\pi}{\lambda_1} \dfrac{\cos\theta_1 \sin^2\theta_1}{\mu_m{}^2 - \mu_1{}^2} \dfrac{\displaystyle\int \frac{(\mu_m{}^2 - \mu^2)(\mu^2 - \mu_1{}^2)}{\mu}\, dx}{\cos^2\theta_1 - \dfrac{\mu_1{}^2}{\mu_m{}^2}\sin^2\theta_1}, \quad \dots\dots(46)$

the difference of phase vanishing, as it ought to do, when $\mu = \mu_1$, or μ_m, or again, when $\theta_1 = 0$.

It should not escape notice that the expressions (10) and (11) have different signs when θ_1 and θ_2 are small. This anomaly, as it must appear from an optical point of view, should be corrected when we consider the significance of $\delta'' - \delta'$. The origin of it lies in the circumstance that, in our application of the boundary conditions, we have, in effect, used different vectors as dependent variables to express light of the two polarisations. For further explanation reference may be made to former writings, e.g. "On the Dynamical Theory of Gratings*."

If throughout the range of integration, μ is intermediate between the terminal values μ_1, μ_m, the reflection is of the kind called positive by Jamin. The transition may well be of this character when there is no contamination. On the other hand, the reflection is negative, if μ has throughout a value outside the range between μ_1 and μ_m. It is probable that something of this kind occurs when water has a greasy surface.

The formulæ required in Optics, viz. (43), (44), (45), (46), are due, in substance, to Lorenz and Van Ryn. The more general expressions (41), (42) do not seem to have been given.

There is no particular difficulty in pursuing the approximation from (32), etc. At the next stage the second term in the expansion of the c's

---

* *Roy. Soc. Proc.* A, 1907, Vol. LXXIX. p. 413.

must be retained, while the s's are still sufficiently represented by the first terms. The result, analogous to (37), (38), is

$$\frac{H_1}{K_1} = -\frac{\sigma_1 a_m}{\sigma_m a_1} \frac{1 - \int_0^d \sigma \cdot \int_0^x \frac{a^2 dx}{\sigma} \cdot dx + i \frac{\sigma_m}{a_m} \int_0^d \frac{a^2}{\sigma} dx}{1 - \int_0^d \frac{a^2}{\sigma} \cdot \int_0^x \sigma dx \cdot dx + i \frac{a_m}{\sigma_m} \int_0^d \sigma dx}, \quad \ldots\ldots\ldots(47)$$

in which the terminal abscissæ of the variable layer are taken to be 0 and d, instead of x_1 and x_{m-1}. I do not follow out the application to particular cases such as $\sigma = $ constant, or $\sigma \sin^2 \theta = $ constant. For this reference may be made to Maclaurin, who, however, uses a different method.

The second case which allows of a simple approximate expression for the reflection arises when all the partial reflections are small. It is then hardly necessary to appeal to the general equations: the method usually employed in Optics suffices. The assumptions are that at each surface of transition the incident waves may be taken to be the same as in the first medium, merely retarded by the appropriate amount, and that each partial reflection reaches the first medium no otherwise modified than by such retardation. This amounts to the neglect of waves three times reflected. Thus

$$\frac{B_1}{A_1} = \frac{\beta_1 - \alpha_1}{\beta_1 + \alpha_1} + \frac{\beta_2 - \alpha_2}{\beta_2 + \alpha_2} e^{-2i a_2 (x_2 - x_1)} + \frac{\beta_3 - \alpha_3}{\beta_3 + \alpha_3} e^{-2i[a_2(x_2-x_1)+a_3(x_3-x_2)]} + \ldots \quad (48)$$

An interesting question suggests itself as to the manner in which the transition from one uniform medium to another must be effected in order to obviate reflection, and especially as to the least thickness of the layer of transition consistent with this result. If there be two transitions only, the least thickness of the layer is obtained by supposing in (48)

$$\frac{\beta_1 - \alpha_1}{\beta_1 + \alpha_1} = \frac{\beta_2 - \alpha_2}{\beta_2 + \alpha_2} \quad \ldots\ldots\ldots\ldots\ldots\ldots\ldots\ldots\ldots(49)$$

and

$$2a_2 (x_2 - x_1) = \pi ; \quad \ldots\ldots\ldots\ldots\ldots\ldots\ldots\ldots(50)$$

and this conclusion, as we have seen already, is not limited to the case of small differences of quality. In its application to perpendicular incidence, (50) expresses that the thickness of the layer is one-quarter of the wavelength proper to the layer. The two partial reflections are equal in magnitude and sign. It is evident that nothing better than this can be done so long as the reflections are of one sign, however numerous the surfaces of transition may be.

If we allow the partial reflections to be of different signs, some reduction of the necessary thickness is possible. For example, suppose that there are two intermediate layers of equal thickness, of which the first is similar to the final uniform medium, and the second similar to the initial uniform medium. Of the three partial reflections the first and third are similar, but the second is of the opposite sign. If three vectors of equal numerical value compensate one another, they must be at angles of 120°. The necessary conditions are satisfied (in the case of perpendicular transmission) if the total thickness (2*l*) is $\frac{1}{6}\lambda$, in accordance with

$$1 - e^{-4\pi i l/\lambda} + e^{-8\pi i l/\lambda} = 0.$$

The total thickness of the layer of transition is thus somewhat reduced, but only by a very artificial arrangement, such as would not usually be contemplated when a layer of transition is spoken of. If the progress from the first to the second uniform quality be always *in one direction*, reflection cannot be obviated unless the layer be at least $\frac{1}{4}\lambda$ thick.

The general formula (48) may be adapted to express the result appropriate to continuous variation of the medium. Suppose, for example, that σ is constant, making $\beta = 1$, and corresponding to the continuity of both ϕ and $d\phi/dx$.* It is convenient to suppose that the variation commences at $x = 0$. Then (48) may be written

$$\frac{B_1}{A_1} = -\int \frac{da}{2a} e^{-2i\int_0^x a\,dx}, \qquad \dots\dots\dots\dots\dots\dots(51)$$

a at any point x being connected with the angle of propagation by the usual relation (3). In the special case of perpendicular propagation, $a = 2\pi\mu/\lambda_1\mu_1$, μ being refractive index and λ_1, μ_1 relating to the first medium.

A curious example, theoretically possible even if unrealizable in experiment, arises when the variable medium is constituted in such a manner that the velocity of propagation is everywhere constant, so that there is no refraction. Then a is constant, $\alpha = 1$, and (48) gives

$$\frac{B_1}{A_1} = \int \frac{d\sigma}{2\sigma} e^{-2ia x}. \qquad \dots\dots\dots\dots\dots\dots\dots(52)$$

Some of the questions relating to the propagation of waves in a variable medium are more readily treated on the basis of the appropriate differential equation. As in (1), we suppose that the waves are plane, and that the medium is stratified in plane strata perpendicular to x, and we usually omit the exponential factors involving t and y, which may be supposed to run through. In the case of perpendicular propagation, y would not appear at all.

Consider the differential equation

$$\frac{d^2\phi}{dx^2} + k^2\phi = 0, \qquad \dots\dots\dots\dots\dots\dots\dots(53)$$

in which (unless k^2 can be infinite) it is necessary to suppose that ϕ and $d\phi/dx$ are continuous; k^2 is a function of x, which must be everywhere

---

* These would be the conditions appropriate to a stretched string of variable longitudinal density vibrating transversely.

positive when the transmission is perpendicular, as, for example, in the case of a stretched string. When the transmission is oblique to the strata, k^2 may become negative, corresponding to "total reflection," but in most of what follows we shall assume that this does not happen. The continuity of ϕ and $d\phi/dx$, even though k^2 be discontinuous, appears to limit the application of (53) to certain kinds of waves, although, as a matter of analysis, the general differential equation of the second order may always be reduced to this form *.

In the theory of a uniform medium, we may consider stationary waves or progressive waves. The former may be either

$$\phi = A \cos k_0 x \cos pt, \quad \text{or} \quad \phi = B \sin k_0 x \sin pt;$$

and, if $B = \pm A$, the two may be combined, so as to constitute progressive waves

$$\phi = A \cos (pt \pm k_0 x).$$

Conversely, progressive waves, travelling in opposite directions, may be combined so as to constitute stationary waves. When we pass to variable media, no ambiguity arises respecting stationary waves; they are such that the phase is the same at all points. But is there such a thing as a progressive wave? In the full sense of the phrase there is not. In general, if we contemplate the wave forms at two different times, the difference between them cannot be represented by a mere shift of position proportional to the interval of time which has elapsed.

The solution of (53) may be taken to be

$$\phi = A'\psi(x) + B'\chi(x), \quad \ldots\ldots\ldots\ldots\ldots\ldots(54)$$

where $\psi(x)$, $\chi(x)$ are real oscillatory functions of x; A', B', arbitrary constants as regards x. If we introduce the time-factor, writing p in place of the less familiar c of (1), we may take

$$\phi = A \cos pt \cdot \psi(x) + B \sin pt \cdot \chi(x); \quad \ldots\ldots\ldots\ldots(55)$$

and this may be put into the form

$$\phi = H \cos(pt - \theta), \quad \ldots\ldots\ldots\ldots\ldots\ldots(56)$$

where
$$H \cos \theta = A\psi(x), \quad H \sin \theta = B\chi(x), \quad \ldots\ldots\ldots\ldots(57)$$

or
$$H^2 = A^2 [\psi(x)]^2 + B^2 [\chi(x)]^2, \quad \ldots\ldots\ldots\ldots\ldots(58)$$

$$\theta = \tan^{-1} \frac{B \cdot \chi(x)}{A \cdot \psi(x)}. \quad \ldots\ldots\ldots\ldots\ldots\ldots(59)$$

But the expression for ϕ in (56) cannot be said to represent in general a progressive wave. We may illustrate this even from the case of the uniform medium where $\psi(x) = \cos kx$, $\chi(x) = \sin kx$. In this case (56) becomes

$$\phi = \{A^2 \cos^2 kx + B^2 \sin^2 kx\}^{\frac{1}{2}} \cos \left\{ pt - \tan^{-1} \left(\frac{B}{A} \tan kx \right) \right\}. \quad \ldots(60)$$

---

* Forsyth's *Differential Equations*, § 59.

6

If $B = \pm A$, reduction ensues to the familiar positive or negative progressive wave. But if B be not equal to $\pm A$, (55), taking the form

$$\phi = \tfrac{1}{2}(A + B)\cos(pt - kx) + \tfrac{1}{2}(A - B)\cos(pt + kx),$$

clearly does not represent a progressive wave. The mere possibility of reduction to the form (57) proves little, without an examination of the character of H and θ.

It may be of interest to consider for a moment the character of θ in (60). If B/A, or, say, m, is positive, θ may be identified with kx at the quadrants but elsewhere they differ, unless $m = 1$. Introducing the imaginary expressions for tangents, we find

$$\theta = kx + M\sin 2kx + \tfrac{1}{2}M^2\sin 4kx + \tfrac{1}{3}M^3\sin 6kx + \ldots, \quad \ldots\ldots(61)$$

where

$$M = \frac{m-1}{m+1}. \quad \ldots\ldots\ldots\ldots\ldots\ldots\ldots\ldots(62)$$

When k is constant, one of the solutions of (53) makes ϕ proportional to e^{-ikx}. Acting on this suggestion, and following out optical ideas, let us assume in general

$$\phi = \eta e^{-i\int a\, dx}, \quad \ldots\ldots\ldots\ldots\ldots\ldots\ldots(63)$$

where the amplitude η and a are real functions of x, which, for the purpose of approximations, may be supposed to vary slowly. Substituting in (53), we find

$$\frac{d^2\eta}{dx^2} + (k^2 - a^2)\eta - 2ia^{\frac{1}{2}}\frac{d}{dx}(a^{\frac{1}{2}}\eta) = 0. \quad \ldots\ldots\ldots\ldots(64)$$

For a first approximation, we neglect $d^2\eta/dx^2$. Hence

$$k = a, \qquad k^{\frac{1}{2}}\eta = C, \ldots\ldots\ldots\ldots\ldots\ldots\ldots\ldots(65)$$

so that

$$\phi = Ck^{-\frac{1}{2}}e^{ipt}e^{-i\int k\, dx}, \quad \ldots\ldots\ldots\ldots\ldots\ldots(66)$$

or in real form,

$$\phi = Ck^{-\frac{1}{2}}\cos(pt - \int k\, dx). \quad \ldots\ldots\ldots\ldots\ldots(67)$$

If we hold rigorously to the suppositions expressed in (65), the satisfaction of (64) requires that $d^2\eta/dx^2 = 0$, or $d^2k^{-\frac{1}{2}}/dx^2 = 0$. With omission of arbitrary constants affecting merely the origin and the scale of x, this makes $k^2 = x^{-4}$, corresponding to the differential equation

$$x^4\frac{d^2\phi}{dx^2} + \phi = 0, \quad \ldots\ldots\ldots\ldots\ldots\ldots(68)$$

whose accurate solution is accordingly

$$\phi = Cxe^{i(pt-1/x)}. \quad \ldots\ldots\ldots\ldots\ldots\ldots(69)$$

In (69) the imaginary part may be rejected. The solution (69) is, of course, easily verified. In all other cases (67) is only an approximation.

As an example, the case where $k^2 = n^2/x^2$ may be referred to. Here $\int k\,dx = n \log x - \epsilon$, and (67) gives

$$\phi = Cx^{\frac{1}{2}} \cos\left(pt - n \log x + \epsilon\right) \quad\ldots\ldots\ldots\ldots\ldots(70)$$

as an approximate solution. We shall see presently that a slight change makes it accurate.

Reverting to (64), we recognize that the first and second terms are real, while the third is imaginary. The satisfaction of the equation requires therefore that

$$a^{\frac{1}{2}}\eta = C, \quad\ldots\ldots\ldots\ldots\ldots\ldots\ldots\ldots(71)$$

and that

$$k^2 = C^4\eta^{-4} - \frac{1}{\eta}\frac{d^2\eta}{dx^2}; \quad\ldots\ldots\ldots\ldots\ldots\ldots(72)$$

while (63) becomes

$$\phi = \eta e^{-iC^2\int\eta^{-2}dx}. \quad\ldots\ldots\ldots\ldots\ldots\ldots\ldots(73)$$

Let us examine in what cases η may take the form Dx^r. From (72),

$$k^2 = C^4 D^{-4} x^{-4r} - r(r-1)x^{-2}. \quad\ldots\ldots\ldots\ldots\ldots(74)$$

If $r = 0$, k^2 is constant. If $r = 1$, $k^2 = C^4 D^{-4} x^{-4}$, already considered in (68). The only other case in which k^2 is a simple power of x occurs when $r = \frac{1}{2}$, making

$$k^2 = (C^4 D^{-4} + \tfrac{1}{4})x^{-2} = n^2 x^{-2} \text{ (say)}. \quad\ldots\ldots\ldots\ldots(75)$$

Here $\eta = Dx^{\frac{1}{2}}$, $C^2\int\eta^{-2}\,dx = C^2/D^2 . \log x - \epsilon$, and the realized form of (73) is

$$\phi = Dx^{\frac{1}{2}}\cos\left\{pt - \sqrt{(n^2 - \tfrac{1}{4})}\log x + \epsilon\right\}, \quad\ldots\ldots\ldots\ldots(76)$$

which is the exact form of the solution obtained by approximate methods in (70). For a discussion of (76) reference may be made to *Theory of Sound*, second edition, § 148 b.

The relation between a and η in (71) is the expression of the energy condition, as appears readily if we consider the application to waves along a stretched string. From (53), with restoration of e^{ipt},

$$\frac{d\phi}{dt} = e^{ipt} e^{-i\int a\,dx} . ip\eta, \qquad \frac{d\phi}{dx} = e^{ipt} e^{-i\int a\,dx}\left\{\frac{d\eta}{dx} - ia\eta\right\}.$$

If the common phase factors be omitted, the parts of $d\phi/dt$ and $d\phi/dx$ which are in the same phase are as $p\eta$ and $a\eta$, and thus the mean work transmitted at any place is as $a\eta^2$. Since there is no accumulation of energy between two places, $a\eta^2$ must be constant.

When the changes are gradual enough, a may be identified with k, and then $\eta \propto k^{-\frac{1}{2}}$, as represented in (67).

If we regard η as a given function of x, a follows when C has been chosen, and also k^2 from (72). In the case of perpendicular propagation k^2 cannot be negative, but this is the only restriction. When η is constant, k^2 is constant;

and thus if we suppose η to pass from one constant value to another through a finite transitional layer, the transition is also from one uniform k^2 to another; and (73) shows that there is no reflection back into the first medium.   If the terminal values of η and therefore of k^2 be given, and the transitional layer be thick enough, it will always be possible, and that in an infinite number of ways, to avoid a negative k^2, and thus to secure complete transmission without reflection back ; but if with given terminal values the layer be too much reduced, k^2 must become negative.   In this case reflection cannot be obviated.

It may appear at first sight as if this argument proved too much, and that there should be no reflection in any case so long as k^2 is positive throughout. But although a constant η requires a constant k^2, it does not follow conversely that a constant k^2 requires a constant η, and, in fact, this is not true. *One* solution of (72), when k^2 is constant, certainly is $\eta^2 = C^2/k$; but the complete solution necessarily includes two arbitrary constants, of which C is not one.   From (60) it may be anticipated that a solution of (72) may be

$$\eta^2 = A^2\cos^2 kx + B^2\sin^2 kx = \tfrac{1}{2}(A^2 + B^2) + \tfrac{1}{2}(A^2 - B^2)\cos 2kx. \quad....(77)$$

From this we find on differentiation

$$\eta^3\frac{d^2\eta}{dx^2} + k^2\eta^4 = k^2 A^2 B^2 ;$$

and thus (72) is satisfied, provided that

$$k^2 A^2 B^2 = C^2. \quad..............................(78)$$

It appears then that (77) subject to (78) is a solution of (72).   The second arbitrary constant evidently takes the form of an arbitrary addition to x, and η will not be constant unless $A^2 = B^2$.

On the supposition that η and a are slowly varying functions, the approximations of (65) may be pursued.   We find

$$\eta = Ck^{-\frac{1}{2}}\left\{1 - \tfrac{1}{4}k^{-\frac{3}{2}}\frac{d^2 k^{-\frac{1}{2}}}{dx^2}\right\}, \quad....................(79)$$

$$a = k + \tfrac{1}{2}k^{-\frac{1}{2}}\frac{d^2 k^{-\frac{1}{2}}}{dx^2}. \quad............................(80)$$

The retardation, as usually reckoned in optics, is $\int k\,dx$.   The additional retardation according to (80) is

$$\tfrac{1}{2}\int k^{-\frac{1}{2}}\frac{d^2 k^{-\frac{1}{2}}}{dx^2}\,dx = \tfrac{1}{2}\left[k^{-\frac{1}{2}}\frac{dk^{-\frac{1}{2}}}{dx}\right] - \tfrac{1}{2}\int\left(\frac{dk^{-\frac{1}{2}}}{dx}\right)^2 dx.$$

As applied to the transition from one uniform medium to another, the retardation is *less* than according to the first approximation by

$$\tfrac{1}{2}\int\left(\frac{dk^{-\frac{1}{2}}}{dx}\right)^2 dx. \quad........................(81)$$

The supposition that η varies slowly excludes more than a very small reflection.

Equations (79), (80) may be tested on the particular case already referred to where $k = n/x$. We get

$$\eta = Cn^{-\frac{1}{2}}x^{\frac{1}{2}}\left(1 + \frac{1}{16n^2}\right), \qquad a = \frac{1}{x}\left(n - \frac{1}{8n}\right);$$

so that

$$\int a\,dx = \left(n - \frac{1}{8n}\right)\log x.$$

When n^{-4} is neglected in comparison with unity, $n - \frac{1}{8}n^{-1}$ may be identified with $\surd(n^2 - \frac{1}{4})$.

Let us now consider what are the possibilities of avoiding reflection when the transition layer $(x_2 - x_1)$ between two uniform media is reduced. If $\eta_1, k_1; \eta_2, k_2$ are the terminal values, (79) requires that

$$k_1^2 = C^4\eta_1^{-4}, \qquad k_2^2 = C^4\eta_2^{-4}.$$

We will suppose that $\eta_2 > \eta_1$. If the transition from η_1 to η_2 be made too quickly, viz., in too short a space, $d^2\eta/dx^2$ will become somewhere so large as to render k^2 negative. The same consideration shows that at the beginning of the layer of transition (x_1), $d\eta/dx$ must vanish. The quickest admissible rise of η will ensue when the curve of rise is such as to make k^2 vanish. When η attains the second prescribed value η_2, it must suddenly become constant, notwithstanding that this makes k^2 positively infinite.

From (72) it appears that the curve of rise thus defined satisfies

$$\frac{d^2\eta}{dx^2} = C^4\eta^{-3}. \quad\dots\dots\dots\dots\dots\dots\dots\dots(82)$$

The solution of (82), subject to the conditions that $\eta = \eta_1$, $d\eta/dx = 0$, when $x = x_1$, is

$$\eta^2 - \eta_1^2 = C^4\eta_1^{-2}(x - x_1)^2 = k_1^2\eta_1^2(x - x_1)^2. \quad\dots\dots\dots\dots(83)$$

Again, when $\eta = \eta_2$, $x = x_2$, so that

$$k_1^2(x_2 - x_1)^2 = \frac{\eta_2^2 - \eta_1^2}{\eta_1^2} = \frac{k_1 - k_2}{k_2}, \quad\dots\dots\dots\dots\dots(84)$$

giving the minimum thickness of the layer of transition.

It will be observed that the minimum thickness of the layer of transition necessary to avoid reflection diminishes without limit with $k_1 - k_2$, that is, as the difference between the two media diminishes. However, the arrangement under discussion is very artificial. In the case of the string, for example, it is supposed that the density drops suddenly from the first uniform value to zero, at which it remains constant for a time. At the end of this it becomes momentarily infinite, before assuming the second uniform value. The infinite longitudinal density at x_2 is equivalent to a finite load there attached. In the layer of transition, if so it may be called, the string remains straight during the passage of the waves.

If, as in the more ordinary use of the term, we require the transition to be such that k^2 moves always in one direction from the first terminal value to the second, the problem is one already considered. The minimum thickness is such that k^2 has throughout it a constant intermediate value, so chosen as to make the reflections equal at the two faces.

It would be of interest to consider a particular case in which k^2 varies continuously and always in the one direction. As appears at once from (72), $d^2\eta/dx^2$, as well as $d\eta/dx$, must vanish at both ends of the layer, and there must also be a third point of inflection between. If the layer be from $x = 0$ to $x = \beta$, we may take

$$\frac{d^2\eta}{dx^2} = Ax(x - \alpha)(x - \beta). \quad\dots\dots\dots\dots\dots(85)$$

We find that $\beta = 2\alpha$, and that

$$\frac{\eta - \eta_1}{\eta_2 - \eta_1} = \frac{15x^3}{4\alpha^5}\left\{\frac{x^2}{20} - \frac{\alpha x}{4} + \frac{\alpha^2}{3}\right\}, \quad\dots\dots\dots\dots(86)$$

$$\frac{1}{\eta_2 - \eta_1}\frac{d^2\eta}{dx^2} = \frac{15x}{4\alpha^5}(x - \alpha)(x - 2\alpha). \quad\dots\dots\dots(87)$$

From these k^2 would have to be calculated by means of (72), and one question would be to find how far α might be reduced without interfering with the prescribed character of k^2. But to discuss this in detail would lead us too far.

If the differences of quality in the variable medium are small, (72) simplifies. If η_0, k_0 be corresponding values, subject to $k_0^2 = C^4\eta_0^{-4}$, we may take

$$\eta = \eta_0 + \eta', \quad k^2 = k_0^2 + \delta k^2, \quad\dots\dots\dots\dots\dots(88)$$

where η' and δk^2 are small, and (72) becomes approximately

$$\frac{d^2\eta'}{dx^2} + 4k_0^2\eta' = -\eta_0\delta k^2. \quad\dots\dots\dots\dots\dots(89)$$

Replacing x by t, representing *time*, we see that the problem is the same as that of a pendulum upon which displacing forces act; see *Theory of Sound*, § 66. The analogue of the transition from one uniform medium to another is that of the pendulum initially at rest in the position of equilibrium, upon which at a certain time a displacing force acts. The force may be variable at first, but ultimately assumes a constant value. If there is to be no reflection in the original problem, the force must be of such a character that when it becomes constant the pendulum is left *at rest* in the new position. If the object be to effect the transition between the two states in the shortest possible time, but with forces which are restricted never to exceed the final value, it is pretty evident that the force must immediately assume the maximum admissible value, and retain it for such a time that the pendulum, then left free, will just reach the new position of equilibrium, after which the force is reimposed. The present solution is excluded, if it be required that the force never decrease in value. Under this restriction the best we can do is to make the force assume at once *half* its final value, and remain constant for a time equal to one-half of the free period. Under this force the pendulum will just swing out to the new position of equilibrium, where it is held on arrival by doubling the force. These cases have already been considered, but the analogue of the pendulum is instructive.

Kelvin[*] has shown that the equation of the second order

$$\frac{d}{dx}\left(\frac{1}{P}\frac{dy_1}{dx}\right) = y_1 \quad\quad\quad\quad\quad\quad (90)$$

can be solved by a machine. It is worth noting that an equation of the form (53) is solved at the same time. In fact, if we make

$$y_1 = \frac{dy_2}{dx}, \quad\quad Py_2 = \frac{dy_1}{dx}, \quad\quad\quad\quad\quad (91)$$

we get on elimination either (90) for y_1, or

$$\frac{d^2y_2}{dx^2} = Py_2 \quad\quad\quad\quad\quad\quad\quad (92)$$

for y_2. Equations (91) are those which express directly the action of the machine.

It now remains to consider more in detail some cases where total reflection occurs. When there is merely a simple transition from one medium (1) to another (2), the transmitted wave is

$$\phi_2 = A_2 e^{-ia_2(x-x_1)} e^{i(ct+by)}. \quad\quad\quad\quad\quad (93)$$

If there is total reflection, a_2 becomes imaginary, say $-ia_2'$; the transmitted wave is then no longer a wave in the ordinary sense, but there remains some disturbance, not conveying energy, and rapidly diminishing as we recede from the surface of transition according to the factor $e^{-a_2'(x-x_1)}$. From (2)

$$a_2{}^2 = \frac{4\pi^2}{\lambda_2{}^2}\cos^2\theta_2 = \frac{4\pi^2}{\lambda_1{}^2}\left(\frac{V_1{}^2}{V_2{}^2} - \sin^2\theta_1\right),$$

or

$$a_2' = \frac{2\pi}{\lambda_1}\sqrt{(\sin^2\theta_1 - V_1{}^2/V_2{}^2)}. \quad\quad\quad\quad (94)$$

It appears that soon after the critical angle is passed, the disturbance in the second medium extends sensibly to a distance of only a few wave-lengths.

The circumstances of total reflection at a sudden transition are thus very simple; but total reflection itself does not require a sudden transition, and takes place however gradual the passage may be from the first medium to the second, the only condition being that when the second is reached the angle of refraction becomes imaginary. From this point of view total reflection is more naturally regarded as a sort of refraction, reflection proper depending on some degree of abruptness of transition. Phenomena of this kind are familiar in Optics under the name of *mirage*.

In the province of acoustics the vagaries of fog-signals are naturally referred to irregular refraction and reflection in the atmosphere, due to temperature or wind differences; but the difficulty of verifying a suggested explanation on these lines is usually serious, owing to our ignorance of the state of affairs overhead *.

The penetration of vibrations into a medium where no regular waves can be propagated is a matter of considerable interest; but, so far as I am aware, there is no discussion of such a case, beyond that already sketched, relating to a sudden transition between two uniform media. It might have been supposed that oblique propagation through a variable medium would involve too many difficulties, but we have already had opportunity to see that, in reality, obliquity need not add appreciably to the complication of the problem.

To fix ideas, let us suppose that we are dealing with waves in a membrane uniformly stretched with tension T, and of superficial density ρ, which is a function of x only. The equation of vibration is (*Theory of Sound*, § 194)

$$T\left(\frac{d^2\phi}{dx^2}+\frac{d^2\phi}{dy^2}\right)=\rho\,\frac{d^2\phi}{dt^2},$$

or, if ϕ be proportional to $e^{i(ct+by)}$, as in (1),

$$\frac{d^2\phi}{dx^2}+(c^2\rho/T-b^2)\,\phi=0, \quad\dots\dots\dots\dots\dots\dots(95)$$

agreeing with (53) if
$$k^2=c^2\rho/T-b^2. \quad\dots\dots\dots\dots\dots\dots\dots(96)$$

The waves originally move towards the less dense parts, and total reflection will ensue when a place is reached, at and after which k^2 is negative. The case which best lends itself to analytical treatment is when ρ is a linear function of x. k^2 is then also a linear function; and, by suitable choice of the origin and scale of x, (95) takes the form

$$\frac{d^2\phi}{dx^2}+\tfrac{1}{3}x\phi=0. \quad\dots\dots\dots\dots\dots\dots\dots(97)$$

---

* An observation during the exceptionally hot weather of last summer recalled my attention to this subject. A train passing at high speed at a distance of not more than 150 yards was almost inaudible. The wheels were in full view, but the situation was such that the line of vision passed for most of its length pretty close to the highly heated ground. It seemed clear that the sound rays which should have reached the observers were deflected upwards over their heads, which were left in a kind of shadow.

The waves are now supposed to come from the positive side and are totally reflected at $x = 0$. The coefficient and sign of x are chosen so as to suit the formulæ about to be quoted.

The solution of (97), appropriate to the present problem, is exactly the integral investigated by Airy to express the intensity of light in the neighbourhood of a caustic*. The line $x = 0$ is, in fact, a caustic in the optical sense, being touched by all the *rays*. Airy's integral is

$$W = \int_0^\infty \cos \tfrac{1}{2}\pi \, (w^3 - mw) \, dw. \qquad\qquad\qquad (98)$$

It was shown by Stokes† to satisfy (97), if

$$x \text{ (in his notation } n) = (\tfrac{1}{2}\pi)^{2/3} \, m. \qquad\qquad (99)$$

Calculating by quadratures and from series proceeding by ascending powers of m, Airy tabulated W for values of m lying between $m = \pm 5\cdot6$. For larger numerical values of m another method is necessary, for which Stokes gave the necessary formulæ. Writing

$$\phi_+^+ = 2 \, (\tfrac{1}{3}x)^{3/2} = \pi \, (\tfrac{1}{3}m)^{3/2}, \qquad\qquad (100)$$

where the numerical values of m and x are supposed to be taken when these quantities are negative, he found when m is positive

$$W = 2^{\frac{1}{4}} (3m)^{-\frac{1}{4}} \{R \cos (\phi - \tfrac{1}{4}\pi) + S \sin (\phi - \tfrac{1}{4}\pi)\}, \qquad (101)$$

where

$$R = 1 - \frac{1 \cdot 5 \cdot 7 \cdot 11}{1 \cdot 2 \, (72\phi)^2} + \frac{1 \cdot 5 \cdot 7 \cdot 11 \cdot 13 \cdot 17 \cdot 19 \cdot 23}{1 \cdot 2 \cdot 3 \cdot 4 \, (72\phi)^4} - \ldots, \quad \ldots(102)$$

$$S = \frac{1 \cdot 5}{1 \cdot 72\phi} - \frac{1 \cdot 5 \cdot 7 \cdot 11 \cdot 13 \cdot 17}{1 \cdot 2 \cdot 3 \, (72\phi)^3} + \ldots \qquad\qquad (103)$$

When m is negative, so that W is the integral expressed by writing $-m$ for m in (98),

$$W = 2^{-\frac{1}{2}} (3m)^{-\frac{1}{4}} e^{-\phi} \left\{ 1 - \frac{1 \cdot 5}{1 \cdot 72\phi} + \frac{1 \cdot 5 \cdot 7 \cdot 11}{1 \cdot 2 \, (72\phi)^3} - \ldots \right\}. \qquad (104)$$

The first form (101) is evidently fluctuating. The roots of $W = 0$ are given by

$$\phi/\pi = i - 0\cdot25 + \frac{0\cdot028145}{4i - 1} - \frac{0\cdot026510}{(4i - 1)^3} + \ldots, \qquad\qquad (105)$$

i being a positive integer, so that for $i = 2, 3, 4$, etc., we get

$$m = 4\cdot3631, \; 5\cdot8922, \; 7\cdot2436, \; 8\cdot4788, \text{ etc.}$$

For $i = 1$, Airy's calculation gave $m = 2\cdot4955$.

---

* *Camb. Phil. Trans.* 1838, Vol. VI. p. 379; 1849, Vol. VIII. p. 595.
† *Camb. Phil. Trans.* 1850, Vol. IX.; *Math. and Phys. Papers*, Vol. II. p. 328.
‡ Here used in another sense.

The complete solution of (97) in series of ascending powers of x is to be obtained in the usual way, and the arbitrary constants are readily determined by comparison with (98).   Lommel* showed that these series are expressible by means of the Bessel's functions $J_{\frac{1}{3}}$, $J_{-\frac{1}{3}}$.   The connection between the complete solutions of (97), as expressed by ascending or by descending semi-convergent series, is investigated in a second memoir by Stokes†.   A reproduction of the most important part of Airy's table will be found in Mascart's *Optics* (Vol. I. p. 397).

As total reflection requires, the waves in our problem are stationary as regards x.   The realized solution of (95) may be written

$$\phi = W \cos (ct + by), \qquad \ldots\ldots\ldots\ldots\ldots\ldots(106)$$

W being the function of x already discussed.   On the negative side, when x numerically exceeds a moderate value, the disturbance becomes insensible.

* *Studien über die Bessel'schen Functionen*, Leipzig, 1868.
† *Camb. Phil. Trans.* 1857, Vol. x. p. 106.

361.

SPECTROSCOPIC METHODS.

[*Nature*, Vol. LXXXVIII. p. 377, 1912.]

IN his interesting address on spectroscopic methods, Prof. Michelson falls into a not uncommon error when he says that, in order to obtain a pure spectrum, "two important modifications must be made in Newton's arrangement. First, the light must be allowed to pass through a very *narrow* aperture, and, secondly, a sharp *image* of this aperture must be formed by a lens or mirror."

Both these modifications were made by Newton himself, and with a clear understanding of their advantages. In *Opticks*, Exper. 11, we read:—"In the Sun's Light let into my darkened Chamber through a small round hole in my Window—shut, at about 10 or 12 feet from the Window, I placed a Lens, by which the Image of the hole might be distinctly cast upon a sheet of white Paper, placed at the distance of six, eight, ten, or twelve Feet from the Lens....For in this case the circular Images of the hole which comprise that Image...were terminated most distinctly without any Penumbra, and therefore extended into one another the least that they could, and by consequence the mixture of the Heterogeneous Rays was now the least of all."

And further on:—

"Yet instead of the circular hole F, 'tis better to substitute an oblong hole shaped like a long Parallelogram with its length Parallel to the Prism ABC. For if this hole be an Inch or two long, and but a tenth or twentieth part of an Inch broad or narrower: the Light of the Image *pt* will be as Simple as before or simpler [*i.e.* as compared with a correspondingly narrow circular hole], and the Image will become much broader, and therefore more fit to have Experiments tried in its Light than before."

Again, it was not Bunsen and Kirchhoff who first introduced the collimator into the spectroscope. Swan employed it in 1847, and fully described its use in *Edin. Trans.* Vol. XVI. p. 375, 1849. See also *Edin. Trans.* Vol. XXI. p. 411, 1857; *Pogg. Ann.* C, p. 306, 1857.

These are very minor matters as compared with what Prof. Michelson has to tell of his own achievements and experiences, but it seems desirable that they should be set right.

362.

ON DEPARTURES FROM FRESNEL'S LAWS OF REFLEXION.

[*Philosophical Magazine*, Vol. XXIII. pp. 431—439, 1912.]

IN the summer of 1907, in connexion with my experiments upon reflexion from glass at the polarizing angle[*], I made observations also upon the diamond, a subject in which Kelvin had expressed an interest. It was known from the work of Jamin and others that the polarization of light reflected from this substance is very far 'from complete at any angle of incidence, and my first experiments were directed to ascertain whether this irregularity could be plausibly attributed to superficial films of foreign matter, such as so greatly influence the corresponding phenomena in the case of water[†]. The arrangements were of the simplest. The light from a paraffin flame seen edgeways was reflected from the diamond and examined with a nicol, the angle being varied until the reflexion was a minimum.

In one important respect the diamond offers advantages, in comparison, for instance, with glass, where the surface is the field of rapid chemical changes due presumably to atmospheric influences. On the other hand, the smallness of the available surfaces is an inconvenience which, however, is less felt than it would be, were high precision necessary in the measurements. Two diamonds were employed—one, kindly lent me by Sir W. Crookes, mounted at the end of a bar of lead, the other belonging to a lady's ring. No particular difference in behaviour revealed itself.

The results of repeated observations seemed to leave it improbable that any process of cleaning would do more than reduce the reflexion at the polarizing angle. Potent chemicals, such as hot chromic acid, may be employed, but there is usually a little difficulty in the subsequent preparation. After copious rinsing, at first under the tap and then with distilled water from a wash-bottle, the question arises how to dry the surface. Any ordinary wiping may be expected to nullify the chemical treatment; but if

* *Phil. Mag.* Vol. XVI. p. 444 (1908); *Scientific Papers*, Vol. V. p. 489.
† *Phil. Mag.* Vol. XXXIII. p. 1 (1892); *Scientific Papers*, Vol. III. p. 496.

drops are allowed to dry on, the effect is usually bad. Sometimes it is possible to shake the drops away sufficiently. After a successful operation of this sort wiping with an ordinarily clean cloth usually increases the minimum reflexion, and of course a touch with the finger, however prepared, is much worse. As the result of numerous trials I got the impression that the reflexion could not be reduced below a certain standard which left the flame still easily visible. Rotation of the diamond surface in its own plane seemed to be without effect.

During the last few months I have resumed these observations, using the same diamonds, but with such additions to the apparatus as are necessary for obtaining measures of the residual reflexion. Besides the polarizing nicol, there is required a quarter-wave mica plate and an analysing nicol, to be traversed successively by the light after reflexion, as described in my former papers. The analysing nicol is set alternately at angles $\beta = \pm 45°$. At each of these angles extinction may be obtained by a suitable rotation of the polarizing nicol; and the observation consists in determining the angle $\alpha' - \alpha$ between the two positions. Jamin's k, representing the ratio of reflected amplitudes for the two principal planes when light incident at the angle $\tan^{-1} \mu$ is polarized at 45° to these planes, is equal to $\tan \frac{1}{2} (\alpha' - \alpha)$. The sign of $\alpha' - \alpha$ is reversed when the mica is rotated through a right angle, and the absolute sign of k must be found independently.

Wiped with an ordinarily clean cloth, the diamond gave at first $\alpha' - \alpha = 2°\cdot3$. By various treatments this angle could be much reduced. There was no difficulty in getting down to 1°. On the whole the best results were obtained when the surface was finally wiped, or rather pressed repeatedly, upon sheet asbestos which had been ignited a few minutes earlier in the blowpipe flame; but they were not very consistent. The lowest reading was $0°\cdot4$; and we may, I think, conclude that with a clean surface $\alpha' - \alpha$ would not exceed $0°\cdot5$. No more than in the case of glass, did the effect seem sensitive to moisture, no appreciable difference being observable when chemically dried air played upon the surface. It is impossible to attain absolute certainty, but my impression is that the angle cannot be reduced much further. So long as it exceeds a few tenths of a degree, the paraffin flame is quite adequate as a source of light.

If we take for diamond $\alpha' - \alpha = 30'$, we get

$$k = \tan \tfrac{1}{2} (\alpha' - \alpha) = \cdot0044.$$

Jamin's value for k is $\cdot019$, corresponding more nearly with what I found for a merely wiped surface.

Similar observations have been made upon the face of a small dispersing prism which has been in my possession some 45 years. When first examined, it gave $\alpha' - \alpha = 9°$, or thereabouts. Treatment with rouge on a piece of calico, stretched over a glass plate, soon reduced the angle to 4° or 3°, but further progress seemed more difficult. Comparisons were rendered somewhat uncertain by the fact that different parts of the surface gave varying numbers. After a good deal of rubbing, $\alpha' - \alpha$ was reduced to such figures as 2°, on one occasion apparently to $1\frac{1}{2}$°. Sometimes the readings were taken without touching the surface after removal from the rouge, at others the face was breathed upon and wiped. In general, the latter treatment seemed to increase the angle. Strong sulphuric acid was also tried, but without advantage, as also putty-powder in place of or in addition to rouge. The behaviour did not appear to be sensitive to moisture, or to alter appreciably when the surface stood for a few days after treatment.

Thinking that possibly changes due to atmospheric influences might in nearly half a century have penetrated somewhat deeply into the glass, I re-ground and polished (sufficiently for the purpose) one of the originally unpolished faces of the prism, but failed even with this surface to reduce $\alpha' - \alpha$ below 2°. As in the case of the diamond, it is impossible to prove absolutely that $\alpha' - \alpha$ cannot be reduced to zero, but after repeated trials I had to despair of doing so. It may be well to record that the refractive index of the glass for yellow rays is 1·680.

These results, in which k (presumably positive) remained large in spite of all treatment, contrast remarkably with those formerly obtained on less refractive glasses, one of which, however, appears to contain lead. It was then found that by re-polishing it was possible to carry k down to zero and to the negative side, somewhat as in the observations upon water it was possible to convert the negative k of ordinary (greasy) water into one with a small positive value, when the surface was purified to the utmost.

There is another departure from Fresnel's laws which is observed when a piece of plate glass is immersed in a liquid of equal index*. Under such circumstances the reflexion ought to vanish.

The liquid may consist of benzole and bisulphide of carbon, of which the first is less and the second more refractive than the glass. If the adjustment is for the yellow, more benzole or a higher temperature will take the ray of equal index towards the blue and *vice versâ*. "For a closer examination the plate was roughened behind (to destroy the second reflexion), and was mounted in a bottle prism in such a manner that the incidence could be rendered grazing. When the adjustment of indices was for the yellow the appearances observed were as follows: if the incidence is pretty oblique, the reflexion is total for the violet and blue; scanty, but not evanescent, for the yellow; more copious again in the red. As the incidence becomes more and more nearly grazing, the region of total reflexion advances from the blue

---

* "On the Existence of Reflexion when the relative Refractive Index is Unity," *Brit. Assoc. Report*, p. 585 (1887); *Scientific Papers*, Vol. III. p. 15.

end closer and closer upon the ray of equal index, and ultimately there is a very sharp transition between this region and the band which now looks very dark. On the other side the reflexion revives, but more gradually, and becomes very copious in the orange and red. On this side the reflexion is not technically total. If the prism be now turned so that the angle of incidence is moderate, it is found that, in spite of the equality of index for the most luminous part of the spectrum, there is a pretty strong reflexion of a candle-flame, and apparently without colour. With the aid of sunlight it was proved that in the reflexion at moderate incidences there was no marked chromatic selection, and in all probability the blackness of the band in the yellow at grazing incidences is a matter of contrast only. Indeed, calculation shows that according to Fresnel's formulæ, the reflexion would be nearly insensible for all parts of the spectrum when the index is adjusted for the yellow." It was further shown that the reflexion could be reduced, but not destroyed, by re-polishing or treatment of the surface with hydrofluoric acid.

I have lately thought it desirable to return to these experiments under the impression that formerly I may not have been sufficiently alive to the irregular behaviour of glass surfaces which are in contact with the atmosphere. I wished also to be able to observe the transmitted as well as the reflected light. A cell was prepared from a tin-plate cylinder 3 inches long and 2 inches in diameter by closing the ends with glass plates cemented on with glue and treacle. Within was the glass plate to be experimented on, of similar dimensions, so as to be nearly a fit. A hole in the cylindrical wall allowed the liquid to be poured in and out. Although the plate looked good and had been well wiped, I was unable to reproduce the old effects; or, for a time, even to satisfy myself that I could attain the right composition of the liquid. Afterwards a clue was found in the spectra formed by the edges of the plate (acting as prisms) when the cell was slewed round. The subject of observation was a candle placed at a moderate distance. When the adjustment of indices is correct for any ray, the corresponding part of the spectrum is seen in the same direction as is the undispersed candle-flame by rays which have passed outside the plate. Either spectrum may be used, but the best for the purpose is that formed by the edge nearer the eye. There was now no difficulty in adjusting the index for the yellow ray, and the old effects ought to have manifested themselves; but they did not. The reflected image showed little deficiency in the yellow, although the incidence was nearly grazing, while at moderate angles it was fairly bright and without colour. This considerable departure from Fresnel's laws could only be attributed to a not very thin superficial modification of the glass rendering it optically different from the interior.

In order to allow of the more easy removal and replacement of the plate under examination, an altered arrangement was introduced, in which the aperture at the top of the cell extended over the whole length. The general dimensions being the same as before, the body of the cell was formed by bending round a rectangular piece of tin-plate A (fig. 1) and securing the ends, to which the glass faces B were to be cemented, by enveloping copper wire. The plate C could then be removed for cleaning or polishing without breaking a joint. In emptying the cell it is necessary to employ a large funnel, as the liquid pours badly.

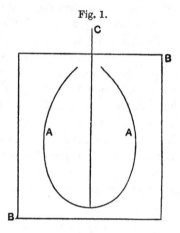

Fig. 1.

The plate tried behaved much as the one just spoken of. In the reflected light, whether at moderate angles or nearly grazing, the yellow-green ray of equal index did not appear to be missing. A line or rather band of polish, by putty-powder applied with the finger, showed a great alteration. Near grazing there was now a dark band in the spectrum of the reflected light as formerly described, and the effect was intensified when the polish affected *both* faces. In the transmitted light the spectrum was shorn of blue and green, the limit coming down as grazing is approached—a consequence of the total reflexion of certain rays which then sets in. But at incidences far removed from grazing the place of equal index in the spectrum of the reflected light showed little weakening. A few days' standing (after polishing) in the air did not appear to alter the behaviour materially. On the same plate other bands were treated with hydrofluoric acid—commercial acid diluted to one-third. This seemed more effective than the putty-powder. At about 15° off grazing, the spectrum of the reflected light still showed some weakening in the ray of equal index.

In the cell with parallel faces it is not possible to reduce the angle of incidence (reckoned from the normal) sufficiently, a circumstance which led me to revert to the 60° bottle-prism. A strip of glass half an inch wide could be inserted through the neck, and this width suffices for the observation of the reflected light. But I experienced some trouble in finding the light until I had made a calculation of the angles concerned. Supposing the plane of the reflecting surface to be parallel to the base of the prism, let us call the angle of incidence upon it χ, and let θ, ϕ be the angles which the ray makes with the normal to the faces, externally and internally, measured in each case *towards* the refracting angle of the prism. Then

$$\chi = 60° - \phi, \quad \phi = \sin^{-1}(\tfrac{2}{3}\sin\theta).$$

The smallest χ occurs when $\theta = 90°$, in which case $\chi = 18° \ 10'$. This value cannot be actually attained, since the emergence would be grazing. If $\chi = 90°$, giving grazing reflexion, $\theta = -48° \ 36'$. Again, if $\theta = 0$, $\chi = 60°$;

and if $\chi = 45°$, $\theta = 22° \ 51'$. We can thus deal with all kinds of reflexion from $\chi = 90°$ down to nearly 18°, and this suffices for the purpose.

The strip employed was of plate glass and was ground upon the back surface. The front reflecting face was treated for about 30^s with hydrofluoric acid. It was now easy to trace the effects all the way from grazing incidence down to an incidence of 45° or less. The ray of equal index was in the yellow-green, as was apparent at once from the spectrum of the reflected light near grazing. There was a very dark band in this region, and total reflection reaching nearly down to it from the blue end. The light was from a paraffin flame, at a distance of about two feet, seen edgeways. As grazing incidence is departed from, the flame continues at first to show a purple colour, and the spectrum shows a weakened, but not totally absent, green. As the angle of incidence χ still further decreases, the reflected light weakens both in intensity and colour. When $\chi = 45°$, or thereabouts, the light was weak and the colour imperceptible. After two further treatments with hydrofluoric acid and immediate examination, the light seemed further diminished, but it remained bright enough to allow the absence of colour to be ascertained, especially when the lamp was temporarily brought nearer. An ordinary candle-flame at the same (2 feet) distance was easily visible.

In order to allow the use of the stopper, the strip was removed from the bottle-prism when the observations were concluded, and it stood for four days exposed to the atmosphere. On re-examination it seemed that the reflexion at $\chi = 45°$ had sensibly increased, a conclusion confirmed by a fresh treatment with hydrofluoric acid.

It remains to consider the theoretical bearing of the two anomalies which manifest themselves (i) at the polarizing angle, and (ii) at other angles when both media have the same index, at any rate for a particular ray. Evidently the cause may lie in a skin due either to contamination or to the inevitable differences which must occur in the neighbourhood of the surface of a solid or fluid body. Such a skin would explain both anomalies and is certainly a part of the true explanation, but it remains doubtful whether it accounts for everything. Under these circumstances it seems worth while to inquire what would be the effect of less simple boundary conditions than those which lead to Fresnel's formulæ.

On the electromagnetic theory, if θ, θ_1 are respectively the angles of incidence and refraction, the ratio of the reflected to the incident vibration is, for the two principal polarizations,

$$\frac{\tan \theta_1 / \tan \theta - \mu/\mu_1}{\tan \theta_1 / \tan \theta + \mu/\mu_1} \quad \dots\dots\dots\dots\dots\dots\dots(A)$$

and

$$\frac{\tan \theta_1 / \tan \theta - K/K_1}{\tan \theta_1 / \tan \theta + K/K_1}, \quad \dots\dots\dots\dots\dots\dots(B)$$

in which K, μ are the electric and magnetic constants for the first medium, K_1, μ_1 for the second*. The relation between θ and θ_1 is

$$K_1\mu_1 : K\mu = \sin^2\theta : \sin^2\theta_1. \quad\ldots\ldots\ldots\ldots\ldots\ldots\ldots(C)$$

It is evident that mere absence of refraction will not secure the evanescence of reflexion for both polarizations, unless we assume *both* $\mu_1 = \mu$ and $K_1 = K$. In the usual theory μ_1 is supposed equal to μ in all cases. (A) then identifies itself with Fresnel's sine-formula, and (B) with the tangent-formula, and both vanish when $K_1 = K$ corresponding to no refraction. Further, (B) vanishes at the Brewsterian angle, even though there be refraction. A slight departure from these laws would easily be accounted for by a difference between μ_1 and μ, such as in fact occurs in some degree (diamagnetism). But the effect of such a departure is not to interfere with the complete evanescence of (B), but merely to displace the angle at which it occurs from the Brewsterian value. If $\mu_1/\mu = 1 + h$, where h is small, calculation shows that the angle of complete polarization is changed by the amount

$$\delta\theta = -\frac{hn^3}{(n^2 + 1)(n^2 - 1)}, \quad\ldots\ldots\ldots\ldots\ldots\ldots\ldots(D)$$

n being the refractive index. The failure of the diamond and dense glass to polarize completely at some angle of incidence is not to be explained in this way.

As I formerly suggested, the anomalies may perhaps be connected with the fact that one at least of the media is *dispersive*. A good deal depends upon the cause of the dispersion. In the case of a stretched string, vibrating transversely and endowed with a moderate amount of stiffness, the boundary conditions would certainly be such as would entail a reflexion in spite of equal velocity of wave-propagation. All optical dispersion is now supposed to be of the same nature as what used to be called *anomalous* dispersion, *i.e.* to be due to resonances lying beyond the visible range. In the simplest form of this theory, as given by Maxwell† and Sellmeier, the resonating bodies take their motion from those parts of the æther with which they are directly connected, but they do not influence one another. In such a case the boundary conditions involve merely the continuity of the displacement and its first derivative, and no complication ensues. When there is no refraction, there is also no reflexion. By introducing a mutual reaction between the resonators, and probably in other ways, it would be possible to modify the situation in such a manner that the boundary conditions would involve higher derivatives, as in the case of the stiff string, and thus to allow reflexion in spite of equality of wave-velocities for a given ray.

* "On the Electromagnetic Theory of Light," *Phil. Mag.* Vol. XII. p. 81 (1881); *Scientific Papers*, Vol. I. p. 521.

† *Cambridge Calendar* for 1869. See *Phil. Mag.* Vol. XLVIII. p. 151 (1899); *Scientific Papers*, Vol. IV. p. 413.

P.S. Jan. 15.—Some later observations upon a surface of *fused quartz* are of interest. The plate, prepared by Messrs Hilger, was ½ inch square, and the surfaces were inclined at a few degrees so as to separate the reflexions. From these surfaces the reflexion at the polarizing angle *sensibly disappears.* The image of the paraffin flame could be quenched by the operation of the polarizing nicol alone. When the quarter wave-plate and analysing nicol were introduced, α' and α could not be distinguished, the difference probably not exceeding ·05°, *i.e.* 3 minutes of angle.

In order to examine the reflexion when the quartz was in contact with a liquid of equal index, I had to mix alcohol with the benzole. The behaviour was then much the same as with glass of which the surface had been renewed by hydrofluoric acid. No precise measures could be taken, but the reflexion at 45° incidence seemed less than from the glass, though still easily visible. In spite of repeated trials with intermediate cleanings, it was difficult to feel sure that the residual effect might not be due to foreign matter, the more as differences could sometimes be detected between various parts of the surface*. Even if the surface could be regarded as clean on immersion, there is no certainty that a capillary film of some sort might not be deposited upon it from the liquid. The cause of the small residual reflexion must remain for the present an open question.

---

* At the top of the plate, where it was attached to a handle, a slight invasion of gelatine (used as a cement) gave rise to a copious reflexion; but this film was easily visible in the air.

363.

THE PRINCIPLE OF REFLECTION IN SPECTROSCOPES.

[*Nature*, Vol. LXXXIX. p. 167, 1912.]

THE application of a reflector to pass light back through a prism, or prisms, is usually ascribed to Littrow. Thus Kayser writes (*Handbuch der Spectroscopie*, Bd. I. p. 513), "Der Erste, der Rückkehr der Strahlen zur Steigerung der Dispersion verwandte, war Littrow" (O. v. Littrow, *Wien. Ber.* XLVII. ii. pp. 26–32, 1863). But this was certainly not the first use of the method. I learned it myself from Maxwell (*Phil. Trans.* Vol. CL. p. 78, 1860), who says, "The principle of reflecting light, so as to pass twice through the same prism, was employed by me in an instrument for combining colours made in 1856, and a reflecting instrument for observing the spectrum has been constructed by M. Porro."

I have not been able to find the reference to Porro; but it would seem that both Maxwell and Porro antedated Littrow. As to the advantages of the method there can be no doubt.

364.

ON THE SELF-INDUCTION OF ELECTRIC CURRENTS IN A THIN ANCHOR-RING.

[*Proceedings of the Royal Society*, A, Vol. LXXXVI. pp. 562—571, 1912.]

IN their useful compendium of "Formulæ and Tables for the Calculation of Mutual and Self-Inductance*," Rosa and Cohen remark upon a small discrepancy in the formulæ given by myself† and by M. Wien‡ for the self-induction of a coil of circular cross-section over which the current is *uniformly distributed*. With omission of n, representative of the number of windings, my formula was

$$L = 4\pi a \left[\log \frac{8a}{\rho} - \frac{7}{4} + \frac{\rho^2}{8a^2} \left(\log \frac{8a}{\rho} + \frac{1}{3} \right) \right], \quad \ldots\ldots\ldots\ldots(1)$$

where ρ is the radius of the section and a that of the circular axis. The first two terms were given long before by Kirchhoff§. In place of the fourth term within the bracket, viz., $+\frac{1}{24}\rho^2/a^2$, Wien found $-\cdot0083\rho^2/a^2$. In either case a correction would be necessary in practice to take account of the space occupied by the insulation. Without, so far as I see, giving a reason, Rosa and Cohen express a preference for Wien's number. The difference is of no great importance, but I have thought it worth while to repeat the calculation and I obtain the same result as in 1881. A confirmation after 30 years, and without reference to notes, is perhaps almost as good as if it were independent. I propose to exhibit the main steps of the calculation and to make extension to some related problems.

The starting point is the expression given by Maxwell‖ for the mutual induction M between two neighbouring co-axial circuits. For the present

---

* *Bulletin of the Bureau of Standards*, Washington, 1908, Vol. III. No. 1.

† *Roy. Soc. Proc.* 1881, Vol. XXXII. p. 104; *Scientific Papers*, Vol. II. p. 15.

‡ *Ann. d. Physik*, 1894, Vol. LIII. p. 934; it would appear that Wien did not know of my earlier calculation.

§ *Pogg. Ann.* 1864, Vol. CXXI. p. 551.

‖ *Electricity and Magnetism*, § 705.

purpose this requires transformation, so as to express the inductance in terms of the situation of the elementary circuits relatively to the circular axis. In the figure, O is the centre of the circular axis, A the centre of a section B through the axis of symmetry, and the position of any point P of the section is given by polar coordinates relatively to A, viz., by $PA\,(\rho)$ and by the angle $PAC(\phi)$. If ρ_1, ϕ_1; ρ_2, ϕ_2 be the coordinates of two points of the section P_1, P_2, the mutual induction between the two circular circuits represented by P_1, P_2 is approximately

$$
\begin{aligned}
\frac{M_{12}}{4\pi a} = &\left\{ 1 + \frac{\rho_1\cos\phi_1 + \rho_2\cos\phi_2}{2a} + \frac{\rho_1^2 + \rho_2^2 + 2\rho_1^2\sin^2\phi_1 + 2\rho_2^2\sin^2\phi_2}{16a^2} \right. \\
&\left. - \frac{2\rho_1\rho_2\cos(\phi_1-\phi_2) + 4\rho_1\rho_2\sin\phi_1\sin\phi_2}{16a^2} \right\} \log\frac{8a}{r} \\
&- 2 - \frac{\rho_1\cos\phi_1 + \rho_2\cos\phi_2}{2a} \\
&+ \frac{3(\rho_1^2 + \rho_2^2) - 4(\rho_1^2\sin^2\phi_1 + \rho_2^2\sin^2\phi_2) + 2\rho_1\rho_2\cos(\phi_1-\phi_2)}{16a^2}, \quad (2)
\end{aligned}
$$

in which r, the distance between P_1 and P_2, is given by

$$ r^2 = \rho_1^2 + \rho_2^2 - 2\rho_1\rho_2\cos(\phi_1-\phi_2). \quad\ldots\ldots\ldots\ldots\ldots\ldots(3) $$

Further details will be found in Wien's memoir; I do not repeat them because I am in complete agreement so far.

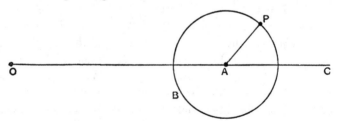

For the problem of a current uniformly distributed we are to integrate (2) twice over the area of the section. Taking first the integrations with respect to ϕ_1, ϕ_2, let us express

$$ \frac{1}{4\pi^2} \int_{-\pi}^{+\pi} \int_{-\pi}^{+\pi} \frac{M_{12}}{4\pi a}\, d\phi_1\, d\phi_2, \quad \ldots\ldots\ldots\ldots\ldots\ldots\ldots(4) $$

of which we can also make another application. The integration of the terms which do not involve $\log r$ is elementary. For those which do involve $\log r$ we may conveniently replace ϕ_2 by $\phi_1 + \phi$, where $\phi = \phi_2 - \phi_1$, and take first the integration with respect to ϕ, ϕ_1 being constant. Subsequently we integrate with respect to ϕ_1.

It is evident that the terms in (2) which involve the first power of ρ vanish in the integration. For a change of ϕ_1, ϕ_2 into $\pi - \phi_1$, $\pi - \phi_2$

respectively reverses $\cos \phi_1$ and $\cos \phi_2$, while it leaves r unaltered. The definite integrals required for the other terms are *

$$\int_{-\pi}^{+\pi} \log (\rho_1^2 + \rho_2^2 - 2\rho_1\rho_2 \cos \phi)\, d\phi = \text{greater of } 4\pi \log \rho_2 \text{ and } 4\pi \log \rho_1, \quad (5)$$

$$\int_{-\pi}^{+\pi} \cos m\phi \, \log (\rho_1^2 + \rho_2^2 - 2\rho_1\rho_2 \cos \phi)\, d\phi$$
$$= -\frac{2\pi}{m} \times \text{smaller of } \left(\frac{\rho_2}{\rho_1}\right)^m \text{ and } \left(\frac{\rho_1}{\rho_2}\right)^m, \quad \ldots\ldots(6)$$

m being an integer.   Thus

$$\frac{1}{4\pi^2} \int\!\!\int \log r \, d\phi_1 \, d\phi_2 = \frac{1}{4\pi^2} \int_{-\pi}^{+\pi} d\phi_1 \int_{-\pi}^{+\pi} d\phi \, \log r = \text{greater of} \log \rho_2 \text{ and } \log \rho_1. \quad (7)$$

So far as the more important terms in (4)—those which do not involve ρ as a factor—we have at once

$$\log (8a) - 2 - \text{greater of } \log \rho_2 \text{ and } \log \rho_1. \quad \ldots\ldots\ldots\ldots(8)$$

If ρ_2 and ρ_1 are equal, this becomes

$$\log (8a/\rho) - 2. \quad \ldots\ldots\ldots\ldots\ldots\ldots\ldots\ldots(9)$$

We have now to consider the terms of the second order in (2). The contribution which these make to (4) may be divided into two parts. The first, not arising from the terms in $\log r$, is easily found to be

$$\frac{\rho_1^2 + \rho_2^2}{8a^2} [\log (8a) + \tfrac{1}{2}]. \quad \ldots\ldots\ldots\ldots\ldots\ldots(10)$$

The difference between Wien's number and mine arises from the integration of the terms in $\log r$, so that it is advisable to set out these somewhat in detail.   Taking the terms in order, we have as in (7)

$$\frac{1}{4\pi^2} \int_{-\pi}^{+\pi}\!\! \int_{-\pi}^{+\pi} \log r \, d\phi_1 \, d\phi_2 = \text{greater of } \log \rho_2 \text{ and } \log \rho_1. \quad \ldots\ldots(11)$$

In like manner

$$\frac{1}{4\pi^2} \int\!\!\int \sin^2 \phi_1 \log r \, d\phi_1 \, d\phi_2 = \tfrac{1}{2} [\text{greater of } \log \rho_2 \text{ and } \log \rho_1], \quad \ldots(12)$$

and $\dfrac{1}{4\pi^2} \displaystyle\int\!\!\int \sin^2 \phi_2 \log r \, d\phi_1 \, d\phi_2$ has the same value.   Also by (6), with $m = 1$,

$$\frac{1}{4\pi^2} \int\!\!\int \cos (\phi_2 - \phi_1) \log r \, d\phi_1 d\phi_2 = -\tfrac{1}{2} [\text{smaller of } \rho_2/\rho_1 \text{ and } \rho_1/\rho_2]. \quad \ldots (13)$$

Finally $\dfrac{1}{4\pi^2} \displaystyle\int\!\!\int \sin \phi_1 \sin \phi_2 \log r \, d\phi_1 \, d\phi_2$

$$= \frac{1}{4\pi^2} \int_{-\pi}^{+\pi} d\phi_1 \sin \phi_1 \int_{-\pi}^{+\pi} (\sin \phi_1 \cos \phi + \cos \phi_1 \sin \phi) \log r \, d\phi$$

$$= -\tfrac{1}{4} [\text{smaller of } \rho_2/\rho_1 \text{ and } \rho_1/\rho_2]. \quad \ldots\ldots\ldots\ldots\ldots(14)$$

* Todhunter's *Int. Calc.* §§ 287, 289.

Thus altogether the terms in (2) of the second order involving $\log r$ yield in (4)

$$-\frac{\rho_1^2 + \rho_2^2}{8a^2} [\text{greater of } \log \rho_2 \text{ and } \log \rho_1] - \frac{\rho_1 \rho_2}{8a^2} \left[\text{smaller of } \frac{\rho_2}{\rho_1} \text{ and } \frac{\rho_1}{\rho_2} \right]. \quad \dots(15)$$

The complete value of (4) to this order of approximation is found by addition of (8), (10), and (15).

By making ρ_2 and ρ_1 equal we obtain at once for the self-induction of a current limited to the circumference of an anchor-ring, and uniformly distributed over that circumference,

$$L = 4\pi a \left[\left(1 + \frac{\rho^2}{4a^2} \right) \log \frac{8a}{\rho} - 2 \right], \quad \dots\dots\dots\dots(16)$$

ρ being the radius of the circular section. The value of L for this case, when ρ^2 is neglected, was virtually given by Maxwell*.

When the current is uniformly distributed over the area of the section, we have to integrate again with respect to ρ_1 and ρ_2 between the limits 0 and ρ in each case. For the more important terms we have from (8)

$$\frac{1}{\rho^4} \iint d\rho_1^2 \, d\rho_2^2 \, [\log 8a - 2 - \text{greater of } \log \rho_2 \text{ and } \log \rho_1]$$

$$= \log 8a - 2 - \frac{1}{2\rho^4} \int d\rho_1^2 \, [\log \rho_1^2 \cdot \rho_1^2 + \rho^2 (\log \rho^2 - 1) - \rho_1^2 (\log \rho_1^2 - 1)]$$

$$= \log 8a - 2 - \log \rho + \frac{1}{4} = \log \frac{8a}{\rho} - \frac{7}{4}. \quad \dots\dots\dots\dots\dots(17)$$

A similar operation performed upon (10) gives

$$\frac{\log(8a) + \frac{1}{2}}{8a^2 \rho^4} \iint (\rho_1^2 + \rho_2^2) \, d\rho_1^2 \, d\rho_2^2 = \frac{\log(8a) + \frac{1}{2}}{8} \frac{\rho^2}{a^2}. \quad \dots\dots\dots(18)$$

In like manner, the first part of (15) yields

$$-\frac{\rho^2}{16a^2} (\log \rho^2 - \tfrac{1}{3}).$$

For the second part we have

$$-\frac{1}{8a^2 \rho^4} \iint d\rho_1^2 d\rho_2^2 \, [\text{smaller of } \rho_2^2, \, \rho_1^2] = -\frac{\rho^2}{24a^2};$$

thus altogether from (15)

$$-\frac{\rho^2}{8a^2} (\log \rho + \tfrac{1}{6}). \quad \dots\dots\dots\dots\dots\dots(19)$$

The terms of the second order are accordingly, by addition of (18) and (19),

$$\frac{\rho^2}{8a^2} \left(\log \frac{8a}{\rho} + \frac{1}{3} \right). \quad \dots\dots\dots\dots\dots(20)$$

* *Electricity and Magnetism*, §§ 692, 706.

To this are to be added the leading terms (17); whence, introducing $4\pi a$, we get finally the expression for L already stated in (1).

It must be clearly understood that the above result, and the corresponding one for a *hollow* anchor-ring, depend upon the assumption of a uniform distribution of current, such as is approximated to when the coil consists of a great number of windings of wire insulated from one another. If the conductor be solid and the currents due to induction, the distribution will, in general, not be uniform. Under this head Wien considers the case where the currents are due to the variation of a homogeneous magnetic field, parallel to the axis of symmetry, and where the distribution of currents is governed by *resistance*, as will happen in practice when the variations are slow enough. In an elementary circuit the electromotive force varies as the square of the radius and the resistance as the first power. Assuming as before that the whole current is unity, we have merely to introduce into (4) the factors

$$\frac{(a + \rho_1 \cos \phi_1)(a + \rho_2 \cos \phi_2)}{a^2}, \quad \dots\dots\dots\dots\dots(21)$$

M_{12} retaining the value given in (2).

The leading term in (21) is unity, and this, when carried into (14), will reproduce the former result. The term of the first order in ρ in (21) is $(\rho_1 \cos \phi_1 + \rho_2 \cos \phi_2)/a$, and this must be combined with the terms of order ρ^0 and ρ^1 in (2). The former, however, contributes nothing to the integral. The latter yield in (4)

$$\frac{\rho_1^2 + \rho_2^2}{4a^2} \{\log 8a - 1 - \text{greater of } \log \rho_1 \text{ and } \log \rho_2\} + \frac{\text{smaller of } \rho_1^2 \text{ and } \rho_2^2}{4a^2}. \quad (22)$$

The term of the second order in (21), viz., $\rho_1 \rho_2 / a^2 . \cos \phi_1 \cos \phi_2$, needs to be combined only with the leading term in (2). It yields in (4)

$$\frac{\text{smaller of } \rho_1^2 \text{ and } \rho_2^2}{4a^2}. \quad \dots\dots\dots\dots\dots(23)$$

If ρ_1 and ρ_2 are equal (ρ), the additional terms expressed by (22), (23) become

$$\frac{\rho^2}{2a^2} \log \frac{8a}{\rho}. \quad \dots\dots\dots\dots\dots(24)$$

If (24), multiplied by $4\pi a$, be added to (16), we shall obtain the self-induction for a shell (of uniform infinitesimal thickness) in the form of an anchor-ring, the currents being excited in the manner supposed. The result is

$$L = 4\pi a \left\{ \left(1 + \frac{3\rho^2}{4a^2}\right) \log \frac{8a}{\rho} - 2 \right\}. \quad \dots\dots\dots\dots(25)$$

We now proceed to consider the solid ring. By (22), (23) the terms, additional to those previously obtained on the supposition that the current was uniformly distributed, are

$$\frac{1}{\rho^4}\int\int d\rho_1{}^2 d\rho_2{}^2 \left[\frac{\text{smaller of } \rho_1{}^2 \text{ and } \rho_2{}^2}{2a^2}\right.$$

$$\left. +\frac{\rho_1{}^2+\rho_2{}^2}{4a^2}\{\log 8a -1 - \text{greater of } \log \rho_1 \text{ and } \log \rho_2\}\right]. \ \dots (26)$$

The first part of this is $\rho^2/6a^2$, and the second is $\dfrac{\rho^2}{4a^2}\{\log 8a -1 - \log \rho + \tfrac{1}{6}\}$. The additional terms are accordingly

$$\frac{\rho^2}{4a^2}\left\{\log\frac{8a}{\rho}-\frac{1}{6}\right\}. \quad\dots\dots\dots\dots\dots\dots\dots\dots(27)$$

These multiplied by $4\pi a$ are to be added to (1). We thus obtain

$$L = 4\pi a\left[\log\frac{8a}{\rho}-\frac{7}{4}+\frac{3\rho^2}{8a^2}\log\frac{8a}{\rho}\right] \quad\dots\dots\dots\dots(28)$$

for the self-induction of the solid ring when currents are slowly generated in it by uniform magnetic forces parallel to the axis of symmetry. In Wien's result for this case there appears an additional term within the bracket equal to $-0{\cdot}092\,\rho^2/a^2$.

A more interesting problem is that which arises when the alternations in the magnetic field are rapid instead of slow. Ultimately the distribution of current becomes independent of *resistance*, and is determined by induction alone. A leading feature is that the currents are *superficial*, although the ring itself may be solid. They remain, of course, symmetrical with respect to the straight axis, and to the plane which contains the circular axis.

The magnetic field may be supposed to be due to a current x_1 in a circuit at a distance, and the whole energy of the field may be represented by

$$T = \tfrac{1}{2}M_{11}x_1{}^2 + \tfrac{1}{2}M_{22}x_2{}^2 + \tfrac{1}{2}M_{33}x_3{}^2 + \dots + M_{12}x_1x_2 + M_{13}x_1x_3 + \dots$$

$$+ M_{23}x_2x_3 + \dots, \quad\dots\dots(29)$$

x_2, x_3, etc., being currents in other circuits where no independent electromotive force acts. If x_1 be regarded as given, the corresponding values of x_2, x_3, ... are to be found by making T a minimum. Thus

$$\left.\begin{array}{l} M_{12}x_1 + M_{22}x_2 + M_{32}x_3 + \dots = 0, \\ M_{13}x_1 + M_{23}x_2 + M_{33}x_3 + \dots = 0, \end{array}\right\} \quad\dots\dots\dots\dots(30)$$

and so on, are the equations by which x_2, etc., are to be found in terms of x_1. What we require is the corresponding value of T', formed from T by omission of the terms containing x_1.

The method here sketched is general. It is not necessary that x_2, etc., be currents in particular circuits. They may be regarded as generalized coordinates, or rather velocities, by which the kinetic energy of the system is defined.

For the present application we suppose that the distribution of current round the circumference of the section is represented by

$$\{\alpha_0 + \alpha_1 \cos \phi_1 + \alpha_2 \cos 2\phi_1 + \ldots\} \frac{d\phi_1}{2\pi}, \qquad \ldots\ldots\ldots\ldots(31)$$

so that the total current is α_0. The doubled energy, so far as it depends upon the interaction of the ring currents, is

$$\frac{1}{4\pi^2} \iint (\alpha_0 + \alpha_1 \cos \phi_1 + \alpha_2 \cos 2\phi_1 + \ldots)(\alpha_0 + \alpha_1 \cos \phi_2 + \ldots) M_{12} d\phi_1 d\phi_2, \quad (32)$$

where M_{12} has the value given in (2), simplified by making ρ_1 and ρ_2 both equal to ρ. To this has to be added the double energy arising from the interaction of the ring currents with the primary current. For each element of the ring currents (31) we have to introduce a factor proportional to the area of the circuit, viz., $\pi (a + \rho \cos \phi_1)^2$. This part of the double energy may thus be taken to be

$$H \int d\phi_1 (a + \rho \cos \phi_1)^2 (\alpha_0 + \alpha_1 \cos \phi_1 + \alpha_2 \cos 2\phi_1 + \ldots),$$

that is
$$2\pi H \{(a^2 + \tfrac{1}{2}\rho^2) \alpha_0 + a\rho\alpha_1 + \tfrac{1}{4}\rho^2\alpha_2\}, \qquad \ldots\ldots\ldots\ldots(33)$$

α_3, etc., not appearing. The sum of (33) and (32) is to be made a minimum by variation of the α's.

We have now to evaluate (32). The coefficient of α_0^2 is the quantity already expressed in (16). For the other terms it is not necessary to go further than the first power of ρ in (2). We get

$$4\pi a \left[\alpha_0^2 \left\{ \log \frac{8a}{\rho} \left(1 + \frac{\rho^2}{4a^2}\right) - 2 \right\} + \tfrac{1}{4}(\alpha_1^2 + \tfrac{1}{2}\alpha_2^2 + \tfrac{1}{3}\alpha_3^2 + \ldots) \right.$$

$$\left. + \frac{\rho}{a} \left\{ \frac{\alpha_0\alpha_1}{2} \left(\log \frac{8a}{\rho} - 1\right) + \frac{\alpha_1}{8}(2\alpha_0 + \alpha_2) + \frac{\alpha_2}{2.8}(\alpha_1 + \alpha_3) + \frac{\alpha_3}{3.8}(\alpha_2 + \alpha_4) + \ldots \right\} \right].$$

$$\ldots\ldots\ldots(34)$$

Differentiating the sum of (33), (34), with respect to α_0, α_1, etc., in turn, we find

$$H(a^2 + \tfrac{1}{2}\rho^2) + 4a\alpha_0 \left\{ \log \frac{8a}{\rho} \left(1 + \frac{\rho^2}{4a^2}\right) - 2 \right\} + \rho\alpha_1 \left(\log \frac{8a}{\rho} - \frac{1}{2}\right) = 0, \quad (35)$$

$$H\rho + \alpha_1 + \frac{\rho}{a} \left\{ \alpha_0 \left(\log \frac{8a}{\rho} - \frac{1}{2}\right) + \frac{3\alpha_2}{8} \right\} = 0, \qquad \ldots\ldots\ldots\ldots\ldots\ldots(36)$$

$$H\rho^2 + 2a\alpha_2 + \rho \left(\frac{3\alpha_1}{2} + \frac{5\alpha_3}{6}\right) = 0. \qquad \ldots\ldots\ldots\ldots\ldots\ldots(37)$$

The leading term is, of course, α_0. Relatively to this, α_1 is of order ρ, α_2 of order ρ^2, and so on. Accordingly, α_2, α_3, etc., may be omitted entirely from (34), which is only expected to be accurate up to ρ^2 inclusive. Also, in α_1 only the leading term need be retained.

The ratio of α_1 to α_0 is to be found by elimination of H between (35), (36). We get

$$\frac{\alpha_1}{\alpha_0} = \frac{\rho}{a} \left\{ 3 \log \frac{8a}{\rho} - \frac{15}{2} \right\}. \quad \text{.....................(38)}$$

Substituting this in (34), we find as the coefficient of self-induction

$$L = 4\pi a \left[\log \frac{8a}{\rho} \left(1 + \frac{\rho^2}{4a^2} \right) - 2 + \frac{\rho^2}{4a^2} \left(3 \log \frac{8a}{\rho} - \frac{15}{2} \right) \left(5 \log \frac{8a}{\rho} - \frac{17}{2} \right) \right]. \quad (39)$$

The approximate value of α_0 in terms of H is

$$\alpha_0 = -\frac{Ha}{4 \left(\log \dfrac{8a}{\rho} - 2 \right)}. \quad \text{.......................(40)}$$

A closer approximation can be found by elimination of α_1 between (35), (36).

In (39) the currents are supposed to be induced by the variation (in time) of an unlimited uniform magnetic field. A problem, simpler from the theoretical point of view, arises if we suppose the uniform field to be limited to a cylindrical space co-axial with the ring, and of diameter less than the smallest diameter of the ring $(2a - 2\rho)$. Such a field may be supposed to be due to a cylindrical current sheet, the length of the cylinder being infinite. The ring currents to be investigated are those arising from the instantaneous abolition of the current sheet and its conductor.

If πb^2 be the area of the cylinder, (33) is replaced simply by

$$H \int d\phi_1 \, b^2 \left(\alpha_0 + \alpha_1 \cos \phi_1 + \ldots \right) = 2\pi H b^2 \alpha_0. \quad \text{...............(41)}$$

The expression (34) remains unaltered and the equations replacing (35), (36) are thus

$$H b^2 + 4a\alpha_0 \left\{ \log \frac{8a}{\rho} \left(1 + \frac{\rho^2}{4a^2} \right) - 2 \right\} + \rho \alpha_1 \left(\log \frac{8a}{\rho} - \frac{1}{2} \right) = 0, \quad \text{....(42)}$$

$$\alpha_1 + \frac{\rho}{a} \left(\log \frac{8a}{\rho} - \frac{1}{2} \right) \alpha_0 = 0. \quad \text{.....................(43)}$$

The introduction of (43) into (34) gives for the coefficient of self-induction in this case—

$$L = 4\pi a \left[\log \frac{8a}{\rho} \left(1 + \frac{\rho^2}{4a^2} \right) - 2 - \frac{\rho^2}{4a^2} \left(\log \frac{8a}{\rho} - \frac{1}{2} \right)^2 \right]. \quad \text{........(44)}$$

It will be observed that the sign of α_1/α_0 is different in (38) and (43).

The peculiarity of the problem last considered is that the primary current occasions no magnetic force at the surface of the ring. The consequences were set out 40 years ago by Maxwell in a passage* whose significance was very slowly appreciated. "In the case of a current sheet of no resistance, the surface integral of magnetic induction remains constant at every point of the current sheet.

"If, therefore, by the motion of magnets or variations of currents in the neighbourhood, the magnetic field is in any way altered, electric currents will be set up in the current sheet, such that their magnetic effect, combined with that of the magnets or currents in the field, will maintain the normal component of magnetic induction at every point of the sheet unchanged. If at first there is no magnetic action, and no currents in the sheet, then the normal component of magnetic induction will always be zero at every point of the sheet.

"The sheet may therefore be regarded as impervious to magnetic induction, and the lines of magnetic induction will be deflected by the sheet exactly in the same way as the lines of flow of an electric current in an infinite and uniform conducting mass would be deflected by the introduction of a sheet of the same form made of a substance of infinite resistance.

"If the sheet forms a closed or an infinite surface, no magnetic actions which may take place on one side of the sheet will produce any magnetic effect on the other side."

All that Maxwell says of a current sheet is, of course, applicable to the surface of a perfectly conducting solid, such as our anchor-ring may be supposed to be. The currents left in the ring after the abolition of the primary current must be such that the magnetic force due to them is *wholly tangential* to the surface of the ring. Under this condition $\int_{-\pi}^{+\pi} M_{12} d\phi_2$ must be independent of ϕ_1, and we might have investigated the problem upon this basis.

In Maxwell's notation α, β, γ denote the components of magnetic force, and the whole energy of the field T is given by

$$T = \frac{1}{8\pi} \iiint (\alpha^2 + \beta^2 + \gamma^2)\, dx\, dy\, dz = \tfrac{1}{2} L a_0^2. \quad \ldots\ldots\ldots\ldots(45)$$

Moreover a_0, the total current, multiplied by 4π is equal to the "circulation" of magnetic force round the ring. In this form our result admits of immediate application to the hydrodynamical problem of the circulation of

---

* *Electricity and Magnetism*, §§ 654, 655. Compare my "Acoustical Observations," *Phil. Mag.* 1882, Vol. XIII. p. 340; *Scientific Papers*, Vol. II. p. 99.

incompressible frictionless fluid round a solid having the form of the ring; for the components of velocity u, v, w are subject to precisely the same conditions as are α, β, γ. If the density be unity, the kinetic energy T of the motion has the expression

$$T = \frac{L}{8\pi} \times (\text{circulation})^2, \quad \dots\dots\dots\dots\dots\dots\dots(46)$$

L having the value given in (44).

P.S. March 4.—Sir W. D. Niven, who in 1881 verified some other results for self-induction—those numbered (11), (12) in the paper referred to—has been good enough to confirm the formulæ (1), (28) of the present communication, in which I differ from M. Wien.

ELECTRICAL VIBRATIONS ON A THIN ANCHOR-RING.

[*Proceedings of the Royal Society*, A, Vol. LXXXVII. pp. 193—202, 1912.]

ALTHOUGH much attention has been bestowed upon the interesting subject of electric oscillations, there are comparatively few examples in which definite mathematical solutions have been gained. These problems are much simplified when conductors are supposed to be perfect, but even then the difficulties usually remain formidable. Apart from cases where the propagation may be regarded as being in one dimension*, we have Sir J. Thomson's solutions for electrical vibrations upon a conducting sphere or cylinder†. But these vibrations have so little persistence as hardly to deserve their name. A more instructive example is afforded by a conductor in the form of a circular ring, whose circular section is supposed small. There is then in the neighbourhood of the conductor a considerable store of energy which is more or less entrapped, and so allows of vibrations of reasonable persistence. This problem was very ably treated by Pocklington‡ in 1897, but with deficient explanations§. Moreover, Pocklington limits his detailed conclusions to one particular mode of free vibration. I think I shall be doing a service in calling attention to this investigation, and in exhibiting the result for the radiation of vibrations in the higher modes. But I do not attempt a complete re-statement of the argument.

Pocklington starts from Hertz's formulæ for an elementary vibrator at the origin of coordinates ξ, η, ζ,

$$P = \frac{d^2\Pi}{d\xi\,d\zeta}, \quad Q = \frac{d^2\Pi}{d\eta\,d\zeta}, \quad R = \frac{d^2\Pi}{d\zeta^2} + \alpha^2\Pi, \quad \text{............(1)}$$

where

$$\Pi = e^{i\alpha\rho}\,e^{ipt}/\rho, \quad \text{...............................(2)}$$

---

* *Phil. Mag.* 1897, Vol. XLIII. p. 125; 1897, Vol. XLIV. p. 199; *Scientific Papers*, Vol. IV. pp. 276, 327.

† *Recent Researches*, 1893, §§ 301, 312. [1913. There is also Abraham's solution for the ellipsoid.]

‡ *Camb. Proceedings*, 1897, Vol. IX. p. 324.

§ Compare W. M^cF. Orr, *Phil. Mag.* 1903, Vol. VI. p. 667.

in which P, Q, R denote the components of electromotive intensity, $2\pi/p$ is the period of the disturbance, and $2\pi/\alpha$ the wave-length corresponding in free æther to this period. At a great distance ρ from the source, we have from (1)

$$P, Q, R = \frac{\alpha^2 e^{i\alpha\rho}}{\rho} \left(-\frac{\xi\zeta}{\rho^2}, \; -\frac{\eta\zeta}{\rho^2}, \; 1 - \frac{\zeta^2}{\rho^2} \right). \quad\text{.............}(3)$$

The resultant is perpendicular to ρ, and in the plane containing ρ and ζ. Its magnitude is

$$-\frac{\alpha^2 e^{i\alpha\rho}}{\rho} \sin \chi, \quad\text{............................}(4)$$

where χ is the angle between ρ and ζ.

The required solution is obtained by a distribution of elementary vibrators of this kind along the circular axis of the ring, the axis of the vibrator being everywhere tangential to the axis of the ring and the coefficient of intensity proportional to $\cos m\phi'$, where m is an integer and ϕ' defines a point upon the axis. The calculation proceeds in terms of semi-polar coordinates z, ϖ, ϕ, the axis of symmetry being that of z, and the origin being at the centre of the circular axis. The radius of the circular axis is a, and the radius of the circular section is ϵ, ϵ being very small relatively to a. The condition to be satisfied is that at every point of the surface of the ring, where $(\varpi - a)^2 + z^2 = \epsilon^2$, the tangential component of (P, Q, R) shall vanish. It is not satisfied absolutely by the above specification; but Pocklington shows that to the order of approximation required the specification suffices, provided α be suitably chosen. The equation determining α expresses the evanescence of that tangential component which is parallel to the circular axis, and it takes the form

$$\int_0^{\pi} d\phi \, \Pi_0 \cos m\phi \, (m^2 - \alpha^2 a^2 \cos \phi) = 0, \quad\text{.................}(5)$$

where

$$\Pi_0 = \frac{e^{i\alpha[\epsilon^2 + 4\varpi a \sin^2 \frac{1}{2}\phi]}}{\sqrt{[\epsilon^2 + 4\varpi a \sin^2 \frac{1}{2}\phi]}}. \quad\text{......................}(6)$$

In (5) we are to retain the large term, arising in the integral when ϕ is small, and the finite term, but we may reject *small* quantities. Thus Pocklington finds

$$\int_0^{\pi} \frac{(a^2\alpha^2 \cos \phi - m^2) \cos m\phi \, d\phi}{\sqrt{\{\epsilon^2 + 4a^2 \sin^2 \frac{1}{2}\phi\}}}$$

$$+ \int_0^{\pi} \frac{(e^{2i\alpha a \sin \frac{1}{2}\phi} - 1)(a^2\alpha^2 \cos \phi - m^2) \cos m\phi \, d\phi}{2a \sin \frac{1}{2}\phi} = 0, \quad\text{.........}(7)$$

the condition being to this order of approximation the same at all points of a cross-section.

The first integral in (7) may be evaluated for any (integral) value of m. Writing $\frac{1}{2}\phi = \psi$, we have

$$\int_0^{\frac{1}{2}\pi} \frac{(a^2\alpha^2 \cos 2\psi - m^2) \cos 2m\psi \, d\psi}{a\sqrt{\{\epsilon^2/4a^2 + \sin^2 \psi\}}}. \qquad\qquad (8)$$

The large part of the integral arises from small values of ψ. We divide the range of integration into two parts, the first from 0 to ψ where ψ, though small, is large compared with $\epsilon/2a$, and the second from ψ to $\frac{1}{2}\pi$. For the first part we may replace $\cos 2\psi$, $\cos 2m\psi$ by unity, and $\sin^2 \psi$ by ψ^2. We thus obtain

$$\frac{a^2\alpha^2 - m^2}{a} \log\{\psi + \sqrt{(\epsilon^2/4a^2 + \psi^2)}\}_0^\psi = \frac{a^2\alpha^2 - m^2}{a}(\log 4a/\epsilon + \log \psi). \ \ ...(9)$$

Thus to a first approximation $\alpha a = \pm m$. In the second part of the range of integration we may neglect $\epsilon^2/4a^2$ in comparison with $\sin^2 \psi$, thus obtaining

$$\int_\psi^{\frac{1}{2}\pi} \frac{(a^2\alpha^2 \cos 2\psi - m^2) \cos 2m\psi \, d\psi}{a \sin \psi}. \qquad\qquad (10)$$

The numerator may be expressed as a sum of terms such as $\cos^{2n} \psi$, and for each of these the integral may be evaluated by taking $\cos \psi = z$, in virtue of

$$\int \frac{z^{2n} \, dz}{z^2 - 1} = z + \frac{z^3}{3} + \frac{z^5}{5} + \ldots\ldots + \frac{z^{2n-1}}{2n-1} + \frac{1}{2}\log\frac{1-z}{1+z}.$$

Accordingly

$$\int_\psi^{\frac{1}{2}\pi} \frac{\cos^{2n} \psi \, d\psi}{\sin \psi} = -\cos \psi - \frac{\cos^3 \psi}{3} - \ldots\ldots - \frac{\cos^{2n-1} \psi}{2n-1} - \log \tan \tfrac{1}{2}\psi$$

$$= -1 - \tfrac{1}{3} - \ldots\ldots - \frac{1}{2n-1} - \log \tfrac{1}{2}\psi, \qquad\qquad (11)$$

when small quantities are neglected. For example,

$$\int_\psi^{\frac{1}{2}\pi} \frac{\cos^2 \psi \, d\psi}{\sin \psi} = -1 - \log \tfrac{1}{2}\psi, \qquad \int_\psi^{\frac{1}{2}\pi} \frac{\cos^4 \psi \, d\psi}{\sin \psi} = -\tfrac{4}{3} - \log \tfrac{1}{2}\psi.$$

The sum of the coefficients in the series of terms (analogous to $\cos^{2n} \psi$) which represents the numerator of (10) is necessarily $a^2\alpha^2 - m^2$, since this is the value of the numerator itself when $\psi = 0$. The particular value of ψ chosen for the division of the range of integration thus disappears from the sum of (9) and (10), as of course it ought to do.

When $m = 1$, corresponding to the gravest mode of vibration specially considered by Pocklington, the numerator in (10) is

$$4a^2\alpha^2 \cos^4 \psi - (4a^2\alpha^2 + 2) \cos^2 \psi + a^2\alpha^2 + 1,$$

8

and the value of the integral is accordingly

$$\frac{1}{a}\left[2 - \frac{4a^2\alpha^2}{3} - (a^2\alpha^2 - 1)\log\tfrac{1}{2}\psi\right].$$

To this is to be added from (9)

$$\frac{a^2\alpha^2 - 1}{a}\left[\log\frac{4a}{\epsilon} + \log\psi\right],$$

making altogether for the value of (8)

$$\frac{1}{a}\left[(a^2\alpha^2 - 1)\log\frac{8a}{\epsilon} + 2 - \frac{4a^2\alpha^2}{3}\right]. \quad\dots\dots\dots\dots(12)$$

The second integral in (7) contributes only finite terms, but it is important as determining the imaginary part of α and thus the rate of dissipation. We may write it

$$\frac{m^2}{2a}\int_0^{\frac{1}{2}\pi} d\psi\,\frac{e^{ix\sin\psi} - 1}{\sin\psi}\{\cos(2m + 2)\,\psi + \cos(2m - 2)\,\psi - 2\cos 2m\psi\},\dots(13)$$

where $\qquad\qquad x^2 = 4a^2\alpha^2 = 4m^2$ approximately.

Pocklington shows that the imaginary part of (13) can be expressed by means of Bessel's functions.   We may take

$$\frac{2}{\pi}\int_0^{\frac{1}{2}\pi} d\psi\,\cos 2n\psi\,e^{ix\sin\psi} = J_{2n}(x) + i\,K_{2n}(x), \quad\dots\dots\dots(14)*$$

whence $\quad\int_0^{\frac{1}{2}\pi} d\psi\,\cos 2n\psi\,\frac{e^{ix\sin\psi} - 1}{\sin\psi} = \frac{i\pi}{2}\int_0^x \{J_{2n}(x) + i\,K_{2n}(x)\}\,dx. \quad\dots..(15)$

Accordingly, (13) may be replaced by

$$\frac{im^2\pi}{4a}\int_0^x dx\,\{J_{2m+2}(x) - 2J_{2m}(x) + J_{2m-2}(x) + i\,(K_{2m+2} - 2K_{2m} + K_{2m-2})\}. \quad(16)$$

Now $\dagger$ $\qquad\qquad J_{2m+2} - 2J_{2m} + J_{2m-2} = 4J''_{2m},$

so that $\quad\int_0^x dx\,\{J_{2m+2} - 2J_{2m} + J_{2m-2}\} = 4J'_{2m} = 2J_{2m-1} - 2J_{2m+1}. \quad\dots..(17)$

The imaginary part of (13) is thus simply

$$\frac{im^2\pi}{2a}\{J_{2m-1}(x) - J_{2m+1}(x)\}. \quad\dots\dots\dots\dots\dots(18)$$

A corresponding theory for the K functions does not appear to have been developed.

When $m = 1$, our equation becomes

$$\left(\frac{x^2}{4} - 1\right)\log\frac{8a}{\epsilon} = -\frac{i\pi}{2}\{J_1(x) - J_3(x)\} + \frac{x^2}{3} - 2$$

$$-\int_0^{\frac{1}{2}\pi} d\psi\,\frac{\cos(x\sin\psi) - 1}{2\sin\psi}(1 - 2\cos 2\psi + \cos 4\psi), \quad\dots\dots(19)$$

---

* Compare *Theory of Sound*, § 302.   † Gray and Mathews, *Bessel's Functions*, p. 13.

and on the right we may replace x by its first approximate value. Referring to (2) we see that the negative sign must be chosen for α and x, so that $x = -2$. The imaginary term on the right is thus

$$\frac{i\pi}{2} \{J_1(2) - J_3(2)\} = 0.70336i.$$

For the real term Pocklington calculates 0.485, so that, L being written for $\log(8a/\epsilon)$,

$$-\alpha = \frac{1}{a} \{1 + (0.243 + 0.352i)/L\}. \quad\ldots\ldots\ldots\ldots\ldots(20)$$

"Hence the period of the oscillation is equal to the time required for a free wave to traverse a distance equal to the circumference of the circle multiplied by $1 - 0.243/L$, and the ratio of the amplitudes of consecutive vibrations is $1 : e^{-2.21/L}$ or $1 - 2.21/L$."

For the general value of m (19) is replaced by

$$(a^2\alpha^2 - m^2) L = \frac{im^2\pi}{2} \{J_{2m-1}(2m) - J_{2m+1}(2m)\} + R, \quad\ldots\ldots(21)$$

where R is a real finite number, and finally

$$-\alpha = \frac{m}{a} \left[1 + \frac{R}{2m^2L} + \frac{i\pi}{4L} \{J_{2m-1}(2m) - J_{2m+1}(2m)\} \right]. \quad\ldots\ldots(22)$$

The ratio of the amplitudes of successive vibrations is thus

$$1 : 1 - \pi^2 \{J_{2m-1}(2m) - J_{2m+1}(2m)\}/2L, \quad\ldots\ldots\ldots\ldots(23)$$

in which the values of $J_{2m-1}(2m) - J_{2m+1}(2m)$ can be taken from the tables (see Gray and Mathews). We have as far as m equal to 12:

m	$J_{2m-1}(2m) - J_{2m+1}(2m)$	m	$J_{2m-1}(2m) - J_{2m+1}(2m)$
1	0.448	7	0.136
2	0.298	8	0.125
3	0.232	9	0.116
4	0.194	10	0.108
5	0.169	11	0.102
6	0.150	12	0.096

It appears that the damping during a *single vibration* diminishes as m increases, viz., the greater the number of subdivisions of the circumference.

An approximate expression for the tabulated quantity when m is large may be at once derived from a formula due to Nicholson[*], who shows that

* *Phil. Mag.* 1908, Vol. XVI. pp. 276, 277.

when n and z are large and nearly equal, $J_n(z)$ is related to Airy's integral. In fact,

$$J_n(z) = \frac{1}{\pi} \left(\frac{6}{z}\right)^{\frac{1}{3}} \int_0^\infty \cos\left\{w^3 + (n-z)\left(\frac{6}{z}\right)^{\frac{1}{3}} w\right\} dw$$

$$= \frac{1}{\pi} \left(\frac{6}{z}\right)^{\frac{1}{3}} \left[\frac{\Gamma\left(\frac{1}{3}\right)}{2\sqrt{3}} - (n-z)\left(\frac{6}{z}\right)^{\frac{1}{3}} \frac{\Gamma\left(\frac{2}{3}\right)}{2\sqrt{3}}\right], \quad \ldots\ldots(24)$$

so that

$$J_{2m-1}(2m) - J_{2m+1}(2m) = \left(\frac{3}{m}\right)^{\frac{2}{3}} \frac{\Gamma\left(\frac{2}{3}\right)}{\pi\sqrt{3}}. \quad \ldots\ldots\ldots\ldots(25)$$

If we apply this formula to $m = 10$, we get 0·111 as compared with the tabular 0·108*.

It follows from (25) that the damping in each vibration diminishes without limit as m increases. On the other hand, the damping in a *given time* varies as $m^{\frac{1}{3}}$ and increases indefinitely, if slowly, with m.

We proceed to examine more in detail the character at a great distance of the vibration radiated from the ring. For this purpose we choose axes of x and y in the plane of the ring, and the coordinates (x, y, z) of any point may also be expressed as $r \sin\theta \cos\phi$, $r \sin\theta \sin\phi$, $r \cos\theta$. The contribution of an element $a\,d\phi'$ at ϕ' is given by (4). The direction cosines of this element are $\sin\phi'$, $-\cos\phi'$, 0; and those of the disturbance due to it are taken to be l, m, n. The direction of this disturbance is perpendicular to r and in the plane containing r and the element of arc $a\,d\phi'$. The first condition gives $lx + my + nz = 0$, and the second gives

$$l \cdot z \cos\phi' + m \cdot z \sin\phi' - n(x\cos\phi' + y\sin\phi') = 0;$$

so that

$$\frac{l}{(z^2+y^2)\sin\phi' + xy\cos\phi'} = \frac{-m}{(z^2+x^2)\cos\phi' + xy\sin\phi'} = \frac{n}{zy\cos\phi' - zx\sin\phi'}.$$

$$\ldots\ldots\ldots\ldots(26)$$

The sum of the squares of the denominators in (26) is

$$r^2\{z^2 - (y\sin\phi' + x\cos\phi')^2\}.$$

Also in (4)

$$\sin^2\chi = 1 - \frac{(x\sin\phi' - y\cos\phi')^2}{r^2} = \frac{z^2 + (x\cos\phi' + y\sin\phi')^2}{r^2}; \quad \ldots(27)$$

and thus

$$\left.\begin{aligned}
r^2 \cdot l \ \sin\chi &= (z^2+y^2)\sin\phi' + xy\cos\phi', \\
-r^2 \cdot m \sin\chi &= (z^2+x^2)\cos\phi' + xy\sin\phi', \\
r^2 \cdot n \ \sin\chi &= zy\cos\phi' - zx\sin\phi'.
\end{aligned}\right\} \quad \ldots\ldots\ldots\ldots(28)$$

To these quantities the components P, Q, R due to the element $a\,d\phi'$ are proportional.

* $\log_{10}\Gamma\left(\frac{2}{3}\right) = 0\cdot13166$.

Before we can proceed to an integration there are two other factors to be regarded. The first relates to the intensity of the source situated at $ad\phi'$. To represent this we must introduce $\cos m\phi'$. Again, there is the question of phase. In $e^{ia\rho}$ we have

$$\rho = r - a \sin \theta \cos (\phi' - \phi);$$

and in the denominator of (4) we may neglect the difference between ρ and r. Thus, as the components due to $ad\phi'$, we have

$$P = -\frac{a^2 a e^{iar}}{r} d\phi' e^{-iaa \sin \theta \cos(\phi'-\phi)} \cos m\phi' \frac{(z^2 + y^2) \sin \phi' + xy \cos \phi'}{r^2}, \quad \ldots(29)$$

with similar expressions for Q and R corresponding to the right-hand members of (28). The integrals to be considered may be temporarily denoted by S, C, where

$$S,\ C = \int_{-\pi}^{+\pi} d\phi' \cos m\phi' e^{-i\zeta \cos(\phi'-\phi)} (\sin \phi', \cos \phi'), \quad \ldots\ldots\ldots(30)$$

ζ being written for $aa \sin \theta$. Here

$$S = \tfrac{1}{2} \int_{-\pi}^{+\pi} d\phi' e^{-i\zeta\cos(\phi'-\phi)} \{\sin (m + 1) \phi' - \sin (m - 1) \phi'\},$$

and in this, if we write ψ for $\phi' - \phi$,

$$\sin (m + 1) \phi' = \sin (m + 1) \psi . \cos (m + 1) \phi + \cos (m + 1) \psi . \sin (m + 1) \phi.$$

We thus find

$$S = \Theta_{m+1} \sin (m + 1) \phi - \Theta_{m-1} \sin (m - 1) \phi, \quad \ldots\ldots\ldots(31)$$

where

$$\Theta_n = \int_0^{\pi} d\psi \cos n\psi\, e^{-i\zeta\cos\psi}. \quad \ldots\ldots\ldots\ldots\ldots\ldots(32)$$

In like manner,

$$C = \Theta_{m+1} \cos (m + 1) \phi + \Theta_{m-1} \cos (m - 1) \phi. \quad \ldots\ldots\ldots(33)$$

Now

$$\Theta_n = \int_0^{\pi} d\psi \cos n\psi \{\cos (\zeta \cos \psi) - i \sin (\zeta \cos \phi)\}.$$

When n is even, the imaginary part vanishes, and

$$\Theta_n = \frac{\pi J_n(\zeta)}{\cos \tfrac{1}{2} n\pi}. \quad \ldots\ldots\ldots\ldots\ldots\ldots\ldots\ldots(34)$$

On the other hand, when n is odd, the real part vanishes, and

$$\Theta_n = -\frac{i\pi J_n(\zeta)}{\sin \tfrac{1}{2} n\pi}. \quad \ldots\ldots\ldots\ldots\ldots\ldots\ldots(35)$$

Thus, when m is even, $m + 1$ and $m - 1$ are both odd and S and C are both pure imaginaries. But when m is odd, S and C are both real.

As functions of direction we may take P, Q, R to be proportional to

$$S \frac{z^2 + y^2}{r^2} + C \frac{xy}{r^2}, \qquad -C \frac{z^2 + x^2}{r^2} - S \frac{xy}{r^2}, \qquad C \frac{zy}{r^2} - S \frac{zx}{r^2}.$$

Whether m be odd or even, the three components are in the same phase. On the same scale the intensity of disturbance, represented by $P^2 + Q^2 + R^2$, is in terms of θ, ϕ

$$\cos^2 \theta \, (S^2 + C^2) + \sin^2 \theta \, (C \cos \phi + S \sin \phi)^2, \quad \ldots\ldots\ldots\ldots(36)$$

an expression whose sign should be changed when m is even. Introducing the values of C and S in terms of Θ from (31), (33), we find that $P^2 + Q^2 + R^2$ is proportional to

$$\cos^2 \theta \, \{\Theta_{m+1}{}^2 + \Theta_{m-1}{}^2 + 2\Theta_{m+1} \Theta_{m-1} \cos 2m\phi\} + \sin^2 \theta \cos^2 m\phi \, \{\Theta_{m+1} + \Theta_{m-1}\}^2.$$
$$\ldots\ldots(37)$$

From this it appears that for directions lying in the plane of the ring ($\cos \theta = 0$) the radiation vanishes with $\cos m\phi$. The expression (37) may also be written

$$\Theta_{m+1}{}^2 + \Theta_{m-1}{}^2 + 2\Theta_{m+1} \Theta_{m-1} \cos 2m\phi - \tfrac{1}{2} \sin^2 \theta \, (\Theta_{m+1} - \Theta_{m-1})^2 \, (1 - \cos 2m\phi),$$
$$\ldots\ldots(38)$$

or, in terms of J's, by (34), (35),

$$\pi^2 \left[J_{m+1}{}^2 + J_{m-1}{}^2 - 2J_{m+1} J_{m-1} \cos 2m\phi - \tfrac{1}{2} \sin^2 \theta \, (J_{m+1} + J_{m-1})^2 \, (1 - \cos 2m\phi) \right],$$
$$\ldots\ldots(39)$$

and this whether m be odd or even. The argument of the J's is $\alpha a \sin \theta$.

Along the axis of symmetry ($\theta = 0$) the expression (39) should be independent of ϕ. That this is so is verified when we remember that $J_n(0)$ vanishes except $n = 0$. The expression (39) thus vanishes altogether with θ unless $m = 1$, when it reduces to π^2 simply*. In the neighbourhood of the axis the intensity is of the order θ^{2m-2}.

In the plane of the ring ($\sin \theta = 1$) the general expression reduces to

$$\pi^2 \, (J_{m+1} - J_{m-1})^2 \cos^2 m\phi, \quad \text{or} \quad 4\pi^2 J_m'^2 \cos^2 m\phi. \quad \ldots\ldots\ldots(40)$$

It is of interest to consider also the *mean* value of (39) reckoned over angular space. The mean with respect to ϕ is evidently

$$\pi^2 \left[J_{m+1}{}^2 + J_{m-1}{}^2 + \tfrac{1}{2} \sin^2 \theta \, (J_{m+1} + J_{m-1})^2 \right]. \quad \ldots\ldots\ldots\ldots(41)$$

By a known formula in Bessel's functions

$$\{J_{m+1}(\zeta) + J_{m-1}(\zeta)\}^2 = \frac{4m^2}{\zeta^2} J_m{}^2(\zeta). \quad \ldots\ldots\ldots\ldots\ldots(42)$$

For the present purpose

$$\zeta^2 = a^2 \alpha^2 \sin^2 \theta = m^2 \sin^2 \theta \,;$$

and (41) becomes

$$\pi^2 \left[J_{m+1}{}^2(\zeta) + J_{m-1}{}^2(\zeta) - 2J_m{}^2(\zeta) \right]. \quad \ldots\ldots\ldots\ldots(43)$$

---

* [*June* 20. Reciprocally, plane waves, travelling parallel to the axis of symmetry and incident upon the ring, excite none of the higher modes of vibration.]

To obtain the mean over angular space we have to multiply this by $\sin\theta\,d\theta$, and integrate from 0 to $\frac{1}{2}\pi$. For this purpose we require

$$\int_0^{\frac{1}{2}\pi} J_n{}^2(m\sin\theta)\sin\theta\,d\theta, \quad\dots\dots\dots\dots\dots(44)$$

an integral which does not seem to have been evaluated.

By a known expansion* we have

$$J_0(2m\sin\theta\sin\tfrac{1}{2}\beta) = J_0{}^2(m\sin\theta) + 2J_1{}^2(m\sin\theta)\cos\beta + 2J_2{}^2(m\sin\theta)\cos 2\beta$$
$$+\dots\dots,$$

whence

$$\int_0^{\frac{1}{2}\pi} J_0(2m\sin\theta\sin\tfrac{1}{2}\beta)\sin\theta\,d\theta$$

$$= \int_0^{\frac{1}{2}\pi} J_0{}^2(m\sin\theta)\sin\theta\,d\theta + 2\cos\beta\int_0^{\frac{1}{2}\pi} J_1{}^2(m\sin\theta)\sin\theta\,d\theta + \dots\dots$$

$$+ 2\cos n\beta\int_0^{\frac{1}{2}\pi} J_n{}^2(m\sin\theta)\sin\theta\,d\theta. \quad\dots\dots\dots\dots\dots\dots(45)$$

Now† for the integral on the left

$$\int_0^{\frac{1}{2}\pi} J_0(2m\sin\theta\sin\tfrac{1}{2}\beta)\sin\theta\,d\theta = \frac{\sin(2m\sin\tfrac{1}{2}\beta)}{2m\sin\tfrac{1}{2}\beta};$$

and thus

$$\int_0^{\frac{1}{2}\pi} J_n{}^2(m\sin\theta)\sin\theta\,d\theta = \frac{1}{2\pi m}\int_0^{\pi} d\beta\cos n\beta\,\frac{\sin(2m\sin\tfrac{1}{2}\beta)}{2m\sin\tfrac{1}{2}\beta}$$

$$= \frac{1}{\pi m}\int_0^{\frac{1}{2}\pi} d\psi\cos 2n\psi\,\frac{\sin(2m\sin\psi)}{\sin\psi} = \frac{1}{2m}\int_0^{2m} J_{2n}(x)\,dx, \quad\dots\dots(46)$$

as in (15). Thus the mean value of (43) is

$$\frac{\pi^2}{2m}\int_0^{2m} dx\,\{J_{2m+2}(x) + J_{2m-2}(x) - 2J_{2m}(x)\} = \frac{2\pi^2}{m}J_{2m}{}'(2m)$$

$$= \frac{\pi^2}{m}\{J_{2m-1}(2m) - J_{2m+1}(2m)\}, \quad\dots\dots(47)$$

as before.

In order to express fully the mean value of $P^2 + Q^2 + R^2$ at distance r, we have to introduce additional factors from (29). If $\alpha = -\alpha_1 - i\alpha_2$, $e^{i\alpha r} = e^{-i\alpha_1 r}e^{\alpha_2 r}$, and these factors may be taken to be $\alpha^4 a^2 e^{2\alpha_2 r}/r^2$. The occurrence of the factor $e^{2\alpha_2 r}$, where α_2 is positive, has a strange appearance; but, as Lamb has shown‡, it is to be expected in such cases as the present, where the vibrations to be found at any time at a greater distance correspond to an earlier vibration at the nucleus.

* Gray and Mathews, p. 28.

† *Enc. Brit.* "Wave Theory of Light," Equation (43), 1888; *Scientific Papers*, Vol. III. p. 98.

‡ *Proc. Math. Soc.* 1900, Vol. XXXII. p. 208.

The calculations just effected afford an independent estimate of the dissipation. The rate at which energy is propagated outwards away from the sphere of great radius r, is

$$- \frac{dE}{dt} = V \cdot 4\pi r^2 \cdot \frac{\alpha^4 a^2 e^{2\alpha_2 r}}{r^2} \frac{\pi^2}{m} \{J_{2m-1} - J_{2m+1}\}, \quad \ldots\ldots\ldots(48)$$

or, since τ (the period) $= 2\pi a/mV$, the loss of energy in one complete vibration is given by

$$- \frac{dE \cdot \tau}{dt} = \frac{8\pi^4 \alpha^4 a^3 e^{2\alpha_2 r}}{m^2} \{J_{2m-1} - J_{2m+1}\}. \quad \ldots\ldots\ldots\ldots(49)$$

With this we have to compare the total energy to be found within the sphere. The occurrence of the factor $e^{2\alpha_2 r}$ is a complication from which we may emancipate ourselves by choosing r great in comparison with a, but still small enough to justify the omission of $e^{2\alpha_2 r}$, conditions which are reconcilable when ϵ is sufficiently small. The mean value of $P^2 + Q^2 + R^2$ at a small distance ρ from the circular axis is $2m^2/a^2\rho^2$. This is to be multiplied by $2\pi a \cdot 2\pi\rho d\rho$, and integrated from ϵ to a value of ρ comparable with a, which need not be further specified. Thus

$$E = \frac{8m^2\pi^2}{a} \int \frac{d\rho}{\rho} = - \frac{8m^2\pi^2}{a} \log \epsilon; \quad \ldots\ldots\ldots\ldots(50)$$

and

$$- \frac{dE \cdot \tau}{E dt} = \frac{\pi^2 \{J_{2m-1}(2m) - J_{2m+1}(2m)\}}{- \log \epsilon}, \quad \ldots\ldots\ldots\ldots(51)$$

in agreement with (23).

366.

COLOURED PHOTOMETRY.

[*Philosophical Magazine*, Vol. XXIV. pp. 301, 302, 1912.]

In his recent paper on the Photometry of Lights of Different Colours* Mr H. Ives remarks:—"No satisfactory theory of the action of the flicker photometer can be said to exist. What does it actually measure? We may assume the existence of a 'luminosity sense' distinct from the colour sense....If, for instance, there exists a physiological process called into action both by coloured and uncoloured light, a measure of this would be a measure of a common property."

Very many years ago it occurred to me that the adjustment of the iris afforded just such a "physiological process"†. The iris contracts when the eye is exposed to a bright red or to a bright green light. There must therefore be some relative brightness of the two lights which tends *equally* to close the iris, and this may afford the measure required. The flicker adjustment is complete when the iris has no tendency to alter under the alternating illumination.

This question was brought home to me very forcibly, when in 1875 I fitted the whole area of the window of a small room with revolving sectors after the manner of Talbot. The intention was to observe, more conveniently than when the eye is at a small hole, the movements of vibrating bodies. The apparatus served this purpose well enough; but incidentally I was much struck with the remarkably disagreeable and even painful sensations experienced when at the beginning or end of operations the slits were revolving slowly so as to generate flashes at the rate of perhaps 3 or 4 per second. I soon learned in self-defence to keep my eyes closed during this phase; and I attributed the discomfort to a vain attempt on the part of the iris to adjust itself to fluctuating conditions.

---

* *Phil. Mag.* Vol. XXIV. p. 178.

† If my memory serves me, I have since read somewhere a similar suggestion, perhaps in Helmholtz.

It is clear, I think, that we have here a common element in variously coloured lights, such as might serve as the basis of coloured photometry. I suppose that there would be no particular difficulty in observing the movements of an iris, and I would suggest that experiments be undertaken to ascertain whether in fact the flicker match coincides with quiescence of the iris. Should this prove to be the case, the view suggested would be amply confirmed; otherwise, it would be necessary to turn to some of the other possibilities discussed by Mr 'Ives.

[1913. Mr H. C. Stevens (*Phil. Mag.* Vol. XXVI. p. 180, 1912), in connexion with the above suggestion, describes an experiment in which the *musculus sphincter pupillae* was paralysed with atropine, without changing " in any observable particular " the appearance of flicker. This observation may prove that an actual movement of the iris is not necessary to the sensation of flicker, but it can hardly be said that the iris has no *tendency* to alter because it is prevented from doing so by the paralysis of the muscle. There must be more than one step between the impression upon the retina which initiates a message to close the iris and the actual closing thereof. The flicker adjustment may, so far as appears, correspond to the absence of such messages.]

367.

ON SOME IRIDESCENT FILMS*.

[*Philosophical Magazine*, Vol. XXIV. pp. 751—755, 1912.]

THE experiments now to be described originated in an accidental observation. Some old lantern-plates, from which the gelatine films had been cleaned off a few years before (probably with nitric acid), being required for use, were again placed in dilute nitric acid to ensure cleanliness. From these plates a gas-flame burning over the dish was seen reflected with colour, of which the cause was not obvious. On examination in daylight a dry plate was observed to be iridescent, but so slightly that the fact might easily escape attention. But when the plate was under water and suitably illuminated, the brilliancy was remarkably enhanced. Upon this question of illumination almost everything depends. The window-shutter of one of the rooms in my laboratory has an aperture about 4 inches square. In front of this the dish of water is placed and at the bottom of the dish a piece of dark-coloured glass. In the water the plate under observation is *tilted*, so as to separate the reflexions of the sky as given by the plate and by the glass underneath. In this way a dark background is ensured. At the corners and edges of the plate the reflected light is white, then follow dark bands, and afterwards the colours which suggest reflexion from a thin plate. On this view it is necessary to suppose that the iridescent film is thinnest at the outside and thickens towards the interior, and further, that the material constituting the film has an index intermediate between those of the glass and of the water. In this way the general behaviour is readily explained, the fact that the colours are so feeble in air being attributed to the smallness of the optical difference between the film and the glass underneath. In the water there would be a better approach to equality between the reflexions at the outer and inner surfaces of the film.

From the first I formed the opinion that the films were due to the use of a silicate substratum in the original preparation, but as the history of the

---

* Read before the British Association at Dundee.

plates was unknown this conjecture could not be satisfactorily confirmed. No ordinary cleaning or wiping had any effect; to remove the films recourse must be had to hydrofluoric acid, or to a polishing operation. My friend Prof. T. W. Richards, after treating one with strong acids and other chemicals, pronounced it to be what chemists would call " very insoluble." The plates first encountered manifested (in the air) a brilliant glassy surface, but afterwards I found others showing in the water nearly or quite as good colours, but in the air presenting a smoky appearance.

Desirous of obtaining the colours as perfectly as possible, I endeavoured to destroy the reflexion from the back surface of the plate, which would, I supposed, dilute the colours due to the iridescent film. But a coating of black sealing-wax, or marine glue, did not do so much good as had been expected. The most efficient procedure was to grind the back of the plate, as is very easily done with carborundum. The colours seemed now to be as good as such colours can ever be, the black also being well developed. Doubtless the success was due in great measure to the special localized character of the illumination. The substitution of strong brine for water made no perceptible improvement.

At this stage I found a difficulty in understanding fully the behaviour of the unground plates. In some places the black would occasionally be good, while in others it had a washed-out appearance, a difference not easily accounted for. A difficulty had already been experienced in deciding upon which side of a plate the film was, and had been attributed to the extreme thinness of the plates. But a suspicion now arose that there were films upon *both* sides, and this was soon confirmed. The best proof was afforded by grinding away half the area upon one side of the plate and the other half of the area upon the other side. Whichever face was uppermost, the unground half witnessed the presence of a film by brilliant coloration.

Attempts to produce silicate films on new glass were for some time an almost complete failure. I used the formula given by Abney (*Instruction in Photography*, 11th edition, p. 342):—

Albumen	1 part.
Water	20 parts.
Silicate of Soda solution of syrupy consistency	1 part.

But whether the plates (coated upon one side) were allowed to drain and dry in the cold, or were more quickly dried off over a spirit flame or before a fire, the resulting films washed away under the tap with the slightest friction or even with no friction at all. Occasionally, however, more adherent patches were observed, which could not so easily be cleaned off. Although it did not seem probable that the photographic film proper played any part, I tried without success a superposed coat of gelatine. In view of these failures

I could only suppose that the formation of a permanent film was the work of time, and some chemical friends were of the same opinion. Accordingly a number of plates were prepared and set aside duly labelled.

Examination at intervals proved that time acted but slowly. After six months the films seemed more stable, but nothing was obtained comparable with the old iridescent plates. It is possible that the desired result might eventually be achieved in this way, but the prospect of experimenting under such conditions is not alluring. Luckily an accidental observation came to my aid. In order to prevent the precipitation of lime in the observing-dish a few drops of nitric acid were sometimes added to the water, and I fancied that films tested in this acidified water showed an advantage. A special experiment confirmed the idea. Two plates, coated similarly with silicate and dried a few hours before, were immersed, one in ordinary tap water, the other in the same water moderately acidified with nitric acid. After some 24 hours' soaking the first film washed off easily, but the second had much greater fixity. There was now no difficulty in preparing films capable of showing as good colours as those of the old plates. The best procedure seems to be to dry off the plates before a fire after coating with recently-filtered silicate solution. In order to obtain the most suitable thickness, it is necessary to accommodate the rapidity of drying to the strength of the solution. If heat is not employed the strength of the above given solution may be doubled. When dry the plates may be immersed for some hours in (much) diluted nitric acid. They are then fit for optical examination, but are best not rubbed at this stage. If the colours are suitable the plates may now be washed and allowed to dry. The full development of the colour effects requires that the back of the plates be treated. In my experience grinding gives the best results when the lighting is favourable, but an opaque varnish may also be used with good effect. The comparative failure of such a treatment of the old plates was due to the existence of films upon both sides. A sufficiently opaque glass, e.g. stained with cobalt or copper, may also be employed. After the films have stood some time subsequently to the treatment with acid, they may be rubbed vigorously with a cloth even while wet; but one or two, which probably had been rubbed prematurely, showed scratches.

The surfaces of the new films are not quite as glassy as the best of the old ones, nor so inconspicuous in the air, but there is, I suppose, no doubt that they are all composed of silica. But I am puzzled to understand how the old plates were manipulated. The films cover both sides without interruption, and are thinner at all the four corners than in the interior.

The extraordinary development of the colours in water as compared with what can be seen in air led me to examine in the same way other thin films deposited on glass. A thin coat of albumen (without silicate) is inconspicuous in air. As in photography it may be rendered insoluble by nitrate of silver acidified with acetic acid, and then exhibits good colours when examined under water with favourable illumination. Filtered gelatine, with which a little bichromate has been mixed beforehand, may also be employed. In this case the dry film should be well exposed to light before washing. Ready-made varnishes also answer well, provided they are capable of withstanding the action of water, at least for a time. I have used amber in chloroform, a "crystal" (benzole) varnish such as is, or was, used by photographers, and bitumen dissolved in benzole. The last is soon disintegrated under water, but the crystal varnish gives very good films. The varnish as sold may probably require dilution in order that the film may be thin enough.

Another varnish which gives interesting results is celluloid in pear-oil. All these films show little in air, but display beautiful colours in water when the reflexion from the back of the glass is got rid of as already described. The advantage from the water depends, of course, upon its mitigating the inequality of the reflexion from the two sides of the film by diminishing the front reflexion. A similar result may be arrived at by another road if we can increase the back reflexion, with the further advantage of enhanced illumination. For this purpose we may use silvering. A glass is coated with a very thin silver film and then with celluloid varnish of suitable consistency. Magnificent colours are then seen without the aid of water, and the only difficulty is to hit off the right thickness for the silver. Other methods of obtaining similar displays are described in Wood's *Physical Optics* (Macmillan, 1905, p. 142).

368.

BREATH FIGURES*.

[*Nature*, Vol. xc. pp. 436, 437, 1912.]

At intervals during the past year I have tried a good many experiments in the hope of throwing further light upon the origin of these figures, especially those due to the passage of a small blow-pipe flame, or of hot sulphuric acid, across the surface of a glass plate on which, before treatment, the breath deposits evenly. The even deposit consists of a multitude of small lenses easily seen with a hand magnifier. In the track of the flame or sulphuric acid the lenses are larger, often passing into flat masses which, on evaporation, show the usual colours of thin plates. When the glass is seen against a dark ground, and is so held that regularly reflected light does not reach the eye, the general surface shows bright, while the track of the flame or acid is by comparison dark or black. It will be convenient thus to speak of the deposit as bright or dark—descriptive words implying no doubtful hypothesis. The question is what difference in the glass surface determines the two kinds of deposit.

In Aitken's view (*Proc. Ed. Soc.* p. 94, 1893; *Nature*, June 15, 1911), the flame acts by the deposit of numerous fine particles constituting nuclei of aqueous condensation, and in like manner he attributes the effect of sulphuric (or hydrofluoric) acid to a water-attracting residue remaining in spite of washing. On the other hand, I was disposed to refer the dark deposit to a greater degree of freedom from grease or other water-repelling contamination (*Nature*, May 25, 1911), supposing that a clean surface of glass would everywhere attract moisture. It will be seen that the two views are sharply contrasted.

My first experiments were directed to improving the washing after hot sulphuric or hydrofluoric acid. It soon appeared that rinsing and soaking prolonged over twenty-four hours failed to abolish the dark track; but probably Mr Aitken would not regard this as at all conclusive. It was more to the point that dilute sulphuric acid (1/10) left no track, even after perfunctory washing. Rather to my surprise, I found that even strong

* See p. 26 of this volume.

sulphuric acid fails if employed cold. A few drops were poured upon a glass (¼-plate photographic from which the film had been removed), and caused to form an elongated pool, say, half an inch wide. After standing level for about five minutes—longer than the time required for the treatment with hot acid—the plate was rapidly washed under the tap, soaked for a few minutes, and finally rinsed with distilled water, and dried over a spirit lamp. Examined when cold by breathing, the plate showed, indeed, the form of the pool, but mainly by the darkness of the *edge*. The interior was, perhaps, not quite indistinguishable from the ground on which the acid had not acted, but there was no approach to darkness. This experiment may, I suppose, be taken to prove that the action of the hot acid is not attributable to a residue remaining after the washing.

I have not found any other treatment which will produce a dark track without the aid of heat. Chromic acid, *aqua regia*, and strong potash are alike ineffective. These reagents do undoubtedly exercise a cleansing action, so that the result is not entirely in favour of the grease theory as ordinarily understood.

My son, Hon. R. J. Strutt, tried for me an experiment in which part of an ordinarily cleaned glass was exposed for three hours to a stream of strongly ozonised oxygen, the remainder being protected. On examination with the breath, the difference between the protected and unprotected parts was scarcely visible.

It has been mentioned that the edges of pools of strong cold sulphuric acid and of many other reagents impress themselves, even when there is little or no effect in the interior. To exhibit this action at its best, it is well to employ a minimum of liquid; otherwise a creeping of the edge during the time of contact may somewhat obscure it. The experiment succeeds about equally well even when distilled water from a wash-bottle is substituted for powerful reagents. On the grease theory the effect may be attributed to the cleansing action of a pure free surface, but other interpretations probably could be suggested.

Very dark deposits, showing under suitable illumination the colours of thin plates, may be obtained on freshly-blown bulbs of soft glass. It is convenient to fill the interior with water, to which a little ink may be added. From this observation no particular conclusion can be deduced, since the surface, though doubtless very clean, has been exposed to the blow-pipe flame. In my former communication, I mentioned that no satisfactory result was obtained when a glass plate was strongly heated *on the back* by a long Bunsen burner; but I am now able to bring forward a more successful experiment.

A test-tube of thin glass, about ½ inch in diameter, was cleaned internally until it gave an even bright deposit. The breath is introduced through a tube of smaller diameter, previously warmed slightly with the hand. The closed end of the test-tube was then heated in a gas flame urged with a foot blow-pipe until there were signs of incipient softening. After cooling, the breath deposit showed interesting features, best brought out by transmitted light under a magnifier. The greater part of the length showed, as before, the usual fine dew. As the closed end was approached the drops became gradually larger, until at about an inch from the end they disappeared, leaving the glass covered with a nearly uniform film. One advantage of the tube is that evaporation of dew, once formed, is slow, unless promoted by suction through the mouth-tube. As the film evaporated, the colours of thin plates were seen by reflected light. Since it is certain that the flame had no access to the internal surface, it seems proved that dark deposits can be obtained on surfaces treated by heat alone.

In some respects a tube of thin glass, open at *both* ends, is more convenient than the test-tube. It is easier to clean, and no auxiliary tube is required to introduce or abstract moisture. I have used one of 3/10 in. diameter. Heated locally over a simple spirit flame to a point *short of softening*, it exhibited similar effects. This easy experiment may be recommended to anyone interested in the subject.

One of the things that I have always felt as a difficulty is the comparative permanence of the dark tracts. On flat plates they may survive in some degree rubbing by the finger, with subsequent rinsing and wiping. Practically the easiest way to bring a plate back to its original condition is to rub it with soapy water. But even this does not fully succeed with the test-tube, probably on account of the less effective rubbing and wiping near the closed end. But what exactly is involved in rubbing and wiping? I ventured to suggest before that possibly grease may penetrate the glass somewhat. From such a situation it might not easily be removed, or, on the other hand, introduced.

There is another form of experiment from which I had hoped to reap decisive results. The interior of a mass of glass cannot be supposed to be greasy, so that a surface freshly obtained by fracture should be clean, and give the dark deposit. One difficulty is that the character of the deposit on the irregular surface is not so easily judged. My first trial on a piece of plate glass $\frac{3}{8}$ in. thick, broken into two pieces with a hammer, gave anomalous results. On part of each new surface the breath was deposited in thin laminæ capable of showing colours, but on another part the water masses were decidedly smaller, and the deposit could scarcely be classified as black. The black and less black parts of the two surfaces were those which had been contiguous before fracture. That there should be a well-marked difference in this respect between parts both inside a rather small piece of glass is very surprising. I have not again met with this anomaly; but further trials on thick glass have revealed deposits which may be considered dark, though I was not always satisfied that they were so dark as those obtained on flat surfaces with the blow-pipe or hot sulphuric acid. Similar experiments with similar results may be made upon the edges of ordinary glass plates (such as are used in photography), cut with a diamond. The breath deposit is best held pretty close to a candle-flame, and is examined with a magnifier.

In conclusion, I may refer to two other related matters in which my experience differs from that of Mr Aitken. He mentions that with an alcohol flame he "could only succeed in getting very slight indications of any action." I do not at all understand this, as I have nearly always used an alcohol flame (with a mouth blow-pipe) and got black deposits. Thinking that perhaps the alcohol which I generally use was contaminated, I replaced it by pure alcohol, but without any perceptible difference in the results.

Again, I had instanced the visibility of a gas flame through a dewed plate as proving that part of the surface was uncovered. I have improved the experiment by using a curved tube through which to blow upon a glass plate already in position between the flame and the eye. I have not been able to find that the flame becomes invisible (with a well-defined outline) at any stage of the deposition of dew. Mr Aitken mentions results pointing in the opposite direction. Doubtless, the highly localized light of the flame is favourable.

[1913. Mr Aitken returned to the subject in a further communication to *Nature*, Vol. xc. p. 619, 1912, to which the reader should refer.]

369.

REMARKS CONCERNING FOURIER'S THEOREM AS APPLIED TO PHYSICAL PROBLEMS.

[*Philosophical Magazine*, Vol. XXIV. pp. 864—869, 1912.]

FOURIER'S theorem is of great importance in mathematical physics, but difficulties sometimes arise in practical applications which seem to have their origin in the aim at too great a precision. For example, in a series of observations extending over time we may be interested in what occurs during seconds or years, but we are not concerned with and have no materials for a remote antiquity or a distant future; and yet these remote times determine whether or not a period precisely defined shall be present. On the other hand, there may be no clearly marked limits of time indicated by the circumstances of the case, such as would suggest the other form of Fourier's theorem where everything is ultimately periodic. Neither of the usual forms of the theorem is exactly suitable. Some method of taking off the edge, as it were, appears to be called for.

The considerations which follow, arising out of a physical problem, have cleared up my own ideas, and they may perhaps be useful to other physicists.

A train of waves of length λ, represented by

$$\psi = e^{2\pi i(ct+\xi)/\lambda}, \quad \dots\dots\dots\dots\dots\dots\dots\dots\dots(1)$$

advances with velocity c in the negative direction. If the medium is absolutely uniform, it is propagated without disturbance; but if the medium is subject to small variations, a reflexion in general ensues as the waves pass any place x. Such reflexion reacts upon the original waves; but if we suppose the variations of the medium to be extremely small, we may neglect the reaction and calculate the aggregate reflexion as if the primary waves were undisturbed. The partial reflexion which takes place at x is represented by

$$d\psi = e^{2\pi i(ct-\xi)/\lambda} \phi(x)\, dx \cdot e^{4\pi ix/\lambda}, \quad \dots\dots\dots\dots\dots\dots(2)$$

9—2

in which the first factor expresses total reflexion supposed to originate at $x = 0$, $\phi(x)\,dx$ expresses the actual reflecting power at x, and the last factor gives the alteration of phase incurred in traversing the distance $2x$. The aggregate reflexion follows on integration with respect to x; with omission of the first factor it may be taken to be

$$C + iS, \dots\dots\dots\dots\dots\dots\dots\dots\dots(3)$$

where $$C = \int_{-\infty}^{+\infty} \phi(v) \cos uv\,dv, \quad S = \int_{-\infty}^{+\infty} \phi(v) \sin uv\,dv, \dots\dots\dots(4)$$

with $u = 4\pi/\lambda$. When ϕ is given, the reflexion is thus determined by (3). It is, of course, a function of λ or u.

In the converse problem we regard (3)—the reflexion—as given for all values of u and we seek thence to determine the form of ϕ as a function of x. By Fourier's theorem we have at once

$$\phi(x) = \frac{1}{\pi} \int_0^\infty du \, \{C \cos ux + S \sin ux\}. \dots\dots\dots\dots(5)$$

It will be seen that we require to know C and S separately. A knowledge of the *intensity* merely, viz. $C^2 + S^2$, does not suffice.

Although the general theory, above sketched, is simple enough, questions arise as soon as we try to introduce the approximations necessary in practice. For example, in the optical application we could find by observation the values of C and S for a finite range only of u, limited indeed in eye observations to less than an octave. If we limit the integration in (5) to correspond with actual knowledge of C and S, the integral may not go far towards determining ϕ. It may happen, however, that we have some independent knowledge of the form of ϕ. For example, we may know that the medium is composed of strata each uniform in itself, so that within each ϕ vanishes. Further, we may know that there are only two kinds of strata, occurring alternately. The value of $\int \phi\,dx$ at each transition is then numerically the same but affected with signs alternately opposite. This is the case of chlorate of potash crystals in which occur repeated twinnings*. Information of this kind may supplement the deficiency of (5) taken by itself. If it be for high values only of u that C and S are not known, the curve for ϕ first obtained may be subjected to any alteration which leaves $\int \phi\,dx$, taken over any small range, undisturbed, a consideration which assists materially where ϕ is known to be discontinuous.

If observation indicates a large C or S for any particular value of u, we infer of course from (5) a correspondingly important periodic term in ϕ. If the large value of C or S is limited to a very small range of u, the periodicity of ϕ extends to a large range of x; otherwise the interference of

---

* *Phil. Mag.* Vol. XXVI. p. 256 (1888); *Scientific Papers*, Vol. III. p. 204.

components with somewhat different values of u may limit the periodicity to a comparatively small range. Conversely, a prolonged periodicity is associated with an approach to discontinuity in the values of C or S.

The complete curve representing $\phi(x)$ will in general include features of various lengths reckoned along x, and a feature of any particular length is associated with values of u grouped round a corresponding centre. For some purposes we may wish to *smooth* the curve by eliminating small features. One way of effecting this is to substitute everywhere for $\phi(x)$ the mean of the values of $\phi(x)$ in the neighbourhood of x, viz.

$$\frac{1}{2a}\int_{x-a}^{x+a}\phi(x)\,dx, \quad\dots\dots\dots\dots\dots\dots\dots(6)$$

the range ($2a$) of integration being chosen suitably. With use of (5) we find for (6)

$$\frac{1}{2a}\int_{x-a}^{x+a}\phi(x)\,dx = \frac{1}{\pi}\int_0^\infty du\,\frac{\sin ua}{ua}\{C\cos ux + S\sin ux\}, \quad\dots\dots(7)$$

differing from the right-hand member of (5) merely by the introduction of the factor $\sin ua \div ua$. The effect of this factor under the integral sign is to diminish the importance of values of u which exceed π/a and gradually to annul the influence of still larger values. If we are content to speak very roughly, we may say that the process of averaging on the left is equivalent to the omission in Fourier's integral of the values of u which exceed $\pi/2a$.

We may imagine the process of averaging to be repeated once or more times upon (6). At each step a new factor $\sin ua \div ua$ is introduced under the integral sign. After a number of such operations the integral becomes practically independent of all values of u for which ua is not small.

In (6) the average is taken in the simplest way with respect to x, so that every part of the range $2a$ contributes equally (fig. 1). Other and perhaps

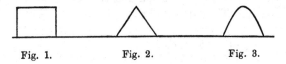

Fig. 1.　　　　　Fig. 2.　　　　　Fig. 3.

better methods of smoothing may be proposed in which a preponderance is given to the central parts. For example we may take (fig. 2)

$$\frac{1}{a^2}\int_0^a (a-\xi)\{\phi(x+\xi)+\phi(x-\xi)\}\,d\xi. \quad\dots\dots\dots\dots(8)$$

From (5) we find that (8) is equivalent to

$$\frac{2}{\pi}\int_0^\infty du\,\frac{1-\cos ua}{u^2a^2}\{C\cos ux + S\sin ux\}, \quad\dots\dots\dots\dots(9)$$

reducing to (5) again when a is made infinitely small. In comparison with (7) the higher values of ua are eliminated more rapidly. Other kinds of averaging over a finite range may be proposed. On the same lines as above the formula next in order is (fig. 3)

$$\frac{3}{4a}\int_{-a}^{+a}\left(1-\frac{\xi^2}{a^2}\right)\phi(x+\xi)\,d\xi$$

$$=\frac{1}{\pi}\int_0^\infty du\,\frac{\sin au - au\cos au}{\frac{1}{3}a^3u^3}\{C\cos ux + S\sin ux\}\,dx. \quad\ldots(10)$$

In the above processes for smoothing the curve representing $\phi(x)$, ordinates which lie at distances exceeding a from the point under consideration are without influence. This may or may not be an advantage. A formula in which the integration extends to infinity is

$$\frac{1}{a\sqrt{\pi}}\int_{-\infty}^{+\infty}\phi(x+\xi)e^{-\xi^2/a^2}\,d\xi=\frac{1}{\pi}\int_0^\infty du\,e^{-u^2a^2/4}\{C\cos ux + S\sin ux\}. \quad\ldots(11)$$

In this case the values of ua which exceed 2 make contributions to the integral whose importance very rapidly diminishes.

The intention of the operation of smoothing is to remove from the curve features whose length is small. For some purposes we may desire on the contrary to eliminate features of *great* length, as for example in considering the record of an instrument whose zero is liable to slow variation from some extraneous cause. In this case (to take the simplest formula) we may subtract from $\phi(x)$—the uncorrected record—the average over a length b relatively large, so obtaining

$$\phi(x)-\frac{1}{2b}\int_{x-b}^{x+b}\phi(x)\,dx=\frac{1}{\pi}\int_0^\infty du\left\{1-\frac{\sin ub}{ub}\right\}\{C\cos ux + S\sin ux\}. \quad\ldots(12)$$

Here, if ub is much less than π, the corresponding part of the range of integration is approximately cancelled and features of great length are eliminated.

There are cases where this operation and that of smoothing may be combined advantageously. Thus if we take

$$\frac{1}{2a}\int_{x-a}^{x+a}\phi(x)\,dx-\frac{1}{2b}\int_{x-b}^{x+b}\phi(x)\,dx$$

$$=\frac{1}{\pi}\int_0^\infty du\left\{\frac{\sin ua}{ua}-\frac{\sin ub}{ub}\right\}\{C\cos ux + S\sin ux\}, \quad\ldots(13)$$

we eliminate at the same time the features whose length is small compared with a and those whose length is large compared with b. The same method may be applied to the other formulæ (9), (10), (11).

A related question is one proposed by Stokes[*], to which it would be interesting to have had Stokes' own answer. What is in common and what

* Smith's Prize Examination, Feb. 1, 1882; *Math. and Phys. Papers*, Vol. v. p. 367.

is the difference between C and S in the two cases (i) where $\phi(x)$ fluctuates between $-\infty$ and $+\infty$ and (ii) where the fluctuations are nearly the same as in (i) between finite limits $\pm a$ but outside those limits tends to zero? When x is numerically great, $\cos ux$ and $\sin ux$ fluctuate rapidly with u; and inspection of (5) shows that $\phi(x)$ is then small, unless C or S are themselves rapidly variable as functions of u. Case (i) therefore involves an approach to discontinuity in the forms of C or S. If we eliminate these discontinuities, or rapid variations, by a smoothing process, we shall annul $\phi(x)$ at great distances and at the same time retain the former values near the origin. The smoothing may be effected (as before) by taking

$$\frac{1}{2a}\int_{u-a}^{u+a} C\,du, \quad \frac{1}{2a}\int_{u-a}^{u+a} S\,du$$

in place of C and S simply. C then becomes

$$\int_{-\infty}^{+\infty} dv\,\phi(v)\cos uv\,\frac{\sin av}{av},$$

$\phi(v)$ being replaced by $\phi(v)\sin av \div av$. The effect of the added factor disappears when av is small, but when av is large, it tends to annul the corresponding part of the integral. The new form for $\phi(x)$ is thus the same as the old one near the origin but tends to vanish at great distances on either side. Case (ii) is thus deducible from case (i) by the application of a smoothing process to C and S, whereby fluctuations of small length are removed.

We may sum up by saying that a smoothing of $\phi(x)$ annuls C and S for large values of u, while a smoothing of C and S (as functions of u) annuls $\phi(x)$ for values of x which are numerically great.

370.

SUR LA RÉSISTANCE DES SPHÈRES DANS L'AIR EN MOUVEMENT.

[*Comptes Rendus*, t. CLVI. p. 109, 1913.]

DANS les *Comptes rendus* du 30 décembre 1912, M. Eiffel donne des résultats très intéressants pour la résistance rencontrée, à vitesse variable, par trois sphères de 16·2, 24·4 et 33 cm. de diamètre. Dans la première figure, ces résultats sont exprimés par les valeurs d'un coefficient K, égal à R/SV^2, où R est la résistance totale, S la surface diamétrale et V la vitesse. En chaque cas, il y a une *vitesse critique*, et M. Eiffel fait remarquer que la *loi de similitude* n'est pas toujours vraie; en effet, les trois sphères donnent des vitesses critiques tout à fait différentes.

D'après la loi de similitude dynamique, précisée par Stokes[*] et Reynolds pour les liquides visqueux, K est une fonction d'une *seule* variable ν/VL, où ν est la *viscosité cinématique*, constante pour un liquide donné, et L est la dimension linéaire, proportionnelle à $S^{\frac{1}{2}}$. Ainsi les vitesses critiques ne doivent pas être les mêmes dans les trois cas, mais inversement proportionnelles à L. En vérité, si nous changeons l'échelle des vitesses suivant cette loi, nous trouvons les courbes de M. Eiffel presque identiques, au moins que ces vitesses ne sont pas très petites.

Je ne sais si les écarts résiduels sont réels ou non. La théorie simple admet que les sphères sont polies, sinon que les inégalités sont proportionnelles aux diamètres, que la compressibilité de l'air est négligeable et que la viscosité cinématique est absolument constante. Si les résultats de l'expérience ne sont pas complètement d'accord avec la théorie, on devra examiner ces hypothèses de plus près.

J'ai traité d'autre part et plus en détail de la question dont il s'agit ici[†].

---

[*] [*Camb. Trans.* 1850; *Math. and Phys. Papers*, Vol. III. p. 17.]

[†] Voir *Scientific Papers*, t. v. 1910, pp. 532—534.

371.

THE EFFECT OF JUNCTIONS ON THE PROPAGATION OF ELECTRIC WAVES ALONG CONDUCTORS.

[Proceedings of the Royal Society, A, Vol. LXXXVIII. pp. 103—110, 1913.]

SOME interesting problems in electric wave propagation are suggested by an experiment of Hertz[*]. In its original form waves of the simplest kind travel in the positive direction (fig. 1), outside an infinitely thin conducting cylindrical shell, AA, which comes to an end, say, at the plane $z = 0$. Co-axial with the cylinder a rod or wire BB (of less diameter) extends to infinity in both directions. The conductors being supposed perfect, it is required to determine the waves propagated onwards beyond the cylinder on the positive side of z, as well as those reflected back outside the cylinder and in the annular space between the cylinder and the rod.

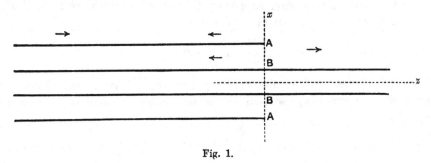

Fig. 1.

So stated, the problem, even if mathematically definite, is probably intractable; but if we modify it by introducing an external co-axial conducting sheath CC (fig. 2), extending to infinity in both directions, and if we further suppose that the diameter of this sheath is small in comparison with the wave-length (λ) of the vibrations, we shall bring it within the scope of approximate methods. It is under this limitation that I propose here to

---

[*] "Ueber die Fortleitung electrischer Wellen durch Drähte," *Wied. Ann.* 1889, Vol. XXXVII. p. 395.

consider the present and a few analogous problems.   Some considerations of a more general character are prefixed.

If P, Q, R be components of electromotive intensity, a, b, c those of magnetisation, Maxwell's general circuital relations* for the dielectric give

$$\frac{da}{dt} = \frac{dQ}{dz} - \frac{dR}{dy} \; , \quad \dots\dots\dots\dots\dots\dots\dots(1)$$

and two similar equations, and

$$\frac{dP}{dt} = V^2\left(\frac{dc}{dy} - \frac{db}{dz}\right), \quad \dots\dots\dots\dots\dots\dots\dots(2)$$

also with two similar equations, V being the velocity of propagation.   From (1) and (2) we may derive

$$\frac{da}{dx} + \frac{db}{dy} + \frac{dc}{dz} = 0, \qquad \frac{dP}{dx} + \frac{dQ}{dy} + \frac{dR}{dz} = 0 ; \quad \dots\dots\dots\dots(3)$$

and, further, that $\left(\dfrac{d^2}{dt^2} - V^2\nabla^2\right)(P, Q, R, a, b, c) = 0, \quad \dots\dots\dots\dots\dots(4)$

where $\nabla^2 = d^2/dx^2 + d^2/dy^2 + d^2/dz^2. \quad \dots\dots\dots\dots\dots(5)$

At any point upon the surface of a conductor, regarded as perfect, the condition to be satisfied is that the vector (P, Q, R) be there *normal*.   In what follows we shall have to deal only with simple vibrations in which all the quantities are proportional to e^{ipt}, so that d/dt may be replaced by ip.

It may be convenient to commence with some cases where the waves are in two dimensions (x, z) only, supposing that a, c, Q vanish, while b, P, R are independent of y.   From (1) and (2) we have

$$P\frac{db}{dx} + R\frac{db}{dz} = 0.$$

At the surface of a conductor P, Q are proportional to the direction cosines of the normal (n); so that the surface condition may be expressed simply by

$$\frac{db}{dn} = 0, \quad \dots\dots\dots\dots\dots\dots\dots\dots\dots\dots(6)$$

which, with $\left(\dfrac{d^2}{dx^2} + \dfrac{d^2}{dz^2} + k^2\right)b = 0, \quad \dots\dots\dots\dots\dots\dots(7)$

suffices to determine b.   In (7) $k = p/V$.   It will be seen that equations (6), (7) are identical with those which apply in two dimensions to aërial vibrations executed in spaces bounded by fixed walls, b then denoting velocity-potential.   When b is known, the remaining functions follow at once.

* *Phil. Trans.* 1868 ; Maxwell's *Scientific Papers*, Vol. II. p. 128.

It may be remarked by the way that the above analogy throws light upon the question under what circumstances electric waves are *guided* by conductors. Some high authorities, it would seem, regard such guidance as ensuing in all cases as a consequence of the boundary condition fixing the direction of the electric force. But in Acoustics, though a similar condition holds good, there is no guidance of aërial waves round convex surfaces, and it follows that there is none in the two-dimensional electric vibrations under consideration. Near the *concave* surface of walls there is in both cases a whispering gallery effect*. The peculiar guidance of electric waves by wires depends upon the conductor being encircled by the magnetic force. No such circulation, for example, could ensue from the incidence of plane waves upon a wire which lies entirely in the plane containing the direction of propagation and that of the magnetic force.

Our first special application is to the extreme form of Hertz's problem (as modified) which occurs when all the radii of the cylindrical surfaces concerned become infinite, while the *differences* CA, AB remain finite and indeed small in comparison with λ. In fig. 2, A, B, C then represent

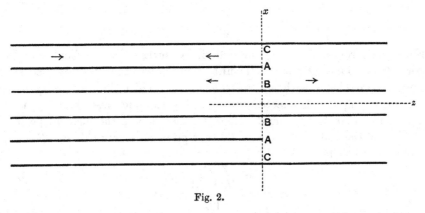

Fig. 2.

planes perpendicular to the plane of the paper and the problem is in two dimensions. The two halves, corresponding to *plus* and *minus* values of x, are isolated, and we need only consider one of them. Availing ourselves of the acoustical analogy, we may at once transfer the solution given (after Poisson) in *Theory of Sound*, § 264. If the incident wave in CA be represented by f_{CA} and that therein reflected by F, while the waves propagated along CB, AB be denoted by f_{CB}, f_{AB}, we have

$$f'_{CB} = f'_{AB} = \frac{2CA}{CB + AB + CA} f'_{CA} = \frac{CA}{CB} f'_{CA} \quad \dots\dots\dots\dots(8)$$

and

$$F' = \frac{AB}{CB} f'_{CA}. \quad \dots\dots\dots\dots\dots\dots(9)$$

* *Phil. Mag.* 1910, Vol. **xx.** p. 1001; *Scientific Papers*, Vol. **v.** p. 617.

The wave in AB is to be regarded as propagated onwards round the corner at A rather than as reflected. As was to be anticipated, the reflected wave F' is smaller, the smaller is AB. It will be understood that the validity of these results depends upon the assumption that the region round A through which the waves are irregular has dimensions which are negligible in comparison with λ.

An even simpler example is sketched in fig. 3, where for the present the

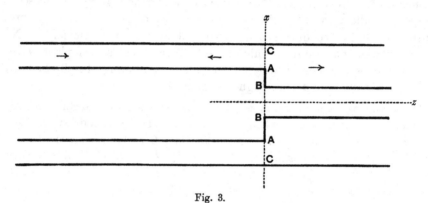

Fig. 3.

various lines represent planes or cylindrical surfaces perpendicular to the paper. One bounding plane C is unbroken. The other boundary consists mainly of two planes with a transition at AB, which, however, may be of any form so long as it is effected within a distance much less than λ. With a notation similar to that used before, f_{CA} may denote the incident positive wave and F the reflected wave, while that propagated onwards in CB is f_{CB}. We obtain in like manner

$$f'_{CB} = \frac{2CA}{CB + CA} f'_{CA}, \quad \dots\dots\dots\dots\dots\dots\dots\dots(10)$$

$$F' = \frac{AB}{CB + CA} f'_{CA}. \quad \dots\dots\dots\dots\dots\dots\dots\dots(11)$$

When AB vanishes we have, of course, $f'_{CB} = f'_{CA}$, $F' = 0$. A little later we shall consider the problem of fig. 3 when the various surfaces are of revolution round the axis of z.

Leaving the two-dimensional examples, we find that the same general method is applicable, always under the condition that the region occupied by irregular waves has dimensions which are small in comparison with λ. Within this region a simplified form of the general equations avails, and thus the difficulty is turned.

An increase in λ means a decrease in p. When this goes far enough, it justifies the omission of d/dt in equations (1), (2), (3), (4). Thus P, Q, R become the derivatives of a simple potential function ϕ, which itself satisfies

$\nabla^2 \phi = 0$; that is, the electric forces obey the laws of electrostatics. Similarly a, b, c are derivatives of another function ψ satisfying the same equation. The only difference is that ψ may be multivalued. The magnetism is that due to steady electric currents. If several wires meet in a point, the total current is zero. This expresses itself in terms of a, b, c as a relation between the "circulations." The method then consists in forming the solutions which apply to the parts at a distance on the two sides from the region of irregularity, and in accommodating them to one another by the conditions which hold good at the margins of this region in virtue of the fact that it is small.

In the application to the problem of fig. 3 we will suppose that the conductors are of revolution round z, though this limitation is not really imposed by the method itself. The problem of the regular waves (whatever may be form of section) was considered in a former paper*. All the dependent variables expressing the electric conditions being proportional to $e^{i(pt \pm kz)}$, d^2/dt^2 in (4) compensates $V^2 d^2/dz^2$, so that

$$\left(\frac{d^2}{dx^2} + \frac{d^2}{dy^2} \right) (P, Q, R, a, b, c) = 0 ; \quad \dots\dots\dots\dots(12)$$

also R and c vanish. In the present case we have for the negative side, where there is both a direct and a reflected wave,

$$P, Q, R = e^{ipt} (H_1 e^{-ikz} + K_1 e^{ikz}) \left(\frac{d}{dx}, \frac{d}{dy}, 0 \right) \log r, \quad \dots\dots(13)$$

where r is the distance of any point from the axis of symmetry, and H_1, K_1 are arbitrary constants. Corresponding to (13),

$$V (a, b, c) = e^{ipt} (- H_1 e^{-ikz} + K_1 e^{ikz}) \left(\frac{d}{dy}, - \frac{d}{dx}, 0 \right) \log r. \quad \dots\dots(14)$$

In the region of regular waves on the positive side there is supposed to be no wave propagated in the negative direction. Here accordingly

$$P, Q, R = H_2 e^{i(pt-kz)} \left(\frac{d}{dx}, \frac{d}{dy}, 0 \right) \log r, \quad \dots\dots\dots\dots(15)$$

$$V (a, b, c) = H_2 e^{i(pt-kz)} \left(- \frac{d}{dy}, \frac{d}{dx}, 0 \right) \log r, \quad \dots\dots\dots(16)$$

H_2 being another constant. We have now to determine the relations between the constants H_1, K_1, H_2, hitherto arbitrary, in terms of the remaining data.

For this purpose consider cross-sections on the two sides both near the origin and yet within the regions of regular waves. The electric force as expressed in (13), (15) is purely radial. On the positive side its integral

* Phil. Mag. 1897, Vol. XLIV. p. 199; Scientific Papers, Vol. IV. p. 327.

between r_2 the radius of the inner and r' that of the outer conductor is, with omission of e^{ipt},

$$H_2 e^{-ikz} \log (r'/r_2),$$

z having the value proper to the section. On the negative side the corresponding integral is

$$(H_1 e^{-ikz} + K_1 e^{ikz}) \log (r'/r_1),$$

r_1 being the radius of the inner conductor at that place. But when we consider the intermediate region, where electrostatical laws prevail, we recognize that these two integrals must be equal; and further that the exponentials may be identified with unity. Accordingly, the first relation is

$$(H_1 + K_1) \log (r'/r_1) = H_2 \log (r'/r_2). \quad \dots\dots\dots\dots\dots\dots(17)$$

In like manner the magnetic force in (14), (16) is purely circumferential. And the circulations at the two sections are as $H_1 - K_1$ and H_2. But since these circulations, representing electric currents which may be treated as steady, are equal, we have as the second relation—

$$H_1 - K_1 = H_2. \quad \dots\dots\dots\dots\dots\dots\dots\dots\dots(18)$$

The two relations (17), (18) determine the wave propagated onwards H_2 and that reflected K_1 in terms of the incident wave H_1. If $r_2 = r_1$, we have of course, $H_2 = H_1$, $K_1 = 0$.

If we suppose r_1, r_2, r' all great and nearly equal and expand the logarithms, we fall back on the solution for the two-dimensional case already given.

In the above the radius of the outer sheath is supposed uniform throughout. If in the neighbourhood of the origin the radius of the sheath changes from r_1' to r_2', while (as before) that of the inner conductor changes from r_1 to r_2, we have instead of (17),

$$(H_1 + K_1) \log (r_1'/r_1) = H_2 \log (r_2'/r_2), \quad \dots\dots\dots\dots\dots(19)$$

while (18) remains undisturbed.

In (19) the logarithmic functions are proportional to the reciprocals of the electric capacities of the system on the two sides, reckoned in each case per unit of length. From the general theory given in the paper referred to we may infer that this substitution suffices to liberate us from the restriction to symmetry round the axis hitherto imposed. The more general functions which then replace $\log r$ on the two sides must be chosen with such coefficients as will make the circulations of magnetic force equal. The generalization here indicated applies equally in the other problems of this paper.

In Hertz's problem, fig. 2, the method is similar. In the region of regular waves on the left in CA we may retain (13), (14), and for the regular waves on the right in CB we retain (15), (16). But now in addition for the regular waves on the left in AB, we have

$$P,\ Q,\ R = K_3 e^{i(pt+kz)} \left(\frac{d}{dx},\ \frac{d}{dy},\ 0 \right) \log r, \quad \dots\dots\dots\dots(20)$$

$$V(a,\ b,\ c) = K_3 e^{i(pt+kz)} \left(\frac{d}{dy},\ -\frac{d}{dx},\ 0 \right) \log r. \quad \dots\dots\dots(21)$$

Three conditions are now required to determine K_1, H_2, K_3 in terms of H_1. We shall denote the radii taken in order, viz. $\frac{1}{2}BB$, $\frac{1}{2}AA$, $\frac{1}{2}CC$, by r_1, r_2, r_3 respectively. As in (17), the electric forces give

$$(H_1 + K_1) \log \frac{r_3}{r_2} + K_3 \log \frac{r_2}{r_1} = H_2 \log \frac{r_3}{r_1}. \quad \dots\dots\dots\dots(22)$$

The magnetic forces yield two equations, which may be regarded as expressing that the currents are the same on the two sides along BB, and that, since the section is at a negligible distance from the insulated end, there is no current in AA. Thus

$$H_1 - K_1 = -K_3 = H_2. \quad \dots\dots\dots\dots\dots(23)$$

From (22) and (23)

$$\frac{K_1}{H_1} = \frac{\log r_2 - \log r_1}{\log r_3 - \log r_1}, \quad \dots\dots\dots\dots\dots(24)$$

$$H_2 = -K_3 = \frac{\log r_3 - \log r_2}{\log r_3 - \log r_1}. \quad \dots\dots\dots\dots(25)$$

If r_2 exceeds r_1 but little, K_1 tends to vanish, while H_2 and $-K_3$ approach unity. Again, if the radii are all great, (24), (25) reduce to

$$\frac{K_1}{H_1} = \frac{r_2 - r_1}{r_3 - r_1}, \qquad H_2 = -K_3 = \frac{r_3 - r_2}{r_3 - r_1}, \quad \dots\dots\dots(26)$$

as already found in (8), (9).

The same method applies with but little variation to the more general problem where waves between one wire and sheath $(r_1,\ r_1')$ divide so as to pass along several wires and sheaths $(r_2,\ r_2')$, $(r_3,\ r_3')$, etc., always under the condition that the whole region of irregularity is negligible in comparison with the wave-length*. The various wires and sheaths are, of course, supposed to be continuous. With a similar notation the direct and reflected waves along the first wire are denoted by H_1, K_1, and those propagated

---

* This condition will usually suffice. But extreme cases may be proposed where, in spite of the smallness of the intermediate region, its shape is such as to entail natural resonances of frequency agreeing with that of the principal waves. The method would then fail.

onwards along the second, third, and other wires by H_2, H_3, etc. The equations are—

$$(H_1 + K_1) \log \frac{r_1'}{r_1} = H_2 \log \frac{r_2'}{r_2} = H_3 \log \frac{r_3'}{r_3} = \ldots\ldots, \quad \ldots\ldots\ldots(27)$$

$$H_1 - K_1 = H_2 = H_3 = \ldots\ldots$$

It is hardly necessary to detail obvious particular cases.

The success of the method used in these problems depends upon the assumption of a great wave-length. This, of course, constitutes a limitation; but it has the advantage of eliminating the irregular motion at the junctions. In the two-dimensional examples it might be possible to pursue the approximation by determining the character of the irregular waves, at least to a certain extent, somewhat as in the question of the correction for the open end of an organ pipe.

372.

THE CORRECTION TO THE LENGTH OF TERMINATED RODS IN ELECTRICAL PROBLEMS.

[*Philosophical Magazine*, Vol. xxv. pp. 1—9, 1913.]

In a short paper "On the Electrical Vibrations associated with thin terminated Conducting Rods"* I endeavoured to show that the difference between the half wave-length of the gravest vibration and the length (*l*) of the rod (of uniform section) tends to vanish relatively when the section is reduced without limit, in opposition to the theory of Macdonald which makes $\lambda = 2\cdot53\, l$. Understanding that the argument there put forward is not considered conclusive, I have tried to treat the question more rigorously, but the difficulties in the way are rather formidable. And this is not surprising in view of the discontinuities presented at the edges where the flat ends meet the cylindrical surface.

The problem assumes a shape simpler in some respects if we suppose that the rod of length *l* and radius *a* surrounded by a cylindrical coaxial conducting case of radius *b* extending to infinity in both directions. One advantage is that the vibrations are now *permanently maintained*, for no waves can escape to infinity along the tunnel, seeing that *l* is supposed great compared with *b*†. The greatness of *l* secures also the independence of the two ends, so that the whole correction to the length, whatever it is, may be regarded as simply the double of that due to the end of a rod infinitely long.

At an interior node of an infinitely long rod the electric forces, giving rise (we may suppose) to potential energy, are a maximum, while the magnetic forces representing kinetic energy are evanescent. The end of a terminated rod corresponds, approximately at any rate, to a node. The complications

---

* *Phil. Mag.* Vol. VIII. p. 105 (1904); *Scientific Papers*, Vol. V. p. 198.

† *Phil. Mag.* Vol. XLIII. p. 125 (1897); *Scientific Papers*, Vol. IV. p. 276. The conductors are supposed to be *perfect*.

due to the end thus tell mainly upon the electric forces*, and the problem is reduced to the electrostatical one of finding the *capacity* of the terminated rod as enclosed in the infinite cylindrical case at potential zero. But this simplified form of the problem still presents difficulties.

Taking cylindrical coordinates z, r, we identify the axis of symmetry with that of z, supposing also that the origin of z coincides with the flat end of the interior conducting rod which extends from $-\infty$ to 0. The enclosing case on the other hand extends from $-\infty$ to $+\infty$. At a distance from the end on the negative side the potential V, which is supposed to be unity on the rod and zero on the case, has the form

$$V_0 = \frac{\log b/r}{\log b/a}, \quad\dots\dots\dots\dots\dots\dots\dots(1)$$

and the *capacity* per unit length is $1/(2 \log b/a)$.

On the plane $z = 0$ the value of V from $r = 0$ to $r = a$ is unity. If we knew also the value of V from $r = a$ to $r = b$, we could treat separately the problems arising on the positive and negative sides. On the positive side we could express the solution by means of the functions appropriate to the complete cylinder $r < b$, and on the negative side by those appropriate to the annual cylindrical space $b > r > a$. If we assume an arbitrary value for V over the part in question of the plane $z = 0$, the criterion of its suitability may be taken to be the equality of the resulting values of dV/dz on the two sides.

We may begin by supposing that (1) holds good on the negative side throughout; and we have then to form for the positive side a function which shall agree with this at $z = 0$. The general expression for a function which shall vanish when $r = b$ and when $z = +\infty$, and also satisfy Laplace's equation, is

$$A_1 J_0(k_1 r) e^{-k_1 z} + A_2 J_0(k_2 r) e^{-k_2 z} + \dots, \quad\dots\dots\dots\dots\dots(2)$$

where k_1, k_2, &c. are the roots of $J_0(kb) = 0$; and this is to be identified when $z = 0$ with (1) from a to b and with unity from 0 to a. The coefficients A are to be found in the usual manner by multiplication with $J_0(k_n r)$ and integration over the area of the circle $r = b$. To this end we require

$$\int_0^a J_0(kr) r\, dr = -\frac{a}{k} J_0'(ka), \quad\dots\dots\dots\dots\dots\dots\dots\dots\dots(3)$$

$$\int_a^b J_0(kr) r\, dr = -\frac{1}{k} \{b J_0'(kb) - a J_0'(ka)\}, \quad\dots\dots\dots\dots\dots\dots(4)$$

$$\int_a^b \log r J_0(kr) r\, dr = -\frac{1}{k} \{b \log b J_0'(kb) - a \log a J_0'(ka)\} - \frac{1}{k^2} J_0(ka). \;\dots (5)$$

---

* Compare the analogous acoustical questions in *Theory of Sound*, §§ 265, 317.

Thus altogether

$$\frac{J_0(ka)}{k^2 \log b/a} = A \int_a^b J_0^2(kr)\, r\, dr = \tfrac{1}{2} b^2 A J_0'^2(kb). \quad \dots\dots\dots(6)$$

For $J_0'^2$ we may write J_1^2; so that if in (2) we take

$$A = \frac{2J_0(ka)}{k^2 b^2 J_1^2(kb) \log b/a}, \quad \dots\dots\dots\dots(7)$$

we shall have a function which satisfies the necessary conditions, and at $z = 0$ assumes the value 1 from 0 to a and that expressed in (1) from a to b. But the values of dV/dz are not the same on the two sides.

If we call the value, so determined on the positive as well as upon the negative side, V_0, we may denote the true value of V by $V_0 + V'$. The conditions for V' will then be the satisfaction of Laplace's equation throughout the dielectric (except at $z = 0$), that on the negative side it make $V' = 0$ both when $r = a$ and when $r = b$, and vanish at $z = -\infty$, and on the positive side $V' = 0$ when $r = b$ and when $z = +\infty$, and that when $z = 0$ V' assume the same value on the two sides between a and b and on the positive side the value zero from 0 to a. A further condition for the exact solution is that dV/dz, or $dV_0/dz + dV'/dz$, shall be the same on the two sides from $r = a$ to $r = b$ when $z = 0$.

Now whatever may be in other respects the character of V' on the negative side, it can be expressed by the series

$$V' = H_1 \phi(h_1 r) e^{h_1 z} + H_2 \phi(h_2 r) e^{h_2 z} + \dots, \quad \dots\dots\dots(8)$$

where $\phi(h_1 r)$, &c. are the normal functions appropriate to the symmetrical vibrations of an annular membrane of radii a and b, so that $\phi(hr)$ vanishes for $r = a$, $r = b$. In the usual notation we may write

$$\phi(hr) = \frac{J_0(hr)}{J_0(ha)} - \frac{Y_0(hr)}{Y_0(ha)}, \quad \dots\dots\dots\dots(9)$$

with the further condition

$$Y_0(ha) J_0(hb) - J_0(ha) Y_0(hb) = 0, \quad \dots\dots\dots(10)$$

determining the values of h. The function ϕ satisfies the same differential equation as do J_0 and Y_0.

Considering for the present only one term of the series (8), we have to find for the positive side a function which shall satisfy the other necessary conditions and when $z = 0$ make $V' = 0$ from 0 to a, and $V' = H\phi(hr)$ from a to b. As before, such a function may be expressed by

$$V' = B_1 J_0(k_1 r) e^{-k_1 z} + B_2 J_0(k_2 r) e^{-k_2 z} + \dots, \quad \dots\dots\dots(11)$$

and the only remaining question is to find the coefficients B. For this purpose we require to evaluate

$$\int_b^a \phi(hr) J_0(kr)\, r\, dr.$$

10—2

From the differential equation satisfied by J_0 and ϕ we get

$$k^2 \int_a^b J_0(kr)\,\phi(hr)\,r\,dr = -\left[r.\phi.\frac{dJ_0}{dr}\right]_a^b + \int_a^b \frac{d\phi}{dr}\frac{dJ_0}{dr}\,r\,dr,$$

and

$$h^2 \int_a^b J_0(kr)\,\phi(hr)\,r\,dr = -\left[r.J_0.\frac{d\phi}{dr}\right]_a^b + \int_a^b \frac{d\phi}{dr}\frac{dJ_0}{dr}\,r\,dr\,;$$

so that

$$(k^2 - h^2)\int_a^b J_0(kr)\,\phi(hr)\,r\,dr = \left[rJ_0\frac{d\phi}{dr} - r\frac{dJ_0}{dr}\phi\right]_a^b$$

$$= -haJ_0(ka)\,\phi'(ha), \quad \ldots\ldots\ldots(12)$$

since here $\phi(ha) = \phi(hb) = 0$, and also $J_0(kb) = 0$. Thus in (11), corresponding to a single term of (8),

$$B = \frac{2haHJ_0(ka)\,\phi'(ha)}{(h^2 - k^2)\,b^2 J_1^2(kb)}. \quad \ldots\ldots\ldots\ldots\ldots(13)$$

The exact solution demands the inclusion in (8) of all the admissible values of h, with addition of (1) which in fact corresponds to a zero value of h. And each value of h contributes a part to each of the infinite series of coefficients B, needed to express the solution on the positive side.

But although an exact solution would involve the whole series of values of h, approximate methods may be founded upon the use of a limited number of them. I have used this principle in calculations relating to the potential from 1870 onwards*. A potential V, given over a closed surface, makes

$$\frac{1}{4\pi}\iiint\left\{\left(\frac{dV}{dx}\right)^2 + \left(\frac{dV}{dy}\right)^2 + \left(\frac{dV}{dz}\right)^2\right\} dx\,dy\,dz, \quad \ldots\ldots\ldots(14)$$

reckoned over the whole included volume, a minimum. If an expression for V, involving a finite or infinite number of coefficients, is proposed which satisfies the surface condition and is such that it necessarily includes the true form of V, we may approximate to the value of (14), making it a minimum by variation of the coefficients, even though only a limited number be included. Every fresh coefficient that is included renders the approximation closer, and as near an approach as we please to the truth may be arrived at by continuing the process. The true value of (14) is equal by Green's theorem to

$$\frac{1}{4\pi}\iint V\frac{dV}{dn}\,dS, \quad \ldots\ldots\ldots\ldots\ldots\ldots(15)$$

the integration being over the surface, so that at all stages of the approximation the calculated value of (14) exceeds the true value of (15). In the application to a *condenser*, whose armatures are at potentials 0 and 1,

* *Phil. Trans.* Vol. CLXI. p. 77 (1870); *Scientific Papers*, Vol. I. p. 33. *Phil. Mag.* Vol. XLIV. p. 328 (1872); *Scientific Papers*, Vol. I. p. 140. Compare also *Phil. Mag.* Vol. XLVII. p. 566 (1899), Vol. XXII. p. 225 (1911).

(15) represents the *capacity*. A calculation of capacity founded upon an approximate value of V in (14) is thus always an overestimate.

In the present case we may substitute (15) for (14), if we consider the positive and negative sides separately, since it is only at $z = 0$ that Laplace's equation fails to receive satisfaction. The complete expression for V on the right is given by combination of (2) and (11), and the surface of integration is composed of the cylindrical wall $r = b$ from $z = 0$ to $z = \infty$, and of the plane $z = 0$ from $r = 0$ to $r = b$*. The cylindrical wall contributes nothing, since V vanishes along it. At $z = 0$

$$V = \Sigma (A + B) J_0 (kr), \quad -dV/dz = \Sigma k (A + B) J_0 (kr);$$

and
$$(15) = \tfrac{1}{4} b^2 \Sigma k (A + B)^2 J_1^2 (kb). \quad \dots \dots \dots \dots (16)$$

On the left the complete value of V includes (1) and (8). There are here two cylindrical surfaces, but $r = b$ contributes nothing for the same reason as before. On $r = a$ we have $V = 1$ and

$$-\frac{dV}{dr} = \frac{1}{a \log b/a} - \Sigma h H \phi' (ha) e^{hz};$$

so that this part of the surface, extending to a great distance $z = -l$, contributes to (15)

$$\frac{1}{2 \log b/a} - \frac{a}{2} \Sigma H \phi' (ha). \quad \dots \dots \dots \dots (17)$$

There remains to be considered the annular area at $z = 0$. Over this

$$V = \frac{\log b/r}{\log b/a} + \Sigma H \phi (hr), \quad \dots \dots \dots \dots (18)$$

$$dV/dz = \Sigma h H \phi (hr). \quad \dots \dots \dots \dots (19)$$

The integrals required are

$$\int_a^b \phi (hr) r \, dr = - h^{-1} \{ b \phi' (hb) - a \phi' (ha) \}, \quad \dots \dots \dots \dots (20)$$

$$\int_0^b \log r \, \phi (hr) r \, dr = - h^{-1} \{ b \log b \, \phi' (hb) - a \log a \, \phi' (ha) \}, \quad \dots (21)$$

$$\int_a^b \{ \phi (hr) \}^2 r \, dr = \tfrac{1}{2} b^2 \{ \phi' (hb) \}^2 - \tfrac{1}{2} a^2 \{ \phi' (ha) \}^2; \quad \dots \dots \dots \dots (22)$$

and we get for this part of the surface

$$\tfrac{1}{2} a \Sigma H \phi' (ha) + \tfrac{1}{4} \Sigma h H^2 [b^2 \{ \phi' (hb) \}^2 - a^2 \{ \phi' (ha) \}^2]. \quad \dots \dots (23)$$

Thus for the whole surface on the left

$$(15) = \frac{1}{2 \log b/a} + \tfrac{1}{4} \Sigma h H^2 [b^2 \phi'^2 (hb) - a^2 \phi'^2 (ha)], \quad \dots \dots \dots (24)$$

---

* The surface at $z = +\infty$ may evidently be disregarded.

the simplification arising from the fact that (1) is practically a member of the series ϕ.

The calculated capacity, an overestimate unless all the coefficients H are correctly assigned, is given by addition of (16) and (24). The first approximation is obtained by omitting all the quantities H, so that the B's vanish also. The additional capacity, derived entirely from (16), is then $\frac{1}{4}b^2\Sigma kA^2J_1^2(kb)$, or on introduction of the value of A,

$$\frac{b}{\log^2 b/a}\Sigma\frac{J_0^2(ka)}{k^3b^3J_1^2(kb)}, \quad\ldots\ldots\ldots\ldots(25)$$

the summation extending to all the roots of $J_0(kb)=0$. Or if we express the result in terms of the correction δl to the length (for one end), we have

$$\delta l=\frac{2b}{\log b/a}\Sigma\frac{J_0^2(ka)}{k^2b^3J_1^2(kb)}, \quad\ldots\ldots\ldots\ldots(26)$$

as the first approximation to δl and an overestimate.

The series in (26) converges sufficiently. $J_0^2(ka)$ is less than unity. The mth root of $J_0(x)=0$ is $x=(m-\frac{1}{4})\pi$ approximately, and $J_1^2(x)=2/\pi x$, so that when m is great

$$\frac{1}{x^3J_1^2(x)}=\frac{8}{\pi(4m-1)^2}. \quad\ldots\ldots\ldots\ldots(27)$$

The values of the reciprocals of $x^3J_1^2(x)$ for the earlier roots can be calculated from the tables* and for the higher roots from (27). I find

m	x	$\pm J_1(x)$	$x^{-3}\div J_1^2(x)$
1	2·4048	·51915	·2668
2	5·5201	·34027	·0513
3	8·6537	·27145	·0209
4	11·7915	·23245	·0113
5	14·9309	·20655	·0070

The next five values are ·0048, ·0035, ·0026, ·0021, ·0017. Thus for any value of a the series in (26) is

$$·2668 J_0^2(2·405 a/b)+·0513 J_0^2(5·520 a/b)+\ldots; \quad\ldots\ldots(28)$$

it can be calculated without difficulty when a/b is given. When a/b is very small, the J's in (28) may be omitted, and we have simply to sum the numbers in the fourth column of the table and its continuation. The first ten roots give ·3720. The remainder I estimate at ·015, making in all ·387. Thus in this case

$$\delta l=\frac{·774b}{\log b/a}. \quad\ldots\ldots\ldots\ldots(29)$$

* Gray and Mathews, *Bessel's Functions*, pp. 244, 247.

It is particularly to be noticed that although (29) is an overestimate, it vanishes when a tends to zero.

The next step in the approximation is the inclusion of H_1 corresponding to the first root h_1 of $\phi(hb) = 0$. For a given k, B has only one term, expressed by (13) when we write h_1, H_1 for h, H. In (16) when we expand $(A + B)^2$, we obtain three series of which the first involving A^2 is that already dealt with. It does not depend upon H_1. Constant factors being omitted, the second series depends upon

$$\Sigma \frac{J_0^2(ka)}{k(h_1^2 - k^2) J_1^2(kb)}, \quad \dots\dots\dots\dots\dots(30)$$

and the third upon

$$\Sigma \frac{k J_0^2(ka)}{(h_1^2 - k^2)^2 J_1^2(kb)}, \quad \dots\dots\dots\dots\dots(31)$$

the summations including all admissible values of k. In (24) we have under Σ merely the single term corresponding to H_1, h_1. The sum of (16) and (24) is a quadratic expression in H_1, and is to be made a minimum by variation of that quantity.

The application of this process to the case of a very small leads to a rather curious result. It is known (*Theory of Sound*, § 213 a) that k_1^2 and h_1^2 are then nearly equal, so that the first terms of (30) and (31) are relatively large, and require a special evaluation. For this purpose we must revert to (10) in which, since ha is small,

$$Y_0(ha) = \log ha J_0(ha) + 2J_2(ha), \quad \dots\dots\dots\dots(32)$$

so that nearly enough

$$J_0(hb) = (h - k) b J_0'(kb) = \frac{Y_0(hb)}{\log ha} = \frac{Y_0(kb)}{\log ka},$$

and

$$k - h = \frac{Y_0(kb)}{b J_1(kb) \log ka}. \quad \dots\dots\dots\dots\dots(33)$$

Thus, when a is small enough, the first terms of (30) and (31) dominate the others, and we may take simply

$$(30) = -\frac{b \log k_1 a}{2 k_1^2 Y_0(k_1 b) J_1(k_1 b)}, \quad \dots\dots\dots\dots(34)$$

$$(31) = \frac{b^2 \log^2 k_1 a}{4 k_1 Y_0^2(k_1 b)}. \quad \dots\dots\dots\dots\dots(35)$$

Also

$$\phi'(k_1 a) = -\frac{1}{k_1 a \log k_1 a}, \qquad \phi'(k_1 b) = \frac{Y_1(k_1 b)}{\log k_1 a}. \quad \dots\dots\dots(36)$$

Using these, we find from (16) and (24)

$$\frac{b}{\log^2 b/a} \Sigma \frac{1}{k^3 b^3 J_1^2(kb)} + \frac{H_1}{k_1^2 b \log b/a \cdot Y_0(k_1 b) J_1(k_1 b)} + \frac{H_1^2}{4 k_1 Y_0^2(k_1 b)}$$

$$+ \frac{l}{2 \log b/a} + \frac{k_1 H_1^2}{4 \log^2 k_1 a} \{ b^2 Y_1^2(k_1 b) - k_1^{-2} \}, \quad \dots\dots(37)$$

as the expression for the capacity which is to be made a minimum. Comparing the terms in H_1^2, we see that the two last, corresponding to the negative side, vanish in comparison with the other in virtue of the large denominator $\log^2 k_1 a$. Hence approximately

$$H_1 = -\frac{2 Y_0 (k_1 b)}{k_1 b \log b/a \,.\, J_1 (k_1 b)}, \quad \dots\dots\dots\dots\dots(38)$$

and (37) becomes

$$\frac{l}{2 \log b/a} + \frac{b}{\log^2 b/a} \Sigma \frac{1}{k^3 b^3 J_1^2 (kb)} - \frac{b}{\log^2 b/a} \frac{1}{k_1^3 b^3 J_1^2 (k_1 b)} \quad \dots\dots (39)$$

when made a minimum by variation of H_1. Thus the effect of the correction depending on the introduction of H_1 is simply to wipe out the initial term of the series which represents the first approximation to the correction.

After this it may be expected that the remaining terms of the first approximation to the correction will also disappear. On examination this conjecture will be found to be verified. Under each value of k in (16) only that part of B is important for which h has the particular value which is nearly equal to k. Thus each new H annuls the corresponding member of the series in (39), so that the continuation of the process leaves us with the first term of (39) isolated. The inference is that the correction to the capacity vanishes in comparison with $b \div \log^2 b/a$, or that δl vanishes in comparison with $b \div \log b/a$. It would seem that δl is of the order $b \div \log^2 b/a$, but it would not be easy to find the numerical coefficient by the present method.

In any case the correction δl to the length of the rod vanishes in the electrostatical problem when the radius of the rod is diminished without limit—a conclusion which I extend to the vibrational problem specified in the earlier portion of this paper.

373.

ON CONFORMAL REPRESENTATION FROM A MECHANICAL POINT OF VIEW.

[*Philosophical Magazine*, Vol. XXV. pp. 698—702, 1913.]

In what is called conformal representation the coordinates of one point x, y in a plane are connected with those of the corresponding point ξ, η by the relation

$$x + iy = f(\xi + i\eta), \quad \dots\dots\dots\dots\dots\dots\dots(1)$$

where f denotes an arbitrary function. In this transformation angles remain unaltered, and corresponding infinitesimal figures are similar, though not in general similarly situated. If we attribute to ξ, η values in arithmetical progression with the same small common difference, the simple square network is represented by two sets of curves crossing one another at right angles so as to form what are ultimately squares when the original common difference is made small enough. For example, as a special case of (1), if

$$x + iy = c \sin (\xi + i\eta), \quad \dots\dots\dots\dots\dots\dots(2)$$

$$x = c \sin \xi \cosh \eta, \quad y = c \cos \xi \sinh \eta \,;$$

and the curves corresponding to $\eta = \text{constant}$ are

$$\frac{x^2}{c^2 \cosh^2 \eta} + \frac{y^2}{c^2 \sinh^2 \eta} = 1, \quad \dots\dots\dots\dots\dots(3)$$

and those corresponding to $\xi = \text{constant}$ are

$$\frac{x^2}{c^2 \sin^2 \xi} - \frac{y^2}{c^2 \cos^2 \xi} = 1, \quad \dots\dots\dots\dots\dots(4)$$

a set of confocal ellipses and hyperbolas.

It is usual to refer x, y and ξ, η to separate planes and, as far as I have seen, no *transition* from the one position to the other is contemplated. But of course there is nothing to forbid the two sets of coordinates being taken in the same plane and measured on the same axes. We may then regard the angular points of the network as moving from the one position to the other.

Some fifteen or twenty years ago I had a model made for me illustrative of these relations. The curves have their material embodiment in wires of hard steel. At the angular points the wires traverse small and rather thick brass disks, bored suitably so as to impose the required perpendicularity, the

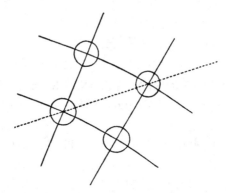

Fig. 1.

two sets of wires being as nearly as may be in the same plane. But something more is required in order to secure that the rectangular element of the network shall be *square*. To this end a third set of wires (shown dotted in fig. 1) was introduced, traversing the corner pieces through borings making 45° with the previous ones. The model answered its purpose to a certain extent, but the manipulation was not convenient on account of the friction entailed as the wires slip through the closely-fitting corner pieces. Possibly with the aid of rollers an improved construction might be arrived at.

The material existence of the corner pieces in the model suggests the consideration of a continuous two-dimensional medium, say a lamina, whose deformation shall represent the transformation. The lamina must be of such a character as absolutely to preclude *shearing*. On the other hand, it must admit of expansion and contraction equal in all (two-dimensional) directions, and if the deformation is to persist without the aid of applied forces, such expansion must be unresisted.

Since the deformation is now regarded as taking place continuously, f in (1) must be supposed to be a function of the time t as well as of $\xi + i\eta$. We may write

$$x + iy = f(t, \xi + i\eta). \quad \dots\dots\dots\dots\dots\dots(5)$$

The component velocities u, v of the particle which at time t occupies the position x, y are given by dx/dt, dy/dt, so that

$$u + iv = \frac{d}{dt} f(t, \xi + i\eta). \quad \dots\dots\dots\dots\dots(6)$$

Between (5) and (6) $\xi + i\eta$ may be eliminated; $u + iv$ then becomes a function of t and of $x + iy$, say

$$u + iv = F(t, x + iy). \quad\ldots\ldots\ldots\ldots\ldots\ldots(7)$$

The equation with which we started is of what is called in Hydrodynamics the Lagrangian type. We follow the motion of an individual particle. On the other hand, (7) is of the Eulerian type, expressing the velocities to be found at any time at a specified place. Keeping t fixed, *i.e.* taking, as it were, an instantaneous view of the system, we see that u, v, as given by (7), satisfy

$$(d^2/dx^2 + d^2/dy^2)(u, v) = 0, \quad\ldots\ldots\ldots\ldots\ldots\ldots(8)$$

equations which hold also for the irrotational motion of an incompressible liquid.

It is of interest to compare the present motion with that of a highly viscous two-dimensional fluid, for which the equations are[*]

$$\rho \frac{Du}{Dt} = \rho X - \frac{dp}{dx} + \mu' \frac{d\theta}{dx} + \mu \left(\frac{d^2 u}{dx^2} + \frac{d^2 u}{dy^2} \right),$$

$$\rho \frac{Dv}{Dt} = \rho Y - \frac{dp}{dy} + \mu' \frac{d\theta}{dy} + \mu \left(\frac{d^2 v}{dx^2} + \frac{d^2 v}{dy^2} \right),$$

where
$$\theta = \frac{du}{dx} + \frac{dv}{dy}.$$

If the pressure is independent of density and if the inertia terms are neglected, these equations are satisfied provided that

$$\rho X + \mu' d\theta/dx = 0, \quad \rho Y + \mu' d\theta/dy = 0.$$

In the case of real viscous fluids, there is reason to think that $\mu' = \frac{1}{3}\mu$. Impressed forces are then required so long as the fluid is moving. The supposition that p is constant being already a large departure from the case of nature, we may perhaps as well suppose $\mu' = 0$, and then no impressed bodily forces are called for either at rest or in motion.

If we suppose that the motion in (7) is *steady* in the hydrodynamical sense, $u + iv$ must be independent of t, so that the elimination of $\xi + i\eta$ between (5) and (6) must carry with it the elimination of t. This requires that df/dt in (6) be a function of f, and not otherwise of t and $\xi + i\eta$; and it follows that (5) must be of the form

$$x + iy = F_1\{t + F_2(\xi + i\eta)\}, \quad\ldots\ldots\ldots\ldots\ldots\ldots(9)$$

---

[*] Stokes, *Camb. Trans.* 1850; *Mathematical and Physical Papers*, Vol. IV. p. 11. It does not seem to be generally known that the laws of dynamical similarity for viscous fluids were formulated in this memoir. Reynolds's important application was 30 years later.

where F_1, F_2 denote arbitrary functions. Another form of (9) is

$$F_3(x + iy) = t + F_2(\xi + i\eta). \quad \ldots\ldots\ldots\ldots\ldots\ldots(10)$$

For an individual particle $F_2(\xi + i\eta)$ is constant, say $a + ib$. The equation of the stream-line followed by this particle is obtained by equating to ib the imaginary part of $F_3(x + iy)$.

As an example of (9), suppose that

$$x + iy = c \sin \{it + \xi + i\eta\}, \quad \ldots\ldots\ldots\ldots\ldots\ldots(11)$$

so that $\quad x = c \sin \xi \cdot \cosh(\eta + t), \quad y = c \cos \xi \cdot \sinh(\eta + t), \quad \ldots\ldots(12)$

whence on elimination of t we obtain (4) as the equation of the stream-lines.

It is scarcely necessary to remark that the law of flow along the stream-lines is entirely different from that with which we are familiar in the flow of incompressible liquids. In the latter case the motion is rapid at any place where neighbouring stream-lines approach one another closely. Here, on the contrary, the motion is exceptionally slow at such a place.

374.

ON THE APPROXIMATE SOLUTION OF CERTAIN PROBLEMS RELATING TO THE POTENTIAL.—II.

[*Philosophical Magazine*, Vol. XXVI. pp. 195—199, 1913.]

THE present paper may be regarded as supplementary to one with the same title published a long while ago*. In two dimensions, if ϕ, ψ be potential and stream-functions, and if (*e.g.*) ψ be zero along the line $y = 0$, we may take

$$\phi = \int f \, dx - \frac{y^2}{1 \cdot 2} f' + \frac{y^4}{1 \cdot 2 \cdot 3 \cdot 4} f''' - \dots , \quad \dots\dots\dots\dots(1)$$

$$\psi = yf - \frac{y^3}{1 \cdot 2 \cdot 3} f'' + \frac{y^5}{1 \cdot 2 \cdot 3 \cdot 4 \cdot 5} f^{\text{iv}} - \dots , \quad \dots\dots\dots(2)$$

f being a function of x so far arbitrary. These values satisfy the general conditions for the potential and stream-functions, and when $y = 0$ make

$$d\phi/dx = f, \qquad \psi = 0.$$

Equation (2) may be regarded as determining the lines of flow (any one of which may be supposed to be the boundary) in terms of f. Conversely, if y be supposed known as a function of x and ψ be constant (say unity), we may find f by successive approximation. Thus

$$f = \frac{1}{y} + \frac{y^2}{6} \frac{d^2}{dx^2} \left(\frac{1}{y} \right) + \frac{y^2}{36} \frac{d^2}{dx^2} \left\{ y^2 \frac{d^2}{dx^2} \left(\frac{1}{y} \right) \right\} - \frac{y^4}{120} \frac{d^4}{dx^4} \left(\frac{1}{y} \right). \quad \dots\dots(3)$$

We may use these equations to investigate the stream-lines for which ψ has a value intermediate between 0 and 1. If η denote the corresponding value of y, we have to eliminate f between

$$1 = yf - \frac{y^3}{6} f'' + \frac{y^5}{120} f^{\text{iv}} - \dots ,$$

and

$$\psi = \eta f - \frac{\eta^3}{6} f'' + \frac{\eta^5}{120} f^{\text{iv}} - \dots ;$$

whence

$$\eta = \psi y + \frac{f''}{6} (y\eta^3 - \eta y^3) - \frac{f^{\text{iv}}}{120} (y\eta^5 - \eta y^5),$$

* *Proc. Lond. Math. Soc.* Vol. VII. p. 75 (1876); *Scientific Papers*, Vol. I. p. 272.

or by use of (3)

$$\eta = \psi y + \frac{y^4 (\psi^3 - \psi)}{6} \frac{d^2}{dx^2}\left(\frac{1}{y}\right) + \frac{y^7 (\psi^2 - 1)(3\psi^2 - \psi)}{36}\left\{\frac{d^2}{dx^2}\left(\frac{1}{y}\right)\right\}^2$$

$$+ \frac{y^4 (\psi^3 - \psi)}{36} \frac{d^2}{dx^2}\left\{y^2 \frac{d^2}{dx^2}\left(\frac{1}{y}\right)\right\} - \frac{y^6 (\psi^5 - \psi)}{120} \frac{d^4}{dx^4}\left(\frac{1}{y}\right). \quad \text{......(4)}$$

The evanescence of ψ when $y = 0$ may arise from this axis being itself a boundary, or from the second boundary being a symmetrical curve situated upon the other side of the axis. In the former paper expressions for the "resistance" and "conductivity" were developed.

We will now suppose that $\psi = 0$ along a *circle* of radius a, in substitution for the axis of x. Taking polar coordinates $a + r$ and θ, we have as the general equation

$$(a + r)^2 \frac{d^2\psi}{dr^2} + (a + r)\frac{d\psi}{dr} + \frac{d^2\psi}{d\theta^2} = 0. \quad \text{..................(5)}$$

Assuming $\psi = R_1 r + R_2 r^2 + R_3 r^3 + \dots , \quad \text{......................(6)}$

where R_1, R_2, &c., are functions of θ, we find on substitution in (5)

$$\left.\begin{array}{l} 2a^2 R_2 + aR_1 = 0, \\ 6a^2 R_3 + 6aR_2 + R_1 + R_1'' = 0 ; \end{array}\right\} \quad \text{..................(7)}$$

so that $\psi = R_1 r - \dfrac{R_1 r^2}{2a} + \dfrac{(2R_1 - R_1'')r^3}{6a^2} \quad \text{..................(8)}$

is the form corresponding to (2) above.

If $\psi = 1$, (8) yields

$$R_1 = \frac{1}{r} + \frac{1}{2a} - \frac{r^2}{12a^2} + \frac{r^2}{6a^2}\frac{d^2}{d\theta^2}\left(\frac{1}{r}\right), \quad \text{..................(9)}$$

expressing R_1 as a function of θ, when r is known as such. To interpolate a curve for which ρ takes the place of r, we have to eliminate R_1 between

$$1 = R_1 r - \frac{R_1 r^2}{2a} + \frac{(2R_1 - R_1'')r^2}{6a^2},$$

and $\psi = R_1 \rho - \dfrac{R_1 \rho^2}{2a} + \dfrac{(2R_1 - R_1'')\rho^2}{6a^2}.$

Thus $\rho = r\psi - \dfrac{R_1}{2a}(\rho r^2 - r\rho^2) + \dfrac{2R_1 - R_1''}{6a^2}(\rho r^3 - r\rho^3),$

and by successive approximation with use of (9)

$$\rho = r\psi + \frac{r^2}{a}\frac{\psi(\psi - 1)}{1.2} + \frac{r^3}{a^2}\frac{\psi(\psi - 1)(\psi - 2)}{1.2.3} + \frac{r^4}{a^2}\frac{\psi(\psi^2 - 1)}{6}\frac{d^2}{d\theta^2}\left(\frac{1}{r}\right). \quad (10)$$

The significance of the first three terms is brought out if we suppose that r is constant (a), so that the last term vanishes. In this case the exact solution is

$$\log \frac{a+\rho}{a} = \psi \log \frac{a+a}{a}, \quad \dots\dots\dots\dots\dots(11)$$

whence

$$\frac{\rho}{a} = \left(\frac{a+a}{a}\right)^{\psi} - 1 = \psi\frac{a}{a} + \frac{\psi(\psi-1)}{1.2}\frac{a^2}{a^2} + \frac{\psi(\psi-1)(\psi-2)}{1.2.3}\frac{a^3}{a^3} + \dots \ \dots(12)$$

in agreement with (10).

In the above investigation ψ is supposed to be zero exactly upon the circle of radius a. If the circle whose centre is taken as origin of coordinates be merely the circle of curvature of the curve $\psi = 0$ at the point ($\theta = 0$) under consideration, ψ will not vanish exactly upon it, but only when r has the approximate value $c\theta^3$, c being a constant. In (6) an initial term R_0 must be introduced, whose approximate value is $-c\theta^3 R_1$. But since R_0'' vanishes with θ, equation (7) and its consequences remain undisturbed and (10) is still available as a formula of interpolation. In all these cases, the success of the approximation depends of course upon the degree of slowness with which y, or r, varies.

Another form of the problem arises when what is given is not a pair of neighbouring curves along each of which (*e.g.*) the stream-function is constant, but *one* such curve together with the variation of potential along it. It is then required to construct a neighbouring stream-line and to determine the distribution of potential upon it, from which again a fresh departure may be made if desired. For this purpose we regard the rectangular coordinates x, y as functions of ξ (potential) and η (stream-function), so that

$$x + iy = f(\xi + i\eta), \quad \dots\dots\dots\dots\dots(13)$$

in which we are supposed to know $f(\xi)$ corresponding to $\eta = 0$, *i.e.*, x and y are there known functions of ξ. Take a point on $\eta = 0$, at which without loss of generality ξ may be supposed also to vanish, and form the expressions for x and y in the neighbourhood. From

$$x + iy = A_0 + iB_0 + (A_1 + iB_1)(\xi + i\eta) + (A_2 + iB_2)(\xi + i\eta)^2 + \dots,$$

we derive

$$x = A_0 + A_1\xi - B_1\eta + A_2(\xi^2 - \eta^2) - 2B_2\xi\eta$$
$$+ A_3(\xi^3 - 3\xi\eta^2) - B_3(3\xi^2\eta - \eta^3)$$
$$+ A_4(\xi^4 - 6\xi^2\eta^2 + \eta^4) - 4B_4(\xi^3\eta - \xi\eta^3) + \dots,$$

$$y = B_0 + B_1\xi + A_1\eta + 2A_2\xi\eta + B_2(\xi^2 - \eta^2)$$
$$+ A_3(3\xi^2\eta - \eta^3) + B_3(\xi^3 - 3\xi\eta^2)$$
$$+ 4A_4(\xi^3\eta - \xi\eta^3) + B_4(\xi^4 - 6\xi^2\eta^2 + \eta^4) + \dots.$$

When $\eta = 0$,
$$x = A_0 + A_1\xi + A_2\xi^2 + A_3\xi^3 + A_4\xi^4 + \dots,$$
$$y = B_0 + B_1\xi + B_2\xi^2 + B_3\xi^3 + B_4\xi^4 + \dots.$$

Since x and y are known as functions of ξ when $\eta = 0$, these equations determine the A's and the B's, and the general values of x and y follow. When $\xi = 0$, but η undergoes an increment,

$$x = A_0 - B_1\eta - A_2\eta^2 + B_3\eta^3 + A_4\eta^4 - \dots, \qquad \dots\dots\dots(14)$$

$$y = B_0 + A_1\eta - B_2\eta^2 - A_3\eta^3 + B_4\eta^4 + \dots, \qquad \dots\dots\dots(15)$$

in which we may suppose $\eta = 1$.

The A's and B's are readily determined if we know the values of x and y for $\eta = 0$ and for equidistant values of ξ, say $\xi = 0$, $\xi = \pm 1$, $\xi = \pm 2$. Thus, if the values of x be called x_0, x_{-1}, x_1, x_2, x_{-2}, we find

$$A_0 = x_0, \qquad \text{and}$$

$$A_1 = \frac{2}{3}(x_1 - x_{-1}) - \frac{1}{12}(x_2 - x_{-2}), \quad A_3 = \frac{x_2 - x_{-2}}{12} - \frac{x_1 - x_{-1}}{6},$$

$$A_2 = \frac{2(x_1 + x_{-1} - 2x_0)}{3} - \frac{x_2 + x_{-2} - 2x_0}{24},$$

$$A_4 = \frac{x_2 + x_{-2} - 2x_0}{24} - \frac{x_1 + x_{-1} - 2x_0}{6}.$$

The B's are deduced from the A's by merely writing y for x throughout. Thus from (14) when $\xi = 0$, $\eta = 1$,

$$x = x_0 - \frac{5}{6}(x_1 + x_{-1} - 2x_0) + \frac{1}{12}(x_2 + x_{-2} - 2x_0)$$

$$- \frac{5}{6}(y_1 - y_{-1}) + \frac{1}{6}(y_2 - y_{-2}). \qquad \dots\dots\dots\dots\dots(16)$$

Similarly $\qquad y = y_0 - \frac{5}{6}(y_1 + y_{-1} - 2y_0) + \frac{1}{12}(y_2 + y_{-2} - 2y_0)$

$$+ \frac{5}{6}(x_1 - x_{-1}) - \frac{1}{6}(x_2 - x_{-2}). \qquad \dots\dots\dots\dots\dots(17)$$

By these formulæ a point is found upon a new stream-line ($\eta = 1$) corresponding to a given value of ξ. And there would be no difficulty in carrying the approximation further if desired.

As an example of the kind of problem to which these results might be applied, suppose that by observation or otherwise we know the form of the upper stream-line constituting part of the free surface when liquid falls steadily over a two-dimensional weir. Since the velocity is known at every point of the free surface, we are in a position to determine ξ along this stream-line, and thus to apply the formulæ so as to find interior stream-lines in succession.

Again (with interchange of ξ and η) we could find what forms are admissible for the second coating of a two-dimensional condenser, in order that the charge upon the first coating, given in size and shape, may have a given value at every point.

[*Sept.* 1916. As another example permanent wave-forms may be noticed.]

375.

ON THE PASSAGE OF WAVES THROUGH FINE SLITS IN THIN OPAQUE SCREENS.

[*Proceedings of the Royal Society*, A, Vol. LXXXIX. pp. 194—219, 1913.]

IN a former paper* I gave solutions applicable to the passage of light through very narrow slits in infinitely thin perfectly opaque screens, for the two principal cases where the polarisation is either parallel or perpendicular to the length of the slit. It appeared that if the width ($2b$) of the slit is very small in comparison with the wave-length (λ), there is a much more free passage when the electric vector is perpendicular to the slit than when it is parallel to the slit, so that unpolarised light incident upon the screen will, after passage, appear polarised in the former manner. This conclusion is in accordance with the observations of Fizeau† upon the very narrowest slits. Fizeau found, however, that somewhat wider slits (scratches upon silvered glass) gave the opposite polarisation; and I have long wished to extend the calculations to slits of width comparable with λ. The subject has also a practical interest in connection with observations upon the Zeeman effect‡.

The analysis appropriate to problems of this sort would appear to be by use of elliptic coordinates; but I have not seen my way to a solution on these lines, which would, in any case, be rather complicated. In default of such a solution, I have fallen back upon the approximate methods of my former paper. Apart from the intended application, some of the problems which present themselves have an interest of their own. It will be convenient to repeat the general argument almost in the words formerly employed

---

* "On the Passage of Waves through Apertures in Plane Screens and Allied Problems," *Phil. Mag.* 1897, Vol. XLIII. p. 259; *Scientific Papers*, Vol. IV. p. 283.

† *Annales de Chimie*, 1861, Vol. LXIII. p. 385; Mascart's *Traité d'Optique*, § 645. See also *Phil. Mag.* 1907, Vol. XIV. p. 350; *Scientific Papers*, Vol. V. p. 417.

‡ Zeeman, *Amsterdam Proceedings*, October, 1912.

Plane waves of simple type impinge upon a parallel screen. The screen is supposed to be infinitely thin and to be perforated by some kind of aperture. Ultimately, one or both dimensions of the aperture will be regarded as small, or, at any rate, as not large, in comparison with the wavelength (λ); and the investigation commences by adapting to the present purpose known solutions concerning the flow of incompressible fluids.

The functions that we require may be regarded as velocity-potentials ϕ, satisfying

$$d^2\phi/dt^2 = V\nabla^2\phi, \quad\dots\dots\dots\dots\dots\dots(1)$$

where $\nabla^2 = d^2/dx^2 + d^2/dy^2 + d^2/dz^2,$

and V is the velocity of propagation. If we assume that the vibration is everywhere proportional to e^{int}, (1) becomes

$$(\nabla^2 + k^2)\,\phi = 0, \quad\dots\dots\dots\dots\dots\dots(2)$$

where $k = n/V = 2\pi/\lambda. \quad\dots\dots\dots\dots\dots\dots(3)$

It will conduce to brevity if we suppress the factor e^{int}. On this understanding the equation of waves travelling parallel to x in the positive direction, and accordingly incident upon the negative side of the screen situated at $x = 0$, is

$$\phi = e^{-ikx}.\quad\dots\dots\dots\dots\dots\dots(4)$$

When the solution is complete, the factor e^{int} is to be restored, and the imaginary part of the solution is to be rejected. The realised expression for the incident waves will therefore be

$$\phi = \cos(nt - kx). \quad\dots\dots\dots\dots\dots\dots(5)$$

There are two cases to be considered corresponding to two alternative boundary conditions. In the first (i) $d\phi/dn = 0$ over the unperforated part of the screen, and in the second (ii) $\phi = 0$. In case (i) dn is drawn outwards normally, and if we take the axis of z parallel to the length of the slit, ϕ will represent the magnetic component parallel to z, usually denoted by c, so that this case refers to vibrations for which the electric vector is perpendicular to the slit. In the second case (ii) ϕ is to be identified with the component parallel to z of the electric vector R, which vanishes upon the walls, regarded as perfectly conducting. We proceed with the further consideration of case (i).

If the screen be complete, the reflected waves under condition (i) have the expression $\phi = e^{ikx}$. Let us divide the actual solution into two parts, χ and ψ; the first, the solution which would obtain were the screen complete; the second, the alteration required to take account of the aperture; and let us distinguish by the suffixes m and p the values applicable upon the negative (*minus*), and upon the positive side of the screen. In the present case we have

$$\chi_m = e^{-ikx} + e^{ikx}, \qquad \chi_p = 0. \quad\dots\dots\dots\dots(6)$$

This χ-solution makes $d\chi_m/dn = 0$, $d\chi_p/dn = 0$ over the whole plane $x = 0$, and over the same plane $\chi_m = 2$, $\chi_p = 0$.

For the supplementary solution, distinguished in like manner upon the two sides, we have

$$\psi_m = \iint \Psi_m \frac{e^{-ikr}}{r} dS, \qquad \psi_p = \iint \Psi_p \frac{e^{-ikr}}{r} dS, \quad \ldots \ldots (7)$$

where r denotes the distance of the point at which ψ is to be estimated from the element dS of the aperture, and the integration is extended over the whole of the area of aperture. Whatever functions of position Ψ_m, Ψ_p may be, these values on the two sides satisfy (2), and (as is evident from symmetry) they make $d\psi_m/dn$, $d\psi_p/dn$ vanish over the wall, viz., the unperforated part of the screen, so that the required condition over the wall for the complete solution is already satisfied. It remains to consider the further conditions that ϕ and $d\phi/dx$ shall be continuous across the *aperture*. These conditions require that on the aperture

$$2 + \psi_m = \psi_p, \qquad d\psi_m/dx = d\psi_p/dx. \quad \ldots \ldots \ldots (8)*$$

The second is satisfied if $\Psi_p = -\Psi_m$; so that

$$\psi_m = \iint \Psi_m \frac{e^{-ikr}}{r} dS, \qquad \psi_p = -\iint \Psi_m \frac{e^{-ikr}}{r} dS, \quad \ldots \ldots (9)$$

making the values of ψ_m, ψ_p equal and opposite at all corresponding points, viz., points which are images of one another in the plane $x = 0$. In order further to satisfy the first condition, it suffices that over the area of aperture

$$\psi_m = -1, \qquad \psi_p = 1, \quad \ldots \ldots \ldots \ldots (10)$$

and the remainder of the problem consists in so determining Ψ_m that this shall be the case.

It should be remarked that Ψ in (9) is closely connected with the normal velocity at dS. In general,

$$\frac{d\psi}{dx} = \iint \Psi \frac{d}{dx}\left(\frac{e^{-ikr}}{r}\right) dS. \quad \ldots \ldots \ldots \ldots (11)$$

At a point (x) infinitely close to the surface, only the neighbouring elements contribute to the integral, and the factor e^{-ikr} may be omitted. Thus

$$\frac{d\psi}{dx} = -\iint \Psi \frac{x}{r^2} dS = -2\pi x \int_x^\infty \Psi \frac{r\,dr}{r^3} = -2\pi \Psi; \quad \text{or} \quad \Psi = -\frac{1}{2\pi}\frac{d\psi}{dn}, \ldots (12)$$

$d\psi/dn$ being the normal velocity at the point of the surface in question.

---

* The use of dx implies that the variation is in a fixed direction, while dn may be supposed to be drawn outwards from the screen in both cases.

In the original paper these results were applied to an aperture, especially of elliptical form, whose dimensions are small in comparison with λ. For our present purpose we may pass this over and proceed at once to consider the case where the aperture is an infinitely long slit with parallel edges, whose width is small, or at the most comparable with λ.

The velocity-potential of a point-source, viz., e^{-ikr}/r, is now to be replaced by that of a linear source, and this, in general, is much more complicated. If we denote it by $D(kr)$, r being the distance from the line of the point where the potential is required, the expressions are *

$$
D(kr) = -\left(\frac{\pi}{2ikr}\right)^{\frac{1}{2}} e^{-ikr} \left\{ 1 - \frac{1^2}{1 \cdot 8ikr} + \frac{1^2 \cdot 3^2}{1 \cdot 2 \cdot (8ikr)^2} - \cdots \right\}
$$
$$
= \left(\gamma + \log \frac{ikr}{2}\right) \left\{ 1 - \frac{k^2 r^2}{2^2} + \frac{k^4 r^4}{2^2 \cdot 4^2} - \cdots \right\}
$$
$$
+ \frac{k^2 r^2}{2^2} S_1 - \frac{k^4 r^4}{2^2 \cdot 4^2} S_2 + \frac{k^6 r^6}{2^2 \cdot 4^2 \cdot 6^2} S_3 - \cdots , \quad \dots\dots\dots(13)
$$

where γ is Euler's constant (0·577215), and

$$
S_m = 1 + \tfrac{1}{2} + \tfrac{1}{3} + \ldots + 1/m. \quad \dots\dots\dots\dots(14)
$$

Of these the first is "semi-convergent" and is applicable when kr is large; the second is fully convergent and gives the form of the function when kr is moderate. The function D may be regarded as being derived from e^{-ikr}/r by integration over an infinitely long and infinitely narrow strip of the surface S.

As the present problem is only a particular case, equations (6) and (10) remain valid, while (9) may be written in the form

$$
\psi_m = \int \Psi_m D(kr)\, dy, \qquad \psi_p = -\int \Psi_m D(kr)\, dy, \quad \dots\dots(15)
$$

the integrations extending over the width of the slit from $y = -b$ to $y = +b$. It remains to determine Ψ_m, so that on the aperture $\psi_m = -1$, $\psi_p = +1$.

At a sufficient distance from the slit, supposed for the moment to be very narrow, $D(kr)$ may be removed from under the integral sign and also be replaced by its limiting form given in (13). Thus

$$
\psi_m = -\left(\frac{\pi}{2ikr}\right)^{\frac{1}{2}} e^{-ikr} \int \Psi_m\, dy. \quad \dots\dots\dots\dots(16)
$$

If the slit be not very narrow, the partial waves arising at different parts of the width will arrive in various phases, of which due account must be taken. The disturbance is no longer circularly symmetrical as in (16). But if, as is usual in observations with the microscope, we restrict ourselves to

* See *Theory of Sound*, § 341.

the direction of original propagation, equality of phase obtains, and (16) remains applicable even in the case of a wide slit. It only remains to determine Ψ_m as a function of y, so that for all points upon the aperture

$$\int_{-b}^{+b} \Psi_m D\,(kr)\,dy = -1, \quad\dots\dots\dots\dots(17)$$

where, since kr is supposed moderate throughout, the second form in (13) may be employed.

Before proceeding further it may be well to exhibit the solution, as formerly given, for the case of a very narrow slit. Interpreting ϕ as the velocity-potential of aerial vibrations and having regard to the known solution for the flow of incompressible fluid through a slit in an infinite plane wall, we may infer that Ψ_m will be of the form $A\,(b^2 - y^2)^{-\frac{1}{2}}$, where A is some constant. Thus (17) becomes

$$A\left[(\gamma + \log \tfrac{1}{2} ik)\, \pi + \int_{-b}^{+b} \frac{\log r\,.\,dy}{\sqrt{(b^2 - y^2)}} \right] = -1. \quad\dots\dots\dots(18)$$

In this equation the first part is obviously independent of the position of the point chosen, and if the form of Ψ_m has been rightly taken the second integral must also be independent of it. If its coordinate be η, lying between $\pm b$,

$$\int_{-b}^{+b} \frac{\log r\,.\,dy}{\sqrt{(b^2 - y^2)}} = \int_{-b}^{\eta} \frac{\log(\eta - y)\,dy}{\sqrt{(b^2 - y^2)}} + \int_{\eta}^{b} \frac{\log(y - \eta)\,dy}{\sqrt{(b^2 - y^2)}} \quad\dots\dots(19)$$

must be independent of η. To this we shall presently return; but merely to determine A in (18) it suffices to consider the particular case of $\eta = 0$. Here

$$\int_{-b}^{+b} \frac{\log r\,.\,dy}{\sqrt{(b^2 - y^2)}} = 2\int_{0}^{b} \frac{\log y\,.\,dy}{\sqrt{(b^2 - y^2)}} = 2\int_{0}^{\frac{1}{2}\pi} \log(b\cos\theta)\,d\theta = \pi \log(\tfrac{1}{2}b).$$

Thus $\qquad A\,(\gamma + \log \tfrac{1}{4} ikb)\,\pi = -1, \quad$ and $\quad \int_{-b}^{+b} \Psi_m dy = \pi A\,;$

so that (16) becomes $\qquad \psi_m = \dfrac{e^{-ikr}}{\gamma + \log(\tfrac{1}{4}ikb)}\left(\dfrac{\pi}{2ikr}\right)^{\frac{1}{2}}. \quad\dots\dots\dots\dots(20)$

From this, ψ_p is derived by simply prefixing a negative sign.

The realised solution is obtained from (20) by omitting the imaginary part after introduction of the suppressed factor e^{int}. If the imaginary part of $\log(\tfrac{1}{4}ikb)$ be neglected, the result is

$$\psi_m = \left(\frac{\pi}{2kr}\right)^{\frac{1}{2}} \frac{\cos(nt - kr - \tfrac{1}{4}\pi)}{\gamma + \log(\tfrac{1}{4}kb)}, \quad\dots\dots\dots\dots(21)$$

corresponding to $\qquad \chi_m = 2\cos nt \cos kx. \quad\dots\dots\dots\dots\dots(22)$

Perhaps the most remarkable feature of the solution is the very limited dependence of the transmitted vibration on the *width* ($2b$) of the aperture.

We will now verify that (19) is independent of the special value of η. Writing $y = b \cos \theta$, $\eta = b \cos \alpha$, we have

$$\int_{-b}^{+b} \frac{\log r \cdot dy}{\sqrt{(b^2 - y^2)}} = \int_0^\pi \log (\tfrac{1}{2} b) \, d\theta + \int_0^a \log 2 \, (\cos \theta - \cos \alpha) \, d\theta$$

$$+ \int_a^\pi \log 2 \, (\cos \alpha - \cos \theta) \, d\theta = \pi \log (\tfrac{1}{2} b)$$

$$+ \int_0^\pi \log \left\{ 2 \sin \frac{\alpha + \theta}{2} \right\} d\theta + \int_0^a \log \left\{ 2 \sin \frac{\alpha - \theta}{2} \right\} d\theta + \int_a^\pi \log \left\{ 2 \sin \frac{\theta - \alpha}{2} \right\} d\theta$$

$$= \pi \log \tfrac{1}{2} b + 2 \int_{\frac{1}{2}a}^{\frac{1}{2}\pi + \frac{1}{2}a} \log (2 \sin \phi) \, d\phi + 2 \int_0^{\frac{1}{2}a} \log (2 \sin \phi) \, d\phi$$

$$+ 2 \int_0^{\frac{1}{2}\pi - \frac{1}{2}a} \log (2 \sin \phi) \, d\phi$$

$$= \pi \log \tfrac{1}{2} b + 2 \int_0^{\frac{1}{2}\pi} \log (2 \sin \phi) \, d\phi + 2 \int_{\frac{1}{2}\pi}^{\frac{1}{2}\pi + \frac{1}{2}a} \log (2 \sin \phi) \, d\phi$$

$$+ 2 \int_0^{\frac{1}{2}\pi - \frac{1}{2}a} \log (2 \sin \phi) \, d\phi$$

$$= \pi \log \tfrac{1}{2} b + 4 \int_0^{\frac{1}{2}\pi} \log (2 \sin \phi) \, d\phi,$$

as we see by changing ϕ into $\pi - \phi$ in the second integral. Since α has disappeared, the original integral is independent of η. In fact*

$$\int_0^{\frac{1}{2}\pi} \log (2 \sin \phi) \, d\phi = 0,$$

and we have $$\int_{-b}^{+b} \frac{\log r \cdot dy}{\sqrt{(b^2 - y^2)}} = \pi \log \tfrac{1}{2} b, \dots\dots\dots\dots\dots(23)$$

as in the particular case of $\eta = 0$.

The required condition (17) can thus be satisfied by the proposed form of Ψ, provided that kb be small enough. When kb is greater, the resulting value of ψ in (15) will no longer be constant over the aperture, but we may find what the actual value is as a function of η by carrying out the integration with inclusion of more terms in the series representing D. As a preliminary, it will be convenient to discuss certain definite integrals which present themselves. The first of the series, which has already occurred, we will call h_0, so that

$$h_0 = \int_0^{\frac{1}{2}\pi} \log (2 \sin \theta) \, d\theta = \int_0^{\frac{1}{2}\pi} \log (2 \cos \theta) \, d\theta = \tfrac{1}{2} \int_0^{\frac{1}{2}\pi} \log (2 \sin 2\theta) \, d\theta$$

$$= \tfrac{1}{4} \int_0^\pi \log (2 \sin \phi) \, d\phi = \tfrac{1}{2} \int_0^{\frac{1}{2}\pi} \log (2 \sin \phi) \, d\phi = \tfrac{1}{2} h_0.$$

* See below.

Accordingly, $h_0 = 0$. More generally we set, n being an even integer,

$$h_n = \int_0^{\frac{1}{2}\pi} \sin^n \theta \log (2 \sin \theta)\, d\theta, \dots\dots\dots\dots\dots(24)$$

or, on integration by parts,

$$h_n = \int_0^{\frac{1}{2}\pi} \cos \theta \left\{ (n-1) \sin^{n-2} \theta \cos \theta \log (2 \sin \theta) + \sin^{n-2} \theta \cos \theta \right\} d\theta$$

$$= (n-1)(h_{n-2} - h_n) + \int_0^{\frac{1}{2}\pi} (\sin^{n-2} \theta - \sin^n \theta)\, d\theta.$$

Thus
$$h_n = \frac{n-1}{n} h_{n-2} + \frac{1}{n^2} \frac{n-3,\, n-5 \dots 1}{n-2,\, n-4 \dots 2} \frac{\pi}{2}, \quad \dots\dots\dots(25)$$

by which the integrals h_n can be calculated in turn. Thus

$$h_2 = \pi/8,$$

$$h_4 = \frac{3}{4} h_2 + \frac{1}{4^2} \cdot \frac{1}{2} \cdot \frac{\pi}{2} = \frac{\pi}{2} \frac{3 \cdot 1}{4 \cdot 2} \left(\frac{1}{1 \cdot 2} + \frac{1}{3 \cdot 4} \right),$$

$$h_6 = \frac{5 \cdot 3 \cdot 1}{6 \cdot 4 \cdot 2} \frac{\pi}{2} \left(\frac{1}{1 \cdot 2} + \frac{1}{3 \cdot 4} \right) + \frac{1}{6^2} \frac{3 \cdot 1}{4 \cdot 2} \frac{\pi}{2}$$

$$= \frac{\pi}{2} \frac{5 \cdot 3 \cdot 1}{6 \cdot 4 \cdot 2} \left(\frac{1}{1 \cdot 2} + \frac{1}{3 \cdot 4} + \frac{1}{5 \cdot 6} \right).$$

Similarly
$$h_8 = \frac{\pi}{2} \frac{7 \cdot 5 \cdot 3 \cdot 1}{8 \cdot 6 \cdot 4 \cdot 2} \left(\frac{1}{1 \cdot 2} + \frac{1}{3 \cdot 4} + \frac{1}{5 \cdot 6} + \frac{1}{7 \cdot 8} \right), \quad \text{and so on.}$$

It may be remarked that the series within brackets, being equal to

$$1 - \tfrac{1}{2} + \tfrac{1}{3} - \tfrac{1}{4} + \dots,$$

approaches ultimately the limit $\log 2$. A tabulation of the earlier members of the series of integrals will be convenient :—

TABLE I.

$$2h_0/\pi = 0$$

$2h_2/\pi$	$= 1/4$	$= 0\cdot25$
$2h_4/\pi$	$= 7/32$	$= 0\cdot21875$
$2h_6/\pi$	$= 37/192$	$= 0\cdot19271$
$2h_8/\pi$	$= 533/3072$	$= 0\cdot17350$
$2h_{10}/\pi$	$= 1627/10240$	$= 0\cdot15889$
$2h_{12}/\pi$	$= 18107/122880$	$= 0\cdot14736$
$2h_{14}/\pi$	$= \dots\dots\dots\dots$	$= 0\cdot13798$
$2h_{16}/\pi$	$= \dots\dots\dots\dots$	$= 0\cdot13018$
$2h_{18}/\pi$	$= \dots\dots\dots\dots$	$= 0\cdot12356$
$2h_{20}/\pi$	$= \dots\dots\dots\dots$	$= 0\cdot11784$

The last four have been calculated in sequence by means of (25).

In (24) we may, of course, replace $\sin \theta$ by $\cos \theta$ throughout. If both $\sin \theta$ and $\cos \theta$ occur, as in

$$\int_0^{\frac{1}{2}\pi} \sin^n \theta \cos^m \theta \log (2 \sin \theta)\, d\theta, \quad \ldots\ldots\ldots\ldots\ldots(26)$$

where n and m are even, we may express $\cos^m \theta$ by means of $\sin \theta$, and so reduce (26) to integrals of the form (24). The particular case where $m = n$ is worthy of notice. Here

$$\int_0^{\frac{1}{2}\pi} \sin^n \theta \cos^n \theta \log (2 \sin \theta)\, d\theta = \int_0^{\frac{1}{2}\pi} \sin^n \theta \cos^n \theta \log (2 \cos \theta)\, d\theta$$

$$= \tfrac{1}{2} \int_0^{\frac{1}{2}\pi} \frac{\sin^n 2\theta}{2^n} \log (2 \sin 2\theta)\, d\theta = \frac{h_n}{2^{n+1}} . \quad \ldots\ldots\ldots\ldots(27)$$

A comparison of the two treatments gives a relation between the integrals h. Thus, if $n = 4$,

$$h_4 - 2h_6 + h_8 = h_4/2^5.$$

We now proceed to the calculation of the left-hand member of (17) with $\Psi = (b^2 - y^2)^{-\frac{1}{2}}$, or, as it may be written,

$$\int_{-b}^{+b} \frac{dy}{\sqrt{(b^2 - y^2)}} \left[\left(\gamma + \log \frac{ikr}{2}\right) J_0(kr) + \frac{k^2 r^2}{2^2} - \frac{k^4 r^4}{2^2 . 4^2} S_2 + \frac{k^6 r^6}{2^2 . 4^2 . 6^2} S_3 - \ldots \right]. (28)$$

The leading term has already been found to be

$$\pi \left(\gamma + \log \frac{ikb}{4}\right) \ldots\ldots\ldots\ldots\ldots\ldots\ldots\ldots(29)$$

In (28) r is equal to $\pm (y - \eta)$. Taking, as before,

$$y = b \cos \theta, \quad \eta = b \cos \alpha,$$

we have

$$\int_0^\pi d\theta \left[\left\{ \gamma + \log \frac{ikb}{4} + \log \pm 2 (\cos \theta - \cos \alpha) \right\} J_0 \{kb (\cos \theta - \cos \alpha)\} \right.$$

$$\left. + \frac{k^2 b^2 (\cos \theta - \cos \alpha)^2}{2^2} - \frac{k^4 b^4 (\cos \theta - \cos \alpha)^4}{2^2 . 4^2} \cdot \frac{3}{2} + \frac{k^6 b^6 (\cos \theta - \cos \alpha)^6}{2^2 . 4^2 . 6^6} \cdot \frac{11}{6} - \ldots \right].$$

$$\ldots\ldots\ldots\ldots(30)$$

As regards the terms which do not involve $\log (\cos \theta - \cos \alpha)$, we have to deal merely with

$$\int_0^\pi (\cos \theta - \cos \alpha)^n\, d\theta, \quad \ldots\ldots\ldots\ldots\ldots\ldots(31)$$

where n is an even integer, which, on expansion of the binomial and integration by a known formula, becomes

$$\pi \left[\frac{n - 1 . n - 3 . n - 5 \ldots 1}{n . n - 2 . n - 4 \ldots 2} + \frac{n . n - 1}{1 . 2} \frac{n - 3 . n - 5 \ldots 1}{n - 2 . n - 4 \ldots 2} \cos^2 \alpha \right.$$

$$\left. + \frac{n . n - 1 . n - 2 . n - 3}{1 . 2 . 3 . 4} \frac{n - 5 . n - 7 \ldots 1}{n - 4 . n - 6 \ldots 2} \cos^4 \alpha + \ldots + \cos^n \alpha \right]. \quad \ldots (32)$$

Thus, if $n = 2$, we get $\pi \left[\frac{1}{2} + \cos^2 \alpha\right]$. If $n = 4$,

$$\pi \left[\frac{3 \cdot 1}{4 \cdot 2} + \frac{4 \cdot 3}{1 \cdot 2} \frac{1}{2} \cos^2 \alpha + \cos^4 \alpha\right], \quad \text{and so on.}$$

The coefficient of (31), or (32), in (30) is

$$(-1)^{\frac{1}{2}n} \frac{k^n b^n}{2^2 \cdot 4^2 \ldots n^2} \left[\gamma + \log \frac{ikb}{4} - S_{\frac{1}{2}n}\right]. \quad \ldots\ldots\ldots\ldots(33)$$

At the centre of the aperture where $\eta = 0$, $\cos \alpha = 0$, (32) reduces to its first term. At the edges where $\cos \alpha = \pm 1$, we may obtain a simpler form directly from (31). Thus

$$(31) = \int_0^\pi (1 \pm \cos \theta)^n \, d\theta = 2^n \pi \frac{2n - 1 \cdot 2n - 3 \ldots 1}{2n \cdot 2n - 2 \ldots 2} = \pi \frac{2n - 1 \cdot 2n - 3 \ldots 1}{n \cdot n - 1 \cdot n - 2 \ldots 1}.$$
$$\ldots\ldots\ldots\ldots(34)$$

For example, if $n = 6$,

$$(34) = \pi \frac{11 \cdot 9 \cdot 7 \cdot 5 \cdot 3 \cdot 1}{6 \cdot 5 \cdot 4 \cdot 3 \cdot 2 \cdot 1} = \frac{231\pi}{16}.$$

We have also in (30) to consider (n even)

$$2^{-n} \int_0^\pi d\theta \, (\cos \theta - \cos \alpha)^n \log \{\pm 2 (\cos \theta - \cos \alpha)\}$$

$$= \int_0^\alpha d\theta \sin^n \frac{\theta + \alpha}{2} \sin^n \frac{\theta - \alpha}{2} \log \left\{4 \sin \frac{\theta + \alpha}{2} \sin \frac{\alpha - \theta}{2}\right\}$$

$$+ \int_\alpha^\pi d\theta \sin^n \frac{\theta + \alpha}{2} \sin^n \frac{\theta - \alpha}{2} \log \left\{4 \sin \frac{\theta + \alpha}{2} \sin \frac{\theta - \alpha}{2}\right\}$$

$$= \int_0^\pi d\theta \sin^n \frac{\theta + \alpha}{2} \sin^n \frac{\theta - \alpha}{2} \log \left\{2 \sin \frac{\theta + \alpha}{2}\right\}$$

$$+ \int_0^\alpha d\theta \sin^n \frac{\theta + \alpha}{2} \sin^n \frac{\theta - \alpha}{2} \log \left\{2 \sin \frac{\alpha - \theta}{2}\right\}$$

$$+ \int_\alpha^\pi d\theta \sin^n \frac{\theta + \alpha}{2} \sin^n \frac{\theta - \alpha}{2} \log \left\{2 \sin \frac{\theta - \alpha}{2}\right\}$$

$$= 2 \int_0^{\frac{1}{2}\pi + \frac{1}{2}\alpha} d\phi \sin^n \phi \sin^n (\phi - \alpha) \log (2 \sin \phi)$$

$$+ 2 \int_0^{\frac{1}{2}\pi - \frac{1}{2}\alpha} d\phi \sin^n \phi \sin^n (\phi + \alpha) \log (2 \sin \phi)$$

$$= 2 \int_0^{\frac{1}{2}\pi} d\phi \sin^n \phi \left\{\sin^n (\phi - \alpha) + \sin^n (\phi + \alpha)\right\} \log (2 \sin \phi)$$

$$+ 2 \int_{\frac{1}{2}\pi}^{\frac{1}{2}\pi + \frac{1}{2}\alpha} d\phi \sin^n \phi \sin^n (\phi - \alpha) \log (2 \sin \phi)$$

$$- 2 \int_{\frac{1}{2}\pi - \frac{1}{2}\alpha}^{\frac{1}{2}\pi} d\phi \sin^n \phi \sin^n (\phi + \alpha) \log (2 \sin \phi)$$

$$= 2 \int_0^{\frac{1}{2}\pi} d\phi \sin^n \phi \left\{\sin^n (\phi - \alpha) + \sin^n (\phi + \alpha)\right\} \log (2 \sin \phi), \ldots\ldots(35)$$

since the last two integrals cancel, as appears when we write $\pi - \psi$ for ϕ, n being even.

In (35)

$$\tfrac{1}{2}\sin^n(\phi + \alpha) + \tfrac{1}{2}\sin^n(\phi - \alpha) = \sin^n \phi \, \cos^n \alpha$$

$$+ \frac{n \cdot n - 1}{1 \cdot 2} \sin^{n-2} \phi \, \cos^2 \phi \, \sin^2 \alpha \, \cos^{n-2} \alpha$$

$$+ \frac{n \cdot n - 1 \cdot n - 2 \cdot n - 3}{1 \cdot 2 \cdot 3 \cdot 4} \sin^{n-4} \phi \, \cos^4 \phi \, \sin^4 \alpha \, \cos^{n-4} \alpha + \ldots + \cos^n \phi \, \sin^n \alpha, \quad (36)$$

and thus the result may be expressed by means of the integrals h. Thus if $n = 2$,

$$(35) = 4 \int_0^{\frac{1}{2}\pi} d\phi \, \sin^2 \phi \, \{\sin^2 \phi \, \cos^2 \alpha + \cos^2 \phi \, \sin^2 \alpha\} \log(2 \sin \phi)$$

$$= 4 \{(\cos^2 \alpha - \sin^2 \alpha) h_4 + \sin^2 \alpha \, h_2\}. \quad \ldots\ldots\ldots\ldots\ldots\ldots\ldots(37)$$

If $n = 4$,

$$(35) = 4 \int_0^{\frac{1}{2}\pi} d\phi \, \sin^4 \phi \, \{\sin^4 \phi \, \cos^4 \alpha + 6 \sin^2 \phi \, \cos^2 \phi \, \sin^2 \alpha \, \cos^2 \alpha$$

$$+ \cos^4 \phi \, \sin^4 \alpha\} \log(2 \sin \phi)$$

$$= 4 \{(\cos^4 \alpha - 6 \sin^2 \alpha \cos^2 \alpha + \sin^4 \alpha) h_8$$

$$+ (6 \sin^2 \alpha \cos^2 \alpha - 2 \sin^4 \alpha) h_6 + \sin^4 \alpha \, h_4\}. \quad \ldots\ldots\ldots(38)$$

If $n = 6$,

$$(35) = 4 \{(\cos^6 \alpha - 15 \cos^4 \alpha \sin^2 \alpha + 15 \cos^2 \alpha \sin^4 \alpha - \sin^6 \alpha) h_{12}$$

$$+ (15 \cos^4 \alpha \sin^2 \alpha - 30 \cos^2 \alpha \sin^4 \alpha + 3 \sin^6 \alpha) h_{10}$$

$$+ (15 \cos^2 \alpha \sin^4 \alpha - 3 \sin^6 \alpha) h_8 + \sin^6 \alpha \, h_6\}. \quad \ldots\ldots\ldots\ldots\ldots(39)$$

It is worthy of remark that if we neglect the small differences between the h's in (39), it reduces to $4 \cos^6 \alpha \, h_{12}$, and similarly in other cases.

When n is much higher than 6, the general expressions corresponding to (37), (38), (39) become complicated. If, however, $\cos \alpha$ be either 0, or ± 1, (36) reduces to a single term, viz., $\cos^n \phi$ or $\sin^n \phi$. Thus at the centre ($\cos \alpha = 0$) from either of its forms

$$(35) = 2^{-n} \cdot 2h_n. \quad \ldots\ldots\ldots\ldots\ldots\ldots\ldots\ldots\ldots(40)$$

On the other hand, at the edges ($\cos \alpha = \pm 1$)

$$(35) = 4 \int_0^{\frac{1}{2}\pi} d\phi \, \sin^{2n} \phi \, \log(2 \sin \phi) = 4h_{2n}. \quad \ldots\ldots\ldots\ldots(41)$$

In (30), the object of our quest, the integral (35) occurs with the coefficient

$$(-1)^{\frac{1}{2}n} \frac{2^n k^n b^n}{2^2 \cdot 4^2 \cdot 6^2 \ldots n^2}. \quad \ldots\ldots\ldots\ldots\ldots\ldots\ldots(42)$$

Thus, expanded in powers of kb, (28) or (30) becomes

$$\pi\left(\gamma + \log\frac{ikb}{4}\right) - \frac{\pi k^2 b^2}{2^2}\left[\left\{\gamma + \log\frac{ikb}{4} - 1\right\}\{\tfrac{1}{2} + \cos^2\alpha\}\right.$$

$$+ \frac{2^3 \cdot 2h_4}{\pi}(\cos^2\alpha - \sin^2\alpha) + \frac{2^3 \cdot 2h_2}{\pi}\sin^2\alpha\bigg]$$

$$+ \frac{\pi k^4 b^4}{2^2 \cdot 4^2}\left[\left\{\gamma + \log\frac{ikb}{4} - \frac{3}{2}\right\}\left\{\frac{3}{8} + 3\cos^2\alpha + \cos^4\alpha\right\}\right.$$

$$+ \frac{2^5 \cdot 2h_8}{\pi}(\cos^4\alpha - 6\cos^2\alpha\,\sin^2\alpha + \sin^4\alpha)$$

$$+ \frac{2^5 \cdot 2h_6}{\pi}(6\cos^2\alpha\,\sin^2\alpha - 2\sin^4\alpha) + \frac{2^5 \cdot 2h_4}{\pi}\sin^4\alpha\bigg]$$

$$- \frac{\pi k^6 b^6}{2^2 \cdot 4^2 \cdot 6^2}\left[\left\{\gamma + \log\frac{ikb}{4} - \frac{11}{6}\right\}\left\{\frac{5}{16} + \frac{45}{8}\cos^2\alpha + \frac{15}{2}\cos^4\alpha + \cos^6\alpha\right\}\right.$$

$$+ \frac{2^7 \cdot 2h_{12}}{\pi}(\cos^6\alpha - 15\cos^4\alpha\,\sin^2\alpha + 15\cos^2\alpha\,\sin^4\alpha - \sin^6\alpha)$$

$$+ \frac{2^7 \cdot 2h_{10}}{\pi}(15\cos^4\alpha\,\sin^2\alpha - 30\cos^2\alpha\,\sin^4\alpha + 3\sin^6\alpha)$$

$$+ \frac{2^7 \cdot 2h_8}{\pi}(15\cos^2\alpha\,\sin^4\alpha - 3\sin^6\alpha) + \frac{2^7 \cdot 2h_6}{\pi}\sin^6\alpha\bigg] + \dots \quad \dots\dots(43)$$

At the centre of the aperture ($\cos\alpha = 0$), in virtue of (40), a simpler form is available. We have

$$\pi\left(\gamma + \log\frac{ikb}{4}\right) - \frac{\pi k^2 b^2}{2^2}\left[\frac{1}{2}\left(\gamma + \log\frac{ikb}{4} - 1\right) + \frac{2h_2}{\pi}\right]$$

$$+ \frac{\pi k^4 b^4}{2^2 \cdot 4^2}\left[\frac{3 \cdot 1}{4 \cdot 2}\left(\gamma + \log\frac{ikb}{4} - \frac{3}{2}\right) + \frac{2h_4}{\pi}\right]$$

$$- \frac{\pi k^6 b^6}{2^2 \cdot 4^2 \cdot 6^2}\left[\frac{5 \cdot 3 \cdot 1}{6 \cdot 4 \cdot 2}\left(\gamma + \log\frac{ikb}{4} - \frac{11}{6}\right) + \frac{2h_6}{\pi}\right]$$

$$+ \frac{\pi k^8 b^8}{2^2 \cdot 4^2 \cdot 6^2 \cdot 8^2}\left[\frac{7 \cdot 5 \cdot 3 \cdot 1}{8 \cdot 6 \cdot 4 \cdot 2}\left(\gamma + \log\frac{ikb}{4} - \frac{25}{12}\right) + \frac{2h_8}{\pi}\right] - \dots \quad (44)$$

Similarly at the edges, by (34), (41), we have

$$\pi\left(\gamma + \log\frac{ikb}{4}\right) - \frac{\pi k^2 b^2}{2^2}\left[\frac{3 \cdot 1}{2 \cdot 1}\left(\gamma + \log\frac{ikb}{4} - 1\right) + 2^3\frac{2h_4}{\pi}\right]$$

$$+ \frac{\pi k^4 b^4}{2^2 \cdot 4^2}\left[\frac{7 \cdot 5 \cdot 3 \cdot 1}{4 \cdot 3 \cdot 2 \cdot 1}\left(\gamma + \log\frac{ikb}{4} - \frac{3}{2}\right) + 2^5\frac{2h_8}{\pi}\right]$$

$$- \frac{\pi k^6 b^6}{2^2 \cdot 4^2 \cdot 6^2}\left[\frac{11 \cdot 9 \cdot 7 \cdot 5 \cdot 3 \cdot 1}{6 \cdot 5 \cdot 4 \cdot 3 \cdot 2 \cdot 1}\left(\gamma + \log\frac{ikb}{4} - \frac{11}{6}\right) + 2^7\frac{2h_{12}}{\pi}\right] + \dots \quad (45)$$

For the general value of α, (43) is perhaps best expressed in terms of $\cos\alpha$, equal to η/b. With introduction of the values of h, we have

$$\pi\left(\gamma+\log\frac{ikb}{4}\right)-\frac{\pi k^2 b^2}{2^2}\left[\left(\gamma+\log\frac{ikb}{4}\right)\left(\cos^2\alpha+\frac{1}{2}\right)+\frac{1}{2}\cos^2\alpha-\frac{1}{4}\right]$$

$$+\frac{\pi k^4 b^4}{2^2.4^2}\left[\left(\gamma+\log\frac{ikb}{4}\right)\left(\cos^4\alpha+3\cos^2\alpha+\frac{3}{8}\right)+\frac{7}{12}\cos^4\alpha-\frac{5}{4}\cos^2\alpha-\frac{11}{32}\right]$$

$$-\frac{\pi k^6 b^6}{2^2.4^2.6^2}\left[\left(\gamma+\log\frac{ikb}{4}\right)\left(\cos^6\alpha+\frac{15}{2}\cos^4\alpha+\frac{45}{8}\cos^2\alpha+\frac{5}{16}\right)\right.$$

$$\left.+\frac{37}{60}\cos^6\alpha-\frac{23}{8}\cos^4\alpha-\frac{159}{32}\cos^2\alpha-\frac{73}{192}\right]+\dots\quad\dots\dots\dots\dots\dots(46)$$

These expressions are the values of

$$\int_{-b}^{+b}\frac{D(kr)\,dy}{\sqrt{(b^2-y^2)}},\quad\dots\dots\dots\dots\dots\dots\dots(47)$$

for the various values of η.

We now suppose that $kb=1$. The values for other particular cases, such as $kb=\frac{1}{2}$, may then easily be deduced. For $\cos\alpha=0$, from (44) we have

$$\pi\left(\gamma+\log\frac{i}{4}\right)\left[1-\frac{1}{2^2}\frac{1}{2}+\frac{1}{2^2.4^2}\frac{3.1}{4.2}-\frac{1}{2^2.4^2.6^2}\frac{5.3.1}{6.4.2}+\dots\right]$$

$$+\pi\left[\frac{1}{2^2}\frac{1}{4}-\frac{1}{2^2.4^2}\frac{11}{32}+\frac{1}{2^2.4^2.6^2}\frac{73}{192}-\dots\right]$$

$$=\pi\left(\gamma+\log\frac{i}{4}\right)[1-0.12500+0.00586+0.00013]$$

$$+\pi[0.06250-0.00537+0.00016]$$

$$=\pi\left(\gamma+\log\frac{i}{4}\right)\times0.88073+\pi\times0.05729$$

$$=\pi[-0.65528+1.3834\,i],\quad\dots\dots\dots\dots\dots\dots\dots(48)$$

since $\gamma=0.577215,\qquad\log 2=0.693147,\qquad\log i=\frac{1}{2}\pi i.$

In like manner, if $kb=\frac{1}{2}$, we get still with $\cos\alpha=0$,

$$\pi\left(\gamma+\log\frac{i}{8}\right)[1-0.03125+0.00037]+\pi[0.01562-0.00033]$$

$$=\pi[-1.4405+1.5223\,i]\quad\dots\dots\dots\dots(49)$$

If $kb=2$, we have

$$\pi\left(\gamma+\log\frac{i}{2}\right)[1-0.5+0.0938-0.0087+0.0005]$$

$$+\pi[0.25-0.0859+0.0102-0.0006]$$

$$=\pi[+0.1058+0.9199\,i].\quad\dots\dots\dots\dots\dots(50)$$

If $kb = 1$ and $\cos\alpha = \pm 1$, we have from (45)

$$\pi\left(\gamma + \log\frac{i}{4}\right)\left[1 - \frac{1}{2^2}\frac{3}{2} + \frac{1}{2^2.4^2}\frac{35}{8} - \frac{1}{2^2.4^2.6^2}\frac{231}{16}\right.$$

$$\left. + \frac{1}{2^2.4^2.6^2.8^2}\frac{6435}{128} - \frac{1}{2^2.4^2.6^2.8^2.10^2}\frac{19.17.6435}{10.9.128} + \cdots\right]$$

$$- \pi\left[\frac{1}{2^2}\frac{1}{4} + \frac{1}{2^2.4^2}\frac{97}{96} - \frac{1}{2^2.4^2.6^2}\frac{7303}{960} + \frac{38.084}{2^2.4^2.6^2.8^2} - \frac{170.64}{2^2.4^2.6^2.8^2.10^2} + \cdots\right]$$

$$= \pi\left(\gamma + \log\frac{i}{4}\right)[1 - 0.375 + 0.068359 - 0.006266 + 0.000341 - 0.000012]$$

$$- \pi[0.0625 + 0.015788 - 0.003302 + 0.000258 + 0.000012]$$

$$= \pi[-0.63141 + 1.0798\,i]. \quad\text{..}(51)$$

Similarly, if $kb = \frac{1}{2}$, we have

$$\pi\left(\gamma + \log\frac{i}{8}\right)[1 - 0.09375 + 0.00427 - 0.00010]$$

$$- \pi[0.01562 + 0.00099 - 0.00005]$$

$$= \pi[-1.3842 + 1.4301\,i]. \quad\text{...}(52)$$

And if $kb = 2$, with diminished accuracy,

$$\pi\left(\gamma + \log\frac{i}{2}\right)[1 - 1.5 + 1.094 - 0.401 + 0.087 - 0.012 + 0.001]$$

$$- \pi[0.25 + 0.253 - 0.211 + 0.066 - 0.012 + 0.001]$$

$$= \pi[-0.378 + 0.422\,i]. \quad\text{...}(53)$$

As an intermediate value of α we will select $\cos^2\alpha = \frac{1}{2}$. For $kb = 1$, from (46)

$$\pi\left(\gamma + \log\frac{i}{4}\right)[1 - 0.25 + 0.03320 - 0.00222 + \cdots]$$

$$+ \pi[0 - 0.01286 + 0.001522 + \cdots]$$

$$= \pi[-0.6432 + 1.2268\,i]. \quad\text{...}(54)$$

Also, when $kb = \frac{1}{2}$,

$$\pi[-1.4123 + 1.4759\,i]. \quad\text{.........................}(55)$$

When $kb = 2$, only a rough value is afforded by (46), viz.,

$$\pi[-0.16 + 0.61\,i]. \quad\text{...........................}(56)$$

The accompanying table exhibits the various numerical results, the factor π being omitted.

TABLE II.

	$kb = \frac{1}{2}$	$kb = 1$	$kb = 2$
$\cos a = 0$	$-1.4405 + 1.5223\,i$	$-0.65528 + 1.3834\,i$	$+0.1058 + 0.9199\,i$
$\cos^2 a = \frac{1}{2}$	$-1.4123 + 1.4759\,i$	$-0.6432 + 1.2268\,i$	$-0.16 \quad +0.61\,i$
$\cos^2 a = 1$	$-1.3842 + 1.4301\,i$	$-0.63141 + 1.0798\,i$	$-0.378 \quad +0.422\,i$

As we have seen already, the tabulated quantity when kb is very small takes the form $\gamma + \log (ikb/4)$, or $\log kb - 0\cdot8091 + 1\cdot5708i$, whatever may be the value of α. In this case the condition (17) can be completely satisfied with $\Psi = A (b^2 - y^2)^{-\frac{1}{2}}$, A being chosen suitably. When kb is finite, (17) can no longer be satisfied for all values of α. But when $kb = \frac{1}{2}$, or even when $kb = 1$, the tabulated number does not vary greatly with α and we may consider (17) to be approximately satisfied if we make in the first case

$$\pi(-1\cdot4123 + 1\cdot4759\,i)\,A = -1, \quad\dots\dots\dots\dots\dots(57)$$

and in the second,

$$\pi(-0\cdot6432 + 1\cdot2268\,i)\,A = -1. \quad\dots\dots\dots\dots(58)$$

The value of ψ, applicable to a point at a distance directly in front of the aperture, is then, as in (16),

$$\psi = -\pi A \left(\frac{\pi}{2\,ikr}\right)^{\frac{1}{2}} e^{-ikr}. \quad\dots\dots\dots\dots\dots\dots(59)$$

In order to obtain a better approximation we require the aid of a second solution with a different form of Ψ. When this is introduced, as an addition to the first solution and again with an arbitrary constant multiplier, it will enable us to satisfy (17) for two distinct values of α, that is of η, and thus with tolerable accuracy over the whole range from $\cos\alpha = 0$ to $\cos\alpha = \pm 1$. Theoretically, of course, the process could be carried further so as to satisfy (17) for any number of assigned values of $\cos\alpha$.

As the second solution we will take simply $\Psi = 1$, so that the left-hand member of (17) is

$$\int_0^{b+\eta} D\,(kr)\,dr + \int_0^{b-\eta} D\,(kr)\,dr. \quad\dots\dots\dots\dots\dots(60)$$

If we omit k, which may always be restored by consideration of homogeneity, we have

$$(60) = \left(\gamma + \log\frac{i}{2}\right)\left[b + \eta - \frac{(b+\eta)^3}{2^2\cdot3} + \frac{(b+\eta)^5}{2^2\cdot4^2\cdot5} - \cdots\right]$$

$$+ \frac{(b+\eta)^3}{2^2\cdot3} - \frac{(b+\eta)^5}{2^2\cdot4^2\cdot5}\,S_2 + \frac{(b+\eta)^7}{2^2\cdot4^2\cdot6^2\cdot7}\,S_3 - \cdots$$

$$+ (b+\eta)\,\{\log(b+\eta) - 1\} - \frac{(b+\eta)^3}{2^2\cdot3}\left\{\log(b+\eta) - \frac{1}{3}\right\}$$

$$+ \frac{(b+\eta)^5}{2^2\cdot4^2\cdot5}\left\{\log(b+\eta) - \frac{1}{5}\right\} - \cdots$$

+ the same expression with the sign of η changed.

The leading term in (60) is thus

$$2b\,(\gamma - 1 + \log\tfrac{1}{2}i) + (b+\eta)\log(b+\eta) + (b-\eta)\log(b-\eta). \dots(61)$$

At the centre of the aperture ($\eta = 0$),

$$(61) = 2b \left\{ \gamma - 1 + \log \tfrac{1}{2} ib \right\},$$

and at the edges ($\eta = \pm b$),

$$(61) = 2b \left\{ \gamma - 1 + \log ib \right\}.$$

It may be remarked that in (61), the real part varies with η, although the imaginary part is independent of that variable.

The complete expression (60) naturally assumes specially simple forms at the centre and edges of the aperture. Thus, when $\eta = 0$,

$$(60) \div 2b = \left(\gamma + \log \frac{ib}{2} \right) \left[1 - \frac{b^2}{2^2 \cdot 3} + \frac{b^4}{2^2 \cdot 4^2 \cdot 5} - \cdots \right]$$

$$- 1 + \frac{b^2}{2^2 \cdot 3} \left(1 + \frac{1}{3} \right) - \frac{b^4}{2^2 \cdot 4^2 \cdot 5} \left(1 + \frac{1}{2} + \frac{1}{5} \right) + \frac{b^6}{2^2 \cdot 4^2 \cdot 6^2 \cdot 7} \left(1 + \frac{1}{2} + \frac{1}{3} + \frac{1}{7} \right) - \cdots ;$$

$$\cdots \cdots (62)$$

and, similarly, when $\eta = \pm b$,

$$(60) \div 2b = (\gamma + \log ib) \left[1 - \frac{(2b)^2}{2^2 \cdot 3} + \frac{(2b)^4}{2^2 \cdot 4^2 \cdot 5} - \cdots \right]$$

$$- 1 + \frac{(2b)^2}{2^2 \cdot 3} \left(1 + \frac{1}{3} \right) - \frac{(2b)^4}{2^2 \cdot 4^2 \cdot 5} \left(1 + \frac{1}{2} + \frac{1}{5} \right) + \frac{(2b)^6}{2^2 \cdot 4^2 \cdot 6^2 \cdot 7} \left(1 + \frac{1}{2} + \frac{1}{3} + \frac{1}{7} \right) - \cdots .$$

$$\cdots \cdots (63)$$

To restore k we have merely to write kb for b in the *right-hand members* of (62), (63).

The calculation is straightforward. For the same values as before of kb and of $\cos^2 \alpha$, equal to η^2 / b^2, we get for $(60) \div 2b$

TABLE III.

η^2/b^2	$kb = \tfrac{1}{2}$	$kb = 1$	$kb = 2$
0	$-1 \cdot 7649 + 1 \cdot 5384\,i$	$-1 \cdot 0007 + 1 \cdot 4447\,i$	$-0 \cdot 2167 + 1 \cdot 1198\,i$
$\tfrac{1}{2}$	$-1 \cdot 4510 + 1 \cdot 4912\,i$	$-0 \cdot 6740 + 1 \cdot 2771\,i$	$-0 \cdot 1079 + 0 \cdot 7166\,i$
1	$-1 \cdot 0007 + 1 \cdot 4447\,i$	$-0 \cdot 2217 + 1 \cdot 1198\,i$	$+0 \cdot 1394 + 0 \cdot 4024\,i$

We now proceed to combine the two solutions, so as to secure a better satisfaction of (17) over the width of the aperture. For this purpose we determine A and B in

$$\Psi = A (b^2 - y^2)^{-\frac{1}{2}} + B, \quad \cdots \cdots \cdots (64)$$

so that (17) may be exactly satisfied at the centre and edges ($\eta = 0$, $\eta = \pm b$). The departure from (17) when $\eta^2/b^2 = \frac{1}{2}$ can then be found. If for any value of kb and $\eta = 0$ the first tabular (complex) number is p and the second q, and for $\eta = \pm b$ the first is r and the second s, the equations of condition from (17) are

$$\pi A \cdot p + 2bB \cdot q = -1, \qquad \pi A \cdot r + 2bB \cdot s = -1. \quad \cdots \cdots (65)$$

When A and B are found, we have in (16)

$$\int_{-b}^{+b} \Psi \, dy = \pi A + 2bB.$$

From (65) we get

$$\pi A = \frac{q - s}{ps - qr}, \qquad 2bB = \frac{r - p}{ps - qr}, \quad \dots\dots\dots\dots(66)$$

so that

$$\int_{-b}^{+b} \Psi \, dy = \frac{q + r - s - p}{ps - qr}. \quad \dots\dots\dots\dots\dots(67)$$

Thus for $kb = 1$ we have

$$p = -0.65528 + 1.3834 \, i, \qquad q = -1.0007 + 1.4447 \, i,$$

$$r = -0.63141 + 1.0798 \, i, \qquad s = -0.2217 + 1.1198 \, i,$$

whence

$$\pi A = +0.60008 + 0.51828 \, i, \qquad 2bB = -0.2652 + 0.1073 \, i,$$

and

$$(67) = +0.3349 + 0.6256 \, i.$$

The above values of πA and $2bB$ are derived according to (17) from the values at the centre and edges of the aperture. The success of the method may be judged by substitution of the values for $\eta^2/b^2 = \frac{1}{4}$. Using these in (17) we get $-0.9801 - 0.0082 \, i$, for what should be -1, a very fair approximation.

In like manner, for $kb = 2$

$$(67) = +0.259 + 1.2415 \, i \,;$$

and for $kb = \frac{1}{2}$

$$(67) = +0.3378 + 0.3526 \, i.$$

As appears from (16), when k is given, the modulus of (67) may be taken to represent the amplitude of disturbance at a distant point immediately in front, and it is this with which we are mainly concerned. The following table gives the values of Mod. and Mod.^{2} for several values of kb. The first three have been calculated from the simple formula, see (20).

TABLE IV.

kb	Mod.^{2}	Mod.
0·01	0·0174	0·1320
0·05	0·0590	0·2429
0·25	0·1372	0·3704
0·50	0·2384	0·4883
1·00	0·5035	0·7096
2·00	1·608	1·268

The results are applicable to the problem of aerial waves, or shallow water waves, transmitted through a slit in a thin fixed wall, and to electric

(luminous) waves transmitted by a similar slit in a thin perfectly opaque
screen, provided that the electric vector is *perpendicular* to the length of
the slit.

In curve *A*, fig. 1, the value of the modulus from the third column of
Table IV is plotted against *kb*.

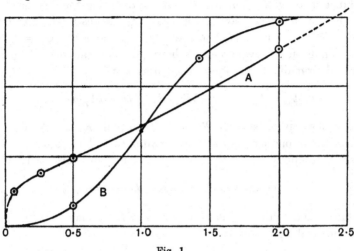

Fig. 1.

When *kb* is large, the limiting form of (67) may be deduced from a
formula, analogous to (12), connecting Ψ and $d\phi/dn$. As in (11),

$$\frac{d\psi}{dx} = \int \Psi \frac{dD}{dx}\, dy,$$

in which, when x is very small, we may take $D = \log r$. Thus

$$\frac{d\psi}{dx} = \Psi \int_{-\infty}^{+\infty} \frac{x\,dy}{x^2 + y^2} = \Psi \left[\tan^{-1}\frac{y}{x} \right]_{-\infty}^{+\infty} = \pi\Psi, \quad \text{or} \quad \Psi = \frac{1}{\pi}\frac{d\psi}{dn}. \quad \text{....(68)}$$

Now, when *kb* is large, $d\psi/dn$ tends, except close to the edges, to assume
the value ik, and ultimately

$$(67) = \int_{-b}^{+b} \Psi\, dy = \frac{2\,ikb}{\pi}, \qquad \text{.......................(69)}$$

of which the modulus is $2kb/\pi$ simply, *i.e.* $0\cdot637\,kb$.

We now pass on to consider case (ii), where the boundary condition to be
satisfied over the wall is $\phi = 0$. Separating from ϕ the solution (χ) which
would obtain were the wall unperforated, we have

$$\chi_m = e^{-ikx} - e^{ikx}, \qquad \chi_p = 0, \quad \text{.....................(70)}$$

giving over the whole plane ($x = 0$),

$$\chi_m = 0, \qquad \chi_p = 0, \qquad d\chi_m/dx = -2ik, \qquad d\chi_p/dx = 0.$$

The supplementary solutions ψ, equal to $\phi - \chi$, may be written

$$\psi_m = \int \frac{dD}{dx} \Psi_m dy, \qquad \psi_p = \int \frac{dD}{dx} \Psi_p dy, \quad \ldots\ldots\ldots\ldots(71)$$

where Ψ_m, Ψ_p are functions of y, and the integrations are over the aperture. D as a function of r is given by (13), and r, denoting the distance between dy and the point (x, η), at which ψ_m, ψ_p are estimated, is equal to $\sqrt{\{x^2 + (y - \eta)^2\}}$. The form (71) secures that on the *walls* $\psi_m = \psi_p = 0$, so that the condition of evanescence there, already satisfied by χ, is not disturbed. It remains to satisfy over the *aperture*

$$\psi_m = \psi_p, \qquad -2ik + d\psi_m/dx = d\psi_p/dx. \quad \ldots\ldots\ldots\ldots(72)$$

The first of these is satisfied if $\Psi_m = -\Psi_p$, so that ψ_m and ψ_p are equal at any pair of corresponding points on the two sides. The values of $d\psi_m/dx$, $d\psi_p/dx$ are then opposite, and the remaining condition is also satisfied if

$$d\psi_m/dx = ik, \qquad d\psi_p/dx = -ik. \quad \ldots\ldots\ldots\ldots(73)$$

At a distance, and if the slit is very narrow, dD/dx may be removed from under the integral sign, so that

$$\psi_p = \frac{dD}{dx} \int_{-b}^{+b} \Psi_p dy, \quad \ldots\ldots\ldots\ldots\ldots\ldots(74)$$

in which

$$\frac{dD}{dx} = \frac{ikx}{r} \left(\frac{\pi}{2\,ikr}\right)^{\frac{1}{2}} e^{-ikr}. \quad \ldots\ldots\ldots\ldots\ldots\ldots(75)$$

And even if kb be not small, (74) remains applicable if the distant point be directly in front of the slit, so that $x = r$. For such a point

$$\psi_p = ik \left(\frac{\pi}{2\,ikr}\right)^{\frac{1}{2}} e^{-ikr} \int_{-b}^{+b} \Psi_p dy. \quad \ldots\ldots\ldots\ldots(76)$$

There is a simple relation, analogous to (68), between the value of Ψ_p at any point (η) of the aperture and that of ψ_p at the same point. For in the application of (71) only those elements of the integral contribute which lie infinitely near the point where ψ_p is to be estimated, and for these $dD/dx = x/r^2$. The evaluation is effected by considering in the first instance a point for which x is finite and afterwards passing to the limit. Thus

$$\psi_p = \Psi_p \int \frac{x\,dy}{x^2 + (y - \eta)^2} = \pi \Psi_p. \quad \ldots\ldots\ldots\ldots(77)$$

It remains to find, if possible, a form for Ψ_p, or ψ_p, which shall make $d\psi_p/dx$ constant over the aperture, as required by (73). In my former paper, dealing with the case where kb is very small, it was shown that known theorems relating to the flow of incompressible fluids lead to the desired conclusion. It appeared that (74), (75) give

$$\psi_p = -\frac{k^2 b^2 x}{2r}\left(\frac{\pi}{2\,ikr}\right)^{\frac{1}{2}} e^{-ikr}, \quad\ldots\ldots\ldots\ldots\ldots(78)$$

showing that when b is small the transmission falls off greatly, much more than in case (i), see (20). The realised solution from (78) is

$$\psi_p = -\frac{k^2 b^2 x}{2r}\left(\frac{\pi}{2\,kr}\right)^{\frac{1}{2}} \cos{(nt - kr - \tfrac{1}{4}\pi)}, \quad\ldots\ldots\ldots(79)$$

corresponding to $\qquad\qquad \chi_m = 2\sin nt\sin kx. \quad\ldots\ldots\ldots\ldots\ldots(80)$

The former method arrived at a result by assuming certain hydrodynamical theorems. For the present purpose we have to go further, and it will be appropriate actually to verify the constancy of $d\psi/dx$ over the aperture as resulting from the assumed form of Ψ, when kb is small. In this case we may take $D = \log r$, where $r^2 = x^2 + (y - \eta)^2$. From (71), the suffix p being omitted,

$$\frac{d\psi}{dx} = \int_{-b}^{+b}\frac{d^2 D}{dx^2}\,\Psi\,dy\,;$$

and herein $\qquad \dfrac{d^2 D}{dx^2} = -\dfrac{d^2 D}{d\eta^2} = -\dfrac{d^2 D}{dy^2}\ (\eta\ \text{const.}).$

Thus, on integration by parts,

$$\frac{d\psi}{dx} = -\left[\Psi\frac{dD}{dy}\right] + \int_{-b}^{+b}\frac{dD}{dy}\frac{d\Psi}{dy}\,dy. \quad\ldots\ldots\ldots(81)$$

In (81) $\qquad \dfrac{dD}{dy} = \dfrac{dD}{dr}\dfrac{dr}{dy} = \dfrac{y - \eta}{(y - \eta)^2 + x^2},$

and so long as η is not equal to $\pm b$, it does not become infinite at the limits $(y = \pm b)$, even though $x = 0$. Thus, if Ψ vanish at the limits, the integrated terms in (81) disappear. We now assume for trial

$$\Psi = \sqrt{(b^2 - y^2)}, \quad\ldots\ldots\ldots\ldots\ldots(82)$$

which satisfies the last-mentioned condition. Writing

$$y = b\cos\theta, \qquad \eta = b\cos\alpha, \qquad x' = x/b,$$

we have $\qquad -\dfrac{d\psi}{dx} = \displaystyle\int_0^{\pi}\dfrac{(\cos\theta - \cos\alpha)^2 + \cos\alpha\,(\cos\theta - \cos\alpha)}{(\cos\theta - \cos\alpha)^2 + x'^2}\,d\theta. \quad\ldots\ldots(83)$

Of the two parts of the integral on the right in (83) the first yields π when $x' = 0$. For the second we have to consider

$$\int_0^{\pi}\frac{\cos\theta - \cos\alpha}{(\cos\theta - \cos\alpha)^2 + x'^2}\,d\theta, \quad\ldots\ldots\ldots\ldots(84)$$

in which $\cos\theta - \cos\alpha$ passes through zero within the range of integration. It will be shown that (84) vanishes ultimately when $x' = 0$. To this end the range of integration is divided into three parts: from 0 to α_1, where $\alpha_1 < \alpha$, from α_1 to α_2, where $\alpha_2 > \alpha$, and lastly from α_2 to π. In evaluating the first and third parts we may put $x' = 0$ at once. And if $z = \tan\frac{1}{2}\theta$

$$\int \frac{d\theta}{\cos\theta - \cos\alpha} = \frac{1}{\sin\alpha} \int \left\{ \frac{dz}{\tan\frac{1}{2}\alpha + z} + \frac{dz}{\tan\frac{1}{2}\alpha - z} \right\}.$$

Sin α being omitted, the first and third parts together are thus

$$\log\frac{z+t}{z-t} + \log\frac{t+t_1}{t-t_1} + \log\frac{t_2-t}{t_2+t},$$

where $t = \tan\frac{1}{2}\alpha$, $t_1 = \tan\frac{1}{2}\alpha_1$, $t_2 = \tan\frac{1}{2}\alpha_2$, and z is to be made infinite.

It appears that the two parts taken together vanish, provided t_1, t_2 are so chosen that $t^2 = t_1 t_2$.

It remains to consider the second part, viz.,

$$\int_{\alpha_1}^{\alpha_2} \frac{d\theta\,(\cos\theta - \cos\alpha)}{(\cos\theta - \cos\alpha)^2 + x'^2}, \quad\dotsfill(85)$$

in which we may suppose the range of integration $\alpha_2 - \alpha_1$ to be very small. Thus

$$(85) = \int_{\alpha_1}^{\alpha_2} \frac{d\theta\,.\,2\sin\frac{1}{2}(\theta+\alpha)\sin\frac{1}{2}(\alpha-\theta)}{4\sin^2\frac{1}{2}(\theta+\alpha)\sin^2\frac{1}{2}(\alpha-\theta) + x'^2}$$

$$= -\frac{1}{2\sin\alpha}\log\frac{\sin^2\alpha\,(\alpha_2-\alpha)^2 + x'^2}{\sin^2\alpha\,(\alpha-\alpha_1)^2 + x'^2},$$

and this also vanishes if $\alpha_2 - \alpha = \alpha - \alpha_1$, a condition consistent with the former to the required approximation. We infer that in (83)

$$-\frac{d\psi}{dx} = \pi, \quad\dotsfill(86)$$

so that, with the aid of a suitable multiplier, (73) can be satisfied. Thus if $\Psi = A\sqrt{(b^2 - y^2)}$, (73) gives $A = ik/\pi$, and the introduction of this into (74) gives (78). We have now to find what departure from (86) is entailed when kb is no longer very small.

Since, in general,

$$d^2D/dx^2 + d^2D/dy^2 + k^2D = 0,$$

we find, as in (81),

$$-\frac{d\psi}{dx} = k^2\int\Psi D\,dy - \int\frac{d\Psi}{dy}\frac{dD}{dy}\,dy, \quad\dotsfill(87)$$

and for the present Ψ has the value defined in (82). The first term on the right of (87) may be treated in the same way as (28) of the former problem, the difference being that $\sqrt{(b^2 - y^2)}$ occurs now in the numerator instead of the denominator. In (30) we are to introduce under the integral sign the additional factor $k^2b^2 \sin^2 \theta$. As regards the second term of (87) we have

$$-\int \frac{d\Psi}{dy} \frac{dD}{dy} dy = \int_{-b}^{+b} \frac{y(y-\eta)\,dy}{\sqrt{(b^2-y^2)}} \frac{1}{r} \frac{dD}{dr},$$

where in $\dfrac{1}{r} \dfrac{dD}{dr}$ we are to replace r by $\pm (y - \eta)$. We then assume as before $y = b \cos \theta$, $\eta = b \cos \alpha$, and the same definite integrals h_n suffice; but the calculations are more complicated.

We have seen already that the leading term in (87) is π. For the next term we have

$$D = \gamma + \log \frac{ikr}{2}, \qquad \frac{1}{r} \frac{dD}{dr} = \frac{k^2}{4} - \frac{k^2}{2} \left(\gamma + \log \frac{ikr}{2} \right),$$

and thus

$$-\frac{1}{k^2b^2} \frac{d\psi}{dx} = \frac{\pi}{4} \left(\gamma + \log \frac{ikb}{4} + \frac{1}{2} \right)$$

$$+ \int_0^\pi d\theta \, (1 - \tfrac{3}{2} \cos^2 \theta + \tfrac{1}{2} \cos \alpha \cos \theta) \log \pm 2 \, (\cos \theta - \cos \alpha). \; \ldots (88)$$

The latter integral may be transformed into

$$2 \int_0^{\frac{1}{2}\pi} d\phi \, \{ 1 - \tfrac{3}{2} \cos^2 (2\phi - \alpha) + \tfrac{1}{2} \cos \alpha \cos (2\phi - \alpha)$$
$$+ 1 - \tfrac{3}{2} \cos^2 (2\phi + \alpha) + \tfrac{1}{2} \cos \alpha \cos (2\phi + \alpha) \} \log (2 \sin \phi),$$

and this by means of the definite integrals h is found to be

$$-\frac{\pi}{8} (1 + 2 \sin^2 \alpha).$$

To this order of approximation the complete value is

$$-\frac{d\psi}{dx} = \pi + \tfrac{1}{4} \pi k^2 b^2 (\gamma - \sin^2 \alpha + \log \tfrac{1}{4} ikb). \quad \ldots \ldots \ldots (89)$$

For the next two terms I find

$$+ \frac{\pi k^4 b^4}{512} [(1 + 4 \cos^2 \alpha)(1 - 4\gamma - 4 \log \tfrac{1}{4} ikb)$$
$$+ 3 \sin^4 \alpha + \tfrac{1}{3} \cos^4 \alpha + 6 \sin^2 \alpha \cos^2 \alpha]$$

$$+ \frac{\pi k^6 b^6}{2^2 . 4^2 . 6} \left[(\tfrac{1}{16} + \tfrac{3}{4} \cos^2 \alpha + \tfrac{1}{2} \cos^4 \alpha)(\gamma + \log \tfrac{1}{4} ikb - \tfrac{2}{3}) \right.$$

$$+ \frac{157}{8^2 . 15} \cos^6 \alpha - \frac{13}{8^2 . 3} \cos^4 \alpha \sin^2 \alpha - \frac{15}{8^2} \cos^2 \alpha \sin^4 \alpha - \frac{7}{8^2 . 3} \sin^6 \alpha \left. \right] \ldots (90)$$

When $\cos \alpha = 0$, or ± 1, the calculation is simpler. Thus, when $\cos \alpha = 0$,

$$-\frac{1}{\pi} \frac{d\psi}{dx} = 1 + \frac{k^2 b^2}{4} \left(\gamma + \log \frac{ikb}{4} - 1 \right) - \frac{k^4 b^4}{128} \left(\gamma + \log \frac{ikb}{4} - 1 \right)$$

$$+ \frac{k^6 b^6}{6 . 4^5} \left(\gamma + \log \frac{ikb}{4} - \frac{5}{4} \right) - \frac{5 k^8 b^8}{9 . 4^9} \left(\gamma + \log \frac{ikb}{4} - \frac{22}{15} \right); \quad \ldots \ldots \ldots (91)$$

and when $\cos \alpha = \pm 1$,

$$-\frac{1}{\pi}\frac{d\psi}{dx} = 1 + \frac{k^2 b^2}{4}\left(\gamma + \log\frac{ikb}{4}\right)$$

$$-\frac{k^4 b^4}{512}\left\{20\left(\gamma + \log\frac{ikb}{4}\right) - \frac{16}{3}\right\} + \frac{k^6 b^6}{6 \cdot 4^5}\left\{21\left(\gamma + \log\frac{ikb}{4}\right) - \frac{683}{60}\right\}$$

$$-\frac{k^8 b^8}{9 \cdot 4^9}\left\{429\left(\gamma + \log\frac{ikb}{4}\right) - 329\right\}, \quad\ldots\ldots\ldots\ldots\ldots\ldots\ldots\ldots(92)$$

the last term, deduced from h_{14}, h_{16}, being approximate.

For the values of $-\pi^{-1}d\psi/dx$ we find from (91), (90), (92) for $kb = \frac{1}{2}$, 1, $\sqrt{2}$, 2:

TABLE V.

	$kb = \frac{1}{2}$	$kb = 1$	$kb = \sqrt{2}$	$kb = 2$
$\cos \alpha = 0$	$0{\cdot}8448 + 0{\cdot}0974\,i$	$0{\cdot}5615 + 0{\cdot}3807\,i$	$0{\cdot}3123 + 0{\cdot}7383\,i$	$0{\cdot}0102 + 1{\cdot}3899\,i$
$\cos^2 \alpha = \frac{1}{2}$	$0{\cdot}8778 + 0{\cdot}0958\,i$	$0{\cdot}6998 + 0{\cdot}3583\,i$	—	$0{\cdot}518 + 1{\cdot}129\,i$
$\cos^2 \alpha = 1$	$0{\cdot}9103 + 0{\cdot}0944\,i$	$0{\cdot}8353 + 0{\cdot}3364\,i$	$0{\cdot}8587 + 0{\cdot}5783\,i$	$1{\cdot}020 + 0{\cdot}861\,i$

These numbers correspond to the value of Ψ expressed in (82).

We have now, in pursuance of our method, to seek a second solution with another form of Ψ. The first which suggests itself with $\Psi = 1$ does not answer the purpose. For (81) then gives as the leading term

$$-\frac{d\psi}{dx} = \left[\frac{y - \eta}{(y - \eta)^2 + x^2}\right]_{-b}^{b} = \frac{2b}{b^2 - \eta^2}, \quad\ldots\ldots\ldots\ldots\ldots(93)$$

becoming infinite when $\eta = \pm b$.

A like objection is encountered if $\Psi = b^2 - y^2$. In this case

$$-\frac{d\psi}{dx} = 2\int\{(y - \eta) + \eta\}\frac{(y - \eta)\,dy}{(y - \eta)^2 + x^2}.$$

The first part gives $4b$ simply when x becomes zero. And

$$2\int\frac{(y - \eta)\,dy}{(y - \eta)^2 + x^2} = \log\frac{(b - \eta)^2 + x^2}{(b + \eta)^2 + x^2};$$

so that

$$-\frac{d\psi}{dx} = 4b + 2\eta\log\frac{b - \eta}{b + \eta}, \quad\ldots\ldots\ldots\ldots\ldots\ldots(94)$$

becoming infinite when $\eta = \pm b$.

So far as this difficulty is concerned we might take $\Psi = (b^2 - y^2)^2$, but another form seems preferable, that is

$$\Psi = b^{-2}(b^2 - y^2)^{3/2}. \quad\ldots\ldots\ldots\ldots\ldots\ldots(95)$$

With the same notation as was employed in the treatment of (82) we have

$$-\frac{d\psi}{dx} = 3\int_0^\pi \frac{\cos\theta\,(\cos\theta - \cos\alpha)\,d\theta}{(\cos\theta - \cos\alpha)^2 + x'^2} - 3\int_0^\pi \frac{\cos^3\theta\,(\cos\theta - \cos\alpha)}{(\cos\theta - \cos .)^2 + x'^2}\,d\theta.$$

The first of these integrals is that already considered in (83). It yields 3π. In the second integral we replace $\cos^3\theta$ by $\{(\cos\theta - \cos\alpha) + \cos\alpha\}^3$, and we find, much as before, that when $x' = 0$

$$\int_0^\pi \frac{\cos^3\theta\,(\cos\theta - \cos\alpha)\,d\theta}{(\cos\theta - \cos\alpha)^2 + x'^2} = \pi\,(\tfrac{1}{2} + \cos^2\alpha). \quad\ldots\ldots\ldots\ldots(96)$$

Thus altogether for the leading term we get

$$-\frac{d\psi}{dx} = 3\pi\,(\tfrac{1}{2} - \cos^2\alpha) = 3\pi\,(\tfrac{1}{2} - \eta^2/b^2). \quad\ldots\ldots\ldots\ldots(97)$$

This is the complete solution for a fluid regarded as incompressible. We have now to pursue the approximation, using a more accurate value of D than that ($\log r$) hitherto employed.

In calculating the next term, we have the same values of D and $r^{-1}dD/dr$ as for (88); and in place of that equation we now have

$$-\frac{1}{k^2 b^2}\frac{d\psi}{dx} = \frac{3\pi}{16}\left(\gamma + \log\frac{ikb}{4} + \frac{1}{2}\right)$$

$$+ \int_0^\pi d\theta\,[\tfrac{5}{2}\sin^4\theta - \tfrac{3}{2}\sin^2\theta + \tfrac{3}{2}\sin^2\theta\,\cos\theta\,\cos\alpha]\log\{\pm\,2\,(\cos\theta - \cos\alpha)\}. \quad(98)$$

The integral may be transformed as before, and it becomes

$$4\int_0^{\tfrac{1}{2}\pi} d\phi\,\log\,(2\sin\phi)\,[\tfrac{5}{2}\,(\sin^4 2\phi\,\cos^4\alpha + 6\sin^2 2\phi\,\cos^2 2\phi\,\sin^2\alpha\,\cos^2\alpha$$

$$+\cos^4 2\phi\,\sin^4\alpha) - \tfrac{3}{2}\,(\sin^2 2\phi\,\cos^2\alpha + \cos^2 2\phi\,\sin^2\alpha)$$

$$+\tfrac{3}{2}\cos\alpha\,\cos 2\phi\,\{\sin^2\alpha\,\cos\alpha + \sin^2 2\phi\,(\cos^3\alpha - 3\sin^2\alpha\,\cos\alpha)\}]. \quad(99)$$

The evaluation could be effected by expressing the square bracket in terms of powers of $\sin^2\phi$, but it may be much facilitated by use of two lemmas.

If $f(\sin^2 2\phi, \cos^2 2\phi)$ denote an integral function of $\sin 2\phi$, $\cos^2 2\phi$,

$$\int_0^{\tfrac{1}{2}\pi} d\phi\,\log\,(2\sin\phi)\,f(\sin^2 2\phi, \cos^2 2\phi) = \int_0^{\tfrac{1}{2}\pi} d\phi\,\log\,(2\cos\phi)\,f(\sin^2 2\phi, \cos^2 2\phi)$$

$$= \tfrac{1}{2}\int_0^{\tfrac{1}{2}\pi} d\phi\,\log\,(2\sin 2\phi)\,f(\sin^2 2\phi, \cos^2 2\phi) = \tfrac{1}{2}\int_0^{\tfrac{1}{2}\pi} d\phi\,\log\,(2\sin\phi)\,f(\sin\phi, \cos^2\phi),$$

$$\ldots\ldots\ldots\ldots\ldots(100)$$

in which the doubled angles are got rid of.

Again, if m be integral,

$$\int_0^{\frac{1}{2}\pi} d\phi \, \sin^{2m} 2\phi \cos 2\phi \log (2 \sin \phi)$$

$$= \frac{1}{4m + 2} \int \log (2 \sin \phi) \, d \sin^{2m+1} 2\phi$$

$$= -\frac{1}{4m + 2} \int_0^{\frac{1}{2}\pi} \sin^{2m} 2\phi \, (1 + \cos 2\phi) \, d\phi$$

$$= -\frac{1}{4m + 2} \int_0^{\frac{1}{2}\pi} \sin^{2m} 2\phi \, d\phi = -\frac{1}{4m + 2} \int_0^{\frac{1}{2}\pi} \sin^{2m} \phi \, d\phi$$

$$= -\frac{1}{4m + 2} \frac{2m - 1 \cdot 2m - 3 \dots 1}{2m \cdot 2m - 2 \dots 2} \frac{\pi}{2} \quad \dots\dots\dots\dots\dots(101)$$

For example, if $m = 0$,

$$\int_0^{\frac{1}{2}\pi} d\phi \cos 2\phi \log (2 \sin \phi) = -\frac{\pi}{4}, \quad \dots\dots\dots\dots(102)$$

and $(m = 1)$ $\quad \displaystyle\int_0^{\frac{1}{2}\pi} d\phi \sin^2 2\phi \cos 2\phi \log (2 \sin \phi) = -\frac{\pi}{24}. \quad \dots\dots\dots(103)$

Using these lemmas, we find

$$(99) = 5h_4 (\cos^4 \alpha - 6 \cos^2 \alpha \, \sin^2 \alpha + \sin^4 \alpha)$$
$$+ h_2 (30 \cos^2 \alpha \, \sin^2 \alpha - 10 \sin^4 \alpha - 3 \cos^2 \alpha + 3 \sin^2 \alpha)$$
$$- \tfrac{1}{4}\pi \cos^2 \alpha \, (\cos^2 \alpha + 3 \sin^2 \alpha);$$

and thence, on introduction of the values of h_2, h_4, for the complete value to this order of approximation,

$$-\frac{d\psi}{dx} = 3\pi \left(\frac{1}{2} - \cos^2 \alpha \right) + \pi k^2 b^2 \left[\frac{3}{16} \left(\gamma + \frac{1}{2} + \log \frac{ikb}{4} \right) \right.$$
$$\left. - \frac{1}{64} (5 \cos^4 \alpha + 18 \cos^2 \alpha \, \sin^2 \alpha + 21 \sin^4 \alpha) \right]. \quad \dots\dots(104)$$

To carry out the calculations to a sufficient approximation with the general value of α would be very tedious. I have limited myself to the extreme cases $\cos \alpha = 0$, $\cos \alpha = \pm 1$. For the former, we have

$$-\frac{1}{\pi} \frac{d\psi}{dx} = \frac{3}{2} + \left(\gamma + \log \frac{ikb}{4} \right) \left\{ \frac{3k^2 b^2}{16} - \frac{k^4 b^4}{256} + \frac{k^6 b^6}{2^2 \cdot 4^2 \cdot 256} \right\}$$
$$- \frac{15k^2 b^2}{64} + \frac{7k^4 b^4}{6 \cdot 256} - \frac{11k^6 b^6}{4^3 \cdot 256 \cdot 8}, \quad \dots\dots\dots(105)$$

and for the latter

$$-\frac{1}{\pi} \frac{d\psi}{dx} = -\frac{3}{2} + \left(\gamma + \log \frac{ikb}{4} \right) \left\{ \frac{3k^2 b^2}{16} - \frac{7k^4 b^4}{16 \cdot 16} + \frac{33k^6 b^6}{4 \cdot 16 \cdot 16 \cdot 16} - \frac{143k^8 b^8}{24 \cdot 16^4} \right\}$$
$$- \frac{5k^2 b^2}{64} + \frac{41k^4 b^4}{16 \cdot 64 \cdot 15} + \frac{1069k^6 b^6}{16 \cdot 3 \cdot 70 \cdot 64 \cdot 64} - \frac{41309k^8 b^8}{16^5 \cdot 9 \cdot 420}$$
$$+ \frac{3k^2 b^2}{32} + \frac{7k^4 b^4}{4 \cdot 16 \cdot 16} - \frac{11k^6 b^6}{2 \cdot 16^3} + \frac{3289k^8 b^8}{16^5 \cdot 36}. \quad \dots\dots\dots\dots(106)$$

From these formulæ the following numbers have been calculated for the value of $-\pi^{-1}d\psi/dx$:

TABLE VI.

	$kb=\frac{1}{2}$	$kb=1$	$kb=\sqrt{2}$	$kb=2$
$\cos a=0$	$1\cdot3716+0\cdot0732i$	$1\cdot1215+0\cdot2885i$	$0\cdot8824+0\cdot5653i$	$0\cdot5499+1\cdot0860i$
$\cos a=\pm1$	$-1\cdot5634+0\cdot0710i$	$-1\cdot6072+0\cdot2546i$	$-1\cdot5693+0\cdot4401i$	$-1\cdot3952+0\cdot6567i$

They correspond to the value of Ψ formulated in (95).

Following the same method as in case (i), we now combine the two solutions, assuming
$$\Psi = A\sqrt{(b^2-y^2)} + Bb^{-2}(b^2-y^2)^{3/2}, \quad\ldots\ldots\ldots\ldots(107)$$
and determining A and B so that for $\cos a = 0$ and for $\cos a = \pm1$, $d\psi/dx$ shall be equal to $-ik$. The value of ψ at a distance in front is given by (76), in which
$$ik\int\Psi\,dy = \frac{\pi\cdot ikb^2}{2}\left(A+\frac{3}{4}B\right). \quad\ldots\ldots\ldots\ldots(108)$$

We may take the modulus of (108) as representing the transmitted vibration, in the same way as the modulus of (67) represented the transmitted vibration in case (i).

Using p, q, r, s, as before, to denote the tabulated complex numbers, we have as the equations to determine A and B,
$$Ap + Bq = Ar + Bs = ik/\pi, \quad\ldots\ldots\ldots\ldots(109)$$
so that
$$ik\int\Psi\,dy = -\frac{k^2b^2}{2}\frac{s-q+\frac{3}{4}(p-r)}{ps-qr} \quad\ldots\ldots\ldots\ldots(110)$$

For the second fraction on the right of (110) and for its modulus we get in the various cases

$kb = \frac{1}{2}$,	$1\cdot1470 - 0\cdot1287\,i$,	$1\cdot1542$,
$kb = 1$,	$1\cdot1824 - 0\cdot6986\,i$,	$1\cdot3733$,
$kb = \sqrt{2}$,	$0\cdot6362 - 1\cdot0258\,i$,	$1\cdot2070$,
$kb = 2$,	$0\cdot1239 - 0\cdot7303\,i$,	$0\cdot7407$.

And thence (on introduction of the value of kb) for the modulus of (110) representing the vibration on the same scale as in case (i)

TABLE VII.

kb	Modulus
$\frac{1}{2}$	$0\cdot1443$
1	$0\cdot6866$
$\sqrt{2}$	$1\cdot2070$
2	$1\cdot4814$

These are the numbers used in the plot of curve B, fig. 1. When kb is much smaller than $\frac{1}{2}$, the modulus may be taken to be $\frac{1}{2}k^2b^2$. When kb is large, the modulus approaches the same limiting form as in case (i).

This curve is applicable to electric, or luminous, vibrations incident upon a thin perfectly conducting screen with a linear perforation when the electric vector is *parallel* to the direction of the slit.

It appears that if the incident light be unpolarised, vibrations perpendicular to the slit preponderate in the transmitted light when the width of the slit is very small, and the more the smaller this width. In the neighbourhood of $kb = 1$, or $2b = \lambda/\pi$, the curves cross, signifying that the transmitted light is unpolarised. When $kb = 1\frac{1}{2}$, or $2b = 3\lambda/2\pi$, the polarisation is reversed, vibrations parallel to the slit having the advantage, but this advantage is not very great. When $kb > 2$, our calculations would hardly succeed, but there seems no reason for supposing that anything distinctive would occur. It follows that if the incident light were white and if the width of the slit were about one-third of the wave-length of yellow-green, there would be distinctly marked opposite polarisations at the ends of the spectrum.

These numbers are in good agreement with the estimates of Fizeau: "Une ligne polarisée perpendiculairement à sa direction a paru être de $\frac{1}{1000}$ de millimètre; une autre, beaucoup moins lumineuse, polarisée parallèlement à sa direction, a été estimée à $\frac{1}{10000}$ de millimètre. Je dois ajouter que ces valeurs ne sont qu'une approximation; elles peuvent être en réalité plus faibles encore, mais il est peu probable qu'elles soient plus fortes. Ce qu'il y a de certain, c'est que la polarisation parallèle n'apparaît que dans les fentes les plus fines, et alors que leur largeur est bien moindre que la longueur d'une ondulation qui est environ de $\frac{1}{2000}$ de millimètre." It will be remembered that the "plane of polarisation" is perpendicular to the electric vector.

It may be well to emphasize that the calculations of this paper relate to an aperture in an *infinitely thin* perfectly conducting screen. We could scarcely be sure beforehand that the conditions are sufficiently satisfied even by a scratch upon a silver deposit. The case of an ordinary spectroscope slit is quite different. It seems that here the polarisation observed with the finest practicable slits corresponds to that from the less fine scratches on silver deposits.

376.

ON THE MOTION OF A VISCOUS FLUID.

[*Philosophical Magazine*, Vol. XXVI. pp. 776—786, 1913.]

IT has been proved by Helmholtz[*] and Korteweg[†] that when the velocities at the boundary are given, the slow steady motion of an incompressible viscous liquid satisfies the condition of making F, the dissipation, an absolute minimum. If u_0, v_0, w_0 be the velocities in one motion M_0, and u, v, w those of another motion M satisfying the same boundary conditions, the difference of the two u', v', w', where

$$u' = u - u_0, \quad v' = v - v_0, \quad w' = w - w_0, \ldots\ldots\ldots\ldots\ldots(1)$$

will constitute a motion M' such that the boundary velocities vanish. If F_0, F, F' denote the dissipation-functions for the three motions M_0, M, M' respectively, all being of necessity positive, it is shown that

$$F = F_0 + F' - 2\mu \int (u' \nabla^2 u_0 + v' \nabla^2 v_0 + w' \nabla^2 w_0) \, dx \, dy \, dz, \ldots\ldots\ldots(2)$$

the integration being over the whole volume. Also

$$F' = -\mu \int (u' \nabla^2 u' + v' \nabla^2 v' + w' \nabla^2 w') \, dx \, dy \, dz$$

$$= \mu \int \left[\left(\frac{dw'}{dy} - \frac{dv'}{dz} \right)^2 + \left(\frac{du'}{dz} - \frac{dw'}{dx} \right)^2 + \left(\frac{dv'}{dx} - \frac{du'}{dy} \right)^2 \right] dx \, dy \, dz. \ \ldots\ldots(3)$$

These equations are purely kinematical, if we include under that head the incompressibility of the fluid. In the application of them by Helmholtz and Korteweg the motion M_0 is supposed to be that which would be steady if small enough to allow the neglect of the terms involving the second powers of the velocities in the dynamical equations. We then have

$$\mu \nabla^2 (u_0, v_0, w_0) = \left(\frac{d}{dx}, \ \frac{d}{dy}, \ \frac{d}{dz} \right) (V\rho + p_0). \ \ldots\ldots\ldots\ldots(4)$$

[*] *Collected Works*, Vol. I. p. 223 (1869).
[†] *Phil. Mag.* Vol. XVI. p. 112 (1883).

where V is the potential of impressed forces.   In virtue of (4)

$$\int (u'\nabla^2 u_0 + v'\nabla^2 v_0 + w'\nabla^2 w_0)\, dx\, dy\, dz = 0, \dots\dots\dots\dots(5)$$

if the space occupied by the fluid be simply connected, or in any case if V be single-valued.   Hence

$$F = F_0 + F', \dots\dots\dots\dots\dots\dots\dots(6)$$

or since F' is necessarily positive, the motion M_0 makes F an absolute minimum.   It should be remarked that F' can vanish only for a motion such as can be assumed by a solid body (Stokes), and that such a motion could not make the boundary velocities vanish.   The motion M_0 determined by (4) is thus unique.

The conclusion expressed in (6) that M_0 makes F an ·absolute minimum is not limited to the supposition of a slow motion.   All that is required to ensure the fulfilment of (5), on which (6) depends, is that $\nabla^2 u_0$, $\nabla^2 v_0$, $\nabla^2 w_0$ should be the derivatives of some single-valued function.   Obviously it would suffice that $\nabla^2 u_0$, $\nabla^2 v_0$, $\nabla^2 w_0$ vanish, as will happen if the motion have a velocity-potential.   Stokes* remarked long ago that when there is a velocity-potential, not only are the ordinary equations of fluid motion satisfied, but the equations obtained when friction is taken into account are satisfied likewise.   A motion with a velocity-potential can always be found which shall have prescribed *normal* velocities at the boundary, and the tangential velocities are thereby determined.   If these agree with the prescribed tangential velocities of a viscous fluid, all the conditions are satisfied by the motion in question.   And since this motion makes F an absolute minimum, it cannot differ from the motion determined by (4) with the same boundary conditions.   We may arrive at the same conclusion by considering the general equation of motion

$$\rho\left(\frac{du}{dt} + u\frac{du}{dx} + v\frac{du}{dy} + w\frac{du}{dz}\right) = \mu\nabla^2 u - \frac{d(\rho V + p)}{dx}. \quad \dots\dots(7)$$

If there be a velocity-potential ϕ, so that $u = d\phi/dx$, &c.,

$$u\frac{du}{dx} + v\frac{du}{dy} + w\frac{du}{dz} = \frac{1}{2}\frac{d}{dx}\left\{\left(\frac{d\phi}{dx}\right)^2 + \left(\frac{d\phi}{dy}\right)^2 + \left(\frac{d\phi}{dz}\right)^2\right\}; \quad \dots\dots(8)$$

and then (7) and its analogues reduce practically to the form (4) if the motion be steady.

Other cases where F is an absolute minimum are worthy of notice.   It suffices that

$$\nabla^2 u_0 = \frac{dH}{dx}, \quad \nabla^2 v_0 = \frac{dH}{dy}, \quad \nabla^2 w_0 = \frac{dH}{dz}, \quad \dots\dots\dots(9)$$

---

* *Camb. Trans.* Vol. IX. (1850); *Math. and Phys. Papers*, Vol. III. p. 73.

where H is a single-valued function, subject to $\nabla^2 H = 0$. If ξ_0, η_0, ζ_0 be the rotations,

$$2\nabla^2\xi_0 = \nabla^2\left(\frac{dw_0}{dy} - \frac{dv_0}{dz}\right) = \frac{d}{dy}\nabla^2 w_0 - \frac{d}{dz}\nabla^2 v_0 = 0 ;$$

and thus (9) requires that

$$\nabla^2\xi_0 = 0, \quad \nabla^2\eta_0 = 0, \quad \nabla^2\zeta_0 = 0. \dots\dots\dots\dots\dots(10)$$

In two dimensions the dynamical equation reduces to $D\zeta_0/Dt = 0$*, so that ζ_0 is constant along a stream-line. Among the cases included are the motion between two planes

$$u_0 = A + By + Cy^2, \quad v_0 = 0, \quad w_0 = 0, \dots\dots\dots\dots(11)$$

and the motion in circles between two coaxal cylinders ($\zeta_0 = $ constant). Also, without regard to the form of the boundary, the uniform rotation, as of a solid body, expressed by

$$u_0 = Cy, \quad v_0 = -Cx. \dots\dots\dots\dots\dots(12)$$

In all these cases F is an absolute minimum.

Conversely, if the conditions (9) be not satisfied, it will be possible to find a motion for which $F < F_0$. To see this choose a place as origin of coordinates where $d\nabla^2 u_0/dy$ is not equal to $d\nabla^2 v_0/dx$. Within a small sphere described round this point as centre let $u' = Cy$, $v' = -Cx$, $w' = 0$, and let $u' = 0$, $v' = 0$, $w' = 0$ outside the sphere, thus satisfying the prescribed boundary conditions. Then in (2)

$$\int (u'\nabla^2 u_0 + v'\nabla^2 v_0 + w'\nabla^2 w_0)\,dx\,dy\,dz = C\int (y\nabla^2 u_0 - x\nabla^2 v_0)\,dx\,dy\,dz, \dots(13)$$

the integration being over the sphere. Within this small region we may take

$$\nabla^2 u_0 = (\nabla^2 u_0)_0 + \frac{d\nabla^2 u_0}{dx_0}x + \frac{d\nabla^2 u_0}{dy_0}y + \frac{d\nabla^2 u_0}{dz_0}z,$$

$$\nabla^2 v_0 = (\nabla^2 v_0)_0 + \frac{d\nabla^2 v_0}{dx_0}x + \frac{d\nabla^2 v_0}{dy_0}y + \frac{d\nabla^2 v_0}{dz_0}z ;$$

so that (13) reduces to

$$C\left(\frac{d\nabla^2 u_0}{dy_0} - \frac{d\nabla^2 v_0}{dx_0}\right)\int (y^2 \text{ or } x^2)\,dx\,dy\,dz.$$

Since the sign of C is at disposal, this may be made positive or negative at pleasure. Also F' in (2) may be neglected as of the second order when u', v', w' are small enough. It follows that F is not an absolute minimum for u_0, v_0, w_0, unless the conditions (9) are satisfied.

Korteweg has also shown that the slow motion of a viscous fluid denoted by M_0 is *stable*. "When in a given region occupied by viscous

---

* Where $D/Dt = d/dt + u\,d/dx + v\,d/dy + w\,d/dz$.

incompressible fluid there exists at a certain moment a mode of motion M which does not satisfy equation (4), then, the velocities along the boundary being maintained constant, the change which must occur in the mode of motion will be such (neglecting squares and products of velocities) that the dissipation of energy by internal friction is constantly decreasing till it reaches the value F_0 and the mode of motion becomes identical with M_0."

This theorem admits of instantaneous proof. If the terms of the second order are omitted, the equations of motion, such as (7), are linear, and any two solutions may be superposed. Consider two solutions, both giving the same velocities at the boundary. Then the difference of these is also a solution representing a possible motion with zero velocities at the boundary. But such a motion necessarily comes to rest. Hence with flux of time the two original motions tend to become and to remain identical. If one of these is the steady motion, the other must tend to become coincident with it.

The stability of the *slow* steady motion of a viscous fluid, or (as we may put it) the steady motion of a *very* viscous fluid, is thus ensured. When the circumstances are such that the terms of the second order must be retained, there is but little definite knowledge as to the character of the motion in respect of stability. Viscous fluid, contained in a vessel which rotates with uniform velocity, would be expected to acquire the same rotation and ultimately to revolve as a solid body, but the expectation is perhaps founded rather upon observation than upon theory. We might, however, argue that any other event would involve perpetual dissipation which could only be met by a driving force applied to the vessel, since the kinetic energy of the motion could not for ever diminish. And such a maintained driving couple would generate angular momentum without limit—a conclusion which could not be admitted. But it may be worth while to examine this case more closely.

We suppose as before that u_0, v_0, w_0 are the velocities in the steady motion M_0 and u, v, w those of the motion M, both motions satisfying the dynamical equations, and giving the prescribed boundary velocities; and we consider the expression for the kinetic energy T' of the motion (1) which is the difference of these two, and so makes the velocities vanish at the boundary. The motion M' with velocities u', v', w' does not in general satisfy the dynamical equations. We have

$$\frac{1}{\rho}\frac{dT'}{dt} = \int \left\{ u'\frac{du'}{dt} + v'\frac{dv'}{dt} + w'\frac{dw'}{dt} \right\} dx\,dy\,dz. \quad \ldots\ldots\ldots\ldots(14)$$

In equations (7) which are satisfied by the motion M we substitute $u = u_0 + u'$, &c.; and since the solution M_0 is steady we have

$$\frac{du_0}{dt} = \frac{dv_0}{dt} = \frac{dw_0}{dt} = 0. \quad \ldots\ldots\ldots\ldots\ldots\ldots\ldots\ldots\ldots(15)$$

We further suppose that $\nabla^2 u_0$, $\nabla^2 v_0$, $\nabla^2 w_0$ are derivatives of a function H, as in (9). This includes the case of uniform rotation expressed by

$$u_0 = y, \quad v_0 = -x, \quad w_0 = 0, \dots\dots\dots\dots\dots\dots(16)$$

as well as those where there is a velocity-potential. Thus (7) becomes

$$\frac{du'}{dt} = \nu \nabla^2 u' - \frac{d\varpi}{dx} - (u_0 + u')\left(\frac{du_0}{dx} + \frac{du'}{dx}\right)$$

$$- (v_0 + v')\left(\frac{du_0}{dy} + \frac{du'}{dy}\right) - (w_0 + w')\left(\frac{du_0}{dz} + \frac{du'}{dz}\right), \dots\dots(17)$$

with two analogous equations, where

$$\varpi = V + p/\rho - \nu H, \quad \nu = \mu/\rho. \dots\dots\dots\dots\dots(18)$$

These values of du'/dt, &c., are to be substituted in (14).

In virtue of the equation of continuity to which u', v', w' are subject, the terms in ϖ contribute nothing to dT'/dt, as appears at once on integration by parts. The remaining terms in dT'/dt are of the first, second, and third degree in u', v', w'. Those of the first degree contribute nothing, since u_0, v_0, w_0 satisfy equations such as

$$u_0 \frac{du_0}{dx} + v_0 \frac{du_0}{dy} + w_0 \frac{du_0}{dz} = -\frac{d\varpi_0}{dx}. \dots\dots\dots\dots(19)$$

The terms of the third degree are

$$-\int \left[u'\left\{ u'\frac{du'}{dx} + v'\frac{du'}{dy} + w'\frac{du'}{dz} \right\} \right.$$

$$+ v'\left\{ u'\frac{dv'}{dx} + v'\frac{dv'}{dy} + w'\frac{dv'}{dz} \right\}$$

$$\left. + w'\left\{ u'\frac{dw'}{dx} + v'\frac{dw'}{dy} + w'\frac{dw'}{dz} \right\} \right] dx\, dy\, dz,$$

which may be written

$$-\frac{1}{2}\int \left[u'\frac{d(u'^2 + v'^2 + w'^2)}{dx} + v'\frac{d(u'^2 + v'^2 + w'^2)}{dy} \right.$$

$$\left. + w'\frac{d(u'^2 + v'^2 + w'^2)}{dz} \right] dx\, dy\, dz;$$

and this vanishes for the same reason as the terms in ϖ.

We are left with the terms of the second degree in u', v', w'. Of these the part involving ν is

$$\nu \int [u'\nabla^2 u' + v'\nabla^2 v' + w'\nabla^2 w']\, dx\, dy\, dz. \dots\dots\dots\dots(20)$$

So far as this part is concerned, we see from (3) that

$$dT'/dt = -F', \dots\dots\dots\dots\dots\dots(21)$$

F' being the dissipation-function calculated from u', v', w'.

Of the remaining 18 terms of the second degree, 9 vanish as before when integrated, in virtue of the equation of continuity satisfied by u_0, v_0, w_0. Finally we have[*]

$$\frac{dT'}{dt} = - F' - \rho \int \left[u' \left\{ u' \frac{du_0}{dx} + v' \frac{du_0}{dy} + w' \frac{du_0}{dz} \right\} \right.$$

$$+ v' \left\{ u' \frac{dv_0}{dx} + v' \frac{dv_0}{dy} + w' \frac{dv_0}{dz} \right\}$$

$$\left. + w' \left\{ u' \frac{dw_0}{dx} + v' \frac{dw_0}{dy} + w' \frac{dw_0}{dz} \right\} \right] dx\, dy\, dz. \quad \dots\dots(22)$$

If the motion u_0, v_0, w_0 be in two dimensions, so that $w_0 = 0$, while u and v_0 are independent of z, (22) reduces to

$$\frac{dT'}{dt} = - F' - \rho \int \left[u'^2 \frac{du_0}{dx} + v'^2 \frac{dv_0}{dy} + u'v' \left(\frac{du_0}{dy} + \frac{dv_0}{dx} \right) \right] dx\, dy\, dz. \quad \dots(23)$$

Under this head comes the case of uniform rotation expressed in (16), for which

$$\frac{du_0}{dx} = 0, \quad \frac{dv_0}{dy} = 0, \quad \frac{du_0}{dy} + \frac{dv_0}{dx} = 0.$$

Here then $dT'/dt = - F'$ simply, that is T' continually diminishes until it becomes insensible. Any motion superposed upon that of uniform rotation gradually dies out.

When the motion u_0, v_0, w_0 has a velocity-potential ϕ, (22) may be written

$$\frac{dT'}{dt} = - F' - \rho \int \left[u'^2 \frac{d^2\phi}{dx^2} + v'^2 \frac{d^2\phi}{dy^2} + w'^2 \frac{d^2\phi}{dz^2} \right.$$

$$\left. + 2u'v' \frac{d^2\phi}{dx\,dy} + 2v'w' \frac{d^2\phi}{dy\,dz} + 2w'u' \frac{d^2\phi}{dz\,dx} \right] dx\, dy\, dz. \quad \dots\dots(24)$$

So far as I am aware, no case of complete stability for all values of μ is known, other than the motion possible to a solid body above considered. It may be doubted whether such cases exist. Under the head of (24) a simple example occurs when $\phi = \tan^{-1}(y/x)$, the irrotational motion taking place in concentric circles. Here if $r^2 = x^2 + y^2$,

$$\frac{dT'}{dt} = - F' - 2\rho \int \left[\frac{xy}{r^4} (u'^2 - v'^2) + \frac{y^2 - x^2}{r^4} u'v' \right] dx\, dy\, dz. \quad \dots\dots(25)$$

---

* Compare O. Reynolds, *Phil. Trans.* 1895, Part i. p. 146. In Lorentz's deduction of a similar equation (*Abhandlungen*, Vol. i. p. 46) the additional motion is assumed to be small. This memoir, as well as that of Orr referred to below, should be consulted by those interested. See also Lamb's *Hydrodynamics*, § 346.

If the superposed motion also be two-dimensional, it may be expressed by means of a stream-function ψ. We have in terms of polar coordinates

$$u' = \frac{d\psi}{dy} = \frac{d\psi}{dr} \sin\theta + \frac{1}{r} \frac{d\psi}{d\theta} \cos\theta,$$

$$-v' = \frac{d\psi}{dx} = \frac{d\psi}{dr} \cos\theta - \frac{1}{r} \frac{d\psi}{d\theta} \sin\theta,$$

so that

$$u'^2 - v'^2 = (\cos^2\theta - \sin^2\theta) \left\{ \frac{1}{r^2} \left(\frac{d\psi}{d\theta}\right)^2 - \left(\frac{d\psi}{dr}\right)^2 \right\} + \frac{4\sin\theta\cos\theta}{r} \frac{d\psi}{dr} \frac{d\psi}{d\theta},$$

$$-u'v' = \cos\theta\sin\theta \left\{ \left(\frac{d\psi}{dr}\right)^2 - \frac{1}{r^2} \left(\frac{d\psi}{d\theta}\right)^2 \right\} + \frac{\cos^2\theta - \sin^2\theta}{r} \frac{d\psi}{dr} \frac{d\psi}{d\theta}.$$

Thus

$$\cos\theta\sin\theta (u'^2 - v'^2) - (\cos^2\theta - \sin^2\theta) u'v' = \frac{1}{r} \frac{d\psi}{dr} \frac{d\psi}{d\theta}, \quad\ldots\ldots(26)$$

and (25) becomes

$$\frac{dT'}{dt} = -F' - 2\rho \iiint \frac{1}{r^2} \frac{d\psi}{dr} \frac{d\psi}{d\theta} \, dr \, d\theta \, dz, \quad\ldots\ldots\ldots\ldots(27)$$

T', F', as well as the last integral, being proportional to z.

We suppose the motion to take place in the space between two coaxal cylinders which revolve with appropriate velocities. If the additional motion be also symmetrical about the axis, the stream-lines are circles, and ψ is a function of r only. The integral in (27) then disappears and dT'/dt reduces to $-F'$, so that under this restriction * the original motion is stable. The experiments of Couette† and of Mallock‡, made with revolving cylinders, appear to show that when u', v', w' are not specially restricted the motion is unstable. It may be of interest to follow a little further the indications of (27).

The general value of ψ is

$$\psi = C_0 + C_1 \cos\theta + S_1 \sin\theta + \ldots + C_n \cos n\theta + S_n \sin n\theta, \quad\ldots\ldots(28)$$

C_n, S_n being functions of r, whence

$$\int \frac{d\psi}{dr} \frac{d\psi}{d\theta} \, d\theta = \pi \Sigma n \left(S_n \frac{dC_n}{dr} - C_n \frac{dS_n}{dr} \right), \quad\ldots\ldots\ldots\ldots(29)$$

n being 1, 2, 3, &c. If S_n, C_n differ only by a constant multiplier, (29) vanishes. This corresponds to

$$\psi = R_0 + R_1 \cos(\theta + \epsilon_1) + \ldots + R_n \cos n(\theta + \epsilon_n) + \ldots, \quad\ldots\ldots(30)$$

* We may imagine a number of thin, coaxal, freely rotating cylinders to be interposed between the extreme ones whose motion is prescribed.

† *Ann. d. Chimie*, t. XXI. p. 433 (1890).

‡ *Proc. Roy. Soc.* Vol. LIX. p. 38 (1895).

where R_0, R_1, &c. are functions of r, while ϵ_1, ϵ_2, &c. are constants. If ψ can be thus limited, dT'/dt reduces to $-F'$, and the original motion is stable.

In general
$$\frac{dT'}{dt} = -F' - 2\pi\rho z \int \Sigma n \left(S_n \frac{dC_n}{dr} - C_n \frac{dS_n}{dr} \right) \frac{dr}{r^2} \quad \ldots\ldots\ldots(31)$$

C_n, S_n must be such as to give at the boundaries
$$C_n = 0, \quad dC_n/dr = 0, \quad S_n = 0, \quad dS_n/dr = 0 ; \quad\ldots\ldots\ldots\ldots(32)$$

otherwise they are arbitrary functions of r. It may be noticed that the sign of any term in (29) may be altered at pleasure by interchange of C_n and S_n.

When μ is great, so that the influence of F preponderates, the motion is stable. On the other hand when μ is small, the motion is probably unstable, unless special restrictions can be imposed.

A similar treatment applies to the problem of the uniform shearing motion of a fluid between two parallel plane walls, defined by
$$u_0 = A + By, \quad v_0 = 0, \quad w_0 = 0. \quad\ldots\ldots\ldots\ldots\ldots(33)$$

From (23)
$$\frac{dT'}{dt} = -F' - \rho B \iint u' v' \, dx \, dy. \quad\ldots\ldots\ldots\ldots\ldots(34)$$

If in the superposed motion $v' = 0$, the double integral vanishes and the original motion is stable. More generally, if the stream-function of the superposed motion be
$$\psi = C \cos kx + S \sin kx, \quad\ldots\ldots\ldots\ldots\ldots(35)$$

where C, S are functions of y, we find
$$\frac{dT'}{dt} = -F' + \rho B \iint \frac{d\psi}{dy} \frac{d\psi}{dx} \, dx \, dy$$
$$= -F' + \frac{\rho B . kx}{2} \int \left(S \frac{dC}{dy} - C \frac{dS}{dy} \right) dy. \quad\ldots\ldots\ldots(36)$$

Here again if the motion can be such that C and S differ only by a constant multiplier, the integral would vanish. When μ is small and there is no special limitation upon the disturbance, instability probably prevails. The question whether μ is to be considered great or small depends of course upon the other data of the problem. If D be the distance between the planes, we have to deal with BD^2/ν (Reynolds).

In an important paper * Orr, starting from equation (34), has shown that if BD^2/ν is less than 177 "every disturbance must automatically decrease, and that (for a higher value than 177) it is possible to prescribe a disturbance which will increase for a time." We must not infer that when

* Proc. Roy. Irish Acad. 1907.

$BD^2/\nu > 177$ the regular motion is necessarily unstable. As the fluid moves under the laws of dynamics, the initial increase of certain disturbances may after a time be exchanged for a decrease, and this decrease may be without limit.

At the other extreme when ν is very small, observation shows that the tangential traction on the walls, moving (say) with velocities $\pm U$, tends to a statistical uniformity and to become proportional, no longer to U, but to U^2. If we assume this law to be absolute in the region of high velocity, the principle of dynamical similarity leads to rather remarkable conclusions. For the tangential traction, having the dimensions of a pressure, must in general be of the form

$$\rho U^2 \cdot f\left(\frac{\nu}{UD}\right), \quad\dots\dots\dots\dots\dots\dots(37)$$

D being the distance between the walls, and f an arbitrary function. In the regular motion (z large) $f(z) = 2z$, and (37) is proportional to U. If (37) is proportional to U^2, f must be a constant and the traction becomes independent not only of μ, but *also* of D.

If the velocity be not quite so great as to reduce f to constancy, we may take

$$f(z) = a + bz,$$

where a and b are numerical constants, so that (37) becomes

$$a\rho U^2 + b\mu\, U/D. \quad\dots\dots\dots\dots\dots\dots(38)$$

It could not be assumed without further proof that b has the value (2) appropriate to a large z; nevertheless, Korteweg's equation (6) suggests that such may be the case.

From data given by Couette I calculate that in c.g.s. measure

$$a = \cdot000027.$$

The tangential traction is thus about a twenty thousandth part of the pressure $(\frac{1}{2}\rho U^2)$ due to the normal impact of the fluid moving with velocity U.

Even in cases where the steady motion of a viscous fluid satisfying the dynamical equations is certainly unstable, there is a distinction to be attended to which is not without importance. It may be a question of the *time* during which the fluid remains in an unstable condition. When fluid moves between two coaxal cylinders, the instability has an indefinite time in which to develop itself. But it is otherwise in many important problems. Suppose that fluid has to move through a narrow place, being guided for example by hyperbolic surfaces, either in two dimensions, or in three with symmetry about an axis. If the walls have suitable tangential velocities, the motion may be irrotational. This irrotational motion is that which would be initiated from rest by propellent impulses acting at a distance. If the viscosity were great, the motion would be steady and stable; if the viscosity is less, it still satisfies the dynamical equations, but is (presumably) unstable. But the instability, as it affects any given portion of the fluid, has a.very short duration. Only as it approaches the narrows has the fluid any considerable velocity, and as soon as the narrows are passed the velocity falls off again. Under these circumstances it would seem probable that the instability in the narrows would be of little consequence, and that the irrotational motion would practically hold its own. If this be so, the tangential movement of the walls exercises a profound influence, causing the fluid to follow the walls on the down stream side, instead of shooting onwards as a jet—the behaviour usually observed when fluid is invited to follow fixed divergent walls, unless indeed the expansion is very gradual.

377.

ON THE STABILITY OF THE LAMINAR MOTION OF AN INVISCID FLUID.

[*Philosophical Magazine*, Vol. XXVI. pp. 1001—1010, 1913.]

THE equations of motion of an inviscid fluid are satisfied by a motion such that U, the velocity parallel to x, is an arbitrary function of y only, while the other component velocities V and W vanish. The motion may be supposed to be limited by two fixed plane walls for each of which y has a constant value. In order to investigate the stability of the motion, we superpose upon it a two-dimensional disturbance u, v, where u and v are regarded as small. If the fluid is incompressible,

$$\frac{du}{dx} + \frac{dv}{dy} = 0 ; \qquad \dots\dots\dots\dots\dots\dots\dots\dots\dots\dots(1)$$

and if the squares and products of small quantities are neglected, the hydro-dynamical equations give*

$$\left(\frac{d}{dt} + U\frac{d}{dx}\right)\left(\frac{du}{dy} - \frac{dv}{dx}\right) + \frac{d^2U}{dy^2}\,v = 0. \qquad \dots\dots\dots\dots(2)$$

From (1) and (2), if we assume that as functions of t and x, u and v are proportional to $e^{i\,(nt+kx)}$, where k is real and n may be real or complex,

$$\left(\frac{n}{k} + U\right)\left(\frac{d^2v}{dy^2} - k^2v\right) + \frac{d^2U}{dy^2}\,v = 0. \qquad \dots\dots\dots\dots(3)$$

In the paper quoted it was shown that under certain conditions n could not be complex; and it may be convenient to repeat the argument. Let

$$n/k = p + iq, \quad v = \alpha + i\beta,$$

---

* *Proceedings of London Mathematical Society*, Vol. XI. p. 57 (1880); *Scientific Papers*, Vol. I. p. 485. Also Lamb's *Hydrodynamics*, § 345.

where p, q, α, β are real. Substituting in (3) and equating separately to zero the real and imaginary parts, we get

$$\frac{d^2\alpha}{dy^2} = k^2\alpha + \frac{d^2U}{dy^2}\frac{(p+U)\alpha + q\beta}{(p+U)^2 + q^2},$$

$$\frac{d^2\beta}{dy^2} = k^2\beta + \frac{d^2U}{dy^2}\frac{-q\alpha + (p+U)\beta}{(p+U)^2 + q^2};$$

whence if we multiply the first by β and the second by α and subtract,

$$\frac{d}{dy}\left(\beta\frac{d\alpha}{dy} - \alpha\frac{d\beta}{dy}\right) = \frac{d^2U}{dy^2}\frac{q(\alpha^2 + \beta^2)}{(p+U)^2 + q^2}. \quad\ldots\ldots\ldots\ldots(4)$$

At the limits, corresponding to finite or infinite values of y, we suppose that v, and therefore both α and β, vanish. Hence when (4) is integrated with respect to y between these limits, the left-hand member vanishes and we infer that q also must vanish unless d^2U/dy^2 changes sign. Thus in the motion between walls if the velocity curve, in which U is ordinate and y abscissa, be of one curvature throughout, n must be wholly real; otherwise, so far as this argument shows, n may be complex and the disturbance exponentially unstable.

Two special cases at once suggest themselves. If the motion be that which is possible to a viscous fluid moving steadily between two fixed walls under external pressure or impressed force, so that for example $U = y^2 - b^2$, d^2U/dy^2 is a finite constant, and complex values of n are clearly excluded. In the case of a simple shearing motion, exemplified by $U = y$, $d^2U/dy^2 = 0$, and no inference can be drawn from (4). But referring back to (3), we see that in this case if n be complex,

$$\frac{d^2v}{dy^2} - k^2v = 0 \quad\ldots\ldots\ldots\ldots\ldots\ldots\ldots\ldots\ldots\ldots\ldots(5)$$

would have to be satisfied over the whole range between the limits where $v = 0$. Since such satisfaction is not possible, we infer that here too a complex n is excluded.

It may appear at first sight as if real, as well as complex, values of n were excluded by this argument. But if n be such that $n/k + U$ vanishes anywhere within the range, (5) need not there be satisfied. In other words, the arbitrary constants which enter into the solution of (5) may there change values, subject only to the condition of making v continuous. The terminal conditions can then be satisfied. Thus any value of $-n/k$ is admissible which coincides with a value of U to be found within the range. But other real values of n are excluded.

Let us now examine how far the above argument applies to real values of n, when d^2U/dy^2 in (3) does not vanish throughout. It is easy to recognize that here also any value of $- kU$ is admissible, and for the same reason as before, viz., that when $n + kU = 0$, dv/dy may be discontinuous. Suppose, for example, that there is but one place where $n + kU = 0$. We may start from either wall with $v = 0$ and with an arbitrary value of dv/dy and gradually build up the solutions inwards so as to satisfy (3)*. The process is to be continued on both sides until we come to the place where $n + kU = 0$. The two values there found for v and for dv/dy will presumably disagree. But by suitable choice of the relative initial values of dv/dy, v may be made continuous, and (as has been said) a discontinuity in dv/dy does not interfere with the satisfaction of (3). If there are other places where U has the same value, dv/dy may there be either continuous or discontinuous. Even when there is but one place where $n + kU = 0$ with the proposed value of n, it may happen that dv/dy is there continuous.

The argument above employed is not interfered with even though U is such that dU/dy is here and there discontinuous, so as to make d^2U/dy^2 infinite. At any such place the necessary condition is obtained by integrating (3) across the discontinuity. As was shown in my former paper (*loc. cit.*), it is

$$\left(\frac{n}{k} + U\right) . \Delta \left(\frac{dv}{dy}\right) - \Delta \left(\frac{dU}{dy}\right) . v = 0, \quad \dots\dots\dots\dots(6)$$

Δ being the symbol of finite differences; and by (6) the corresponding sudden change in dv/dy is determined.

It appears then that any value of $- kU$ is a possible value of n. Are other real values admissible? If so, $n + kU$ is of one sign throughout. It is easy to see that if d^2U/dy^2 has throughout the same sign as $n + kU$, no solution is possible. I propose to prove that no solution is possible in any case if $n + kU$, being real, is of one sign throughout.

If U' be written for $U + n/k$, our equation (3) takes the form

$$U' \frac{d^2v}{dy^2} - v \frac{d^2U'}{dy^2} = k^2 U'v, \quad \dots\dots\dots\dots\dots(7)$$

or on integration with respect to y,

$$U' \frac{dv}{dy} - v \frac{dU'}{dy} = K + k^2 \int_0^y U'v \, dy, \quad \dots\dots\dots\dots (8)$$

where K is an arbitrary constant. Assume $v = U'v'$; then

$$\frac{dv'}{dy} = \frac{K}{U'^2} + \frac{k^2}{U'^2} \int_0^y v' U'^2 \, dy ; \quad \dots\dots\dots\dots(9)$$

---

* Graphically, the equation directs us with what curvature to proceed at any point already reached.

whence, on integration and replacement of v,

$$v = HU' + KU' \int_0^y \frac{dy}{U'^2} + k^2 U' \int_0^y \frac{dy}{U'^2} \int_0^y U'v\, dy, \quad \ldots\ldots(10)$$

H denoting a second arbitrary constant.

In (10) we may suppose y measured from the first wall, where $v = 0$. Hence, unless U' vanish with y, $H = 0$. Also from (8) when $y = 0$,

$$\left(U' \frac{dv}{dy}\right)_0 = K. \quad \ldots\ldots\ldots\ldots\ldots\ldots\ldots(11)$$

Let us now trace the course of v as a function of y, starting from the wall where $y = 0$, $v = 0$; and let us suppose first that U' is everywhere positive. By (11) K has the same sign as $(dv/dy)_0$, that is the same sign as the early values of v. Whether this sign be positive or negative, v as determined by (10) cannot again come to zero. If, for example, the initial values of v are positive, both (remaining) terms in (10) necessarily continue positive; while if v begins by being negative, it must remain finitely negative. Similarly, if U' be everywhere negative, so that K has the opposite sign to that of the early values of v, it follows that v cannot again come to zero. No solution can be found unless U' somewhere vanishes, that is unless n coincides with some value of $-kU$.

In the above argument U', and therefore also n, is supposed to be *real*, but the formula (10) itself applies whether n be real or complex. It is of special value when k is very small, that is when the wave-length along x of the disturbance is very great; for it then gives v explicitly in the form

$$v = K(U + n/k) \int_0^y \frac{dy}{(U + n/k)^2}. \quad \ldots\ldots\ldots\ldots(12)$$

When k is small, but not so small as to justify (12), a second approximation might be found by substituting from (12) in the last term of (10).

If we suppose in (12) that the second wall is situated at $y = l$, n is determined by

$$\int_0^l \frac{dy}{(U + n/k)^2} = 0. \quad \ldots\ldots\ldots\ldots\ldots\ldots(13)$$

The integrals (12), (13) must not be taken through a place where $U + n/k = 0$, as appears from (8). We have already seen that any value of n for which this can occur is admissible. But (13) shows that no other real value of n is admissible; and it serves to determine any complex values of n.

In (13) suppose (as before) that $n/k = p + iq$; then separating the real and imaginary parts, we get

$$\int_0^l \frac{(p + U)^2 - q^2}{\{(p + U)^2 + q^2\}^2}\, dy = 0, \quad \int_0^l \frac{q(p + U)\, dy}{\{(p + U)^2 + q^2\}^2} = 0, \quad \ldots\ldots(14)$$

from the second of which we may infer that if q be finite, $p + U$ must change sign, as we have already seen that it must do when $q = 0$. In every case then, when k is small, the *real part* of n must equal some value of $-kU$*.

It may be of interest to show the application of (13) to a case formerly treated† in which the velocity-curve is made up of straight portions and is anti-symmetrical with respect to the point lying midway between the two walls, now taken as origin of y. Thus on the positive side

from $\qquad y = 0$ to $y = \tfrac{1}{2}b'$, $\qquad U = \dfrac{Vy}{\tfrac{1}{2}b'}$;

from $\qquad y = \tfrac{1}{2}b'$ to $y = \tfrac{1}{2}b' + b$, $\quad U = \dfrac{Vy}{\tfrac{1}{2}b'} + \mu V(y - \tfrac{1}{2}b')$;

while on the negative side U takes symmetrically the opposite values. Then if we write $n/kV = n'$, (13) becomes

$$0 = \int_0^{\tfrac{1}{2}b'} \frac{dy}{(2y/b + n')^2} + \int_{\tfrac{1}{2}b'}^{\tfrac{1}{2}b' + b} \frac{dy}{\{2y/b + \mu(y - \tfrac{1}{2}b') + n'\}^2}$$

$\qquad$ + same with n' reversed.

Effecting the integrations, we find after reduction

$$n'^2 = \frac{n^2}{k^2 V^2} = \frac{2b + b' + 2\mu b\,(b + b') + \mu^2 b^2 b'}{2b + b'}, \quad \ldots\ldots\ldots(15)$$

in agreement with equation (23) of the paper referred to when k is there made small. Hence n, if imaginary at all, is a pure imaginary, and it is imaginary only when μ lies between $-1/b$ and $-1/b - 2/b'$. The regular motion is then exponentially unstable.

In the only unstable cases hitherto investigated the velocity-curve is made up of straight portions meeting at finite angles, and it may perhaps be thought that the instability has its origin in this discontinuity. The method now under discussion disposes of any doubt. For obviously in (13) it can make no important difference whether dU/dy is discontinuous or not. If a motion is definitely unstable in the former case, it cannot become stable merely by easing off the finite angles in the velocity-curve. There exist, therefore, exponentially unstable motions in which both U and dU/dy are continuous. And it is further evident that any proposed velocity-curve may be replaced approximately by straight lines as in my former papers.

---

* By the method of a former paper "On the question of the Stability of the Flow of Fluids" (*Phil. Mag.* Vol. xxxiv. p. 59 (1892); *Scientific Papers*, Vol. iii. p. 579) the conclusion that $p + U$ must change sign may be extended to the problem of the simple shearing motion between two parallel walls of a *viscous* fluid, and this whatever may be the value of k.

† *Proc. Lond. Math. Soc.* Vol. xix. p. 67 (1887); *Scientific Papers*, Vol. iii. p. 20, figs. (3), (4), (5).

The fact that n in equation (15) appears only as n^2 is a simple consequence of the anti-symmetrical character of U. For if in (13) we measure y from the centre and integrate between the limits $\pm \frac{1}{2}l$, we obtain in that case

$$\int_0^{\frac{1}{2}l} \frac{n^2/k^2 + U^2}{(n^2/k^2 - U^2)^2}\, dy = 0, \qquad \ldots\ldots\ldots\ldots\ldots\ldots\ldots(16)$$

in which only n^2 occurs. But it does not appear that n^2 is necessarily real, as happens in (15).

Apart from such examples as were treated in my former papers in which d^2U/dy^2 vanishes except at certain definite places, there are very few cases in which (3) can be solved analytically. If we suppose that $v = \sin(\pi y/l)$, vanishing when $y = 0$ and when $y = l$, and seek what is then admissible for U, we get

$$U + n/k = A \cos \{k^2 + \pi^2/l^2\}^{\frac{1}{2}} y + B \sin \{k^2 + \pi^2/l^2\}^{\frac{1}{2}} y, \quad \ldots\ldots(17)$$

in which A and B are arbitrary and n may as well be supposed to be zero. But since U varies with k, the solution is of no great interest.

In estimating the significance of our results respecting stability, we must of course remember that the disturbance has been assumed to be and to remain infinitely small. Where stability is indicated, the magnitude of the admissible disturbance may be very restricted. It was on these lines that Kelvin proposed to explain the apparent contradiction between theoretical results for an inviscid fluid and observation of what happens in the motion of real fluids which are all more or less viscous. Prof. McF. Orr has carried this explanation further *. Taking the case of a simple shearing motion between two walls, he investigates a composite disturbance, periodic with respect to x but not with respect to t, given initially as

$$v = B \cos lx \cos my, \qquad \ldots\ldots\ldots\ldots\ldots\ldots(18)$$

and he finds, equation (38), that when m is large the disturbance may increase very much, though ultimately it comes to zero. Stability in the mathematical sense (B infinitely small) may thus be not inconsistent with a practical instability. A complete theoretical proof of instability requires not only a method capable of dealing with finite disturbances but also a definition, not easily given, of what is meant by the term. In the case of stability we are rather better situated, since by absolute stability we may understand complete recovery from disturbances of any kind however large, such as Reynolds showed to occur in the present case when viscosity is paramount †. In the absence of dissipation, stability in this sense is not to be expected.

---

* *Proc. Roy. Irish Academy*, Vol. xxvii. Section A, No. 2, 1907. Other related questions are also treated.

† See also Orr, *Proc. Roy. Irish Academy*, 1907, p. 124.

Another manner of regarding the present problem of the shearing motion of an inviscid fluid is instructive. In the original motion the vorticity is constant throughout the whole space between the walls. The disturbance is represented by a superposed vorticity, which may be either positive or negative, and this vorticity everywhere *moves with the fluid*. At any subsequent time the same vorticities exist as initially; the only question is as to their distribution. And when this distribution is known, the whole motion is determined. Now it would seem that the added vorticities will produce most effect if the positive parts are brought together, and also the negative parts, as much as is consistent with the prescribed periodicity along x, and that even if this can be done the effect cannot be out of proportion to the magnitude of the additional vorticities. If this view be accepted, the temporary large increase in Prof. Orr's example would be attributed to a specially unfavourable distribution initially in which (m large) the positive and negative parts of the added vorticities are closely intermingled. We may even go further and regard the subsequent tendency to evanescence, rather than the temporary increase, as the normal phenomenon. The difficulty in reconciling the observed behaviour of actual fluids with the theory of an inviscid fluid still seems to me to be considerable, unless indeed we can admit a distinction between a fluid of infinitely small viscosity and one of none at all.

At one time I thought that the instability suggested by observation might attach to the stages through which a viscous liquid must pass in order to acquire a uniform shearing motion rather than to the final state itself. Thus in order to find an explanation of " skin friction " we may suppose the fluid to be initially at rest between two infinite fixed walls, one of which is then suddenly made to move in its own plane with a uniform velocity. In the earlier stages the other wall has no effect and the problem is one considered by Fourier in connexion with the conduction of heat. The velocity U in the laminar motion satisfies generally an equation of the form

$$\frac{dU}{dt} = \frac{d^2U}{dy^2}, \qquad \dots\dots\dots\dots\dots\dots\dots\dots\dots\dots(19)$$

with the conditions that initially ($t = 0$) $U = 0$, and that from $t = 0$ onwards $U = 1$ when $y = 0$, and (if we please) $U = 0$ when $y = l$. We might employ Fourier's solution, but all that we require follows at once from the differential equation itself. It is evident that dU/dt, and therefore d^2U/dy^2, is everywhere positive and accordingly that a non-viscous liquid, moving laminarly as the viscous fluid moves in any of these stages, is stable. It would appear then that no explanation is to be found in this direction.

Hitherto we have supposed that the disturbance is periodic as regards x, but a simple example, not coming under this head, may be worthy of notice. It is that of the disturbance due to a single vortex filament in which the vorticity differs from the otherwise uniform vorticity of the neighbouring fluid. In the figure the lines AA, BB represent the situation of the walls and AM the velocity-curve of the original shearing motion rising from zero at A to a finite value at M. For the present purpose, however, we suppose material walls to be absent, but that the same effect (of prohibiting normal motion) is arrived at by suitable suppositions as to the fluid lying outside and now imagined infinite. It is only necessary to continue the velocity-curve in the manner shown $AMCN...$, the vorticities in the alternate layers of equal width being equal and opposite. Symmetry then shows that under the operation of these vorticities the fluid moves as if AA, BB, &c. were material walls.

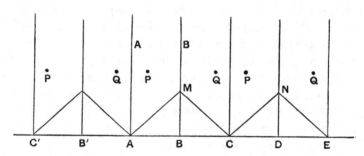

We have now to trace the effect of an additional vorticity, supposed positive, at a point P. If the wall AA were alone concerned, its effect would be imitated by the introduction of an opposite vorticity at the point Q which is the image of P in AA. Thus P would move under the influence of the original vorticities, already allowed for, and of the negative vorticity at Q. Under the latter influence it would move parallel to AA with a certain velocity, and for the same reason Q would move similarly, so that PQ would remain perpendicular to AA. To take account of both walls the more complicated arrangement shown in the figure is necessary, in which the points P represent equal positive vorticities and Q equal negative vorticities. The conditions at both walls are thus satisfied; and as before all the vortices P, Q move under each other's influence so as to remain upon a line perpendicular to AA. Thus, to go back to the original form of the problem, P moves parallel to the walls with a constant velocity, and no change ensues in the character of the motion—a conclusion which will appear the more remarkable when we remember that there is no limitation upon the magnitude of the added vorticity.

The same method is applicable—in imagination at any rate—whatever be the distribution of vorticities between the walls, and the corresponding velocity at any point is determined by quadratures on Helmholtz's principle. The new positions of all the vorticities after a short time are thus found, and then a new departure may be taken, and so on indefinitely.

378.

REFLECTION OF LIGHT AT THE CONFINES OF A DIFFUSING MEDIUM.

[*Nature*, Vol. XCII. p. 450, 1913.]

I SUPPOSE that everyone is familiar with the beautifully graded illumination of a paraffin candle, extending downwards from the flame to a distance of several inches. The thing is seen at its best when there is but one candle in an otherwise dark room, and when the eye is protected from the direct light of the flame. And it must often be noticed when a candle is broken across, so that the two portions are held together merely by the wick, that the part below the fracture is much darker than it would otherwise be, and the part above brighter, the contrast between the two being very marked. This effect is naturally attributed to reflection, but it does not at first appear that the cause is adequate, seeing that at perpendicular incidence the reflection at the common surface of wax and air is only about 4 per cent.

A little consideration shows that the efficacy of the reflection depends upon the incidence not being limited to the neighbourhood of the perpendicular. In consequence of diffusion* the propagation of light within the wax is not specially along the length of the candle, but somewhat approximately equal in *all* directions. Accordingly at a fracture there is a good deal of "total reflection." The general attenuation downwards is doubtless partly due to defect of transparency, but also, and perhaps more, to the lateral escape of light at the surface of the candle, thereby rendered visible. By hindering this escape the brightly illuminated length may be much increased.

The experiment may be tried by enclosing the candle in a reflecting tubular envelope. I used a square tube composed of four rectangular pieces of mirror glass, 1 in. wide, and 4 or 5 in. long, held together by strips of

---

* To what is the diffusion due? Actual cavities seem improbable. Is it chemical heterogeneity, or merely varying orientation of chemically homogeneous material operative in virtue of double refraction?

pasted paper. The tube should be lowered over the candle until the whole of the flame projects, when it will be apparent that the illumination of the candle extends decidedly lower down than before.

In imagination we may get quit of the lateral loss by supposing the diameter of the candle to be increased without limit, the source of light being at the same time extended over the whole of the horizontal plane.

To come to a definite question, we may ask what is the proportion of light reflected when it is incident equally in all directions upon a surface of transition, such as is constituted by the candle fracture. The answer depends upon a suitable integration of Fresnel's expression for the reflection of light of the two polarisations, viz.

$$S^2 = \frac{\sin^2(\theta - \theta')}{\sin^2(\theta + \theta')}, \qquad T^2 = \frac{\tan^2(\theta - \theta')}{\tan^2(\theta + \theta')}, \quad \dots\dots\dots\dots(1)$$

where θ, θ' are the angles of incidence and refraction. We may take first the case where $\theta > \theta'$, that is, when the transition is from the less to the more refractive medium.

The element of solid angle is $2\pi \sin\theta\, d\theta$, and the area of cross-section corresponding to unit area of the refracting surface is $\cos\theta$; so that we have to consider

$$2 \int_0^{\frac{1}{2}\pi} \sin\theta \cos\theta \, (S^2 \text{ or } T^2) \, d\theta, \quad \dots\dots\dots\dots\dots(2)$$

the multiplier being so chosen as to make the integral equal to unity when S^2 or T^2 has that value throughout. The integral could be evaluated analytically, at any rate in the case of S^2, but the result would scarcely repay the trouble. An estimate by quadratures in a particular case will suffice for our purposes, and to this we shall presently return.

In (2) θ varies from 0 to $\frac{1}{2}\pi$ and θ' is always real. If we suppose the passage to be in the other direction, viz. from the more to the less refractive medium, S^2 and T^2, being symmetrical in θ and θ', remain as before, and we have to integrate

$$2 \sin\theta' \cos\theta' \, (S^2 \text{ or } T^2) \, d\theta'.$$

The integral divides itself into two parts, the first from 0 to α, where α is the critical angle corresponding to $\theta = \frac{1}{2}\pi$. In this S^2, T^2 have the values given in (1). The second part of the range from $\theta' = \alpha$ to $\theta' = \frac{1}{2}\pi$ involves "total reflection," so that S^2 and T^2 must be taken equal to unity. Thus altogether we have

$$2 \int_0^{\alpha} \sin\theta' \cos\theta' \, (S^2 \text{ or } T^2) \, d\theta' + 2 \int_{\alpha}^{\frac{1}{2}\pi} \sin\theta' \cos\theta' \, d\theta', \quad \dots\dots(3)$$

in which $\sin \alpha = 1/\mu$, μ (greater than unity) being the refractive index. In (3)

$$2 \sin \theta' \cos \theta' \, d\dot{\theta}' = d \sin^2 \theta' = \mu^{-2} d \sin^2 \theta,$$

and thus

$$(3) = \mu^{-2} \times (2) + 1 - \mu^{-2} = \frac{1}{\mu^2} \left\{ \mu^2 - 1 + \int_0^{\frac{1}{2}\pi} \sin 2\theta \, (S^2 \text{ or } T^2) \, d\theta \right\}, \dots (4)$$

expressing the proportion of the uniformly diffused incident light reflected in this case.

Much the more important part is the light totally reflected. If $\mu = 1\cdot 5$, this amounts to 5/9 or 0·5556.

With the same value of μ, I find by Weddle's rule

$$\int_0^{\frac{1}{2}\pi} \sin 2\theta \, . \, S^2 d\theta = 0\cdot1460, \quad \int_0^{\frac{1}{2}\pi} \sin 2\theta \, . \, T^2 d\theta = 0\cdot0339.$$

Thus for light vibrating perpendicularly to the plane of incidence

$$(4) = 0\cdot5556 + 0\cdot0649 = 0\cdot6205;$$

while for light vibrating in the plane of incidence

$$(4) = 0\cdot5556 + 0\cdot0151 = 0\cdot5707.$$

The increased reflection due to the diffusion of the light is thus abundantly explained, by far the greater part being due to the total reflection which ensues when the incidence in the denser medium is somewhat oblique.

379.

THE PRESSURE OF RADIATION AND CARNOT'S PRINCIPLE.

[*Nature*, Vol. XCII. pp. 527, 528, 1914.]

As is well known, the pressure of radiation, predicted by Maxwell, and since experimentally confirmed by Lebedew and by Nichols and Hull, plays an important part in the theory of radiation developed by Boltzmann and W. Wien. The existence of the pressure according to electromagnetic theory is easily demonstrated*, but it does not appear to be generally remembered that it could have been deduced with some confidence from thermodynamical principles, even earlier than in the time of Maxwell. Such a deduction was, in fact, made by Bartoli in 1876, and constituted the foundation of Boltzmann's work †. Bartoli's method is quite sufficient for his purpose; but, mainly because it employs irreversible operations, it does not lend itself to further developments. It may therefore be of service to detail the elementary argument on the lines of Carnot, by which it appears that in the absence of a pressure of radiation it would be possible to raise heat from a lower to a higher temperature.

The imaginary apparatus is, as in Boltzmann's theory, a cylinder and piston formed of perfectly reflecting material, within which we may suppose the radiation to be confined. This radiation is always of the kind characterised as complete (or black), a requirement satisfied if we include also a very small black body with which the radiation is in equilibrium. If the operations are slow enough, the size of the black body may be reduced without limit, and then the whole energy at a given temperature is that of the radiation and proportional to the volume occupied. When we have occasion to introduce or abstract heat, the communication may be supposed

---

* See, for example, J. J. Thomson, *Elements of Electricity and Magnetism* (Cambridge, 1895, § 241); Rayleigh, *Phil. Mag.* Vol. XLV. p. 222 (1898); *Scientific Papers*, Vol. IV. p. 354.

† *Wied. Ann.* Vol. XXXII. pp. 31, 291 (1884). It is only through Boltzmann that I am acquainted with Bartoli's reasoning.

in the first instance to be with the black body. The operations are of two kinds: (1) compression (or rarefaction) of the kind called *adiabatic*, that is, without communication of heat. If the volume increases, the temperature must fall, even though in the absence of pressure upon the piston no work is done, since the same energy of complete radiation now occupies a larger space. Similarly a rise of temperature accompanies adiabatic contraction. In the second kind of operation (2) the expansions and contractions are *isothermal*—that is, without change of temperature. In this case heat must pass, into the black body when the volume expands and out of it when the volume contracts, and at a given temperature the amount of heat which must pass is proportional to the change of volume.

The cycle of operations to be considered is the same as in Carnot's theory, the only difference being that here, in the absence of pressure, there is no question of external work. Begin by isothermal expansion at the lower temperature during which heat is taken in. Then compress adiabatically until a higher temperature is reached. Next continue the compression isothermally until the same amount of heat is given out as was taken in during the first expansion. Lastly, restore the original volume adiabatically. Since no heat has passed upon the whole in either direction, the final state is identical with the initial state, the temperature being recovered as well as the volume. The sole result of the cycle is that heat is raised from a lower to a higher temperature. Since this is assumed to be impossible, the supposition that the operations can be performed without external work is to be rejected—in other words, we must regard the radiation as exercising a pressure upon the moving piston. Carnot's principle and the absence of a pressure are incompatible.

For a further discussion it is, of course, desirable to employ the general formulation of Carnot's principle, as in a former paper*. If p be the pressure, θ the absolute temperature,

$$\theta \frac{dp}{d\theta} = M, \quad\dots\dots\dots\dots\dots\dots\dots\dots\dots\dots(29)$$

where $M\,dv$ represents the heat that must be communicated, while the volume alters by dv and $d\theta = 0$. In the application to radiation M cannot vanish, and therefore p cannot. In this case clearly

$$M = U + p, \quad\dots\dots\dots\dots\dots\dots\dots\dots\dots(30)$$

where U denotes the volume-density of the energy—a function of θ only. Hence

$$\theta \frac{dp}{d\theta} = U + p. \quad\dots\dots\dots\dots\dots\dots\dots\dots(31)$$

---

* "On the Pressure of Vibrations," *Phil. Mag.* Vol. III. p. 338, 1902; *Scientific Papers*, Vol. v. p. 47.

14

If we assume from electromagnetic theory that

$$p = \tfrac{1}{3}U, \quad \dotfill (32)$$

it follows at once that

$$U \propto \theta^4, \quad \dotfill (33)$$

the well-known law of Stefan.

In (31) if p be known as a function of θ, U as a function of θ follows immediately. If, on the other hand, U be known, we have

$$d\left(\frac{p}{\theta}\right) = \frac{U}{\theta^2}\, d\theta,$$

and thence

$$\frac{p}{\theta} = \int_0^\theta \frac{U}{\theta^2}\, d\theta + C. \quad \dotfill (34)$$

380.

FURTHER APPLICATIONS OF BESSEL'S FUNCTIONS OF HIGH ORDER TO THE WHISPERING GALLERY AND ALLIED PROBLEMS.

[*Philosophical Magazine*, Vol. XXVII. pp. 100—109, 1914.]

IN the problem of the Whispering Gallery* waves in two dimensions, of length small in comparison with the circumference, were shown to run round the concave side of a wall with but little tendency to spread themselves inwards. The wall was supposed to be perfectly reflecting for all kinds of waves. But the question presents itself whether the sensibly perfect reflexion postulated may not be attained on the principle of so-called "total reflexion," the wall being merely the transition between two uniform media of which the outer is the less refracting. It is not to be expected that absolutely no energy should penetrate and ultimately escape to an infinite distance. The analogy is rather with the problem treated by Stokes† of the communication of vibrations from a vibrating solid, such as a bell or wire, to a surrounding gas, when the wave-length in the gas is somewhat large compared with the dimensions of the vibrating segments. The energy radiated to a distance may then be extremely small, though not mathematically evanescent.

A comparison with the simple case where the surface of the vibrating body is *plane* ($x = 0$) is interesting, especially as showing how the partial

---

* *Phil. Mag.* Vol. XX. p. 1001 (1910); *Scientific Papers*, Vol. V. p. 619. But the numbers there given require some correction owing to a slip in Nicholson's paper from which they were derived, as was first pointed out to me by Prof. Macdonald. Nicholson's table should be interpreted as relating to the values, not of $2 \cdot 1123 \, (n-z)/z^{\frac{1}{3}}$, but of $1 \cdot 3447 \, (n-z)/z^{\frac{1}{3}}$, see Nicholson, *Phil. Mag.* Vol. XXV. p. 200 (1913). Accordingly, in my equation (5) $1 \cdot 1814 \, n^{\frac{1}{3}}$ should read $1 \cdot 8558 \, n^{\frac{1}{3}}$, and in equation (8) $\cdot 51342 \, n^{\frac{1}{3}}$ should read $\cdot 8065 \, n^{\frac{1}{3}}$. [1916. Another error should be noticed. In (3), $= \int_0^\infty \cos n \, (\omega - \sin \omega) \, d\omega/\pi$ must be omitted, the integrand being periodic. See Watson, *Phil. Mag.* Vol. XXXII. p. 233, 1916.]

† *Phil. Trans.* 1868. See *Theory of Sound*, Vol. II. § 324.

escape of energy is connected with the curvature of the surface. If V be the velocity of propagation, and $2\pi/k$ the wave-length of plane waves of the given period, the time-factor is e^{ikVt}, and the equation for the velocity-potential in two dimensions is

$$\frac{d^2\phi}{dx^2} + \frac{d^2\phi}{dy^2} + k^2\phi = 0. \qquad \qquad (1)$$

If ϕ be also proportional to $\cos my$, (1) reduces to

$$\frac{d^2\phi}{dx^2} + (k^2 - m^2)\,\phi = 0, \qquad \qquad (2)$$

of which the solution changes its form when m passes through the value k. For our purpose m is to be supposed greater than k, viz. the wave-length of plane waves is to be greater than the linear period along y. That solution of (1) on the positive side which does not become infinite with x is proportional to $e^{-x\sqrt{(m^2-k^2)}}$, so that we may take

$$\phi = \cos kVt \,.\, \cos my \,.\, e^{-x\sqrt{(m^2-k^2)}}. \qquad \qquad (3)$$

However the vibration may be generated at $x = 0$, provided only that the linear period along y be that assigned, it is limited to relatively small values of x and, since no energy can escape, no work is done on the whole at $x = 0$. And this is true by however little m may exceed k.

The reason of the difference which ensues when the vibrating surface is curved is now easily seen. Suppose, for example, that in two dimensions ϕ is proportional to $\cos n\theta$, where θ is a vectorial angle. Near the surface of a cylindrical vibrator the conditions may be such that (3) is approximately applicable, and ϕ rapidly diminishes as we go outwards. But when we reach a radius vector r which is sensibly different from the initial one, the conditions may change. In effect the linear dimension of the vibrating compartment increases proportionally to r, and ultimately the equation (2) changes its form and ϕ oscillates, instead of continuing an exponential decrease. *Some* energy always escapes, but the amount must be very small if there is a sufficient margin to begin with between m and k.

It may be well before proceeding further to follow a little more closely what happens when there is a transition at a plane surface $x = 0$ from a more to a less refractive medium. The problem is that of total reflexion when the incidence is grazing, in which case the usual formulæ* become nugatory. It will be convenient to fix ideas upon the case of sonorous waves, but the results are of wider application. The general differential equation is of the form

$$\frac{d^2\phi}{dt^2} = V^2 \left(\frac{d^2\phi}{dx^2} + \frac{d^2\phi}{dy^2} \right), \qquad \qquad (4)$$

---

* See for example *Theory of Sound*, Vol. II. § 270.

which we will suppose to be adapted to the region where x is negative. On the right (x positive) V is to be replaced by V_1, where $V_1 > V$, and ϕ by ϕ_1. In optical notation $V_1/V = \mu$, where μ (greater than unity) is the refractive index. We suppose ϕ and ϕ_1 to be proportional to $e^{i(by+ct)}$, b and c being the same in both media. Further, on the left we suppose b and c to be related as they would be for simple plane waves propagated parallel to y. Thus (4) becomes, with omission of $e^{i(by+ct)}$,

$$\frac{d^2\phi}{dx^2} = 0, \quad \frac{d^2\phi_1}{dx^2} = b^2(\mu^2 - 1)/\mu^2, \quad \ldots\ldots\ldots\ldots\ldots(5)$$

of which the solutions are

$$\phi = A + Bx, \qquad \phi_1 = C e^{-bx\sqrt{(\mu^2-1)}/\mu}, \quad \ldots\ldots\ldots\ldots(6)$$

A, B, C denoting constants so far arbitrary. The boundary conditions require that when $x = 0$, $d\phi/dx = d\phi_1/dx$ and that $\rho\phi = \rho_1\phi_1$, ρ, ρ_1 being the densities. Hence discarding the imaginary part, and taking $A = 1$, we get finally

$$\phi = \left\{1 - \frac{\rho bx\sqrt{(\mu^2-1)}}{\rho_1\mu}\right\} \cos(by + ct), \quad \ldots\ldots\ldots\ldots(7)$$

$$\phi_1 = \frac{\rho}{\rho_1} e^{-bx\sqrt{(\mu^2-1)}/\mu} \cos(by + ct). \quad \ldots\ldots\ldots\ldots(8)$$

It appears that while nothing can escape on the positive side, the amplitude on the negative side increases rapidly as we pass away from the surface of transition.

If $\mu < 1$, a wave of the ordinary kind is propagated into the second medium, and energy is conveyed away.

In proceeding to consider the effect of curvature it will be convenient to begin with Stokes' problem, taking advantage of formulæ relating to Bessel's and allied functions of high order developed by Lorenz, Nicholson, and Macdonald*. The motion is supposed to take place in two dimensions, and ideas may be fixed upon the case of aerial vibrations. The velocity-potential ϕ is expressed by means of polar coordinates r, θ, and will be assumed to be proportional to $\cos n\theta$, attention being concentrated upon the case where n is a large integer. The problem is to determine the motion at a distance due to the normal vibration of a cylindrical surface at $r = a$, and it turns upon the character of the function of r which represents a disturbance propagated outwards. If $D_n(kr)$ denote this function, we have

$$\phi = e^{ikVt} \cos n\theta . D_n(kr), \quad \ldots\ldots\ldots\ldots\ldots(9)$$

and $D_n(z)$ satisfies Bessel's equation

$$D_n'' + \frac{1}{z} D_n' + \left(1 - \frac{n^2}{z^2}\right) D_n = 0. \quad \ldots\ldots\ldots\ldots(10)$$

* Compare also Debye, *Math. Ann.* Vol. LXVII. (1909).

It may be expressed in the form

$$D_n = \frac{J_{-n} - e^{in\pi} J_n}{\sin n\pi}, \qquad \ldots\ldots\ldots\ldots\ldots(11)$$

which, however, requires a special evaluation when n is an integer. Using Schläfli's formula

$$J_n(z) = \frac{1}{\pi} \int_0^\pi \cos(z \sin\theta - n\theta)\, d\theta - \frac{\sin n\pi}{\pi} \int_0^\infty e^{-n\theta - z \sinh\theta}\, d\theta, \ldots(12)$$

n being positive or negative, and z positive, we find

$$D_n(z) = \frac{1}{\pi} \int_0^\infty e^{n\theta - z \sinh\theta}\, d\theta + \frac{\cos n\pi}{\pi} \int_0^\infty e^{-n\theta - z \sinh\theta}\, d\theta$$

$$- \frac{1}{\pi} \int_0^\pi \sin(z \sin\theta - n\theta)\, d\theta - \frac{i}{\pi} \int_0^\pi \cos(z \sin\theta - n\theta)\, d\theta, \ldots\ldots(13)$$

the imaginary part being $- iJ_n(z)$ simply. This holds good for any integral value of n. The present problem requires the examination of the form assumed by D_n when n is very great and the ratio z/n decidedly greater, or decidedly less, than unity.

In the former case we set $n = z \sin\alpha$, and the important part of D_n arises from the two integrals last written. It appears* that

$$D_n = \left(\frac{2}{\pi z \cos\alpha}\right)^{\frac{1}{2}} e^{-i\rho}, \qquad \ldots\ldots\ldots\ldots\ldots(14)$$

where

$$\rho = \tfrac{1}{4}\pi + z \{\cos\alpha - (\tfrac{1}{2}\pi - \alpha) \sin\alpha\}, \ldots\ldots\ldots\ldots(15)$$

or when z is extremely large ($\alpha = 0$)

$$D_n(z) = \left(\frac{2}{\pi z}\right)^{\frac{1}{2}} e^{-i(\frac{1}{4}\pi + z)}. \qquad \ldots\ldots\ldots\ldots\ldots(16)$$

At a great distance the value of ϕ in (9) thus reduces to

$$\phi = \left(\frac{2}{\pi k r}\right)^{\frac{1}{2}} \cos n\theta \cdot e^{i \{k(Vt - r) - \frac{1}{4}\pi\}}, \qquad \ldots\ldots\ldots\ldots(17)$$

from which finally the imaginary part may be omitted.

When on the other hand z/n is decidedly less than unity, the most important part of (13) arises from the first and last integrals. We set $n = z \cosh\beta$, and then, n being very great,

$$D_n(z) = \left(\frac{\coth\beta}{2n\pi}\right)^{\frac{1}{2}} \{2 e^{-t} - i e^t\}, \qquad \ldots\ldots\ldots\ldots(18)$$

where

$$t = n(\tanh\beta - \beta). \qquad \ldots\ldots\ldots\ldots\ldots(19)$$

---

* Nicholson, B. A. Report, Dublin, 1908, p. 595; Phil. Mag. Vol. xix. p. 240 (1910); Macdonald, Phil. Trans. Vol. ccx. p. 135 (1909).

Also, the most important part of the real and imaginary terms being retained,

$$D_n'(z) = -\left(\frac{\sinh\beta\cosh\beta}{2n\pi}\right)^{\frac{1}{2}}\{2e^{-t} + ie^t\}. \ldots\ldots\ldots(20)$$

The application is now simple. From (9) with introduction of an arbitrary coefficient

$$\frac{d\phi}{dr} = kA\,e^{ikVt}\cos n\theta\,.\,D_n'(kr). \ldots\ldots\ldots\ldots(21)$$

If we suppose that the normal velocity of the vibrating cylindrical surface $(r = a)$ is represented by $e^{ikVt}\cos n\theta$, we have

$$kAD_n'(ka) = 1, \ldots\ldots\ldots\ldots\ldots(22)$$

and thus at distance r

$$\phi = e^{ikVt}\cos n\theta\,\frac{D_n(kr)}{k\,D_n'(ka)}, \ldots\ldots\ldots\ldots(23)$$

or when r is very great

$$\phi = \cos n\theta\left(\frac{2}{\pi kr}\right)^{\frac{1}{2}}\frac{e^{i\{k(Vt-r)-\frac{1}{4}\pi\}}}{k\,D_n'(ka)}. \ldots\ldots\ldots\ldots(24)$$

We may now, following Stokes, compare the actual motion at a distance with that which would ensue were lateral motion prevented, as by the insertion of a large number of thin plane walls radiating outwards along the lines $\theta = \text{constant}$, the normal velocity at $r = a$ being the same in both cases. In the altered problem we have merely in (23) to replace D_n, D_n' by D_0, D_0'. When z is great enough, $D_n(z)$ has the value given in (16), independently of the particular value of n. Accordingly the ratio of velocity-potentials at a distance in the two cases is represented by the symbolic fraction

$$\frac{D_0'(ka)}{D_n'(ka)}, \ldots\ldots\ldots\ldots\ldots\ldots(25)$$

in which

$$D_0'(ka) = -i\left(\frac{2}{\pi ka}\right)^{\frac{1}{2}}e^{-i(\frac{1}{4}\pi + ka)}. \ldots\ldots\ldots\ldots(26)$$

We have now to introduce the value of $D_n'(ka)$. When n is very great, and ka/n decidedly less than unity, t is negative in (20), and e^t is negligible in comparison with e^{-t}. The modulus of (25) is therefore

$$\left(\frac{n/ka}{\sinh\beta\cosh\beta}\right)^{\frac{1}{2}}e^t, \quad\text{or}\quad \frac{e^{-n(\beta - \tanh\beta)}}{\sinh^{\frac{1}{2}}\beta}. \ldots\ldots\ldots\ldots(27)$$

For example, if $n = 2ka$, so that the linear period along the circumference of the vibrating cylinder $(2\pi a/n)$ is half the wave-length,

$$\cosh\beta = 2, \quad \beta = 1\cdot317, \quad \sinh\beta = 1\cdot7321, \quad \tanh\beta = \cdot8660,$$

and the numerical value of (27) is

$$e^{-\cdot4510\,n} \div \sqrt{(1\cdot732)}.$$

When n is great, the vibration at a distance is extraordinarily small in comparison with what it would have been were lateral motion prevented. As another example, let $n = 1.1\, ka$. Then $(27) = e^{-.027\, n} \div \sqrt{(.4587)}$. Here n would need to be about 17 times larger for the same sort of effect.

The extension of Stokes' analysis to large values of n only emphasizes his conclusion as to the insignificance of the effect propagated to a distance when the vibrating segments are decidedly smaller than the wave-length.

We now proceed to the case of the whispering gallery supposed to act by "total reflexion." From the results already given, we may infer that when the refractive index is moderate, the escape of energy must be very small, and accordingly that the vibrations inside have long persistence. There is, however, something to be said upon the other side. On account of the concentration near the reflecting wall, the store of energy to be drawn upon is diminished. At all events the problem is worthy of a more detailed examination.

Outside the surface of transition $(r = a)$ we have the same expression (9) as before for the velocity-potential, k and V having values proper to the outer medium. Inside k and V are different, but the product kV is the same. We will denote the altered k by h. In accordance with our suppositions $h > k$, and h/k represents the refractive index (μ) of the inside medium relatively to that outside. On account of the damping k and h are complex, though their ratio is real; but the imaginary part is relatively small. Thus, omitting the factors $e^{ikVt} \cos n\theta$, we have $(r > a)$

$$\phi = A D_n (kr), \quad\dots\dots\dots\dots\dots\dots\dots(28)$$

and inside $(r < a)$
$$\phi = B J_n (hr). \quad\dots\dots\dots\dots\dots\dots\dots(29)$$

The boundary conditions to be satisfied when $r = a$ are easily expressed. The equality of normal motions requires that

$$k A D_n{}' (ka) = h B J_n{}' (ha); \quad\dots\dots\dots\dots\dots(30)$$

and the equality of pressures requires that

$$\sigma A D_n (ka) = \rho B J_n (ha), \quad\dots\dots\dots\dots\dots(31)$$

σ, ρ being the densities of the outer and inner media respectively. The equation for determining the values of ha, ka (in addition to $h/k = \mu$) is accordingly

$$\frac{k D_n{}' (ka)}{\sigma D_n (ka)} = \frac{h J_n{}' (ha)}{\rho J_n (ha)}. \quad\dots\dots\dots\dots\dots(32)$$

Equation (32) cannot be satisfied exactly by real values of h and k; for, although $J_n{}'/J_n$ is then real, $D_n{}'/D_n$ includes an imaginary part. But since the imaginary part is relatively small, we may conclude that *approximately* h and k are real, and the first step is to determine these real values.

Since ka is supposed to be decidedly less than n, D_n and D_n' are given by (18), (20); and, if we neglect the imaginary part,

$$\frac{D_n'(ka)}{D_n(ka)} = -\sinh\beta. \qquad\qquad\dots\dots\dots\dots\dots\dots(33)$$

Thus (32) becomes
$$\frac{J_n'(ha)}{J_n(ha)} = -\frac{\rho k}{\sigma h}\sinh\beta, \qquad\dots\dots\dots\dots\dots(34)$$

the right-hand member being real and negative. Of this a solution can always be found in which $ha = n$ very nearly. For* $J_n(z)$ increases with z from zero until $z = n + \cdot8065\,n^{\frac{1}{3}}$, when $J_n'(z) = 0$, and then decreases until it vanishes when $z = n + 1\cdot8558\,n^{\frac{1}{3}}$. Between these limits for z, J_n'/J_n assumes all possible negative values. Substituting n for ha on the right in (34), we get

$$-\frac{\rho ka}{\sigma n}\sinh\beta, \quad \text{or} \quad -\frac{\rho}{\sigma}\tanh\beta, \qquad\dots\dots\dots\dots(35)$$

while $\cosh\beta = \mu$. The approximate real value of ha is thus n simply, while that of ka is n/μ.

These results, though stated for aerial vibrations, have as in all such (two-dimensional) cases a wider application, for example to electrical vibrations, whether the electric force be in or perpendicular to the plane of r, θ. For ordinary gases, of which the compressibility is the same,

$$\rho/\sigma = h^2/k^2 = \mu^2.$$

Hitherto we have neglected the small imaginary part of D_n'/D_n. By (18), (20), when z is real,

$$\frac{D_n'(z)}{D_n(z)} = -\sinh\beta\,\frac{2e^{-t} + i\,e^t}{2e^{-t} - i\,e^t} = -\sinh\beta\cdot(1 + i\,e^{2t}) \dots\dots\dots(36)$$

approximately, with $\cosh\beta = n/z$. We have now to determine what small imaginary additions must be made to ha, ka in order to satisfy the complete equation.

Let us assume $ha = x + iy$, where x and y are real, and y is small. Then approximately

$$\frac{J_n'(x + iy)}{J_n(x + iy)} = \frac{J_n'(x) + iy\,J_n''(x)}{J_n(x) + iy\,J_n'(x)},$$

and
$$J_n''(x) = -\frac{1}{x}J_n'(x) - \left(1 - \frac{n^2}{x^2}\right)J_n(x).$$

Since the approximate value of x is n, J_n'' is small compared with J_n or J_n', and we may take

$$\frac{J_n'(x + iy)}{J_n(x + iy)} = \frac{J_n'(x)}{J_n(x)}\left\{1 - iy\,\frac{J_n'(x)}{J_n(x)}\right\}. \qquad\dots\dots\dots\dots(37)$$

* See paper quoted on p. 211 and correction.

Similarly, if we write $ka = x' + iy'$, where $x' = x/\mu$, $y' = y/\mu$,

$$\frac{D_n'(x' + iy')}{D_n(x' + iy')} = \frac{D_n'(x') + iy'\,D_n''(x')}{D_n(x') + iy'\,D_n'(x')},$$

and in virtue of (10)

$$D_n''(x') = -\frac{\cosh\beta}{n}\,D_n'(x') + \sinh^2\beta\,D_n(x'),$$

where $\cosh\beta = n/x'$. Thus

$$\frac{D_n'(x' + iy')}{D_n(x' + iy')} = \frac{D_n'(x')}{D_n(x')}\left\{1 + iy'\left(-\frac{\cosh\beta}{n} + \sinh^2\beta\,\frac{D_n}{D_n'} - \frac{D_n'}{D_n}\right)\right\}.$$

Accordingly with use of (36)

$$\frac{D_n'(x' + iy')}{D_n(x' + iy')} = -\sinh\beta\left\{1 + i\,e^{2t} + iy'\left(-\frac{\cosh\beta}{n} + \sinh^2\beta\,\frac{D_n}{D_n'} - \frac{D_n'}{D_n}\right)\right\}. \quad (38)$$

Equation (32) asserts the equality of the expressions on the two sides of (38) with

$$\frac{h\sigma}{k\rho}\frac{J_n'(x)}{J_n(x)}\left\{1 - iy\,\frac{J_n'(x)}{J_n(x)}\right\}. \quad\quad\ldots\ldots\ldots\ldots\ldots(37)$$

If we neglect the imaginary terms in (38), (37), we fall back on (34). The imaginary terms themselves give a second equation. In forming this we notice that the terms in y' vanish in comparison with that in y. For in the coefficient of y' the first part, viz. $-n^{-1}\cosh\beta$, vanishes when n is made infinite, while the second and third parts compensate one another in virtue of (33). Accordingly (32) gives with regard to (34)

$$y = \frac{\sigma h}{\rho k}\frac{e^{2t}}{\sinh\beta} = \frac{\mu\sigma}{\rho}\frac{e^{-2n\,(\beta-\tanh\beta)}}{\sinh\beta}, \quad\quad\ldots\ldots\ldots\ldots(39)$$

in which

$$\cosh\beta = \mu. \quad\quad\ldots\ldots\ldots\ldots\ldots\ldots(40)$$

In (39) iy is the imaginary increment of ha, of which the principal real part is n. In the time-factor e^{ikVt}, the exponent

$$ikVt = \frac{iha\,Vt}{\mu a} = \frac{in\,Vt}{\mu a}\left\{1 + \frac{i\,(39)}{n}\right\}.$$

In one complete period τ, $nVt/\mu a$ undergoes the increment 2π. The exponential factor giving the decrement in one period is thus

$$e^{-2\pi\,(39)/n}, \quad\quad\ldots\ldots\ldots\ldots\ldots\ldots(41)$$

or with regard to the smallness of (39)

$$1 - \frac{2\pi\mu\sigma}{n\rho}\frac{e^{-2n\,(\beta-\tanh\beta)}}{\sinh\beta}. \quad\quad\ldots\ldots\ldots\ldots(42)$$

This is the factor by which the amplitude is reduced after each complete period.

In the case of ordinary gases $\rho/\sigma = \mu^2$. As an example, take $\mu = \cosh \beta = 1\cdot3$; then (42) gives

$$1 - \cdot581 n^{-1} e^{-\cdot236 n}. \quad\quad\quad\quad\quad\dots\dots\dots\dots\dots\dots(43)$$

When n rises beyond 10, the damping according to (43) becomes small; and when n is at all large, the vibrations have very great persistence.

In the derivation of (42) we have spoken of stationary vibrations. But the damping is, of course, the same for vibrations which progress round the circumference, since these may be regarded as compounded of two sets of stationary vibrations which differ in phase by 90°.

Calculation thus confirms the expectation that the whispering gallery effect does not require a perfectly reflecting wall, but that the main features are reproduced in transparent media, provided that the velocity of waves is moderately larger outside than inside the surface of transition. And further, the less the curvature of this surface, the smaller is the refractive index (greater than unity) which suffices.

381.

ON THE DIFFRACTION OF LIGHT BY SPHERES OF SMALL * RELATIVE INDEX.

[*Proceedings of the Royal Society*, A, Vol. XC. pp. 219—225, 1914.]

IN a short paper "On the Diffraction of Light by Particles Comparable with the Wave-length†," Keen and Porter describe curious observations upon the intensity and colour of the light transmitted through small particles of precipitated sulphur, while still in a state of suspension, when the size of the particles is comparable with, or decidedly larger than, the wave-length of the light. The particles principally concerned in their experiments appear to have decidedly exceeded those dealt with in a recent paper‡, where the calculations were pushed only to the point where the circumference of the sphere is $2 \cdot 25 \lambda$. The authors cited give as the size of the particles, when the intensity of the light passing through was a minimum, $6\,\mu$ to $10\,\mu$, that is over 10 wave-lengths of yellow light, and they point out the desirability of extending the theory to larger spheres.

The calculations referred to related to the particular case where the (relative) refractive index of the spherical obstacles is $1\cdot5$. This value was chosen in order to bring out the peculiar polarisation phenomena observed in the diffracted light at angles in the neighbourhood of 90°, and as not inappropriate to experiments upon particles of high index suspended in water. I remarked that the extension of the calculations to greater particles would be of interest, but that the arithmetical work would rapidly become heavy.

There is, however, another particular case of a more tractable character, viz., when the relative refractive index is *small* *; and although it may not be the one we should prefer, its discussion is of interest and would be expected

---

* [1914. It would have been in better accordance with usage to have said "of Relative Index differing little from Unity."]

† *Roy. Soc. Proc.* A, Vol. LXXXIX. p. 370 (1913).

‡ *Roy. Soc. Proc.* A, Vol. LXXXIV. p. 25 (1910); *Scientific Papers*, Vol. V. p. 547.

to throw some light upon the general course of the phenomenon. It has already been treated up to a certain point, both in the paper cited and the earlier one * in which experiments upon precipitated sulphur were first described. It is now proposed to develop the matter further.

The specific inductive capacity of the general medium being unity, that of the sphere of radius R is supposed to be K, where $K-1$ is very small. Denoting electric displacements by f, g, h, the primary wave is taken to be

$$h_0 = e^{int} e^{ikx}, \quad \dots \dots \dots \dots \dots \dots \dots \dots \dots (1)$$

so that the direction of propagation is along x (negatively), and that of vibration parallel to z. The electric displacements (f_1, g_1, h_1) in the scattered wave, so far as they depend upon the first power of $(K-1)$, have at a great distance the values

$$f_1, g_1, h_1 = \frac{k^2 P}{4\pi r} \left(\frac{\alpha\gamma}{r^2}, \frac{\beta\gamma}{r^2}, -\frac{\alpha^2 + \beta^2}{r^2} \right), \quad \dots \dots \dots \dots (2)$$

in which

$$P = -(K-1) \cdot e^{int} \iiint e^{ik(x-r)} \, dx \, dy \, dz. \dots \dots \dots \dots (3)$$

In these equations r denotes the distance between the point (α, β, γ) where the disturbance is required to be estimated, and the element of volume $(dx\,dy\,dz)$ of the obstacle. The centre of the sphere R will be taken as the origin of coordinates. It is evident that, so far as the secondary ray is concerned, P depends only upon the angle (χ) which this ray makes with the primary ray. We will suppose that $\chi = 0$ in the direction backwards along the primary ray, and that $\chi = \pi$ along the primary ray continued. The integral in (3) may then be found in the form

$$\frac{2\pi R^2 e^{-ikr}}{k \cos \frac{1}{2}\chi} \int_0^{\frac{1}{2}\pi} J_1(2kR \cos \tfrac{1}{2}\chi \cdot \cos \phi) \cos^2 \phi \, d\phi, \quad \dots \dots \dots (4)$$

r now denoting the distance of the point of observation from the centre of the sphere. Expanding the Bessel's function, we get

$$P = -\frac{4\pi R^3 (K-1) e^{i(nt-kr)}}{3} \left\{ 1 - \frac{m^2}{2.5} + \frac{m^4}{2.4.5.7} - \frac{m^6}{2.4.6.5.7.9} \right.$$

$$\left. + \frac{m^8}{2.4.6.8.5.7.9.11} - \dots \right\}, \quad \dots \dots \dots (5)$$

in which m is written for $2kR \cos \frac{1}{2}\chi$. It is to be observed that in this solution there is no limitation upon the value of R if $(K-1)^2$ is neglected absolutely. In practice it will suffice that $(K-1) R/\lambda$ be small, λ (equal to $2\pi/k$) being the wave-length.

---

* Phil. Mag. Vol. XII. p. 81 (1881); Scientific Papers, Vol. I. p. 518.

These are the formulæ previously given. I had not then noticed that the integral in (4) can be expressed in terms of circular functions. By a general theorem due to Hobson *

$$\int_0^{\frac{1}{2}\pi} J_1 (m \cos \phi) \cos^2 \phi \, d\phi = \sqrt{\left(\frac{\pi}{2m}\right)} J_{\frac{3}{2}} (m) = \frac{\sin m}{m^2} - \frac{\cos m}{m}, \dots \dots (6)$$

so that $$P = -(K-1). \, 4\pi R^3 . \, e^{i \, (nt-kr)} \left(\frac{\sin m}{m^3} - \frac{\cos m}{m^2}\right), \dots \dots \dots (7)$$

in agreement with (5). The secondary disturbance vanishes with P, viz., when $\tan m = m$, or

$$m = 2kR \cos \tfrac{1}{2}\chi = \pi \, (1\cdot4303, \ 2\cdot4590, \ 3\cdot4709, \ 4\cdot4774, \ 5\cdot4818, \ \text{etc.})\dagger. \, \dots (8)$$

The smallest value of kR for which P vanishes occurs when $\chi = 0$, i.e. in the direction *backwards* along the primary ray. In terms of λ the diameter is

$$2R = 0\cdot715\lambda. \, \dots \dots \dots \dots \dots \dots \dots (9)$$

In directions nearly along the primary ray *forwards*, $\cos \frac{1}{2}\chi$ is small, and evanescence of P requires much larger ratios of R to λ. As was formerly fully discussed, the secondary disturbance vanishes, independently of P, in the direction of primary vibration ($a = 0$, $\beta = 0$).

In general, the intensity of the secondary disturbance is given by

$$f_1^2 + g_1^2 + h_1^2 = \left(\frac{k^2 P_0}{4\pi r}\right)^2 \left(1 - \frac{\gamma^2}{r^2}\right), \, \dots \dots \dots \dots (10)$$

in which P_0 denotes P with the factor $e^{i \, (nt-kr)}$ omitted, and is a function of χ, the angle between the secondary ray and the axis of x. If we take polar coordinates (χ, ϕ) round the axis of x,

$$1 - \frac{\gamma^2}{r^2} = 1 - \sin^2 \chi \, \cos^2 \phi \, ; \dots \dots \dots \dots \dots \dots (11)$$

and the intensity at distance r and direction (χ, ϕ) may be expressed in terms of these quantities. In order to find the effect upon the transmitted light, we have to integrate (10) over the whole surface of the sphere r. Thus

$$r^2 \iint \sin \chi \, d\chi \, d\phi \, (f_1^2 + g_1^2 + h_1^2) = \pi \int_0^\pi \sin \chi \, d\chi \left(\frac{k^2 P_0}{4\pi}\right)^2 (1 + \cos^2 \chi)$$

$$= \pi k^4 (K-1)^2 \, R^6 \int_0^\pi \sin \chi \, d\chi \, (1 + \cos^2 \chi) \frac{(\sin m - m \cos m)^2}{m^6}$$

$$= \tfrac{1}{2}\pi k^2 R^4 (K-1)^2 \int_0^{2kR} \frac{dm}{m^5} \left\{2 - \frac{m^2}{k^2 R^2} + \frac{m^4}{4k^4 R^4}\right\}$$

$$\times \{1 + m^2 + (m^2 - 1) \cos 2m - 2m \sin 2m\}. \, \dots \dots (12)$$

* *Lond. Math. Soc. Proc.* Vol. xxv. p. 71 (1893).
† See *Theory of Sound*, Vol. ii. § 207.

The integral may be expressed by means of functions regarded as known. Thus on integration by parts

$$\int_0^m \{1 + m^2 + (m^2 - 1)\cos 2m - 2m \sin 2m\} \frac{dm}{m^5}$$

$$= -\frac{1 - \cos 2m}{4m^4} + \frac{\sin 2m}{2m^3} - \frac{1}{2m^2} + \frac{1}{2},$$

$$\int_0^m \{1 + m^2 + (m^2 - 1)\cos 2m - 2m \sin 2m\} \frac{dm}{m^3}$$

$$= -\frac{1}{2m^2} + \int_0^m \frac{1 - \cos 2m}{m} \, dm + \frac{\cos 2m}{2m^2} + \frac{\sin 2m}{m} - 1,$$

$$\int_0^m \{1 + m^2 + (m^2 - 1)\cos 2m - 2m \sin 2m\} \frac{dm}{m}$$

$$= \int_0^m \frac{1 - \cos 2m}{m} \, dm + \frac{m^2}{2} + \frac{m \sin 2m}{2} + \frac{5 \cos 2m}{4} - \frac{5}{4}.$$

Accordingly, if m now stand for $2kR$, we get

$$r^2 \iint \sin \chi \, d\chi \, d\phi \, (f_1^2 + g_1^2 + h_1^2) = \frac{\pi R^2 (K-1)^2}{8} \left\{ -\frac{7(1 - \cos 2m)}{2m^2} \right.$$

$$\left. - \frac{\sin 2m}{m} + 5 + m^2 + \left(\frac{4}{m^2} - 4\right) \int_0^m \frac{1 - \cos 2m}{m} \, dm \right\}. \quad \ldots\ldots(13)$$

If m is small, the { } in (13) reduces to

$$0 + 0 \times m^2 + \tfrac{4}{27} m^4,$$

so that ultimately

$$(13) = \tfrac{8}{27} \pi k^4 R^6 (K-1)^2, \quad\ldots\ldots\ldots\ldots\ldots(14)$$

in agreement with the result which may be obtained more simply from (5). If we include another term, we get

$$(13) = \tfrac{8}{27} \pi k^4 R^6 (K-1)^2 \left(1 - \frac{2k^2 R^2}{5}\right). \quad\ldots\ldots\ldots(15)$$

As regards the definite integral, still written as such, in (13), we have

$$\int_0^m \frac{1 - \cos 2m}{m} \, dm = \int_0^{2m} \left\{ \frac{x}{1.2} - \frac{x^3}{4!} + \frac{x^5}{6!} - \ldots \right\} dx = \gamma + \log(2m) - \mathrm{Ci}(2m), \quad (16)$$

where γ is Euler's constant (0·5772156) and Ci is the cosine-integral, defined by

$$\mathrm{Ci}(x) = \int_\infty^x \frac{\cos u}{u} \, du\ldots\ldots\ldots\ldots\ldots\ldots(17)$$

As in (16), when x is moderate, we may use

$$\mathrm{Ci}(x) = \gamma + \log x - \tfrac{1}{2} \frac{x^2}{1.2} + \tfrac{1}{4} \frac{x^4}{1.2.3.4} - \ldots, \quad\ldots\ldots\ldots(18)$$

which is always convergent.   When x is great, we have the semi-convergent series

$$\mathrm{Ci}\,(x) = \sin x \left\{\frac{1}{x} - \frac{1\,.\,2}{x^3} + \frac{1\,.\,2\,.\,3\,.\,4}{x^5} - \dots\right\}$$
$$- \cos x \left\{\frac{1}{x^2} - \frac{1\,.\,2\,.\,3}{x^4} + \frac{1\,.\,2\,.\,3\,.\,4\,.\,5}{x^6} - \dots\right\} \dots\dots\dots(19)$$

Fairly complete tables of $\mathrm{Ci}\,(x)$, as well as of related integrals, have been given by Glaisher*.

When m is large, $\mathrm{Ci}\,(2m)$ tends to vanish, so that ultimately

$$\int_0^m \frac{1 - \cos 2m}{m}\,dm = \gamma + \log\,(2m).$$

Hence, when kR is large, (13) tends to the form

$$(13) = \tfrac{1}{2}\pi k^2 R^4\,(K-1)^2. \quad\dots\dots\dots\dots\dots\dots(20)$$

Glaisher's Table XII gives the maxima and minima values of the cosine-integral, which occur when the argument is an odd multiple of $\tfrac{1}{2}\pi$.   Thus:

n	$\mathrm{Ci}\,(n\pi/2)$	n	$\mathrm{Ci}\,(n\pi/2)$
1	$+0\cdot4720007$	7	$-0\cdot0895640$
3	$-0\cdot1984076$	9	$+0\cdot0700653$
5	$+0\cdot1237723$	11	$-0\cdot0575011$

These values allow us to calculate the $\{\ \}$ in (13), viz.,

$$-\frac{7\,(1 - \cos 2m)}{2m^2} - \frac{\sin 2m}{m} + 5 + m^2 + \left(\frac{4}{m^2} - 4\right)[\gamma + \log 2m - \mathrm{Ci}\,(2m)], \quad (21)$$

when $2m = n\pi/2$, and n is an odd integer.   In this case $\cos 2m = 0$ and $\sin 2m = \pm 1$, so that (21) reduces to

$$-\frac{56}{n^2\pi^2} \pm \frac{4}{n\pi} + 5 + \frac{n^2\pi^2}{16} + \left(\frac{64}{n^2\pi^2} - 4\right)[\gamma + \log\,(n\pi/2) - \mathrm{Ci}\,(n\pi/2)]. \quad (22)$$

We find

n	(22)	n	(22)
1	$0\cdot0530$	7	$23\cdot440$
3	$2\cdot718$	9	$42\cdot382$
5	$10\cdot534$	11	$65\cdot958$

* *Phil. Trans.* Vol. CLX. p. 367 (1870).

For values of n much greater, (22) is sufficiently represented by $n^2\pi^2/16$, or m^2, simply. It appears that there is no tendency to a falling-off in the scattering, such as would allow an increased transmission.

In order to make sure that the special choice of values for m has not masked a periodicity, I have calculated also the results when n is even. Here $\sin 2m = 0$ and $\cos 2m = \pm 1$, so that (21) reduces to

$$-\frac{112\,(1 \text{ or } 0)}{n^2\pi^2} + 5 + \frac{n^2\pi^2}{16} + \left(\frac{64}{n^2\pi^2} - 4\right)[\gamma + \log\,(n\pi/2) - \mathrm{Ci}\,(n\pi/2)]. \quad (23)$$

The following are required:

n	$\mathrm{Ci}\,(n\pi/2)$	n	$\mathrm{Ci}\,(n\pi/2)$
2	$+0{\cdot}0738$	8	$-0{\cdot}0061$
4	$-0{\cdot}0224$	10	$+0{\cdot}0040$
6	$+0{\cdot}0106$		

of which the first is obtained by interpolation from Glaisher's Table VI, and the remainder directly from (19). Thus:

n	(23)	n	(23)
2	$0{\cdot}7097$	8	$32{\cdot}336$
4	$6{\cdot}1077$	10	$53{\cdot}477$
6	$16{\cdot}156$		

The better to exhibit the course of the calculation, the actual values of the several terms of (23) when $n = 10$ may be given. We have

$$-\frac{112}{n^2\pi^2} = -0{\cdot}11348, \quad \frac{n^2\pi^2}{16} = 61{\cdot}685,$$

$$4 - \frac{64}{n^2\pi^2} = 4 - 0{\cdot}06485 = 3{\cdot}93515,$$

$$\gamma + \log\,(\pi/2) + \log n - \mathrm{Ci}\,(n\pi/2) = 0{\cdot}57722 + 0{\cdot}45158 + 2{\cdot}30259 - 0{\cdot}0040$$
$$= 13{\cdot}094,$$

so that $\qquad \left(4 - \frac{64}{n^2\pi^2}\right)\{\gamma + \log\,(n\pi/2) - \mathrm{Ci}\,(n\pi/2)\} = 13{\cdot}094.$

It will be seen that from this onwards the term $n^2\pi^2/16$, viz., m^2, greatly preponderates; and this is the term which leads to the limiting form (20).

The values of $2R/\lambda$ concerned in the above are very moderate. Thus, $n = 10$, making $m = 4\pi R/\lambda = 10\pi/4$, gives $2R/\lambda = 5/4$ only. Neither below this point, nor beyond it, is there anything but a steady rise in the value of (13) as λ diminishes when R is constant. *A fortiori* is this the case when R increases and λ is constant.

An increase in the light scattered from a single spherical particle implies, of course, a decrease in the light directly transmitted through a suspension containing a given number of particles in the cubic centimetre. The calculation is detailed in my paper " On the Transmission of Light through an Atmosphere containing Small Particles in Suspension *," and need not be repeated. It will be seen that no explanation is here arrived at of the augmentation of transparency at a certain stage observed by Keen and Porter. The discrepancy may perhaps be attributed to the fundamental supposition of the present paper, that the relative index is very small [or rather very near unity], a supposition not realised when sulphur and water are in question. But I confess that I should not have expected so wide a difference, and, indeed, the occurrence of anything special at so great diameters as 10 wave-lengths is surprising.

One other matter may be alluded to. It is not clear from the description that the light observed was truly transmitted in the technical sense. This light was much attenuated—down to only 5 per cent. Is it certain that it contained no sensible component of scattered light, but slightly diverted from its original course? If such admixture occurred, the question would be much complicated.

* *Phil. Mag.* Vol. xlvii. p. 375 (1899) ; *Scientific Papers*, Vol. iv. p. 397.

382.

SOME CALCULATIONS IN ILLUSTRATION OF FOURIER'S THEOREM.

[*Proceedings of the Royal Society*, A, Vol. XC. pp. 318—323, 1914.]

ACCORDING to Fourier's theorem a curve whose ordinate is arbitrary over the whole range of abscissæ from $x = -\infty$ to $x = +\infty$ can be compounded of harmonic curves of various wave-lengths. If the original curve contain a discontinuity, infinitely small wave-lengths must be included, but if the discontinuity be eased off, infinitely small wave-lengths may not be necessary. In order to illustrate this question I commenced several years ago calculations relating to a very simple case. These I have recently resumed, and although the results include no novelty of principle they may be worth putting upon record.

The case is that where the ordinate is constant (π) between the limits ± 1 for x and outside those limits vanishes.

In general

$$\phi(x) = \frac{1}{\pi} \int_0^\infty dk \int_{-\infty}^{+\infty} dv \, \phi(v) \cos k(v - x). \quad \dots\dots\dots\dots(1)$$

Here

$$\int_{-\infty}^{+\infty} dv \, \phi(v) \cos k(v - x) = 2\pi \cos kx \int_0^1 dv \cos kv = 2\pi \cos kx \, \frac{\sin k}{k}$$

$$= \frac{\pi}{k} \{\sin k(x + 1) - \sin k(x - 1)\},$$

and

$$\phi(x) = \int_0^\infty \frac{dk}{k} \{\sin k(x + 1) - \sin k(x - 1)\}. \quad \dots\dots\dots\dots(2)$$

As is well known, each of the integrals in (2) is equal to $\pm \frac{1}{2}\pi$; so that, as was required, $\phi(x)$ vanishes outside the limits ± 1 and between those limits takes the value π. It is proposed to consider what values are assumed by $\phi(x)$ when in (2) we omit that part of the range of integration in which k exceeds a finite value k_1.

The integrals in (2) are at once expressible by what is called the *sine-integral*, defined by

$$\text{Si}(\theta) = \int_0^\theta \frac{\sin\theta}{\theta}\,d\theta. \qquad\qquad\qquad\dots\dots\dots\dots(3)$$

Thus
$$\phi(x) = \text{Si}\,k_1(x+1) - \text{Si}\,k_1(x-1), \qquad\dots\dots\dots(4)$$

and if the sine-integral were thoroughly known there would be scarcely anything more to do. For moderate values of θ the integral may be calculated from an ascending series which is always convergent. For larger values this series becomes useless; we may then fall back upon a descending series of the semi-convergent class, viz.,

$$\text{Si}(\theta) = \frac{\pi}{2} - \cos\theta\left\{\frac{1}{\theta} - \frac{1.2}{\theta^3} + \frac{1.2.3.4}{\theta^5} - \dots\right\}$$

$$-\sin\theta\left\{\frac{1}{\theta^2} - \frac{1.2.3}{\theta^4} + \frac{1.2.3.4.5}{\theta^6} - \dots\right\}. \qquad\dots\dots(5)$$

Dr Glaisher* has given very complete tables extending from $\theta = 0$ to $\theta = 1$, and also from 1 to 5 at intervals of 0·1. Beyond this point he gives the function for integer values of θ from 5 to 15 inclusive, and afterwards only at intervals of 5 for 20, 25, 30, 35, &c. For my purpose these do not suffice, and I have calculated from (5) the values for the missing integers up to $\theta = 60$. The results are recorded in the Table below. In each case, except those quoted from Glaisher, the last figure is subject to a small error.

For the further calculation, involving merely subtractions, I have selected the special cases $k_1 = 1, 2, 10$. For $k_1 = 1$, we have

$$\phi(x) = \text{Si}(x+1) - \text{Si}(x-1). \qquad\dots\dots\dots\dots(6)$$

θ	Si (θ)	θ	Si (θ)	θ	Si (θ)	θ	Si (θ)
16	1·63130	28	1·60474	39	1·56334	50	1·55162
17	1·59013	29	1·59731	40	1·58699	51	1·55600
18	1·53662	30	1·56676	41	1·59494	52	1·57357
19	1·51863	31	1·54177	42	1·58083	53	1·58798
20	1·54824	32	1·54424	43	1·55836	54	1·58634
21	1·59490	33	1·57028	44	1·54808	55	1·57072
22	1·61609	34	1·59525	45	1·55871	56	1·55574
23	1·59546	35	1·59692	46	1·57976	57	1·55490
24	1·55474	36	1·57512	47	1·59184	58	1·56845
25	1·53148	37	1·54861	48	1·58445	59	1·58368
26	1·54487	38	1·54549	49	1·56507	60	1·58675
27	1·58029						

In every case $\phi(x)$ is an even function, so that it suffices to consider x positive.

* *Phil. Trans.* Vol. CLX. p. 367 (1870).

$$k_1 = 1.$$

x	$\phi(x)$	x	$\phi(x)$	x	$\phi(x)$
0·0	+1·8922	2·5	+0·5084	6·0	−0·0953
0·5	1·8178	3·0	+0·1528	7·0	+0·1495
1·0	1·6054	3·5	−0·1244	8·0	+0·2104
1·5	1·2854	4·0	−0·2987	9·0	+0·0842
2·0	0·9026	5·0	−0·3335	10·0	−0·0867

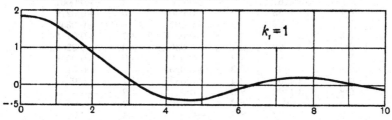

When $k_1 = 2$, $\phi(x) = \text{Si}(2x+2) - \text{Si}(2x-2)$, ...(7)

and we find

$$k_1 = 2.$$

x	$\phi(x)$	x	$\phi(x)$	x	$\phi(x)$
0·0	+3·2108	0·9	+1·9929	3·0	−0·1840
0·1	3·1934	1·0	1·7582	3·5	+0·1151
0·2	3·1417	1·1	1·5188	4·0	+0·2337
0·3	3·0566	1·2	1·2794	4·5	+0·1237
0·4	2·9401	1·3	1·0443	5·0	−0·0692
0·5	2·7947	1·4	0·8179	5·5	−0·1657
0·6	2·6235	1·5	+0·6038	6·0	−0·1021
0·7	2·4300	2·0	−0·1807		
0·8	2·2184	2·5	−0·3940		

Both for $k_1 = 1$ and for $k_1 = 2$ all that is required for the above values of $\phi(x)$ is given in Glaisher's tables.

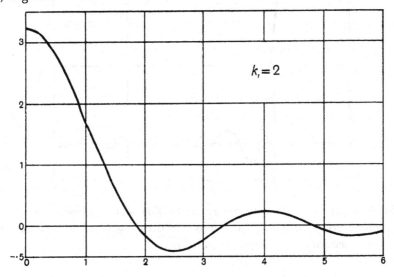

When $k_1 = 10$, $\phi(x) = \text{Si}(10x + 10) - \text{Si}(10x - 10)$.   ....(8)

We find

$$k_1 = 10.$$

x	$\phi(x)$	x	$\phi(x)$	x	$\phi(x)$
0·0	+3·3167	1·7	+0·1257	3·4	−0·0067
0·1	3·2433	1·8	+0·0305	3·5	+0·0272
0·2	3·0792	1·9	−0·0677	3·6	+0·0349
0·3	2·9540	2·0	−0·0916	3·7	+0·0115
0·4	2·9809	2·1	−0·0365	3·8	−0·0203
0·5	3·1681	2·2	+0·0393	3·9	−0·0322
0·6	3·3895	2·3	+0·0709	4·0	−0·0151
0·7	3·4388	2·4	+0·0390	4·1	+0·0142
0·8	3·1420	2·5	−0·0213	4·2	+0·0293
0·9	2·4647	2·6	−0·0562	4·3	+0·0178
1·0	1·5482	2·7	−0·0415	4·4	−0·0089
1·1	0·6488	2·8	+0·0089	4·5	−0·0262
1·2	+0·0107	2·9	+0·0447	4·6	−0·0194
1·3	−0·2532	3·0	+0·0387	4·7	+0·0063
1·4	−0·2035	3·1	+0·0000	4·8	+0·0230
1·5	−0·0184	3·2	−0·0353	4·9	+0·0203
1·6	+0·1202	3·3	−0·0371	5·0	−0·0002

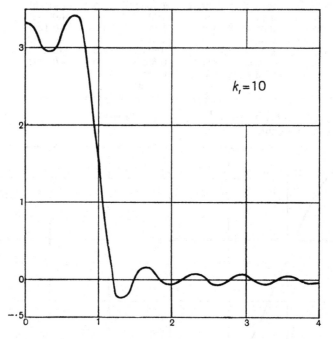

$k_1 = 10$

The same set of values of Si up to Si(60) would serve also for the calculation of $\phi(x)$ for $k_1 = 20$ and from $x = 0$ to $x = 2$ at intervals of 0·05. It is hardly necessary to set this out in detail.

An inspection of the curves plotted from the above tables shows the approximation towards discontinuity as k_1 increases.

That the curve remains undulatory is a consequence of the sudden stoppage of the integration at $k = k_1$. If we are content with a partial suppression only of the shorter wave-lengths, a much simpler solution is open to us. We have only to introduce into (1) the factor e^{-ak}, where a is positive, and to continue the integration up to $x = \infty$. In place of (2), we have

$$\phi(x) = \int_0^\infty \frac{dk\, e^{-ak}}{k} \{\sin k\,(x+1) - \sin k\,(x-1)\} = \tan^{-1}\left(\frac{x+1}{a}\right) - \tan^{-1}\left(\frac{x-1}{a}\right).$$

$$\ldots\ldots(9)$$

The discontinuous expression corresponds, of course, to $a = 0$. If a is merely small, the discontinuity is eased off. The following are values of $\phi(x)$, calculated from (9) for $a = 1,\ 0\cdot5,\ 0\cdot05$:

$$a = 1.$$

x	$\phi(x)$	x	$\phi(x)$	x	$\phi(x)$
0·0	1·571	2·0	0·464	4·0	0·124
0·5	1·446	2·5	0·309	5·0	0·080
1·0	1·107	3·0	0·219	6·0	0·055
1·5	0·727				

$$a = 0\cdot5.$$

x	$\phi(x)$	x	$\phi(x)$	x	$\phi(x)$
0·00	2·214	1·00	1·326	2·00	0·298
0·25	2·173	1·25	0·888	2·50	0·180
0·50	2·111	1·50	0·588	3·00	0·120
0·75	1·756	1·75	0·408	3·50	0·087

$$a = 0\cdot05.$$

x	$\phi(x)$	x	$\phi(x)$	x	$\phi(x)$
0·00	3·041	0·90	2·652	1·20	0·222
0·20	3·037	0·95	2·331	1·40	0·103
0·40	3·023	1·00	1·546	1·60	0·064
0·60	2·986	1·05	0·761	1·80	0·045
0·80	2·869	1·10	0·440	2·00	0·033

As is evident from the form of (9), $\phi(x)$ falls continuously as x increases whatever may be the value of a.

383.

FURTHER CALCULATIONS CONCERNING THE MOMENTUM OF PROGRESSIVE WAVES.

[*Philosophical Magazine*, Vol. XXVII. pp. 436—440, 1914.]

THE question of the momentum of waves in fluid is of interest and has given rise to some difference of opinion. In a paper published several years ago* I gave an approximate treatment of some problems of this kind. For a fluid moving in one dimension for which the relation between pressure and density is expressed by

$$p = f(\rho), \dots\dots\dots\dots\dots\dots(1)$$

it appeared that the momentum of a progressive wave of mean density equal to that of the undisturbed fluid is given by

$$\left\{ \frac{\rho_0 f''(\rho_0)}{4a^3} + \frac{1}{2a} \right\} \times \text{total energy}, \dots\dots\dots\dots(2)$$

in which ρ_0 is the undisturbed density and a the velocity of propagation. The momentum is reckoned positive when it is in the direction of wave-propagation.

For the "adiabatic" law, viz.:

$$p/p_0 = (\rho/\rho_0)^\gamma, \dots\dots\dots\dots\dots\dots(3)$$

$$f'(\rho_0) = \frac{\gamma p_0}{\rho_0} = a^2, \qquad f''(\rho_0) = \frac{p_0 \gamma(\gamma-1)}{\rho_0{}^2}; \dots\dots\dots\dots(4)$$

so that

$$\frac{\rho_0 f''(\rho_0)}{4a^3} + \frac{1}{2a} = \frac{\gamma+1}{4a}. \dots\dots\dots\dots(5)$$

In the case of Boyle's law we have merely to make $\gamma = 1$ in (5).

For ordinary gases $\gamma > 1$ and the momentum is positive; but the above argument applies to all positive values of γ. If γ be negative, the pressure would increase as the density decreases, and the fluid would be essentially unstable.

* *Phil. Mag.* Vol. x. p. 364 (1905); *Scientific Papers*, Vol. v. p. 265.

However, a slightly modified form of (3) allows the exponent to be negative. If we take

$$p/p_0 = 2 - (\rho/\rho_0)^{-\beta} \quad \dots\dots\dots\dots\dots\dots\dots(6)$$

with β positive, we get as above

$$f'(\rho_0) = \frac{\beta p_0}{\rho_0} = a^2, \quad f''(\rho_0) = -\frac{(\beta + 1)\, a^2}{\rho_0}, \quad \dots\dots\dots(7)$$

and accordingly

$$\frac{\rho_0 f''(\rho_0)}{4a^3} + \frac{1}{2a} = \frac{1 - \beta}{4a}. \quad \dots\dots\dots\dots\dots(8)$$

If $\beta = 1$, the law of pressure is that under which waves can be propagated without a change of type, and we see that the momentum is zero. In general, the momentum is positive or negative according as β is less or greater than 1.

In the above formula (2) the calculation is approximate only, powers of the disturbance above the *second* being neglected. In the present note it is proposed to determine the sign of the momentum under the laws (3) and (6) more generally and further to extend the calculations to waves in a liquid moving in two dimensions under gravity.

It should be clearly understood that the discussion relates to *progressive* waves. If this restriction be dispensed with, it would always be possible to have a disturbance (limited if we please to a finite length) without momentum, as could be effected very simply by beginning with displacements unaccompanied by velocities. And the disturbance, considered as a whole, can never acquire (or lose) momentum. In order that a wave may be progressive in one direction only, a relation must subsist between the velocity and density at every point. In the case of Boyle's law this relation, first given by De Morgan[*], is

$$u = a \log(\rho/\rho_0), \quad \dots\dots\dots\dots\dots\dots\dots(9)$$

and more generally [†]

$$u = \int \sqrt{\left(\frac{dp}{d\rho}\right)} \cdot \frac{d\rho}{\rho}. \quad \dots\dots\dots\dots\dots\dots(10)$$

Wherever this relation is violated, a wave emerges travelling in the negative direction.

For the adiabatic law (3), (10) gives

$$u = \frac{2a}{\gamma - 1} \left\{ \left(\frac{\rho}{\rho_0}\right)^{\frac{\gamma - 1}{2}} - 1 \right\}, \quad \dots\dots\dots\dots(11)$$

---

[*] Airy, *Phil. Mag.* Vol. xxxiv. p. 402 (1849).
[†] Earnshaw, *Phil. Trans.* 1859, p. 146.

a being the velocity of infinitely small disturbances, and this reduces to (9) when $\gamma = 1$. Whether γ be greater or less than 1, u is positive when ρ exceeds ρ_0. Similarly if the law of pressure be that expressed in (6),

$$u = \frac{2a}{\beta + 1} \left\{ 1 - \left(\frac{\rho}{\rho_0} \right)^{-\frac{\beta+1}{2}} \right\}. \quad \text{...................(12)}$$

Since β is positive, values of ρ greater than ρ_0 are here also accompanied by positive values of u.

By definition the momentum of the wave, whose length may be supposed to be limited, is per unit of cross-section

$$\int \rho u \, dx, \quad \text{................................(13)}$$

the integration extending over the whole length of the wave. If we introduce the value of u given in (11), we get

$$(13) = \frac{2\rho_0 a}{\gamma - 1} \int \left\{ \left(\frac{\rho}{\rho_0} \right)^{\frac{\gamma+1}{2}} - \frac{\rho}{\rho_0} \right\} dx; \quad \text{.................(14)}$$

and the question to be examined is the sign of (14). For brevity we may write unity in place of ρ_0, and we suppose that the wave is such that its mean density is equal to that of the undisturbed fluid, so that $\int \rho \, dx = l$, where l is the length of the wave. If l be divided into n equal parts, then when n is great enough the integral may be represented by the sum

$$\left\{ \rho_1^{\frac{\gamma+1}{2}} + \rho_2^{\frac{\gamma+1}{2}} + \rho_3^{\frac{\gamma+1}{2}} + \ldots - \rho_1 - \rho_2 - \ldots \right\} \frac{l}{n}, \quad \text{............(15)}$$

in which all the ρ's are positive. Now it is a proposition in Algebra that

$$\frac{\rho_1^{\frac{\gamma+1}{2}} + \rho_2^{\frac{\gamma+1}{2}} + \ldots}{n} > \left(\frac{\rho_1 + \rho_2 + \ldots}{n} \right)^{\frac{\gamma+1}{2}}$$

when $\frac{1}{2}(\gamma + 1)$ is negative, or positive and greater than unity; but that the reverse holds when $\frac{1}{2}(\gamma + 1)$ is positive and less than unity. Of course the inequality becomes an equality when all the n quantities are equal. In the present application the sum of the ρ's is n, and under the adiabatic law (3), γ and $\frac{1}{2}(\gamma + 1)$ are positive. Hence (15) is positive or negative according as $\frac{1}{2}(\gamma + 1)$ is greater or less than unity, viz., according as γ is greater or less than unity. In either case the momentum represented by (13) is *positive*, and the conclusion is not limited to the supposition of small disturbances.

In like manner if the law of pressure be that expressed in (6), we get from (12)

$$(13) = \frac{2\rho_0 a}{\beta + 1} \int \left\{ \frac{\rho}{\rho_0} - \left(\frac{\rho}{\rho_0} \right)^{-\frac{\beta-1}{2}} \right\} dx, \quad \text{.................(16)}$$

from which we deduce almost exactly as before that the momentum (13) is positive if β (being positive) is less than 1 and negative if β is greater than 1. If $\beta = 1$, the momentum vanishes. The conclusions formerly obtained on the supposition of small disturbances are thus extended.

We will now discuss the momentum in certain cases of fluid motion under gravity. The simplest is that of *long* waves in a uniform canal. If η be the (small) elevation at any point x measured in the direction of the length of the canal and u the corresponding fluid velocity parallel to x, which is uniform over the section, the dynamical equation is[*]

$$\frac{du}{dt} = -g\frac{d\eta}{dx}. \quad\quad\quad\quad\quad\quad\quad\quad (17)$$

As is well known, long waves of small elevation are propagated without change of form. If c be the velocity of propagation, a positive wave may be represented by

$$\eta = F(ct - x), \quad\quad\quad\quad\quad\quad\quad\quad (18)$$

where F denotes an arbitrary function, and c is related to the depth h_0 according to

$$c^2 = gh_0. \quad\quad\quad\quad\quad\quad\quad\quad (19)$$

From (17), (18)

$$u = \frac{g\eta}{c} = \sqrt{\left(\frac{g}{h_0}\right)}.\eta \quad\quad\quad\quad\quad\quad\quad\quad (20)$$

is the relation obtaining between the velocity and elevation at any place in a positive progressive wave of small elevation.

Equation (20), however, does not suffice for our present purpose. We may extend it by the consideration that in a long wave of finite disturbance the elevation and velocity may be taken as relative to the neighbouring parts of the wave. Thus, writing du for u and h for h_0, so that $\eta = dh$, we have

$$du = \sqrt{\left(\frac{g}{h}\right)}\,dh,$$

and on integration

$$u = 2\sqrt{g}\,\{h^{\frac{1}{2}} + C\}.$$

The arbitrary constant of integration is determined by the fact that outside the wave $u = 0$ when $h = h_0$, whence and replacing h by $h_0 + \eta$, we get

$$u = 2\sqrt{g}\,\{\sqrt{(h_0 + \eta)} - \sqrt{h_0}\}, \quad\quad\quad\quad\quad (21)$$

as the generalized form of (20). It is equivalent to a relation given first in another notation by De Morgan[†], and it may be regarded as the condition

---

[*] Lamb's *Hydrodynamics*, § 168.

[†] Airy, *Phil. Mag.* Vol. xxxiv. p. 402 (1849).

which must be satisfied if the emergence of a negative wave is to be obviated.

We are now prepared to calculate the momentum. For a wave in which the mean elevation is zero, the momentum corresponding to unit horizontal breadth is

$$\rho \int u\,(h_0 + \eta)\,dx = \tfrac{3}{4}\rho \,\sqrt{(g/h_0)} \int \eta^2 dx, \dots\dots\dots\dots\dots(22)$$

when we omit cubes and higher powers of η. We may write (22) also in the form

$$\text{Momentum} = \frac{3}{4}\,\frac{\text{Total Energy}}{c}, \dots\dots\dots\dots\dots(23)$$

c being the velocity of propagation of waves of small elevation.

As in (14), with γ equal to 2, we may prove that the momentum is positive without restriction upon the value of η.

As another example, periodic waves moving on the surface of deep water may also be referred to. The momentum of such waves has been calculated by Lamb*, on the basis of Stokes' second approximation. It appears that the momentum per wave-length and per unit width perpendicular to the plane of motion is

$$\pi\rho a^2 c, \dots\dots\dots\dots\dots\dots\dots\dots\dots(24)$$

where c is the velocity of propagation of the waves in question and the wave form is approximately

$$\eta = a \cos \frac{2\pi}{\lambda}\,(ct - x). \dots\dots\dots\dots\dots(25)$$

The forward velocity of the surface layers was remarked by Stokes. For a simple view of the matter reference may be made also to *Phil. Mag.* Vol. I. p. 257 (1876); *Scientific Papers*, Vol. I. p. 263.

---

* *Hydrodynamics*, § 246.

384.

FLUID MOTIONS.

[*Proc. Roy. Inst.* March, 1914; *Nature*, Vol. XCIII. p. 364, 1914.]

THE subject of this lecture has received the attention of several generations of mathematicians and experimenters. Over a part of the field their labours have been rewarded with a considerable degree of success. In all that concerns small vibrations, whether of air, as in sound, or of water, as in waves and tides, we have a large body of systematized knowledge, though in the case of the tides the question is seriously complicated by the fact that the rotation of the globe is actual and not merely relative to the sun and moon, as well as by the irregular outlines and depths of the various oceans. And even when the disturbance constituting the vibration is not small, some progress has been made, as in the theory of sound waves in one dimension, and of the tidal *bores*, which are such a remarkable feature of certain estuaries and rivers.

The general equations of fluid motion, when friction or viscosity is neglected, were laid down in quite early days by Euler and Lagrange, and in a sense they should contain the whole theory. But, as Whewell remarked, it soon appeared that these equations by themselves take us a surprisingly little way, and much mathematical and physical talent had to be expended before the truths hidden in them could be brought to light and exhibited in a practical shape. What was still more disconcerting, some of the general propositions so arrived at were found to be in flagrant contradiction with observation, even in cases where at first sight it would not seem that viscosity was likely to be important. Thus a solid body, submerged to a sufficient depth, should experience no resistance to its motion through water. On this principle the screw of a submerged boat would be useless, but, on the other hand, its services would not be needed. It is little wonder that practical men should declare that theoretical hydrodynamics has nothing at all to do with real fluids. Later we will return to some of these difficulties, not yet fully surmounted, but for the moment I will call your attention to simple phenomena of which theory can give a satisfactory account.

Considerable simplification attends the supposition that the motion is always the same at the same place—is *steady*, as we say—and fortunately this covers many problems of importance.   Consider the flow of water along a pipe whose section varies.   If the section were uniform, the pressure would vary along the length only in consequence of friction, which now we are neglecting.   In the proposed pipe how will the pressure vary?   I will not prophesy as to a Royal Institution audience, but I believe that most unsophisticated people suppose that a contracted place would give rise to an *increased* pressure.   As was known to the initiated long ago, nothing can be further from the fact.   The experiment is easily tried, either with air or water, so soon as we are provided with the right sort of tube.   A suitable shape is shown in fig. 1, but it is rather troublesome to construct in metal.

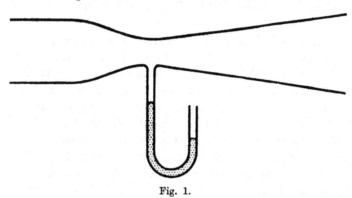

Fig. 1.

W. Froude found paraffin-wax the most convenient material for ship models, and I have followed him in the experiment now shown.   A brass tube is filled with candle-wax and bored out to the desired shape, as is easily done with templates of tin plate.   When I blow through, a *suction* is developed at the narrows, as is witnessed by the rise of liquid in a manometer connected laterally.

In the laboratory, where dry air from an acoustic bellows or a gas-holder is available, I have employed successfully tubes built up of cardboard, for a circular cross-section is not necessary.   Three or more precisely similar pieces, cut for example to the shape shown in fig. 2 and joined together

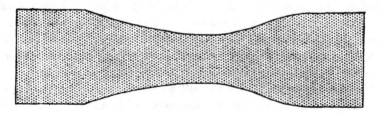

Fig. 2.

closely along the edges, give the right kind of tube, and may be made air-tight with pasted paper or with sealing-wax. Perhaps a square section requiring four pieces is best. It is worth while to remark that there is no stretching of the cardboard, each side being merely *bent* in one dimension. A model is before you, and a study of it forms a simple and useful exercise in solid geometry.

Another form of the experiment is perhaps better known, though rather more difficult to think about. A tube (fig. 3) ends in a flange. If I blow through the tube, a card presented to the flange is drawn up pretty closely, instead of being blown away as might be expected. When we consider the

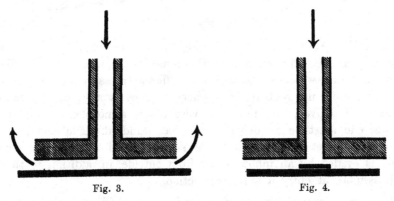

Fig. 3.   Fig. 4.

matter, we recognize that the channel between the flange and the card through which the air flows after leaving the tube is really an expanding one, and thus that the inner part may fairly be considered as a contracted place. The suction here developed holds the card up.

A slight modification enhances the effect. It is obvious that immediately opposite the tube there will be pressure upon the card and not suction. To neutralize this a sort of cap is provided, attached to the flange, upon which the objectionable pressure is taken (fig. 4). By blowing smartly from the mouth through this little apparatus it is easy to lift and hold up a penny for a short time.

The facts then are plain enough, but what is the explanation? It is really quite simple. In steady motion the quantity of fluid per second passing any section of the tube is everywhere the same. If the fluid be incompressible, and air in these experiments behaves pretty much as if it were, this means that the product of the velocity and area of cross-section is constant, so that at a narrow place the velocity of flow is necessarily increased. And when we enquire how the additional velocity in passing from a wider to a narrower place is to be acquired, we are compelled to recognize that it can only be in consequence of a fall of pressure. The suction at the narrows is the only result consistent with the great principle of conservation of energy;

but it remains rather an inversion of ordinary ideas that we should have to deduce the forces from the motion, rather than the motion from the forces.

The application of the principle is not always quite straightforward. Consider a tube of slightly conical form, open at both ends, and suppose that we direct upon the narrower end a jet of air from a tube having the same (narrower) section (fig. 5). We might expect this jet to enter the

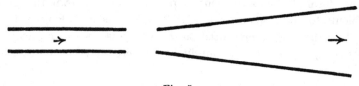

Fig. 5.

conical tube without much complication. But if we examine more closely a difficulty arises. The stream in the conical tube would have different velocities at the two ends, and therefore different pressures. The pressures at the ends could not both be atmospheric. Since at any rate the pressure at the wider delivery end must be very nearly atmospheric, that at the narrower end must be decidedly below that standard. The course of the events at the inlet is not so simple as supposed, and the apparent contradiction is evaded by an inflow of air from outside, in addition to the jet, which assumes at entry a narrower section.

If the space surrounding the free jet is enclosed (fig. 6), suction is there developed and ultimately when the motion has become steady the jet enters the conical tube without contraction. A model shows the effect, and the principle is employed·in a well-known laboratory instrument arranged for working off the water-mains.

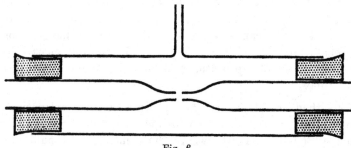

Fig. 6.

I have hitherto dealt with air rather than water, not only because air makes no mess, but also because it is easier to ignore gravitation. But there is another and more difficult question. You will have noticed that in our expanding tubes the section changes only gradually. What happens when the expansion is more sudden—in the extreme case when the diameter of a previously uniform tube suddenly becomes infinite? (fig. 3) without card. Ordinary experience teaches that in such a case the flow does not follow the walls round the corner, but shoots across as a jet, which for a time preserves its individuality and something like its original section. Since the velocity is not lost, the pressure which would replace it is not developed. It is instructive to compare this case with another, experimented on by Savart* and W. Froude †, in which a free jet is projected through a short cone, or a mere hole in a thin wall, into a vessel under a higher pressure. The apparatus consists of two precisely similar vessels with apertures, in which the fluid (water) may be at different levels (fig. 7, copied from Froude). Savart found that not a single drop of liquid was spilt so long as the pressure in the recipient vessel did not exceed one-sixth of that under which the jet issues. And Froude reports that so long as the head in the discharge cistern is maintained at a moderate height above that in the

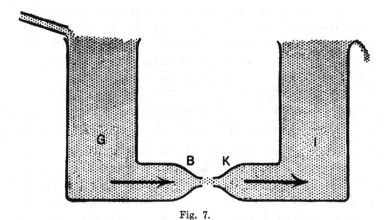

Fig. 7.

recipient cistern, the whole of the stream enters the recipient orifice, and there is "no waste, except the small sprinkling which is occasioned by inexactness of aim, and by want of exact circularity in the orifices." I am disposed to attach more importance to the small spill, at any rate when the conoids are absent or very short. For if there is no spill, the jet (it would seem) might as well be completely enclosed; and then it would propagate itself into the recipient cistern without sudden expansion and consequent recovery of pressure. In fact, the pressure at the narrows would never fall below that of the recipient cistern, and the discharge would be correspondingly lessened. When a decided spill occurs, Froude explains it as due to the retardation by friction of the outer layers which are thus unable to force themselves against the pressure in front.

Evidently it is the behaviour of these outer layers, especially at narrow places, which determines the character of the flow in a large variety of cases.

* *Ann. de Chimie*, Vol. LV. p. 257, 1833.
† *Nature*, Vol. XIII. p. 93, 1875.

They are held back, as Froude pointed out, by friction acting from the walls; but, on the other hand, when they lag, they are pulled forward by layers farther in which still retain their velocity. If the latter prevail, the motion in the end may not be very different from what would occur in the absence of friction; otherwise an entirely altered motion may ensue. The situation as regards the rest of the fluid is much easier when the layers upon which the friction tells most are allowed to escape. This happens in instruments of the injector class, but I have sometimes wondered whether full advantage is taken of it. The long gradually expanding cones are overdone, perhaps, and the friction which they entail must have a bad effect.

Similar considerations enter when we discuss the passage of a solid body through a large mass of fluid otherwise at rest, as in the case of an airship or submarine boat. I say a submarine, because when a ship moves upon the surface of the water the formation of waves constitutes a complication, and one of great importance when the speed is high. In order that the water in its relative motion may close in properly behind, the after-part of the ship must be suitably shaped, fine lines being more necessary at the stern than at the bow, as fish found out before men interested themselves in the problem. In a well-designed ship the whole resistance (apart from wave-making) may be ascribed to *skin friction*, of the same nature as that which is encountered when the ship is replaced by a thin plane moving edgeways.

At the other extreme we may consider the motion of a thin disk or blade flatways through the water. Here the actual motion differs altogether from that prescribed by the classical hydrodynamics, according to which the character of the motion should be the same behind as in front. The liquid refuses to close in behind, and a region of more or less "dead water" is developed, entailing a greatly increased resistance. To meet this Helmholtz, Kirchhoff, and their followers have given calculations in which the fluid behind is supposed to move strictly with the advancing solid, and to be separated from the remainder of the mass by a surface at which a finite slip takes place. Although some difficulties remain, there can be no doubt that this theory constitutes a great advance. But the surface of separation is unstable, and in consequence of fluid friction it soon loses its sharpness, breaking up into more or less periodic eddies, described in some detail by Mallock (fig. 8). It is these eddies which cause the whistling of the wind in trees and the more musical notes of the æolian harp.

The obstacle to the closing-in of the lines of flow behind the disk is doubtless, as before, the layer of liquid in close proximity to the disk, which at the edge has insufficient velocity for what is required of it. It would be an interesting experiment to try what would be the effect of allowing a small "spill." For this purpose the disk or blade would be made double, with a suction applied to the narrow interspace. Relieved of the slowly

moving layer, the liquid might then be able to close in behind, and success
would be witnessed by a greatly diminished resistance.

Fig. 8.

When a tolerably fair-shaped body moves through fluid, the relative
velocity is greatest at the maximum section of the solid which is the minimum
section for the fluid, and consequently the pressure is there least. Thus the
water-level is depressed at and near the midship section of an advancing
steamer, as is very evident in travelling along a canal. On the same principle
may be explained the stability of a ball sustained on a vertical jet as in a
well-known toy (shown). If the ball deviate to one side, the jet in bending
round the surface develops a suction pulling the ball back. As Mr Lanchester
has remarked, the effect is aided by the rotation of the ball. That a convex
surface is attracted by a jet playing obliquely upon it was demonstrated by
T. Young more than 100 years ago by means of a model, of which a copy is
before you (fig. 9).

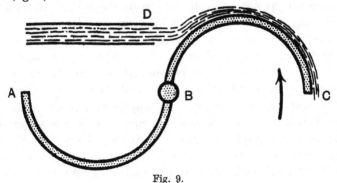

Fig. 9.

A plate, bent into the form *ABC*, turning on centre *B*, is
impelled by a stream of air *D* in the direction shown.

It has been impossible in dealing with experiments to keep quite clear
of friction, but I wish now for a moment to revert to the ideal fluid of hydrodynamics, in which pressure and inertia alone come into account. The
possible motions of such a fluid fall into two great classes—those which do
and those which do not involve *rotation*. What exactly is meant by rotation
is best explained after the manner of Stokes. If we imagine any spherical

portion of the fluid in its motion to be suddenly solidified, the resulting solid may be found to be rotating. If so, the original fluid is considered to possess rotation. If a mass of fluid moves irrotationally, no spherical portion would revolve on solidification. The importance of the distinction depends mainly upon the theorem, due to Lagrange and Cauchy, that the irrotational character is permanent, so that any portion of fluid at any time destitute of rotation will always remain so. Under this condition fluid motion is comparatively simple, and has been well studied. Unfortunately many of the results are very unpractical.

As regards the other class of motions, the first great step was taken in 1858, by Helmholtz, who gave the theory of the vortex-ring. In a perfect fluid a vortex-ring has a certain permanence and individuality, which so much impressed Kelvin that he made it the foundation of a speculation as to the nature of matter. To him we owe also many further developments in pure theory.

On the experimental side, the first description of vortex-rings that I have come across is that by W. B. Rogers*, who instances their production during the bursting of bubbles of phosphuretted hydrogen, or the escape of smoke from cannon and from the lips of expert tobacconists. For private observation nothing is simpler than Helmholtz's method of drawing a partially immersed spoon along the surface, for example, of a cup of tea. Here half a ring only is developed, and the places where it meets the surface are shown as dimples, indicative of diminished pressure. The experiment, made on a larger scale, is now projected upon the screen, the surface of the liquid and its motion being made more evident by powder of lycopodium or sulphur scattered over it. In this case the ring is generated by the motion of a half-immersed circular disk, withdrawn after a travel of two or three inches. In a modified experiment the disk is replaced by a circular or semi-circular *aperture* cut in a larger plate, the level of the water coinciding with the horizontal diameter of the aperture. It may be noticed that while the first forward motion of the plate occasions a ring behind, the stoppage of the plate gives rise to a second ring in front. As was observed by Reusch†, the same thing occurs in the more usual method of projecting smoke-rings from a box; but in order to see it the box must be transparent.

In a lecture given here in 1877, Reynolds showed that a Helmholtz ring can push the parent disk before it, so that for a time there appears to be little resistance to its motion.

For an explanation of the origin of these rings we must appeal to friction, for in a perfect fluid no rotation can develop. It is easy to recognize that friction against the wall in which the aperture is perforated, or against the

---

* *Amer. J. Sci.* Vol. XXVI. p. 246, 1858.
† *Pogg. Ann.* Vol. CX. p. 309, 1860.

face of the disk in the other form of experiment, will start a rotation which, in a viscous fluid, such as air or water actually is, propagates itself to a finite distance inwards. But although a general explanation is easy, many of the details remain obscure.

It is apparent that in dealing with a large and interesting class of fluid motions we cannot go far without including fluid friction, or *viscosity* as it is generally called, in order to distinguish it from the very different sort of friction encountered by solids, unless well lubricated. In order to define it, we may consider the simplest case where fluid is included between two parallel walls, at unit distance apart, which move steadily, each in its own plane, with velocities which differ by unity. On the supposition that the fluid also moves in plane strata, the viscosity is measured by the tangential force per unit of area exercised by each stratum upon its neighbours. When we are concerned with internal motions only, we have to do rather with the so-called "kinematic viscosity," found by dividing the quantity above defined by the density of the fluid. On this system the viscosity of water is much *less* than that of air.

Viscosity varies with temperature; and it is well to remember that the viscosity of air increases while that of water decreases as the temperature rises. Also that the viscosity of water may be greatly increased by admixture with alcohol. I used these methods in 1879 during investigations respecting the influence of viscosity upon the behaviour of such fluid jets as are sensitive to sound and vibration.

Experimentally the simplest case of motion in which viscosity is paramount is the flow of fluid through capillary tubes. The laws of such motion are simple, and were well investigated by Poiseuille. This is the method employed in practice to determine viscosities. The apparatus before you is arranged to show the diminution of viscosity with rising temperature. In the cold the flow of water through the capillary tube is slow, and it requires sixty seconds to fill a small measuring vessel. When, however, the tube is heated by passing steam through the jacket surrounding it, the flow under the same head is much increased, and the measure is filled in twenty-six seconds. Another case of great practical importance, where viscosity is the leading consideration, relates to lubrication. In admirably conducted experiments Tower showed that the solid surfaces moving over one another should be separated by a complete film of oil, and that when this is attended to there is no wear. On this basis a fairly complete theory of lubrication has been developed, mainly by O. Reynolds. But the capillary nature of the fluid also enters to some extent, and it is not yet certain that the whole character of a lubricant can be expressed even in terms of both surface tension and viscosity.

It appears that in the extreme cases, when viscosity can be neglected and again when it is paramount, we are able to give a pretty good account of what passes. It is in the intermediate region, where both inertia and viscosity are of influence, that the difficulty is greatest. But even here we are not wholly without guidance. There is a general law, called the law of dynamical similarity, which is often of great service. In the past this law has been unaccountably neglected, and not only in the present field. It allows us to infer what will happen upon one scale of operations from what has been observed at another. On the present occasion I must limit myself to viscous fluids, for which the law of similarity was laid down in all its completeness by Stokes as long ago as 1850. It appears that similar motions may take place provided a certain condition be satisfied, viz. that the product of the linear dimension and the velocity, divided by the kinematic viscosity of the fluid, remain unchanged. Geometrical similarity is presupposed. An example will make this clearer. If we are dealing with a single fluid, say air under given conditions, the kinematic viscosity remains of course the same. When a solid sphere moves uniformly through air, the character of the motion of the fluid round it may depend upon the size of the sphere and upon the velocity with which it travels. But we may infer that the motions remain *similar*, if only the product of diameter and velocity be given. Thus, if we know the motion for a particular diameter and velocity of the sphere, we can infer what it will be when the velocity is halved and the diameter doubled. The fluid velocities also will everywhere be halved at the *corresponding* places. M. Eiffel found that for any sphere there is a velocity which may be regarded as critical, *i.e.* a velocity at which the law of resistance changes its character somewhat suddenly. It follows from the rule that these critical velocities should be inversely proportional to the diameters of the spheres, a conclusion in pretty good agreement with M. Eiffel's observations*. But the principle is at least equally important in effecting a comparison between different fluids. If we know what happens on a certain scale and at a certain velocity in *water*, we can infer what will happen in *air* on any other scale, provided the velocity is chosen suitably. It is assumed here that the compressibility of the air does not come into account, an assumption which is admissible so long as the velocities are small in comparison with that of sound.

But although the principle of similarity is well established on the theoretical side and has met with some confirmation in experiment, there has been much hesitation in applying it, due perhaps to certain discrepancies with observation which stand recorded. And there is another reason. It is rather difficult to understand how viscosity can play so large a part as it seems to do, especially when we introduce numbers, which make it appear that the viscosity of air, or water, is very small in relation to the other data occurring in practice. In order to remove these doubts it is very desirable to experiment with different viscosities, but this is not easy to do on a

* *Comptes Rendus*, Dec. 30, 1912, Jan. 13, 1913. [This volume, p. 136.]

moderately large scale, as in the wind channels used for aeronautical purposes. I am therefore desirous of bringing before you some observations that I have recently made with very simple apparatus.

When liquid flows from one reservoir to another through a channel in which there is a contracted place, we can compare what we may call the *head* or driving pressure, *i.e.* the difference of the pressures in the two reservoirs, with the *suction, i.e.* the difference between the pressure in the recipient vessel and that lesser pressure to be found at the narrow place. The ratio of head to suction is a purely numerical quantity, and according to the principle of similarity it should for a given channel remain unchanged, provided the velocity be taken proportional to the kinematic viscosity of the fluid. The use of the same material channel throughout has the advantage that no question can arise as to geometrical similarity, which in principle should extend to any roughnesses upon the surface, while the necessary changes of velocity are easily attained by altering the head and those of viscosity by altering the temperature.

The apparatus consisted of two aspirator bottles (fig. 10) containing water and connected below by a passage bored in a cylinder of lead, 7 cm.

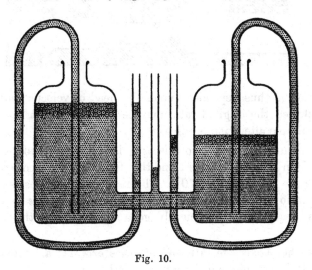

Fig. 10.

long, fitted water-tight with rubber corks. The form of channel actually employed is shown in fig. 11. On the up-stream side it contracts pretty suddenly from full bore (8 mm.) to the narrowest place, where the diameter is 2·75 mm. On the down-stream side the expansion takes place in four or five steps, corresponding to the drills available. It had at first been intended to use a smooth curve, but preliminary trials showed that this was unnecessary, and the expansion by steps has the advantage of bringing before the mind the dragging action of the jets upon the thin layers of fluid between them and the walls. The three pressures concerned are indicated on manometer tubes as shown, and the two differences of level representing head and suction can be taken off with compasses and referred to a milli-metre scale. In starting an observation the water is drawn up in the discharge vessel, as far as may be required, with the aid of an air-pump. The rubber cork at the top of the discharge vessel necessary for this purpose is not shown.

As the head falls during the flow of the water, the ratio of head to suction increases. For most of the observations I contented myself with recording the head for which the ratio of head to suction was exactly 2 : 1, as indicated by proportional compasses. Thus on January 23, when the temperature of the water was 9° C., the 2 : 1 ratio occurred on four trials at 120, 130, 123, 126, mean 125 mm. head. The temperature was then raised with precaution by pouring in warm water with passages backwards and forwards. The occurrence of the 2 : 1 ratio was now much retarded, the mean head being only 35 mm., corresponding to a mean temperature of 37° C. The ratio of

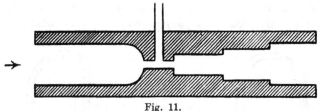

Fig. 11.

head to suction is thus dependent upon the head or velocity, but when the velocity is altered the original ratio may be recovered if at the same time we make a suitable alteration of viscosity.

And the required alteration of viscosity is about what might have been expected. From Landolt's tables I find that for 9° C. the viscosity of water is ·01368, while for 37° C. it is ·00704. The ratio of viscosities is accordingly 1·943. The ratio of heads is 125 : 35. The ratio of *velocities* is the square-root of this or 1·890, in sufficiently good agreement with the ratio of viscosities.

In some other trials the ratio of velocities *exceeded* a little the ratio of viscosities. It is not pretended that the method would be an accurate one for the comparison of viscosities. The change in the ratio of head to suction is rather slow, and the measurement is usually somewhat prejudiced by unsteadiness in the suction manometer. Possibly better results would be obtained in more elaborate observations by several persons, the head and suction being recorded separately and referred to a time scale so as to facilitate interpolation. But as they stand the results suffice for my purpose, showing directly and conclusively the influence of viscosity as compensating a change in the velocity.

In conclusion, I must touch briefly upon a part of the subject where theory is still at fault, and I will limit myself to the simplest case of all— the uniform shearing motion of a viscous fluid between two parallel walls, one of which is at rest, while the other moves tangentially with uniform velocity. It is easy to prove that a uniform shearing motion of the fluid satisfies the dynamical equations, but the question remains: Is this motion stable? Does a small departure from the simple motion tend of itself to die out? In the case where the viscosity is relatively great, observation suggests an affirmative answer; and O. Reynolds, whose illness and comparatively early death were so great a loss to science, was able to deduce the same conclusion from theory. Reynolds' method has been improved, more especially by Professor Orr of Dublin. The simple motion is thoroughly stable if the viscosity exceed a certain specified value relative to the velocity of the moving plane and the distance between the planes; while if the viscosity is less than this, it is possible to propose a kind of departure from the original motion which will increase *for a time*. It is on this side of the question that there is a deficiency. When the viscosity is very small, observation appears to show that the simple motion is unstable, and we ought to be able to derive this result from theory. But even if we omit viscosity altogether, it does not appear possible to prove instability *à priori*, at least so long as we regard the walls as mathematically plane. We must confess that at the present we are unable to give a satisfactory account of skin-friction, in order to overcome which millions of horse-power are expended in our ships. Even in the older subjects there are plenty of problems left !

385.

ON THE THEORY OF LONG WAVES AND BORES.

[*Proceedings of the Royal Society*, A, Vol. xc. pp. 324—328, 1914.]

In the theory of long waves in two dimensions, which we may suppose to be reduced to a "steady" motion, it is assumed that the length is so great in proportion to the depth of the water that the velocity in a vertical direction can be neglected, and that the horizontal velocity is uniform across each section of the canal. This, it should be observed, is perfectly distinct from any supposition as to the height of the wave. If l be the undisturbed depth, and h the elevation of the water at any point of the wave, u_0, u the velocities corresponding to l, $l+h$ respectively, we have, as the equation of continuity,

$$u = \frac{lu_0}{l+h}. \quad \ldots\ldots\ldots\ldots\ldots\ldots\ldots(1)$$

By the principles of hydrodynamics, the increase of pressure due to retardation will be

$$\tfrac{1}{2}\rho\,(u_0{}^2 - u^2) = \frac{\rho u_0{}^2}{2} \cdot \frac{2lh + h^2}{(l+h)^2}. \quad \ldots\ldots\ldots\ldots(2)$$

On the other hand, the loss of pressure (at the surface) due to height will be $g\rho h$; and therefore the total gain of pressure over the undisturbed parts is

$$\left(\frac{\rho u_0{}^2}{l} \cdot \frac{1 + h/2l}{(1 + h/l)^2} - g\rho\right) h. \quad \ldots\ldots\ldots\ldots(3)$$

If, now, the ratio h/l be very small, the coefficient of h becomes

$$\rho\,(u_0{}^2/l - g), \quad \ldots\ldots\ldots\ldots\ldots\ldots\ldots(4)$$

and we conclude that the condition of a free surface is satisfied, provided $u_0{}^2 = gl$. This determines the rate of flow u_0, in order that a stationary wave may be possible, and gives, of course, at the same time the velocity of a wave in still water.

Unless h^2 can be neglected, it is impossible to satisfy the condition of a free surface for a stationary long wave—which is the same as saying that it is impossible for a long wave of finite height to be propagated in still water without change of type.

Although a constant gravity is not adequate to compensate the changes of pressure due to acceleration and retardation in a long wave of finite height, it is evident that complete compensation is attainable if gravity be made a suitable function of height; and it is worth while to enquire what the law of force must be in order that long waves of unlimited height may travel with type unchanged. If f be the force at height h, the condition of constant surface pressure is

$$\tfrac{1}{2}\rho u_0{}^2 \left\{ 1 - \frac{l^2}{(l+h)^2} \right\} = \rho \int_0^h f\,dh ; \quad \dots\dots\dots\dots(5)$$

whence

$$f = -\frac{u_0{}^2}{2} \cdot \frac{d}{dh}\frac{l^2}{(l+h)^2} = u_0{}^2 \frac{l^2}{(l+h)^3}, \quad \dots\dots\dots(6)$$

which shows that the force must vary inversely as the cube of the distance from the bottom of the canal. Under this law the waves may be of any height, and they will be propagated unchanged with the velocity $\surd(f_1 l)$, where f_1 is the force at the undisturbed level*.

It may be remarked that we are concerned only with the values of f at water-levels which actually occur. A change in f below the lowest water-level would have no effect upon the motion, and thus no difficulty arises from the law of inverse cube making the force infinite at the bottom of the canal.

When a wave is limited in length, we may speak of its velocity relatively to the undisturbed water lying beyond it on the two sides, and it is implied that the uniform levels on the two sides are the same. But the theory of long waves is not thus limited, and we may apply it to the case where the uniform levels on the two sides of the variable region are different, as, for example, to *bores*. This is a problem which I considered briefly on a former occasion†, when it appeared that the condition of conservation of energy could not be satisfied with a constant gravity. But in the calculation of the loss of energy a term was omitted, rendering the result erroneous, although the general conclusions are not affected. The error became apparent in applying the method to the case above considered of a gravity varying as the inverse cube of the depth. But, before proceeding to the calculation of energy, it may be well to give the generalised form of the relation between velocity and height which must be satisfied in a *progressive* wave‡, whether or not the type be permanent.

* *Phil. Mag.* Vol. i. p. 257 (1876); *Scientific Papers*, Vol. i. p. 254.

† *Roy. Soc. Proc.* A, Vol. lxxxi. p. 448 (1908); *Scientific Papers*, Vol. v. p. 495.

‡ Compare *Scientific Papers*, Vol. i. p. 253 (1899).

In a small positive progressive wave, the relation between the particle-velocity u at any point (now reckoned relatively to the parts outside the wave) and the elevation h is

$$u = \sqrt{(f/l)} \cdot h. \qquad \qquad (7)$$

If this relation be violated anywhere, a wave will emerge, travelling in the negative direction. In applying (7) to a wave of finite height, the appropriate form of (7) is

$$du = \sqrt{\left(\frac{f}{l+h}\right)} dh, \qquad \qquad (8)$$

where f is a known function of $l + h$, or on integration

$$u = \int_0^h \sqrt{\left(\frac{f}{l+h}\right)} dh. \qquad \qquad (9)$$

To this particle-velocity is to be added the wave-velocity

$$\sqrt{\{(l+h)f\}}, \qquad \qquad (10)$$

making altogether for the velocity of, *e.g.*, the crest of a wave relative to still water

$$\int_0^h \sqrt{\left(\frac{f}{l+h}\right)} dh + \sqrt{\{(l+h)f\}}. \qquad \qquad (11)$$

Thus if f be constant, say g, (9) gives De Morgan's formula

$$u = 2\sqrt{g}\{(l+h)^{\frac{1}{2}} - l^{\frac{1}{2}}\}, \qquad \qquad (12)$$

and (11) becomes

$$3\sqrt{g}\sqrt{(l+h)} - 2\sqrt{(gl)}. \qquad \qquad (13)$$

If, again,
$$f = \frac{f_1 l^3}{(l+h)^3}, \qquad \qquad (14)$$

(11) gives as the velocity of a crest

$$\frac{f_1^{\frac{1}{2}} l^{\frac{1}{2}} h}{l+h} + \frac{f_1^{\frac{1}{2}} l^{\frac{3}{2}}}{l+h} = \sqrt{(f_1 l)}, \qquad \qquad (15)$$

which is independent of h, thus confirming what was found before for this law of force.

As regards the question of a bore, we consider it as the transition from a uniform velocity u and depth l to a uniform velocity u' and depth l', l' being greater than l. The first relation between these four quantities is that given by continuity, viz.,

$$lu = l'u'. \qquad \qquad (16)$$

The second relation arises from a consideration of momentum. It may be convenient to take first the usual case of a constant gravity g. The mean pressures at the two sections are $\frac{1}{2}gl$, $\frac{1}{2}gl'$, and thus the equation of momentum is

$$lu(u - u') = \frac{1}{2}g(l'^2 - l^2). \qquad \qquad (17)$$

By these equations u and u' are determined in terms of l, l' :

$$u^2 = \tfrac{1}{2}g\,(l + l')\,.\,l'/l, \qquad u'^2 = \tfrac{1}{2}g\,(l + l')\,.\,l/l'. \quad\ldots\ldots\ldots\ldots(18)$$

We have now to consider the question of energy. The difference of work done by the pressure at the two ends (reckoned per unit of time and per unit of breadth) is $lu\,(\tfrac{1}{2}gl - \tfrac{1}{2}gl')$. And the difference between the *kinetic* energies entering and leaving the region is $lu\,(\tfrac{1}{2}u^2 - \tfrac{1}{2}u'^2)$, the density being taken as unity. But this is not all. The *potential* energies of the liquid leaving and entering the region are different. The centre of gravity rises through a height $\tfrac{1}{2}(l' - l)$, and the gain of potential energy is therefore $lu\,.\,\tfrac{1}{2}g\,(l' - l)$. The whole *loss* of energy is accordingly

$$lu\,\{\tfrac{1}{2}gl - \tfrac{1}{2}gl' + \tfrac{1}{2}u^2 - \tfrac{1}{2}u'^2 - \tfrac{1}{2}g\,(l' - l)\} = lu\left\{gl - gl' + \tfrac{1}{4}g\,(l + l')\left(\frac{l'}{l} + \frac{l}{l'}\right)\right\}$$

$$= lu\,.\,\frac{g\,(l' - l)^3}{4ll'}\,. \qquad\ldots\ldots\ldots\ldots\ldots\ldots(19)$$

This is much smaller than the value formerly given, but it remains of the same sign. "That there should be a loss of energy constitutes no difficulty, at least in the presence of viscosity; but the impossibility of a gain of energy shows that the motions here contemplated cannot be reversed."

We now suppose that the constant gravity is replaced by a force f, which is a function of y, the distance from the bottom. The pressures p, p' at the two sections are also functions of y, such that

$$p = \int_y^l f\,dy, \qquad p' = \int_y^{l'} f\,dy. \quad\ldots\ldots\ldots\ldots\ldots(20)$$

The equation of momentum replacing (17) is now

$$lu\,(u - u') = \int_0^{l'} p'\,dy - \int_0^l p\,dy = \left[p'y\right]_0^{l'} - \left[py\right]_0^l - \int_0^{l'} y\,\frac{dp'}{dy}\,dy + \int_0^l y\,\frac{dp}{dy}\,dy$$

$$= \int_0^{l'} yf\,dy - \int_0^l yf\,dy = \int_l^{l'} yf\,dy, \quad\ldots\ldots\ldots\ldots\ldots\ldots(21)$$

the integrated terms vanishing at the limits. This includes, of course, all special cases, such as $f = \text{constant}$, or $f \propto y^{-3}$.

As regards the reckoning of energy, the first two terms on the left of (19) are replaced by

$$lu\left\{\frac{1}{l}\int_0^l p\,dy - \frac{1}{l'}\int_0^{l'} p'\,dy\right\}. \quad\ldots\ldots\ldots\ldots\ldots(22)$$

The third and fourth terms representing kinetic energy remain as before. For the potential energy we have to consider that a length u and depth l is converted into a length u' and depth l'. If we reckon from the bottom, the potential energy is in the first case

$$u\int_0^l dy \int_0^y f\,dy,$$

in which

$$\int_0^y f\,dy = \int_0^l f\,dy - \int_y^l f\,dy = p_0 - p,$$

p_0 denoting the pressure at the bottom, so that the potential energy is

$$ul\left\{p_0 - \frac{1}{l}\int_0^l p\,dy\right\}.$$

The difference of potential energies, corresponding to the fifth and sixth terms of (19), is thus

$$lu\left\{p_0 - p_0' - \frac{1}{l}\int_0^l p\,dy + \frac{1}{l'}\int_0^{l'} p'\,dy\right\}. \qquad \dots\dots\dots\dots(23)$$

The integrals in (23) compensate those of (22), and we have finally as the loss of energy

$$lu\left\{p_0 - p_0' + \tfrac{1}{2}u^2 - \tfrac{1}{2}u'^2\right\} = lu\left\{\tfrac{1}{2}u^2 - \tfrac{1}{2}u'^2 - \int_l^{l'} f\,dy\right\}. \qquad \dots\dots(24)$$

It should be remarked that it is only for values of y between l and l' that f is effectively involved.

In the special case where $f = \mu y^{-3}$, equations (16), (21) give

$$u^2 l^2 = \mu, \qquad u'^2 l'^2 = \mu, \qquad \dots\dots\dots\dots\dots\dots\dots(25)$$

the introduction of which into (24) shows that, in this case, the loss of energy vanishes; all the conditions can be satisfied, even though there be no dissipation. The reversed motion is then equally admissible.

Experimental.

The formation of bores is illustrated by a very ordinary observation, probably not often thought of in this connection. Something of the kind may usually be seen whenever a stream of water from a tap strikes a horizontal surface [or when water from a can is poured into a flat bath]. The experiment is best made by directing a vertically falling stream into a flat and shallow dish from which the water overflows*. The effective depth may be varied by holding a glass plate in a horizontal position under the water surface. Where the jet strikes, it expands into a thin sheet which diverges for a certain distance, and this distance diminishes as the natural depth of the water over the plate is made greater. The circular boundary where the transition from a small to a greater depth takes place constitutes a bore on a small scale. The flow may be made two-dimensional by limiting it with two battens held in contact with the glass. I have not attempted measures. On the smallest scale surface-tension doubtless plays a considerable part, but this may be minimised by increasing the stream, and correspondingly the depth of the water over the plate, so far as may be convenient.

* The tap that I employed gives a jet whose diameter is 6 mm. A much larger tap may need to be fitted with a special nozzle.—May 14, [1914].

386.

THE SAND-BLAST.

[*Nature*, Vol. XCIII. p. 188, 1914.]

AMONG the many remarkable anticipations contained in T. Young's *Lectures on Natural Philosophy* (1807) is that in which he explains the effect of what is now commonly known as the sand-blast. On p. 144 he writes:—
"There is, however, a limit beyond which the velocity of a body striking another cannot be increased without overcoming its resilience, and breaking it, however small the bulk of the first body may be, and this limit depends on the inertia of the parts of the second body, which must not be disregarded when they are impelled with a considerable velocity. For it is demonstrable that there is a certain velocity, dependent on the nature of a substance, with which the effect of any impulse or pressure is transmitted through it; a certain portion of time, which is shorter accordingly as the body is more elastic, being required for the propagation of the force through any part of it; and if the actual velocity of any impulse be in a greater proportion to this velocity than the extension or compression, of which the substance is capable, is to its whole length, it is obvious that a separation must be produced, since no parts can be extended or compressed which are not yet affected by the impulse, and the length of the portion affected at any instant is not sufficient to allow the required extension or compression. Thus if the velocity with which an impression is transmitted by a certain kind of wood be 15,000 ft. in a second, and it be susceptible of compression to the extent of 1/200 of its length, the greatest velocity that it can resist will be 75 ft. in a second, which is equal to that of a body falling from a height of about 90 ft."

Doubtless this passage was unknown to O. Reynolds when, with customary penetration, in his paper on the sand-blast (*Phil. Mag.* Vol. XLVI. p. 337, 1873) he emphasises that "the intensity of the pressure between bodies on first impact is independent of the size of the bodies."

After his manner, Young was over-concise, and it is not clear precisely what circumstances he had in contemplation. Probably it was the longitudinal impact of bars, and at any rate this affords a convenient example. We may begin by supposing the bars to be of the same length, material, and section, and before impact to be moving with equal and opposite velocities v. At impact, the impinging faces are reduced to rest, and remain at rest so long as the bars are in contact at all. This condition of rest is propagated in each bar as a wave moving with a velocity a, characteristic of the material. In such a progressive wave there is a general relation between the particle-velocity (estimated relatively to the parts outside the wave) and the compression (e), viz., that the velocity is equal to ae. In the present case the relative particle-velocity is v, so that $v = ae$. The limit of the strength of the material is reached when e has a certain value, and from this the greatest value of v (*half* the original relative velocity) which the bars can bear is immediately inferred.

But the importance of the conclusion depends upon an extension now to be considered. It will be seen that the length of the bars does not enter into the question. Neither does the equality of the lengths. However short one of them may be, we may contemplate an interval after first impact so short that the wave will not have reached the further end, and then the argument remains unaffected. However short one of the impinging bars, the above calculated relative velocity is the highest which the material can bear without undergoing disruption.

As more closely related to practice, the case of two spheres of radii r, r', impinging directly with relative velocity v, is worthy of consideration. According to ordinary elastic theory the only remaining data of the problem are the densities ρ, ρ', and the elasticities. The latter may be taken to be the Young's moduli q, q', and the Poisson's ratios, σ, σ', of which the two last are purely numerical. The same may be said of the ratios q'/q, ρ'/ρ, and r'/r. So far as dimensional quantities are concerned, any maximum strain e may be regarded as a function of r, v, q, and ρ. The two last can occur only in the combination q/ρ, since strain is of no dimensions. Moreover, $q/\rho = a^2$, where a is a velocity. Regarding e as a function of r, v, and a, we see that v and a can occur only as the ratio v/a, and that r cannot appear at all. The maximum strain then is independent of the linear scale; and if the rupture depends only on the maximum strain, it is as likely to occur with small spheres as with large ones. The most interesting case occurs when one sphere is very large relatively to the other, as when a grain of sand impinges upon a glass surface. If the velocity of impact be given, the glass is as likely to be broken by a small grain as by a much larger one. It may be remarked that this conclusion would be upset if rupture depends upon the *duration* of a strain as well as upon its *magnitude*.

The general argument from dynamical similarity that the maximum strain during impact is independent of linear scale, is, of course, not limited to the case of spheres, which has been chosen merely for convenience of statement.

387.

THE EQUILIBRIUM OF REVOLVING LIQUID UNDER CAPILLARY FORCE.

[*Philosophical Magazine*, Vol. XXVIII. pp. 161—170, 1914.]

THE problem of a mass of homogeneous incompressible fluid revolving with uniform angular velocity (ω) and held together by capillary tension (T) is suggested by well-known experiments of Plateau. If there is no rotation, the mass assumes a spherical form. Under the influence of rotation the sphere flattens at the poles, and the oblateness increases with the angular velocity. At higher rotations Plateau's experiments suggest that an annular form may be one of equilibrium. The earlier forms, where the liquid still meets the axis of rotation, have been considered in some detail by Beer*, but little attention seems to have been given to the equilibrium in the form of a ring. A general treatment of this case involves difficulties, but if we assume that the ring is *thin*, viz. that the diameter of the section is small compared with the diameter of the circular axis, we may prove that the form of the section is approximately circular and investigate the small departures from that figure. It is assumed that in the cases considered the surface is one of revolution about the axis of rotation.

Fig. 1 represents a section by a plane through the axis Oy, O being the point where the axis meets the equatorial plane. One of the principal

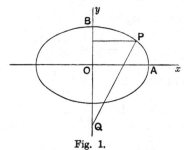

Fig. 1.

* *Pogg. Ann.* Vol. XCVI. p. 210 (1855); compare Poincaré's *Capillarité*, 1895.

curvatures of the surface at P is that of the meridianal curve, the radius of the other principal curvature is PQ—the normal as terminated on the axis. The pressure due to the curvature is thus

$$T\left(\frac{1}{\rho} + \frac{1}{PQ}\right),$$

and the equation of equilibrium may be written

$$\frac{1}{\rho} + \frac{1}{PQ} = \frac{\sigma\omega^2x^2}{2T} + \frac{p_0}{T}, \quad \dots\dots\dots\dots\dots\dots(1)$$

where p_0 is the pressure at points lying upon the axis, and σ is the density of the fluid.

The curvatures may most simply be expressed by means of s, the length of the arc of the curve measured say from A. Thus

$$\frac{1}{PQ} = \frac{1}{x}\frac{dy}{ds}, \quad \frac{1}{\rho} = \frac{d^2y/ds^2}{dx/ds}, \quad \dots\dots\dots\dots\dots\dots(2)$$

so that (1) becomes

$$\frac{dy}{ds}\frac{dx}{ds} + x\frac{d^2y}{ds^2} = \frac{\sigma\omega^2x^3}{2T}\frac{dx}{ds} + \frac{p_0x}{T}\frac{dx}{ds},$$

or on integration

$$x\frac{dy}{ds} = \frac{\sigma\omega^2x^4}{8T} + \frac{p_0x^2}{2T} + \text{const.} \quad \dots\dots\dots\dots(3)$$

Thus dy/ds is a function of x of known form, say X, and we get for y in terms of x

$$y = \pm\int\frac{X\,dx}{\sqrt{(1 - X^2)}}, \quad \dots\dots\dots\dots\dots\dots(4)$$

as given by Beer.

If, as in fig. 1, the curve meets the axis, (3) must be satisfied by $x = 0$, $dy/ds = 0$. The constant accordingly disappears, and we have the much simplified form

$$\frac{dy}{ds} = \frac{\sigma\omega^2x^3}{8T} + \frac{p_0x}{2T}. \quad \dots\dots\dots\dots\dots\dots(5)$$

At the point A on the equator $dy/ds = 1$. If $OA = a$,

$$1 = \frac{\sigma\omega^2a^3}{8T} + \frac{p_0a}{2T};$$

whence eliminating p_0 and writing

$$\Omega = \frac{\sigma\omega^2a^3}{8T}, \quad \dots\dots\dots\dots\dots\dots(6)$$

we get

$$\frac{dy}{ds} = \Omega\frac{x^3}{a^3} + (1 - \Omega)\frac{x}{a}. \quad \dots\dots\dots\dots\dots\dots(7)$$

In terms of y and x from (7)

$$\pm \frac{dy}{dx} = \frac{x\left(\Omega \frac{x^2}{a^2} + 1 - \Omega\right)}{\sqrt{\left\{a^2 - x^2\left(\Omega \frac{x^2}{a^2} + 1 - \Omega\right)^2\right\}}}, \quad \dotsc\dotsc\dotsc\dotsc(8)$$

or if we write

$$x^2/a^2 = 1 - z, \quad \dotsc\dotsc\dotsc\dotsc\dotsc\dotsc\dotsc(9)$$

$$-\frac{2\,dy}{a\,dz} = \frac{1 - \Omega z}{\sqrt{z}\,.\,\sqrt{\{1 + 2(1-z)\,\Omega - z(1-z)\,\Omega^2\}}}$$

$$= (1 - \Omega + \tfrac{3}{2}\Omega^2)\,z^{-\frac{1}{2}} - \tfrac{3}{2}\Omega^2\,z^{\frac{1}{2}}, \quad \dotsc\dotsc\dotsc\dotsc(10)$$

when we neglect higher powers of Ω than Ω^2. Reverting to x, we find for the integral of (10)

$$\pm\frac{y}{a} = (1 - \Omega + \tfrac{3}{2}\Omega^2)\left(1 - \frac{x^2}{a^2}\right)^{\frac{1}{2}} - \frac{\Omega^2}{2}\left(1 - \frac{x^2}{a^2}\right)^{\frac{3}{2}}, \quad \dotsc\dotsc\dotsc(11)$$

no constant being added since $y = 0$ when $x = a$.

If we stop at Ω, we have

$$\frac{x^2}{a^2} + \frac{y^2}{a^2(1 - \Omega)^2} = 1 \quad \dotsc\dotsc\dotsc\dotsc\dotsc(12)$$

representing an ellipse whose minor axis OB is $a(1 - \Omega)$.

When Ω^2 is retained,

$$OB = (1 - \Omega + \Omega^2)\,a. \quad \dotsc\dotsc\dotsc\dotsc\dotsc(13)$$

The approximation in powers of Ω could of course be continued if desired.

So long as $\Omega < 1$, p_0 is positive and the (equal) curvatures at B are convex. When $\Omega = 1$, $p_0 = 0$ and the surface at B is flat. In this case (8) gives

$$\pm\frac{dy}{dx} = \frac{x^3}{\sqrt{\{a^6 - x^6\}}}, \quad \dotsc\dotsc\dotsc\dotsc\dotsc(14)$$

or if we set $x = a\sin^{\frac{1}{3}}\phi$,

$$\pm\frac{dy}{d\phi} = \frac{a}{3}\sin^{\frac{1}{3}}\phi. \quad \dotsc\dotsc\dotsc\dotsc\dotsc(15)$$

Here $x = a$ corresponds to $\phi = \tfrac{1}{2}\pi$, and $x = 0$ corresponds to $\phi = 0$. Hence

$$OB = \frac{a}{3}\int_0^{\frac{1}{2}\pi}\sin^{\frac{1}{3}}\phi\,d\phi. \quad \dotsc\dotsc\dotsc\dotsc(16)$$

The integral in (16) may be expressed in terms of gamma functions and we get

$$OB = a\sqrt{\pi}\,.\,\Gamma(\tfrac{2}{3}) \div \Gamma(\tfrac{1}{6}) = \cdot 4312a. \quad \dotsc\dotsc\dotsc\dotsc(17)$$

When $\Omega > 1$, the curvature at B is concave and p_0 is negative, as is quite permissible.

In order to trace the various curves we may calculate by quadratures from (4) the position of a sufficient number of points. This, as I understand, was the procedure adopted by Beer. An alternative method is to trace the curves by direct use of the radius of curvature at the point arrived at. Starting from (7) we find

$$\frac{d^2y}{ds^2} = \left(\Omega \frac{3x^2}{a^3} + \frac{1-\Omega}{a}\right) \frac{dx}{ds},$$

and thence

$$\frac{a}{\rho} = a \frac{d^2y/ds^2}{dx/ds} = \Omega \frac{3x^2}{a^2} + 1 - \Omega. \quad \dots\dots\dots\dots\dots(18)$$

From (18) we see at once that $\Omega = 0$ makes $\rho = a$ throughout, and that when $\Omega = 1$, $x = 0$ makes $\rho = \infty$.

In tracing a curve we start from the point A in a known direction and with $\rho = a/(2\Omega + 1)$, and at every point arrived at we know with what curvature to proceed. If, as has been assumed, the curve meets the axis, it must do so at right angles, and a solution is then obtained.

The method is readily applied to the case $\Omega = 1$ with the advantage that we know where the curve should meet the axis of y. From (18) with $\Omega = 1$ and $a = 5$,

$$\frac{1}{\rho} = \frac{24x^2}{1000}. \quad \dots\dots\dots\dots\dots\dots\dots(19)$$

Starting from $x = 5$ we draw small portions of the curve corresponding to decrements of x equal to ·2, thus arriving in succession at the points for which $x = 4\cdot8$, $4\cdot6$, $4\cdot4$, &c. For these portions we employ the *mean* curvatures, corresponding to $x = 4\cdot9$, $4\cdot7$, &c. calculated from (19). It is convenient to use squared paper and fair results may be obtained with the ordinary ruler and compasses. There is no need actually to draw the normals. But for such work the procedure recommended by Boys* offers great advantages. The ruler and compasses are replaced by a straight scale divided upon a strip of semi-transparent celluloid. At one point on the scale a fine pencil point protrudes through a small hole and describes the diminutive circular arc. Another point of the scale at the required distance occupies the centre of the circle and is held temporarily at rest with the aid of a small brass tripod standing on sharp needle points. After each step the celluloid is held firmly to the paper and the tripod is moved to the point of the scale required to give the next value of the curvature. The ordinates of the curve so drawn are given in the second and fifth columns of the annexed table. It will be seen that from $x = 0$ to $x = 2$ the curve is very flat. Fig. (1).

* *Phil. Mag.* Vol. xxxvi. p. 75 (1893). I am much indebted to Mr Boys for the loan of suitable instruments. The use is easy after a little practice.

Another case of special interest is the last figure reaching the axis of symmetry at all, which occurs at the point $x = 0$. We do not know beforehand to what value of Ω this corresponds, and curves must be drawn tentatively. It appears that $\Omega = 2\cdot4$ approximately, and the values of y obtained from this curve are given in columns 3 and 6 of the table. Fig. (2)*.

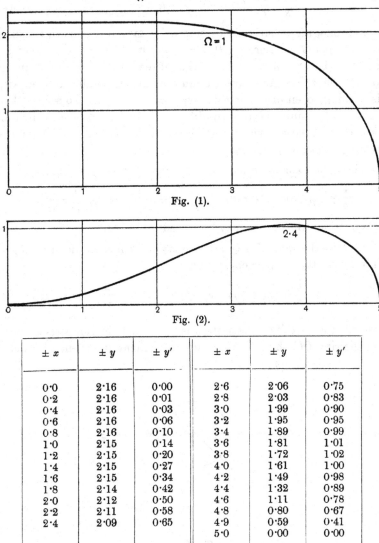

Fig. (1).

Fig. (2).

$\pm x$	$\pm y$	$\pm y'$	$\pm x$	$\pm y$	$\pm y'$
0·0	2·16	0·00	2·6	2·06	0·75
0·2	2·16	0·01	2·8	2·03	0·83
0·4	2·16	0·03	3·0	1·99	0·90
0·6	2·16	0·06	3·2	1·95	0·95
0·8	2·16	0·10	3·4	1·89	0·99
1·0	2·15	0·14	3·6	1·81	1·01
1·2	2·15	0·20	3·8	1·72	1·02
1·4	2·15	0·27	4·0	1·61	1·00
1·6	2·15	0·34	4·2	1·49	0·98
1·8	2·14	0·42	4·4	1·32	0·89
2·0	2·12	0·50	4·6	1·11	0·78
2·2	2·11	0·58	4·8	0·80	0·67
2·4	2·09	0·65	4·9	0·59	0·41
			5·0	0·00	0·00

There is a little difficulty in drawing the curve through the point of zero curvature. I found it best to begin at both ends ($x = 0, y = 0$) and ($x = 5, y = 0$) with an assumed value of Ω and examine whether the two parts could be made to fit.

* [1916. These figures were omitted in the original memoir.]

When $\Omega > 2\cdot4$ and the curve does not meet the axis at all, the constant in (3) must be retained, and the difficulty is much increased. If we suppose that $dy/ds = +1$ when $x = a_2$ and $dy/ds = -1$ when $x = a_1$, we can determine p_0 as well as the constant of integration, and (3) becomes

$$x\frac{dy}{ds} = \frac{\sigma\omega^2}{8T}(x^2 - a_1{}^2)(x^2 - a_2{}^2) + \frac{x^2 - a_1 a_2}{a_2 - a_1}. \quad\ldots\ldots\ldots\ldots(20)$$

We may imagine a curve to be traced by means of this equation. We start from the point A where $y = 0$, $x = a_2$ and in the direction perpendicular to OA, and (as before) we are told in what direction to proceed at any point reached. When $x = a_1$, the tangent must again be parallel to the axis, but there is nothing to ensure that this occurs when $y = 0$. To secure this end and so obtain an annular form of equilibrium, $\sigma\omega^2/T$ must be chosen suitably, but there is no means apparent of doing this beforehand. The process of curve tracing can only be tentative.

If we form the expression for the curvature as before, we obtain

$$\frac{1}{\rho} = \frac{\sigma\omega^2}{8T}\left(3x^2 - a_1{}^2 - a_2{}^2 - \frac{a_1{}^2 a_2{}^2}{x^2}\right) + \frac{1}{a_2 - a_1} + \frac{a_1 a_2}{x^2(a_2 - a_1)} \quad\ldots\ldots(21)$$

by means of which the curves may be traced tentatively.

If we retain the normal PQ, as we may conveniently do in using Boys' method, we have the simpler expression

$$\frac{1}{\rho} + \frac{1}{PQ} = \frac{\sigma\omega^2}{4T}(2x^2 - a_1{}^2 - a_2{}^2) + \frac{2}{a_2 - a_1}. \quad\ldots\ldots\ldots\ldots(22)$$

When the radius CP of the section is very small in comparison with the radius of the ring OC, the conditions are approximately satisfied by a circular

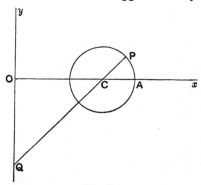

Fig. 2.

form. We write $CP = r$, $OC = a$, $PCA = \theta$. Then, r being supposed constant, the principal radii of curvature are r and $a\sec\theta + r$, so that the equation of equilibrium is

$$\frac{p_0}{T} = \frac{1}{r} + \frac{\cos\theta}{a + r\cos\theta} - \frac{\omega^2}{2T}(a + r\cos\theta)^2, \quad\ldots\ldots\ldots\ldots(23)$$

in which p_0 should be constant as θ varies. In this

$$\frac{\cos \theta}{a + r \cos \theta} = \frac{1}{a} \left\{ -\frac{r}{2a} + \left(1 + \frac{3r^2}{4a^2} \right) \cos \theta - \frac{r}{2a} \cos 2\theta + \frac{r^2}{4a^2} \cos 3\theta \right\},$$

$$\left(1 + \frac{r}{a} \cos \theta \right)^2 = 1 + \frac{r^2}{2a^2} + \frac{2r}{a} \cos \theta + \frac{r^2}{2a^2} \cos 2\theta.$$

Thus approximately

$$\frac{ap_0}{T} = \frac{a}{r} - \frac{r}{2a} - \frac{\omega^2 a^3}{2T} \left(1 + \frac{r^2}{2a^2} \right) + \cos \theta \left\{ 1 + \frac{3r^2}{4a^2} - \frac{\omega^2 a^3}{2T} \frac{2r}{a} \right\}$$

$$+ \cos 2\theta \left\{ -\frac{r}{2a} - \frac{\omega^2 a}{2T} \frac{r^2}{2a^2} \right\} + \cos 3\theta \cdot \frac{r^2}{4a^2}. \quad \dots (24)$$

The term in $\cos \theta$ will vanish if we take ω so that

$$\frac{\omega^2 a^2}{T} = \frac{1}{r} \left(1 - \frac{3r^2}{a^2} \right). \quad \dots\dots\dots\dots\dots\dots (25)$$

The coefficient of $\cos 2\theta$ then becomes

$$-\frac{3r}{4a} + \text{cubes of } \frac{r}{a}. \quad \dots\dots\dots\dots\dots\dots (26)$$

If we are content to neglect r/a in comparison with unity, the condition of equilibrium is satisfied by the circular form; otherwise there is an inequality of pressure of this order in the term proportional to $\cos 2\theta$. From (25) it is seen that if a and T be given, the necessary angular velocity increases as the radius of the section decreases.

In order to secure a better fulfilment of the pressure equation it is necessary to suppose r variable, and this of course complicates the expressions for the curvatures. For that in the meridianal plane we have

$$\frac{1}{\rho} = \frac{r^2 - r \frac{d^2 r}{d\theta^2} + 2 \left(\frac{dr}{d\theta} \right)^2}{\left\{ r^2 + \left(\frac{dr}{d\theta} \right)^2 \right\}^{\frac{3}{2}}},$$

or with sufficient approximation

$$\frac{1}{\rho} = \frac{1}{r} \left\{ 1 - \frac{1}{r} \frac{d^2 r}{d\theta^2} + \frac{1}{2r^2} \left(\frac{dr}{d\theta} \right)^2 \right\}. \quad \dots\dots\dots\dots (27)$$

For the curvature in the perpendicular plane we have to substitute PQ', measured along the normal, for PQ, whose expression remains as before (fig. 3). Now

$$\frac{PQ}{PQ'} = \frac{\sin Q'}{\sin Q} = \cos QPQ' - \tan \theta \sin QPQ'$$

in which

$$\cos QPQ' = \frac{CN}{r} = \left\{ 1 + \frac{1}{r^2} \left(\frac{dr}{d\theta} \right)^2 \right\}^{-\frac{1}{2}} = 1 - \frac{1}{2r^2} \left(\frac{dr}{d\theta} \right)^2$$

approximately,

$$\sin QPQ' = -\frac{1}{r}\frac{dr}{d\theta}\left\{1 - \frac{1}{2r^2}\left(\frac{dr}{d\theta}\right)^2\right\}.$$

Thus

$$\frac{1}{PQ} = \frac{\cos\theta}{a + r\cos\theta}\left\{1 - \frac{1}{2r^2}\left(\frac{dr}{d\theta}\right)^2\right\} + \frac{\sin\theta}{a + r\cos\theta}\frac{1}{r}\frac{dr}{d\theta}\left\{1 - \frac{1}{2r^2}\left(\frac{dr}{d\theta}\right)^2\right\}.$$

$$\ldots\ldots\ldots(28)$$

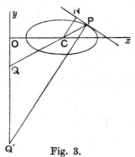

Fig. 3.

It will be found that it is unnecessary to retain $(dr/d\theta)^2$, and thus the pressure equation becomes

$$\frac{ap_0}{T} = \frac{a}{r}\left\{1 - \frac{1}{r}\frac{d^2r}{d\theta^2}\right\} + \frac{a\cos\theta}{a + r\cos\theta} + \frac{a\sin\theta}{a + r\cos\theta}\frac{1}{r}\frac{dr}{d\theta} - \frac{\omega^2 a^3}{2T}\left(1 + \frac{r\cos\theta}{a}\right)^2.$$

$$\ldots\ldots\ldots(29)$$

It is proposed to satisfy this equation so far as terms of the order r^2/a^2 inclusive.

As a function of θ, r may be taken to be

$$r = r_0 + \delta r = r_0 + r_1\cos\theta + r_2\cos 2\theta + \ldots, \qquad\ldots\ldots\ldots(30)$$

where r_1, r_2, &c. are constants small relatively to r_0. It will appear that to our order of approximation $(\delta r/r_0)^2$ may be neglected and that it is unnecessary to include the r's beyond r_3 inclusive. We have

$$\frac{a}{r} = \frac{a}{r_0}\left\{1 - \frac{\delta r}{r_0} + \left(\frac{\delta r}{r_0}\right)^2\right\},$$

$$\frac{a\cos\theta}{a + r\cos\theta} = -\frac{r_0}{2a} - \frac{r_2}{4a} + \cos\theta\left\{1 + \frac{3r_0^2}{4a^2} - \frac{3r_1}{4a} - \frac{r_3}{4a}\right\}$$

$$- \cos 2\theta\left\{\frac{r_0}{2a} + \frac{r_2}{2a}\right\} + \cos 3\theta\left\{\frac{r_0^2}{4a^2} - \frac{r_1}{4a} - \frac{r_3}{2a}\right\},$$

$$\left(1 + \frac{r\cos\theta}{a}\right)^2 = 1 + \frac{r_1}{a^2} + \frac{r_0^2}{2a^2} + \frac{r_0 r_2}{2a^2} + \cos\theta\left\{\frac{2r_0}{a} + \frac{r_2}{a} + \frac{3r_0 r_1}{2a^2} + \frac{r_0 r_3}{2a^2}\right\}$$

$$+ \cos 2\theta\left\{\frac{r_1}{a} + \frac{r_3}{a} + \frac{r_0^2}{2a^2} + \frac{r_0 r_2}{a^2}\right\} + \cos 3\theta\left\{\frac{r_2}{a} + \frac{r_0 r_3}{a^2} + \frac{r_0 r_1}{2a^2}\right\},$$

$$-\frac{a}{r^2}\frac{d^2r}{d\theta^2} = \frac{a}{r_0^2}\left\{r_1\cos\theta + 4r_2\cos 2\theta + 9r_3\cos 3\theta\right\},$$

$$\frac{a\sin\theta}{a+r\cos\theta}\frac{1}{r}\frac{dr}{d\theta} = -\frac{r_1}{2r_0} + \frac{r_2}{2a} + \cos\theta\left\{\frac{r_1}{4a} - \frac{r_2}{r_0} + \frac{3r_3}{4a}\right\}$$

$$+\cos 2\theta\left\{\frac{r_1}{2r_0} - \frac{3r_3}{2r_0}\right\} + \cos 3\theta\left\{\frac{r_2}{r_0} - \frac{r_1}{4a}\right\}.$$

Thus altogether for the coefficient of $\cos\theta$ on the right of (29) we get

$$1 + \frac{3r_0^2}{4a^2} - \frac{r_1}{2a} - \frac{r_2}{r_0} - \frac{\omega^2 a^3}{2T}\left\{\frac{2r_0}{a} + \frac{r_2}{a}\right\}.$$

This will be made to vanish if we take ω such that

$$\frac{\omega^2 a^2 r_0}{T} = 1 + \frac{3r_0^2}{4a^2} - \frac{r_1}{2a} - \frac{3r_2}{2r_0}. \quad\dots\dots\dots\dots\dots(31)$$

The coefficient of $\cos 2\theta$ is

$$\frac{3ar_2}{r_0^2} - \frac{r_0}{2a} + \frac{r_1}{2r_0} - \frac{3r_3}{2r_0} - \frac{\omega^2 a^3}{2T}\left\{\frac{r_1}{a} + \frac{r_3}{a} + \frac{r_0^2}{2a^3}\right\},$$

or when we introduce the value of ω from (31)

$$\frac{3ar_2}{r_0^2} - \frac{3r_0}{4a} - \frac{2r_3}{r_0}. \quad\dots\dots\dots\dots\dots\dots(32)$$

The coefficient of $\cos 3\theta$ is in like manner

$$\frac{8ar_3}{r_0^2} + \frac{r_0^2}{4a^2} + \frac{r_2}{2r_0}. \quad\dots\dots\dots\dots\dots\dots(33)$$

These coefficients are annulled and ap_0/T is rendered constant so far as the second order of r_0/a inclusive, when we take r_4, r_5, &c. equal to zero and

$$r_2/r_0 = r_0^2/4a^2, \qquad r_3/r_0 = -3r_0^3/64a^3. \quad\dots\dots\dots(34)$$

We may also suppose that $r_1 = 0$.

The solution of the problem is accordingly that

$$r = r_0\left\{1 + \frac{r_0^2}{4a^2}\cos 2\theta - \frac{3r_0^3}{64a^3}\cos 3\theta\right\} \quad\dots\dots\dots(35)$$

gives the figure of equilibrium, provided ω be such that

$$\frac{\omega^2 a^2 r_0}{T} = 1 + \frac{3r_0^2}{8a^2}. \quad\dots\dots\dots\dots\dots(36)$$

The form of a thin ring of equilibrium is thus determined; but it seems probable that the equilibrium would be unstable for disturbances involving a departure from symmetry round the axis of revolution.

388.

FURTHER REMARKS ON THE STABILITY OF VISCOUS FLUID MOTION.

[*Philosophical Magazine*, Vol. XXVIII. pp. 609—619, 1914.]

AT an early date my attention was called to the problem of the stability of fluid motion in connexion with the acoustical phenomena of sensitive jets, which may be ignited or unignited. In the former case they are usually referred to as sensitive *flames*. These are naturally the more conspicuous experimentally, but the theoretical conditions are simpler when the jets are unignited, or at any rate not ignited until the question of stability has been decided.

The instability of a surface of separation in a non-viscous liquid, *i.e.* of a surface where the velocity is discontinuous, had already been remarked by Helmholtz, and in 1879 I applied a method, due to Kelvin, to investigate the character of the instability more precisely. But nothing very practical can be arrived at so long as the original steady motion is treated as discontinuous, for in consequence of viscosity such a discontinuity in a real fluid must instantly disappear. A nearer approach to actuality is to suppose that while the *velocity* in a laminated steady motion is continuous, the *rotation* or vorticity changes suddenly in passing from one layer of finite thickness to another. Several problems of this sort have been treated in various papers*. The most general conclusion may be thus stated. The steady motion of a non-viscous liquid in two dimensions between fixed parallel plane walls is stable provided that the velocity U, everywhere parallel to the walls and a function of y only, is such that d^2U/dy^2 is of one sign throughout, y being the coordinate measured perpendicularly to the walls. It is here assumed that the disturbance is in two dimensions and *infinitesimal*. It involves

---

. * *Proc. Lond. Math. Soc.* Vol. x. p. 4 (1879); xi. p. 57 (1880); xix. p. 67 (1887); xxvii. p. 5 (1895); *Phil. Mag.* Vol. xxxiv. p. 59 (1892); xxvi. p. 1001 (1913); *Scientific Papers*, Arts. 58, 66, 144, 216, 194. [See also Art. 377.]

a slipping at the walls, but this presents no inconsistency so long as the fluid is regarded as absolutely non-viscous.

The steady motions for which stability in a non-viscous fluid may be inferred include those assumed by a viscous fluid in two important cases, (i) the simple shearing motion between two planes for which $d^2U/dy^2 = 0$, and (ii) the flow (under suitable forces) between two fixed plane walls for which d^2U/dy^2 is a finite constant. And the question presented itself whether the effect of viscosity upon the disturbance could be to introduce instability. An affirmative answer, though suggested by common experience and the special investigations of O. Reynolds[*], seemed difficult to reconcile with the undoubted fact that great viscosity makes for stability.

It was under these circumstances that "the Criterion of the Stability and Instability of the Motion of a Viscous Fluid," with special reference to cases (i) and (ii) above, was proposed as the subject of an Adams Prize essay[†], and shortly afterwards the matter was taken up by Kelvin[‡] in papers which form the foundation of much that has since been written upon the subject. His conclusion was that in both cases the steady motion is wholly stable for infinitesimal disturbances, whatever may be the value of the viscosity (μ); but that when the disturbances are finite, the limits of stability become narrower and narrower as μ diminishes. Two methods are employed: the first a special method applicable only to case (i) of a simple shear, the second (ii) more general and applicable to both cases. In 1892 (l.c.) I had occasion to take exception to the proof of stability by the second method, and Orr[§] has since shown that the same objection applies to the special method. Accordingly Kelvin's proof of stability cannot be considered sufficient, even in case (i). That Kelvin himself (partially) recognized this is shown by the following interesting and characteristic letter, which I venture to give in full.

July 10 (? 1895).

" On Saturday I saw a splendid illustration by Arnulf Mallock of our ideas regarding instability of water between two parallel planes, one kept moving and the other fixed. (Fig. 1) Coaxal cylinders, nearly enough planes for our illustration. The rotation of the outer can was kept very accurately uniform at whatever speed the governor was set for, when left to itself. At one of the speeds he shewed me, the water came to regular regime, *quite smooth*. I dipped a disturbing rod an inch or two down into the water and immediately the torque increased largely. *Smooth* regime could only be

[*] *Phil. Trans.* 1883, Part III. p. 935.

[†] *Phil. Mag.* Vol. XXIV. p. 142 (1887). The suggestion came from me, but the notice was (I think) drawn up by Stokes.

[‡] *Phil. Mag.* Vol. XXIV. pp. 188, 272 (1887); *Collected Papers*, Vol. IV. p. 321.

[§] Orr, *Proc. Roy. Irish Acad.* Vol. XXVII. (1907).

re-established by slowing down and bringing up to speed again, gradually enough.

"Without the disturbing rod at all, I found that by resisting the outer can by hand somewhat suddenly, but not very much so, the torque increased suddenly and the motion became visibly turbulent at the lower speed and remained so.

"I have no doubt we should find with higher and higher speeds, very gradually reached, stability of laminar or non-turbulent motion, but with narrower and narrower limits as to magnitude of disturbance; and so find through a large range of velocity, a confirmation of *Phil. Mag.* 1887, 2, pp. 191—196. The experiment would, at high velocities, fail to prove the stability which the mathematical investigation proves for every velocity however high.

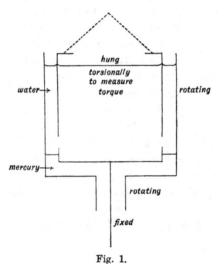

Fig. 1.

"As to *Phil. Mag.* 1887, 2, pp. 272—278, I admit that the mathematical proof is not complete, and withdraw [temporarily?] the words 'virtually inclusive' (p. 273, line 3). I still think it probable that the laminar motion is stable for this case also. In your (*Phil. Mag.* July 1892, pp. 67, 68) refusal to admit that stability is proved you don't distinguish the case in which my proof was complete from the case in which it seems, and therefore is, not complete.

"Your equation (24) of p. 68 is only valid for infinitely small motion, in which the squares of the total velocities are everywhere negligible; and in this case the motion is manifestly periodic, for any stated periodic conditions of the boundary, and comes to rest according to the logarithmic law if the boundary is brought to rest at any time.

"In your p. 62, lines 11 and 12 are 'inaccurate.' Stokes limits his investigation to the case in which the squares of the velocities can be neglected

$$(i.e. \ \frac{\text{radius of globe} \times \text{velocity}}{\text{diffusivity}} \ \text{very small}),$$

in which it is manifest that the steady motion is the same whatever the viscosity; but it is manifest that when the squares cannot be neglected, the steady motion is very different (and horribly difficult to find) for different degrees of viscosity.

"In your p. 62, near the foot, it is not explained what V is; and it disappears henceforth.—Great want of explanation here—Did you not want your paper to be understandable without Basset in hand? I find your two papers of July/92, pp. 61—70, and Oct./93, pp. 355—372, very difficult reading, in every page, and in some ∞ ly difficult.

"Pp. 366, 367 very mysterious. The elastic problem is not defined. It is impossible that there can be the rectilineal motion of the fluid asserted in p. 367, lines 17—19 from foot, in circumstances of motion, quite undefined, but of some kind making the lines of motion on the right side different from those on the left. The conditions are not explained for either the elastic-solid*, or the hydraulic case.

"See p. 361, lines 19, 20, 21 from foot. The formation of a backwater depends essentially on the non-negligibility of squares of velocities; and your p. 367, lines 1—4, and line 17 from foot, are not right.

"If you come to the R. S. Library Committee on Thursday we may come to agreement on some of these questions."

Although the main purpose in Kelvin's papers of 1887 was not attained, his special solution for a disturbed vorticity in case (i) is not without interest. The general dynamical equation for the vorticity in two dimensions is

$$\frac{D\zeta}{Dt} = \frac{d\zeta}{dt} + u\frac{d\zeta}{dx} + v\frac{d\zeta}{dy} = \nu \nabla^2\zeta, \quad \dots\dots\dots\dots\dots(1)$$

where $\nu \ (= \mu/\rho)$ is the kinematic viscosity and $\nabla^2 = d^2/dx^2 + d^2/dy^2$. In this hydrodynamical equation ζ is itself a feature of the motion, being connected with the velocities u, v by the relation

$$\zeta = \tfrac{1}{2}\left(\frac{du}{dy} - \frac{dv}{dx}\right), \quad \dots\dots\dots\dots\dots\dots\dots(2)$$

while u, v themselves satisfy the "equation of continuity"

$$\frac{du}{dx} + \frac{dv}{dy} = 0. \quad \dots\dots\dots\dots\dots\dots\dots(3)$$

---

* I think Kelvin did not understand that the analogous elastic problem referred to is that of a thin *plate*. See words following equation (5) of my paper.

In other applications of (1), *e.g.* to the diffusion of heat or dissolved matter in a moving fluid, ζ is a new dependent variable, not subject to (2), and representing temperature or salinity. We may then regard the motion as known while ζ remains to be determined. In any case $\frac{1}{2} D\zeta^2/Dt = \nu \zeta \nabla^2 \zeta$. If the fluid move within fixed boundaries, or extend to infinity under suitable conditions, and we integrate over the area included,

$$\iint \frac{D\zeta^2}{Dt} dx\, dy = \frac{d}{dt} \iint \zeta^2 dx\, dy,$$

so that

$$\frac{1}{2} \frac{d}{dt} \iint \zeta^2 dx\, dy = \nu \iint \zeta \nabla^2 \zeta\, dx\, dy = \nu \int \zeta \frac{d\zeta}{dn} ds - \nu \iint \left\{ \left(\frac{d\zeta}{dx}\right)^2 + \left(\frac{d\zeta}{dy}\right)^2 \right\} dx\, dy,$$

$$\dots\dots(4)$$

by Green's theorem. The boundary integral disappears, if either ζ or $d\zeta/dn$ there vanishes, and then the integral on the left necessarily diminishes as time progresses*. The same conclusion follows if ζ and $d\zeta/dn$ have all along the boundary contrary signs. Under these conditions ζ tends to zero over the whole of the area concerned. The case where at the boundary ζ is required to have a constant finite value Z is virtually included, since if we write $Z + \zeta'$ for ζ, Z disappears from (1), and ζ everywhere tends to the value Z.

In the hydrodynamical problem of the simple shearing motion, ζ is a constant, say Z, u is a linear function of y, say U, and $v = 0$. If in the disturbed motion the vorticity be $Z + \zeta$, and the components of velocity be $U + u$ and v, equation (1) becomes

$$\frac{d\zeta}{dt} + (U + u) \frac{d\zeta}{dx} + v \frac{d\zeta}{dy} = \nu \nabla^2 \zeta, \quad\dots\dots\dots(5)$$

in which ζ, u, and v relate to the *disturbance*. If the disturbance be treated as infinitesimal, the terms of the second order are to be omitted and we get simply

$$\frac{d\zeta}{dt} + U \frac{d\zeta}{dx} = \nu \nabla^2 \zeta. \quad\dots\dots\dots(6)$$

In (6) the motion of the fluid, represented by U simply, is given independently of ζ, and the equation is the same as would apply if ζ denoted the temperature, or salinity, of the fluid moving with velocity U. Any conclusions that we may draw have thus a widened interest.

In Kelvin's solution of (6) the disturbance is supposed to be periodic in x, proportional to e^{ikx}, and U is taken equal to βy. He assumes for trial

$$\zeta = T e^{i\{kx + (n - k\beta t) y\}}, \quad\dots\dots\dots\dots(7)$$

* Compare Orr, *l.c.* p. 115.

where T is a function of t. On substitution in (6) he finds

$$\frac{dT}{dt} = -\nu\left\{k^2 + (n - k\beta t)^2\right\} T,$$

whence

$$T = Ce^{-\nu t\{k^2 + n^2 - nk\beta t + \frac{1}{3}k^2\beta^2 t^2\}}, \quad\ldots\ldots\ldots\ldots\ldots\ldots(8)$$

and comes ultimately to zero. Equations (7) and (8) determine ζ and so suffice for the heat and salinity problems in an infinitely extended fluid. As an example, if we suppose $n = 0$ and take the real part of (7),

$$\zeta = T \cos k\,(x - \beta t \,.\, y), \quad\ldots\ldots\ldots\ldots\ldots\ldots(9)$$

reducing to $\zeta = C \cos kx$ simply when $t = 0$. At this stage the lines of constant ζ are parallel to y. As time advances, T diminishes with increasing rapidity, and the lines of constant ζ tend to become parallel to x. If x be constant, ζ varies more and more rapidly with y. This solution gives a good idea of the course of events when a liquid of unequal salinity is stirred.

In the hydrodynamical problem we have further to deduce the small velocities u, v corresponding to ζ. From (2) and (3), if u and v are proportional to e^{ikx},

$$\zeta = \frac{i}{2k}\left(\frac{d^2v}{dy^2} - k^2v\right) = \frac{i}{2k}\,\nabla^2 v. \quad\ldots\ldots\ldots\ldots\ldots(10)$$

Thus, corresponding to (9),

$$v = -\frac{2T}{k\,(1 + \beta^2 t^2)}\sin k\,(x - \beta t \,.\, y). \quad\ldots\ldots\ldots\ldots(11)$$

No complementary terms satisfying $d^2v/dy^2 - k^2v = 0$ are admissible, on account of the assumed periodicity with x. It should be mentioned that in Kelvin's treatment the disturbance is not limited to be two-dimensional.

Another remarkable solution for an unlimited fluid of Kelvin's equation (6) with $U = \beta y$ has been given by Oseen[*]. In this case the initial value of ζ is concentrated at one point (ξ, η), and the problem may naturally be regarded as an extension of one of Fourier relating to the conduction of heat. Oseen finds

$$\zeta\,(x, y, t) = \frac{Ce^{-\frac{\{\xi - x + \frac{1}{2}\beta t\,(\eta + y)\}^2}{4\nu t\,(1 + \frac{1}{12}\beta^2 t^2)} - \frac{(\eta - y)^2}{4\nu t}}}{4\pi\nu t\,\sqrt{(1 + \frac{1}{12}\beta^2 t^2)}}, \quad\ldots\ldots\ldots\ldots(12)$$

where

$$C = \iint \zeta(\xi, \eta, 0)\,d\xi\,d\eta\,; \quad\ldots\ldots\ldots\ldots\ldots\ldots(13)$$

and the result may be verified by substitution.

* *Arkiv för Matematik, Astronomi och Fysik*, Upsala, Bd. vii. No. 15 (1911).

"The curves $\zeta = $ const. constitute a system of coaxal and similar ellipses, whose centre at $t = 0$ coincides with the point ξ, η, and then moves with the velocity $\beta \eta$ parallel to the x-axis. For very small values of t the eccentricity of the ellipse is very small and the angle which the major axis makes with the x-axis is about $45°$. With increasing t this angle becomes smaller. At the same time the eccentricity becomes larger. For infinitely great values of t, the angle becomes infinitely small and the eccentricity infinitely great."

When $\beta = 0$ in (12), we fall back on Fourier's solution. Without loss of generality we may suppose $\xi = 0$, $\eta = 0$, and then ($r^2 = x^2 + y^2$)

$$\zeta (x, y, t) = \frac{Ce^{-r^2/4\nu t}}{4\pi \nu t}, \quad \dots\dots\dots\dots\dots(14)$$

representing the diffusion of heat, or vorticity, in two dimensions. It may be worth while to notice the corresponding tangential velocity in the hydrodynamical problem. If ψ be the stream-function,

$$2\zeta = \frac{d^2\psi}{dx^2} + \frac{d^2\psi}{dy^2} = \frac{1}{r} \frac{d}{dr} \left(r \frac{d\psi}{dr} \right),$$

so that

$$r \frac{d\psi}{dr} = \frac{C}{\pi} (1 - e^{-r^2/4\nu t}), \quad \dots\dots\dots\dots\dots(15)$$

the constant of integration being determined from the known value of $d\psi/dr$ when $r = \infty$. When r is small (15) gives

$$\frac{d\psi}{dr} = \frac{Cr}{4\pi \nu t}, \quad \dots\dots\dots\dots\dots(16)$$

becoming finite when $r = 0$ so soon as t is finite.

At time t the greatest value of $d\psi/dr$ occurs when

$$r^2 = 1\cdot256 \times 4\nu t. \quad \dots\dots\dots\dots\dots(17)$$

On the basis of his solution Oseen treats the problem of the stability of the shearing motion between two parallel planes and he arrives at the conclusion, in accordance with Kelvin, that the motion is stable for infinitesimal disturbances. For this purpose he considers "the specially unfavourable case" where the distance between the planes is infinitely great. I cannot see myself that Oseen has proved his point. It is doubtless true that a great distance between the planes is unfavourable to stability, but to arrive at a sure conclusion there must be no limitation upon the character of the infinitesimal disturbance, whereas (as it appears to me) Oseen assumes that the disturbance does not sensibly reach the walls. The simultaneous evanescence at the walls of both velocity-components of an otherwise sensible disturbance would seem to be of the essence of the question.

It may be added that Oseen is disposed to refer the instability observed in practice not merely to the square of the disturbance neglected in (6), but also to the inevitable unevenness of the walls.

We may perhaps convince ourselves that the infinitesimal disturbances of (6), with $U = \beta y$, tend to die out by an argument on the following lines, in which it may suffice to consider the operation of a single wall. The argument could, I think, be extended to both walls, but the statement is more complicated. When there is but one wall, we may as well fix ideas by supposing that the wall is at rest (at $y = 0$).

The difficulty of the problem arises largely from the circumstance that the operation of the wall cannot be imitated by the introduction of imaginary vorticities on the further side, allowing the fluid to be treated as uninterrupted. We may indeed in this way satisfy *one* of the necessary conditions. Thus if corresponding to every real vorticity at a point on the positive side we introduce the opposite vorticity at the image of the point in the plane $y = 0$, we secure the annulment in an unlimited fluid of the velocity-component v parallel to y, but the component u, parallel to the flow, remains finite. In order further to annul u, it is in general necessary to introduce new vorticity at $y = 0$. The vorticities on the positive side are not wholly arbitrary.

Let us suppose that initially the only (additional) vorticity in the interior of the fluid is at A, and that this vorticity is clockwise, or positive, like that of the undisturbed motion (fig. 2). If this existed alone, there would be of necessity a finite velocity u along the wall in its neighbourhood. In order

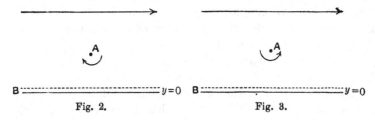

Fig. 2.   Fig. 3.

to satisfy the condition $u = 0$, there must be instantaneously introduced at the wall a negative vorticity of an amount sufficient to give compensation. To this end the local intensity must be inversely as the distance from A and as the sine of the angle between this distance and the wall (Helmholtz). As we have seen these vorticities tend to diffuse and in addition to move with the velocity of the fluid, those near the wall slowly and those arising from A more quickly. As A is carried on, new negative vorticities are developed at those parts of the wall which are being approached. At the other end the vorticities near the wall become excessive and must be compensated. To effect this, new *positive* vorticity must be developed at the wall, whose diffusion over short distances rapidly annuls the negative so far as may be required.   After a time, dependent upon its distance, the vorticity arising from A loses its integrity by coming into contact with the negative diffusing from the wall and thus suffers diminution.   It seems evident that the end can only be the annulment of all the additional vorticity and restoration of the undisturbed condition.   So long as we adhere to the suppositions of equation (6), the argument applies equally well to an original negative vorticity at A, and indeed to any combination of positive and negative vorticities, however distributed.

It is interesting to inquire how this argument would be affected by the retention in (5) of the additional velocities u, v, which are omitted in (6), though a definite conclusion is hardly to be expected.   In fig. 2 the negative vorticity which diffuses inwards is subject to a backward motion due to the vorticity at A in opposition to the slow forward motion previously spoken of. And as A passes on, this negative vorticity in addition to the diffusion is also convected inwards in virtue of the component velocity v due to A. The effect is thus a continued passage inwards behind A of negative vorticity, which tends to neutralize in this region the original constant vorticity (Z). When the additional vorticity at A is negative (fig. 3), the convection behind A acts in opposition to diffusion, and thus the positive developed near the wall remains closer to it, and is more easily absorbed as A passes on.   It is true that in front of A there is a convection of positive inwards; but it would seem that this would lead to a more rapid annulment of A itself; and that upon the whole the tendency is for the effect of fig. 2 to preponderate.   If this be admitted, we may perhaps see in it an explanation of the diminution of vorticity as we recede from a wall observed in certain circumstances.   But we are not in a position to decide whether or not a disturbance dies down.   By other reasoning (Reynolds, Orr) we know that it will do so if β be small enough in relation to the other elements of the problem, viz. the distance between the walls and the kinematic viscosity ν.

A precise formulation of the problem for free infinitesimal disturbances was made by Orr (1907).   We suppose that ζ and v are proportional to $e^{int} e^{i\beta x}$, where $n = p + iq$.   If $\nabla^2 v = S$, we have from (6) and (10)

$$\frac{d^2 S}{dy^2} = \left\{ k^2 - \frac{q}{\nu} + \frac{i}{\nu} (p + k\beta y) \right\} S, \quad \dots\dots\dots\dots(18)$$

and

$$\frac{d^2 v}{dy^2} - k^2 v = S, \quad \dots\dots\dots\dots\dots\dots(19)$$

with the boundary conditions that $v = 0$, $dv/dy = 0$ at the walls.   Orr easily shows that the period-equation takes the form

$$\int S_1 e^{ky} \, dy \cdot \int S_2 e^{-ky} \, dy - \int S_1 e^{-ky} \, dy \cdot \int S_2 e^{ky} \, dy = 0, \quad \dots\dots(20)$$

where S_1, S_2 are any two independent solutions of (18), and the integrations are extended over the interval between the walls. An equivalent equation was given a little later (1908) independently by Sommerfeld *.

Stability requires that for no value of k shall any of the q's determined by (20) be negative. In his discussion Orr arrives at the conclusion that this condition is satisfied, though he does not claim that his method is rigorous. Another of Orr's results may be mentioned here. He shows that $p + k\beta y$ necessarily changes sign in the interval between the walls.

The stability problem has further been skilfully treated by v. Mises† and by Hopf ‡, the latter of whom worked at the suggestion of Sommerfeld, with the result of confirming the conclusions of Kelvin and Orr. Doubtless the reasoning employed was sufficient for the writers themselves, but the statements of it put forward hardly carry conviction to the mere reader. The problem is indeed one of no ordinary difficulty. It may, however, be simplified in one respect, as has been shown by v. Mises. It suffices to prove that q can never be zero, inasmuch as it is certain that in some cases $(\beta = 0)$ q is positive.

In this direction it may be possible to go further. When $\beta = 0$, it is easy to show that not merely q, but $q - k^2\nu$, is positive§. According to Hopf, this is true generally. Hence it should suffice to omit $k^2 - q/\nu$ in (18), and then to prove that the S-solutions obtained from the equation so simplified cannot satisfy (20). The functions S_1 and S_2, satisfying the simplified equation

$$\frac{d^2S}{d\eta^2} = i\eta S, \dots\dots\dots\dots\dots\dots\dots\dots(21)$$

where η is *real*, being a linear function of y with real coefficients, could be completely tabulated by the combined use of ascending and descending series, as explained by Stokes in his paper of 1857 ‖. At the walls η takes opposite signs.

Although a simpler demonstration is desirable, there can remain (I suppose) little doubt but that the shearing motion is stable for infinitesimal disturbances. It has not yet been proved theoretically that the stability can fail for finite disturbances on the supposition of perfectly smooth walls; but such failure seems probable. We know from the work of Reynolds, Lorentz, and Orr that no failure of stability can occur unless $\beta D^2/\nu > 177$, where D is the distance between the walls, so that βD represents their relative motion.

* *Atti del IV. Congr. intern. dei Math.* Roma (1909).

† *Festschrift H. Weber*, Leipzig (1912), p. 252; *Jahresber. d. Deutschen Math. Ver.* Bd. xxi. p. 241 (1913). The mathematics has a very wide scope.

‡ *Ann. der Physik*, Bd. xliv. p. 1 (1914).

§ *Phil. Mag.* Vol. xxxiv. p. 69 (1892); *Scientific Papers*, Vol. iii. p. 583.

‖ *Camb. Phil. Trans.* Vol. x. p. 106; *Math. and Phys. Papers*, Vol. iv. p. 77. This appears to have long preceded the work of Hankel. I may perhaps pursue the line of inquiry here suggested.

389.

NOTE ON THE FORMULA FOR THE GRADIENT WIND.

[*Advisory Committee for Aeronautics. Reports and Memoranda.*
No. 147. January, 1915.]

AN instantaneous derivation of the formula for the "gradient wind" has been given by Gold*. "For the steady horizontal motion of air along a path whose radius of curvature is r, we may write directly the equation

$$\frac{(\omega r \sin \lambda \pm v)^2}{r} = \frac{1}{\rho} \frac{dp}{dr} + \frac{(\omega r \sin \lambda)^2}{r},$$

expressing the fact that the part of the centrifugal force arising from the motion of the wind is balanced by the effective gradient of pressure.

"In the equation p is atmospheric pressure, ρ density, v velocity of moving air, λ is latitude, and ω is the angular velocity of the earth about its axis." Gold deduces interesting consequences relating to the motion and pressure of air in anti-cyclonic regions†.

But the equation itself is hardly obvious without further explanations, unless we limit it to the case where $\sin \lambda = 1$ (at the pole) and where the relative motion of the air takes place about the same centre as the earth's rotation. I have thought that it may be worth while to take the problem avowedly in two dimensions, but without further restriction upon the character of the relative steady motion.

The axis of rotation is chosen as axis of z. The axes of x and y being supposed to rotate in their own plane with angular velocity ω, we denote by u, v, the velocities at time t, relative to these axes, of the particle which then occupies the position x, y. The actual velocities of the same particle, parallel to the instantaneous positions of the axes, will be $u - \omega y$, $v + \omega x$, and the accelerations in the same directions will be

$$\frac{du}{dt} + u \frac{du}{dx} + v \frac{du}{dy} - 2\omega v - \omega^2 x$$

---

* *Proc. Roy. Soc.* Vol. LXXX A. p. 436 (1908).
† See also Shaw's *Forecasting Weather*, Chapter II.

and

$$\frac{dv}{dt} + u\frac{dv}{dx} + v\frac{dv}{dy} + 2\omega u - \omega^2 y *$$

Since the relative motion is supposed to be steady, du/dt, dv/dt disappear, and the dynamical equations are

$$\frac{1}{\rho}\frac{dp}{dx} = \omega^2 x + 2\omega v - u\frac{du}{dx} - v\frac{du}{dy}, \quad\dots\dots\dots\dots(1)$$

$$\frac{1}{\rho}\frac{dp}{dy} = \omega^2 y - 2\omega u - u\frac{dv}{dx} - v\frac{dv}{dy}. \quad\dots\dots\dots\dots(2)$$

The velocities u, v may be expressed by means of the relative stream-function ψ:

$$u = d\psi/dy, \qquad v = -d\psi/dx.$$

Equations (1), (2) then become

$$\frac{1}{\rho}\frac{dp}{dx} = \omega^2 x - 2\omega\frac{d\psi}{dx} - \frac{1}{2}\frac{d}{dx}\left\{\left(\frac{d\psi}{dx}\right)^2 + \left(\frac{d\psi}{dy}\right)^2\right\} + \nabla^2\psi\cdot\frac{d\psi}{dx}, \quad\dots\dots(3)$$

$$\frac{1}{\rho}\frac{dp}{dy} = \omega^2 y - 2\omega\frac{d\psi}{dy} - \frac{1}{2}\frac{d}{dy}\left\{\left(\frac{d\psi}{dx}\right)^2 + \left(\frac{d\psi}{dy}\right)^2\right\} + \nabla^2\psi\cdot\frac{d\psi}{dy}; \quad\dots\dots(4)$$

and on integration, if we leave out the part of p independent of the relative motion,

$$\frac{p}{\rho} = -2\omega\psi - \frac{1}{2}\left\{\left(\frac{d\psi}{dx}\right)^2 + \left(\frac{d\psi}{dy}\right)^2\right\} + \int\nabla^2\psi\,d\psi, \quad\dots\dots\dots\dots(5)$$

in which by a known theorem $\nabla^2\psi$ is a function of ψ only. If ω be omitted, (5) coincides with the equation given long ago by Stokes[†]. It expresses p in terms of ψ; but it does not directly allow of the expression of ψ in terms of p, as is required if the data relate to a barometric chart.

We may revert to the more usual form, if in (1) or (3) we take the axis of x perpendicular to the direction of (relative) motion at any point. Then $u = 0$, and

$$\frac{1}{\rho}\frac{dp}{dx} = 2\omega v + \frac{d\psi}{dx}\frac{d^2\psi}{dy^2}. \quad\dots\dots\dots\dots\dots\dots(6)$$

But since $d\psi/dy = 0$, the curvature at this place of the stream-line ($\psi =$ const.) is

$$\pm\frac{1}{r} = \frac{d^2\psi}{dy^2} \div \frac{d\psi}{dx},$$

and thus

$$\frac{1}{\rho}\frac{dp}{dx} = 2\omega v \pm \frac{v^2}{r}, \quad\dots\dots\dots\dots\dots\dots\dots(7)$$

---

* Lamb's *Hydrodynamics*, § 206.
† *Camb. Phil. Trans.* Vol. VII. 1842; *Math. and Phys. Papers*, Vol. I. p. 9.

giving the velocity v in terms of the barometric gradient dp/dx by means of a quadratic.   As is evident from the case $\omega = 0$, the positive sign in the alternative is to be taken when x and r are drawn in opposite directions.

In (7) r is not derivable from the barometric chart, nor can ψ be determined strictly by means of p.   But in many cases it appears that the more important part of p, at any rate in moderate latitudes, is that which depends upon ω, so that approximately from (5)

$$\psi = - p/2\rho\omega. \quad\dotfill(8)$$

Substituting this value of ψ in the smaller terms, we get as a second approximation

$$2\rho\omega \,.\, \psi = - p - \frac{1}{8\omega^2\rho} \left\{ \left(\frac{dp}{dx}\right)^2 + \left(\frac{dp}{dy}\right)^2 \right\} + \frac{1}{4\omega^2\rho} \int \nabla^2 p \, dp. \quad\dots\dots(9)$$

With like approximation we may identify r in (7) with the radius of curvature of the *isobaric* curve which passes through the point in question.

The interest of these formulæ depends largely upon the fact that the velocity calculated as above from the barometric gradient represents fairly well the wind actually found at a moderate elevation.   At the surface the discrepancy is larger, especially over the land, owing doubtless to friction.

390.

SOME PROBLEMS CONCERNING THE MUTUAL INFLUENCE OF RESONATORS EXPOSED TO PRIMARY PLANE WAVES.

[*Philosophical Magazine*, Vol. XXIX. pp. 209—222, 1915.]

RECENT investigations, especially the beautiful work of Wood on "Radiation of Gas Molecules excited by Light"*, have raised questions as to the behaviour of a cloud of resonators under the influence of plane waves of their own period. Such questions are indeed of fundamental importance. Until they are answered we can hardly approach the consideration of *absorption*, viz. the conversion of radiant into thermal energy. The first action is upon the molecule. We may ask whether this can involve on the average an increase of translatory energy. It does not seem likely. If not, the transformation into thermal energy must await collisions.

The difficulties in the way of answering the questions which naturally arise are formidable. In the first place we do not understand what kind of vibration is assumed by the molecule. But it seems desirable that a beginning should be made; and for this purpose I here consider the case of the simple aerial resonator vibrating symmetrically. The results cannot be regarded as even roughly applicable in a quantitative sense to radiation, inasmuch as this type is inadmissible for transverse vibrations. Nevertheless they may afford suggestions.

The action of a simple resonator under the influence of suitably tuned primary aerial waves was considered in *Theory of Sound*, § 319 (1878). The primary waves were supposed to issue from a simple source at a finite distance c from the resonator. With suppression of the time-factor, and at a distance r from their source, they are represented† by the potential

$$\phi = \frac{e^{-ikr}}{r}, \dots\dots\dots\dots\dots\dots\dots\dots\dots\dots\dots\dots(1)$$

---

* A convenient summary of many of the more important results is given in the Guthrie Lecture, *Proc. Phys. Soc.* Vol. XXVI. p. 185 (1914).

† A slight change of notation is introduced.

in which $k = 2\pi/\lambda$, and λ is the wave-length; and it appeared that the potential of the secondary waves diverging from the resonator is

$$\psi = \frac{e^{-ikc}}{ikc} \frac{e^{-ikr'}}{r'} , \quad \dots\dots\dots\dots\dots(2)$$

so that

$$4\pi r'^2 \operatorname{Mod}^2 \psi = 4\pi/k^2 c^2 . \quad \dots\dots\dots(3)$$

The left-hand member of (3) may be considered to represent the energy dispersed. At the distance of the resonator

$$\operatorname{Mod}^2 \phi = 1/c^2.$$

If we inquire what area S of primary wave-front propagates the same energy as is dispersed by the resonator, we have

$$S/c^2 = 4\pi/k^2 c^2,$$

or

$$S = 4\pi/k^2 = \lambda^2/\pi. \quad \dots\dots\dots\dots\dots(4)$$

Equation (4) applies of course to *plane* primary waves, and is then a particular case of a more general theorem established by Lamb*.

It will be convenient for our present purpose to start *de novo* with plane primary waves, still supposing that the resonator is simple, so that we are concerned only with symmetrical terms, of zero order in spherical harmonics.

Taking the place of the resonator as origin and the direction of propagation as initial line, we may represent the primary potential by

$$\phi = e^{ikr\cos\theta} = 1 + ikr\cos\theta - \tfrac{1}{2}k^2 r^2 \cos^2\theta + \dots \quad \dots\dots(5)$$

The potential of the symmetrical waves issuing from the resonator may be taken to be

$$\psi = \frac{a e^{-ikr}}{r} = \frac{a}{r}(1 - ikr + \dots). \quad \dots\dots\dots\dots(6)$$

Since the resonator is supposed to be an ideal resonator, concentrated in a point, r is to be treated as infinitesimal in considering the conditions to be there satisfied. The first of these is that no *work* shall be done at the resonator, and it requires that total pressure and total radial velocity shall be in quadrature. The total pressure is proportional to $d(\phi + \psi)/dt$, or to $i(\phi + \psi)$, and the total radial velocity is $d(\phi + \psi)/dr$. Thus $(\phi + \psi)$ and $d(\phi + \psi)/dr$ must be in the same (or opposite) phases, in other words their *ratio* must be *real*. Now, with sufficient approximation,

$$\phi + \psi = 1 + \frac{a}{r}(1 - ikr), \quad \frac{d(\phi + \psi)}{dr} = -\frac{a}{r^2};$$

so that

$$a^{-1} - ik = \text{real}. \quad \dots\dots\dots\dots(7)$$

---

* *Camb. Trans.* Vol. XVIII. p. 348 (1899); *Proc. Math. Soc.* Vol. XXXII. p. 11 (1900). The resonator is no longer limited to be *simple*. See also Rayleigh, *Phil. Mag.* Vol. III. p. 97 (1902); *Scientific Papers*, Vol. V. p. 8.

If we write

$$a = A e^{ia}, \quad 1/a = A^{-1} e^{-ia}, \quad \dots\dots\dots(8)$$

then

$$A = - k^{-1} \sin \alpha. \quad \dots\dots\dots\dots(9)$$

So far α is arbitrary, since we have used no other condition than that no work is being done at the resonator. For instance, (9) applies when the source of disturbance is merely the presence at the origin of a small quantity of gas of varied character. The peculiar action of a *resonator* is to make A a maximum, so that $\sin \alpha = \pm 1$, say -1. Then

$$A = 1/k, \quad a = -i/k, \quad \dots\dots\dots(10)$$

and

$$\psi = - \frac{i e^{-ikr}}{kr}. \quad \dots\dots\dots\dots(11)$$

As in (3),

$$4\pi r^2 \, \mathrm{Mod}^2 \, \psi = 4\pi / k^2 = \lambda^2 / \pi, \dots\dots\dots(12)$$

and the whole energy dispersed corresponds to an area of primary wave-front equal to λ^2 / π.

The condition of resonance implies a definite relation between $(\phi + \psi)$ and $d(\phi + \psi)/dr$. If we introduce the value of a from (10), we see that this is

$$\frac{\phi + \psi}{d(\phi + \psi)/dr} = \frac{1/a + 1/r - ik}{-1/r^2} = -r; \quad \dots\dots\dots(13)$$

and this is the relation which must hold at a resonator so tuned as to respond to the primary waves, when isolated from all other influences.

The above calculation relates to the case of a single resonator. For many purposes, especially in Optics, it would be desirable to understand the operation of a company of resonators. A strict investigation of this question requires us to consider each resonator as under the influence, not only of the primary waves, but also of the secondary waves dispersed by its neighbours, and in this many difficulties are encountered. If, however, the resonators are not too near one another, or too numerous, they may be supposed to act independently. From (11) it will be seen that the standard of distance is the wave-length.

The action of a number (n) of similar and irregularly situated centres of secondary disturbance has been considered in various papers on the light from the sky*. The phase of the disturbance from a single centre, as it reaches a distant point, depends of course upon this distance and upon the situation of the centre along the primary rays. If all the circumstances are accurately prescribed, we can calculate the aggregate effect at a distant point, and the resultant intensity may be anything between 0 and that corresponding to complete agreement of phase among all the components. But such a calculation would have little significance for our present purpose.

---

* Compare also "Wave Theory of Light," *Enc. Brit.* Vol. xxiv. (1888), § 4; *Scientific Papers*, Vol. iii. pp. 53, 54.

Owing to various departures from ideal simplicity, *e.g.* want of homogeneity in the primary vibrations, movement of the disturbing centres, the impossibility of observing what takes place at a mathematical point, we are in effect only concerned with the average, and the average intensity is n times that due to a single centre.

In the application to a cloud of acoustic resonators the restriction was necessary that the resonators must not be close compared with λ; otherwise they would react upon one another too much. This restriction may appear to exclude the case of the light from the sky, regarded as due mainly to the molecules of air; but these molecules are not resonators—at any rate as regards visible radiations. We can most easily argue about an otherwise uniform medium disturbed by numerous small obstacles composed of a medium of different quality. There is then no difficulty in supposing the obstacles so small that their mutual reaction may be neglected, even although the average distance of immediate neighbours is much less than a wavelength. When the obstacles are small enough, the whole energy dispersed may be trifling, but it is well to observe that there must be some. No medium can be fully transparent in all directions to plane waves, which is not itself quite uniform. Partial exceptions may occur, *e.g.* when the want of uniformity is a stratification in plane strata. The dispersal then becomes a regular reflexion, and this may vanish in certain cases, even though the changes of quality are sudden (black in Newton's rings)*. But such transparency is limited to certain directions of propagation.

To return to resonators: when they may be close together, we have to consider their mutual reaction. For simplicity we will suppose that they all lie on the same primary wave-front, so that as before in the neighbourhood of each resonator we may take

$$\phi = 1, \quad d\phi/dr = 0. \quad \dots\dots\dots\dots\dots\dots\dots(14)$$

Further, we suppose that all the resonators are similarly situated as regards their neighbours, *e.g.*, that they lie at the angular points of a regular polygon. The waves diverging from each have then the same expression, and altogether

$$\psi = a \left\{ \frac{e^{-ikr_1}}{r_1} + \frac{e^{-ikr_2}}{r_2} + \dots \right\}, \quad \dots\dots\dots\dots\dots(15)$$

where $r_1, r_2, \dots$ are the distances of the point where ψ is measured from the various resonators, and a is a coefficient to be determined. The whole potential is $\phi + \psi$, and it suffices to consider the state of things at the first resonator. With sufficient approximation

$$\phi + \psi = 1 + \frac{a}{r_1} (1 - ikr_1) + a \Sigma \frac{e^{-ikR}}{R}, \quad \dots\dots\dots\dots(16)$$

* See *Proc. Roy. Soc.* Vol. LXXXVI A, p. 207 (1912); [This volume, p. 77].

R being the distance of any other resonator from the first, while (as before)

$$\frac{d(\phi + \psi)}{dr} = -\frac{a}{r_1^2}. \qquad \ldots \ldots \ldots \ldots \ldots \ldots (17)$$

We have now to distinguish two cases. In the first, which is the more important, the tuning of the resonators is such that each singly would respond as much as possible to the primary waves. The ratio of (16) to (17) must then, as we have seen, be equal to $-r_1$, when r_1 is indefinitely diminished. Accordingly

$$\frac{1}{a} = ik - \Sigma \frac{e^{-ikR}}{R}, \qquad \ldots \ldots \ldots \ldots \ldots \ldots (18)$$

which, of course, includes (10). If we write $a = Ae^{i\alpha}$, then

$$A^2 = \frac{1/k^2}{\left[\Sigma \dfrac{\cos kR}{kR}\right]^2 + \left[1 + \Sigma \dfrac{\sin kR}{kR}\right]^2}. \qquad \ldots \ldots \ldots \ldots (19)$$

The other case arises when the resonators are so tuned that the *aggregate* responds as much as possible to the primary waves. We may then proceed as in the investigation for a single resonator. In order that no work may be done at the disturbing centres, $(\phi + \psi)$ and $d(\phi + \psi)/dr$ must be in the same phase, and this requires that

$$\frac{1}{a} + \frac{1}{r_1} - ik + \Sigma \frac{e^{-ikR}}{R} = \text{real},$$

or

$$\frac{1}{a} = \text{real} + ik + i\Sigma \frac{\sin kR}{R}. \qquad \ldots \ldots \ldots \ldots \ldots \ldots (20)$$

The condition of maximum resonance is that the real part in (20) shall vanish, so that

$$\frac{1}{a} = ik \left\{ 1 + \Sigma \frac{\sin kR}{kR} \right\} \qquad \ldots \ldots \ldots \ldots \ldots \ldots (21)$$

or

$$A = \frac{1/k}{1 + \Sigma \dfrac{\sin kR}{kR}}. \qquad \ldots \ldots \ldots \ldots \ldots \ldots (22)$$

The present value of A^2 is greater than that in (19), as was of course to be expected. In either case the disturbance is given by (15) with the value of a determined by (18), or (21).

The simplest example is when there are only two resonators and the sign of summation may be omitted in (18). In order to reckon the energy dispersed, we may proceed by either of two methods. In the first we consider the value of ψ and its modulus at a great distance r from the resonators. It is evident that ψ is symmetrical with respect to the line R joining the resonators, and if θ be the angle between r and R, $r_1 - r_2 = R \cos \theta$. Thus

$$r^2 \cdot \text{Mod}^2 \psi = A^2 \{2 + 2 \cos(kR \cos \theta)\};$$

and on integration over angular space,

$$2\pi r^2 \int_0^\pi \mathrm{Mod}^2\,\psi\,.\,\sin\theta\,d\theta = 8\pi A^2 \left\{1 + \frac{\sin kR}{kR}\right\}. \qquad\ldots\ldots\ldots(23)$$

Introducing the value of A^2 from (19), we have finally

$$2\pi r^2 \int_0^\pi \mathrm{Mod}^2\,\psi\,.\,\sin\theta\,d\theta = \frac{8\pi k^{-2}\left(1 + \dfrac{\sin kR}{kR}\right)}{1 + \dfrac{1}{k^2R^2} + 2\,\dfrac{\sin kR}{kR}}. \qquad\ldots\ldots\ldots(24)$$

If we suppose that kR is large, but still so that R is small compared with r, (24) reduces to $8\pi k^{-2}$ or $2\lambda^2/\pi$. The energy dispersed is then the double of that which would be dispersed by each resonator acting alone; otherwise the mutual reaction complicates the expression.

The greatest interference naturally occurs when kR is small. (24) then becomes $2k^2R^2\,.\,2\lambda^2/\pi$, or $16\pi R^2$, in agreement with *Theory of Sound*, § 321. The whole energy dispersed is then much *less* than if there were only one resonator.

It is of interest to trace the influence of distance more closely. If we put $kR = 2\pi m$, so that $R = m\lambda$, we may write (24)

$$S = (2\lambda^2/\pi)\,.\,F, \qquad\ldots\ldots\ldots\ldots\ldots\ldots\ldots\ldots\ldots(25)$$

where S is the area of primary wave-front which carries the same energy as is dispersed by the two resonators and

$$F = \frac{2\pi m + \sin(2\pi m)}{2\pi m + (2\pi m)^{-1} + 2\sin(2\pi m)}. \qquad\ldots\ldots\ldots\ldots(26)$$

If $2m$ is an integer, the sine vanishes and

$$F = \frac{1}{1 + (2\pi m)^{-2}}, \qquad\ldots\ldots\ldots\ldots\ldots\ldots\ldots(27)$$

not differing much from unity even when $2m = 1$; and whenever $2m$ is great, F approaches unity.

The following table gives the values of F for values of $2m$ not greater than 2:

$2m$	F	$2m$	F	$2m$	F
0·05	0·0459	0·70	0·7042	1·40	1·266
0·10	0·1514	0·80	0·7588	1·50	1·269
0·20	0·3582	0·90	0·8256	1·60	1·226
0·30	0·4836	1·00	0·9080	1·70	1·159
0·40	0·5583	1·10	1·006	1·80	1·088
0·50	0·6110	1·20	1·113	1·90	1·026
0·60	0·6569	1·30	1·208	2·00	0·975

In the case of two resonators the integration in (23) presents no difficulty; but when there are a larger number, it is preferable to calculate the emission of energy in the dispersed waves from the work which would have to be done to generate them at the resonators (in the absence of primary waves)—a method which entails no integration. We continue to suppose that all the resonators are similarly situated, so that it suffices to consider the work done at one of them—say the first. From (15)

$$\psi = a \left\{ \frac{1 - ikr}{r} + \Sigma \frac{e^{-ikR}}{R} \right\}, \qquad \frac{d\psi}{dr} = -\frac{a}{r^2}.$$

The pressure is proportional to $i\psi$, and the part of it which is in the same phase as $d\psi/dr$ is proportional to

$$a \left\{ k + \Sigma \frac{\sin kR}{R} \right\}.$$

Accordingly the work done at each source is proportional to

$$A^2 \left\{ 1 + \Sigma \frac{\sin kR}{kR} \right\}. \quad\dotfill(28)$$

Hence altogether by (19) the energy dispersed by n resonators is that carried by an area S of primary wave-front, where

$$S = \frac{n\lambda^2}{\pi} \frac{1 + \Sigma \dfrac{\sin kR}{kR}}{\left[\Sigma \dfrac{\cos kR}{kR} \right]^2 + \left[1 + \Sigma \dfrac{\sin kR}{kR} \right]^2}, \quad\dotfill(29)$$

the constant factor being determined most simply by a comparison with the case of a single resonator, for which $n = 1$ and the Σ's vanish. We fall back on (24) by merely putting $n = 2$, and dropping the signs of summation, as there is then only one R.

If the tuning is such as to make the effect of the *aggregate* of resonators a maximum, the cosines in (29) are to be dropped, and we have

$$S = \frac{n\lambda^2/\pi}{1 + \Sigma \dfrac{\sin kR}{kR}}. \quad\dotfill(30)$$

As an example of (29), we may take 4 résonators at the angular points of a square whose side is b. There are then 3 R's to be included in the summation, of which two are equal to b and one to $b\sqrt{2}$, so that (28) becomes

$$A^2 \left\{ 1 + 2 \frac{\sin kb}{kb} + \frac{\sin (kb\sqrt{2})}{kb\sqrt{2}} \right\}. \quad\dotfill(31)$$

A similar result may be arrived at from the value of ψ at an infinite distance, by use of the definite integral*

$$\int_0^{\frac{1}{2}\pi} J_0\,(x \sin \theta) \sin \theta\, d\theta = \frac{\sin x}{x}. \quad \dots\dots\dots\dots(32)$$

As an example where the company of resonators extends to infinity, we may suppose that there is a row of them, equally spaced at distance R. By (18)

$$\frac{1}{a} = ik - 2 \left\{ \frac{e^{-ikR}}{R} + \frac{e^{-2ikR}}{2R} + \frac{e^{-3ikR}}{3R} + \dots \right\}. \quad \dots\dots\dots(33)$$

The series may be summed.   If we write

$$\Sigma = e^{-ix} + \frac{he^{-2ix}}{2} + \frac{h^2\, e^{-3ix}}{3} + \dots, \quad \dots\dots\dots\dots(34)$$

where h is real and less than unity, we have

$$\frac{d\Sigma}{dx} = -\frac{i\,e^{-ix}}{1 - h\,e^{-ix}},$$

and

$$\Sigma = -\frac{1}{h} \log\,(1 - h\,e^{-ix}), \quad \dots\dots\dots\dots\dots(35)$$

no constant of integration being required, since

$$\Sigma = -h^{-1} \log\,(1 - h) \quad \text{when } x = 0.$$

If now we put $h = 1$,

$$\Sigma = -\log\,(1 - e^{-ix}) = -\log\left(2 \sin \frac{x}{2}\right) + \tfrac{1}{2}i\,(x - \pi) + 2i\,n\pi. \dots\dots(36)$$

Thus

$$\frac{1}{ka} = i - \frac{2}{kR}\left\{-\log\left(2 \sin \frac{kR}{2}\right) + \tfrac{1}{2}i\,(kR - \pi) + 2in\pi\right\}. \dots\dots(37)$$

If $kR = 2m\pi$, or $R = m\lambda$, where m is an integer, the logarithm becomes infinite and a tends to vanish †.

When R is very small, a is also very small, tending to

$$a = R \div 2 \log\,(kR). \dots\dots\dots\dots\dots\dots(38)$$

The longitudinal density of the now approximately linear source may be considered to be a/R, and this tends to vanish.   The multiplication of resonators ultimately annuls the effect at a distance.   It must be remembered that the tuning of each resonator is supposed to be as for itself alone.

In connexion with this we may consider for a moment the problem in two dimensions of a linear resonator parallel to the primary waves, which responds symmetrically.   As before, we may take at the resonator

$$\phi = 1, \quad d\phi/dr = 0.$$

---

* *Enc. Brit.* l. c. equation (43); *Scientific Papers*, Vol. III. p. 98.

† *Phil. Mag.* Vol. XIV. p. 60 (1907); *Scientific Papers*, Vol. V. p. 409.

As regards ψ, the potential of the waves diverging in two dimensions, we must use different forms when r is small (compared with λ) and when r is large*. When r is small

$$\psi/a = \left(\gamma + \log \frac{ikr}{2}\right) \left\{1 - \frac{k^2 r^2}{2^2} + \frac{k^4 r^4}{2^2 \cdot 4^2} - \dots\right\}$$

$$+ \frac{k^2 r^2}{2^2} - \frac{k^4 r^4}{2^2 \cdot 4^2}(1 + \tfrac{1}{2}) + \frac{k^6 r^6}{2^2 \cdot 4^2 \cdot 6^2}(1 + \tfrac{1}{2} + \tfrac{1}{3}) - \dots; \quad \dots\dots(39)$$

and when r is large,

$$\psi/a = -\left(\frac{\pi}{2ikr}\right)^{\frac{1}{2}} e^{-ikr} \left\{1 - \frac{1^2}{1 \cdot 8ikr} + \frac{1^2 \cdot 3^2}{1 \cdot 2 \cdot (8ikr)^2} - \dots\right\}. \quad \dots(40)$$

By the same argument as for a point resonator we find, as the condition that no work is done at $r = 0$, that the imaginary part of $1/a$ is $-i\pi/2$. For maximum resonance

$$a = 2i/\pi, \quad \dots\dots\dots\dots\dots\dots\dots\dots\dots(41)$$

so that at a distance ψ approximates to

$$\psi = -\frac{\sqrt{\lambda}}{\pi\sqrt{r}} e^{-i(kr - \frac{1}{4}\pi)}. \quad \dots\dots\dots\dots\dots(42)$$

Thus
$$2\pi r \cdot \text{Mod}^2 \psi = \frac{2\lambda}{\pi}, \quad \dots\dots\dots\dots\dots\dots(43)$$

which expresses the width of primary wave-front carrying the same energy as is dispersed by the linear resonator tuned to maximum resonance.

A subject which naturally presents itself for treatment is the effect of a distribution of point resonators over the whole plane of the primary wave-front. Such a distribution may be either regular or haphazard. A regular distribution, e.g. in square order, has the advantage that all the resonators are similarly situated. The whole energy dispersed is then expressed by (29), though the interpretation presents difficulties in general. But even this would not cover all that it is desirable to know. Unless the side of the square (b) is smaller than λ, the waves directly reflected back are accompanied by lateral "spectra" whose directions may be very various. When $b < \lambda$, it seems that these are got rid of. For then not only the infinite lines forming sides of the squares which may be drawn through the points, but a fortiori lines drawn obliquely, such as those forming the diagonals, are too close to give spectra. The whole of the effect is then represented by the specular reflexion.

In some respects a haphazard distribution forms a more practical problem, especially in connexion with resonating vapours. But a precise calculation of the averages then involved is probably not easy.

* *Theory of Sound*, § 341.

If we suppose that the scale (b) of the regular structure is very small compared with λ, we can proceed further in the calculation of the regularly reflected wave. Let Q be one of the resonators, O the point in the plane of the resonators opposite to P, at which ψ is required; $OP = x$, $OQ = y$, $PQ = r$. Then if m be the number of resonators per unit area,

$$\psi = 2\pi ma \int_0^\infty y\, dy\, \frac{e^{-ikr}}{r},$$

or since $y\, dy = r\, dr$,

$$\psi = 2\pi ma \int_x^\infty e^{-ikr}\, dr.$$

The integral, as written, is not convergent; but as in the theory of diffraction we may omit the integral at the upper limit, if we exclude the case of a nearly circular boundary. Thus

$$\psi = \frac{2\pi ma}{ik} e^{-ikx}, \qquad\qquad\qquad\dots\dots\dots\dots\dots(44)$$

and

$$\text{Mod}^2\, \psi = \frac{4\pi^2 m^2 A^2}{k^2}. \qquad\qquad\dots\dots\dots\dots\dots(45)$$

The value of A^2 is given by (19). We find, with the same limitation as above,

$$\Sigma \frac{\cos kR}{R} = 2\pi m \int_0^\infty \cos kR\, dR = 0,$$

$$\Sigma \frac{\sin kR}{R} = 2\pi m \int_0^\infty \sin kR\, dR = 2\pi m/k.$$

Thus $A^2 = 1/(k + 2\pi m/k)^2$

and

$$\text{Mod}^2\, \psi = \frac{4\pi^2 m^2}{(k^2 + 2\pi m)^2}. \qquad\qquad\dots\dots\dots\dots\dots(46)$$

When the structure is very fine compared with λ, k^2 in the denominator may be omitted, and then $\text{Mod}^2\, \psi = 1$, that is the regular reflexion becomes total.

The above calculation is applicable in strictness only to resonators arranged in regular order and very closely distributed. It seems not unlikely that a similar result, viz. a nearly total specular reflexion, would ensue even when there are only a few resonators to the square wave-length, and these are in motion, after the manner of gaseous molecules; but this requires further examination.

In the foregoing investigation we have been dealing solely with forced vibrations, executed in synchronism with primary waves incident upon the resonators, and it has not been necessary to enter into details respecting the constitution of the resonators. All that is required is a suitable adjustment to one another of the virtual mass and spring. But it is also of interest to consider *free* vibrations. These are of necessity subject to damping, owing to the communication of energy to the medium, forthwith propagated away; and their persistence depends upon the nature of the resonator as regards mass and spring, and not merely upon the ratio of these quantities.

Taking first the case of a single resonator, regarded as bounded at the surface of a small sphere, we have to establish the connexion between the motion of this surface and the aerial pressure operative upon it as the result of vibration. We suppose that the vibrations have such a high degree of persistence that we may calculate the pressure as if they were permanent. Thus if ψ be the velocity-potential, we have as before with sufficient approximation

$$\psi/a = \frac{1 - ikr}{r}, \quad \frac{1}{a}\frac{d\psi}{dr} = -\frac{1}{r^2};$$

so that, if ρ be the radial displacement of the spherical surface, $d\rho/dt = -a/r^2$, and

$$\psi = -r(1 - ikr)\,d\rho/dt. \quad\ldots\ldots\ldots\ldots\ldots\ldots(47)$$

Again, if σ be the density of the fluid and δp the variable part of the pressure,

$$\delta p = -\sigma\,d\psi/dt = \sigma r(1 - ikr)\,d^2\rho/dt^2, \quad\ldots\ldots\ldots\ldots(48)$$

which gives the pressure in terms of the displacement ρ at the surface of a sphere of small radius r. Under the circumstances contemplated we may use (48) although the vibration slowly dies down according to the law of e^{int}, where n is not wholly real.

If M denotes the "mass" and μ the coefficient of restitution applicable to ρ, the equation of motion is

$$M\frac{d^2\rho}{dt^2} + \mu\rho + 4\pi\sigma r^3(1 - ikr)\frac{d^2\rho}{dt^2} = 0, \quad\ldots\ldots\ldots\ldots(49)$$

or if we introduce e^{int} and write M' for $M + 4\pi\sigma r^3$,

$$n^2(-M' + 4\pi\sigma kr^4 . i) + \mu = 0. \quad\ldots\ldots\ldots\ldots\ldots(50)$$

Approximately,

$$n = \sqrt{(\mu/M')} . \{1 + i . 2\pi\sigma kr^4/M'\};$$

and if we write $n = p + iq$,

$$p = \sqrt{(\mu/M')}, \quad q = p . 2\pi\sigma kr^4/M'. \quad\ldots\ldots\ldots\ldots(51)$$

If T be the time in which vibrations die down in the ratio of $e : 1$, $T = 1/q$.

If there be a second precisely similar vibrator at a distance R from the first, we have for the potential

$$\psi_2 = -\frac{r^2}{R}e^{-ikR}\frac{d\rho_2}{dt}, \quad\ldots\ldots\ldots\ldots\ldots\ldots(52)$$

and for the pressure due to it at the surface of the first vibrator

$$\delta p = \frac{\sigma r^2}{R} e^{-ikR} \frac{d^2 \rho_2}{dt^2}. \qquad \ldots\ldots\ldots\ldots\ldots\ldots(53)$$

The equation of motion for ρ_1 is accordingly

$$M \frac{d^2 \rho_1}{dt^2} + \mu \rho_1 + 4\pi \sigma r^3 \left\{ (1 - ikr) \frac{d^2 \rho_1}{dt^2} + \frac{r e^{-ikR}}{R} \frac{d^2 \rho_2}{dt^2} \right\} = 0;$$

and that for ρ_2 differs only by the interchange of ρ_1 and ρ_2. Assuming that both ρ_1 and ρ_2 are as functions of the time proportional to e^{int}, we get to determine n

$$n^2 \left\{ M' - 4\pi \sigma r^3 . ikr \right\} - \mu = \pm n^2 . 4\pi \sigma r^4 R^{-1} e^{-ikR},$$

or approximately

$$n = \sqrt{\frac{\mu}{M'}} . \left\{ 1 + \frac{2\pi \sigma r^4}{RM'} (ikR \pm e^{-ikR}) \right\}. \ldots\ldots\ldots\ldots(54)$$

If, as before, we take $n = p + iq$,

$$p = \sqrt{\frac{\mu}{M'}} . \left(1 \pm \frac{2\pi \sigma r^4}{RM'} \cos kR \right), \qquad \ldots\ldots\ldots\ldots(55)$$

$$q = p . \frac{2\pi \sigma r^4}{RM'} (kR \mp \sin kR). \qquad \ldots\ldots\ldots\ldots\ldots\ldots(56)$$

We may observe that the reaction of the neighbour does not disturb the frequency if $\cos kR = 0$, or the damping if $\sin kR = 0$. When kR is small, the damping in one alternative disappears. The two vibrators then execute their movements in opposite phases and nothing is propagated to a distance.

The importance of the disturbance of frequency in (55) cannot be estimated without regard to the damping. The question is whether the two vibrations get out of step *while they still remain considerable*. Let us suppose that there is a relative gain or loss of half a period while the vibration dies down in the ratio of $e : 1$, viz. in the time denoted previously by T, so that

$$(p_1 - p_2) T = \pi.$$

Calling the undisturbed values of p and q respectively P and Q, and supposing kR to be small, we have

$$\frac{P}{Q} \frac{4\pi \sigma r^4}{RM'} = \pi,$$

in which $Q/P = 2\pi \sigma k r^4 / M'$. According to this standard the disturbance of frequency becomes important only when $kR < 1/\pi$, or R less than λ/π^2. It has been assumed throughout that r is much less than R.

391.

ON THE WIDENING OF SPECTRUM LINES.

[*Philosophical Magazine*, Vol. XXIX. pp. 274—284, 1915.]

MODERN improvements in optical methods lend additional interest to an examination of the causes which interfere with the absolute homogeneity of spectrum lines. So far as we know these may be considered under five heads, and it appears probable that the list is exhaustive:

(i) The translatory motion of the radiating particles in the line of sight, operating in accordance with Doppler's principle.

(ii) A possible effect of the rotation of the particles.

(iii) Disturbance depending on collision with other particles either of the same or of another kind.

(iv) Gradual dying down of the luminous vibrations as energy is radiated away.

(v) Complications arising from the multiplicity of sources in the line of sight. Thus if the light from a flame be observed through a similar one, the increase of illumination near the centre of the spectrum line is not so great as towards the edges, in accordance with the principles laid down by Stewart and Kirchhoff; and the line is effectively widened. It will be seen that this cause of widening cannot act alone, but merely aggravates the effect of other causes.

There is reason to think that in many cases, especially when vapours in a highly rarefied condition are excited electrically, the first cause is the most important. It was first considered by Lippich* and somewhat later independently by, myself†. Subsequently, in reply to Ebert, who claimed to have discovered that the high interference actually observed was inconsistent with Doppler's principle and the theory of gases, I gave a more complete

---

* *Pogg. Ann.* Vol. CXXXIX. p. 465 (1870).

† *Nature*, Vol. VIII. p. 474 (1873); *Scientific Papers*, Vol. I. p. 183.

calculation*, taking into account the variable velocity of the molecules as defined by Maxwell's law, from which it appeared that there was really no disagreement with observation. Michelson compared these theoretical results with those of his important observations upon light from vacuum-tubes and found an agreement which was thought sufficient, although there remained some points of uncertainty.

The same ground was traversed by Schönrock†, who made the notable remark that while the agreement was good for the monatomic gases it failed for diatomic hydrogen, oxygen, and nitrogen; and he put forward the suggestion that in these cases the chemical atom, rather than the usual molecule, was to be regarded as the carrier of the emission-centres. By this substitution, entailing an increase of velocity in the ratio $\sqrt{2}:1$, the agreement was much improved.

While I do not doubt that Schönrock's comparison is substantially correct, I think that his presentation of the theory is confused and unnecessarily complicated by the introduction (in two senses) of the "width of the spectrum line," a quantity not usually susceptible of direct observation. Unless I misunderstand, what he calls the observed width is a quantity not itself observed at all but deduced from the visibility of interference bands by arguments which already assume Doppler's principle and the theory of gases. I do not see what is gained by introducing this quantity. Given the nature of the radiating gas and its temperature, we can calculate from known data the distribution of light in the bands corresponding to any given retardation, and from photometric experience we can form a pretty good judgment as to the maximum retardation at which they should still be visible. This theoretical result can then be compared with a purely experimental one, and an agreement will confirm the principles on which the calculation was founded. I think it desirable to include here a sketch of this treatment of the question on the lines followed in 1889, but with a few slight changes of notation.

The phenomenon of interference in its simplest form occurs when two equal trains of waves are superposed, both trains having the same frequency and one being retarded relatively to the other by a linear retardation X‡. Then if λ denote the wave-length, the aggregate may be represented by

$$\cos nt + \cos (nt - 2\pi X/\lambda) = 2 \cos (\pi X/\lambda) . \cos (nt - \pi X/\lambda). \ldots\ldots(1)$$

The intensity is given by

$$I = 4 \cos^2 (\pi X/\lambda) = 2 \{1 + \cos (2\pi X/\lambda)\}. \quad\ldots\ldots\ldots\ldots(2)$$

If we regard X as gradually increasing from zero, I is periodic, the maxima (4) occurring when X is a multiple of λ and the minima (0) when X is an odd

---

* "On the limits to interference when light is radiated from moving molecules," *Phil. Mag.* Vol. xxvii. p. 298 (1889); *Scientific Papers*, Vol. iii. p. 258.

† *Ann. der Physik*, Vol. xx. p. 995 (1906).

‡ In the paper of 1889 the retardation was denoted by 2Δ.

multiple of $\frac{1}{2}\lambda$. If bands are visible corresponding to various values of X, the darkest places are absolutely devoid of light, and this remains true however great X may be, that is however high the order of interference.

The above conclusion requires that the light (duplicated by reflexion or otherwise) should have an absolutely definite frequency, i.e. should be absolutely homogeneous. Such light is not at our disposal; and a defect of homogeneity will usually entail a limit to interference, as X increases. We are now to consider the particular defect arising in accordance with Doppler's principle from the motion of the radiating particles in the line of sight. Maxwell showed that for gases in temperature equilibrium the number of molecules whose velocities resolved in three rectangular directions lie within the range $d\xi\,d\eta\,d\zeta$ must be proportional to

$$e^{-\beta\,(\xi^2+\eta^2+\zeta^2)}\,d\xi\,d\eta\,d\zeta.$$

If ξ be the direction of the line of sight, the component velocities η, ζ are without influence in the present problem. All that we require to know is that the number of molecules for which the component ξ lies between ξ and $\xi + d\xi$ is proportional to

$$e^{-\beta\xi^2}\,d\xi. \quad\dots\dots\dots\dots\dots\dots\dots\dots\dots(3)$$

The relation of β to the mean (resultant) velocity v is

$$v = \frac{2}{\sqrt{(\pi\beta)}}. \quad\dots\dots\dots\dots\dots\dots\dots\dots(4)$$

It was in terms of v that my (1889) results were expressed, but it was pointed out that v needs to be distinguished from the velocity of mean square with which the pressure is more directly connected. If this be called v',

$$v' = \sqrt{\left(\frac{3}{2\beta}\right)}, \quad\dots\dots\dots\dots\dots\dots\dots(5)$$

so that

$$\frac{v}{v'} = \sqrt{\left(\frac{8}{3\pi}\right)}. \quad\dots\dots\dots\dots\dots\dots\dots(6)$$

Again, the relation between the original wave-length Λ and the actual wave-length λ, as disturbed by the motion, is

$$\frac{\Lambda}{\lambda} = 1 + \frac{\xi}{c}, \quad\dots\dots\dots\dots\dots\dots\dots(7)$$

c denoting the velocity of light. The intensity of the light in the interference bands, so far as dependent upon the molecules moving with velocity ξ, is by (2)

$$dI = 2\left\{1 + \cos\frac{2\pi X}{\Lambda}\left(1 + \frac{\xi}{c}\right)\right\}e^{-\beta\xi^2}\,d\xi,\dots\dots\dots\dots(8)$$

and this is now to be integrated with respect to ξ between the limits $\pm\infty$. The bracket in (8) is

$$1 + \cos\frac{2\pi X}{\Lambda}\cos\frac{2\pi X\xi}{\Lambda c} - \sin\frac{2\pi X}{\Lambda}\sin\frac{2\pi X\xi}{\Lambda c}.$$

The third term, being uneven in ξ, contributes nothing. The remaining integrals are included in the well-known formula

$$\int_{-\infty}^{+\infty} e^{-a^2x^2}\cos(2rx)\,dx = \frac{\sqrt{\pi}}{a}e^{-r^2/a^2}.$$

Thus

$$I = \frac{2\sqrt{\pi}}{\sqrt{\beta}}\left[1 + \cos\frac{2\pi X}{\Lambda}.\operatorname{Exp}\left(-\frac{\pi^2X^2}{c^2\beta\Lambda^2}\right)\right].\quad\ldots\ldots(9)$$

The intensity I_1 at the darkest part of the bands is found by making X an odd multiple of $\frac{1}{2}\lambda$, and I_2 the maximum brightness by making X a multiple of λ.

Thus

$$\operatorname{Exp}\left(-\frac{\pi^2X^2}{c^2\beta\Lambda^2}\right) = \frac{I_2-I_1}{I_2+I_1} = V, \quad\ldots\ldots(10)$$

where V denotes the "visibility" according to Michelson's definition. Equation (10) is the result arrived at in my former paper, and β can be expressed in terms of either the mean velocity v, or preferably of the velocity of mean square v'*.

The next question is what is the smallest value of V for which the bands are recognizable. Relying on photometric experience, I estimated that a relative difference of 5 per cent. between I_1 and I_2 would be about the limit in the case of high interference bands, and I took $V = \cdot025$. Shortly afterwards† I made special experiments upon bands well under control, obtained by means of double refraction, and I found that in this very favourable case the bands were still just distinctly seen when the relative difference between I_1 and I_2 was reduced to 4 per cent. It would seem then that the estimate $V = \cdot025$ can hardly be improved upon. On this basis (10) gives in terms of v

$$\frac{X}{\Lambda} = \frac{2c}{\pi^{3/2}v}\sqrt{(\log_e 40)} = \cdot690\frac{c}{v}, \quad\ldots\ldots(11)$$

as before. In terms of v' by (6)

$$\frac{X}{\Lambda} = \frac{\sqrt{3}.c}{\pi\sqrt{2}.v'}\sqrt{(\log_e 40)} = \cdot749\frac{c}{v'}. \quad\ldots\ldots(12)$$

As an example of (12), let us apply it to hydrogen molecules at $0°$ C. Here $v' = 1839\times10^2$ cm./sec.‡, and $c = 3\times10^{10}$. Thus

$$X/\Lambda = 1\cdot222\times10^5. \quad\ldots\ldots(13)$$

* See also *Proc. Roy. Soc.* Vol. LXXVI A. p. 440 (1905); *Scientific Papers*, Vol. v. p. 261.
† *Phil. Mag.* Vol. XXVII. p. 484 (1889); *Scientific Papers*, Vol. III. p. 277.
‡ It seems to be often forgotten that the first published calculation of molecular velocities was that of Joule (*Manchester Memoirs*, Oct. 1848, *Phil. Mag.* ser. 4, Vol. XIV. p. 211).

This is for the hydrogen *molecule*. For the hydrogen *atom* (13) must be divided by $\sqrt{2}$. Thus for absolute temperature T and for radiating centres whose mass is m times that of the hydrogen *atom*, we have

$$\frac{X}{\Lambda} = \frac{1{\cdot}222 \times \sqrt{(273)} \times 10^5}{\sqrt{2}} \sqrt{\left(\frac{m}{T}\right)} = 1{\cdot}427 \times 10^6 \sqrt{\left(\frac{m}{T}\right)}. \quad \dots(14)$$

In Buisson and Fabry's corresponding formula, which appears to be derived from Schönrock, 1·427 is replaced by the appreciably different number 1·22*.

The above value of X is the retardation corresponding to the *limit* of visibility, taken to be represented by $V = {\cdot}025$. In Schönrock's calculation the retardation X_1, corresponding to $V = {\cdot}5$, is considered. In (12), $\sqrt{(\log_e 40)}$ would then be replaced by $\sqrt{(\log_e 2)}$, and instead of (14) we should have

$$\frac{X_1}{\Lambda} = 6{\cdot}186 \times 10^5 \sqrt{\left(\frac{m}{T}\right)}. \quad \dots\dots\dots\dots\dots(15)$$

But I do not understand how $V = {\cdot}5$ could be recognized in practice with any precision.

Although it is not needed in connexion with high interference, we can of course calculate the width of a spectrum line according to any conventional definition. Mathematically speaking, the width is infinite; but if we disregard the outer parts where the intensity is less than *one-half* the maximum the limiting value of ξ by (3) is given by

$$\beta \xi^2 = \log_e 2, \quad \dots\dots\dots\dots\dots\dots\dots(16)$$

and the corresponding value of λ by

$$\frac{\lambda - \Lambda}{\Lambda} = \frac{\xi}{c} = \frac{\sqrt{(\log_e 2)}}{c\sqrt{\beta}}. \quad \dots\dots\dots\dots\dots(17)$$

Thus, if $\delta\lambda$ denote the half-width of the line according to the above definition,

$$\frac{\delta\lambda}{\Lambda} = \frac{\sqrt{({\cdot}6931)}}{c\sqrt{\beta}} = 3{\cdot}57 \times 10^{-7} \sqrt{\left(\frac{T}{m}\right)}, \quad \dots\dots\dots(18)$$

T denoting absolute temperature and m the mass of the particles in terms of that of the hydrogen atom, in agreement with Schönrock.

In the application to particular cases the question at once arises as to what we are to understand by T and m. In dealing with a flame it is natural to take the temperature of the flame as ordinarily understood, but when we pass to the rare vapour of a vacuum-tube electrically excited, the matter is not so simple. Michelson assumed from the beginning that the temperature with which we are concerned is that of the tube itself or not much higher. This view is amply confirmed by the beautiful experiments of Buisson and Fabry†,

---

* [1916. I understand from M. Fabry that the difference between our numbers has its origin in a somewhat different estimate of the minimum value of V. The French authors admit an allowance for the more difficult conditions under which high interference is observed.]

† *Journ. de Physique*, t. II. p. 442 (1912).

who observed the limit of interference when tubes containing helium, neon, and krypton were cooled in liquid air. Under these conditions bands which had already disappeared at room temperature again became distinct, and the ratios of maximum retardations in the two cases (1·66, 1·60, 1·58) were not much less than the theoretical 1·73 calculated on the supposition that the temperature of the gas is that of the tube. The highest value of X/Λ, in their notation N, hitherto observed is 950,000, obtained from krypton in liquid air. With all three gases the agreement at room temperature between the observed and calculated values of N is extremely good, but as already remarked their theoretical numbers are a little lower than mine (14). We may say not only that the observed effects are accounted for almost completely by Doppler's principle and the theory of gases, but that the temperature of the emitting gas is not much higher than that of the containing tube.

As regards m, no question arises for the inert monatomic gases. In the case of hydrogen Buisson and Fabry follow Schönrock in taking the atom rather than the molecule as the moving source, so that $m = 1$; and further they find that this value suits not only the lines of the first spectrum of hydrogen but equally those of the second spectrum whose origin has sometimes been attributed to impurities or aggregations.

In the case of sodium, employed in a vacuum-tube, Schönrock found a fair agreement with the observations of Michelson, on the assumption that the *atom* is in question. It may be worth while to make an estimate for the D lines from soda in a Bunsen flame. Here $m = 23$, and we may perhaps take T at 2500. These data give in (14) as the maximum number of bands

$$X/\Lambda = 137,000.$$

The number of bands actually seen is very dependent upon the amount of soda present. By reducing this Fizeau was able to count 50,000 bands, and it would seem that this number cannot be much increased*, so that observation falls very distinctly behind calculation†. With a large supply of soda the number of bands may drop to two or three thousand, or even further.

The second of the possible causes of loss of homogeneity enumerated above, viz. *rotation* of the emitting centres, was briefly discussed many years ago in a letter to Michelson‡, where it appeared that according to the views then

---

* "Interference Bands and their Applications," *Nature*, Vol. XLVIII. p. 212 (1893); *Scientific Papers*, Vol. IV. p. 59. The parallel plate was a layer of water superposed upon mercury. An enhanced illumination may be obtained by substituting nitro-benzol for water, and the reflexions from the mercury and oil may be balanced by staining the latter with aniline blue. But a thin layer of nitro-benzol takes a surprisingly long time to become level.

† Smithells (*Phil. Mag.* Vol. XXXVII. p. 245, 1894) argues with much force that the actually operative parts of the flame may be at a much higher temperature (if the word may be admitted) than is usually supposed, but it would need an almost impossible allowance to meet the discrepancy. The chemical questions involved are very obscure. The coloration with soda appears to require the presence of oxygen (Mitcherlich, Smithells).

‡ *Phil. Mag.* Vol. XXXIV. p. 407 (1892); *Scientific Papers*, Vol. IV. p. 15.

widely held this cause should be more potent than (i). The transverse vibrations emitted from a luminous source cannot be uniform in all directions, and the effect perceived in a fixed direction from a rotating source cannot in general be simple harmonic. In illustration it may suffice to mention the case of a bell vibrating in four segments and rotating about the axis of symmetry. The sound received by a stationary observer is intermittent and therefore not homogeneous. On the principle of equipartition of energy between translatory and rotatory motions, and from the circumstance that the dimensions of molecules are much less than optical wave-lengths, it followed that the loss of homogeneity from (ii) was much greater than from (i). I had in view diatomic molecules—for at that time mercury vapour was the only known exception; and the specific heats at ordinary temperatures showed that two of the possible three rotations actually occurred in accordance with equipartition of energy. It is now abundantly clear that the widening of spectrum lines at present under consideration does not in fact occur; and the difficulty that might be felt is largely met when we accept Schönrock's supposition that the radiating centres are in all cases monatomic. Still there are questions remaining behind. Do the atoms rotate, and if not, why not? I suppose that the quantum theory would help here, but it may be noticed that the question is not merely of acquiring rotation. A permanent rotation, not susceptible of alteration, should apparently make itself felt. These are problems relating to the constitution of the atom and the nature of radiation, which I do not venture further to touch upon.

The third cause of widening is the disturbance of free vibration due to encounters with other bodies. That something of this kind is to be expected has long been recognized, and it would seem that the widening of the D lines when more than a very little soda is present in a Bunsen flame can hardly be accounted for otherwise. The simplest supposition open to us is that an entirely fresh start is made at each collision, so that we have to deal with a series of regular vibrations limited at both ends. The problem thus arising has been treated by Godfrey[*] and by Schönrock[†]. The Fourier analysis of the limited train of waves of length r gives for the intensity of various parts of the spectrum line

$$k^{-2} \sin^2(\pi r k), \quad\dots\dots\dots\dots\dots\dots\dots\dots\dots(19)$$

where k is the reciprocal of the wave-length, measured from the centre of the line. In the application to radiating vapours, integrations are required with respect to r.

Calculations of this kind serve as illustrations; but it is not to be supposed that they can represent the facts at all completely. There must surely

* *Phil. Trans.* A. Vol. cxcv. p. 346 (1899). See also *Proc. Roy. Soc.* Vol. lxxvi. A. p. 440 (1905); *Scientific Papers*, Vol. v. p. 257.

† *Ann. der Physik*, Vol. xxii. p. 209 (1907).

be encounters of a milder kind where the free vibrations are influenced but yet not in such a degree that the vibrations after the encounter have no relation to the previous ones. And in the case of flames there is another question to be faced : Is there no distinction in kind between encounters first of two sodium atoms and secondly of one sodium atom and an atom say of nitrogen ? The behaviour of soda flames shows that there is. Otherwise it seems impossible to explain the great effect of relatively very small additions of soda in presence of large quantities of other gases. The phenomena suggest that the failure of the least coloured flames to give so high an interference as is calculated from Doppler's principle may be due to encounters with other gases, but that the rapid falling off when the supply of soda is increased is due to something special. This might be of a quasi-chemical character, *e.g.* temporary associations of atoms ; or again to vibrators in close proximity putting one another out of tune. In illustration of such effects a calculation has been given in the previous paper*. It is in accordance with this view that, as Gouy found, the emission of light tends to increase as the square root of the amount of soda present.

We come now to cause (iv). Although it is certain that this cause must operate, we are not able at the present time to point to any experimental verification of its influence. As a theoretical illustration " we may consider the analysis by Fourier's theorem of a vibration in which the amplitude follows an exponential law, rising from zero to a maximum and afterwards falling again to zero. It is easily proved that

$$e^{-a^2 x^2} \cos rx = \frac{1}{2a\sqrt{\pi}} \int_0^\infty du \cos ux \left\{ e^{-(u-r)^2/4a^2} + e^{-(u+r)^2/4a^2} \right\}, \quad ...(20)$$

in which the second member expresses an aggregate of trains of waves, each individual train being absolutely homogeneous. If a be small in comparison with r, as will happen when the amplitude on the left varies but slowly, $e^{-(u+r)^2/4a^2}$ may be neglected, and $e^{-(u-r)^2/4a^2}$ is sensible only when u is very nearly equal to r "†.

An analogous problem, in which the vibration is represented by $e^{-at} \sin bt$, has been treated by Garbasso‡. I presume that the form quoted relates to positive values of t and that for negative values of t it is to be replaced by zero. But I am not able to confirm Garbasso's formula§.

As regards the fifth cause of (additional) widening enumerated at the beginning of this paper, the case is somewhat similar to that of the fourth. It must certainly operate, and yet it does not appear to be important in practice. In such rather rough observations as I have made, it seems to make no

---

* *Phil. Mag. suprà*, p. 209. [This volume, Art. 390.]

† *Phil. Mag.* Vol. xxxiv. p. 407 (1892) ; *Scientific Papers*, Vol. iv. p. 16.

‡ *Ann. der Physik*, Vol. xx. p. 848 (1906).

§ Possibly the sign of a is supposed to change when t passes through zero. But even then what are perhaps misprints would need correction.

great difference whether two surfaces of a Bunsen soda flame (front and back) are in action or only one. If the supply of soda to each be insufficient to cause dilatation, the multiplication of flames in line (3 or 4) has no important effect either upon the brightness or the width of the lines. Actual measures, in which no high accuracy is needed, would here be of service.

The observations referred to led me many years ago to make a very rough comparison between the light actually obtained from a nearly undilated soda line and that of the corresponding part of the spectrum from a black body at the same temperature as the flame. I quote it here rather as a suggestion to be developed than as having much value in itself. Doubtless, better data are now available.

How does the intrinsic brightness of a just undilated soda flame compare with the total brightness of a black body at the temperature of the flame? As a source of light Violle's standard, viz. one sq. cm. of just melting platinum, is equal to about 20 candles. The candle presents about 2 sq. cm. of area, so that the radiating platinum is about 40 times as bright. Now platinum is not a black body and the Bunsen flame is a good deal hotter than the melting metal. I estimated (and perhaps under estimated) that a factor of 5 might therefore be introduced, making the black body at flame temperature 200 times as bright as the candle.

To compare with a candle a soda flame of which the D-lines were just beginning to dilate, I reflected the former nearly perpendicularly from a single glass surface. The soda flame seemed about half as bright. At this rate the intrinsic brightness of the flame was $\frac{1}{2} \times \frac{1}{25} = \frac{1}{50}$ of that of the candle, and accordingly $\frac{1}{10,000}$ of that of the black body.

The black body gives a continuous spectrum. What would its brightness be when cut down to the narrow regions occupied by the D-lines? According to Abney's measures the brightness of that part of sunlight which lies between the D's would be about $\frac{1}{250}$ of the whole. We may perhaps estimate the region actually covered by the soda lines as $\frac{1}{25}$ of this. At this rate we should get

$$\frac{1}{25} \times \frac{1}{250} = \frac{1}{6250},$$

as the fraction of the whole radiation of the black body which has the wavelengths of the soda lines. The actual brightness of a soda flame is thus of the same order of magnitude as that calculated for a black body when its spectrum is cut down to that of the flame, and we may infer that the light of a powerful soda flame is due much more to the widening of the spectrum lines than to an increased brightness of their central parts.

392.

THE PRINCIPLE OF SIMILITUDE.

[*Nature*, Vol. xcv. pp. 66—68, March, 1915.]

I HAVE often been impressed by the scanty attention paid even by original workers in physics to the great principle of similitude. It happens not infrequently that results in the form of " laws " are put forward as novelties on the basis of elaborate experiments, which might have been predicted *a priori* after a few minutes' consideration. However useful verification may be, whether to solve doubts or to exercise students, this seems to be an inversion of the natural order. One reason for the neglect of the principle may be that, at any rate in its applications to particular cases, it does not much interest mathematicians. On the other hand, engineers, who might make much more use of it than they have done, employ a notation which tends to obscure it. I refer to the manner in which gravity is treated. When the question under consideration depends essentially upon gravity, the symbol of gravity (g) makes no appearance, but when gravity does not enter into the question at all, g obtrudes itself conspicuously.

I have thought that a few examples, chosen almost at random from various fields, may help to direct the attention of workers and teachers to the great importance of the principle. The statement made is brief and in some cases inadequate, but may perhaps suffice for the purpose. Some foreign considerations of a more or less obvious character have been invoked in aid. In using the method practically, two cautions should be borne in mind. First, there is no prospect of determining a numerical coefficient from the principle of similarity alone ; it must be found, if at all, by further calculation, or experimentally. Secondly, it is necessary as a preliminary step to specify clearly *all* the quantities on which the desired result may reasonably be supposed to depend, after which it may be possible to drop one or more if further consideration shows that in the circumstances they cannot enter. The following, then, are some conclusions, which may be arrived at by this method :

Geometrical similarity being presupposed here as always, how does the strength of a bridge depend upon the linear dimension and the force of gravity ?

In order to entail the same strains, the force of gravity must be inversely as the linear dimension. Under a given gravity the larger structure is the weaker.

The velocity of propagation of periodic waves on the surface of deep water is as the square root of the wave-length.

The periodic time of liquid vibration under gravity in a deep cylindrical vessel of any section is as the square root of the linear dimension.

The periodic time of a tuning-fork, or of a Helmholtz resonator, is directly as the linear dimension.

The intensity of light scattered in an otherwise uniform medium from a small particle of different refractive index is inversely as the fourth power of the wave-length.

The resolving power of an object-glass, measured by the reciprocal of the angle with which it can deal, is directly as the diameter and inversely as the wave-length of the light.

The frequency of vibration of a globe of liquid, vibrating in any of its modes under its own gravitation, is independent of the diameter and directly as the square root of the density.

The frequency of vibration of a drop of liquid, vibrating under capillary force, is directly as the square root of the capillary tension and inversely as the square root of the density and as the $1\frac{1}{2}$ power of the diameter.

The time-constant (*i.e.* the time in which a current falls in the ratio $e:1$) of a linear conducting electric circuit is directly as the inductance and inversely as the resistance, measured in electro-magnetic measure.

The time-constant of circumferential electric currents in an infinite conducting cylinder is as the square of the diameter.

In a gaseous medium, of which the particles repel one another with a force inversely as the nth power of the distance, the viscosity is as the $(n+3)/(2n-2)$ power of the absolute temperature. Thus, if $n=5$, the viscosity is proportional to temperature.

Eiffel found that the resistance to a sphere moving through air changes its character somewhat suddenly at a certain velocity. The consideration of viscosity shows that the critical velocity is inversely proportional to the diameter of the sphere.

If viscosity may be neglected, the mass (M) of a drop of liquid, delivered slowly from a tube of diameter (a), depends further upon (T) the capillary tension, the density (σ), and the acceleration of gravity (g). If these data suffice, it follows from similarity that

$$M = \frac{Ta}{g} \cdot F\left(\frac{T}{g\sigma a^2}\right),$$

where F denotes an arbitrary function. Experiment shows that F varies but little and that within somewhat wide limits it may be taken to be 3·8. Within these limits Tate's law that M varies as a holds good.

In the Æolian harp, if we may put out of account the compressibility and the viscosity of the air, the pitch (n) is a function of the velocity of the wind (v) and the diameter (d) of the wire. It then follows from similarity that the pitch is directly as v and inversely as d, as was found experimentally by Strouhal. If we include viscosity (ν), the form is

$$n = v/d \cdot f(\nu/vd),$$

where f is arbitrary.

As a last example let us consider, somewhat in detail, Boussinesq's problem of the steady passage of heat from a good conductor immersed in a stream of fluid moving (at a distance from the solid) with velocity v. The fluid is treated as incompressible and for the present as inviscid, while the solid has always the same *shape* and presentation to the stream. In these circumstances the total heat (h) passing in unit time is a function of the linear dimension of the solid (a), the temperature-difference (θ), the stream-velocity (v), the capacity for heat of the fluid per unit volume (c), and the conductivity (κ). The density of the fluid clearly does not enter into the question. We have now to consider the "dimensions" of the various symbols.

Those of a are   (Length)^1,

,,   ,, v ,,   (Length)^1 (Time)$^{-1}$,

,,   ,, θ ,,   (Temperature)^1,

,,   ,, c ,,   (Heat)^1 (Length)$^{-3}$ (Temp.)$^{-1}$,

,,   ,, κ ,,   (Heat)^1 (Length)$^{-1}$ (Temp.)$^{-1}$ (Time)$^{-1}$,

,,   ,, h ,,   (Heat)^1 (Time)$^{-1}$.

Hence if we assume

$$h = a^x \theta^y v^z c^u \kappa^v,$$

we have

by heat $1 = u + v,$

by temperature $0 = y - u - v,$

by length $0 = x + z - 3u - v,$

by time $-1 = -z - v;$

so that

$$h = \kappa a \theta \left(\frac{avc}{\kappa}\right)^z.$$

Since z is undetermined, any number of terms of this form may be combined, and all that we can conclude is that

$$h = \kappa a \theta \cdot F(avc/\kappa),$$

where F is an arbitrary function of the one variable avc/κ. An important particular case arises when the solid takes the form of a cylindrical wire of any section, the length of which is perpendicular to the stream. In strictness similarity requires that the length l be proportional to the linear dimension of the section b; but when l is relatively very great h must become proportional to l and a under the functional symbol may be replaced by b. Thus

$$h = \kappa l\theta \cdot F(bvc/\kappa).$$

We see that in all cases h is proportional to θ, and that for a given fluid F is constant provided v be taken inversely as a or b.

In an important class of cases Boussinesq has shown that it is possible to go further and actually to determine the form of F. When the layer of fluid which receives heat during its passage is very thin, the flow of heat is practically in one dimension and the circumstances are the same as when the plane boundary of a uniform conductor is suddenly raised in temperature and so maintained. From these considerations it follows that F varies as $v^{\frac{1}{2}}$, so that in the case of the wire

$$h \propto l\theta \cdot \sqrt{(bvc/\kappa)},$$

the remaining constant factor being dependent upon the shape and purely numerical. But this development scarcely belongs to my present subject.

It will be remarked that since viscosity is neglected, the fluid is regarded as flowing past the surface of the solid with finite velocity, a serious departure from what happens in practice. If we include viscosity in our discussion, the question is of course complicated, but perhaps not so much as might be expected. We have merely to include another factor, ν^w, where ν is the kinematic viscosity of dimensions $(\text{Length})^2$ $(\text{Time})^{-1}$, and we find by the same process as before

$$h = \kappa a\theta \cdot \left(\frac{avc}{\kappa}\right)^z \cdot \left(\frac{c\nu}{\kappa}\right)^w.$$

Here z and w are both undetermined, and the conclusion is that

$$h = \kappa a\theta \cdot F\left\{\frac{avc}{\kappa}, \frac{c\nu}{\kappa}\right\},$$

where F is an arbitrary function of the *two* variables avc/κ and $c\nu/\kappa$. The latter of these, being the ratio of the two diffusivities (for momentum and for temperature), is of no dimensions; it appears to be constant for a given kind of gas, and to vary only moderately from one gas to another. If we may assume the accuracy and universality of this law, $c\nu/\kappa$ is a merely numerical constant, the same for all gases, and may be omitted, so that h reduces to the forms already given when viscosity is neglected altogether, F being again a function of a single variable, avc/κ or bvc/κ. In any case F is constant for a given fluid, provided v be taken inversely as a or b.

[*Nature*, Vol. xcv. p. 644, Aug. 1915.]

The question raised by Dr Riabouchinsky (*Nature*, July 29, p. 105)* belongs rather to the logic than to the use of the principle of similitude, with which I was mainly concerned. It would be well worthy of further discussion. The conclusion that I gave follows on the basis of the usual Fourier equation for the conduction of heat, in which heat and temperature are regarded as *sui generis*. It would indeed be a paradox if further knowledge of the nature of heat afforded by molecular theory put us in a worse position than before in dealing with a particular problem. The solution would seem to be that the Fourier equations embody something as to the nature of heat and temperature which is ignored in the alternative argument of Dr Riabouchinsky.

[1917.   Reference may be made also to a letter signed J. L. in the same number of *Nature*, and to *Nature*, April 22, 1915. See further Buckingham, *Nature*, Vol. xcvi. p. 396, Dec. 1915. Mr Buckingham had at an earlier date (Oct. 1914) given a valuable discussion of the whole theory (*Physical Review*, Vol. iv. p. 345), and further questions have been raised in the same Review by Tolman.

As a variation of the last example, we may consider the passage of heat between two infinite parallel plane surfaces maintained at fixed temperatures differing by θ, when the intervening space is occupied by a stream of incompressible viscous fluid (*e.g.* water) of mean velocity v. In a uniform regime the heat passing across is proportional to the time and to the area considered; but in many cases the uniformity is not absolute and it is necessary to take the *mean* passage over either a large enough area or a long enough time. On this understanding there is a definite quantity h', representing the passage of heat per unit area and per unit time.

If there be no stream ($v = 0$), or in any case if the kinematic viscosity (ν) is infinite, we have

$$h' = \kappa\theta/a,$$

a being the distance between the surfaces, since then the motion, if any, takes place in plane strata. But when the velocity is high enough, or the viscosity low enough, the motion becomes *turbulent*, and the flow of heat may be greatly augmented. With the same reasoning and with the same notation as before we have

$$h' = \frac{\kappa\theta}{a} . F\left(\frac{avc}{\kappa}, \frac{cv}{\kappa}\right),$$

---

* "In *Nature* of March 18, Lord Rayleigh gives this formula $h = \kappa a\theta . F (avc/\kappa)$, considering heat, temperature, length, and time as four 'independent' units. If we suppose that only three of these quantities are really independent, we obtain a different result. For example, if the temperature is defined as the mean kinetic energy of the molecules, the principle of similarity allows us only to affirm that $h = \kappa a\theta . F (v/\kappa a^2, ca^3)$."

or which comes to the same

$$h' = \frac{\kappa\theta}{a} \cdot F_1\left(\frac{av}{\nu}, \frac{c\nu}{\kappa}\right),$$

F, F_1 being arbitrary functions of two variables.   And, as we have seen, $F(0, c\nu/\kappa) = 1$.

For a given fluid $c\nu/\kappa$ is constant and may be omitted. Dynamical similarity is attained when av is kept constant, so that a complete determination of F, experimentally or otherwise, does not require a variation of *both* a and v. There is advantage in retaining a constant; for if a varies, geometrical similarity demands that any roughnesses shall be in proportion.

It should not be overlooked that in the above argument, c, κ, ν are treated as constants, whereas they would really vary with the temperature. The assumption is completely justified only when the temperature difference θ is very small.

Another point calls for attention. The régime ultimately established may in some cases depend upon the initial condition. Reynolds' observations suggest that with certain values of av/ν the simple stratified motion once established may persist; but that the introduction of disturbances exceeding a certain amount may lead to an entirely different (turbulent) régime. Over part of the range F would have double values.

It would be of interest to know what F becomes when av tends to infinity. It seems probable that F too becomes infinite. but perhaps very slowly.]

393.

DEEP WATER WAVES, PROGRESSIVE OR STATIONARY, TO THE THIRD ORDER OF APPROXIMATION.

[*Proceedings of the Royal Society*, A, Vol. XCI. pp. 345—353, 1915.]

As is well known, the form of periodic waves progressing over deep water *without change of type* was determined by Stokes[*] to a high degree of approximation. The wave-length (λ) in the direction of x being 2π and the velocity of propagation unity, the form of the surface is given by

$$y = a \cos (x - t) - \tfrac{1}{2} a^2 \cos 2\,(x - t) + \tfrac{3}{8} a^3 \cos 3\,(x - t), \quad \ldots\ldots\ldots(1)$$

and the corresponding gravity necessary to maintain the motion by

$$g = 1 - a^2. \quad \ldots\ldots\ldots\ldots\ldots\ldots\ldots\ldots\ldots\ldots\ldots(2)$$

The generalisation to other wave-lengths and velocities follows by "dimensions."

These and further results for progressive waves of permanent type are most easily arrived at by use of the stream-function on the supposition that the waves are reduced to rest by an opposite motion of the water as a whole, when the problem becomes one of steady motion[†]. My object at present is to extend the scope of the investigation by abandoning the initial restriction to progressive waves of permanent type. The more general equations may then be applied to progressive waves as a particular case, or to stationary waves in which the principal motion is proportional to a simple circular function of the time, and further to ascertain what occurs when the conditions necessary for the particular cases are not satisfied. Under these circumstances the use of the stream-function loses much of its advantage, and the method followed is akin to that originally adopted by Stokes.

---

[*] *Camb. Phil. Trans.* Vol. VIII. p. 441 (1847); *Math. and Phys. Papers*, Vol. I. p. 197.

[†] *Phil. Mag.* Vol. I. p. 257 (1876); *Scientific Papers*, Vol. I. p. 262. Also *Phil. Mag.* Vol. XXI. p. 183 (1911); [This volume, p. 11].

The velocity-potential ϕ, being periodic in x, may be expressed by the series

$$\phi = \alpha e^{-y} \sin x - \alpha' e^{-y} \cos x + \beta e^{-2y} \sin 2x$$
$$- \beta' e^{-2y} \cos 2x + \gamma e^{-3y} \sin 3x - \gamma' e^{-3y} \cos 3x + \dots, \quad \dots(3)$$

where α, α', β, etc., are functions of the time only, and y is measured downwards from mean level. In accordance with (3) the component velocities are given by

$$u = d\phi/dx = e^{-y} (\alpha \cos x + \alpha' \sin x) + 2e^{-2y} (\beta \cos 2x + \beta' \sin 2x) + \dots$$

$$- v = d\phi/dy = e^{-y} (\alpha \sin x - \alpha' \cos x) + 2e^{-2y} (\beta \sin 2x - \beta' \cos 2x) + \dots.$$

The density being taken as unity, the pressure equation is

$$p = - d\phi/dt + F + gy - \tfrac{1}{2}(u^2 + v^2), \quad \dots\dots\dots\dots\dots(4)$$

in which F is a function of the time.

In applying (4) we will regard α, α', as small quantities of the first order, while β, β', γ, γ', are small quantities of the second order at most; and for the present we retain only quantities of the second order. β, etc., will then not appear in the expression for $u^2 + v^2$. In fact

$$u^2 + v^2 = e^{-2y} (\alpha^2 + \alpha'^2),$$

and

$$p = - \frac{d\alpha}{dt} e^{-y} \sin x + \frac{d\alpha'}{dt} e^{-y} \cos x - \frac{d\beta}{dt} e^{-2y} \sin 2x + \dots$$
$$+ gy - \tfrac{1}{2} e^{-2y} (\alpha^2 + \alpha'^2) + F. \quad \dots(5)$$

The surface conditions are (i) that p be there zero, and (ii) that

$$\frac{Dp}{Dt} = \frac{dp}{dt} + u \frac{dp}{dx} + v \frac{dp}{dy} = 0. \quad \dots\dots\dots\dots\dots(6)$$

The first is already virtually expressed in (5). For the second

$$\frac{dp}{dt} = - \frac{d^2\alpha}{dt^2} e^{-y} \sin x + \frac{d^2\alpha'}{dt^2} e^{-y} \cos x - \dots - e^{-2y} \left(\alpha \frac{d\alpha}{dt} + \alpha' \frac{d\alpha'}{dt} \right) + F',$$

$$\frac{dp}{dx} = - \frac{d\alpha}{dt} e^{-y} \cos x - \frac{d\alpha'}{dt} e^{-y} \sin x - \dots,$$

$$\frac{dp}{dy} = \frac{d\alpha}{dt} e^{-y} \sin x - \frac{d\alpha'}{dt} e^{-y} \cos x + \dots + g + e^{-2y} (\alpha^2 + \alpha'^2).$$

In forming equation (6) to the second order of small quantities we need to include only the principal term of u, but v must be taken correct to the second order. As the equation of the free surface we assume

$$y = a \cos x + a' \sin x + b \cos 2x + b' \sin 2x + c \cos 3x + c' \sin 3x + \dots \quad \dots(7)$$

in which b, b', c, c', are small compared with a, a'. Thus (6) gives

$$(1 - a \cos x - a' \sin x) \left(-\frac{d^2\alpha}{dt^2} \sin x + \frac{d^2\alpha'}{dt^2} \cos x \right) - \frac{d^2\beta}{dt^2} \sin 2x$$

$$+ \frac{d^2\beta'}{dt^2} \cos 2x - \frac{d^2\gamma}{dt^2} \sin 3x + \frac{d^2\gamma}{dt^2} \cos 3x - a \frac{da}{dt} - a' \frac{da'}{dt} + F''$$

$$- (a \cos x + a' \sin x) \left(\frac{da}{dt} \cos x + \frac{da'}{dt} \sin x \right) - \{(1 - a \cos x - a' \sin x)$$

$$\times (a \sin x - a' \cos x) + 2\beta \sin 2x - 2\beta' \cos 2x + 3\gamma \sin 3x - 3\gamma' \cos 3x\}$$

$$\times \left\{ g + \frac{da}{dt} \sin x - \frac{da'}{dt} \cos x \right\} = 0. \quad\dots\dots\dots\dots\dots\dots\dots(8)$$

This equation is to hold good to the second order for all values of x, and therefore for each Fourier component separately. The terms in $\sin x$ and $\cos x$ give

$$\frac{d^2\alpha}{dt^2} + g\alpha = 0, \qquad \frac{d^2\alpha'}{dt^2} + g\alpha' = 0. \quad\dots\dots\dots\dots\dots(9)$$

The term in $\sin 2x$ gives

$$\frac{d^2\beta}{dt^2} + 2g\beta = \frac{a}{2} \left(\frac{d^2\alpha}{dt^2} + g\alpha \right) - \frac{a'}{2} \left(\frac{d^2\alpha'}{dt^2} + g\alpha' \right) = 0, \quad\dots\dots\dots(10)$$

and, similarly, that in $\cos 2x$ gives

$$\frac{d^2\beta'}{dt^2} + 2g\beta' = 0. \quad\dots\dots\dots\dots\dots\dots(11)$$

In like manner

$$\frac{d^2\gamma}{dt^2} + 3g\gamma = 0, \qquad \frac{d^2\gamma'}{dt^2} + 3g\gamma' = 0, \dots\dots\dots\dots\dots(12)$$

and so on. These are the results of the surface condition $Dp/Dt = 0$. From the other surface condition ($p = 0$) we find in the same way

$$-\frac{da}{dt} + ga' = 0, \qquad \frac{da'}{dt} + ga = 0. \quad\dots\dots\dots\dots(13)$$

$$gb' = \frac{d\beta}{dt} + \frac{a'}{2} \frac{da'}{dt} - \frac{a}{2} \frac{da}{dt} = \frac{d\beta}{dt} - gaa'. \quad\dots\dots\dots\dots(14)$$

$$gb = -\frac{d\beta}{dt} + \frac{a'}{2} \frac{da}{dt} + \frac{a}{2} \frac{da'}{dt} = -\frac{d\beta}{dt} + \tfrac{1}{2}g(a'^2 - a^2). \quad\dots\dots\dots(15)$$

$$-\frac{d\gamma}{dt} + gc' = 0, \qquad \frac{d\gamma'}{dt} + gc = 0. \quad\dots\dots\dots\dots(16)$$

From equations (9) to (16) we see that a, a' satisfy the same equations (9) as do α, α', and also that c, c' satisfy the same equations (12) as do γ, γ'; but that b, b' are not quite so simply related to β, β'.

Let us now suppose that the principal terms represent a progressive wave. In accordance with (9) we may take

$$a = A \cos t', \qquad a' = A \sin t', \dots\dots\dots\dots(17)$$

where $t' = \sqrt{g} \cdot t$. Then if β, β', γ, γ', do not appear, c, c', are zero, and $b = \frac{1}{2}A^2 (\sin^2 t' - \cos^2 t')$, $b' = -A^2 \cos t' \sin t'$; so that

$$y = A \cos (x - t') - \frac{1}{2}A^2 \cos 2 (x - t'), \dots\dots\dots\dots(18)$$

representing a permanent wave-form propagated with velocity $\sqrt{g}$. So far as it goes, this agrees with (1). But now in addition to these terms we may have others, for which b, b' need only to satisfy

$$(d^2/dt'^2 + 2) (b, b') = 0, \dots\dots\dots\dots(19)$$

and c, c' need only to satisfy

$$(d^2/dt'^2 + 3) (c, c') = 0. \dots\dots\dots\dots(20)$$

The corresponding terms in y represent merely such waves, propagated in either direction, and of wave-lengths equal to an aliquot part of the principal wave-length, as might exist alone of infinitesimal height, when there is no primary wave at all. When these are included, the aggregate, even though it be all propagated in the same direction, loses its character of possessing a permanent wave-shape, and further it has no tendency to acquire such a character as time advances.

If the principal wave is *stationary* we may take

$$a = A \cos t', \qquad a' = 0. \dots\dots\dots\dots(21)$$

If β, β', γ, γ', vanish,

$$b = -\frac{1}{2}a^2, \quad b' = 0, \quad c = 0, \quad c' = 0,$$

and

$$y = A \cos x \cdot \cos t' - \frac{1}{2}A^2 \cos 2x \cdot \cos^2 t'. \dots\dots\dots\dots(22)$$

According to (22) the surface comes to its zero position everywhere when $\cos t' = 0$, and the displacement is a maximum when $\cos t' = \pm 1$. Then

$$y = \pm A \cos x - \frac{1}{2}A^2 \cos 2x, \dots\dots\dots\dots(23)$$

so that at this moment the wave-form is the same as for the progressive wave (18). Since y is measured downwards, the maximum elevation above the mean level exceeds numerically the maximum depression below it.

In the more general case (still with β, etc., evanescent) we may write

$$a = A \cos t' + B \sin t', \quad a' = A' \cos t' + B' \sin t',$$

with

$$b' = -aa', \quad b = \frac{1}{2}(a'^2 - a^2), \quad c' = 0, \quad c = 0.$$

When β, β', γ, γ', are finite, waves such as might exist alone, of lengths equal to aliquot parts of the principal wave-length and of corresponding frequencies, are superposed. In these waves the amplitude and phase are arbitrary.

When we retain the third order of small quantities, the equations naturally become more complicated. We now assume that in (3) β, β', are small quantities of the second order, and γ, γ', small quantities of the third order. For p, as an extension of (5), we get

$$p = e^{-y}\left(-\frac{d\alpha}{dt}\sin x + \frac{d\alpha'}{dt}\cos x\right) + e^{-2y}\left(-\frac{d\beta}{dt}\sin 2x + \frac{d\beta'}{dt}\cos 2x\right)$$

$$+ e^{-3y}\left(-\frac{d\gamma}{dt}\sin 3x + \frac{d\gamma'}{dt}\cos 3x\right) + gy + F - \tfrac{1}{2}e^{-2y}(\alpha^2 + \alpha'^2)$$

$$- 2e^{-3y}\{(\alpha\beta + \alpha'\beta')\cos x + (\alpha\beta' - \alpha'\beta)\sin x\}. \quad\ldots\ldots\ldots\ldots\ldots(24)$$

This is to be made to vanish at the surface. Also we find, on reduction,

$$-\frac{Dp}{Dt} = (1 - y + \tfrac{1}{2}y^2)\left\{\left(\frac{d^2\alpha}{dt^2} + g\alpha\right)\sin x - \left(\frac{d^2\alpha'}{dt^2} + g\alpha'\right)\cos x\right\}$$

$$+ (1 - 2y)\left\{\left(\frac{d^2\beta}{dt^2} + 2g\beta\right)\sin 2x - \left(\frac{d^2\beta'}{dt^2} + 2g\beta'\right)\cos 2x\right\}$$

$$+ \left(\frac{d^2\gamma}{dt^2} + 3g\gamma\right)\sin 3x - \left(\frac{d^2\gamma'}{dt^2} + 3g\gamma'\right)\cos 3x - F'$$

$$+ 2(1 - 2y)\left(\alpha\frac{d\alpha}{dt} + \alpha'\frac{d\alpha'}{dt}\right) + 4\sin x\frac{d}{dt}(\alpha\beta' - \alpha'\beta)$$

$$+ 4\cos x\frac{d}{dt}(\alpha\beta' + \alpha'\beta) + (\alpha^2 + \alpha'^2)(\alpha\sin x - \alpha'\cos x); \quad\ldots\ldots(25)$$

and at the surface $Dp/Dt = 0$ for all values of x. In (25) y is of the form (7), where b, b', are of the second order, c, c', of the third order.

Considering the coefficients of $\sin x$, $\cos x$, in (25) when reduced to Fourier's form, we see that $d^2\alpha/dt^2 + g\alpha$, $d^2\alpha'/dt^2 + g\alpha'$, are both of the third order of small quantities, so that in the first line the factor $(1 - y + \tfrac{1}{2}y^2)$ may be replaced by unity. Again, from the coefficients of $\sin 2x$, $\cos 2x$, we see that to the third order inclusive

$$\frac{d^2\beta}{dt^2} + 2g\beta = 0, \qquad \frac{d^2\beta'}{dt^2} + 2g\beta' = 0,\ldots\ldots\ldots\ldots\ldots(26)$$

and from the coefficients of $\sin 3x$, $\cos 3x$ that to the third order inclusive

$$\frac{d^2\gamma}{dt^2} + 3g\gamma = 0, \qquad \frac{d^2\gamma'}{dt^2} + 3g\gamma' = 0. \quad\ldots\ldots\ldots\ldots\ldots(27)$$

And now returning to the coefficients of $\sin x$, $\cos x$, we get

$$\frac{d^2\alpha}{dt^2} + g\alpha - 2\alpha'\frac{d}{dt}(\alpha^2 + \alpha'^2) + 4\frac{d}{dt}(\alpha\beta' - \alpha'\beta) + \alpha(\alpha^2 + \alpha'^2) = 0, \ldots(28)$$

$$\frac{d^2\alpha'}{dt^2} + g\alpha' + 2\alpha\frac{d}{dt}(\alpha^2 + \alpha'^2) - 4\frac{d}{dt}(\alpha\beta' + \alpha'\beta) + \alpha'(\alpha^2 + \alpha'^2) = 0. \quad(29)$$

Passing next to the condition $p = 0$, we see from (24), by considering the coefficients of $\sin x$, $\cos x$, that

$$- \frac{da}{dt} + ga' + \text{terms of 3rd order} = 0,$$

$$\frac{da'}{dt} + ga + \text{terms of 3rd order} = 0.$$

The coefficients of $\sin 2x$, $\cos 2x$, require, as in (14), (15), that

$$b' = \frac{1}{g} \frac{d\beta}{dt} - aa', \qquad b = - \frac{1}{g} \frac{d\beta'}{dt} + \frac{a'^2 - a^2}{2}. \quad\ldots\ldots\ldots\ldots(30)$$

Again, the coefficients of $\sin 3x$, $\cos 3x$, give

$$c' = \frac{1}{g} \frac{d\gamma}{dt} - \tfrac{3}{2}(a'b + ab') + \tfrac{3}{8}a'(a'^2 - 3a^2)$$

$$= \frac{1}{g} \left\{ \frac{d\gamma}{dt} - \frac{3a}{2} \frac{d\beta}{dt} + \frac{3a'}{2} \frac{d\beta'}{dt} \right\} - \frac{3a'}{8}(a'^2 - 3a^2), \ldots\ldots\ldots(31)$$

$$c = - \frac{1}{g} \frac{d\gamma'}{dt} + \tfrac{3}{2}(a'b' - ab) + \frac{3a}{8}(3a'^2 - a^2)$$

$$= \frac{1}{g} \left\{ - \frac{d\gamma'}{dt} + \frac{3a'}{2} \frac{d\beta}{dt} + \frac{3a}{2} \frac{d\beta'}{dt} \right\} - \frac{3a}{8}(3a'^2 - a^2). \ldots\ldots\ldots(32)$$

When β, β', γ, γ', vanish, these results are much simplified. We have

$$b' = -aa', \qquad b = \tfrac{1}{2}(a'^2 - a^2),\ldots\ldots\ldots\ldots\ldots\ldots(33)$$

$$c' = - \frac{3a'}{8}(a'^2 - 3a^2), \qquad c = - \frac{3a}{8}(3a'^2 - a^2). \ldots\ldots\ldots(34)$$

If the principal terms represent a purely progressive wave, we may take, as in (17),

$$a = A \cos nt, \qquad a' = A \sin nt, \ldots\ldots\ldots\ldots\ldots(35)$$

where n is for the moment undetermined. Accordingly

$$b' = - \tfrac{1}{2} A^2 \sin 2nt, \qquad b = - \tfrac{1}{2} A^2 \cos 2nt,$$

$$c' = \tfrac{3}{8} A^3 \sin 3nt, \qquad c = \tfrac{3}{8} A^3 \cos 3nt;$$

so that

$$y = A \cos(x - nt) - \tfrac{1}{2} A^2 \cos 2(x - nt) + \tfrac{3}{8} A^3 \cos 3(x - nt), \ldots\ldots(36)$$

representing a progressive wave of permanent type, as found by Stokes.

To determine n we utilize (28), (29), in the small terms of which we may take

$$\alpha = g \int a' dt = - \frac{gA}{n} \cos nt, \qquad \alpha' = - g \int a\, dt = - \frac{gA}{n} \sin nt,$$

so that

$$\alpha^2 + \alpha'^2 = g^2 A^2 / n^2.$$

Thus

$$\frac{d^2(\alpha, \alpha')}{dt^2} + \left(g + \frac{g^2 A^2}{n^2} \right)(\alpha, \alpha') = 0,$$

and

$$n^2 = g + g^2 A^2 / n^2 = g(1 + A^2), \ldots\ldots\ldots\ldots(37)$$

or, if we restore homogeneity by introduction of $k (= 2\pi/\lambda)$,

$$n^2 = g/k \cdot (1 + k^2 A^2). \quad\dots\dots\dots\dots\dots\dots\dots(38)$$

Let us next suppose that the principal terms represent a stationary, instead of a progressive, wave and take

$$a = A \cos nt, \qquad a' = 0. \quad\dots\dots\dots\dots\dots(39)$$

Then by (33), (34),

$$b' = 0, \qquad b = -\tfrac{1}{2}A^2 \cos^2 nt, \qquad c' = 0, \qquad c = \tfrac{3}{8}A^3 \cos^3 nt;$$

and

$$y = A \cos nt \cos x - \tfrac{1}{2}A^2 \cos^2 nt \cos 2x + \tfrac{3}{8}A^3 \cos^3 nt \cos 3x. \quad\dots(40)$$

When $\cos nt = 0$, $y = 0$ throughout; when $\cos nt = 1$,

$$y = A \cos x - \tfrac{1}{2}A^2 \cos 2x + \tfrac{3}{8}A^3 \cos 3x,$$

so that at this moment of maximum displacement the form is the same as for the progressive wave (36).

We have still to determine n so as to satisfy (28), (29), with evanescent β, β'. The first is satisfied by $\alpha = 0$, since $a' = 0$. The second becomes

$$\frac{d^2\alpha'}{dt^2} + g\alpha' + 4a\alpha'\frac{da'}{dt} + \alpha'^3 = 0.$$

In the small terms we may take $\alpha' = -g \int a\,dt = -\dfrac{gA}{n} \sin nt$, so that

$$\frac{d^2\alpha'}{dt^2} + g\alpha' + \frac{g^2 A^3}{4n}(\sin nt + 5 \sin 3nt) = 0.$$

To satisfy this we assume

$$\alpha' = H \sin nt + K \sin 3nt.$$

Then $\qquad H(g - n^2) + \dfrac{g^2 A^3}{4n} = 0, \qquad K(g - 9n^2) + \dfrac{5g^2 A^3}{4n} = 0,$

from the first of which

$$n^2 = g + \frac{g^2 A^3}{4nH} = g - \frac{gA^2}{4}, \quad\dots\dots\dots\dots\dots\dots(41)$$

or, if we restore homogeneity by introduction of k,

$$n^2 = g/k \cdot (1 - \tfrac{1}{4}k^2 A^2). \quad\dots\dots\dots\dots\dots\dots\dots(42)$$

With this value of n the stationary vibration

$$y = A \cos nt \cos kx - \tfrac{1}{2}kA^2 \cos^2 nt \cos 2kx + \tfrac{3}{8}k^2 A^3 \cos^3 nt \cos 3kx,\dots(43)$$

satisfies all the conditions. It may be remarked that according to (42) the frequency of vibration is diminished by increase of amplitude.

The special cases above considered of purely progressive or purely stationary waves possess an exceptional simplicity. In general, with omission of β, β', equations (28), (29), become

$$\frac{d^2\alpha}{dt^2} + g\alpha - \frac{2}{g}\frac{da}{dt}\frac{d(\alpha^2 + \alpha'^2)}{dt} + \alpha(\alpha^2 + \alpha'^2) = 0, \quad\dots\dots\dots(44)$$

and a like equation in which α and α' are interchanged. In the terms of the third order, we take

$$\alpha = P \cos nt + Q \sin nt, \qquad \alpha' = P' \cos nt + Q' \sin nt, \dots\dots\text{(45)}$$

so that

$$\alpha^2 + \alpha'^2 = \tfrac{1}{2}(P^2 + Q^2 + P'^2 + Q'^2) + \tfrac{1}{2}(P^2 + P'^2 - Q^2 - Q'^2)\cos 2nt$$
$$+ (PQ + P'Q') \sin 2nt.$$

The third order terms in (44) are

$$\tfrac{1}{2}(P^2 + P'^2 + Q^2 + Q'^2)(P \cos nt + Q \sin nt)$$

$$+ 2 \cos nt \cos 2nt \left\{ \tfrac{1}{4}P(P^2 + P'^2 - Q^2 - Q'^2) - \frac{2n^2 Q}{g}(PQ + P'Q') \right\}$$

$$+ 2 \sin nt \sin 2nt \left\{ \tfrac{1}{2}Q(PQ + P'Q') - \frac{n^2 P}{g}(P^2 + P'^2 - Q^2 - Q'^2) \right\}$$

$$+ 2 \sin nt \cos 2nt \left\{ \tfrac{1}{4}Q(P^2 + P'^2 - Q^2 - Q'^2) + \frac{2n^2 P}{g}(PQ + P'Q') \right\}$$

$$+ 2 \cos nt \sin 2nt \left\{ \tfrac{1}{2}P(PQ + P'Q') + \frac{n^2 Q}{g}(P^2 + P'^2 - Q^2 - Q'^2) \right\},$$

of which the part in $\sin nt$ has the coefficient

$$Q\{\tfrac{1}{4}(P^2 + P'^2) + \tfrac{3}{4}(Q^2 + Q'^2)\} + \tfrac{1}{2}P(PQ + P'Q')$$
$$+ n^2/g . \{Q(P^2 + P'^2 - Q^2 - Q'^2) - 2P(PQ + P'Q')\}$$

or, since $n^2 = g$ approximately,

$$Q\{\tfrac{5}{4}(P^2 + P'^2) - \tfrac{1}{4}(Q^2 + Q'^2)\} - \tfrac{3}{2}P(PQ + P'Q'). \dots\dots\text{(46)}$$

In like manner the coefficient of $\cos nt$ is

$$P\{\tfrac{5}{4}(Q^2 + Q'^2) - \tfrac{1}{4}(P^2 + P'^2)\} - \tfrac{3}{2}Q(PQ + P'Q'), \dots\dots\text{(47)}$$

differing merely by the interchange of P and Q.

But when these values are employed in (44), it is not, in general, possible, with constant values of P, Q, P', Q', to annul the terms in $\sin nt$, $\cos nt$. We obtain from the first

$$n^2 = g + \tfrac{5}{4}(P^2 + P'^2) - \tfrac{1}{4}(Q^2 + Q'^2) - \frac{3P}{2Q}(PQ + P'Q'), \dots\dots\text{(48)}$$

and from the second

$$n^2 = g + \tfrac{5}{4}(Q^2 + Q'^2) - \tfrac{1}{4}(P^2 + P'^2) - \frac{3Q}{2P}(PQ + P'Q'); \dots\dots\text{(49)}$$

and these are inconsistent, unless

$$(PP' + QQ')(PQ' - P'Q) = 0. \dots\dots\dots\text{(50)}$$

The latter condition is unaltered by interchange of dashed and undashed letters, and thus it serves equally for the equation in α'.

The two alternatives indicated in (50) correspond to the particular cases already considered.   In the first $(PP' + QQ' = 0)$ we have a purely progressive wave and in the second a purely stationary one.

When the condition (50) does not hold good, it is impossible to satisfy our equations as before with constant values of n, P, Q, P', Q' ; and it is perhaps hardly worth while to pursue the more complicated questions which then arise.   It may suffice to remark that an approximately stationary wave can never pass into an approximately progressive wave, nor *vice versâ*.   The progressive wave has momentum, while the stationary wave has none, and momentum is necessarily conserved.

When β, β', γ, γ', are not zero, additional terms enter.   Equations (26), (30), show that the additions to b, b', vary as the sine and cosine of $\sqrt{(2g)} \cdot t$, and represent waves which might exist in the complete absence of the principal wave.

The additions to c, c', are more complicated.   As regards the parts depending in (31), (32), on $d\gamma/dt$, $d\gamma'/dt$, they are proportional to the sine and cosine of $\sqrt{(3g)} \cdot t$, and represent waves which might exist alone.   But besides these there are other parts, analogous to the combination-tones of Acoustics, resulting from the interaction of the β-waves with the principal wave.   These vary as the sine and cosine of $\sqrt{g} \cdot \{\sqrt{2} \pm 1\}\, t$, thus possessing frequencies differing from the former frequencies.   Similar terms will enter into the expression for n^2 as determined from (28), (29).

In the particular case of β, β', vanishing, even though γ, γ' (assumed still to be of the third order) remain, we recover most of the former simplicity, the only difference being the occurrence in c, c', of terms in $\sqrt{(3g)} \cdot t$, such as might exist alone.

394.

ÆOLIAN TONES.

[*Philosophical Magazine*, Vol. XXIX. pp. 433—444, 195, 1915.]

IN what has long been known as the Æolian Harp, a stretched string, such as a pianoforte wire or a violin string, is caused to vibrate in one of its possible modes by the impact of wind; and it was usually supposed that the action was analogous to that of a violin bow, so that the vibrations were executed in the plane containing the direction of the wind. A closer examination showed, however, that this opinion was erroneous and that in fact the vibrations are transverse to the wind*. It is not essential to the production of sound that the string should take part in the vibration, and the general phenomenon, exemplified in the whistling of wind among trees, has been investigated by Strouhal† under the name of *Reibungstöne*.

In Strouhal's experiments a vertical wire or rod attached to a suitable frame was caused to revolve with uniform velocity about a parallel axis. The pitch of the æolian tone generated by the relative motion of the wire and of the air was found to be independent of the length and of the tension of the wire, but to vary with the diameter (D) and with the speed (V) of the motion. Within certain limits the relation between the frequency of vibration (N) and these data was expressible by

$$N = \cdot 185\, V/D, \quad\dots\dots\dots\dots\dots\dots\dots\dots(1)\ddagger$$

the centimetre and the second being units.

When the speed is such that the æolian tone coincides with one of the proper tones of the wire, supported so as to be capable of free independent vibration, the sound is greatly reinforced, and with this advantage Strouhal found it possible to extend the range of his observations. Under the more extreme conditions then practicable the observed pitch deviated considerably

---

* *Phil. Mag.* Vol. VII. p. 149 (1879); *Scientific Papers*, Vol. I. p. 413.

† *Wied. Ann.* Vol. v. p. 216 (1878).

‡ In (1) V is the velocity of the wire relatively to the walls of the laboratory.

from the value given by (1). He further showed that with a given diameter and a given speed a rise of temperature was attended by a fall in pitch.

If, as appears probable, the compressibility of the fluid may be left out of account, we may regard N as a function of the relative velocity V, D, and ν the kinematic coefficient of viscosity. In this case N is necessarily of the form

$$N = V/D \cdot f(\nu/VD), \quad\quad\quad\quad\quad\quad\quad\quad (2)$$

where f represents an arbitrary function; and there is dynamical similarity, if $\nu \propto VD$. In observations upon air at one temperature ν is constant; and if D vary inversely as V, ND/V should be constant, a result fairly in harmony with the observations of Strouhal. Again, if the temperature rises, ν increases, and in order to accord with observation, we must suppose that the function f diminishes with increasing argument.

"An examination of the actual values in Strouhal's experiments shows that ν/VD was always small; and we are thus led to represent f by a few terms of MacLaurin's series. If we take

$$f(x) = a + bx + cx^2,$$

we get
$$N = a\,\frac{V}{D} + b\,\frac{\nu}{D^2} + c\,\frac{\nu^2}{VD^3}. \quad\quad\quad\quad\quad (3)$$

"If the third term in (3) may be neglected, the relation between N and V is linear. This law was formulated by Strouhal, and his diagrams show that the coefficient b is negative, as is also required to express the observed effect of a rise of temperature. Further,

$$D\,\frac{dN}{dV} = a - \frac{c\nu^2}{V^2 D^2}, \quad\quad\quad\quad\quad\quad (4)$$

so that $D \cdot dN/dV$ is very nearly constant, a result also given by Strouhal on the basis of his measurements.

"On the whole it would appear that the phenomena are satisfactorily represented by (2) or (3), but a dynamical theory has yet to be given. It would be of interest to extend the experiments to liquids*."

Before the above paragraphs were written I had commenced a systematic deduction of the form of f from Strouhal's observations by plotting ND/V against VD. Lately I have returned to the subject, and I find that nearly all his results are fairly well represented by two terms of (3). In C.G.S. measure

$$\frac{ND}{V} = \cdot 195 \left(1 - \frac{3\cdot02}{VD}\right) \quad\quad\quad\quad\quad\quad (5)$$

Although the agreement is fairly good, there are signs that a change of wire introduces greater discrepancies than a change in V—a circumstance

---

* *Theory of Sound*, 2nd ed. Vol. II. § 372 (1896).

which may possibly be attributed to alterations in the character of the surface. The simple form (2) assumes that the wires are smooth, or else that the roughnesses are in proportion to D, so as to secure geometrical similarity.

The completion of (5) from the theoretical point of view requires the introduction of ν. The temperature for the experiments in which ν would enter most was about 20° C., and for this temperature

$$\nu = \frac{\mu}{\rho} = \frac{1806 \times 10^{-7}}{\cdot 00120} = \cdot 1505 \text{ C.G.S.}$$

The generalized form of (5) is accordingly

$$\frac{ND}{V} = \cdot 195 \left(1 - \frac{20 \cdot 1\nu}{VD} \right), \quad \dots \dots \dots \dots \dots (6)$$

applicable now to any fluid when the appropriate value of ν is introduced. For water at 15° C., $\nu = \cdot 0115$, much *less* than for air.

Strouhal's observations have recently been discussed by Krüger and Lauth*, who appear not to be acquainted with my theory. Although they do not introduce viscosity, they recognize that there is probably some cause for the observed deviations from the simplest formula (1), other than the complication arising from the circulation of the air set in motion by the revolving parts of the apparatus. Undoubtedly this circulation marks a weak place in the method, and it is one not easy to deal with. On this account the numerical quantities in (6) may probably require some correction in order to express the true formula when V denotes the velocity of the wire through otherwise undisturbed fluid.

We may find confirmation of the view that viscosity enters into the question, much as in (6), from some observations of Strouhal on the effect of *temperature*. Changes in ν will tell most when VD is small, and therefore I take Strouhal's table XX., where $D = \cdot 0179$ cm. In this there appears

$$t_1 = 11°, \quad V_1 = 385, \quad N_1/V_1 = 6 \cdot 70, \quad \nu_1,$$
$$t_2 = 31°, \quad V_2 = 381, \quad N_2/V_2 = 6 \cdot 48, \quad \nu_2.$$

Introducing these into (6), we get

$$6 \cdot 70 - 6 \cdot 48 = \frac{195}{D} \left(1 - \frac{20 \cdot 1\nu_1}{V_1 D} \right) - \frac{195}{D} \left(1 - \frac{20 \cdot 1\nu_2}{V_2 D} \right),$$

or with sufficient approximation

$$\nu_2 - \nu_1 = \frac{\cdot 52 D^2 V}{\cdot 195 \times 20 \cdot 1} = \cdot 016 \text{ C.G.S.}$$

* "Theorie der Hiebtöne," *Ann. d. Physik*, Vol. XLIV. p. 801 (1914).

We may now compare this with the known values of ν for the temperatures in question. We have

$$\mu_{31} = 1853 \times 10^{-7}, \quad \rho_{31} = \cdot001161,$$

$$\mu_{11} = 1765 \times 10^{-7}, \quad \rho_{11} = \cdot001243;$$

so that $\qquad\qquad \nu_2 = \cdot1596, \qquad \nu_1 = \cdot1420,$

and $\qquad\qquad\qquad \nu_2 - \nu_1 = \cdot018.$

The difference in the values of ν at the two temperatures thus accounts in (6) for the change of frequency both in sign and in order of magnitude.

As regards dynamical explanation it was evident all along that the origin of vibration was connected with the instability of the vortex sheets which tend to form on the two sides of the obstacle, and that, at any rate when a wire is maintained in transverse vibration, the phenomenon must be unsymmetrical. The alternate formation in water of detached vortices on the two sides is clearly described by H. Bénard*. "Pour une vitesse suffisante, au-dessous de laquelle il n'y a pas de tourbillons (cette vitesse limite croît avec la viscosité et decroît quand l'épaisseur transversale des obstacles augmente), *les tourbillons produits périodiquement se détachent alternativement à droite et à gauche du remous d'arrière qui suit le solide; ils gagnent presque immédiatement leur emplacement définitif, de sorte qu'à l'arrière de l'obstacle se forme une double rangée alternée d'entonnoirs stationnaires, ceux de droite dextrogyres, ceux de gauche lévogyres, séparés par des intervalles égaux.*"

The symmetrical and unsymmetrical processions of vortices were also figured by Mallock† from direct observation.

In a remarkable theoretical investigation‡ Kármán has examined the question of the stability of such processions. The fluid is supposed to be incompressible, to be devoid of viscosity, and to move in two dimensions. The vortices are concentrated in points and are disposed at equal intervals (l) along two parallel lines distant h. Numerically the vortices are all equal, but those on different lines have opposite signs.

Apart from stability, steady motion is possible in two arrangements (a) and (b), fig. 1, of which (a) is symmetrical. Kármán shows that (a) is always unstable, whatever may be the ratio of h to l; and further that (b) is usually unstable also. The single exception occurs when $\cosh(\pi h/l) = \sqrt{2}$, or $h/l = 0\cdot283$. With this ratio of h/l, (b) is stable for every kind of displacement except one, for which there is neutrality. The only procession which can possess a practical permanence is thus defined.

* C. R. t. 147, p. 839 (1908).

† Proc. Roy. Soc. Vol. LXXXIV. A, p. 490 (1910).

‡ Göttingen Nachrichten, 1912, Heft 5, S. 547; Kármán and Rubach, *Physik. Zeitschrift*, 1912, p. 49. I have verified the more important results.

The corresponding motion is expressed by the complex potential (ϕ potential, ψ stream-function)

$$\phi + i\psi = \frac{i\zeta}{2\pi} \log \frac{\sin \{\pi (z_0 - z)/l\}}{\sin \{\pi (z_0 + z)/l\}}, \quad \dots\dots\dots\dots(7)$$

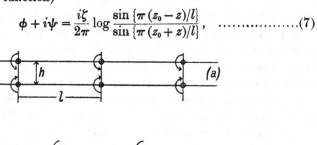

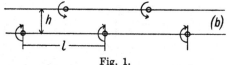

Fig. 1.

in which ζ denotes the strength of a vortex, $z = x + iy$, $z_0 = \frac{1}{4}l + ih$. The x-axis is drawn midway between the two lines of vortices and the y-axis halves the distance between neighbouring vortices with opposite rotation. Kármán gives a drawing of the stream-lines thus defined.

The constant velocity of the processions is given by

$$u = \frac{\zeta}{2l} \tanh \frac{\pi h}{l} = \frac{\zeta}{l\sqrt{8}} \quad \dots\dots\dots\dots\dots(8)$$

This velocity is relative to the fluid at a distance.

The observers who have experimented upon water seem all to have used obstacles not susceptible of vibration. For many years I have had it in my mind to repeat the æolian harp effect with water*, but only recently have brought the matter to a test. The water was contained in a basin, about 36 cm. in diameter, which stood upon a sort of turn-table. The upper part, however, was not properly a table, but was formed of two horizontal beams crossing one another at right angles, so that the whole apparatus resembled rather a turn-*stile*, with four spokes. It had been intended to drive from a small water-engine, but ultimately it was found that all that was needed could more conveniently be done by hand after a little practice. A metronome beat approximate half seconds, and the spokes (which projected beyond the basin) were pushed gently by one or both hands until the rotation was uniform with passage of one or two spokes in correspondence with an assigned number of beats. It was necessary to allow several minutes in order to

---

* From an old note-book. "Bath, Jan. 1884. I find in the baths here that if the spread fingers be drawn pretty quickly through the water (palm foremost was best), they are thrown into transverse vibration and strike one another. This seems like æolian string.... The blade of a flesh-brush about 1½ inch broad seemed to vibrate transversely in its own plane when moved through water broadways forward. It is pretty certain that with proper apparatus these vibrations might be developed and observed."

make sure that the water had attained its ultimate velocity. The axis of
rotation was indicated by a pointer affixed to a small stand resting on the
bottom of the basin and rising slightly above the level of the water.

The pendulum (fig. 2), of which the lower part was immersed, was
supported on two points (A, B) so that the possible vibrations were limited
to one vertical plane. In the usual arrangement the vibrations of the rod
would be radial, *i.e.* transverse to the motion of the water, but it was easy to
turn the pendulum round when it was desired to test whether a circumferential
vibration could be maintained. The rod C itself was of brass tube $8\frac{1}{2}$ mm.
in diameter, and to it was clamped a hollow cylinder of lead D. The time

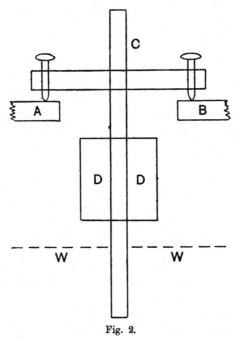

Fig. 2.

of complete vibration (τ) was about half a second. When it was desired to
change the diameter of the immersed part, the rod C was drawn up higher
and prolonged below by an additional piece—a change which did not much
affect the period τ. In all cases the length of the part immersed was
about 6 cm.

Preliminary observations showed that in no case were vibrations generated
when the pendulum was so mounted that the motion of the rod would be
circumferential, viz. in the direction of the stream, agreeably to what had
been found for the æolian harp. In what follows the vibrations, if any, are
radial, that is transverse to the stream.

In conducting a set of observations it was found convenient to begin with
the highest speed, passing after a sufficient time to the next lower, and so on,

with the minimum of intermission. I will take an example relating to the
main rod, whose diameter (D) is $8\frac{1}{2}$ mm., $\tau = 60/106$ sec., beats of metronome
62 in 30 sec. The speed is recorded by the number of beats corresponding
to the passage of *two* spokes, and the vibration of the pendulum (after the
lapse of a sufficient time) is described as small, fair, good, and so on. Thus on
Dec. 21, 1914:

> 2 spokes to 4 beats gave fair vibration,
> . . . . 5 . . . . good . . . . . .
> . . . . 6 . . . . . rather more . : .
> . . . . 7 . . . . good . . . . . .
> . . . . 8 . . . . fair . . . . . .

from which we may conclude that the maximum effect corresponds to 6 beats,
or to a time (T) of revolution of the turn-table equal to $2 \times 6 \times 30/62$ sec.
The distance (r) of the rod from the axis of rotation was 116 mm., and the
speed of the water, supposed to move with the basin, is $2\pi r/T$. The result
of the observations may intelligibly be expressed by the ratio of the distance
travelled by the water during one complete vibration of the pendulum to the
diameter of the latter, viz.

$$\frac{\tau . 2\pi r/T}{D} = \frac{2\pi \times 116 \times 62}{8\cdot5 \times 6 \times 106} = 8\cdot36.$$

Concordant numbers were obtained on other occasions.

In the above calculation the speed of the water is taken as if it were
rigidly connected with the basin, and must be an over estimate. When the
pendulum is away, the water may be observed to move as a solid body after
the rotation has been continued for two or three minutes. For this purpose
the otherwise clean surface may be lightly dusted over with sulphur. But
when the pendulum is immersed, the rotation is evidently hindered, and that
not merely in the neighbourhood of the pendulum itself. The difficulty
thence arising has already been referred to in connexion with Strouhal's
experiments and it cannot easily be met in its entirety. It may be mitigated
by increasing r, or by diminishing D. The latter remedy is easily applied up
to a certain point, and I have experimented with rods 5 mm. and $3\frac{1}{2}$ mm. in
diameter. With a 2 mm. rod no vibration could be observed. The final
results were thus tabulated:

Diameter ...	8·5 mm.	5·0 mm.	3·5 mm.
Ratio ...	8·35	7·5	7·8

from which it would appear that the disturbance is not very serious. The
difference between the ratios for the 5·0 mm. and 3·5 mm. rods is hardly outside the limits of error; and the prospect of reducing the ratio much below 7
seemed remote.

The instinct of an experimenter is to try to get rid of a disturbance, even
though only partially; but it is often equally instructive to increase it. The

observations of Dec. 21 were made with this object in view; besides those already given they included others in which the disturbance due to the vibrating pendulum was augmented by the addition of a similar rod (8½ mm.) immersed to the same depth and situated symmetrically on the same diameter of the basin. The anomalous effect would thus be doubled. The record was as follows:

> 2 spokes to 3 beats gave little or no vibration,
> . . . . 4 . . . . . . fair   . . . . . .
> . . . . 5 . . . . . . large   . . . . . .
> . . . . 6 . . . . . . less   . . . . . .
> . . . . 7 . . . . . . little or no . . . . . .

As the result of this and another day's similar observations it was concluded that the 5 beats with additional obstruction corresponded with 6 beats without it. An approximate correction for the disturbance due to improper action of the pendulum may thus be arrived at by decreasing the calculated ratio in the proportion of 6 : 5; thus

$$\tfrac{5}{6}(8\cdot35) = 7\cdot0$$

is the ratio to be expected in a uniform stream. It would seem that this cannot be far from the mark, as representing the travel at a distance from the pendulum in an otherwise uniform stream during the time of one complete vibration of the latter. Since the correction for the other diameters will be decidedly less, the above number may be considered to apply to all three diameters experimented on.

In order to compare with results obtained from air, we must know the value of ν/VD. For water at 15° C. $\nu = \mu = \cdot0115$ c.g.s.; and for the 8·5 mm. pendulum $\nu/VD = \cdot0011$. Thus from (6) it appears that ND/V should have nearly the full value, say ·190. The reciprocal of this, or 5·3, should agree with the ratio found above as 7·0; and the discrepancy is larger than it should be.

An experiment to try whether a change of viscosity had appreciable influence may be briefly mentioned. Observations were made upon water heated to about 60° C. and at 12° C. No difference of behaviour was detected. At 60° C. $\mu = \cdot0049$, and at 12° C. $\mu = \cdot0124$.

I have described the simple pendulum apparatus in some detail, as apart from any question of measurements it demonstrates easily the general principle that the vibrations are transverse to the stream, and when in good action it exhibits very well the double row of vortices as witnessed by dimples upon the surface of the water.

The discrepancy found between the number from water (7·0) and that derived from Strouhal's experiments on air (5·3) raises the question whether the latter can be in error. So far as I know, Strouhal's work has not been repeated; but the error most to be feared, that arising from the circulation of the air, acts in the wrong direction. In the hope of further light I have remounted my apparatus of 1879. The draught is obtained from a chimney. A structure of wood and paper is fitted to the fire-place, which may prevent all access of air to the chimney except through an elongated horizontal aperture in the front (vertical) wall. The length of the aperture is 26 inches (66 cm.), and the width 4 inches (10·2 cm.); and along its middle a gut string is stretched over bridges.

The draught is regulated mainly by the amount of fire. It is well to have a margin, as it is easy to shunt a part through an aperture at the top of the enclosure, which can be closed partially or almost wholly by a superposed card. An adjustment can sometimes be got by opening a door or window. A piece of paper thrown on the fire increases the draught considerably for about half a minute.

The string employed had a diameter of ·95 mm., and it could readily be made to vibrate (in 3 segments) in unison with a fork of pitch 256. The octave, not difficult to mistake, was verified by a resonator brought up close to the string. That the vibration is transverse to the wind is confirmed by the behaviour of the resonator, which goes out of action when held symmetrically. The sound, as heard in the open without assistance, was usually feeble, but became loud when the ear was held close to the wooden frame. The difficulty of the experiment is to determine the velocity of the wind, where it acts upon the string. I have attempted to do this by a pendulum arrangement designed to determine the wind by its action upon an elongated piece of mirror (10·1 cm. × 1·6 cm.) held perpendicularly and just in front of the string. The pendulum is supported on two points—in this respect like the one used for the water experiments; the mirror is above, and there is a counter-weight below. An arm projects horizontally forward on which a rider can be placed. In commencing observations the wind is cut off by a large card inserted across the aperture and just behind the string. The pendulum then assumes a sighted position, determined in the usual way by reflexion. When the wind operates the mirror is carried with it, but is brought back to the sighted position by use of a rider of mass equal to ·485 gm.

Observations have been taken on several occasions, but it will suffice to record one set whose result is about equal to the average. The (horizontal) distance of the rider from the axis of rotation was 62 mm., and the vertical distance of the centre line of the mirror from the same axis is 77 mm. The force of the wind upon the mirror was thus $62 \times ·485 \div 77$ gms. weight. The mean pressure P is

$$\frac{62 \times ·485 \times 981}{77 \times 16·2} = 23·7 \ \frac{\text{dynes}}{\text{cm.}^2}.$$

The formula connecting the velocity of the wind V with the pressure P may be written

$$P = C\rho V^2,$$

where ρ is the density; but there is some uncertainty as to the constancy of C. It appears that for large plates $C = \cdot 62$, but for a plate 2 inches square Stanton found $C = \cdot 52$. Taking the latter value*, we have

$$V^2 = \frac{23\cdot7}{\cdot52\rho} = \frac{23\cdot7}{\cdot52 \times \cdot00123},$$

on introduction of the value of ρ appropriate to the circumstances of the experiment. Accordingly

$$V = 192 \text{ cm./sec.}$$

The frequency of vibration (τ^{-1}) was nearly enough 256; so that

$$\frac{V\tau}{D} = \frac{192}{256 \times \cdot095} = 7\cdot9.$$

In comparing this with Strouhal, we must introduce the appropriate value of VD, that is 19, into (5). Thus

$$\frac{V}{ND} = \frac{V\tau}{D} = 6\cdot1.$$

Whether judged from the experiments with water or from those just detailed upon air, this (Strouhal's) number would seem to be too low; but the uncertainty in the value of C above referred to precludes any very confident conclusion. It is highly desirable that Strouhal's number should be further checked by some method justifying complete confidence.

When a wire or string exposed to wind does not itself enter into vibration, the sound produced is uncertain and difficult to estimate. No doubt the wind is often different at different parts of the string, and even at the same part it may fluctuate rapidly. A remedy for the first named cause of unsteadiness is to listen through a tube, whose open end is brought pretty close to the obstacle. This method is specially advantageous if we take advantage of our knowledge respecting the mode of action, by using a tube drawn out to a narrow bore (say 1 or 2 mm.) and placed so as to face the processions of vortices behind the wire. In connexion with the fire-place arrangement the drawn out glass tube is conveniently bent round through 180° and continued to the ear by a rubber prolongation. In the wake of the obstacle the sound is well heard, even at some distance (50 mm.) behind; but little or nothing reaches the ear when the aperture is in front or at the side, even though quite close up, unless the wire is itself vibrating. But the special arrangement for

---

* But I confess that I feel doubts as to the diminution of C with the linear dimension. [1917. See next paper.]

a draught, where the observer is on the high pressure side, is not necessary ; in a few minutes any one may prepare a little apparatus competent to show the effect. Fig. 3 almost explains itself. A is the drawn out glass tube

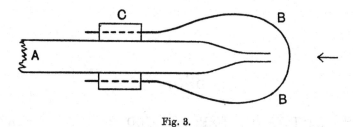

Fig. 3.

B the loop of iron or brass wire (say 1 mm. in diameter), attached to the tube with the aid of a cork C. The rubber prolongation is not shown. Held in the crack of a slightly opened door or window, the arrangement yields a sound which is often pure and fairly steady.

395.

ON THE RESISTANCE EXPERIENCED BY SMALL PLATES EXPOSED TO A STREAM OF FLUID.

[*Philosophical Magazine*, Vol. XXX. pp. 179—181, 1915.]

In a recent paper on Æolian Tones* I had occasion to determine the velocity of wind from its action upon a narrow strip of mirror (10·1 cm. × 1·6 cm.), the incidence being normal. But there was some doubt as to the coefficient to be employed in deducing the velocity from the density of the air and the force per unit area. Observations both by Eiffel and by Stanton had indicated that the resultant pressure (force reckoned per unit area) is less on small plane areas than on larger ones; and although I used provisionally a diminished value of C in the equation $P = C\rho V^2$ in view of the narrowness of the strip, it was not without hesitation†. I had in fact already commenced experiments which appeared to show that no variation in C was to be detected. Subsequently the matter was carried a little further; and I think it worth while to describe briefly the method employed. In any case I could hardly hope to attain finality, which would almost certainly require the aid of a proper wind channel, but this is now of less consequence as I learn that the matter is engaging attention at the National Physical Laboratory.

According to the principle of similitude a departure from the simple law would be most apparent when the kinematic viscosity is large and the stream velocity small. Thus, if the delicacy can be made adequate, the use of *air* resistance and such low speeds as can be reached by walking through a still atmosphere should be favourable. The principle of the method consists in balancing the two areas to be compared by mounting them upon a vertical axis, situated in their common plane, and capable of turning with the minimum of friction. If the areas are equal, their centres must be at the same distance (on opposite sides) from the axis. When the apparatus is carried forward through the air, equality of mean pressures is witnessed by the plane of the obstacles assuming a position of perpendicularity to the line of motion. If in

---

* *Phil. Mag.* Vol. XXIX. p. 442 (1915). [Art. 394.]

† See footnote on p. [324].

this position the mean pressure on one side is somewhat deficient, the plane on that side advances against the relative stream, until a stable balance is attained in an oblique position, in virtue of the displacement (forwards) of the centres of pressure from the centres of figure.

The plates under test can be cut from thin card and of course must be accurately measured. In my experiments the axis of rotation was a sewing-needle held in a U-shaped strip of brass provided with conical indentations. The longitudinal pressure upon the needle, dependent upon the spring of the brass, should be no more than is necessary to obviate shift. The arms connecting the plates with the needle are as slender as possible consistent with the necessary rigidity, not merely in order to save weight but to minimise their resistance. They may be made of wood, provided it be accurately shaped, or of wire, preferably of aluminium. Regard must be paid to the proper balancing of the resistances of these arms, and this may require otherwise superfluous additions. It would seem that a practical solution may be attained, though it must remain deficient in mathematical exactness. The junctions of the various pieces can be effected quite satisfactorily with sealing-wax used sparingly. The brass U itself is mounted at the end of a rod held horizontally in front of the observer and parallel to the direction of motion. I found it best to work indoors in a long room or gallery.

Although in use the needle is approximately vertical, it is necessary to eliminate the possible effect of gravity more completely than can thus be attained. When the apparatus is otherwise complete, it is turned so as to make the needle horizontal, and small balance weights (finally of wax) adjusted behind the plates until equilibrium is neutral. In this process a good opinion can be formed respecting the freedom of movement.

In an experiment, suggested by the case of the mirror above referred to, the comparison was between a rectangular plate 2 inches × $1\frac{1}{2}$ inches and an elongated strip ·51 inch broad, the length of the strip being parallel to the needle, *i.e.* vertical in use. At first this length was a little in excess, but was cut down until the resistance balance was attained. For this purpose it seemed that equal areas were required to an accuracy of about one per cent., nearly on the limit set by the delicacy of the apparatus.

According to the principle of similitude the influence of linear scale (l) upon the mean pressure should enter only as a function of ν/Vl, where ν is the kinematic viscosity of air and V the velocity of travel. In the present case $\nu = ·1505$, V (4 miles per hour) = 180, and l, identified with the width of the strip, = 1·27, all in C.G.S. measure. Thus

$$\nu/Vl = ·00066.$$

In view of the smallness of this quantity, it is not surprising that the influence of linear scale should fail to manifest itself.

In virtue of the more complete symmetry realizable when the plates to be compared are not merely equal in area but also similar in shape, this method would be specially advantageous for the investigation of the possible influence of thickness and of the smoothness of the surfaces.

When the areas to be compared are unequal, so that their centres need to be at different distances from the axis, the resistance balance of the auxiliary parts demands special attention. I have experimented upon circular disks whose areas are as $2:1$. When there was but one smaller disk (6 cm. in diameter) the arms of the lever had to be also as $2:1$ (fig. 1). In another

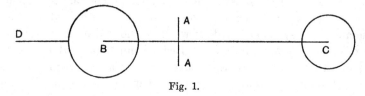

Fig. 1.

experiment *two* small disks (each 4 cm. in diameter) were balanced against a larger one of equal total area (fig. 2). Probably this arrangement is the better. In neither case was any difference of mean pressures detected.

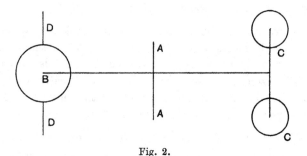

Fig. 2.

In the figures AA represents the needle, B and C the large and small disks respectively, D the extra attachments needed for the resistance balance of the auxiliary parts.

396.

HYDRODYNAMICAL PROBLEMS SUGGESTED BY PITOT'S TUBES.

[*Proceedings of the Royal Society*, A, Vol. XCI. pp. 503—511, 1915.]

THE general use of Pitot's tubes for measuring the velocity of streams suggests hydrodynamical problems. It can hardly be said that these are of practical importance, since the action to be observed depends simply upon Bernoulli's law. In the interior of a long tube of any section, closed at the further end and facing the stream, the pressure must be that due to the velocity (v) of the stream, *i.e.* $\frac{1}{2}\rho v^2$, ρ being the density. At least, this must be the case if viscosity can be neglected. I am not aware that the influence of viscosity here has been detected, and it does not seem likely that it can be sensible under ordinary conditions. It would enter in the combination ν/vl, where ν is the kinematic viscosity and l represents the linear dimension of the tube. Experiments directed to show it would therefore be made with small tubes and low velocities.

In practice a tube of circular section is employed. But, even when viscosity is ignored, the problem of determining the motion in the neighbourhood of a circular tube is beyond our powers. In what follows, not only is the fluid supposed frictionless, but the circular tube is replaced by its two-dimensional analogue, *i.e.* the channel between parallel plane walls. Under this head two problems naturally present themselves.

The first problem proposed for consideration may be defined to be the flow of electricity in two dimensions, when the uniformity is disturbed by the presence of a channel whose infinitely thin non-conducting walls are parallel to the flow. By themselves these walls, whether finite or infinite, would cause no disturbance; but the channel, though open at the finite end, is supposed to be closed at an infinite distance away, so that, on the whole, there is no stream through it. If we suppose the flow to be of liquid instead of electricity, the arrangement may be regarded as an idealized Pitot's tube, although we know that, in consequence of the sharp edges, the electrical law would be widely departed from. In the recesses of the tube there is no motion, and the pressure developed is simply that due to the velocity of the stream.

The problem itself may be treated as a modification of that of Helmholtz*, where flow is imagined to take place within the channel and to come to evanescence outside at a distance from the mouth. If in the usual notation† $z = x + iy$, and $w = \phi + i\psi$ be the complex potential, the solution of Helmholtz's problem is expressed by

$$z = w + e^w, \quad\quad\quad\quad\quad\quad\dots\dots\dots\dots\dots(1)$$

or

$$x = \phi + e^\phi \cos \psi, \quad\quad y = \psi + e^\phi \sin \psi. \quad\dots\dots\dots(2)$$

The walls correspond to $\psi = \pm \pi$, where y takes the same values, and they extend from $x = -\infty$ to $x = -1$. Also the stream-line $\psi = 0$ makes $y = 0$, which is a line of symmetry. In the recesses of the channel ϕ is negative and large, and the motion becomes a uniform stream.

To annul the internal stream we must superpose upon this motion, expressed say by $\phi_1 + i\psi_1$, another of the form $\phi_2 + i\psi_2$ where

$$\phi_2 + i\psi_2 = -x - iy.$$

In the resultant motion,

$$\phi = \phi_1 + \phi_2 = \phi_1 - x, \quad\quad \psi = \psi_1 + \psi_2 = \psi_1 - y,$$

so that

$$\phi_1 = \phi + x, \quad\quad \psi_1 = \psi + y,$$

and we get

$$0 = \phi + e^{\phi+x} \cos (\psi + y), \quad\quad 0 = \psi + e^{\phi+x} \sin (\psi + y), \quad\dots\dots (3)$$

whence

$$x = -\phi + \log \sqrt{(\phi^2 + \psi^2)}, \quad\quad y = -\psi + \tan^{-1} (\psi/\phi) \quad\dots\dots(4)$$

or, as it may also be written,

$$z = -w + \log w. \quad\quad\quad\quad\dots\dots\dots\dots(5)$$

It is easy to verify that these expressions, no matter how arrived at, satisfy the necessary conditions. Since x is an even function of ψ, and y an odd function, the line $y = 0$ is an axis of symmetry. When $\psi = 0$, we see from (3) that $\sin y = 0$, so that $y = 0$ or $\pm \pi$, and that $\cos y$ and ϕ have opposite signs. Thus when ϕ is negative, $y = 0$; and when ϕ is positive, $y = \pm \pi$. Again, when ϕ is negative, x ranges from $+\infty$ to $-\infty$; and when ϕ is positive x ranges from $-\infty$ to -1, the extreme value at the limit of the wall, as appears from the equation

$$dx/d\phi = -1 + 1/\phi = 0,$$

making $\phi = 1$, $x = -1$. The central stream-line may thus be considered to pass along $y = 0$ from $x = \infty$ to $x = -\infty$. At $x = -\infty$ it divides into two

---

* *Berlin Monatsber.* 1868; *Phil. Mag.* Vol. xxxvi. p. 337 (1868). In this paper a new path was opened.

† See Lamb's *Hydrodynamics*, § 66.

branches along $y = \pm \pi$. From $x = -\infty$ to $x = -1$, the flow is along the inner side of the walls, and from $x = -1$ to $x = -\infty$ back again along the outer side. At the turn the velocity is of course infinite.

We see from (4) that when ψ is given the difference in the final values of y, corresponding to infinite positive and negative values of ϕ, amounts to π, and that the smaller is ψ the more rapid is the change in y.

The corresponding values of x and y for various values of ϕ, and for the stream-lines $\psi = -1, -\frac{1}{2}, -\frac{1}{4}$, are given in Table I, and the more important parts are exhibited in the accompanying plots (fig. 1).

TABLE I.

ϕ	$\psi = -\frac{1}{4}$		$\psi = -\frac{1}{2}$		$\psi = -1$	
	x	y	x	y	x	y
-10	12·303	0·2750	12·30	0·550	12·31	1·100
-5	6·610	0·3000	6·614	0·600	6·63	1·198
-3	4·102	0·3333	4·112	0·665	4·15	1·322
-2	2·701	0·3745	2·723	0·745	2·80	1·464
-1	1·030	0·495	1·111	0·964	1·35	1·785
$-0·50$	0·081	0·714	0·153	1·285	—	—
$-0·25$	$-0·790$	1·035	—	—	—	—
$0·00$	$-1·386$	1·821	$-0·693$	2·071	0·00	2·571
$0·25$	$-1·290$	2·606	—	—	—	—
$0·50$	$-1·081$	2·928	$-0·847$	2·881	$-0·388$	3·035
$1·0$	$-0·970$	3·147	$-0·888$	3·178	$-0·653$	3·356
$2·0$	$-1·299$	3·267	$-1·277$	3·397	$-1·195$	3·678
$3·0$	$-1·898$	3·308	$-1·888$	3·477	—	—
$4·0$	—	—	—	—	$-2·584$	3·897
$5·0$	$-3·389$	3·342	$-3·386$	3·542	—	—
$10·0$	$-7·697$	3·367	—	—	$-7·692$	4·042
$20·0$	—	—	—	—	$-17·00$	4·092

In the second form of the problem we suppose, after Helmholtz and Kirchhoff, that the infinite velocity at the edge, encountered when the fluid adheres to the wall, is obviated by the formation of a surface of discontinuity where the condition to be satisfied is that of constant pressure and velocity. It is, in fact, a particular case of one treated many years ago by Prof. Love, entitled "Liquid flowing against a disc with an elevated rim," when the height of the rim is made infinite[*]. I am indebted to Prof. Love for the form into which the solution then degrades. The origin O' (fig. 2) of $x + iy$ or z is taken at one edge. The central stream-line ($\psi = 0$) follows the line of symmetry AB from $y = +\infty$ to $y = -\infty$. At $y = -\infty$ it divides, one half following the inner side of the wall CO' from $y = -\infty$ to $y = 0$, then becomes a free surface $O'D$ from $y = 0$ to $y = -\infty$. The connexion between

---

[*] *Camb. Phil. Proc.* Vol. VII. p. 185 (1891).

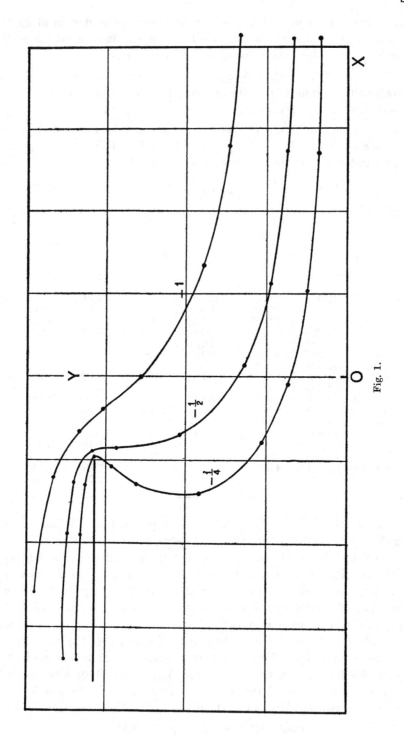

Fig. 1.

z and $w (= \phi + i\psi)$ is expressed with the aid of an auxiliary variable θ. Thus

$$z = \tan\theta - \theta - \tfrac{1}{4}i\tan^2\theta - i\log\cos\theta, \quad\dots\dots\dots\dots(6)$$

$$w = \tfrac{1}{4}\sec^2\theta. \quad\dots\dots\dots\dots\dots\dots(7)$$

If we put $\tan\theta = \xi + i\eta$, we get

$$w = \tfrac{1}{4}(1 + \xi^2 - \eta^2 + 2i\xi\eta),$$

so that $\qquad \phi = \tfrac{1}{4}(1 + \xi^2 - \eta^2), \qquad \psi = \tfrac{1}{2}\xi\eta. \quad\dots\dots\dots(8)$

We find further (Love),

$$z = \xi + i\eta + \tfrac{1}{2}\xi\eta - \frac{i}{4}(\xi^2 - \eta^2) - \tfrac{1}{2}\tan^{-1}\frac{2\xi}{1 - \xi^2 - \eta^2} - \tfrac{1}{2}\tan^{-1}\frac{2\xi\eta}{1 + \xi^2 - \eta^2}$$

$$+ \frac{i}{2}\log\{(1 - \eta)^2 + \xi^2\} \quad\dots\dots(9)$$

so that $\qquad x = \xi + \psi + \tfrac{1}{2}\tan^{-1}\dfrac{2\xi}{\eta^2 + \xi^2 - 1} + \tfrac{1}{2}\tan^{-1}\dfrac{4\psi}{\eta^2 - \xi^2 - 1}, \quad\dots\dots(10)$

$$y = \eta - \tfrac{1}{4}(\xi^2 - \eta^2) + \tfrac{1}{2}\log\{(1 - \eta)^2 + \xi^2\}. \quad\dots\dots\dots\dots(11)$$

The stream-lines, corresponding to a constant ψ, may be plotted from (10), (11), if we substitute $2\psi/\xi$ for η and regard ξ as the variable parameter. Since by (8)

$$\phi = \tfrac{1}{4}(1 + \xi^2) - \psi^2/\xi^2, \qquad d\phi/d\xi = \tfrac{1}{2}\xi + 2\psi^2/\xi^3,$$

there is no occasion to consider negative values of ξ, and ϕ and ξ vary always in the same direction.

As regards the fractions under the sign of $\tan^{-1}$, we see that both vanish when $\xi = 0$, and also when $\xi = \infty$. The former, viz., $2\xi \div (4\psi^2/\xi^2 + \xi^2 - 1)$, at first $+$ when ξ is very small, rises to ∞ when $\xi^2 = \tfrac{1}{2}\{1 \pm \sqrt{(1 - 16\psi^2)}\}$, which happens when $\psi < \tfrac{1}{4}$, but not otherwise. In the latter case the fraction is always positive. When $\psi < \tfrac{1}{4}$, the fraction passes through ∞, there changing sign. The numerically least negative value is reached when $\xi^2 = \tfrac{1}{2}\{\sqrt{(1 + 48\psi^2)} - 1\}$. The fraction then retraces its entire course, until it becomes zero again when $\xi = \infty$. On the other hand the second fraction, at first positive, rises to infinity in all cases when $\xi^2 = \tfrac{1}{2}\{\sqrt{(1 + 16\psi^2)} - 1\}$, after which it becomes negative and decreases numerically to zero, no part of its course being retraced. As regards the ambiguities in the resulting angles, it will suffice to suppose both angles to start from zero with ξ. This choice amounts to taking the origin of x at O, instead of O'.

When ψ is very small the march of the functions is peculiar. The first fraction becomes infinite when $\xi^2 = 4\psi^2$, that is when ξ is still small. The turn occurs when $\xi^2 = 12\psi^2$, and the corresponding least negative value is also small. The first $\tan^{-1}$ thus passes from 0 to π while ξ is still small. The second fraction also becomes infinite when $\xi^2 = 4\psi^2$, there changing sign, and again approaches zero while ξ is of the same order of magnitude.

The second $\tan^{-1}$ thus passes from 0 to π, thereby completing its course, while ξ is still small.

When $\psi = 0$ absolutely, either ξ or η, or both, must vanish, but we must still have regard to the relative values of ψ and ξ. Thus when ξ is small *enough*, $x = 0$, and this part of the stream-line coincides with the axis of symmetry. But while ξ is still small, x changes from 0 to π, the new value representing the inner face of the wall. The transition occurs when $\xi = 2\psi$, $\eta = 1$, making in (11) $y = -\infty$. The point O' at the edge of the wall $(x = \pi, y = 0)$ corresponds to $\xi = 0$, $\eta = 0$.

For the free part of the stream-line we may put $\eta = 0$, so that

$$x = \xi + \tfrac{1}{2}\tan^{-1}\frac{2\xi}{\xi^2 - 1} + \frac{\pi}{2} = \xi - \tan^{-1}\xi + \pi, \quad \dots\dots\dots\dots(12)$$

where $\tan^{-1}\xi$ is to be taken between 0 and $\tfrac{1}{2}\pi$. Also

$$y = -\tfrac{1}{4}\xi^2 + \tfrac{1}{2}\log(1 + \xi^2). \quad \dots\dots\dots\dots\dots(13)$$

When ξ is very great,

$$x = \xi + \tfrac{1}{2}\pi, \qquad y = -\tfrac{1}{4}\xi^2, \quad \dots\dots\dots\dots\dots(14)$$

and the curve approximates to a parabola.

When ξ is small,

$$x - \pi = \tfrac{1}{3}\xi^3, \qquad y = \tfrac{1}{4}\xi^2, \quad \dots\dots\dots\dots\dots(15)$$

so that the ratio $(x - \pi)/y$ starts from zero, as was to be expected.

The upward movement of y is of but short duration. It may be observed that, while $dx/d\xi$ is always positive,

$$\frac{dy}{d\xi} = \frac{\xi(1 - \xi^2)}{2(1 + \xi^2)}, \quad \dots\dots\dots\dots\dots(16)$$

which is positive only so long as $\xi < 1$. And when $\xi = 1$,

$$x - \pi = 1 - \tfrac{1}{4}\pi = 0{\cdot}2146, \qquad y = -\tfrac{1}{4} + \log 2 = 0{\cdot}097.$$

Some values of x and y calculated from (12), (13) are given in Table II and the corresponding curve is shown in fig. 3.

TABLE II.—$\psi = 0$.

ξ	x	y	ξ	x	y
0·0	3·142	0	2·5	4·451	− 0·571
0·5	3·178	+0·050	3·0	4·892	− 1·098
1·0	3·356	+0·097	4·0	5·816	− 2·583
1·5	3·659	+0·027	5·0	6·768	− 4·62
2·0	4·034	−0·195	20·0	21·621	− 97·00

It is easy to verify that the velocity is constant along the curve defined by (12), (13). We have

$$\frac{dx}{d\phi} = \frac{\xi^2}{1+\xi^2}\frac{d\xi}{d\phi}, \qquad \frac{dy}{d\phi} = \frac{\xi}{2}\frac{1-\xi^2}{1+\xi^2}\frac{d\xi}{d\phi};$$

and when $\psi = 0$,

$$\phi = \tfrac14(1+\xi^2), \qquad d\phi/d\xi = \tfrac12\xi.$$

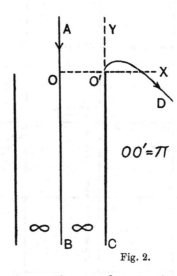

$$OO' = \pi$$

Fig. 2.

Thus

$$\frac{dx}{d\phi} = \frac{2\xi}{1+\xi^2}, \qquad \frac{dy}{d\phi} = \frac{1-\xi^2}{1+\xi^2},$$

and

$$(dx/d\phi)^2 + (dy/d\phi)^2 = 1. \quad\quad\quad\quad\dots\dots\dots\dots\dots(17)$$

The square root of the expression on the left of (17) represents the reciprocal of the resultant velocity.

TABLE III.—$\psi = \tfrac{1}{10}$.

ξ	x	y	ξ	x	y
0	0	∞	0·40	2·9667	+0·076
0·05	0·1667	9·098	0·50	3·0467	0·130
0·10	0·2995	3·008	0·60	3·1089	0·162
0·13	0·4668	1·535	0·80	3·2239	0·198
0·15	0·6725	0·766	1·00	3·3454	0·207
0·17	1·0368	+0·109	1·50	3·6947	+0·125
0·18	1·2977	−0·143	2·00	4·0936	−0·112
0·19	1·5907	−0·304	2·50	4·5234	−0·501
0·20	1·8708	−0·370	3·00	4·9725	−1·032
0·22	2·2828	−0·331	4·00	5·9039	−2·536
0·25	2·5954	−0·195	6·00	7·8305	−7·161
0·30	2·8036	−0·047			

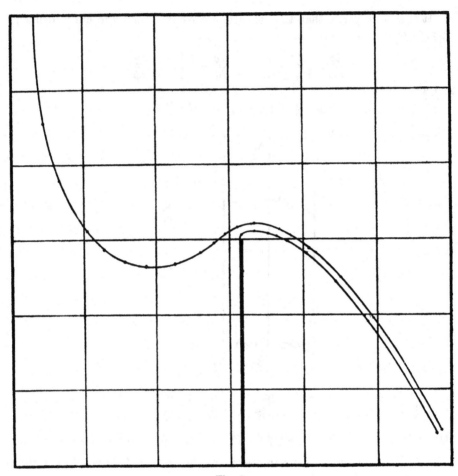

Fig. 3.

When ψ differs from zero, the calculations are naturally more complicated. The most interesting and instructive cases occur when ψ is small. I have chosen $\psi = 1/10$. The corresponding values of ξ, x, and y are given in Table III, calculated from equations (10), (11), and a plot is shown in fig. 3.

As in the former problem, where the liquid is supposed to adhere to the walls notwithstanding the sharp edges, the pressure in the recesses of the tube is simply that due to the velocity at a distance. At other places the pressure can be deduced from the stream-function in the usual way.

397.

ON THE CHARACTER OF THE "S" SOUND.

[*Nature*, Vol. xcv. pp. 645, 646, 1915.]

SOME two years ago I asked for suggestions as to the formation of an artificial hiss, and I remarked that the best I had then been able to do was by blowing through a rubber tube nipped at about half an inch from the open end with a screw clamp, but that the sound so obtained was perhaps more like an *f* than an *s*. "There is reason to think that the ear, at any rate of elderly people, tires rapidly to a maintained hiss. The pitch is of the order of 10,000 per second[*]." The last remark was founded upon experiments already briefly described[†] under the head "Pitch of Sibilants."

"Doubtless this may vary over a considerable range. In my experiments the method was that of nodes and loops (*Phil. Mag.* Vol. VII. p. 149 (1879); *Scientific Papers*, Vol. I. p. 406), executed with a sensitive flame and sliding reflector. A hiss given by Mr Enock, which to me seemed very high and not over audible, gave a wave-length (λ) equal to 25 mm., with good agreement on repetition. A hiss which I gave was graver and less definite, corresponding to $\lambda = 32$ mm. The frequency would be of the order of 10,000 per second, more than 5 octaves above middle C."

Among the replies, publicly or privately given, with which I was favoured, was one from Prof. E. B. Titchener, of Cornell University[‡], who wrote :

"Lord Rayleigh's sound more like an *f* than an *s* is due, according to Köhler's observations, to a slightly too high pitch. A Galton whistle, set for a tone of 8400 v.d., will give a pure *s*."

It was partly in connexion with this that I remarked later[§] that I doubted whether any pure tone gives the full impression of an *s*, having often experimented with bird-calls of about the right pitch. In my published papers I

---

[*] *Nature,* Vol. xci. p. 319, 1913.
[†] *Phil. Mag.* Vol. xvi. p. 235, 1908 ; *Scientific Papers*, Vol. v. p. 486.
[‡] *Nature*, Vol. xci. p. 451, 1913.
[§] *Nature*, Vol. xci. p. 558, 1913.

find references to wave-lengths 31·2 mm., 1·304 in. = 33·1 mm., 1·28 in. = 32·5 mm.*
It is true that these are of a pitch too high for Köhler's optimum, which at
ordinary temperatures corresponds to a wave-length of 40·6 mm., or 1·60 inches;
but they agree pretty well with the pitch found for actual hisses in my observations with Enock.

Prof. Titchener has lately returned to the subject. In a communication
to the American Philosophical Society† he writes :

"It occurred to me that the question might be put to the test of experiment.
The sound of a Galton's whistle set for 8400 v.d. might be imitated by the
mouth, and a series of observations might be taken upon material composed
partly of the natural (mouth) sounds and partly of the artificial (whistle) tones.
If a listening observer were unable to distinguish between the two stimuli,
and if the mouth sound were shown, phonetically, to be a true hiss, then it
would be proved that the whistle also gives an s, and Lord Rayleigh would
be answered.

"The experiment was more troublesome than I had anticipated; but I may
say at once that it has been carried out, and with affirmative result."

A whistle of Edelmann's pattern (symmetrical, like a steam whistle) was
used, actuated by a rubber bulb ; and it appears clear that a practised operator
was able to imitate the whistle so successfully that the observer could not say
with any certainty which was which. More doubt may be felt as to whether
the sound was really a fully developed hiss. Reliance seems to have been
placed almost exclusively upon the position of the lips and tongue of the
operator. I confess I should prefer the opinion of unsophisticated observers
judging of the result simply by ear. The only evidence of this kind mentioned
is in a footnote (p. 328): "Mr Stephens' use of the word 'hiss' was spontaneous, not due to suggestion." I have noticed that sometimes a hiss passes
momentarily into what may almost be described as a whistle, but I do not
think this can be regarded as a normal s.

Since reading Prof. Titchener's paper I have made further experiments
with results that I propose to describe. The pitch of the sounds was determined by the sensitive flame and sliding reflector method, which is abundantly
sensitive for the purpose. The reflector is gradually drawn back from the
burner, and the positions noted in which the flame is unaffected. This phase
occurs when the burner occupies a *node* of the stationary waves. It is a place
where there is no to and fro *motion*. The places of recovery are thus at
distances from the reflector which are (odd or even) multiples of the half
wave-length. The reflector was usually drawn back until there had been five

---

* *Scientific Papers*, Vol. I. p. 407; Vol. II. p. 100.
† *Proceedings*, Vol. LIII. August—December, 1914, p. 323.

recoveries, indicating that the distance from the burner was now $5 \times \frac{1}{2}\lambda$, and this distance was then measured.

The first observations were upon a whistle on Edelmann's pattern of my own construction. The flame and reflector gave $\lambda = 1·7$ in., about a semi-tone flat on Köhler's optimum. As regards the character of the sound, it seemed to me and others to bear some resemblance to an s, but still to be lacking in something essential. I should say that since my own hearing for s's is now distinctly bad, I have always confirmed my opinion by that of other listeners whose hearing is good. That there should be some resemblance to an s at a pitch which is certainly the predominant pitch of an s is not surprising; and it is difficult to describe exactly in what the deficiency consisted. My own impression was that the sound was too nearly a pure tone, and that if it had been quite a pure tone the resemblance to an s would have been less. In subsequent observations the pitch was raised through $\lambda = 1·6$ in., but without modifying the above impressions.

Wishing to try other sources which I thought more likely to give pure tones, I fell back on bird-calls. A new one, with adjustable distance between the perforated plates, gave on different trials $\lambda = 1·8$ in., $\lambda = 1·6$ in. In neither case was the sound judged to be at all a proper s, though perhaps some resemblance remained. The effect was simply that of a high note, like the squeak of a bird or insect. Further trials on another day gave confirmatory results.

The next observations were made with the highest pipe from an organ, gradually raised in pitch by cutting away at the open end. There was some difficulty in getting quite high enough, but measures were taken giving $\lambda = 2·2$ in., $\lambda = 1·9$ in., and eventually $\lambda = 1·6$ in. In no case was there more than the slightest suggestion of an s.

As I was not satisfied that at the highest pitch the organ-pipe was speaking properly, I made another from lead tube, which could be blown from an *adjustable* wind nozzle. Tuned to give $\lambda = 1·6$ in., it sounded faint to my ear, and conveyed no s. Other observers, who heard it well, said it was no s.

In all these experiments the sounds were *maintained*, the various instruments being blown from a loaded bag, charged beforehand with a foot blower. In this respect they are not fully comparable with those of Prof. Titchener, whose whistle was actuated by squeezing a rubber bulb. However, I have also tried a glass tube, 10·4 in. long, supported at the middle and rubbed with a resined leather. This should be of the right pitch, but the squeak heard did not suggest an s. I ought perhaps to add that the thing did not work particularly well,

It will be seen that my conclusions differ a good deal from those of Prof. Titchener, but since these estimates depend upon individual judgment, perhaps not uninfluenced by prepossessions, they are not fully satisfactory. Further independent aural observations are desirable. I fear a record, or ocular observation, of vibrations at so high a pitch is hardly feasible.

I may perhaps be asked if a characteristic s, having a dominant pitch, is not a pure tone, what is it? I am disposed to think that the vibration is irregular. A fairly defined pitch does not necessitate regular sequences of more than a few (say 3—10) vibrations. What is the state of affairs in an organ-pipe which does not speak well, or in a violin string badly bowed? An example more amenable to observation is afforded by the procession of drops into which a liquid jet breaks up. If the jet is well protected from outside influences, the procession is irregular, and yet there is a dominant interval between consecutive drops, giving rise under suitable conditions to a sound having a dominant pitch. Vibrations of this sort deserve more attention than they have received. In the case of the s the pitch is so high that there would be opportunity for interruptions so frequent that they would not be separately audible, and yet not so many as to preclude a fairly defined dominant pitch. I have an impression, too, that the s includes subordinate components decidedly graver than the dominant pitch.

Similar questions naturally arise over the character of the sh, f, and th sounds.

398.

ON THE STABILITY OF THE SIMPLE SHEARING MOTION OF A VISCOUS INCOMPRESSIBLE FLUID.

[*Philosophical Magazine*, Vol. xxx. pp. 329—338, 1915.]

A PRECISE formulation of the problem for free infinitesimal disturbances was made by Orr (1907)*. It is supposed that ζ (the vorticity) and v (the velocity perpendicular to the walls) are proportional to $e^{int} e^{ikx}$, where $n = p + iq$. If $\nabla^2 v = S$, we have

$$\frac{d^2 S}{dy^2} = \left\{ k^2 - \frac{q}{\nu} + \frac{i}{\nu} (p + k\beta y) \right\} S, \quad \dots\dots\dots\dots\dots(1)$$

and
$$d^2 v / dy^2 - k^2 v = S, \quad \dots\dots\dots\dots\dots\dots\dots\dots(2)$$

with the boundary conditions that $v = 0$, $dv/dy = 0$ at the walls where y is constant. Here ν is the kinematic viscosity, and β is proportional to the initial constant vorticity. Orr easily shows that the period-equation takes the form

$$\int S_1 e^{ky} dy \cdot \int S_2 e^{-ky} dy - \int S_1 e^{-ky} dy \cdot \int S_2 e^{ky} dy = 0, \quad \dots\dots\dots(3)$$

where S_1, S_2 are any two independent solutions of (1) and the integrations are extended over the interval between the walls. An equivalent equation was given a little later (1908) independently by Sommerfeld.

Stability requires that for no value of k shall any of the q's determined by (3) be negative. In his discussion Orr arrives at the conclusion that this condition is satisfied. Another of Orr's results may be mentioned. He shows that $p + k\beta y$ necessarily changes sign in the interval between the walls†.

In the paper quoted reference was made also to the work of v. Mises and Hopf, and it was suggested that the problem might be simplified if it could be shown that $q - \nu k^2$ cannot vanish. If so, it will follow that q is always

---

* *Proc. Roy. Irish Acad.* Vol. xxvii.

† *Phil. Mag.* Vol. xxviii. p. 618 (1914).

positive and indeed greater than νk^2, inasmuch as this is certainly the case when $\beta = 0$*. The assumption that $q = \nu k^2$, by which the real part of the { } in (1) disappears, is indeed a considerable simplification, but my hope that it would lead to an easy solution of the stability problem has been disappointed. Nevertheless, a certain amount of progress has been made which it may be desirable to record, especially as the preliminary results may have other applications.

If we take a real η such that

$$p + k\beta y = -(9\nu k^2 \beta^2)^{\frac{1}{3}}\, \eta, \dots\dots\dots\dots\dots(4)$$

we obtain

$$\frac{d^2 S}{d\eta^2} = -9i\eta S. \dots\dots\dots\dots\dots(5)$$

This is the equation discussed by Stokes in several papers†, if we take x in his equation (18) to be the pure imaginary $i\eta$.

The boundary equation (3) retains the same form with $e^{\lambda \eta}\, d\eta$ for $e^{ky}\, dy$, where

$$\lambda^3 = 9\nu k^2/\beta. \dots\dots\dots\dots\dots(6)$$

In (5), (6) η and λ are non-dimensional.

Stokes exhibits the general solution of the equation

$$\frac{d^2 S}{dx^2} - 9xS = 0 \dots\dots\dots\dots\dots(7)$$

in two forms. In ascending series which are always convergent,

$$S = A\left\{1 + \frac{9x^3}{2.3} + \frac{9^2 x^6}{2.3.5.6} + \frac{9^3 x^9}{2.3.5.6.8.9} + \dots\right\}$$

$$+ B\left\{x + \frac{9x^4}{3.4} + \frac{9^2 x^7}{3.4.6.7} + \frac{9^3 x^{10}}{3.4.6.7.9.10} + \dots\right\}. \dots\dots(8)$$

The alternative semi-convergent form, suitable for calculation when x is large, is

$$S = Cx^{-\frac{1}{4}} e^{-2x^{\frac{3}{2}}}\left\{1 - \frac{1.5}{1.144x^{\frac{3}{2}}} + \frac{1.5.7.11}{1.2.144^2 x^3} - \frac{1.5.7.11.13.17}{1.2.3.144^3 x^{\frac{9}{2}}} + \dots\right\}$$

$$+ Dx^{-\frac{1}{4}} e^{2x^{\frac{3}{2}}}\left\{1 + \frac{1.5}{1.144x^{\frac{3}{2}}} + \frac{1.5.7.11}{1.2.144^2 x^3} + \frac{1.5.7.11.13.17}{1.2.3.144^3 x^{\frac{9}{2}}} + \dots\right\}, \dots(9)$$

in which, however, the constants C and D are liable to a discontinuity. When x is real—the case in which Stokes was mainly interested—or a pure imaginary, the calculations are of course simplified.

---

* *Phil. Mag.* Vol. xxxiv. p. 69 (1892); *Scientific Papers*, Vol. iii. p. 583.

† Especially *Camb. Phil. Trans.* Vol. x: p. 106 (1857); *Collected Papers*, Vol. iv. p. 77.

If we take as S_1 and S_2 the two series in (8), the real and imaginary parts of each are readily separated. Thus if

$$S_1 = s_1 + it_1, \qquad S_2 = s_2 + it_2, \quad \ldots\ldots\ldots\ldots\ldots\ldots(10)$$

we have on introduction of $i\eta$

$$s_1 = 1 - \frac{9^2\eta^6}{2.3.5.6} + \frac{9^4\eta^{12}}{2.3.5.6.8.9.11.12} - \ldots, \quad \ldots\ldots(11)$$

$$t_1 = -\frac{9\eta^3}{2.3} + \frac{9^3\eta^9}{2.3.5.6.8.9} - \ldots, \quad \ldots\ldots\ldots\ldots\ldots\ldots(12)$$

$$s_2 = \frac{9\eta^4}{3.4} - \frac{9^3\eta^{10}}{3.4.6.7.9.10} + \ldots, \quad \ldots\ldots\ldots\ldots\ldots\ldots(13)$$

$$t_2 = \eta - \frac{9^2\eta^7}{3.4.6.7} + \frac{9^4\eta^{13}}{3.4.6.7.9.10.12.13} - \ldots, \quad \ldots\ldots(14)$$

in which it will be seen that s_1, s_2 are even in η, while t_1, t_2 are odd.

When $\eta < 2$, these ascending series are suitable. When $\eta > 2$, it is better to use the descending series, but for this purpose it is necessary to know the connexion between the constants A, B and C, D. For $x = i\eta$ these are (Stokes)

$$A = \pi^{-\frac{1}{2}}\Gamma(\tfrac{1}{3})\{C + De^{-i\pi/6}\}, \qquad B = 3\pi^{-\frac{1}{2}}\Gamma(\tfrac{2}{3})\{-C + De^{i\pi/6}\}. \ldots(15)$$

Thus for the first series S_1 ($A = 1$, $B = 0$ in (8))

$$\log D = \bar{1}{\cdot}5820516, \qquad C = De^{i\pi/6}; \ldots\ldots\ldots\ldots(16)$$

and for S_2 ($A = 0$, $B = 1$)

$$\log D' = \bar{1}{\cdot}4012366, \qquad -C' = D'e^{-i\pi/6}, \ldots\ldots\ldots(17)$$

so that if the two functions in (9) be called Σ_1 and Σ_2,

$$S_1 = C\Sigma_1 + D\Sigma_2, \qquad S_2 = C'\Sigma_1 + D'\Sigma_2. \ldots\ldots\ldots\ldots(18)$$

These values may be confirmed by a comparison of results calculated first from the ascending series and secondly from the descending series when $\eta = 2$. Much of the necessary arithmetic has been given already by Stokes[*]. Thus from the ascending series

$$s_1(2) = -13{\cdot}33010, \qquad t_1(2) = 11{\cdot}62838;$$
$$s_2(2) = -2{\cdot}25237, \qquad t_2(2) = -11{\cdot}44664.$$

In calculating from the descending series the more important part is Σ_1, since

$$e^{-2x^{\frac{3}{2}}} = e^{-2i^{\frac{3}{2}}\eta^{\frac{3}{2}}} = e^{\sqrt{2}\,.\,\eta^{\frac{3}{2}}(1-i)}.$$

For $\eta = 2$ Stokes finds

$$\Sigma_1 = -14{\cdot}98520 + 43{\cdot}81046i,$$

of which the log. modulus is $1{\cdot}6656036$, and the phase $+108° 52' 58''{\cdot}99$. When the multiplier C or C' is introduced, there will be an addition of $\pm 30°$ to this phase. Towards the value of S_1 I find

$$-13{\cdot}32487 + 11{\cdot}63096\,i;$$

---

[*] *Loc. cit.* Appendix. It was to take advantage of this that the "9" was introduced in (5).

and towards that of S_2

$$- 2 \cdot 24892 - 11 \cdot 44495 \, i.$$

For the other part involving D or D' we get in like manner

$$- \cdot 00523 - \cdot 00258 \, i,$$

and

$$- \cdot 00345 - \cdot 00170 \, i.$$

TABLE I.

η	s_1	t_1	s_2	t_2
0·0	+ 1·0000	− ·0000	+ ·0000	+ ·0000
0·1	+ 1·0000	− ·0015	+ ·0001	+ ·1000
0·2	+ 1·0000	− ·0120	+ ·0012	+ ·2000
0·3	+ ·9997	− ·0405	+ ·0061	+ ·3000
0·4	+ ·9982	− ·0960	+ ·0192	+ ·3997
0·5	+ ·9930	− ·1874	+ ·0469	+ ·4987
0·6	+ ·9790	·− ·3234	+ ·0971	+ ·5955
0·7	+ ·9393	− ·5485	+ ·1969	+ ·6845
0·8	+ ·8825	− ·7605	+ ·3055	+ ·7663
0·9	+ ·7619	− 1·0717	+ ·4865	+ ·8234
1·0	+ ·554	− 1·444	+ ·734	+ ·840
1·1	+ ·215	− 2·007	+ 1·057	+ ·790
1·2	− ·310	− 2·304	+ 1·456	+ ·634
1·3	− 1·083	− 2·707	+ 1·923	+ ·320
1·4	− 2·173	− 2·979	+ 2·424	− ·221
1·5	− 3·635	− 2·972	+ 2·893	− 1·067
1·6	− 5·493	− 2·466	+ 3·212	− 2·303
1·7	− 7·694	− 1·161	+ 3·191	− 3·998
1·8	− 10·057	+ 1·325	+ 2·550	− 6·173
1·9	− 12·177	+ 5·441	+ ·899	− 8·745
2·0	− 13·330	+ 11·628	− 2·252	− 11·447
2·1	− 12·34	+ 20·19	− 7·46	− 13·70
2·2	− 7·49	+ 31·01	− 15·24	− 14·50
2·3	+ 3·54	+ 43·20	− 25·84	− 12·22
2·4	+ 23·55	+ 54·54	− 38·90	− 4·53
2·5	+ 55·20	+ 60·44	− 52·70	+ 11·59

It appears that with the values of C, D, C', D' defined by (16), (17) the calculations from the ascending and descending series lead to the same results when $\eta = 2$. What is more, and it is for this reason principally that I have detailed the numbers, the second part involving Σ_2 loses its importance when η exceeds 2. Beyond this point the numbers given in the table are calculated from Σ_1 only. Thus $(\eta > 2)$

$$s_1 + i t_1 = D \eta^{-\frac{1}{4}} e^{\sqrt{2} \, . \, \eta^{\frac{3}{2}}} e^{- i \left(\sqrt{2} \, . \, \eta^{\frac{3}{2}} + \pi/8 - \pi/6 \right)}$$

$$\times \left\{ 1 - \frac{1.5}{1 \cdot 144 \, (i\eta)^{\frac{3}{2}}} + \frac{1.5.7.11}{1 \cdot 2 \cdot 144^2 \, (i\eta)^3} - \ldots \right\}, \quad \ldots\ldots\ldots (19)$$

$$s_2 + i t_2 = - D' \eta^{-\frac{1}{4}} e^{\sqrt{2} \, . \, \eta^{\frac{3}{2}}} e^{- i \left(\sqrt{2} \, . \, \eta^{\frac{3}{2}} + \pi/8 + \pi/6 \right)}$$

$$\times \left\{ 1 - \frac{1.5}{1 \cdot 144 \, (i\eta)^{\frac{3}{2}}} + \frac{1.5.7.11}{1 \cdot 2 \cdot 144^2 \, (i\eta)^3} - \ldots \right\}, \quad \ldots\ldots\ldots (20)$$

the only difference being the change from D to $-D'$ and the reversal of sign in $\pi/6$, equivalent to the introduction of a constant (complex) factor.

When η exceeds $2\cdot5$, the second term of the series within $\{\ \}$ in Σ_1 is less than 10^{-2}, so that for rough purposes the $\{\ \}$ may be omitted altogether. We then have

$$s_1 = D\eta^{-\frac{1}{4}} e^{\sqrt{2}\cdot\eta^{\frac{3}{2}}} \cos(\sqrt{2}\cdot\eta^{\frac{3}{2}} - \pi/24), \quad\dots\dots\dots\dots(21)$$

$$t_1 = -D\eta^{-\frac{1}{4}} e^{\sqrt{2}\cdot\eta^{\frac{3}{2}}} \sin(\sqrt{2}\cdot\eta^{\frac{3}{2}} - \pi/24), \quad\dots\dots\dots\dots(22)$$

$$s_2 = D'\eta^{-\frac{1}{4}} e^{\sqrt{2}\cdot\eta^{\frac{3}{2}}} \sin(\sqrt{2}\cdot\eta^{\frac{3}{2}} - \pi/24 - \pi/6), \quad\dots\dots(23)$$

$$t_2 = D'\eta^{-\frac{1}{4}} e^{\sqrt{2}\cdot\eta^{\frac{3}{2}}} \cos(\sqrt{2}\cdot\eta^{\frac{3}{2}} - \pi/24 - \pi/6). \quad\dots\dots(24)$$

Here D and D' are both positive—the logarithms have already been given—and we see that s_1, t_2 are somewhat approximately in the same phase, and t_1, s_2 in approximately opposite phases. When η exceeds a small integer, the functions fluctuate with great rapidity and with correspondingly increasing maxima and minima. When in one period $\sqrt{2}\cdot\eta^{\frac{3}{2}}$ increases by 2π, the exponential factor is multiplied by $e^{2\pi}$, viz. $535\cdot4$. From the approximate expressions applicable when η exceeds a small integer it appears that s_1, t_1 are in quadrature, as also s_2, t_2.

For some purposes it may be more convenient to take Σ_1, Σ_2, or (expressed more correctly) the functions which identify themselves with Σ_1, Σ_2 when η is great, rather than S_1, S_2, as fundamental solutions. When η is small, these functions must be calculated from the ascending series. Thus by (15) $(C=1, D=0)$

$$\Sigma_1 = \pi^{-\frac{1}{2}}\Gamma(\tfrac{1}{3})S_1 - 3\pi^{-\frac{1}{2}}\Gamma(\tfrac{2}{3})S_2, \quad\dots\dots\dots\dots(25)$$

and $(C=0, D=1)$

$$\Sigma_2 = \pi^{-\frac{1}{2}}\Gamma(\tfrac{1}{3})e^{-i\pi/6}S_1 + 3\pi^{-\frac{1}{2}}\Gamma(\tfrac{2}{3})e^{i\pi/6}S_2. \dots\dots\dots(26)$$

Some general properties of the solutions of (5) are worthy of notice. If $S = s + it$, we have

$$d^2s/d\eta^2 = 9\eta t, \qquad d^2t/d\eta^2 = -9\eta s.$$

Let $R = \frac{1}{2}(s^2 + t^2)$; then

$$\frac{dR}{d\eta} = s\frac{ds}{d\eta} + t\frac{dt}{d\eta},$$

and

$$\frac{d^2R}{d\eta^2} = \left(\frac{ds}{d\eta}\right)^2 + \left(\frac{dt}{d\eta}\right)^2 + s\frac{d^2s}{d\eta^2} + t\frac{d^2t}{d\eta^2},$$

of which the two last terms cancel, so that $d^2R/d\eta^2$ is always positive. In the case of S_1, when $\eta = 0$, $s_1(0) = 1$, $t_1(0) = 0$, $s_1'(0) = 0$, so that $R(0) = \frac{1}{2}$, $R'(0) = 0$. Again, when $\eta = 0$, $s_2(0) = 0$, $t_2(0) = 0$, so that $R(0) = 0$, $R'(0) = 0$. In neither case can R vanish for a finite (real) value of η, and the same is true of S_1 and S_2.

Since (5) is a differential equation of the second order, its solutions are connected in a well-known manner. Thus

$$S_1 \frac{d^2 S_2}{d\eta^2} - S_2 \frac{d^2 S_1}{d\eta^2} = 0, \quad \dots \dots \dots \dots \dots (27)$$

and on integration

$$S_1 \frac{dS_2}{d\eta} - S_2 \frac{dS_1}{d\eta} = \text{constant} = i, \quad \dots \dots \dots \dots (28)$$

as appears from the value assumed when $\eta = 0$. Thus

$$\frac{S_2}{S_1} = i \int_0^\eta \frac{d\eta}{S_1^2}, \quad \dots \dots \dots \dots \dots (29)$$

which defines S_2 in terms of S_1.

A similar relation holds for any two particular solutions. For example,

$$\frac{\Sigma_2}{\Sigma_1} = 6i \int_\infty^\eta \frac{d\eta}{\Sigma_1^2}. \quad \dots \dots \dots \dots \dots (30)$$

The difficulty of the stability problem lies in the treatment of the boundary condition

$$\int_{\eta_1}^{\eta_2} S_1 e^{\lambda \eta} d\eta \cdot \int_{\eta_1}^{\eta_2} S_2 e^{-\lambda \eta} d\eta - \int_{\eta_1}^{\eta_2} S_1 e^{-\lambda \eta} d\eta \cdot \int_{\eta_1}^{\eta_2} S_2 e^{\lambda \eta} d\eta = 0, \dots (31)*$$

in which η_2, η_1, and λ are arbitrary, except that we may suppose η_2 and λ to be positive, and η_1 negative. In (31) we may replace $e^{\lambda \eta}$, $e^{-\lambda \eta}$ by $\cosh \lambda \eta$, $\sinh \lambda \eta$ respectively, and the substitution is especially useful when the limits of integration are such that $\eta_1 = -\eta_2$. For in this case

$$\int_{\eta_1}^{\eta_2} S \cosh \lambda \eta \, d\eta = 2 \int_0^{\eta_2} s \cosh \lambda \eta \, d\eta,$$

$$\int_{\eta_1}^{\eta_2} S \sinh \lambda \eta \, d\eta = 2i \int_0^{\eta_2} t \sinh \lambda \eta \, d\eta \,;$$

and the equation reduces to

$$\int_0^{\eta_2} s_1 \cosh \lambda \eta \, d\eta \cdot \int_0^{\eta_2} t_2 \sinh \lambda \eta \, d\eta$$

$$- \int_0^{\eta_2} s_2 \cosh \lambda \eta \, d\eta \cdot \int_0^{\eta_2} t_1 \sinh \lambda \eta \, d\eta = 0, \quad \dots \dots \dots (32)$$

thus assuming a real form, derived, however, from the imaginary term in (31). In general with separation of real and imaginary parts we have by (31) from the real part

$$\int s_1 e^{\lambda \eta} d\eta \cdot \int s_2 e^{-\lambda \eta} d\eta - \int t_1 e^{\lambda \eta} d\eta \cdot \int t_2 e^{-\lambda \eta} d\eta$$

$$- \int s_1 e^{-\lambda \eta} d\eta \cdot \int s_2 e^{\lambda \eta} d\eta + \int t_1 e^{-\lambda \eta} d\eta \cdot \int t_2 e^{\lambda \eta} d\eta = 0, \quad \dots \dots (33)$$

* Rather to my surprise I find this condition already laid down in private papers of Jan. 1893.

and from the imaginary part

$$\int s_1 e^{\lambda\eta}\, d\eta \cdot \int t_2 e^{-\lambda\eta}\, d\eta + \int s_2 e^{-\lambda\eta}\, d\eta \cdot \int t_1 e^{\lambda\eta}\, d\eta$$

$$-\int s_1 e^{-\lambda\eta}\, d\eta \cdot \int t_2 e^{\lambda\eta}\, d\eta - \int s_2 e^{\lambda\eta}\, d\eta \cdot \int t_1 e^{-\lambda\eta}\, d\eta = 0. \dots\dots(34)$$

If we introduce the notation of double integrals, these equations become

$$\iint \sinh \lambda\,(\eta - \eta') \{s_1(\eta) \cdot s_2(\eta') - t_1(\eta) \cdot t_2(\eta')\}\, d\eta\, d\eta' = 0, \ \dots\dots(35)$$

$$\iint \sinh \lambda\,(\eta - \eta') \{s_1(\eta) \cdot t_2(\eta') - s_2(\eta) \cdot t_1(\eta')\}\, d\eta\, d\eta' = 0, \ \dots\dots(36)$$

the limits for η and η' being in both cases η_1 and η_2. In these we see that the parts for which η and η' are nearly equal contribute little to the result.

A case admitting of comparatively simple treatment occurs when λ is so large that the exponential terms $e^{\lambda\eta}$, $e^{-\lambda\eta}$ dominate the integrals. As we may see by integration by parts, (31) then reduces to

$$S_1(\eta_2) \cdot S_2(\eta_1) - S_1(\eta_1) \cdot S_2(\eta_2) = 0, \ \dots\dots\dots\dots(37)$$

or with use of (29)

$$S_1(\eta_2) \cdot S_1(\eta_1) \cdot \int_{\eta_1}^{\eta_2} \frac{d\eta}{S_1^2(\eta)} = 0. \ \dots\dots\dots\dots(38)$$

We have already seen that $S_1(\eta)$ cannot vanish; and it only remains to prove that neither can the integral do so. Owing to the character of S_1, only moderate values of η contribute sensibly to its value. For further examination it conduces to clearness to write $\eta_2 = a$, $\eta_1 = -b$, where a and b are positive. Thus

$$\int_{\eta_1}^{\eta_2} \frac{d\eta}{S_1^2} = \int_0^a \frac{d\eta}{(s_1 + it_1)^2} + \int_0^b \frac{d\eta}{(s_1 - it_1)^2}$$

$$= \int_0^a \frac{(s_1^2 - t_1^2)\, d\eta}{(s_1^2 + t_1^2)^2} + \int_0^b \frac{(s_1^2 - t_1^2)\, d\eta}{(s_1^2 + t_1^2)^2} - 2i \int_0^a \frac{s_1 t_1\, d\eta}{(s_1^2 + t_1^2)^2} + 2i \int_0^b \frac{s_1 t_1\, d\eta}{(s_1^2 + t_1^2)^2}; \ (39)$$

and it suffices to show that $\displaystyle\int_0^a \frac{(s_1^2 - t_1^2)\, d\eta}{(s_1^2 + t_1^2)^2}$ cannot vanish. A short table makes this apparent [see p. 348].

The fifth column represents the sums up to various values of η. The approximate value of $\displaystyle\int_0^\infty \frac{(s_1^2 - t_1^2)\, d\eta}{(s_1^2 + t_1^2)^2}$ is thus $\cdot 2 \times 2\cdot 834$ or $\cdot 567$. The true value of this integral is $(D'/D) \sin 60°$ or $\cdot 571$, as we see from (30) and (19), (20).

We conclude that (37) cannot be satisfied with any values of η_2 and η_1.

When the value of λ is not sufficiently great to justify the substitution of (37) for (31) in the general case, we may still apply the argument in a rough manner to the special case $(\eta_2 + \eta_1 = 0)$ of (32), at any rate when η_2

is moderately great. For, although capable of evanescence, the functions s_1, t_1, s_2, t_2 increase in amplitude so rapidly with η that the extreme value of η may be said to dominate the integrals. The hyperbolic functions then disappear and the equation reduces* to

$$s_1(\eta_2) . t_2(\eta_2) - s_2(\eta_2) . t_1(\eta_2) = 0, \ldots\ldots\ldots\ldots\ldots(40)$$

TABLE II.

η	$s_1^2 - t_1^2$	$(s_1^2 + t_1^2)^2$	$\dfrac{s_1^2 - t_1^2}{(s_1^2 + t_1^2)^2}$	Sums of fourth column
·1	+ 1·000	1·000	+1·000	1·000
·3	+ 0·997	1·002	+ ·995	1·995
·5	+ 0·951	1·042	+ ·913	2·908
·7	+ 0·581	1·399	+ ·415	3·323
·9	− 0·569	2·989	− ·191	3·132
1·1	− 3·982	16·60	− ·240	2·892
1·3	− 6·155	72·25	− ·085	2·807
1·5	+ 4·38	485·8	+ ·009	2·816
1·7	+ 57·9	3660·0	+ ·016	2·832
1·9	+ 119·0	31700·0	+ ·004	2·836
2·1	− 255·0	314000·0	− ·001	2·835
2·3	− 1854·0	353×10^{-4}	− ·001	2·834
2·5	− 616·0	45×10^{-6}	− ·000	2·834

which cannot be satisfied by a moderately large value of η_2. For it appears from the appropriate expressions (21)...(24) that the left-hand member of (40) is then

$$DD' e^{2\sqrt{2} . \eta^{\frac{3}{2}}} \cos(\pi/6),$$

a positive and rapidly increasing quantity. Again, it is evident from Table I that the left-hand member of (32)' remains positive for all values of η_2 from zero up to some value which must exceed 1·1, since up to that point the functions s_1, s_2, t_2 are positive while t_1 is negative. Even without further examination it seems fairly safe to conclude that (32) cannot be satisfied by any values of η_2 and λ.

Another case admitting of simple treatment occurs when η_2 and η_1 are both small, although λ may be great. We have approximately

$$s_1 = 1, \quad t_1 = -\tfrac{3}{2}\eta^3, \quad s_2 = \tfrac{3}{4}\eta^4, \quad t_2 = \eta,$$

the next terms being in each case of 6 higher degrees in η. Thus with omission of terms in η^7 under the integral sign, (31) becomes

$$\int e^{\lambda\eta} d\eta . \int \eta\, e^{-\lambda\eta} d\eta - \int e^{-\lambda\eta} d\eta . \int \eta\, e^{\lambda\eta} d\eta = 0, \ldots\ldots\ldots\ldots(41)$$

---

* Regard being paid to the character of the functions. Needless to say, it is no general proposition that the value of an integral is determined by the greatest value, however excessive, of the integrand.

or on effecting the integrations

$$\lambda (\eta_2 - \eta_1) \sinh \lambda (\eta_2 - \eta_1) + 2 - 2 \cosh \lambda (\eta_2 - \eta_1) = 0. \quad \ldots\ldots(42)$$

It is easy to show that (42) cannot be satisfied. For, writing $\lambda (\eta_2 - \eta_1) = x$,

$$x \sinh x = x^2 + \frac{x^4}{2.3} + \frac{x^6}{2.3.4.5} + \cdots,$$

$$2 (\cosh x - 1) = x^2 + \frac{x^4}{3.4} + \frac{x^6}{3.4.5.6} + \cdots,$$

every term of the first series exceeding the corresponding term of the second series. The left-hand member of (42) is accordingly always positive. This disposes of the whole question when η_2 and η_1 are small enough (numerically), say distinctly less than unity.

399.

ON THE THEORY OF THE CAPILLARY TUBE.

[*Proceedings of the Royal Society*, A, Vol. XCII. pp. 184—195, Oct. 1915.]

A RECENT paper by Richards and Coombs[*] discusses in some detail the determination of surface-tension by the rise of the liquid in capillary tubes, and reflects mildly upon the inadequate assistance afforded by mathematics. It is true that no complete analytical solution of the problem can be obtained, even when the tube is accurately cylindrical. We may have recourse to graphical constructions, or to numerical calculations by the method of Runge[†], who took an example from this very problem. But for experimental purposes all that is really needed is a sufficiently approximate treatment of the two extreme cases of a narrow and of a wide tube. The former question was successfully attacked by Poisson, whose final formula [(18) below] would meet all ordinary requirements. Unfortunately doubts have been thrown upon the correctness of Poisson's results, especially by Mathieu[‡], who rejects them altogether in the only case of much importance, *i.e.* when the liquid wets the walls of the tube—a matter which will be further considered later on. Mathieu also reproaches Poisson's investigation as implying two different values of h, of which the second is really only an improvement upon the first, arising from a further approximation. It must be admitted, however, that the problem is a delicate one, and that Poisson's explanation at a critical point leaves something to be desired. In the investigation which follows I hope to have succeeded in carrying the approximation a stage beyond that reached by Poisson.

In the theory of narrow tubes the lower level from which the height of the meniscus is reckoned is the free plane level. In experiment, the lower level is usually that of the liquid in a wide tube connected below with the narrow one, and the question arises how wide this tube needs to be in order that the inner part of the meniscus may be nearly enough plane. Careful

[*] *Journ. Amer. Chem. Soc.* No. 7, July, 1915.

[†] *Math. Ann.* Vol. XLVI. p. 175 (1895).

[‡] *Théorie de la Capillarité*, Paris, 1883, pp. 46—49.

experiments by Richards and Coombs led to the conclusion that in the case of water the diameter of the wide tube should exceed 33 mm., and that probably 38 mm. suffices. Such smaller diameters as are often employed (20 mm.) involve very appreciable error. Here, again, we should naturally look to mathematics to supply the desired information. The case of a straight wall, making the problem two-dimensional, is easy *, but that of the circular wall is much more complicated.

Some drawings (from theory) given by Kelvin, figs. 24, 26, 28†, indicate clearly that diameters of 1·8 cm. and 2·6 cm. are quite inadequate. I have attempted below an analytical solution, based upon the assumption that the necessary diameter is large, as it will be, if the prescribed error at the axis is small enough. Although this assumption is scarcely justified in practice, the calculation indicates that a diameter of 4·7 cm. may not be too large.

As Richards and Coombs remark, the observed curvature of the lower part of the meniscus may be used as a test. Theory shows that there should be no sensible departure from straightness over a length of about 1 cm.

The Narrow Tube.

For the surface of liquid standing in a vertical tube of circular section, we have

$$x \sin \psi = \frac{x \, dz/dx}{\sqrt{\{1 + (dz/dx)^2\}}} = \frac{1}{a^2} \int_0^x zx \, dx, \quad \ldots\ldots\ldots\ldots(1)$$

in which z is the vertical co-ordinate measured upwards from the free plane level, x is the horizontal co-ordinate measured from the axis, ψ is the angle the tangent at any point makes with the horizontal, and $a^2 = T/g\rho$‡, where T is the surface-tension, g the acceleration of gravity, and ρ the density of the fluid. The equation expresses the equilibrium of the cylinder of liquid of radius x. At the wall, where $x = r$, ψ assumes a given value $(\frac{1}{2}\pi - i)$, and (1) becomes

$$a^2 r \cos i = \int_0^r zx \, dx. \quad \ldots\ldots\ldots\ldots\ldots\ldots(2)$$

If the radius (r) of the tube is *small*, the total curvature is nearly constant, that is, the surface is nearly spherical. We take

$$z = l - \sqrt{(c^2 - x^2)} + u, \quad \ldots\ldots\ldots\ldots\ldots\ldots(3)$$

where l is the height of the centre and c the radius of the sphere, while u represents the correction required for a closer approximation. If we omit u altogether, (2) gives

$$a^2 r \cos i = \frac{1}{2} l r^2 + \frac{1}{3} \{(c^2 - r^2)^{\frac{3}{2}} - c^3\}. \quad \ldots\ldots\ldots\ldots(4)$$

* Compare *Phil. Mag.* Vol. xxxiv. p. 309, Appendix, 1892; *Scientific Papers*, Vol. iv. p. 13.

† The reference is given below.

‡ It may be remarked that a^2 is sometimes taken to denote the double of the above quantity.

Also, if h be the height at the lowest point of the meniscus, the quantity directly measured in experiment,

$$h = l - c. \dots\dots\dots\dots\dots\dots\dots\dots\dots\dots(5)$$

In this approximation $r/c = \cos i$, and thus in terms of c

$$a^2 r^2/c = \tfrac{1}{2}r^2(h + c) + \tfrac{1}{3}(c^2 - r^2)^{\frac{3}{2}} - \tfrac{1}{3}c^3. \dots\dots\dots\dots(6)$$

When the angle of contact (i) is zero, $c = r$, and

$$a^2 = \tfrac{1}{2}r(h + \tfrac{1}{3}r), \dots\dots\dots\dots\dots\dots\dots(7)$$

the well-known formula.

When we include u, it becomes a question whether we should retain the value of c, i.e. $r \sec i$, appropriate when the surface is supposed to be exactly spherical. It appears, however, to be desirable, if not necessary, to leave the precise value of c open. Substituting the value of z from (3) in (1), we get, with neglect of $(du/dx)^3$,

$$1 + \frac{du}{dx}\frac{(c^2 - x^2)^{\frac{3}{2}}}{xc^2} - \frac{3(c^2 - x^2)^2}{2c^4}\left(\frac{du}{dx}\right)^2 = \frac{c}{a^2 x^2}\left[\frac{lx^2}{2} + \frac{(c^2 - x^2)^{\frac{3}{2}} - c^3}{3} + \int_0^x ux\,dx\right]. \dots(8)$$

For the purposes of the next approximation we may omit $(du/dx)^2$ and the integral, which is to be divided by a^2. Thus

$$\frac{du}{dx} = \left(\frac{cl}{2a^2} - 1\right)\frac{c^2 x}{(c^2 - x^2)^{\frac{3}{2}}} - \frac{c^6}{3a^2}\frac{1}{x(c^2 - x^2)^{\frac{3}{2}}} + \frac{c^3}{3a^2 x}, \dots\dots(9)$$

and on integration

$$u = c^2\left(\frac{cl}{2a^2} - \frac{c^2}{3a^2} - 1\right)\frac{1}{\sqrt{(c^2 - x^2)}} + \frac{c^3}{3a^2}\log\{c + \sqrt{(c^2 - x^2)}\} + C. \quad (10)$$

We suppose with Poisson and Mathieu that

$$\frac{cl}{2a^2} - \frac{c^2}{3a^2} - 1 = 0, \dots\dots\dots\dots\dots\dots(11)$$

so that

$$u = \frac{c^3}{3a^2}\log\{c + \sqrt{(c^2 - x^2)}\} + C, \dots\dots\dots\dots(12)$$

corresponding to

$$\frac{du}{dx} = \frac{c^3}{3a^2}\frac{\sqrt{(c^2 - x^2)} - c}{x\sqrt{(c^2 - x^2)}}. \dots\dots\dots\dots\dots(13)$$

To determine c we have the boundary condition

$$\cot i = \left(\frac{dz}{dx}\right)_{x=r} = \frac{r}{\sqrt{(c^2 - r^2)}} + \left(\frac{du}{dx}\right)_{x=r}$$

$$= \frac{r}{\sqrt{(c^2 - r^2)}}\left\{1 - \frac{c^3}{3a^2}\frac{c - \sqrt{(c^2 - r^2)}}{r^2}\right\}, \dots\dots\dots(14)$$

which gives c in terms of i and r. Explicitly

$$c = \frac{r}{\cos i} - \frac{r^3}{3a^2}\frac{\sin^2 i}{(1 + \sin i)\cos^3 i}. \dots\dots\dots\dots(15)$$

These latter equations are given by Mathieu.

We have now to find the value of a^2 to the corresponding approximation. For the observed height of the meniscus

$$h = l - c + u_{x=0} = l - c + C + \frac{c^3}{3a^2} \log(2c); \quad \ldots\ldots\ldots(16)$$

and

$$a^2 r \cos i = \int_0^r zx\,dx = \frac{r^2}{2}(l+C) + \frac{1}{3}\{(c^2 - r^2)^{\frac{3}{2}} - c^3\} + \int_0^r (u - C)\,x\,dx$$

$$= (h + c)\frac{r^2}{2} + \frac{1}{3}\{(c^2 - r^2)^{\frac{3}{2}} - c^3\}$$

$$+ \frac{c^3}{3a^2}\left[\frac{r^2}{2}\log\frac{c + \sqrt{(c^2 - r^2)}}{2c} + \frac{\{c + \sqrt{(c^2 - r^2)}\}^2}{4} - c\sqrt{(c^2 - r^2)}\right]. \quad (17)$$

In the important case where $i = 0$, the liquid *wetting* the walls of the tube, $c = r$ simply, and

$$a^2 = \frac{r}{2}\left(h + \frac{r}{3}\right) - \frac{r^4}{6a^2}\left(\log 2 - \frac{1}{2}\right)$$

$$= \frac{r}{2}\left\{h + \frac{r}{3} - \frac{2r^2}{3h}\left(\log 2 - \frac{1}{2}\right)\right\}$$

$$= \tfrac{1}{2}r\,(h + \tfrac{1}{3}r - 0{\cdot}1288\,r^2/h). \quad \ldots\ldots\ldots\ldots\ldots(18)$$

This is the formula given long since by Poisson[*], the only difference being that his a^2 is the double of the quantity here so denoted.

It is remarkable that Mathieu rejects the above equations as applicable to the case $i = 0$, $c = r$, on the ground that then du/dx in (13) becomes infinite when $x = r$. But $d\sqrt{(r^2 - x^2)}/dx$, with which du/dx comes into comparison, is infinite at the same time; and, in fact, both

$$\frac{du}{dx}(r^2 - x^2)^{\frac{3}{2}} \quad \text{and} \quad \left(\frac{du}{dx}\right)^2 (r^2 - x^2)^2,$$

in equation (8) *vanish* when $x = r$. It is this circumstance which really determines the choice of l in (11).

We may now proceed to a yet closer approximation, introducing approximate values of the terms previously neglected altogether. From (13)

$$(c^2 - x^2)^2\left(\frac{du}{dx}\right)^2 = \frac{c^6}{9a^4 x^2}\{c^2(c^2 - x^2) + (c^2 - x^2)^2 - 2c(c^2 - x^2)^{\frac{3}{2}}\},$$

and from (12)

$$\int_0^x ux\,dx = \tfrac{1}{2}Cx^2 + \frac{c^3}{6a^2}[x^2 \log\{c + \sqrt{(c^2 - x^2)}\} + \tfrac{1}{2}c^2 - c\sqrt{(c^2 - x^2)} + \tfrac{1}{2}(c^2 - x^2)].$$

[*] *Nouvelle Théorie de l'Action Capillaire*, 1831, p. 112.

Thus
$$\frac{du}{dx} = \left\{\frac{c\,(l+C)}{2a^2} - 1\right\}\frac{c^2x}{(c^2-x^2)^{\frac{3}{2}}} - \frac{c^6}{3a^2x\,(c^2-x^2)^{\frac{3}{2}}} + \frac{c^3}{3a^2x}$$

$$+ \frac{c^4}{6xa^4\,(c^2-x^2)^{\frac{3}{2}}}\left[\frac{3c^2\,(c^2-x^2)}{2} + (c^2-x^2)^2 - 2c\,(c^2-x^2)^{\frac{3}{2}}\right.$$

$$\left. + c^2x^2\log\{c + \sqrt{(c^2-x^2)}\} + \tfrac{1}{2}c^4 - c^3\sqrt{(c^2-x^2)}\right], \quad \ldots\ldots(19)$$

$$u = \left\{\frac{c\,(l+C)}{2a^2} - \frac{c^2}{3a^2} - 1\right\}\frac{c^2}{\sqrt{(c^2-x^2)}} + \frac{c^3}{3a^2}\log\{c + \sqrt{(c^2-x^2)}\}$$

$$+ \frac{c^5}{6a^4}\left[-2\log\{c + \sqrt{(c^2-x^2)}\} + \frac{\sqrt{(c^2-x^2)}}{c} - 1 + \frac{c}{2\sqrt{(c^2-x^2)}}\right.$$

$$\left. + \frac{c}{\sqrt{(c^2-x^2)}}\log\{c + \sqrt{(c^2-x^2)}\}\right] + \text{constant}.$$

We have now to choose l, or rather $(l+C)$, and it may appear at first sight as though we might take it almost at pleasure. But this is not the case, at any rate if we wish our results to be applicable when $c = r$. For this purpose it is necessary that $(du/dx)_r \times (r^2 - x^2)$ be a small quantity, and only a particular choice of $(l+C)$ will make it so. For when $x = c = r$,

$$\left(\frac{du}{dx}\right)_r\frac{r^2-x^2}{r^2} = \frac{r}{\sqrt{(r^2-x^2)}}\left\{\frac{r\,(l+C)}{2a^2} - 1 - \frac{r^2}{3a^2} + \frac{r^4}{6a^4}\left(\log r + \frac{1}{2}\right)\right\} - \frac{r^4}{6a^4}$$

$+$ terms vanishing when $x = r$.

We must therefore take

$$\frac{c\,(l+C)}{2a^2} - 1 - \frac{c^2}{3a^2} + \frac{c^4}{6a^4}\left(\log c + \frac{1}{2}\right) = 0, \quad \ldots\ldots\ldots\ldots(20)$$

making

$$u = \frac{c^3}{3a^2}\left(1 - \frac{c^2}{a^2}\right)\log\frac{c + \sqrt{(c^2-x^2)}}{c}$$

$$+ \frac{c^5}{6a^4}\left[\frac{\sqrt{(c^2-x^2)} - c}{c} + \frac{c}{\sqrt{(c^2-x^2)}}\log\frac{c + \sqrt{(c^2-x^2)}}{c}\right] + C'. \quad \ldots(21)$$

It should be noticed that u so determined does not become infinite when $c = r$ and $x = r$. For we have

$$u_r = -\frac{r^5}{6a^4} + \frac{r^6}{6a^4\sqrt{(r^2-x^2)}}\log\left\{1 + \frac{\sqrt{(r^2-x^2)}}{r}\right\} + C' = C'.$$

Also with the general value of c

$$u_0 = \frac{c^3}{3a^2}\left(1 - \frac{c^2}{2a^2}\right)\log 2 + C'. \quad \ldots\ldots\ldots\ldots\ldots(22)$$

As before
$$h = l - c + u_0,$$

and
$$ra^2\cos i = \frac{(l+C')\,r^2}{2} + \frac{(c^2-r^2)^{\frac{3}{2}} - c^3}{3} + \int_0^r (u - C')\,x\,dx$$

$$= \frac{r^2}{2}\left\{h + c - \frac{c^3}{3a^2}\left(1 - \frac{c^2}{2a^2}\right)\log 2\right\} + \frac{(c^2-r^2)^{\frac{3}{2}} - c^3}{3} + \int_0^r (u - C')\,x\,dx. \quad \ldots(23)$$

The integral in (23) can be expressed.

We find

$$\int_0^r (u - C')\, x\, dx = \frac{c^3}{3a^2}\left(1 - \frac{c^2}{a^2}\right)\left[\frac{r^2}{2}\log\frac{c + \surd(c^2 - r^2)}{c} + \frac{c^2}{4} - \frac{c\,\surd(c^2 - r^2)}{2} + \frac{c^2 - r^2}{4}\right]$$

$$+ \frac{c^5}{6a^4}\left[\frac{c^3 - (c^2 - r^2)^{\frac{3}{2}}}{3c} - \frac{r^2}{2} - c\left\{c + \surd(c^2 - r^2)\right\}\left\{\log\frac{c + \surd(c^2 - r^2)}{c} - 1\right\}\right.$$

$$\left. + 2c^2(\log 2 - 1)\right]. \qquad\qquad\qquad\qquad (24)$$

The expression for $ra^2 \cos i$ in terms of c is complicated, and so is the relation between c and i demanded by the boundary condition

$$\cot i = \frac{r}{\surd(c^2 - r^2)} + \left(\frac{du}{dx}\right)_{x=r}. \qquad\qquad (25)$$

But in the particular case of greatest interest ($i = 0$) much simplification ensues. It follows easily from (25) that $c = r$. When we introduce this condition into (24), we get

$$\int_0^r (u - C')\, x\, dx = \frac{r^5}{12a^2} + \frac{r^7}{6a^4}\left(2\log 2 - \frac{5}{3}\right), \qquad (26)$$

and accordingly

$$a^2 = \frac{r}{2}\left(h + \frac{r}{3}\right) - \frac{r^4}{6a^2}\left(\log 2 - \frac{1}{2}\right) + \frac{5r^6}{36a^4}(3\log 2 - 2). \qquad (27)$$

Hence by successive approximations

$$a^2 = \frac{r}{2}\left\{h + \frac{r}{3} - \frac{2r^2}{3h}\left(\log 2 - \frac{1}{2}\right) + \frac{r^3}{9h^2}(32\log 2 - 21)\right\}$$

$$= \tfrac{1}{2} r\left\{h + \tfrac{1}{3} r - 0{\cdot}1288\, r^2/h + 0{\cdot}1312\, r^3/h^2\right\}. \qquad (28)$$

If the ratio of r to h is at all such as should be employed in experiment, this formula will yield a^2, viz., $T/g\rho$, with abundant accuracy.

Our equations give for the whole height of the meniscus in the case $i = 0$, $c = r$,

$$z_r - z_0 = r + u_r - u_0 = r - \frac{r^3}{3a^2}\left(1 - \frac{r^2}{2a^2}\right)\log 2. \qquad (29)$$

Another method of calculating the correction for a small tube, originating apparently with Hagen and Desains, is to assume an elliptical form of surface in place of the circular, the minor axis of the ellipse being vertical. In any case this should allow of a closer approximation, and drawings made for Kelvin* by Prof. Perry suggest that the representation is really a good one.

* *Proc. Roy. Inst.* 1886; "Popular Lectures and Addresses," I. p. 40.

If the semi-axis minor of the ellipse be β, the curvature at the end of this axis is β/r^2, and in our previous notation $\beta = hr^2/2a^2$. Also, i being equal to 0,

$$a^2 r = \int_0^r zx\,dx = \int_0^r \left\{ h + \beta - \beta \sqrt{\left(1 - \frac{x^2}{r^2}\right)} \right\} x\,dx,$$

and
$$a^2 = \tfrac{1}{2}r\,(h + \tfrac{1}{3}\beta) = \tfrac{1}{2}hr\,(1 + r^2/6a^2). \quad\quad\ldots\ldots\ldots\ldots(30)$$

This yields a quadratic in a^2; hence

$$a^2 = \frac{hr}{4} + \frac{hr}{4} \sqrt{\left(1 + \frac{4r}{3h}\right)} = \frac{r}{2}\left\{ h + \frac{r}{3} - \frac{r^2}{9h} + \frac{2r^3}{27h^2}\right\}$$

$$= \tfrac{1}{2}r\,\{h + \tfrac{1}{3}r - 0 \cdot 1111\,r^2/h + 0 \cdot 0741\,r^3/h^2\} \quad\ldots\ldots\ldots\ldots(31)$$

approximately. It will be seen that this differs but little numerically from (28), which, however, professes to be the accurate result so far as the term in r^3/h^2 inclusive.

The Wide Tube.

The equation of the second order for the surface of the liquid, assumed to be of revolution about the axis of z, is well known and may be derived from (1) by differentiation. It is

$$\frac{d^2 z}{dx^2} + \frac{1}{x}\frac{dz}{dx}\left\{1 + \left(\frac{dz^2}{dx}\right)\right\} = \frac{z}{a^2}\left\{1 + \left(\frac{dz^2}{dx}\right)\right\}^{\frac{3}{2}}. \quad\ldots\ldots\ldots(32)$$

If dz/dx be small, (32) becomes approximately

$$\frac{d^2 z}{dx^2} + \frac{1}{x}\frac{dz}{dx} - \frac{z}{a^2} = \frac{3z}{2a^2}\left(\frac{dz}{dx}\right)^2 - \frac{1}{x}\left(\frac{dz}{dx}\right)^3. \quad\ldots\ldots\ldots(33)$$

In the interior part of the surface under consideration $(dz/dx)^2$ may be neglected, and the approximate solution is

$$z = h_0 J_0\,(ix/a) = h_0 I_0\,(x/a) = h_0 \left\{1 + \frac{x^2}{2^2 a^2} + \frac{x^4}{2^2.\,4^2.\,a^4} + \ldots\right\}, \quad\ldots(34)$$

J_0 denoting, as usual, the Bessel's, or rather Fourier's, function of zero order and h_0 being the elevation at the axis above the free absolutely plane level. For the present purpose h_0 is to be so small as to be negligible in experiment, and the question is how large must r be.

When h_0 is small *enough*, x/a may be large while dz/dx still remains small. Eventually dz/dx increases so that the formula fails. But when x is large enough before this occurs, we may if necessary carry on with the two-dimensional solution properly adjusted to fit, as will be further explained later. In the meantime it will be convenient to give some numerical examples of the increase in dz/dx. In the usual notation

$$\frac{dz}{dx} = \frac{h_0}{a} I_1\left(\frac{x}{a}\right), \quad\ldots\ldots\ldots\ldots\ldots\ldots\ldots(35)$$

and the values of I_1, up to $x/a = 6$, are tabulated*.

* *Brit. Assoc. Rep. for* 1889 ; or Gray and Mathews' *Bessel's Functions*, Table VI.

In the case of water $a = 0.27$ cm. If we take $h_0/a = 0.01$, and $x/a = 4$, we have $dz/dx = 0.098$, so that $(dz/dx)^2$ is still fairly small. Here for water $h_0 = 0.0027$ cm. and $2x = 2.2$ cm. A diameter of 2.2 cm. is thus quite insufficient, unless an error exceeding 0.003 cm. be admissible. Again, suppose $h_0/a = 0.001$, and take $x/a = 6$. Then $dz/dx = 0.061$, again small. For water $h_0 = 0.00027$ cm., and $2x = 3.2$ cm. This last value of h_0 is about that (0.003 mm.) given by Richards and Coombs as the maximum admissible error of reading, and we may conclude that a diameter of 3.2 cm. is quite inadequate to take advantage of this degree of refinement.

We may go further in this example without too great a loss of accuracy. Retaining $h_0/a = 0.001$, let us make $x/a = 7$. I find $I_1(7) = 156$ about, so that the extreme value of dz/dx is 0.156, still moderately small. Here $2x = 3.8$ cm., which is thus shown to be inadequate in the case of water.

But apart from the question of the necessary diameter of tube, information sufficient for experimental purposes can be derived in another manner. The initial value of z (on the axis) is h_0; and $z = 2h_0$ when $I_0(x/a) = 2$, i.e. when $x = 1.8a$. For the best work h_0 should be on the limit of what can be detected and then h_0 and $2h_0$ could just be distinguished. The observer may be satisfied if no difference of level can be seen over the range $x = \pm 1.8a$; in the case of water this range is $2 \times 1.8 \times 0.27 = 0.97$ cm., or say 1 cm.

It has already been remarked that when h_0 is small enough x/a may become great within the limits of application of (35). To shorten our expressions we will take a temporarily as the unit of length. Then when x is very great,

$$I_1(x) = I_0(x) = \frac{e^x}{\sqrt{(2\pi x)}} . \qquad \dots\dots\dots\dots\dots\dots(36)$$

Thus if ψ be the angle the tangent to the curve makes with the horizontal,

$$\psi = \tan\psi = \frac{dz}{dx} = \frac{h_0 e^x}{\sqrt{(2\pi x)}}; \qquad \dots\dots\dots\dots\dots(37)$$

an equation which may be employed when h_0 is so small that a large x is consistent with a small ψ.

In order to follow the curve further, up to $\psi = \frac{1}{2}\pi$, we may employ the two-dimensional solution, the assumption being that the region of moderate ψ occupies a range of x small in comparison with its actual value, i.e. a value not much less than r, the radius of the tube. On account of the magnitude of x we have only the one curvature to deal with. For this curvature

$$\frac{1}{R} = \frac{d\psi}{ds} = \frac{d\psi}{dz}\sin\psi = z, \qquad \dots\dots\dots\dots\dots(38)$$

so that $\frac{1}{2}z^2 = C - \cos\psi = 1 - \cos\psi,$

since when $\psi = 0$, z^2 is exceedingly small. Accordingly

$$z = 2 \sin \tfrac{1}{2} \psi. \quad\ldots\ldots\ldots\ldots\ldots\ldots\ldots(39)$$

Also $$dx = \frac{dz}{\tan \psi} = \left(\frac{1}{\sin \tfrac{1}{2}\psi} - 2 \sin\tfrac{1}{2}\psi \right) d\left(\tfrac{1}{2}\psi\right),$$

and $$x = \log \tan \left(\tfrac{1}{4}\psi\right) + 2 \cos \tfrac{1}{2}\psi + C'. \quad\ldots\ldots\ldots(40)$$

The constant is determined by the consideration that at the wall ($x = r$), $\psi = \tfrac{1}{2}\pi$; thus

$$r - x = \log \tan (\pi/8) + \sqrt{2} - \log \tan \left(\tfrac{1}{4}\psi\right) - 2 \cos \left(\tfrac{1}{2}\psi\right)$$

$$= \log \tan (\pi/8) + \sqrt{2} - 2 + 2 \log 2 - \log \psi, \quad\ldots\ldots\ldots(41)$$

since ψ is small.

The value of x is supposed to be the same here as in (37), so that

$$x = \log \psi + \tfrac{1}{2} \log (2\pi x) - \log h_0, \quad\ldots\ldots\ldots\ldots(42)$$

whence on elimination of ψ and restoration of a,

$$r/a = -\log (\sqrt{2} + 1) + \sqrt{2} - 2 + 2 \log 2 + \tfrac{1}{2} \log (2\pi x/a) - \log (h_0/a). \quad\ldots(43)$$

With sufficient approximation, when h_0 is small enough, we may here substitute r for x, and thus

$$r/a - \tfrac{1}{2} \log (r/a) = -\log (\sqrt{2} + 1) + \sqrt{2} - 2 + 2 \log 2 + \tfrac{1}{2} \log (2\pi) - \log (h_0/a)$$

$$= 0{\cdot}8381 + \log (a/h_0). \quad\ldots\ldots\ldots\ldots\ldots\ldots\ldots\ldots\ldots(44)$$

This formula should give the relation between r/a and h_0/a when h_0/a is small enough, but it is only roughly applicable to the case of greatest interest, where $a/h_0 = 1000$, corresponding to the accuracy of reading found by Richards and Coombs. In this case

$$0{\cdot}8381 + \log (a/h_0) = 7{\cdot}746.$$

For this value of r/a we should have $\tfrac{1}{2} \log (r/a) = 1{\cdot}024$. It is true that according to (44) r/a will be somewhat greater, but on the other hand the proper value of x (replaced by r) is less than r. We may fairly take

$$r/a = 7{\cdot}746 + 1{\cdot}024 = 8{\cdot}770,$$

making with $a = 0{\cdot}27$ cm.

$$2r = 4{\cdot}74 \text{ cm.}$$

This calculation indicates that a diameter greater even than those contemplated by Richards and Coombs may be necessary to reduce h_0 to negligibility, but it must be admitted that it is too rough to inspire great confidence in the close accuracy of the final number. Probably it would be feasible to continue the approximation, employing an approximate value for the second curvature in place of neglecting it altogether. But although the integration can be effected, the work is rather long.

[*Added November* 17.—Since this paper was communicated, I have been surprised to find that the problem of the last paragraphs was treated long ago by Laplace in the *Mécanique Céleste** by a similar method, and with a result equivalent to that (44) arrived at above for the relation between the radius of a wide tube and the small elevation at the axis. Laplace uses the definite integral expression for I_0, and obtains the approximate form appropriate to large arguments. In view of Laplace's result, I have been tempted to carry the approximation further, as suggested already.

In the previous notation, the differential equation of the surface may be written

$$\frac{\sin \psi \, d\psi}{dz} + \frac{\sin \psi}{x} = \frac{z}{a^2}. \quad \dots \dots (45)$$

In the first approximation, where the second curvature on the left is omitted, we get

$$\frac{z^2 - z_0^2}{2a^2} = 1 - \cos \psi = 2 \sin^2 \frac{\psi}{2},$$

z_0 being the elevation at the axis, where $\psi = 0$. For the present purpose z_0^2 is to be regarded as exceedingly small, so that we may take at this stage, as in (39),

$$z = 2a \sin \left(\tfrac{1}{2} \psi\right). \quad \dots \dots \dots (46)$$

We now introduce an approximate value for the second curvature in (45), writing $x = r$, where r is the radius of the tube, and making, according to (46),

$$\sin \psi = \frac{z}{a} \sqrt{\left(1 - \frac{z^2}{4a^2}\right)}. \quad \dots \dots (47)$$

On integration

$$C - \cos \psi = \frac{z^2}{2a^2} + \frac{4a}{3r} \left(1 - \frac{z^2}{4a^2}\right)^{\frac{3}{2}} = \frac{z^2}{2a^2} + \frac{4a}{3r} \cos^3 \frac{\psi}{2}, \quad \dots \dots (48)$$

on substitution in the small term of the approximate value of z. When $\psi = 0$, z^2 is very small, so that $C = 1 + 4a/3r$, and

$$\frac{z}{a} = 2 \sin \frac{\psi}{2} + \frac{2a}{3r} \frac{1 - \cos^3 \left(\tfrac{1}{2} \psi\right)}{\sin \tfrac{1}{2} \psi} \quad \dots \dots (49)$$

is the second approximation to z.

From (49)

$$\frac{1}{a} \frac{dz}{d\psi} = \cos \frac{\psi}{2} + \frac{a}{3r} \frac{3 \cos^2 \tfrac{1}{2} \psi . \sin^2 \tfrac{1}{2} \psi - \cos \tfrac{1}{2} \psi \left(1 - \cos^3 \tfrac{1}{2} \psi\right)}{\sin^2 \left(\tfrac{1}{2} \psi\right)}. \quad \dots \dots (50)$$

We are now in a position to find x by the relation

$$x = \int \cot \psi \, (dz/d\psi) \, d\psi, \quad \dots \dots \dots (51)$$

* Supplément au Xe Livre, pp. 60—64, 1805.

the constant of integration being determined by the correspondence of $x = r$, $\psi = \frac{1}{2}\pi$. Thus

$$\frac{r-x}{a} = \log(\sqrt{2}-1) + \sqrt{2} + \frac{a}{3r}\left\{-2 + \frac{\sqrt{2}}{2} + \frac{3}{2}\log(\sqrt{2}-1)\right\} - \log\tan\frac{\psi}{4}$$

$$- 2\cos\frac{\psi}{2} + \frac{a}{3r}\left\{\frac{1}{2(1+\cos\frac{1}{2}\psi)} + 2\sin^2\frac{\psi}{2} - \frac{3}{2}\log\frac{1-\cos\frac{1}{2}\psi}{\sin\frac{1}{2}\psi}\right\}, \quad \dots(52)$$

giving when ψ is small

$$\frac{r-x}{a} = \alpha - \frac{a\beta}{3r} - \left(1 + \frac{a}{2r}\right)\log\psi, \quad \dots\dots\dots\dots(53)$$

where $$\alpha = \log(\sqrt{2}-1) + \sqrt{2} + \log 4 - 2 = -0.0809, \quad \dots\dots\dots(54)$$

$$-\tfrac{1}{3}\beta = \log 2 + \tfrac{1}{2}\log(\sqrt{2}-1) + \tfrac{1}{8}\sqrt{2} - 7/12 = -0.0952. \quad \dots(55)$$

The other equation, derived from the flat part of the surface, is

$$\psi = \frac{dz}{dx} = \frac{h_0}{a}I_1\left(\frac{x}{a}\right) = \frac{h_0/a \cdot e^{x/a}}{\sqrt{(2\pi x/a)}}\left(1 - \frac{3a}{8x}\right), \quad \dots\dots\dots(56)$$

in which x/a is regarded as large ; or

$$\frac{x}{a} = \log\psi + \log\frac{a}{h_0} + \tfrac{1}{2}\log\frac{2\pi x}{a} + \frac{3a}{8x}. \quad \dots\dots\dots\dots(57)$$

In equations (53), (57) x and ψ are to be identified. On elimination of ψ

$$\frac{r}{a} - \log\frac{a}{h_0} = \frac{\alpha - a\beta/3r}{1 + a/2r} + \frac{r-x}{2r+a} + \tfrac{1}{2}\log\frac{2\pi x}{a} + \frac{3a}{8x}, \quad \dots\dots(58)$$

in which we may put

$$\log\frac{2\pi x}{a} = \log\frac{2\pi r}{a} + \log\left(1 - \frac{r-x}{r}\right) = \log\frac{2\pi r}{a} - \frac{r-x}{r},$$

$$\frac{3a}{8x} = \frac{3a}{8r}\left(1 + \frac{r-x}{r}\right).$$

Thus $$\frac{r}{a} - \log\frac{a}{h_0} = \frac{\alpha - a\beta/3r}{1 + a/2r} + \frac{a(r-x)}{8r^2} + \frac{3a}{8r} + \tfrac{1}{2}\log\frac{2\pi r}{a}, \quad \dots\dots(59)$$

in which, since x is nearly equal to r, $a(r-x)/8r^2$ may usually be neglected. Also, in view of the smallness of α and β, it is scarcely necessary to retain the denominator $1 + a/2r$, so that we may write

$$\frac{r}{a} - \log\frac{a}{h_0} = -0.0809 + 0.2798\frac{a}{r} + \tfrac{1}{2}\log\frac{2\pi r}{a}$$

$$= 0.8381 + 0.2798 a/r + \tfrac{1}{2}\log(r/a). \quad \dots\dots\dots(60)$$

The effect of the second approximation is the introduction of the second term on the right of (60).

To take an example, let us suppose as before that $a/h_0 = 1000$, so that $\log(a/h_0) = 6\cdot908$. By successive approximation we find from (60)

$$r/a = 8\cdot869, \quad \dotsi\dotsi\dotsi(61)$$

so that if $a = 0\cdot27$ cm. (as for water),

$$2r = 4\cdot79 \text{ cm.} \quad \dotsi\dotsi\dotsi(62)$$

The correction to Laplace's formula is here unimportant.

The above is the diameter of tube required to render h_0 negligible according to the standard adopted.

It may sometimes be convenient to invert the calculation, and deduce the value of h_0 from the diameter of the tube (not much less than 4 cm.) and an approximate value of a. For this purpose we may use (60), or preferably (59), taking $x = \frac{4}{5}r$ for instance. The calculated value of h_0 would then be used as a correction. The accompanying small Table may be useful for this purpose.

r/a	$-\log_{10}(h_0/a)$	Difference	h_0/a
6	1·8275		0·0149
7	2·2319	0·4044	0·0059
8	2·6399	0·4080	0·0023
9	3·0508	0·4109	0·00089
10	3·4639	0·4131	0·00034

We have supposed throughout that the liquid surface is symmetrical about the axis, as happens when the section of the containing tube is circular. It may be worth remarking that without any restriction to symmetry the differential equation of the nearly flat parts of a large surface may be taken to be

$$\frac{d^2z}{dx^2} + \frac{d^2z}{dy^2} - \frac{z}{a^2} = 0, \quad \dotsi\dotsi\dotsi(63)$$

so that z may be expressed by the series

$$z = A_0 I_0(r/a) + (A_1\cos\theta + B_1\sin\theta)I_1(r/a)$$
$$+ (A_2\cos 2\theta + B_2\sin 2\theta)I_2(r/a) + \dots, \quad \dotsi\dotsi(64)$$

r, θ denoting the usual polar co-ordinates in the horizontal plane.]

400.

THE CONE AS A COLLECTOR OF SOUND.

[*Advisory Committee for Aeronautics,* T. 618, 1915.]

THE action of a cone in collecting sound coming in the direction of the axis may be investigated theoretically. If the diameter of the mouth be small compared with the wave-length (λ) of the sound, the cone may operate as a resonator, and the effect will vary greatly with the precise relation between λ and the length of the cone. On the other hand, the effect will depend very little upon the direction of the sound. It is probably more useful to consider the opposite extreme, where the diameter of the mouth is a large, or at any rate a moderate, multiple of λ, when the effect may be expected to fall off with rapidity as the obliquity of the sound increases.

A simple way of regarding the matter is to suppose the sound, incident axially, to be a *pulse, e.g.* a condensation confined to a narrow stratum bounded by parallel planes. If the angle of the cone be small, the pulse may be supposed to enter without much modification and afterwards to be propagated along. As the area diminishes, the condensation within the pulse must be supposed to increase. Finally the pulse would be reflected, and after emergence from the mouth would retrace its course. But the argument is not satisfactory, seeing that the condition for a progressive wave, *i.e.* of a wave propagated without reflection, is different in a cylindrical and in a conical tube. The usual condition in a cylindrical tube, or in plane waves where there is no tube, viz. $u = as$, where u is the particle velocity, a that of sound, and s the condensation, is replaced in spherical waves by

$$u = as - \frac{a}{r^2} \int sr\,dr,$$

showing that a pulse of condensation alone cannot be propagated without undergoing some reflection. If there is to be no reflection at all, the integral taken over the thickness of the pulse must vanish, and this it cannot do unless the pulse include also a rarefaction.

Apart from what may happen afterwards, there is a preliminary question at the mouth. In the passage from plane to spherical waves there is a phase-disturbance (between the centre and the edge) to be reckoned with, represented by

$$R\,(1 - \cos\theta) = 2R\theta \times \tfrac{1}{4}\theta,$$

where R is the length of the cone, and θ the semi-vertical angle. That this may be a small fraction of λ, itself a small fraction of the diameter of the mouth $(2R\theta)$, it is evident that θ must be very small.

We may now consider the incidence along the axis (x) of plane waves of simple type. Within the cone, supposed to be complete up to the vertex, the vibrations are stationary, and since no energy passes into the cone, the same must be true of the plane waves just outside—at any rate over the greater part of the mouth. The velocity potential just outside may therefore be denoted by

$$\psi = \cos kat \,.\, \cos\,(kx + e),$$

making at the mouth $(x = 0)$

$$\psi = \cos kat \,.\, \cos e, \qquad d\psi/dx = -\,k\cos kat\,.\,\sin e.$$

On the other hand, in the cone

$$\psi = A\,\frac{\sin kr}{kr}\,\cos kat,$$

making at the mouth $(r = R)$

$$\psi = A\,\frac{\sin kR}{kR}\,\cos kat, \qquad \frac{d\psi}{dr} = kA\left\{\frac{\cos kR}{kR} - \frac{\sin kR}{k^2 R^2}\right\}.$$

Equating the two values at the mouth of ψ and $d\psi/dx$ or $d\psi/dr$, we get

$$\cos e = A\,\frac{\sin kR}{kR}, \qquad -\sin e = A\left\{\frac{\cos kR}{kR} - \frac{\sin kR}{k^2 R^2}\right\}$$

and

$$1 = \frac{A^2}{k^2 R^2}\left\{1 - \frac{\sin 2kR}{kR} + \frac{\sin^2 kR}{k^2 R^2}\right\}.$$

When kR is considerable, the second and third terms may be neglected, whatever may be the particular value of kR, so that for a long enough cone

$$A = kR \quad \text{simply,}$$

in which $k = 2\pi/\lambda$. Here A is the maximum value of ψ at the vertex of the cone, and the maximum value of ψ in the stationary waves outside the mouth is unity, the particular *place* where this maximum occurs being variable with the precise value of kR.

The increase of ψ, or of the condensation, at the vertex of the cone as compared with that obtained by simple reflection at a wall is represented by the factor kR, which, under our suppositions, is a large number.

Although the complete fulfilment of the conditions above laid down is hardly realisable in practice with sounds of moderate pitch, one would certainly expect the use of a cone to be of more advantage than appears from the observations at the Royal Aircraft Factory (*Report*, T. 577). In the year 1875, I experimented with a zinc cone 10 inches wide at the mouth and about 9 feet long, but I cannot find any record of the observations. My recollection, however, is that I was disappointed with the results. Perhaps I may find opportunity for further trial, when I propose to use wave-lengths of about 3 inches.

401.

THE THEORY OF THE HELMHOLTZ RESONATOR.

[*Proceedings of the Royal Society*, A, Vol. XCII. pp. 265—275,.1915.]

THE ideal form of Helmholtz resonator is a cavernous space, almost enclosed by a thin, immovable wall, in which there is a small perforation establishing a communication between the interior and exterior gas. An approximate theory, based upon the supposition that the perforation is small, and consequently that the wave-length of the aërial vibration is great, is due to Helmholtz*, who arrived at definite results for perforations whose outline is circular or elliptic. A simplified, and in some respects generalised, treatment was given in my paper on "Resonance†." In the extreme case of a wave-length sufficiently great, the kinetic energy of the vibration is that of the gas near the mouth as it moves in and out, much as an incompressible fluid might do, and the potential energy is that of the almost uniform compressions and rarefactions of the gas in the interior. The latter is a question merely of the volume S of the cavity and of the quantity of gas which has passed, but the calculation of the kinetic energy presents difficulties which have been only partially overcome. In the case of simple apertures in the thin wall (regarded as plane), only circular and elliptic forms admit of complete treatment. The mathematical problem is the same as that of finding the electrostatic *capacity* of a thin conducting plate having the form of the aperture, and supposed to be situated in the open.

The project of a stricter treatment of the problem, in the case of a spherical wall and an aperture of circular outline, has been in my mind more than 40 years, partly with the hope of reaching a closer approximation, and partly because some mathematicians have found the former method unsatisfactory, or, at any rate, difficult to follow. The present paper is on ordinary lines, using the appropriate spherical (Legendre's) functions, much as in a former one, "On the Acoustic Shadow of a Sphere‡."

---

* *Crelle Journ. Math.* Vol. LVII. (1860).

† *Phil. Trans.* Vol. CLXI. p. 77 (1870); *Scientific Papers*, Vol. I. p. 33. Also *Theory of Sound*, ch. XVI.

‡ *Phil. Trans.* A, Vol. CCIII. p. 87 (1904); *Scientific Papers*, Vol. V. p. 149.

The first step is to find the velocity-potential (ψ) due to a normal motion at the surface of the sphere localised at a single point, the normal motion being zero at every other point. This problem must be solved both for the exterior and for the interior of the sphere, but in the end the potential is required only for points lying infinitely near the spherical surface. Then if we assume a normal motion given at every point on the aperture, that is on the portion of the spherical surface not occupied by the walls, we are in a position to calculate ψ upon the two sides of the aperture. If these values are equal at every point of the aperture, it will be a proof that the normal velocity has been rightly assumed, and a solution is arrived at. If the agreement is not sufficiently good—there is no question of more than an approximation—some other distribution of normal velocities must be tried. In what follows, the preliminary work is the same as in the paper last referred to, and the same notation is employed.

The general differential equation satisfied by ψ, and corresponding to a simple vibration, is

$$\frac{d^2\psi}{dx^2} + \frac{d^2\psi}{dy^2} + \frac{d^2\psi}{dz^2} + k^2\psi = 0, \dots\dots\dots\dots\dots(1)$$

where $k = 2\pi/\lambda$, and λ denotes the length of plane waves of the same pitch. For brevity we may omit k; it can always be restored on paying attention to "dimensions." The solution in polar co-ordinates applicable to a wave of the nth order in Laplace's series may be written (with omission of the time-factor)

$$\psi_n = S_n r^n \chi_n(r). \dots\dots\dots\dots\dots\dots(2)$$

The differential equation satisfied by χ_n is

$$\frac{d^2\chi_n}{dr^2} + \frac{2n+2}{r}\frac{d\chi_n}{dr} + \chi_n = 0. \dots\dots\dots\dots(3)$$

The solution of (3) applicable to a wave diverging outwards is

$$\chi_n(r) = \left(-\frac{d}{r\,dr}\right)^n \frac{e^{-ir}}{r}. \dots\dots\dots\dots(4)$$

Putting $n = 0$ and $n = 1$, we have

$$\chi_0(r) = \frac{e^{-ir}}{r}, \qquad \chi_1(r) = \frac{(1+ir)e^{-ir}}{r^3}. \dots\dots\dots\dots(5)$$

It is easy to verify that (4) satisfies (3). For if χ_n satisfies (3), $r^{-1}\chi_n'$ satisfies the corresponding equation for χ_{n+1}. And $r^{-1}e^{-ir}$ satisfies (3) when $n = 0$.

From (3) and (4) the following sequence formulæ may be verified:

$$\chi_n'(r) = -r\chi_{n+1}(r), \dots\dots\dots\dots\dots(6)$$

$$r\chi_n'(r) + (2n+1)\chi_n(r) = \chi_{n-1}(r), \dots\dots\dots\dots(7)$$

$$\chi_{n+1}(r) = \frac{(2n+1)\chi_n(r) - \chi_{n-1}(r)}{r^2}. \dots\dots\dots\dots(8)$$

By means of the last, χ_2, χ_3, etc., may be built up in succession from χ_0 and χ_1.

From (2)

$$d\psi_n/dr = S_n \left(nr^{n-1}\chi_n + r^n \chi_n' \right),$$

or with use of (7)

$$d\psi_n/dr = r^{n-1} S_n \left\{ \chi_{n-1} - (n+1)\chi_n \right\}. \quad \ldots\ldots\ldots\ldots(9)$$

Thus if U_n be the nth component of the normal velocity at the surface of the sphere $(r = c)$

$$U_n = c^{n-1} S_n \left\{ \chi_{n-1}(c) - (n+1)\chi_n(c) \right\}. \quad \ldots\ldots\ldots\ldots(10)$$

When $n = 0$,

$$U_0 = S_0 \chi_0'(c) = - S_0 c \chi_1(c). \quad \ldots\ldots\ldots\ldots\ldots\ldots(11)$$

The introduction of S_n from (10), (11) into (2) gives ψ_n in terms of U_n supposed known.

When r is very great in comparison with the wave-length, we get from (4)

$$\chi_n(r) = \frac{i^n e^{-ir}}{r^{n+1}}, \quad \ldots\ldots\ldots\ldots\ldots\ldots(12)$$

so that

$$\psi_n = S_n \frac{i^n e^{-ir}}{r}. \quad \ldots\ldots\ldots\ldots\ldots\ldots(13)$$

We have now to apply these formulæ to the particular case where U is sensible over an infinitesimal area $d\sigma$, but vanishes over the remainder of the surface of the sphere. If μ be the cosine of the angle (θ) between $d\sigma$ and the point at which U is expressed, $P_n(\mu)$ Legendre's function, we have

$$U_n = \frac{2n+1}{4\pi c^2} U d\sigma . P_n(\mu), \quad \ldots\ldots\ldots\ldots\ldots(14)$$

and accordingly for the velocity-potential at the *surface of the sphere*,

$$\psi = \frac{U d\sigma}{4\pi c} \Sigma \frac{(2n+1)\chi_n(c) . P_n(\mu)}{\chi_{n-1}(c) - (n+1)\chi_n(c)} . \quad \ldots\ldots\ldots\ldots(15)$$

When $n = 0$, $\chi_{n-1} - (n+1)\chi_n$ is to be replaced by $-c^2\chi_1$. Equation (15) gives the value of ψ at a point whose angular distance (θ) from $d\sigma$ is $\cos^{-1}\mu$. If χ_n has the form given by (4), the result applies to the *exterior* surface of the sphere.

We have also to consider the corresponding problem for the interior. The only change required is to replace χ_n as given in (4) by the form appropriate to the interior. For this purpose we might take simply the imaginary part of (4), but since a constant multiplier has no significance, it suffices to make

$$\chi_n(r) = \left(-\frac{d}{r\,dr} \right)^n \frac{\sin r}{r}. \quad \ldots\ldots\ldots\ldots\ldots(16)$$

With this alteration (15) holds good for the interior, U denoting the localised normal velocity at the surface still measured outwards, since $U = d\psi/dr$.

We have now to introduce approximate values of $\chi_{n-1}(c) \div \chi_n(c)$ in (15), having regard to the assumed smallness of c, or rather kc. For this purpose we expand the sine and cosine of $c*$:—

$$\frac{\cos c}{c} = \frac{1}{c} - \frac{c}{1.2} + \frac{c^3}{4!} - \frac{c^5}{6!} + \cdots,$$

$$-\frac{1}{c}\frac{d}{dc}\left(\frac{\cos c}{c}\right) = \frac{1}{c^3} + \frac{1}{1.2c} - \frac{3c}{4!} + \frac{5c^3}{6!} - \frac{7c^5}{8!} + \cdots,$$

$$\left(-\frac{1}{c}\frac{d}{dc}\right)^2 \frac{\cos c}{c} = \frac{3}{c^5} + \frac{1}{1.2c^3} + \frac{3}{c.4!} - \frac{5.3.c}{6!} + \frac{7.5.c^3}{8!} - \cdots,$$

and so on ;

$$\frac{\sin c}{c} = 1 - \frac{c^2}{2.3} + \frac{c^4}{5!} - \frac{c^6}{7!} + \cdots,$$

$$-\frac{1}{c}\frac{d}{dc}\frac{\sin c}{c} = \frac{2}{2.3} - \frac{4c^2}{5!} + \frac{6c^4}{7!} - \cdots,$$

$$\left(-\frac{1}{c}\frac{d}{dc}\right)^2 \frac{\sin c}{c} = \frac{4.2}{5!} - \frac{6.4.c^2}{7!} + \frac{8.6.c^4}{9!} - \cdots,$$

and so on. Thus for the outside

$$\chi_n = \frac{1.3.5\dots(2n-1)}{c^{2n+1}}\left\{1 + \frac{c^2}{2(2n-1)} + \cdots\right\}$$

$$- \frac{i}{1.3.5\dots(2n+1)}\left\{1 - \frac{c^2}{2(2n+3)} + \cdots\right\}. \quad\dots\dots(17)$$

For géneral values of n, we may take

$$\chi_{n-1} \div \chi_n = \frac{c^2}{2n-1}. \quad\dots\dots\dots\dots\dots\dots(18)$$

For $n = 1$

$$\frac{\chi_0}{\chi_1} = \frac{\dfrac{1}{c} - \dfrac{c}{2} - i}{\dfrac{1}{c^3}\left(1 + \dfrac{c^2}{2}\right) - \dfrac{i}{3}} = c^2(1 - c^2 - ic). \quad\dots\dots\dots(19)$$

For $n = 2$

$$\frac{\chi_1}{\chi_2} = \frac{c^2}{3} + \text{terms in } c^4. \quad\dots\dots\dots\dots\dots(20)$$

---

* 1917. In the expansions for the derivative of cos c/c terms (now inserted) were accidentally omitted, as has been pointed out by Mr F. P. White (*Proc. Roy. Soc.* Vol. xcii. p. 549). Equation (17) as originally given was accordingly erroneous. Corresponding corrections have been introduced in (19), (23), (24), (36), (38) which however do not affect the approximation employed in (39). Mr White's main object was to carry the approximation further than is attained in (57) and (60).

Thus in general by (18)

$$\frac{2n+1}{\chi_{n-1}/\chi_n - n - 1} = -2 + \frac{1}{n+1} - \frac{(2n+1)\,c^2}{(n+1)^2\,(2n-1)} \; ; \ldots\ldots(21)$$

while for $n = 1$

$$\frac{3}{\chi_0/\chi_1 - 2} = -2 + \tfrac{1}{2} - \tfrac{3}{4}c^2 + \text{terms in } c^3, \ldots\ldots\ldots(22)$$

in accordance with (21). When $n = 0$

$$\frac{\chi_0}{-c^2\chi_1} = -1 + c^2 + ic + \text{terms in } c^3. \ldots\ldots\ldots\ldots(23)$$

Using these values in (15), we see that, so far as c^2 inclusive,

$$\Sigma \, (\text{outside}) = (-1 + c^2 + ic)\,P_0$$

$$+ \left(-2 + \frac{1}{2} - \frac{3c^2}{4}\right) P_1 + \left(-2 + \frac{1}{3} - \frac{5c^2}{3^2.\,3}\right) P_2 + \ldots$$

$$= -2 \, \{P_0\,(\mu) + P_1\,(\mu) + \ldots + P_n\,(\mu) + \ldots\}$$

$$+ P_0\,(\mu) + \frac{1}{2}\,P_1\,(\mu) + \ldots + \frac{1}{n+1}\,P_n\,(\mu) + \ldots$$

$$+ ic + c^2 - \sum_{1}^{\infty} \frac{(2n+1)\,c^2}{(n+1)^2\,(2n-1)}\,P_n\,(\mu). \ldots\ldots\ldots(24)$$

In like manner for the form of χ_n appropriate to the inside

$$\chi_n\,(c) = \frac{1}{1\,.\,3\,.\,5 \ldots (2n+1)} \left\{1 - \frac{c^2}{2\,(2n+3)}\right\}, \ldots\ldots\ldots(25)$$

so that in general

$$\frac{\chi_{n-1}}{\chi_n} = 2n + 1 - \frac{c^2}{2n+3}, \ldots\ldots\ldots\ldots\ldots(26)$$

and

$$\frac{2n+1}{\chi_{n-1}/\chi_n - n - 1} = 2 + \frac{1}{n} + \frac{(2n+1)\,c^2}{n^2\,(2n+3)}. \ldots\ldots\ldots(27)$$

This suffices for $n = 1$ and onwards. When $n = 0$

$$\frac{\chi_0}{-c^2\chi_1} = -\frac{3}{c^2} \left\{1 - \frac{c^2}{15} - \frac{c^4}{525}\right\}. \ldots\ldots\ldots\ldots(28)$$

Accordingly, so far as c^2 inclusive,

$$\Sigma \, (\text{inside}) = 2 \, \{P_0\,(\mu) + P_1\,(\mu) + \ldots + P_n\,(\mu)\}$$

$$+ P_1\,(\mu) + \frac{1}{2}\,P_2\,(\mu) + \ldots + \frac{1}{n}\,P_n\,(\mu)$$

$$- \frac{3}{c^2} - 1\frac{4}{5} + \frac{c^2}{175} + \sum_{1}^{\infty} \frac{(2n+1)\,c^2}{n^2\,(2n+3)}\,P_n\,(\mu). \ldots\ldots(29)$$

The first two series of P's on the right of (24) and (29) become divergent when $\mu = 1$, or $\theta = 0$. To evaluate them we have

$$\frac{1}{\sqrt{\{1 - 2\alpha \cos \theta + \alpha^2\}}} = 1 + \alpha P_1 + \alpha^2 P_2 + \dots , \quad \dots\dots\dots\dots(30)$$

so that

$$1 + P_1 + P_2 + \dots = \frac{1}{\sqrt{(2 - 2 \cos \theta)}} = \frac{1}{2 \sin \frac{1}{2}\theta}. \quad \dots\dots\dots(31)$$

Again, by integration of (30),

$$\alpha + \frac{1}{2}\alpha^2 P_1 + \frac{1}{3}\alpha^3 P_2 + \dots = \int_0^\alpha \frac{d\alpha}{\sqrt{\{1 - 2\alpha \cos \theta + \alpha^2\}}}$$

$$= \log \left[\alpha - \cos \theta + \sqrt{\{1 - 2\alpha \cos \theta + \alpha^2\}}\right] - \log \left[1 - \cos \theta\right] *,$$

so that

$$1 + \frac{1}{2}P_1 + \frac{1}{3}P_2 + \dots = \log (1 + \sin \frac{1}{2}\theta) - \log \sin \frac{1}{2}\theta. \quad \dots\dots\dots(32)$$

In much the same way we may sum the third series $\Sigma n^{-1} P_n$. We have

$$P_1 + \alpha P_2 + \alpha^2 P_3 + \dots = \frac{1}{\alpha \sqrt{\{1 - 2\alpha\mu + \alpha^2\}}} - \frac{1}{\alpha},$$

$$\alpha P_1 + \frac{1}{2}\alpha^2 P_2 + \frac{1}{3}\alpha^3 P_3 + \dots = \int_0^\alpha \frac{d\alpha}{\alpha \sqrt{\{1 - 2\alpha\mu + \alpha^2\}}} - \int_0^\alpha \frac{d\alpha}{\alpha}.$$

We denote the right-hand member of this equation by I and differentiate it with respect to μ.

Thus

$$\frac{dI}{d\mu} = \int_0^\alpha \frac{d\alpha}{\{(\alpha - \mu)^2 + 1 - \mu^2\}^{\frac{3}{2}}} = \frac{\alpha - \mu}{(1 - \mu^2)\sqrt{\{1 - 2\alpha\mu + \alpha^2\}}} + \frac{\mu}{1 - \mu^2},$$

or when $\alpha = 1$

$$\frac{dI}{d\mu} = \frac{1}{4 \sin \frac{1}{2}\theta \cdot \cos^2 \frac{1}{2}\theta} + \frac{\mu}{1 - \mu^2}. \quad \dots\dots\dots\dots\dots(33)$$

On integration

$$I = \log \tan \frac{1}{4}(\pi - \theta) - \log \sin \theta + C. \quad \dots\dots\dots\dots(34)$$

The constant is to be found by putting $\mu = 0$, $\theta = \frac{1}{2}\pi$. In this case

$$I = \int_0^\alpha \frac{d\alpha}{\alpha \sqrt{(1 + \alpha^2)}} - \int_0^\alpha \frac{d\alpha}{\alpha} = \log 2 - \log \{1 + \sqrt{(1 + \alpha^2)}\}.$$

Thus

$$C = \log \frac{2}{1 + \sqrt{2}} - \log \tan \frac{\pi}{8} = \log 2,$$

---

* If we integrate this equation again with respect to α between the limits 0 and 1, we find

$$\frac{1}{1 \cdot 2} + \frac{P_1}{2 \cdot 3} + \dots + \frac{P_n}{(n+1)(n+2)} = 1 - 2 \sin \frac{1}{2}\theta + 2 \sin^2 \frac{1}{2}\theta [\log (1 + \sin \frac{1}{2}\theta) - \log \sin \frac{1}{2}\theta].$$

When θ is small, the more important part is

$$1 - \theta - \frac{1}{2}\theta^2 \log \theta.$$

and accordingly

$$P_1 + \tfrac{1}{2} P_2 + \tfrac{1}{3} P_3 + \ldots = \log \tan \tfrac{1}{4} (\pi - \theta) - \log (\tfrac{1}{2} \sin \theta). \quad \ldots\ldots(35)$$

For the values of Σ in (15) we now have with restoration of k

$$\Sigma \text{ (outside)} = - \frac{1}{\sin \tfrac{1}{2}\theta} - \log \sin \tfrac{1}{2}\theta + \log (1 + \sin \tfrac{1}{2}\theta)$$

$$+ ikc + k^2 c^2 - \sum_1^\infty \frac{(2n + 1) k^2 c^2}{(n + 1)^2 (2n - 1)} P_n (\mu), \quad \ldots\ldots(36)$$

$$\Sigma \text{ (inside)} = \frac{1}{\sin \tfrac{1}{2}\theta} - \log (\tfrac{1}{2} \sin \theta) + \log \tan \tfrac{1}{4} (\pi - \theta)$$

$$- \frac{3}{k^2 c^2} - \frac{9}{5} + \frac{k^2 c^2}{175} + \sum_1^\infty \frac{(2n + 1) k^2 c^2}{n^2 (2n + 3)} P_n (\mu). \quad \ldots\ldots(37)$$

These equations give the value of ψ at any point of the sphere, either inside or outside, due to a normal velocity at a single point, so far as $k^2 c^2$ inclusive. The inside value is dominated by the term $- 3/k^2 c^2$, except when θ is small. As to the sums in $k^2 c^2$ not evaluated, we may remark that they cannot exceed the values assumed when $\theta = 0$ and $P_n (\mu) = 1$. Approximate calculation of the limiting values is easy. Thus

$$\sum_1^\infty \frac{2n + 1}{(n + 1)^2 (2n - 1)}$$

$$= \sum_1^5 \frac{2n + 1}{(n + 1)^2 (2n - 1)} - \sum_1^5 \{n^{-2} - n^{-3} + \tfrac{3}{2} n^{-4}\} + \sum_1^\infty \{n^{-2} - n^{-3} + \tfrac{3}{2} n^{-4}\}$$

$$= - 0\cdot79040 + 1\cdot64493 - 1\cdot20206 + 1\cdot62348 = 1\cdot2759 \text{ *}.$$

In like manner

$$\sum_1^\infty \frac{2n + 1}{n^2 (2n + 3)} = - 0\cdot9485 + \sum_1^\infty \{n^{-2} - n^{-3} + \tfrac{3}{2} n^{-4}\} = 1\cdot1178 \dagger.$$

* Chrystal's *Algebra*, Part II. p. 343.

† 1917. Mr White has shown that the accurate value of the first sum is

$$\frac{4}{9} + \frac{8}{9} \log 2 + \frac{1}{3} \left(\frac{\pi^2}{6} - 1 \right),$$

and that of the second sum

$$\frac{8}{27} + \frac{8}{9} (1 - \log 2) + \frac{1}{3} \frac{\pi^2}{6};$$

so that for the two taken together as in (38), we have

$$\frac{8}{27} + 1 + \frac{\pi^2}{9} = 2\cdot39292.$$

The coefficient of $k^2 c^2$ in (38) is then

$$\frac{1}{175} - 1 + 2\cdot39292 = 1\cdot39863.$$

Further in this equation

$$\log \frac{1 - \tan \tfrac{1}{4}\theta}{1 + \tan \tfrac{1}{4}\theta} - \log \cos \frac{\theta}{2} - \log \left(1 + \sin \frac{\theta}{2} \right) = - 2 \log \left(1 + \sin \frac{\theta}{2} \right).$$

Our special purpose is concerned with the *difference* in the values of ψ on the two sides of the surface $r = c$, and thus only with the difference of Σ's. We have

$$\Sigma\,(\text{inside}) - \Sigma\,(\text{outside}) = \frac{2}{\sin\frac{1}{2}\theta} - \log\cos\frac{\theta}{2} + \log\frac{1 - \tan\frac{1}{4}\theta}{1 + \tan\frac{1}{4}\theta}$$

$$- \log\left(1 + \sin\frac{\theta}{2}\right) - \frac{3}{k^2c^2} - \frac{9}{5} - ikc$$

$$+ k^2c^2\left\{\frac{1}{175} - 1 + \sum_{1}^{\infty}\frac{(2n+1)\,P_n\,(\mu)}{n^2\,(2n+3)} + \sum_{1}^{\infty}\frac{(2n+1)\,P_n\,(\mu)}{(n+1)^2(2n-1)}\right\}. \quad \dots(38)$$

In the application we have to deal only with small values of θ and we shall omit k^2c^2, so that we take

$$\Sigma\,(\text{in}) - \Sigma\,(\text{out}) = \frac{2}{\sin\frac{1}{2}\theta} - \theta - \frac{3}{k^2c^2} - \frac{9}{5} - ikc; \quad \dots\dots(39)$$

it will indeed appear later that we do not need even the term in θ, since it is of the order k^2c^2.

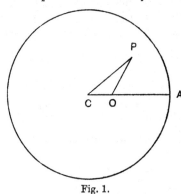

Fig. 1.

In pursuance of our plan we have now to assume a form for U over the circular aperture and examine how far it leads to agreement in the values of ψ on the inside and on the outside. For this purpose we avail ourselves of information derived from the first approximation. If C, fig. 1, be the centre and CA the angular radius of the spherical segment constituting the aperture, P any other point on it, we assume that U at P is proportional to $\{CA^2 - CP^2\}^{-\frac{1}{2}}$, and we require to examine the consequences at another arbitrary point O.

Writing $CA = a$, $CO = b$, $PO = \theta$, $POA = \phi$, we have from the spherical triangle

$$\cos CP = \cos b \cos \theta + \sin b \sin \theta \cos \phi,$$

or when we neglect higher powers than the cube of the small angles,

$$CP^2 = b^2 + \theta^2 + 2b\theta \cos \phi. \quad \dots\dots\dots\dots\dots(40)$$

Thus

$$CA^2 - CP^2 = a^2 - b^2 - \theta^2 - 2b\theta \cos \phi = a^2 - b^2 \sin^2 \phi - (\theta + b \cos \phi)^2, \quad \dots(41)$$

and we wish to make

$$\iint \frac{\sin \theta\, d\theta\, d\phi\,[\Sigma\,(\text{in}) - \Sigma\,(\text{out})]}{\sqrt{\{a^2 - b^2 - \theta^2 - 2b\theta \cos \phi\}}} = 0, \quad \dots\dots\dots\dots(42)$$

as far as possible for all values of b, the integration covering the whole area of aperture. We may write θ for $\sin \theta^*$, since we are content to neglect terms

---

* [Except as regards the product of $\sin \theta$ and the first term on the right of (39), since the term in θ^2 is in point of fact retained in the calculation. W. F. S.]

of order θ^2 in comparison with the principal term.   Reference to (39) shows that as regards the numerator of the integrand we have to deal with terms in θ^0, θ^1, and θ^2.

For the principal term we have

$$4 \iint \frac{d\theta \, d\phi}{\sqrt{\{a^2 - b^2 - \theta^2 - 2b\theta \cos \phi\}}}. \qquad\qquad\dots\dots\dots\dots(43)$$

Now $\quad \displaystyle\int \frac{d\theta}{\sqrt{\{\quad\}}} = \int \frac{d(\theta + b \cos \phi)}{\sqrt{\{\quad\}}} = \sin^{-1} \frac{\theta + b \cos \phi}{\sqrt{\{a^2 - b^2 \sin^2 \phi\}}}.$

For a given ϕ the lower limit of θ is 0 and the upper limit θ_1 is such as to make $a^2 = b^2 + \theta_1^2 + 2b\theta_1 \cos \phi$,

or $\qquad\qquad \theta_1 + b \cos \phi = \sqrt{(a^2 - b^2 \sin^2 \phi)}. \qquad\dots\dots\dots\dots\dots(44)$

Thus $\qquad \displaystyle\int_0^{\theta_1} \frac{d\theta}{\sqrt{\{\quad\}}} = \frac{\pi}{2} - \sin^{-1} \frac{b \cos \phi}{\sqrt{(a^2 - b^2 \sin^2 \phi)}}. \qquad\dots\dots\dots(45)$

When this is integrated with respect to ϕ, the second part disappears, and we are left with π^2 simply, so that the principal term (43) is $4\pi^2$. That this should turn out independent of b, that is the same at all points of the aperture, is only what was to be expected from the known theory respecting the motion of an incompressible fluid.

The term in θ, corresponding to the constant part of $\Sigma\,(\text{in}) - \Sigma\,(\text{out})$, is represented by

$$\iint \frac{\theta \, d\theta \, d\phi}{\sqrt{\{a^2 - CP^2\}}}. \qquad\qquad\dots\dots\dots\dots\dots\dots(46)$$

Here $\theta \, d\theta \, d\phi$ is merely the polar element of area, and the integral is, of course, independent of b. To find its value we may take the centre C as the pole of θ. We get at once

$$2\pi \int_0^a \frac{\theta \, d\theta}{\sqrt{(a^2 - \theta^2)}} = 2\pi a ; \qquad\qquad\dots\dots\dots\dots\dots(47)$$

so that this part of (42) is

$$- 2\pi a \left(\frac{3}{k^2 c^2} + \frac{9}{5} + ikc \right). \qquad\qquad\dots\dots\dots\dots\dots\dots(48)$$

For the third part (in θ^2), we write

$$\theta^2 = - (a^2 - b^2 - 2b\theta \cos \phi - \theta^2) - 2b \cos \phi \,(\theta + b \cos \phi) + a^2 - b^2 + 2b^2 \cos^2 \phi,$$

giving rise to three integrals in θ, of which the first is

$$- \int d\theta \, \sqrt{\{a^2 - b^2 - 2b\theta \cos \phi - \theta^2\}}$$
$$= - \tfrac{1}{2} (\theta + b \cos \phi) \, \sqrt{\{a^2 - b^2 \sin^2 \phi - (\theta + b \cos \phi)^2\}}$$
$$- \frac{a^2 - b^2 \sin^2 \phi}{2} \sin^{-1} \frac{\theta + b \cos \phi}{\sqrt{(a^2 - b^2 \sin^2 \phi)}}. \qquad\dots\dots\dots\dots(49)$$

The second integral is

$$- 2b \cos \phi \int \frac{(\theta + b \cos \phi) \, d\theta}{\sqrt{\{a^2 - CP^2\}}} = 2b \cos \phi \, \sqrt{\{a^2 - b^2 - 2b\theta \cos \phi - \theta^2\}}, \dots(50)$$

and the third is, as for the principal term,

$$(a^2 - b^2 + 2b^2 \cos^2 \phi) \sin^{-1} \frac{\theta + b \cos \phi}{\sqrt{(a^2 - b^2 \sin^2 \phi)}}. \quad \ldots\ldots\ldots\ldots(51)$$

Thus altogether, when the three integrals are taken between the limits 0 and θ_1, we get

$$-\tfrac{2}{3} b \cos \phi \sqrt{(a^2 - b^2)} + [\tfrac{1}{2} a^2 + b^2 (2 \cos^2 \phi + \tfrac{1}{2} \sin^2 \phi - 1)]$$
$$\times \left[\frac{\pi}{2} - \sin^{-1} \frac{b \cos \phi}{\sqrt{(a^2 - b^2 \sin^2 \phi)}} \right],$$

and finally after integration with respect to ϕ

$$\tfrac{1}{2} \pi^2 (a^2 + \tfrac{1}{2} b^2). \quad \ldots\ldots\ldots\ldots\ldots\ldots\ldots\ldots(52)$$

Thus altogether the integral on the left of (42) becomes

$$4\pi^2 - 2\pi a \left(\frac{3}{k^2 c^2} + \frac{9}{5} + ikc \right) - \frac{\pi^2}{2} \left(a^2 + \frac{b^2}{2} \right). \quad \ldots\ldots\ldots\ldots(53)*$$

In consequence of the occurrence of b^2, this expression cannot be made to vanish at all points of the aperture, a sign that the assumed form of U is imperfect. If, however, we neglect the last term, arising from $-\theta$ in $\Sigma \text{(in)} - \Sigma \text{(out)}$, our expression vanishes provided

$$\frac{3}{k^2 c^2} + \frac{9}{5} + ikc = \frac{2\pi}{a}, \quad \ldots\ldots\ldots\ldots\ldots\ldots(54)$$

showing that a is of the order $k^2 c^2$, so that this equation gives the relation between a and kc to a sufficient approximation. Helmholtz's solution corresponds to the neglect of the second and third terms on the left of (54), making

$$\frac{3}{k^2 c^2} = \frac{2\pi}{a} = \frac{2\pi c}{R}, \quad \ldots\ldots\ldots\ldots\ldots\ldots(55)$$

where R denotes the linear radius of the circular aperture. If we introduce $\lambda (= 2\pi/k)$,

$$\lambda = \pi \sqrt{(2S/R)}, \quad \ldots\ldots\ldots\ldots\ldots\ldots\ldots(56)$$

S denoting the capacity of the sphere, the known approximate value.

The third term on the left of (54) represents the decay of the vibration due to the propagation of energy away from the resonator. Omitting this for the moment, we have as the corrected value of λ,

$$\lambda = \pi \sqrt{(2S/R)} . \left\{ 1 - \frac{9}{10} \frac{R}{2\pi c} \right\}. \quad \ldots\ldots\ldots\ldots(57)$$

Let us now consider the term representing decay of the vibrations. The time factor, hitherto omitted, is e^{ikVt}, or if we take $k = k_1 + ik_2$, $e^{-k_2 Vt} e^{ik_1 Vt}$. If $t = \tau$, the period, $k_1 V\tau = 2\pi$, and $e^{-k_2 V\tau} = e^{-2\pi k_2/k_1}$. This is the factor by which the amplitude of vibration is reduced in one period. Now from (55)

$$kc = \sqrt{\left(\frac{3R}{2\pi c} \right)},$$

* [For "$\frac{\pi^2}{2}$" read "$\frac{3\pi^2}{4}$", and three lines below read

"arising from $-\tfrac{2}{3}\theta^2$ in $\sin\theta$ [$\Sigma \text{(in)} - \Sigma \text{(out)}$]" :—

see footnote on p. 372. W. F. S.]

so that (54) becomes

$$\frac{3}{k^2 c^2} + i \sqrt{\left(\frac{3R}{2\pi c}\right)} = \frac{2\pi c}{R}, \quad \dots\dots\dots\dots\dots(58)$$

whence

$$k_1 + i k_2 = \sqrt{\left(\frac{3R}{2\pi c^3}\right)} \cdot \left\{1 + i \frac{\sqrt{3}}{2}\left(\frac{R}{2\pi c}\right)^{\frac{3}{2}}\right\}, \quad \dots\dots\dots(59)$$

and

$$\frac{2\pi k_2}{k_1} = \pi \sqrt{3}\left(\frac{R}{2\pi c}\right)^{\frac{3}{2}}. \quad \dots\dots\dots\dots\dots(60)$$

This gives the reduction of amplitude after one vibration. The decay is least when R is small relatively to c, although it is then estimated for a longer time.

The value found in (60) differs a little from that given in *Theory of Sound*, § 311, where the aperture is supposed to be surrounded by an infinite flange, the effect of which is to favour the propagation of energy away from the resonator.

So far we have supposed the boundary of the aperture to be circular. A comparison with the corresponding process in *Theory of Sound*, § 306 (after Helmholtz), shows that to the degree of approximation here attained the results may be extended to an elliptic aperture provided we replace R by

$$\frac{\pi R_1}{2F(e)}, \quad \dots\dots\dots\dots\dots\dots(61)$$

where R_1 denotes the semi-axis major of the ellipse, e the eccentricity, and F the symbol of the complete elliptic function of the first order. It is there further shown that for any form of aperture not too elongated, the truth is approximately represented if we take $\sqrt{(\sigma/\pi)}$ instead of the radius R of the circle, where σ denotes the *area* of aperture.

It would be of interest to ascertain the electric capacity of a disc of nearly circular outline to the next approximation involving the square of δR, the deviation of the radius in direction ω from the mean value. If $\delta R = \alpha_n \cos n\omega$, α_1 would not appear, and the effect of α_2 is known from the solution for the ellipse. For other values of n further investigation is required.

In the case of the ellipse elongated apertures are not excluded, provided of course that the longer diameter is small enough in comparison with the diameter of the sphere. When e is nearly equal to unity,

$$F(e) = \log\frac{4}{\sqrt{(1 - e^2)}} = \log\frac{4R_1}{R_2}, \quad \dots\dots\dots\dots(62)$$

R_2 being the semi-axis minor. The pitch of the resonator is now comparatively independent of the small diameter of the ellipse, the large diameter being given.

402.

ON THE PROPAGATION OF SOUND IN NARROW TUBES OF VARIABLE SECTION.

[*Philosophical Magazine*, Vol. XXXI. pp. 89—96, 1916.]

UNDER this head there are two opposite extreme cases fairly amenable to analytical treatment, (i) when the changes of section are so slow that but little alteration occurs within a wave-length of the sound propagated and (ii) when any change that may occur is complete within a distance small in comparison with a wave-length.

In the first case we suppose the tube to be of revolution. A very similar analysis would apply to the corresponding problem in two dimensions, but this is of less interest. If the velocity-potential ϕ of the simple sound be proportional to e^{ikat}, the equation governing ϕ is

$$\frac{d^2\phi}{dr^2} + \frac{1}{r}\frac{d\phi}{dr} + \frac{d^2\phi}{dx^2} + k^2\phi = 0, \quad \dots\dots\dots\dots(1)$$

where x is measured along the axis of symmetry and r perpendicular to it. Since there are no *sources* of sound along the axis, the appropriate solution is *

$$\phi = J_0\left\{r\sqrt{(d^2/dx^2 + k^2)}\right\} F(x), \quad \dots\dots\dots\dots(2)$$

in which F, a function of x only, is the value of ϕ when $r = 0$.

At the wall of the tube $r = y$, a known function of x; and the boundary condition, that the motion shall there be tangential, is expressed by

$$-\frac{d\phi}{dx}\frac{dy}{dx} + \frac{d\phi}{dr} = 0, \quad \dots\dots\dots\dots\dots\dots(3)$$

in which $(r = y)$

$$\frac{d\phi}{dx} = \frac{dF}{dx} - \frac{y^2}{2^2}\left(\frac{d^2}{dx^2} + k^2\right)\frac{dF}{dx} + \frac{y^4}{2^2 \cdot 4^2}\left(\frac{d^2}{dx^2} + k^2\right)^2\frac{dF}{dx} - \dots, \quad \dots(4)$$

$$\frac{d\phi}{dr} = -\frac{y}{2}\left(\frac{d^2}{dx^2} + k^2\right)F + \frac{y^3}{2^2 \cdot 4}\left(\frac{d^2}{dx^2} + k^2\right)^2 F - \dots. \quad \dots\dots\dots(5)$$

---

* Compare *Proc. Lond. Math. Soc.* Vol. VII. p. 70 (1876); *Scientific Papers*, Vol. I. p. 275.

Using these in (3), we obtain an equation which may be put into the form

$$\frac{d^2}{dx^2}(yF) + k^2(yF) = \frac{d^2y}{dx^2}F + \frac{y^2}{2}\frac{dy}{dx}\left(\frac{d^2}{dx^2} + k^2\right)\frac{dF}{dx}$$

$$-\frac{y^4}{32}\frac{dy}{dx}\left(\frac{d^2}{dx^2} + k^2\right)^2\frac{dF}{dx} + \dots + \frac{y^3}{8}\left(\frac{d^2}{dx^2} + k^2\right)^2 F - \dots \ \dots(6)$$

As a first approximation we may neglect all the terms on the right of (6), so that the solution is

$$F(x) = \frac{Ae^{-ikx} + Be^{ikx}}{y}, \ \dots\dots\dots\dots\dots\dots(7)$$

where A and B are constants. To the same approximation,

$$\frac{d^2F}{dx^2} + k^2F = -\frac{2}{y}\frac{dy}{dx}\frac{dF}{dx}. \ \dots\dots\dots\dots\dots(8)$$

For a second approximation we retain on the right of (6) all terms of the order d^2y/dx^2, or $(dy/dx)^2$. By means of (8) we find sufficiently for our purpose

$$\left(\frac{d^2}{dx^2} + k^2\right)\frac{dF}{dx} = -\frac{2}{y}\frac{dy}{dx}\frac{d^2F}{dx^2},$$

$$\left(\frac{d^2}{dx^2} + k^2\right)^2 F = 4\frac{d^2F}{dx^2}\left\{\frac{2}{y^2}\left(\frac{dy}{dx}\right)^2 - \frac{1}{y}\frac{d^2y}{dx^2}\right\},$$

$$\left(\frac{d^2}{dx^2} + k^2\right)^2\frac{dF}{dx} = 0, \quad \left(\frac{d^2}{dx^2} + k^2\right)^3 F = 0.$$

Our equation thus becomes

$$\left(\frac{d^2}{dx^2} + k^2\right)(yF) = \frac{d^2y}{dx^2}F - \frac{y^2}{2}\frac{d^2y}{dx^2}\frac{d^2F}{dx^2} = (1 + \tfrac{1}{2}k^2y^2)\frac{d^2y}{dx^2}.F(x), \quad \dots(9)$$

in which on the right the first approximation (7) suffices. Thus

$$yF(x) = \frac{1}{2ik}\left\{e^{ikx}\int Y(B + Ae^{-2ikx})\,dx - e^{-ikx}\int Y(A + Be^{2ikx})\,dx\right\}, \quad (10)$$

where

$$Y = \frac{1 + \tfrac{1}{2}k^2y^2}{y}\frac{d^2y}{dx^2}. \ \dots\dots\dots\dots\dots\dots(11)$$

In (10) the lower limit of the integrals is undetermined; if we introduce arbitrary constants, we may take the integration from $-\infty$ to x.

In order to attack a more definite problem, let us suppose that d^2y/dx^2, and therefore Y, vanishes everywhere except over the finite range from $x = 0$ to $x = b$, b being positive. When x is negative the integrals disappear, only the arbitrary constants remaining; and when x is positive the integrals may be taken from 0 to x.   As regards the values of the constants of integration (10) may be supposed to identify itself with (7) on the negative side.   Thus

$$yF(x) = e^{-ikx}\left\{A - \frac{1}{2ik}\int_0^x Y(A + Be^{2ikx})\,dx\right\}$$
$$+ e^{ikx}\left\{B + \frac{1}{2ik}\int_0^x Y(B + Ae^{-2ikx})\,dx\right\}. \quad\dots(12)$$

The integrals disappear when x is negative, and when x exceeds b they assume constant values.

Let us now further suppose that when x exceeds b there is no negative wave, *i.e.* no wave travelling in the negative direction.   The negative wave on the negative side may then be regarded as the *reflexion* of the there travelling positive wave.   The condition is

$$B\left\{1 + \frac{1}{2ik}\int_0^b Y\,dx\right\} + \frac{A}{2ik}\int_0^b Ye^{-2ikx}\,dx = 0, \quad\dots\dots\dots(13)$$

giving the reflected wave (B) in terms of the incident wave (A).   There is no reflexion if

$$\int_0^b Ye^{-2ikx}\,dx = 0; \quad\dots\dots\dots\dots\dots(14)$$

and then the transmitted wave ($x > b$) is given by

$$F(x) = \frac{Ae^{-ikx}}{y}\left\{1 - \frac{1}{2ik}\int_0^b Y\,dx\right\}. \quad\dots\dots\dots\dots(15)$$

Even when there is reflexion, it is at most of the second order of smallness, since Y is of that order.   For the transmitted wave our equations give ($x > b$)

$$F(x) = \frac{Ae^{-ikx}}{y}\left\{1 - \frac{1}{2ik}\int_0^b Y\,dx - \frac{1}{4k^2}\frac{\int_0^b Ye^{-2ikx}\,dx.\int_0^b Ye^{2ikx}\,dx}{1 + \frac{1}{2ik}\int_0^b Y\,dx}\right\}; \quad(16)$$

but if we stop at the second order of smallness the last part is to be omitted, and (16) reduces to (15).   It appears that to this order of approximation the intensity of the transmitted sound is equal to that of the incident sound, at least if the tube recovers its original diameter.   If the final value of y differs from the initial value, the intensity is changed so as to secure an equal propagation of energy.

The effect of Y in (15) is upon the *phase* of the transmitted wave.   It appears, rather unexpectedly, that there is a linear *acceleration* amounting to

$$\frac{1}{2k^2}\int_0^b Y\,dx, \quad\dots\dots\dots\dots\dots\dots(17)$$

or, since the ends of the disturbed region at 0 and b are cylindrical,

$$\frac{1}{2k^2}\int_0^b \frac{1}{y^2}\left(\frac{dy}{dx}\right)^2 (1 - \tfrac{1}{2}k^2y^2)\, dx, \quad\ldots\ldots\ldots\ldots\ldots(18)$$

from which the term in k^2y^2 may be dropped.

That the reflected wave should be very small when the changes are sufficiently gradual is what might have been expected. We may take (13) in the form

$$\frac{B}{A} = \frac{i}{2k}\int_0^b Ye^{-2ikx}\, dx = \frac{i}{2k}\int_0^b \frac{1}{y}\frac{d^2y}{dx^2}e^{-2ikx}\, dx. \quad\ldots\ldots\ldots(19)$$

As an example let us suppose that from $x = 0$ to $x = b$

$$y = y_0 + \eta\, (1 - \cos mx), \quad\ldots\ldots\ldots\ldots\ldots(20)$$

where y_0 is the constant value of y outside the region of disturbance, and $m = 2\pi/b$. If we suppose further that η is small, we may remove $1/y$ from under the sign of integration, so that

$$\frac{B}{A} = \frac{im^2\eta}{2ky_0}\int_0^b \cos mx\, e^{-2ikx}\, dx = \frac{m^2\eta}{4k^2y_0}\{1 - \cos 2kb + i\sin 2kb\}. \quad\ldots(21)$$

Independently of the last factor (which may vanish in certain cases) B is very small in virtue of the factors m^2/k^2 and η/y_0.

In the second problem proposed we consider the passage of waves proceeding in the positive direction through a tube (not necessarily of revolution) of uniform section σ_1 and impinging on a region of irregularity, whose length is small compared with the wave-length (λ). Beyond this region the tube again becomes regular of section σ_2 (fig. 1). It is convenient to imagine the

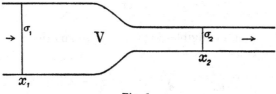

Fig. 1.

axes of the initial and final portions to be coincident, but our principal results will remain valid even when the irregularity includes a bend. We seek to determine the transmitted and reflected waves as proportional to the given incident wave.

The velocity-potentials of the incident and reflected waves on the left of the irregularity and of the transmitted wave on the right are represented respectively by

$$\phi_1 = A e^{-ikx} + Be^{ikx}, \qquad \phi_2 = Ce^{-ikx}; \quad\ldots\ldots\ldots\ldots(22)$$

so that at x_1 and x_2 we have

$$\phi_1 = A\,e^{-ikx_1} + B\,e^{ikx_1}, \qquad \phi_2 = C\,e^{-ikx_2}, \quad \dots\dots\dots\dots(23)$$

$$d\phi_1/dx = ik\,(-\,A\,e^{-ikx_1} + B\,e^{ikx_1}), \qquad d\phi_2/dx = -\,ikC\,e^{-ikx_2}. \quad \dots(24)$$

When λ is sufficiently great we may ignore altogether the space between x_1 and x_2, that is we may suppose that the pressures are the same at these two places and that the total flow is also the same, as if the fluid were incompressible. As there is now no need to distinguish between x_1 and x_2, we may as well suppose both to be zero. The condition $\phi_1 = \phi_2$ gives

$$A + B = C, \quad \dots\dots\dots\dots\dots\dots\dots\dots\dots(25)$$

and the condition $\sigma_1 d\phi_1/dx = \sigma_2 d\phi_2/dx$ gives

$$\sigma_1\,(-\,A + B) = -\,\sigma_2 C. \quad \dots\dots\dots\dots\dots\dots(26)$$

Thus

$$\frac{B}{A} = \frac{\sigma_1 - \sigma_2}{\sigma_1 + \sigma_2}, \qquad \frac{C}{A} = \frac{2\sigma_1}{\sigma_1 + \sigma_2}. \quad \dots\dots\dots\dots\dots(27)$$

These are Poisson's formulæ*. If σ_1 and σ_2 are equal, we have of course $B = 0$, $C = A$. Our task is now to proceed to a closer approximation, still supposing that the region of irregularity is small.

For this purpose both of the conditions just now employed need correction. Since the volume V of the irregular region is to be regarded as sensible and the fluid is really susceptible of condensation (s), we have

$$V\frac{ds}{dt} = \sigma_1\frac{d\phi_1}{dx_1} - \sigma_2\frac{d\phi_2}{dx_2},$$

and since in general $s = -\,a^{-2}d\phi/dt$, we may take

$$\frac{ds}{dt} = -\,a^{-2}\frac{d^2\phi_1}{dt^2} \quad \text{or} \quad -\,a^{-2}\frac{d^2\phi_2}{dt^2},$$

the distinction being negligible in this approximation in virtue of the smallness of V. Thus

$$\sigma_1\frac{d\phi_1}{dx_1} - \sigma_2\frac{d\phi_2}{dx_2} = -\,\frac{V}{a^3}\frac{d^2\phi_2}{dt^2} = k^2 V\phi_2. \quad \dots\dots\dots\dots(28)$$

In like manner, assimilating the flow to that of an incompressible fluid, we have for the second condition

$$\phi_2 - \phi_1 = R\sigma_2\frac{d\phi_2}{dx_2}, \quad \dots\dots\dots\dots\dots\dots(29)$$

where R may be defined in electrical language as the *resistance* between x_1 and x_2, when the material supposed to be bounded by non-conducting walls coincident with the walls of the tube is of unit specific resistance.

---

* Compare *Theory of Sound*, § 264.

In substituting the values of ϕ and $d\phi/dx$ from (23), (24) it will shorten our expressions if for the time we merge the exponentials in the constants, writing

$$A' = Ae^{-ikx_1}, \quad B' = Be^{ikx_1}, \quad C' = Ce^{-ikx_2}. \quad\ldots\ldots\ldots\ldots(30)$$

Thus
$$\sigma_1(-A' + B') + \sigma_2 C' = -ikVC', \quad\ldots\ldots\ldots\ldots(31)$$

$$A' + B' - C' = ik\sigma_2 RC'. \quad\ldots\ldots\ldots\ldots\ldots(32)$$

We may check these equations by applying them to the case where there is really no break in the regularity of the tube, so that

$$\sigma_1 = \sigma_2, \quad V = (x_2 - x_1)\,\sigma, \quad R = (x_2 - x_1)/\sigma.$$

Then (31), (32) give $B' = 0$, or $B = 0$, and

$$\frac{C'}{A'} = \frac{1}{1 + ik\,(x_2 - x_1)} = e^{-ik\,(x_2 - x_1)},$$

with sufficient approximation. Thus

$$C'e^{ikx_2} = A'e^{ikx_1}, \quad \text{or} \quad C = A.$$

The undisturbed propagation of the waves is thus verified.

In general,

$$\frac{B'}{A'} = \frac{\sigma_1 - \sigma_2 + ik\,(\sigma_1\sigma_2 R - V)}{\sigma_1 + \sigma_2 + ik\,(\sigma_1\sigma_2 R + V)}, \quad\ldots\ldots\ldots\ldots(33)$$

$$\frac{C'}{A'} = \frac{2\sigma_1}{\sigma_1 + \sigma_2 + ik\,(\sigma_1\sigma_2 R + V)}. \quad\ldots\ldots\ldots\ldots(34)$$

When $\sigma_1 - \sigma_2$ is finite, the effect of the new terms is only upon the phases of the reflected and transmitted waves. In order to investigate changes of intensity we should need to consider terms of still higher order.

When $\sigma_1 = \sigma_2$, we have

$$C' = A'\left\{1 - \frac{ik}{2\sigma}\,(\sigma^2 R + V)\right\} = A'e^{-ik(\sigma^2 R + V)/2\sigma},$$

$$C = A\,e^{ik(x_2 - x_1 - \frac{1}{2}\sigma R - V/2\sigma)}, \quad\ldots\ldots\ldots\ldots\ldots\ldots(35)$$

making, as before, $C = A$, if there be no interruption. Also, when $\sigma_1 = \sigma_2$ absolutely,

$$\frac{B'}{A'} = \frac{ik\,(\sigma^2 R - V)}{2\sigma}, \quad\ldots\ldots\ldots\ldots\ldots(36)$$

indicating a change of phase of $90°$, and an intensity referred to that of the incident waves equal to

$$\frac{k^2\,(\sigma^2 R - V)^2}{4\sigma^2} \quad\ldots\ldots\ldots\ldots\ldots(37)$$

As an example let us take the case of a tube of revolution for which y, being equal to y_0 over the regular part, becomes $y_0 + \delta y$ between x_1 and x_2. We have

$$\frac{V}{2\sigma} = \frac{1}{2}(x_2 - x_1) + \int \frac{\delta y}{y_0} dx + \frac{1}{2} \int \left(\frac{\delta y}{y_0}\right)^2 dx.$$

Also *

$$\frac{1}{2}\sigma R = \frac{1}{2} y_0^2 \int \frac{dx}{y^2} \left\{1 + \frac{1}{2}\left(\frac{dy}{dx}\right)^2\right\} dx$$

$$= \frac{1}{2}(x_2 - x_1) - \int \frac{\delta y}{y_0} dx + \frac{3}{2} \int \left(\frac{\delta y}{y_0}\right)^2 dx + \frac{1}{4} \int \left(\frac{d\delta y}{dx}\right)^2 dx,$$

and

$$\frac{V}{2\sigma} + \frac{\sigma R}{2} = (x_2 - x_1) + \int_{x_1}^{x_2} \left\{2 \left(\frac{\delta y}{y_0}\right)^2 + \frac{1}{4}\left(\frac{d\delta y}{dx}\right)^2\right\} dx, \quad \text{.........(38)}$$

the terms of the first order in δy disappearing. Thus in the exponent of (35)

$$x_2 - x_1 - \tfrac{1}{2}\sigma R - V/2\sigma = -\int \left\{2 (\delta y/y_0)^2 + \tfrac{1}{4}(d\delta y/dx)^2\right\} dx, \quad \text{...(39)}$$

of which the right-hand member, taken with the positive sign, expresses the retardation of the transmitted wave due to the departure from regularity.

---

* *Theory of Sound*, § 308.

403.

ON THE ELECTRICAL CAPACITY OF APPROXIMATE SPHERES AND CYLINDERS.

[*Philosophical Magazine*, Vol. XXXI. pp. 177—186, March 1916.]

MANY years ago I had occasion to calculate these capacities[*] so far as to include the squares of small quantities, but only the results were recorded. Recently, in endeavouring to extend them, I had a little difficulty in retracing the steps, especially in the case of the cylinder. The present communication gives the argument from the beginning. It may be well to remark at the outset that there is an important difference between the two cases. The capacity of a sphere situated in the open is finite, being equal to the radius. But when we come to the cylinder, supposed to be entirely isolated, we have to recognize that the capacity reckoned per unit length is infinitely small. If a be the radius of the cylinder and b that of a coaxal enveloping case at potential zero, the capacity of a length l is[†]

$$\frac{\frac{1}{2}l}{\log\,(b/a)},$$

which diminishes without limit as b is increased. For clearness it may be well to retain the enveloping case in the first instance.

In the intervening space we may take for the potential in terms of the usual polar coordinates

$$\phi = H_0 \log\,(r/b) + H_1 r^{-1} \cos\,(\theta - \epsilon_1) + K_1 r \cos\,(\theta - \epsilon_1') + \dots$$
$$+ H_n r^{-n} \cos\,(n\theta - \epsilon_n) + K_n r^n \cos\,(n\theta - \epsilon_n').$$

Since $\phi = 0$ when $r = b$,

$$\epsilon_n' = \epsilon_n, \quad K_n = - H_n b^{-2n},$$

and

$$\phi = H_0 \log\,(r/b) + H_1 \left(\frac{1}{r} - \frac{r}{b^2}\right) \cos\,(\theta - \epsilon_1) + H_2 \left(\frac{1}{r^2} - \frac{r^2}{b^4}\right) \cos\,(2\theta - \epsilon_2) + \dots .$$
$$\dots\dots\dots\dots(1)$$

---

[*] "On the Equilibrium of Liquid Conducting Masses charged with Electricity," *Phil. Mag.* Vol. XIV. p. 184 (1882); *Scientific Papers*, Vol. II. p. 130.

[†] Maxwell's *Electricity*, § 126.

At this stage we may suppose b infinite in connexion with H_1, H_2, &c., so that the positive powers of r disappear. For brevity we write $\cos(n\theta - \epsilon_n) = F_n$, and we replace r^{-1} by u. Thus

$$\phi = - H_0 \log(ub) + H_1 u F_1 + H_2 u^2 F_2 + \dots \quad \dots\dots\dots(2)$$

We have now to make $\phi = \phi_1$ at the surface of the approximate cylinder, where ϕ_1 is constant and

$$u = u_0 + \delta u = u_0(1 + C_1 G_1 + C_2 G_2 + \dots).$$

Herein
$$G_n = \cos(n\theta - e_n),$$

and the C's are small constants. So far as has been proved, e_n might differ from ϵ_n, but the approximate identity may be anticipated, and at any rate we may assume for trial that it exists and consider G_n to be the same as F_n, making

$$u = u_0 + \delta u = u_0(1 + C_1 F_1 + C_2 F_2 + \dots). \quad \dots\dots\dots(3)$$

On the cylinder we have

$$\phi_1 = - H_0 \log(u_0 b) + H_1 u_0 F_1 + H_2 u_0^2 F_2 + \dots$$
$$+ \frac{\delta u}{u_0}\{- H_0 + H_1 u_0 F_1 + 2 H_2 u_0^2 F_2 + 3 H_3 u_0^3 F_3 + \dots\}$$
$$+ \left(\frac{\delta u}{u_0}\right)^2 \{\tfrac{1}{2} H_0 + H_2 u_0^2 F_2 + 3 H_3 u_0^3 F_3 + \dots + \tfrac{1}{2} p(p-1) H_p u_0^p F_p\}, \dots(4)$$

and in this

$$\delta u/u_0 = C_1 F_1 + C_2 F_2 + C_3 F_3 + \dots \quad \dots\dots\dots(5)$$

The electric charge Q, reckoned per unit length of the cylinder, is readily found from (2). We have, integrating round an enveloping cylinder of radius r,

$$Q = -\frac{1}{4\pi}\int_0^{2\pi} \frac{d\phi}{dr} r\, d\theta = -\frac{H_0}{2}, \quad \dots\dots\dots(6)$$

and Q/ϕ_1 is the *capacity*.

We now introduce the value of $\delta u/u_0$ from (5) into (4) and make successive approximations. The value of H_n is found by multiplication of (4) by F_n, where $n = 1, 2, 3$, &c., and integration with respect to θ between 0 and 2π, when products such as $F_1 F_2$, $F_2 F_3$, &c., disappear. For the first step, where C^2 is neglected, we have

$$0 = H_n u_0^n \int F_n^2 d\theta - H_0 C_n \int F_n^2 d\theta, \quad \dots\dots\dots(7)$$

or
$$H_n u_0^n = H_0 C_n. \quad \dots\dots\dots(8)$$

Direct integration of (4) gives also

$$\phi_1 = - H_0 \log(u_0 b) + \int \frac{d\theta}{2\pi} \frac{\delta u}{u_0}\{H_1 u_0 F_1 + 2 H_2 u_0^2 F_2$$
$$+ 3 H_3 u_0^3 F_3 + \dots\} + \tfrac{1}{2} H_0 \int \frac{d\theta}{2\pi}\left(\frac{\delta u}{u_0}\right)^2, \quad \dots\dots(9)$$

cubes of C being neglected at this stage. On introduction of the value of H_n from (8) and of δu from (5),

$$\phi_1 = - H_0 \log (u_0 b) + \tfrac{1}{4} H_0 \{3C_1^2 + 5C_2^2 + 7C_3^2 + \ldots\}. \quad \ldots\ldots\ldots(10)$$

Thus
$$\phi_1/Q = 2 \log (u_0 b) - \tfrac{1}{2} \{3C_1^2 + 5C_2^2 + 7C_3^2 = \ldots\}. \quad \ldots\ldots\ldots(11)$$

In the application to an electrified liquid considered in my former paper, it must be remembered that u_0 is not constant during the deformation. If the liquid is incompressible, it is the volume, or in the present case the sectional area (σ), which remains constant. Now

$$2\sigma = \int_0^{2\pi} \frac{d\theta}{(u_0 + \delta u)^2} = \frac{1}{u_0^2} \int d\theta \left\{1 - 2\frac{\delta u}{u_0} + 3\left(\frac{\delta u}{u_0}\right)^2\right\}$$

$$= \frac{2\pi}{u_0^2} \{1 + \tfrac{3}{2} (C_1^2 + C_2^2 + C_3^2 + \ldots)\},$$

so that if a denote the radius of the circle whose area is σ,

$$u_0^2 = a^{-2} \{1 + \tfrac{3}{2} (C_1^2 + C_2^2 + C_3^2 + \ldots). \quad \ldots\ldots\ldots\ldots(12)$$

Accordingly,
$$\log u_0^2 = - 2 \log a + \tfrac{3}{2} (C_1^2 + C_2^2 + C_3^2 + \ldots),$$

and (11) becomes

$$\phi_1/Q = 2 \log (b/a) - C_2^2 - 2C_3^2 - \ldots - (p-1) C_p^2, \quad \ldots\ldots\ldots(13)$$

the term in C_1 disappearing, as was to be expected.

The potential energy of the charge is $\tfrac{1}{2}\phi_1 Q$. If the change of potential energy due to the deformation be called P', we have

$$P' = - \tfrac{1}{2} Q^2 \{C_2^2 + 2C_3^2 + \ldots + (p-1) C_p^2\}, \quad \ldots\ldots\ldots(14)$$

in agreement with my former results.

There are so few forms of surface for which the electric capacity can be calculated that it seems worth while to pursue the approximation beyond that attained in (11), supposing, however, that all the ϵ's vanish, everything being symmetrical about the line $\theta = 0$. Thus from (4), as an extension of (7) with inclusion of C^2,

$$0 = (H_n u_0^n - H_0 C_n) \int \frac{d\theta}{2\pi} F_n^2$$

$$+ \int \frac{d\theta}{2\pi} F_n (C_1 F_1 + C_2 F_2 + \ldots)(H_1 u_0 F_1 + 2H_2 u_0^2 F_2 + 3H_3 u_0^3 F_3 + \ldots)$$

$$+ \frac{H_0}{2} \int \frac{d\theta}{2\pi} F_n (C_1 F_1 + C_2 F_2 + C_3 F_3 + \ldots)^2, \quad \ldots\ldots\ldots\ldots(15)$$

or with use of (8)

$$u_0^n H_n/H_0 = C_n - \int_0^{2\pi} \frac{d\theta}{2\pi} F_n (C_1 F_1 + C_2 F_2 + C_3 F_3 + \ldots)$$

$$(3C_1 F_1 + 5C_2 F_2 + 7C_3 F_3 + \ldots); \quad \ldots\ldots\ldots(16)$$

by which H_n is determined by means of definite integrals of the form

$$\int_0^{2\pi} F_n F_p F_q \, d\theta, \quad \dots\dots\dots\dots\dots\dots(17)$$

n, p, q being positive integers. It will be convenient to denote the integral on the right of (16) by I_n, I_n being of the second order in the C's.

Again, by direct integration of (4) with retention of C^3,

$$\phi_1 = -H_0 \log(u_0 b) + \int \frac{d\theta}{2\pi} (C_1 F_1 + C_2 F_2 + C_3 F_3 + \dots)$$

$$(H_1 u_0 F_1 + 2H_2 u_0^2 F_2 + 3H_3 u_0^3 F_3 + \dots)$$

$$+ \tfrac{1}{2} H_0 \int \frac{d\theta}{2\pi} (C_1 F_1 + C_2 F_2 + C_3 F_3 + \dots)^2$$

$$+ \int \frac{d\theta}{2\pi} (C_1 F_1 + C_2 F_2 + C_3 F_3 + \dots)^2 \{ H_2 u_0^2 F_2 + 3 H_3 u_0^3 F_3$$

$$+ \dots + \tfrac{1}{2} p(p-1) H_p u_0^p F_p \}.$$

In the last integral we may substitute the first approximate value of H_p from (8). Thus in extension of (11)

$$\phi_1/Q = 2 \log(u_0 b) - \tfrac{1}{2} \{ 3C_1^2 + 5C_2^2 + 7C_3^2 + \dots \} + C_1 I_1 + 2C_2 I_2 + 3C_3 I_3 + \dots$$

$$- \int_0^{2\pi} \frac{d\theta}{\pi} (C_1 F_1 + C_2 F_2 + C_3 F_3 + \dots)^2 \{ C_2 F_2 + 3C_3 F_3 + \dots$$

$$+ \tfrac{1}{2} p(p-1) C_p F_p \}. \quad \dots\dots\dots(18)$$

The additional integrals required in (18) are of the same form (17) as those needed for I_n.

As regards the integral (17), it may be written

$$\int_0^{2\pi} d\theta \cos n\theta \cos p\theta \cos q\theta.$$

Now four times the latter integral is equal to the sum of integrals of cosines of $(n - p - q)\theta$, $(n - p + q)\theta$, $(n + p - q)\theta$, and $(n + p + q)\theta$, of which the last vanishes in all cases. We infer that (17) vanishes unless one of the three quantities n, p, q is equal to the sum of the other two. In the excepted cases

$$(17) = \tfrac{1}{2}\pi. \quad \dots\dots\dots\dots\dots\dots(19)$$

If p and q are equal, (17) vanishes unless $n = 2p$; also whenever n, p, q are all odd.

We may consider especially the case in which only C_p occurs, so that

$$u = u_0(1 + C_p \cos p\theta). \quad \dots\dots\dots\dots(20)$$

In (16) $$I_n = (2p+1) C_p^2 \int_0^{2\pi} \frac{d\theta}{2\pi} F_n F_p^2,$$

so that I_n vanishes unless $n = 2p$. But I_{2p} disappears in (18), presenting itself only in association with C_{2p}, which we are supposing not to occur. Also the last integral in (18) makes no contribution, reducing to

$$\tfrac{1}{2}p\,(p-1)\,C_p{}^3 \int_0^{2\pi} \frac{d\theta}{\pi}\cos^3 p\theta,$$

which vanishes.   Thus

$$\phi_1/Q = 2\log{(u_0 b)} - (p + \tfrac{1}{2})\,C_p{}^2, \quad\ldots\ldots\ldots\ldots\ldots(21)$$

the same as in the former approximation, as indeed might have been anticipated, since a change in the sign of C_p amounts only to a shift in the direction from which θ is measured.

The corresponding problem for the approximate sphere, to which we now proceed, is simpler in some respects, though not in others. In the general case u, or r^{-1}, is a function of the two angular polar coordinates θ, ω, and the expansion of δu is in Laplace's functions. When there is symmetry about the axis, ω disappears and the expansion involves merely the Legendre functions $P_n(\mu)$, in which $\mu = \cos\theta$. Then

$$u = u_0 + \delta u = u_0\,\{1 + C_1 P_1(\mu) + C_2 P_2(\mu) + \ldots\}, \quad\ldots\ldots\ldots(22)$$

where C_1, C_2, ... are to be regarded as small. We will assume δu to be of this form, though the restriction to symmetry makes no practical difference in the solution so far as the second order of small quantities.

For the form of the potential (ϕ) outside the surface, we have

$$\phi = H_0 u + H_1 u^2 P_1(\mu) + H_2 u^3 P_2(\mu) + \ldots; \quad\ldots\ldots\ldots\ldots(23)$$

and on the surface

$$\phi_1 = H_0 u_0 + H_1 u_0{}^2 P_1 + H_2 u_0{}^3 P_2 + \ldots$$

$$+ \delta u\,\{H_0 + 2u_0 H_1 P_1 + 3u_0{}^2 H_2 P_2 + \ldots\}$$

$$+ (\delta u)^2\,\{H_1 P_1 + 3u_0 H_2 P_2 + \ldots + \tfrac{1}{2}p\,(p+1)\,u_0{}^{p-1} H_p P_p\}, \quad\ldots(24)$$

in which we are to substitute the values of δu, $(\delta u)^2$ from (22). In this equation ϕ_1 is constant, and H_1, H_2, ... are small in comparison with H_0.

The procedure corresponds closely with that already adopted for the cylinder. We multiply (24) by P_n, where n is a positive integer, and integrate with respect to μ over angular space, i.e. between -1 and $+1$. Thus, omitting the terms of the second order, we get

$$u_0{}^n H_n = - H_0 C_n, \quad\ldots\ldots\ldots\ldots\ldots\ldots(25)$$

as a first approximation to the value of H_n.

Direct integration of (24) gives

$$\phi_1 \int d\mu = H_0 u_0 \int d\mu + u_0 \int \{C_1 P_1 + C_2 P_2 + \ldots\}\{2u_0 H_1 P_1 + 3u_0^2 H_2 P_2 + \ldots\}\, d\mu$$

$$= H_0 u_0 \int d\mu + u_0 \int \{2u_0 H_1 C_1 P_1^2 + 3u_0^2 H_2 C_2 P_2^2 + 4u_0^3 H_3 C_3 P_3^2 + \ldots\}\, d\mu,$$

or on substitution for H_n from (25)

$$\phi_1 = H_0 u_0 \left\{1 - \tfrac{2}{3}C_1^2 - \tfrac{3}{5}C_2^2 - \ldots - \frac{p+1}{2p+1}\,C_p^2\right\}, \qquad \ldots\ldots\ldots(26)$$

inasmuch as
$$\int_{-1}^{+1} P_p^2(\mu)\, d\mu = \frac{2}{2p+1}. \qquad \ldots\ldots\ldots\ldots\ldots\ldots(27)$$

As appears from (23), H_0 is identical with the electric charge upon the sphere, which we may denote by Q, and Q/ϕ_1 is the electrostatic capacity, so that to this order of approximation

$$\text{Capacity} = u_0^{-1}\left\{1 + \tfrac{2}{3}C_1^2 + \ldots + \frac{p+1}{2p+1}\,C_p^2\right\}. \qquad \ldots\ldots\ldots(28)$$

Here, again, we must remember that u_0^{-1} differs from the radius of the true sphere whose volume is equal to that of the approximate sphere under consideration. If that radius be called a

$$u_0^{-1} = a\left\{1 - \frac{2C_1^2}{3} - \frac{2C_2^2}{5} - \ldots - \frac{2C_p^2}{2p+1}\right\}, \qquad \ldots\ldots\ldots(29)$$

and
$$\text{Capacity} = a\left\{1 + \frac{C_2^2}{5} + \ldots + \frac{p-1}{2p+1}\,C_p^2\right\}, \qquad \ldots\ldots\ldots(30)$$

in which C_1 does not appear.

The potential energy of the charge is $\tfrac{1}{2}Q^2 \div$ Capacity. Reckoned from the initial configuration $(C = 0)$, it is

$$P' = -\frac{Q^2}{2a}\left\{\frac{C_2^2}{5} + \ldots + \frac{p-1}{2p+1}\,C_p^2\right\}. \qquad \ldots\ldots\ldots\ldots(31)$$

It has already been remarked that to this order of approximation the restriction to symmetry makes little difference. If we take

$$\delta u / u_0 = F_1 + F_2 + \ldots + F_p, \qquad \ldots\ldots\ldots\ldots\ldots(32)$$

where the F's are Laplace's functions,

$$\frac{1}{4\pi} \iint F_p^2\, d\mu\, d\omega \quad \text{corresponds to} \quad \frac{C_p^2}{2p+1}.$$

This substitution suffices to generalize (30), (31), and the result is in harmony with that formerly given.

The expression for the capacity (30) may be tested on the case of the planetary ellipsoid of revolution for which the solution is known*. Here

---
* Maxwell's *Electricity*, § 151.

$C_2 = \frac{1}{3}e^2$, e being the eccentricity. It must be remembered that a in (30) is not the semi-axis major, but the spherical radius of equal volume. In terms of the semi-axis major (α), the accurate value of the capacity is $\alpha e/\sin^{-1} e$.

We may now proceed to include the terms of the next order in C. The extension of (25) is

$$u_0{}^n H_n \; H_0 = - C_n + \tfrac{1}{2}(2n+1) \int_{-1}^{+1} d\mu \, P_n \, \{C_1 P_1 + \ldots + C_p P_p\}$$
$$\{2C_1 P_1 + \ldots + (q+1) C_q P_q\}, \quad \ldots\ldots(33)$$

where in the small term the approximate value of H_n from (25) has been substituted. We set

$$\int_{-1}^{+1} d\mu \, P_n \, \{C_1 P_1 + \ldots + C_p P_p\} \{2C_1 P_1 + \ldots + (q+1) C_q P_q\} = J_n, \; \ldots(34)$$

where J_n is of order C^2 and depends upon definite integrals of the form

$$\int_{-1}^{+1} P_n P_p P_q \, d\mu, \; \ldots\ldots\ldots\ldots\ldots\ldots\ldots\ldots\ldots(35)$$

n, p, q being positive integers.

In like manner the extension of (26) is

$$\phi_1/Qu_0 = 1 - \tfrac{2}{3}C_1{}^2 - \tfrac{3}{5}C_2{}^2 - \ldots - \frac{p+1}{2p+1} C_p{}^2 + \tfrac{1}{2}\{2C_1 J_1 + 3C_2 J_2 + 4C_3 J_3 + \ldots\}$$

$$- \tfrac{1}{2}\int_{-1}^{+1} d\mu \, (C_1 P_1 + C_2 P_2 + \ldots)^2 \, \{C_1 P_1 + 3C_2 P_2 + \ldots + \tfrac{1}{2}p \, (p+1) C_p P_p\}. \quad (36)$$

Here, again, the definite integrals required are of the form (35).

These definite integrals have been evaluated by Ferrers* and Adams†. In Adams' notation $n + p + q = 2s$, and

$$(35) = \frac{2}{2s+1} \cdot \frac{A\,(s-n) \cdot A\,(s-p) \cdot A\,(s-q)}{A\,(s)}, \quad \ldots\ldots(37)$$

where
$$A\,(n) = \frac{1 \cdot 3 \cdot 5 \ldots (2n-1)}{1 \cdot 2 \cdot 3 \ldots n}. \; \ldots\ldots\ldots\ldots\ldots(38)\ddagger$$

In order that the integral may be finite, no one of the quantities n, p, q must be greater than the sum of the other two, and $n + p + q$ must be an even integer. The condition in order that the integral may be finite is less severe than we found before in the two dimensional problem, and this, in general, entails a greater complication.

But the case of a single term in δu, say $C_p P_p (\mu)$, remains simple. In (36) J_n occurs only when multiplied by C_n, so that only J_p appears, and

$$J_p = (p+1) C_p{}^2 \int P_p{}^3 \, d\mu. \; \ldots\ldots\ldots\ldots\ldots(39)$$

---

* *Spherical Harmonics*, London, 1877, p. 156.
† *Proc. Roy. Soc.* Vol. xxvii. p. 63 (1878).
‡ [Following Adams, $A\,(o)$ must be taken as equal to unity. W. F. S.]

Thus (36) becomes

$$\phi_1/Qu_0 = 1 - \frac{p+1}{2p+1}\, C_p{}^2 + \frac{(p+1)(p+2)}{4}\, C_p{}^3 \int_{-1}^{+1} P_p{}^3\, d\mu. \quad .....(40)$$

When p is odd, the integral vanishes, and we fall back upon the former result; when p is even, by (37), (38),

$$\int_{-1}^{+1} P_p{}^3\, d\mu = \frac{2}{3p+1}\, \frac{\{A\,(\tfrac{1}{2}p)\}^3}{A\,(\tfrac{3}{2}p)}. \quad ....(41)$$

For example, if $p = 2$,

$$\int_{-1}^{+1} P_2{}^3\, d\mu = \tfrac{4}{35},$$

and

$$\phi_1/Qu_0 = 1 - \tfrac{3}{5}C_2{}^2 + \tfrac{12}{35}C_2{}^3. \quad ....(42)$$

Again, if two terms with coefficients C_p, C_q occur in δu, we have to deal only with J_p, J_q. The integrals to be evaluated are limited to

$$\int P_p{}^3\, d\mu, \quad \int P_p{}^2 P_q\, d\mu, \quad \int P_p P_q{}^2\, d\mu, \quad \int P_q{}^3\, d\mu.$$

If p be odd, the first and third of these vanish, and if q be odd the second and fourth. If p and q are both odd, the terms of the third order in C disappear altogether.

As appears at once from (34), (36), the last statement may be generalized. However numerous the components may be, if only odd suffixes occur, the terms of the third order disappear and (36) reduces to (26).

[1917. *Conf. Cisotti, R. Ist. Lombardo Rend.* Vol. XLIX. May, 1916.

In his Kelvin lecture (*Journ. Inst. El. Eng.* Vol. XXXV. Dec. 1916), Dr A. Russell quotes K. Aichi as pointing out that the capacity of an ellipsoidal conductor is given very approximately by $(S/4\pi)^{\frac{1}{2}}$, where S is the *surface* of the ellipsoid, and he further shows that this expression gives approximate values for the capacity in a variety of other calculable cases. As applied to an ellipsoid of revolution, his equation (6) gives

$$\text{Capacity} = \sqrt{\left(\frac{S}{4\pi}\right)} \cdot \left\{ 1 \pm \frac{2e^6}{945} \right\}, \quad ....(43)$$

where e is the eccentricity of the generating ellipse, the *plus* sign relating to the *prolatum* and the *minus* to the *oblatum*. It may thus be of interest to obtain the formula by which u_0 in (28) is expressed in terms of S rather than, as in (29), (30), by the *volume* of the conductor. For a reason which will presently appear it is desirable to include the *cube* of the particular coefficient C_2.

In terms of u, equal to $1/r$, the general formula for S is

$$S = 2\pi \int_0^{\pi} \frac{\sin\theta\, d\theta}{u^2} \sqrt{\left\{1 + \frac{1}{u^2}\left(\frac{du}{d\theta}\right)^2\right\}}. \quad \dots\dots\dots(44)$$

By (22) $\dfrac{1}{u^2}\left(\dfrac{du}{d\theta}\right)^2 = \sin^2\theta\,(C_1 P_1' + C_2 P_2' + \dots)^2(1 - 2C_2 P_2)$,

and hence with regard to well-known properties of Legendre's functions we find

$$S = \frac{4\pi}{u_0^2}\left[1 + \tfrac{4}{3}C_1^2 + \tfrac{6}{5}C_2^2 + \dots + \frac{p(p+1)+6}{2(2p+1)}C_p^2\right.$$
$$\left. - C_2^3 \int_{-1}^{+1} d\mu\,\{(1-\mu^2)\,P_2 P_2'^2 + 2P_2^3\}\right].$$

By (41) $\displaystyle\int P_2^3\, d\mu = 4/35$,

and by use of the particular form of P_2 we readily find

$$\int_{-1}^{+1} d\mu\,(1-\mu^2)\,P_2 P_2'^2 = 12/35.$$

Accordingly

$$\sqrt{\left(\frac{S}{4\pi}\right)} = u_0^{-1}\left\{1 + \tfrac{2}{3}C_1^2 + \tfrac{3}{5}C_2^2 + \frac{p(p+1)+6}{4(2p+1)}C_p^2 - \tfrac{2}{7}C_2^3\right\}. \quad \dots(45)$$

If we omit C_2^3 and combine (45) with (28), we get

$$\text{Capacity} = \sqrt{\left(\frac{S}{4\pi}\right)} \cdot \left\{1 - \frac{C_3^2}{14} - \dots - \frac{(p-1)(p-2)}{4(2p+1)}C_p^2\right\}, \quad \dots(46)$$

the terms in C_1 and C_2 disappearing. When the cubes of the C's are neglected, the capacity is *less* than $\sqrt{(S/4\pi)}$, the radius of the sphere of equal surface. If the surface be symmetrical with respect to the equatorial plane, as in the case of ellipsoids, the C's of odd order do not occur, so that the earliest in (46) is C_4.

For a *prolatum* of minor axis $2b$ and eccentricity e,

$$u^2 = b^{-2}(1 - \mu^2 e^2),$$

whence $u = u_0(1 - \tfrac{1}{3}e^2 P_2 + \text{terms in } e^4)$,

so that $C_2 = -\tfrac{1}{3}e^2$, C_4 is of order e^4, &c.

In like manner for an *oblatum*

$$C_2 = +\tfrac{1}{3}e^2, \quad C_4 \text{ is of order } e^4, \text{ &c.}$$

In both cases the corrections according to (46) would be of order e^8, but we obtain a term in e^6 when we retain C_2^3.

By (40), (41) we obtain as an extension of (28),

$$\text{Capacity} = u_0^{-1}\left\{1 + \tfrac{2}{3}C_1^2 + \tfrac{3}{5}C_2^2 + \dots + \frac{p+1}{2p+1}C_p^2 - \tfrac{12}{35}C_2^3\right\}, \dots(47)$$

and by comparison with (43)

$$\text{Capacity} = \sqrt{\left(\frac{S}{4\pi}\right)}.\left\{1 - \frac{C_3^2}{14} - \dots - \frac{(p-1)(p-2)}{4(2p+1)}C_p^2 - \tfrac{2}{35}C_2^3\right\}. \dots(48)$$

In the case of the ellipsoid $C_2 = \mp \dfrac{e^2}{3}$, and as far as e^6 inclusive we get

$$\text{Capacity} = \sqrt{\left(\frac{S}{4\pi}\right)}.\left\{1 \pm \frac{2e^6}{945}\right\},$$

as given by Russell in (43).]

404.

ON LEGENDRE'S FUNCTION $P_n(\theta)$, WHEN n IS GREAT AND θ HAS ANY VALUE*.

[*Proceedings of the Royal Society*, A, Vol. XCII. pp. 433—437, 1916.]

As is well known, an approximate formula for Legendre's function $P_n(\theta)$, when n is very large, was given by Laplace. The subject has been treated with great generality by Hobson†, who has developed the complete series proceeding by descending powers of n, not only for P_n but also for the "associated functions." The generality aimed at by Hobson requires the use of advanced mathematical methods. I have thought that a simpler derivation, sufficient for practical purposes and more within the reach of physicists with a smaller mathematical equipment, may be useful. It had, indeed, been worked out independently.

The series, of which Laplace's expression constitutes the first term, is arithmetically useful only when $n\theta$ is at least moderately large. On the other hand, when θ is small, P_n tends to identify itself with the Bessel's function $J_0(n\theta)$, as was first remarked by Mehler. A further development of this approximation is here proposed. Finally, a comparison of the results of the two methods of approximation with the numbers calculated by A. Lodge for $n = 20$‡ is exhibited.

The differential equation satisfied by Legendre's function P_n is

$$\frac{d^2u}{d\theta^2} + \cot\theta\,\frac{du}{d\theta} + n(n+1)u = 0. \quad\dots\dots\dots\dots\dots(1)$$

If we assume $u = v(\sin\theta)^{-\frac{1}{2}}$, and write m for $n + \frac{1}{2}$, we have

$$\frac{d^2v}{d\theta^2} + m^2v = -\frac{v}{4\sin^2\theta}. \quad\dots\dots\dots\dots\dots\dots(2)$$

---

* [1917. It would be more correct to say $P_n(\cos\theta)$, where $\cos\theta$ lies between ± 1.]

† "On a Type of Spherical Harmonics of Unrestricted Degree, Order, and Argument," *Phil. Trans.* A, Vol. CLXXXVII. (1896).

‡ "On the Acoustic Shadow of a Sphere," *Phil. Trans.* A, Vol. CCIII. (1904); *Scientific Papers*, Vol. V. p. 163.

If we take out a further factor, $e^{im\theta}$, writing

$$u = v \sin^{-\frac{1}{2}}\theta = w\, e^{im\theta} \sin^{-\frac{1}{2}}\theta, \quad \ldots\ldots\ldots\ldots\ldots\ldots(3)$$

of which ultimately only the real part is to be retained, we find

$$\frac{d^2w}{d\theta^2} + 2im\frac{dw}{d\theta} + \frac{w}{4\sin^2\theta} = 0. \quad \ldots\ldots\ldots\ldots\ldots(4)$$

We next change the independent variable to z, equal to $\cot\theta$, thus obtaining

$$\frac{dw}{dz} = \frac{-i}{2m}\left\{(1+z^2)\frac{d^2w}{dz^2} + 2z\frac{dw}{dz} + \frac{w}{4}\right\}. \quad \ldots\ldots\ldots\ldots(5)$$

From this equation we can approximate to the desired solution, treating m as a large quantity and supposing that $w = 1$ when $z = 0$, or $\theta = \frac{1}{2}\pi$.

The second approximation gives

$$\frac{dw}{dz} = -\frac{i}{8m}, \quad \text{whence} \quad w = 1 - \frac{iz}{8m}.$$

After two more steps we find

$$w = 1 - iz\left(\frac{1}{8m} - \frac{9}{128m^3}\right) - \frac{9z^2}{128m^2} + \frac{75iz^3}{1024m^3}. \quad \ldots\ldots\ldots\ldots(6)$$

Thus in realized form a solution of (1) is

$$u = C \sin^{-\frac{1}{2}}\theta\left[\left(1 - \frac{9\cot^2\theta}{128m^2}\right)\cos(m\theta + \gamma)\right.$$

$$\left. + \left\{\frac{\cot\theta}{8m} - \frac{9\cot\theta}{128m^3} - \frac{75\cot^3\theta}{1024m^3}\right\}\sin(m\theta + \gamma)\right]; \quad \ldots\ldots(7)$$

and this may be identified with P_n provided that the constants C, γ, can be so chosen that u and $du/d\theta$ have the correct values when $\theta = \frac{1}{2}\pi$. For this value of θ we must have

$$P_n(\tfrac{1}{2}\pi) = C\cos(\tfrac{1}{2}m\pi + \gamma), \quad \ldots\ldots\ldots\ldots\ldots(8)$$

$$(dP_n/d\theta)_{\frac{1}{2}\pi} = C\left(-m - \frac{1}{8m} + \frac{9}{128m^3}\right)\sin(\tfrac{1}{2}m\pi + \gamma). \quad \ldots\ldots\ldots(9)$$

We may express $(dP_n/d\theta)_{\frac{1}{2}\pi}$ by means of $P_{n+1}(\tfrac{1}{2}\pi)$. In general

$$\sin^2\theta\,\frac{dP_n}{d\cos\theta} = (n+1)(\cos\theta . P_n - P_{n+1}),$$

so that when $\theta = \frac{1}{2}\pi$,

$$dP_n/d\theta = -dP_n/d\cos\theta = (n+1)P_{n+1}. \quad \ldots\ldots\ldots\ldots(10)$$

When n is even, $(dP_n/d\theta)_{\frac{1}{2}\pi}$ vanishes, and, C being still undetermined, we may take to satisfy (9), $\gamma = -\frac{1}{4}\pi$; and then from (8)

$$C\cos(\tfrac{1}{2}n\pi) = P_n(\tfrac{1}{2}\pi) = (-1)^{\frac{1}{2}n}\frac{1.3.5\ldots(n-1)}{2.4.6\ldots\ n},$$

so that

$$C = \frac{1 . 3 . 5 \ldots (n-1)}{2 . 4 . 6 \ldots \quad n}.$$

Here n is even, say $2r$, and it is supposed to be great. Thus

$$C = \frac{1 . 2 . 3 \ldots (2r-1) \; 2r}{2^2 . 4^2 . 6^2 \ldots\ldots\ldots\ldots (2r)^2} = \frac{(2r)!}{2^{2r} (r!)^2};$$

and when r is great,

$$r! = \sqrt{(2\pi r)} . r^r \, e^{-r+1/12r - 1/360r^3}.$$

Thus

$$C = \frac{1}{\sqrt{(\pi r)}} \mathrm{Exp.} \left(\frac{1}{24r} - \frac{1}{360 . 8r^3} - \frac{1}{6r} + \frac{2}{360r^3} \right)$$

$$= \frac{1}{\sqrt{(\pi r)}} \left(1 - \frac{1}{8r} + \frac{1}{128r^2} + \frac{5}{1024r^3} \right)$$

$$= \sqrt{\left(\frac{2}{\pi n} \right)} . \left(1 - \frac{1}{4n} + \frac{1}{32n^2} + \frac{5}{128n^3} \right). \quad\ldots\ldots\ldots\ldots (11)$$

When n is even and with this value of C,

$$P_n = C \sin^{-\frac{1}{2}} \theta \left[\left(1 - \frac{9 \cot^2 \theta}{128m^2} \right) \cos \left\{ (n + \tfrac{1}{2}) \theta - \tfrac{1}{4} \pi \right\} \right.$$

$$\left. + \left(\frac{\cot \theta}{8m} - \frac{9 \cot \theta}{128m^3} - \frac{75 \cot^3 \theta}{1024m^3} \right) \sin \left\{ (n + \tfrac{1}{2}) \theta - \tfrac{1}{4} \pi \right\} \right]. \quad \ldots\ldots (12)$$

When n is odd, the same value of γ, viz. $-\tfrac{1}{4}\pi$, secures the required evanescence in (8), and we may conjecture that the same value of C will also serve. Laplace* indeed was content to determine γ from the case of n odd and C from the case of n even. I suppose it was this procedure that Todhunter† regarded as unsatisfactory. At any rate there is no difficulty in verifying that (9) is satisfied by the same value of C. From that equation and (10),

$$C = (-1)^{\frac{1}{2}(n+1)} \frac{(n+1) P_{n+1}(\tfrac{1}{2}\pi)}{m + 1/8m - 9/128m^3},$$

and

$$(-1)^{\frac{1}{2}(n+1)} P_{n+1}(\tfrac{1}{2}\pi) = \frac{1 . 3 . 5 \ldots \quad n}{2 . 4 . 6 \ldots (n+1)}$$

$$= \sqrt{\left\{ \frac{2}{(n+1)\pi} \right\}} . \left\{ 1 - \frac{1}{4(n+1)} + \frac{1}{32(n+1)^2} + \frac{5}{128(n+1)^3} \right\}.$$

Here, as throughout, $m = n + \tfrac{1}{2}$, and when we expand these expressions in descending powers of n we recover (11). Equations (11) and (12) are thus applicable to odd as well as to even values of n.

---

* *Méc. Cél.* Supplément au Ve volume.
† *Functions of Laplace*, etc. p. 71.

But whether n be even or odd, (12) fails when θ is so small that $n\theta$ is not moderately large. For this case our original equation (1) takes approximately the form

$$\frac{d^2u}{d\theta^2} + \frac{1}{\theta}\frac{du}{d\theta} + a^2u = 0, \quad \ldots\ldots\ldots\ldots\ldots\ldots(13)$$

where a^2 is written for $n(n+1)$; and of this the solution is

$$u = J_0(a\theta). \quad \ldots\ldots\ldots\ldots\ldots\ldots(14)$$

It is evident that the Bessel's function of the second kind, infinite when $\theta = 0$, does not enter, and that no constant multiplier is required, since u is to be unity when $\theta = 0$. For a second approximation we replace (13) by

$$\frac{d^2u}{d\theta^2} + \frac{1}{\theta}\frac{du}{d\theta} + a^2u = \frac{du}{d\theta}\left(\frac{1}{\theta} - \frac{\cos\theta}{\sin\theta}\right) = \frac{\theta}{3}\frac{du}{d\theta} = \frac{a\theta}{3}J_0'(a\theta);$$

or, if $a\theta = z$,

$$\frac{d^2u}{dz^2} + \frac{1}{z}\frac{du}{dz} + u = \frac{z}{3a^2}J_0'(z). \quad \ldots\ldots\ldots\ldots(15)$$

In order to solve (15) we assume as usual

$$u = v \cdot J_0(z). \quad \ldots\ldots\ldots\ldots\ldots(16)$$

This substitution gives

$$\frac{d^2v}{dz^2} + \frac{dv}{dz}\left(\frac{2J_0'}{J_0} + \frac{1}{z}\right) = \frac{z}{3a^2}\frac{J_0'}{J_0}, \quad \ldots\ldots\ldots\ldots(17)$$

a linear equation of the first order in dv/dz. In this

$$\int\left(\frac{2J_0'}{J_0} + \frac{1}{z}\right)dz = \log(zJ_0^2);$$

so that

$$\frac{dv}{dz} = \frac{A}{zJ_0^2} + \frac{1}{3a^2zJ_0^2}\int z^2J_0J_0'\,dz.$$

Here

$$\int z^2J_0J_0'\,dz = \tfrac{1}{2}z^2J_0^2 - \int J_0^2z\,dz = \tfrac{1}{2}z^2J_0^2 - \tfrac{1}{2}z^2(J_0^2 + J_0'^2) = -\tfrac{1}{2}z^2J_0'^2.$$

Thus

$$\frac{dv}{dz} = \frac{A}{zJ_0^2} - \frac{z}{6a^2}\frac{J_0'^2}{J_0^2}, \quad \ldots\ldots\ldots\ldots\ldots(18)$$

which has now to be integrated again.

$$\int\frac{J_0'^2z\,dz}{J_0^2} = -\int zJ_0'\,d\left(\frac{1}{J_0}\right) = -\frac{zJ_0'}{J_0} + \int\frac{zJ_0'' + J_0'}{J_0}\,dz$$

$$= -\frac{zJ_0'}{J_0} - \int z\,dz = -\frac{zJ_0'}{J_0} - \frac{z^2}{2},$$

regard being paid to the differential equation satisfied by J_0.

Thus
$$v = B + A \int \frac{dz}{z J_0^2} + \frac{1}{12a^2}(z^2 + 2z J_0'/J_0), \quad \ldots\ldots\ldots\ldots(19)$$

and
$$u = B J_0 + A J_0 \int \frac{dz}{z J_0^2} + \frac{1}{12a^2}(z^2 J_0 + 2z J_0'). \quad \ldots\ldots\ldots(20)$$

For the present purpose $A = 0$, $B = 1$; so that for P_n, identified with u, we get

$$P_n(\theta) = J_0(z) + \frac{1}{12a^2}\{z^2 J_0(z) + 2z J_0'(z)\}, \quad \ldots\ldots\ldots(21)$$

in which　　　　　　　$z = a\theta$, $\quad a^2 = n(n+1)$.

The functions J_0, $J_0' = -J_1$, are thoroughly tabulated*.

The Table annexed shows in the second column P_{20} calculated from (21) for values of θ ranging from $0°$ to $35°$. The third column gives the results from (11), (12), beginning with $\theta = 10°$.ˑ In the fourth column are the values of P_{20} calculated directly by A. Lodge. It will be seen that for $\theta = 15°$ and $20°$ the discrepancies are small in the fifth place of decimals. For smaller values of θ, the formula involving the Bessel's functions gives the best results, and for larger values of θ the extended form of Laplace's expression. When θ exceeds about $35°$ the latter formula gives P_{20} correct to six places. For n greater than 20 the combined use of the two methods would of course allow a still closer approximation.

Table for P_{20}.

θ	Formula (21)	From (11) and (12)	Calculated by Lodge
°			
0	1·000000	—	1·000000
5	0·346521	—ˑ	0·346521
10	−0·390581	−0·390420	−0·390588
15	−0·052776	−0·052753	−0·052772
20	+0·300174	+0·300191	+0·300203
25	−0·078051	−0·078085	−0·078085
30	−0·216914	−0·216997	−0·216999
35	+0·155472	+0·155635	+0·155636
40	—	+0·127328	+0·127328
45	—	−0·193065	−0·193065

* See Gray and Mathew's *Bessel's Functions*.

405.

MEMORANDUM ON FOG SIGNALS.

[Report to Trinity House, May 1916.]

PROLONGED experience seems to show that, no matter how much power may be employed in the production of sound-in-air signals, their audibility cannot be relied upon much beyond a mile. At a less distance than two miles the most powerful signals may be lost in certain directions when the atmospheric conditions are unfavourable. There is every reason to surmise that in these circumstances the sound goes over the head of the observer, but, so far as I know, there is little direct confirmation of this. It would clear up the question very much could it be proved that when a signal is prematurely lost at the surface of the sea it could still be heard by an observer at a considerable elevation. In these days of airships it might be possible to get a decision.

But for practical purposes the not infrequent failure of sound-in-air signals must be admitted to be without remedy, and the question arises what alternatives are open. I am not well informed as to the success or otherwise of submarine signals, viz. of sounds propagated through water, over long distances. What I wish at present to draw attention to is the probable advantage of so-called " wireless " signals. The waves constituting these signals are indeed for the most part propagated through air, but they are far more nearly independent of atmospheric conditions—temperature and wind—than are ordinary sound waves. With very moderate appliances they can be sent and observed with certainty at distances such as 10 or 20 miles.

As to how they should be employed, it may be remarked that the mere reception of a signal is in itself of no use. The signal must give information as to the *distance*, or *bearing*, or both, of the sending station. The estimation of *distance* would depend upon the intensity of the signals received and would probably present difficulties if any sort of precision was aimed at. On the other hand the *bearing* of the sending station can be determined at the receiving station with fair accuracy, that is to within two or three degrees. The special apparatus required is not complicated, but it is rather cumbrous since coils of large area have to be capable of rotation. I assume that this part of the work would be done at the Shore Station. A ship arriving near the land and desirous of ascertaining her position would make wireless signals at regular short intervals. The operator on land would determine the bearing of the Ship from which the signals came and communicate this bearing to the Ship. In many cases this might suffice; otherwise the Ship could proceed upon her course for a mile or two and then receive another intimation of her bearing from the Shore Station. The two bearings, with the speed and course of the Ship, would fix her position completely.

I do not suppose that much can be done at the present time towards testing this proposal, but I would suggest that it be borne in mind when considering any change in the Shore Stations concerned. I feel some confidence that the requirements of liners making the land will ultimately be met in some such way and that they cannot be met with certainty and under unfavourable conditions in any other.

[1918. Reference may be made to *Phil. Mag.* Vol. XXXVI, p. 1 (1918), where Prof. Joly discusses lucidly and fully the method of "Synchronous signals." In this method it is *distance* which is found in the first instance. It depends upon the use of signals propagated at different speeds and it involves the audibility of sounds reaching the observer through air, or through water, or through both media.]

406.

LAMB'S *HYDRODYNAMICS*.

[*Nature*, Vol. xcvii. p. 318, 1916.]

THAT this work should have already reached a fourth edition speaks well for the study of mathematical physics. By far the greater part of it is entirely beyond the range of the books available a generation ago. And the improvement in the style is as conspicuous as the extension of the matter. My thoughts naturally go back to the books in current use at Cambridge in the early sixties. With rare exceptions, such as the notable one of Salmon's *Conic Sections* and one or two of Boole's books, they were arid in the extreme, with scarcely a reference to the history of the subject treated, or an indication to the reader of how he might pursue his study of it. At the present time we have excellent books in English on most branches of mathematical physics and certainly on many relating to pure mathematics.

The progressive development of his subject is often an embarrassment to the writer of a text-book. Prof. Lamb remarks that his " work has less pretensions than ever to be regarded as a complete account of the science with which it deals. The subject has of late attracted increased attention in various countries, and it has become correspondingly difficult to do justice to the growing literature. Some memoirs deal chiefly with questions of mathematical method and so fall outside the scope of this book; others though physically important hardly admit of a condensed analysis; others, again, owing to the multiplicity of publications, may unfortunately have been overlooked. And there is, I am afraid, the inevitable personal equation of the author, which leads him to take a greater interest in some branches of the subject than in others."

Most readers will be of opinion that the author has held the balance fairly. Formal proofs of " existence theorems " are excluded. Some of these, though demanded by the upholders of mathematical rigour, tell us only what we knew before, as Kelvin used to say. Take, for example, the existence of a possible stationary temperature within a solid when the temperature at the surface is arbitrarily given. A physicist feels that nothing can make this any clearer or more certain. What is strange is that there should be so wide a gap between his intuition and the lines of argument necessary to satisfy the pure mathematician. Apart from this question it may be said that everywhere the mathematical foundation is well and truly laid, and that in not a few cases the author's formulations will be found the most convenient starting point for investigations in other subjects as well as in hydrodynamics. To almost all parts of his subject he has made entirely original contributions; and, even when this could not be claimed, his exposition of the work of others is often so much simplified and improved as to be of not inferior value. As examples may be mentioned the account of Cauchy and Poisson's theory of the waves produced in deep water by a local disturbance of the surface (§ 238) —the first satisfactory treatment of what is called in Optics a dispersive medium—and of Sommerfeld's investigation of the diffraction of plane waves of sound at the edge of a semi-infinite screen (§ 308).

Naturally a good deal of space is devoted to the motion of a liquid devoid of rotation and to the reaction upon immersed solids. When the solids are "fair" shaped, this theory gives a reasonable approximation to what actually occurs; but when a real liquid flows past projecting angles the motion is entirely different, and unfortunately this is the case of greatest practical importance. The author, following Helmholtz, lays stress upon the negative pressure demanded at sharp corners in order to maintain what may be called the electric character of flow. This explanation may be adequate in some cases; but it is now well known that liquids are capable of sustaining negative pressures of several atmospheres. How too does the explanation apply to gases, which form jets under quite low pressure differences?* It seems probable that viscosity must be appealed to. This is a matter which much needs further elucidation. It is one on which Kelvin and Stokes held strongly divergent views.

The later chapters deal with Vortex Motion, Tidal Waves, Surface Waves, Waves of Expansion (Sound), Viscosity, and Equilibrium of Rotating Masses. On all these subjects the reader will find expositions which could hardly be improved, together with references to original writings of the author and others where further developments may be followed.

It would not have accorded with the author's scheme to go into detail upon experimental matters, but one feels that there is room for a supplementary volume which should have regard more especially to the practical side of the subject. Perhaps the time for this has not yet come. During the last few years much work has been done in connexion with artificial flight. We may hope that before long this may be coordinated and brought into closer relation with theoretical hydrodynamics. In the meantime one can hardly deny that much of the latter science is out of touch with reality.

---

* The fact that liquids do not break under moderate negative pressure was known to T. Young. "The magnitude of the cohesion between liquids and solids, as well as of the particles of fluid with each other, is more directly shewn by an experiment on the continuance of a column of mercury, in the tube of a barometer, at a height considerably greater than that at which it usually stands, on account of the pressure of the atmosphere. If the mercury has been well boiled in the tube, it may be made to remain in contact with the closed end, at the height of 70 inches or more" (*Young's Lectures*, p. 626, 1807). If the mercury be wet, boiling may be dispensed with and negative pressures of two atmospheres are easily demonstrated.

407.

ON THE FLOW OF COMPRESSIBLE FLUID PAST AN OBSTACLE.

[*Philosophical Magazine*, Vol. XXXII. pp. 1—6, 1916.]

IT is well known that according to classical Hydrodynamics a steady stream of frictionless incompressible fluid exercises no resultant force upon an obstacle, such as a rigid sphere, immersed in it. The development of a "resistance" is usually attributed to viscosity, or when there is a sharp edge to the negative pressure which may accompany it (Helmholtz). In either case it would seem that resistance involves something of the nature of a *wake*, extending behind the obstacle to an infinite distance. When the system of disturbed velocities, although it may mathematically extend to infinity, remains as it were attached to the obstacle, there can be no resistance.

The absence of resistance is asserted for an *incompressible* fluid; but it can hardly be supposed that a small degree of compressibility, as in water, would affect the conclusion. On the other hand, high relative velocities, exceeding that of sound in the fluid, must entirely alter the conditions. It seems worth while to examine this question more closely, especially as the first effects of compressibility are amenable to mathematical treatment.

The equation of continuity for a compressible fluid in steady motion is in the usual notation

$$u\frac{d\rho}{dx} + v\frac{d\rho}{dy} + w\frac{d\rho}{dz} + \rho\left(\frac{du}{dx} + \frac{dv}{dy} + \frac{dw}{dz}\right) = 0, \ldots\ldots\ldots\ldots(1)$$

or, if there be a velocity-potential ϕ,

$$\frac{d\phi}{dx}\frac{d\log\rho}{dx} + \frac{d\phi}{dy}\frac{d\log\rho}{dy} + \frac{d\phi}{dz}\frac{d\log\rho}{dz} + \nabla^2\phi = 0. \ldots\ldots\ldots(2)$$

In most cases we may regard the pressure p as a given function of the density ρ, dependent upon the nature of the fluid. The simplest is that of Boyle's law where $p = a^2\rho$, a being the velocity of sound. The general equation

$$\int\frac{dp}{\rho} = C - \tfrac{1}{2}q^2, \ldots\ldots\ldots\ldots\ldots\ldots\ldots\ldots(3)$$

where q is the resultant velocity, so that

$$q^2 = (d\phi/dx)^2 + (d\phi/dy)^2 + (d\phi/dz)^2 \quad \dots\dots\dots\dots(4)$$

reduces in this case to

$$a^2 \log \rho = C - \tfrac{1}{2}q^2,$$

or

$$a^2 \log (\rho/\rho_0) = -\tfrac{1}{2}q^2, \quad \dots\dots\dots\dots\dots(5)$$

if ρ_0 correspond to $q = 0$.   From (2) and (5) we get

$$\nabla^2\phi = \frac{1}{2a^2} \left\{ \frac{d\phi}{dx}\frac{dq^2}{dx} + \frac{d\phi}{dy}\frac{dq^2}{dy} + \frac{d\phi}{dz}\frac{dq^2}{dz} \right\}. \quad \dots\dots\dots(6)$$

When q^2 is small in comparison with a^2, this equation may be employed to estimate the effects of compressibility.   Taking a known solution for an incompressible fluid, we calculate the value of the right-hand member and by integration obtain a second approximation to the solution in the actual case.   The operation may be repeated, and if the integrations can be effected, we obtain a solution in series proceeding by descending powers of a^2.   It may be presumed that this series will be convergent so long as q^2 is less than a^2.

There is no difficulty in the first steps for obstacles in the form of spheres or cylinders, and I will detail especially the treatment in the latter case.   If U, parallel to $\theta = 0$, denote the uniform velocity of the stream at a distance, the velocity-potential for the motion of incompressible fluid is known to be

$$\phi = U \left(r + c^2/r\right) \cos \theta, \quad \dots\dots\dots\dots\dots\dots(7)$$

the origin of polar coordinates (r, θ) being at the centre of the cylinder.   At the surface of the cylinder $r = c$, $d\phi/dr = 0$, for all values of θ.

On the right hand of (6)

$$\frac{d\phi}{dx}\frac{dq^2}{dx} + \frac{d\phi}{dy}\frac{dq^2}{dy} = \frac{d\phi}{dr}\frac{dq^2}{dr} + \frac{1}{r^2}\frac{d\phi}{d\theta}\frac{dq^2}{d\theta}; \quad \dots\dots\dots\dots(8)$$

and from (7)

$$\frac{q^2}{U^2} = \frac{1}{U^2}\left\{ \left(\frac{d\phi}{dr}\right)^2 + \frac{1}{r^2}\left(\frac{d\phi}{d\theta}\right)^2 \right\} = 1 + \frac{c^4}{r^4} - \frac{2c^2}{r^2}\cos 2\theta. \quad \dots\dots(9)$$

Also

$$\frac{1}{U}\frac{d\phi}{dr} = \left(1 - \frac{c^2}{r^2}\right)\cos\theta, \qquad \frac{1}{U}\frac{d\phi}{rd\theta} = -\left(1 + \frac{c^2}{r^2}\right)\sin\theta;$$

$$\frac{1}{U^2}\frac{dq^2}{dr} = -\frac{4c^4}{r^5} + \frac{4c^2}{r^3}\cos 2\theta, \qquad \frac{1}{U^2}\frac{dq^2}{rd\theta} = \frac{4c^2}{r^3}\sin 2\theta.$$

Accordingly

$$\nabla^2\phi = \frac{2U^3c^2}{a^2r^3}\left\{ -\frac{c^2}{r^2}\left(2 - \frac{c^2}{r^2}\right)\cos\theta + \cos 3\theta \right\}. \quad \dots\dots\dots(10)$$

The terms on the right of (10) are all of the form $r^p \cos n\theta$, so that for the present purpose we have to solve

$$\nabla^2\phi = \frac{d^2\phi}{dr^2} + \frac{1}{r}\frac{d\phi}{dr} + \frac{1}{r^2}\frac{d^2\phi}{d\theta^2} = r^p \cos n\theta. \quad \dots\dots\dots(11)$$

If we assume that ϕ varies as $r^m \cos n\theta$, we see that $m = p + 2$, and that the complete solution is

$$\phi = \cos n\theta \left\{ A r^n + B r^{-n} + \frac{r^{p+2}}{(p+2)^2 - n^2} \right\}, \quad \ldots\ldots\ldots(12)$$

A and B being arbitrary constants. In (10) we have to deal with $n = 1$ associated with $p = -5$ and -7, and with $n = 3$ associated with $p = -3$. The complete solution as regards terms in $\cos \theta$ and $\cos 3\theta$ is accordingly

$$\phi = (Ar + Br^{-1}) \cos \theta + (Cr^3 + Dr^{-3}) \cos 3\theta$$
$$+ \frac{2 U^3 c^2}{a^2} \left[\cos \theta \left(-\frac{c^2}{4r^3} + \frac{c^4}{24r^5} \right) - \frac{\cos 3\theta}{8r} \right]. \quad \ldots\ldots(13)$$

The conditions to be satisfied at infinity require that, as in (7), $A = U$, and that $C = 0$. We have also to make $d\phi/dr$ vanish when $r = c$. This leads to

$$B = c^2 U + \frac{13 U^3 c^2}{12 a^2}, \quad D = \frac{U^3 c^4}{12 a^2}. \quad \ldots\ldots\ldots\ldots(14)$$

Thus

$$\phi = U \left\{ r + \frac{c^2}{r} + \frac{13 U^2 c^2}{12 a^2 r} \right\} \cos \theta + \frac{U^3 c^4}{12 a^2 r^3} \cos 3\theta$$
$$+ \frac{U^3 c^2}{a^2} \left\{ \cos \theta \left(-\frac{c^2}{2r^3} + \frac{c^4}{12 r^5} \right) - \frac{\cos 3\theta}{4r} \right\} \quad \ldots\ldots(15)$$

satisfies all the conditions and is the value of ϕ complete to the second approximation.

That the motion determined by (15) gives rise to no resultant force in the direction of the stream is easily verified. The pressure at any point is a function of q^2, and on the surface of the cylinder $q^2 = c^{-2} (d\phi/d\theta)^2$. Now $(d\phi/d\theta)^2$ involves θ in the forms $\sin^2 \theta$, $\sin^2 3\theta$, $\sin \theta \sin 3\theta$, and none of these are changed by the substitution of $\pi - \theta$ for θ; the pressures on the cylinder accordingly constitute a balancing system.

There is no particular difficulty in pursuing the approximation so as to include terms involving the square and higher powers of U^2/a^2. The right-hand member of (6) will continue to include only terms in the cosines of odd multiples of θ with coefficients which are simple powers of r, so that the integration can be effected as in (11), (12). And the general conclusion that there is no resultant force upon the cylinder remains undisturbed.

The corresponding problem for the *sphere* is a little more complicated, but it may be treated upon the same lines with use of Legendre's functions $P_n (\cos \theta)$ in place of cosines of multiples of θ. In terms of the usual polar coordinates (r, θ, ω), the last of which does not appear, the first approximation, as for an incompressible fluid, is

$$\phi = U \cos \theta \left(r + \frac{c^3}{2r^2} \right) = U \left(r + \frac{c^3}{2r^2} \right) P_1, \quad \ldots\ldots\ldots\ldots(16)$$

c denoting the radius of the sphere.   As in (8),

$$\Sigma \frac{d\phi}{dx}\frac{dq^2}{dx} = \frac{d\phi}{dr}\frac{dq^2}{dr} + \frac{1}{r^2}\frac{d\phi}{d\theta}\frac{dq^2}{d\theta} = U^3\left\{\left(-\frac{36c^6}{5r^7} + \frac{9c^9}{2r^{10}}\right)P_1\right.$$

$$\left. + \left(\frac{6c^3}{r^4} - \frac{24c^6}{5r^7} + \frac{3c^9}{2r^{10}}\right)P_3\right\}, \quad\ldots\ldots(17)$$

on substitution from (16) of the values of ϕ and q^2.   This gives us the right-hand member of (6).

In the present problem

$$\nabla^2 = \frac{d^2}{dr^2} + \frac{2}{r}\frac{d}{dr} + \frac{1}{r^2\sin\theta}\frac{d}{d\theta}\left(\sin\theta\frac{d}{d\theta}\right), \quad\ldots\ldots\ldots\ldots(18)$$

while P_n satisfies

$$\frac{1}{\sin\theta}\frac{d}{d\theta}\left(\sin\theta\frac{dP_n}{d\theta}\right) + n(n+1)P_n = 0; \quad\ldots\ldots\ldots\ldots(19)$$

so that

$$\nabla^2\phi = r^p P_n \quad\ldots\ldots\ldots\ldots\ldots\ldots\ldots(20)$$

reduces to

$$\frac{d^2\phi}{dr^2} + \frac{2}{r}\frac{d\phi}{dr} - \frac{n(n+1)}{r^2}\phi = r^p P_n. \quad\ldots\ldots\ldots\ldots\ldots(21)$$

The solution, corresponding to the various terms of (17), is thus

$$\phi = \frac{r^{p+2}P_n}{(p+2)(p+3) - n(n+1)}. \quad\ldots\ldots\ldots\ldots(22)$$

With use of (22), (6) gives

$$\phi = \frac{U^3}{a^2}\left\{-\frac{c^6 P_1}{5r^5} + \frac{c^9 P_1}{24r^8} - \frac{3c^3 P_3}{10r^2} - \frac{3c^6 P_3}{10r^5} + \frac{3c^9 P_3}{176r^8}\right\}$$

$$+ ArP_1 + Br^{-2}P_1 + Cr^3 P_3 + Dr^{-4}P_3, \quad\ldots\ldots\ldots\ldots(23)$$

A, B, C, D being arbitrary constants.   The conditions at infinity require $A = U, C = 0$.   The conditions at the surface of the sphere give

$$B = \frac{c^3}{2}\left(U + \frac{2U^3}{3a^2}\right), \quad D = \frac{27c^5 U^3}{55a^2}; \quad\ldots\ldots\ldots\ldots(24)$$

and thus ϕ is completely determined to the second approximation.

The P's which occur in (23) are of *odd* order, and are polynomials in $\mu\ (=\cos\theta)$ of *odd* degree.   Thus $d\phi/dr$ is odd (in μ) and $d\phi/d\theta = \sin\theta \times$ even function of μ.   Further,

$$q^2 = \text{even function} + \sin^2\theta \times \text{even function} = \text{even function},$$

$$dq^2/dr = \text{even function}, \quad dq^2/d\theta = \sin\theta \times \text{odd function}.$$

Accordingly

$$\frac{d\phi}{dr}\frac{dq^2}{dr} + \frac{1}{r^2}\frac{d\phi}{d\theta}\frac{dq^2}{d\theta} = \text{odd function of } \mu,$$

and can be resolved into a series of P's of odd order.   Thus not only is there no resultant force discovered in the second approximation, but this character is preserved however far we may continue the approximations. And since the coefficients of the various P's are simple polynomials in $1/r$, the integrations present no difficulty in principle.

Thus far we have limited ourselves to Boyle's law, but it may be of interest to make extension to the general adiabatic law, of which Boyle's is a particular case. We have now to suppose

$$p/p_0 = (\rho/\rho_0)^\gamma, \quad \dots\dots\dots\dots\dots\dots(25)$$

making

$$\frac{dp}{d\rho} = \frac{\gamma p_0}{\rho_0} \left(\frac{\rho}{\rho_0}\right)^{\gamma-1} = a^2 \left(\frac{\rho}{\rho_0}\right)^{\gamma-1}, \quad \dots\dots\dots\dots(26)$$

if a denote the velocity of sound corresponding to ρ_0. Then by (3)

$$\frac{a^2}{\gamma-1} \left(\frac{\rho}{\rho_0}\right)^{\gamma-1} = C - \tfrac{1}{2}q^2. \quad \dots\dots\dots\dots\dots(27)$$

If we suppose that ρ_0 corresponds to $q = 0$, $C = a^2/(\gamma-1)$, and

$$\left(\frac{\rho}{\rho_0}\right)^{\gamma-1} = 1 - \frac{(\gamma-1)\,q^2}{2a^2}, \quad \dots\dots\dots\dots(28)$$

whence

$$\frac{d \log \rho}{dx} = -\frac{dq^2/dx}{2a^2 - (\gamma-1)\,q^2}. \quad \dots\dots\dots\dots(29)$$

The use of this in (2) now gives

$$\nabla^2\phi = \frac{1}{2a^2 - (\gamma-1)\,q^2} \left\{ \frac{d\phi}{dx}\frac{dq^2}{dx} + \frac{d\phi}{dy}\frac{dq^2}{dy} + \frac{d\phi}{dz}\frac{dq^2}{dz} \right\}, \quad \dots\dots(30)$$

from which we can fall back upon (6) by supposing $\gamma = 1$. So far as the first and second approximations, the substitution of (30) for (6) makes no difference at all.

As regards the general question it would appear that so long as the series are convergent there can be no resistance and no wake as the result of compressibility. But when the velocity U of the stream exceeds that of sound, the system of velocities in front of the obstacle expressed by our equations cannot be maintained, as they would be at once swept away down stream. It may be presumed that the passage from the one state of affairs to the other synchronizes with a failure of convergency. For a discussion of what happens when the velocity of sound is exceeded, reference may be made to a former paper*.

* *Proc. Roy. Soc.* A, Vol. LXXXIV. p. 247 (1910); *Scientific Papers*, Vol. v. p. 608.
[1917. See P.S. to Art. 411 for a reference to the work of Prof. Cisotti.]

408.

ON THE DISCHARGE OF GASES UNDER HIGH PRESSURES.

[*Philosophical Magazine*, Vol. XXXII. pp. 177—187, 1916.]

THE problem of the passage of gas through a small aperture or nozzle from one vessel to another in which there is a much lower pressure has had a curious history. It was treated theoretically and experimentally a long while ago by Saint-Venant and Wantzel[*] in a remarkable memoir, where they point out the absurd result which follows from the usual formula, when we introduce the supposition that the pressure in the escaping jet is the same as that which prevails generally in the recipient vessel. In Lamb's notation[†], if the gas be subject to the adiabatic law ($p \propto \rho^\gamma$),

$$q^2 = 2 \int_p^{p_0} \frac{dp}{\rho} = \frac{2\gamma}{\gamma-1} \frac{p_0}{\rho_0} \left\{ 1 - \left(\frac{p}{p_0} \right)^{\frac{\gamma-1}{\gamma}} \right\} = \frac{2}{\gamma-1} (c_0^2 - c^2), \quad \ldots\ldots(1)$$

where q is the velocity corresponding to pressure p; p_0, ρ_0 the pressure and density in the discharging vessel where $q = 0$; c the velocity of sound in the gas when at pressure p and density ρ; c_0 that corresponding to p_0, ρ_0. According to (1) the velocity increases as p diminishes, but only up to a maximum, equal to $c_0 \sqrt{\{2/(\gamma-1)\}}$, when $p = 0$. If $\gamma = 1\cdot408$, this limiting velocity is $2\cdot214\,c_0$. It is to be observed, however, that in considering the rate of discharge we are concerned with what the authors cited call the "reduced velocity," that is the result of multiplying q by the corresponding *density* ρ. Now ρ diminishes indefinitely with p, so that the reduced velocity corresponding to an evanescent p is zero. Hence if we identify p with the pressure p_1 in the recipient vessel, we arrive at the impossible conclusion that the rate of discharge into a vacuum is zero. From this our authors infer that the *identification cannot be made*; and their experiments showed that from $p_1 = 0$ upwards to $p_1 = \cdot4p_0$ the rate of discharge is sensibly constant. As p_1 still further increases, the discharge falls off, slowly at first,

---

[*] "Mémoire et expériences sur l'écoulement de l'air, déterminé par des différences de pressions considérables," *Journ. de l'École Polyt.* t. XVI. p. 85 (1839).

[†] *Hydrodynamics*, §§ 23, 25 (1916).

afterwards with greater rapidity, until it vanishes when the pressures become equal.

The work of Saint-Venant and Wantzel was fully discussed by Stokes in his Report on Hydrodynamics [*]. He remarks "These experiments show that when the difference of pressure in the first and second spaces is considerable, we can by no means suppose that the mean pressure at the orifice is equal to the pressure at a distance in the second space, nor even that there exists a contracted vein, at which we may suppose the pressure to be the same as at a distance." But notwithstanding this the work of the French writers seems to have remained very little known. It must have been unknown to O. Reynolds when in 1885 he traversed much the same ground[†], adding, however, the important observation that the maximum reduced velocity occurs when the actual velocity coincides with that of sound under the conditions then prevailing. When the actual velocity at the orifice reaches this value, a further reduction of pressure in the recipient vessel does not influence the rate of discharge, as its effect cannot be propagated backwards against the stream. If $\gamma = 1\cdot408$, this argument suggests that the discharge reaches a maximum when the pressure in the recipient vessel falls to $\cdot527\, p_0$, and then remains constant. In the somewhat later work of Hugoniot[‡] on the same subject there is indeed a complimentary reference to Saint-Venant and Wantzel, but the reader would hardly gather that they had insisted upon the difference between the pressure in the jet at the orifice and in the recipient vessel as the explanation of the impossible conclusion deducible from the contrary supposition.

In the writings thus far alluded to there seems to be an omission to consider what becomes of the jet after full penetration into the receiver. The idea appears to have been that the jet gradually widens in section as it leaves the orifice and that in the absence of friction it would ultimately attain the velocity corresponding to the entire fall of pressure. The first to deal with this question seem to have been Mach and Salcher[§], but the most elaborate examination is that of R. Emden[||], who reproduces interesting pictures of the effluent jet obtained by the simple shadow method of Dvorák[¶]. Light from the sun or from an electric spark, diverging from a small aperture as source, falls perpendicularly upon the jet and in virtue of differences of refraction depicts various features upon a screen held at some distance behind. A permanent record can be obtained by photography. Emden thus describes some of his results. When a jet of air, or better of carbonic

[*] B.A. Report for 1846; Math. and Phys. Papers, Vol. I. p. 176.
[†] Phil. Mag. Vol. XXI. p. 185 (1886).
[‡] Ann. de Chim. t. IX. p. 383 (1886).
[§] Wied. Ann. Bd. XLI. p. 144 (1890).
[||] Wied. Ann. Bd. LXIX. pp. 264, 426 (1899).
[¶] Wied. Ann. Bd. IX. p. 502 (1879).

acid or coal-gas, issues from the nozzle into the open under a pressure of a few millimetres, it is seen to rise as a slender column of the same diameter to a height of perhaps 30 or 40 cm. Sometimes the column disappears without visible disturbance of the air; more often it ends in a small vortex column. When the pressure is raised, the column shortens until finally the funnel-shaped vortex attaches itself to the nozzle. At a pressure of about one-fifth of an atmosphere there appears again a jet 2 or 3 cm. long. As the pressure rises still further, the jet becomes longer and more distinct and suddenly exhibits thin, bright, and fairly equidistant disks to the number of perhaps 10 or 12, crossing the jet perpendicularly. The first disks have exactly the diameter of the nozzle, but they diminish as the jet attenuates. Under still higher pressures the interval between the disks increases, and at the same time the jet is seen to swell out between them. These swellings further increase and oblique markings develop which hardly admit of merely verbal description.

Attributing these periodic features to stationary sound waves in the jet, Emden set himself to determine the wave-length (λ), that is the distance between consecutive disks, and especially the pressure at which the waves begin to develop. He employed a variety of nozzles, and thus sums up his principal results:

1. When air, carbonic acid, and hydrogen escape from equal sufficiently high pressures, the length of the sound waves in the jet is the same for the same nozzle and the same pressure.

2. The pressure at which the stationary sound waves begin to develop is the same in air, carbonic acid, and hydrogen, and is equal to ·9 atmosphere.

This is the pressure-*excess* behind the nozzle, so that the whole pressure there is 1·9 atmosphere. The environment of the jet is at one atmosphere pressure.

Emden, comparing his observations with the theory of Saint-Venant and Wantzel, then enunciates the following conclusion: The critical pressure, in escaping from which into the atmosphere the gas at the nozzle's mouth moves with the velocity of sound, is equal to the pressure at which stationary sound waves begin to form in the jet. So far, I think, Emden makes out his case; but he appears to over-shoot the mark when he goes on to maintain that *after* the critical pressure-ratio is exceeded, the escaping jet moves everywhere with the same velocity, viz. the sound-velocity; and that everywhere within it the free atmospheric pressure prevails. He argues from what happens when the motion is strictly in one dimension. It is true that then a wave can be stationary in space only when the stream moves with the velocity of sound; but here the motion is not limited to one dimension, as is shown by the swellings between the disks. Indeed the propagation of any wave at all is inconsistent with uniformity of pressure within the jet.

At the *surface* of the jet, but not within it, the condition is imposed that the pressure must be that of the surrounding atmosphere.

The problem of a jet in which the motion is completely *steady* in the hydrodynamical sense and approximately uniform was taken up by Prandtl*, both for the case of symmetry round the axis (of z) and in two dimensions. In the former, which is the more practical, the velocity component w is supposed to be nearly constant, say W, while u and v are small. We may employ the usual Eulerian equations. Of these the third,

$$\frac{dw}{dt} + u\frac{dw}{dx} + v\frac{dw}{dy} + w\frac{dw}{dz} = -\frac{1}{\rho}\frac{dp}{dz},$$

reduces to

$$W\frac{dw}{dz} = -\frac{1}{\rho}\frac{dp}{dz}, \quad\dots\dots\dots\dots\dots(2)$$

when we introduce the supposition of steady motion and neglect the terms of the second order. In like manner the other equations become

$$W\frac{du}{dz} = -\frac{1}{\rho}\frac{dp}{dx}, \quad W\frac{dv}{dz} = -\frac{1}{\rho}\frac{dp}{dy}. \quad\dots\dots\dots(3)$$

Further, the usual equation of continuity, viz.

$$\frac{d(\rho u)}{dx} + \frac{d(\rho v)}{dy} + \frac{d(\rho w)}{dz} = 0, \quad\dots\dots\dots\dots(4)$$

here reduces to

$$\rho\left(\frac{du}{dx} + \frac{dv}{dy} + \frac{dw}{dz}\right) + W\frac{d\rho}{dz} = 0. \quad\dots\dots\dots(5)$$

If we introduce a velocity-potential ϕ, we have with use of (2)

$$\nabla^2\phi = -\frac{W}{\rho}\frac{d\rho}{dz} = \frac{W^2}{a^2}\frac{dw}{dz} = \frac{W^2}{a^2}\frac{d^2\phi}{dz^2}, \quad\dots\dots\dots(6)$$

where $a, = \sqrt{(dp/d\rho)}$, is the velocity of sound in the jet. In the case we are now considering, where there is symmetry round the axis, this becomes $(r^2 = x^2 + y^2)$

$$\frac{d^2\phi}{dr^2} + \frac{1}{r}\frac{d\phi}{dr} + \left(1 - \frac{W^2}{a^2}\right)\frac{d^2\phi}{dz^2} = 0, \quad\dots\dots\dots(7)$$

and a similar equation holds for w, since $w = d\phi/dz$.

If the periodic part of w is proportional to $\cos\beta z$, we have for this part

$$\frac{d^2w}{dr^2} + \frac{1}{r}\frac{dw}{dr} + \left(\frac{W^2}{a^2} - 1\right)\beta^2 w = 0, \quad\dots\dots\dots(8)$$

and we may take as the solution

$$w = W + H\cos\beta z \cdot J_0\{\sqrt{(W^2 - a^2)}\cdot\beta r/a\}, \quad\dots\dots\dots(9)$$

since the Bessel's function of the second kind, infinite when $r = 0$, cannot here appear. The condition to be satisfied at the boundary $(r = R)$ is that

* *Phys. Zeitschrift*, 5 Jahrgang, p. 599 (1904).

the pressure be constant, equal to that of the surrounding quiescent air, and this requires that the variable part of w vanish, since the pressure varies with the total velocity. Accordingly

$$J_0 \left\{ \surd(W^2 - a^2) \cdot \beta R/a \right\} = 0, \quad \dots\dots\dots\dots\dots(10)$$

which can be satisfied only when $W > a$, that is when the mean velocity of the jet *exceeds* that of sound. The wave-length (λ) of the periodic features along the jet is given by $\lambda = 2\pi/\beta$.

The most important solution corresponds to the first root of (10), viz. 2·405. In this case

$$\lambda = \frac{2\pi R \ \surd(W^2/a^2 - 1)}{2\cdot405} . \quad \dots\dots\dots\dots\dots(11)$$

The problem for the two-dimensional jet is even simpler. If b be the width of the jet, the principal wave-length is given by

$$\lambda = 2b \ \surd(W^2/a^2 - 1). \quad \dots\dots\dots\dots\dots(12)$$

The above is substantially the investigation of Prandtl, who finds a sufficient agreement between (11) and Emden's measurements [*].

It may be observed that the problem can equally well be treated as one of the small vibrations of a stationary column of gas as developed in *Theory of Sound*, §§ 268, 340 (1878). If the velocity-potential, symmetrical about the axis of z, be also proportional to $e^{i\,(kat+\beta z)}$, where k is such that the wave-length of plane waves of the same period is $2\pi/k$, the equation is § 340 (3)

$$\frac{d^2\phi}{dr^2} + \frac{1}{r} \frac{d\phi}{dr} + (k^2 - \beta^2)\, \phi = 0, \quad \dots\dots\dots\dots\dots(13)$$

and if $k > \beta$

$$\phi = e^{i\,(kat+\beta z)} \ J_0 \left\{ \surd(k^2 - \beta^2) \cdot r \right\}. \quad \dots\dots\dots\dots\dots(14)$$

The condition of constant pressure when $r = R$ gives as before for the principal vibration

$$\surd(k^2 - \beta^2) \cdot R = 2\cdot405. \quad \dots\dots\dots\dots\dots(15)$$

The velocity of propagation of the waves is ka/β. If we equate this to W and suppose that a velocity W is superposed upon the vibrations, the motion becomes steady. When we substitute in (15) the value of k, viz. $W\beta/a$, we recover (11). It should perhaps be noticed that it is only after the vibrations have been made stationary that the effect of the surrounding air can be properly represented by the condition of uniformity of pressure. To assume it generally would be tantamount to neglecting the inertia of the outside air.

The above calculation of λ takes account only of the principal vibration. Other vibrations are possible corresponding to higher roots of (10), and if

---

[*] When $W < a$, β must be imaginary. The jet no longer oscillates, but settles rapidly down into complete uniformity. This is of course the usual case of gas escaping from small pressures.

these occur appreciably, strict periodicity is lost. Further, if we abandon the restriction to symmetry, a new term, $r^{-2}d^2\phi/d\theta^2$, enters in (13) and the solution involves a new factor $\cos(n\theta + \epsilon)$ in conjunction with the Bessel's function J_n in place of J_0.

The particular form of the differential equation exhibited in (13) is appropriate only when the section of the stream is circular. In general we have

$$\frac{d^2\phi}{dx^2} + \frac{d^2\phi}{dy^2} + (k^2 - \beta^2)\,\phi = 0, \quad \ldots\ldots\ldots\ldots\ldots(16)$$

the same equation as governs the vibrations of a stretched membrane (*Theory of Sound*, § 194). For example, in the case of a square section of side b, we have

$$\phi = \cos\frac{\pi x}{b} \cdot \cos\frac{\pi y}{b} \cdot e^{i\,(kat+\beta z)}, \quad \ldots\ldots\ldots\ldots\ldots(17)$$

vanishing when $x = \pm\frac{1}{2}b$ and when $y = \pm\frac{1}{2}b$. This represents the principal vibration, corresponding to the gravest tone of a membrane. The differential equation is satisfied provided

$$k^2 - \beta^2 = 2\pi^2/b^2, \quad \ldots\ldots\ldots\ldots\ldots(18)$$

the equation which replaces (15). It is shown in *Theory of Sound* that provided the deviation from the circular form is not great the question is mainly one of the *area* of the section. Thus the difference between (15) and (18) is but moderate when we suppose πR^2 equal to b^2.

It may be worth remarking that when V the wave-velocity exceeds a, the group-velocity U falls short of a. Thus in (15), (18)

$$V = \frac{ka}{\beta}, \quad U = \frac{d\,(\beta V)}{d\beta} = a\,\frac{dk}{d\beta} = \frac{\beta a}{k};$$

so that

$$UV = a^2. \quad \ldots\ldots\ldots\ldots\ldots(19)$$

Returning to the jet of circular section, we may establish the connexion between the variable pressure along the axis and the amount of the swellings observed to take place between the disks. From (9)

$$\phi = \int w\,dz = Wz + H\beta^{-1}\sin\beta z \cdot J_0\{\sqrt{(W^2/a^2 - 1)} \cdot \beta r\},$$

and

$$\left(\frac{d\phi}{dr}\right)_R = H\sqrt{(W^2/a^2 - 1)} \cdot \sin\beta z \cdot J_0'(2\cdot405). \quad \ldots\ldots\ldots(20)$$

The latter equation gives the radial velocity at the boundary. If δR denote the variable part of the radius of the jet,

$$\delta R = \int\frac{1}{W}\left(\frac{d\phi}{dr}\right)_R dz = -\frac{H\cos\beta z}{\beta W}\sqrt{\left(\frac{W^2}{a^2} - 1\right)} \cdot J_0'(2\cdot405). \quad \ldots(21)$$

Again, if δp be the variable part of the pressure at the axis ($r = 0$),

$$\frac{\delta p}{\rho} = C - \tfrac{1}{2}q^2 = C' - \tfrac{1}{2}w^2 = - W\delta w,$$

where ρ is the average density in the jet and δw the variable part of the component velocity parallel to z.   Accordingly

$$\frac{\delta p}{\rho} = - WH \cos \beta z \, ; \qquad\qquad\qquad\ldots\ldots\ldots\ldots\ldots\ldots(22)$$

and

$$\frac{\delta R}{\delta p/\rho} = \frac{J_0' \,(2\cdot405) \,\sqrt{(W^2/a^2 - 1)}}{\beta W^2} \, . \qquad\ldots\ldots\ldots\ldots(23)$$

In (23) we may substitute for β its value, viz.

$$\frac{2\cdot405a}{R \sqrt{(W^2 - a^2)}},$$

and for $J_0'\,(2\cdot405)$ we have from the tables of Bessel's functions $-0\cdot5191$, so that

$$\frac{\delta R}{\delta p/\rho} = - 0\cdot2158 \, R\,(a^{-2} - W^{-2}). \qquad\ldots\ldots\ldots\ldots\ldots(24)$$

As was to be expected, the greatest swelling is to be found where the pressure at the axis is least.

A complete theory of the effects observed by Mach and Emden would involve a calculation of the optical retardation along every ray which traverses the jet.   For the jet of circular section this seems scarcely practicable; but for the jet in two dimensions the conditions are simpler and it may be worth while briefly to consider this case.   As before, we may denote the general thickness of the two-dimensional jet by b, and take $b + \eta$ to represent the actual thickness at the place (z) where the retardation is to be determined.   The retardation is then sufficiently represented by Δ, where

$$\Delta = \int_0^{\frac{1}{2}(b+\eta)} (\rho - \rho_1) \, dy = \int_0^{\frac{1}{2}(b+\eta)} \rho\, dy - \tfrac{1}{2}\rho_1 (b + \eta), \quad \ldots\ldots\ldots(25)$$

ρ being the density in the jet and ρ_1 that of the surrounding gas.   The total stream

$$= \int_0^{\frac{1}{2}(b+\eta)} \rho\,(W + \delta w) \, dy = W \int_0^{\frac{1}{2}(b+\eta)} \rho\, dy + \rho \int_0^{\frac{1}{2}b} \delta w \, dy \, ;$$

and this is constant along the jet.   Thus

$$\Delta = C - \tfrac{1}{2}\rho_1\eta - \frac{\rho}{W} \int_0^{\frac{1}{2}b} \delta w \, dy, \qquad\ldots\ldots\ldots\ldots\ldots(26)$$

C being a constant, and squares of small quantities being omitted.

In analogy with (9), we may here take

$$\delta w = H \cos \beta z \, . \, \cos \{\beta y \,\sqrt{(W^2/a^2 - 1)}\}, \qquad\ldots\ldots\ldots\ldots(27)$$

and for the principal vibration the argument of the cosine is to become $\frac{1}{2}\pi$ when $y = \frac{1}{2}b$. Hence

$$\int_0^{\frac{1}{2}b} \delta w\, dy = \frac{H \cos \beta z}{\beta \sqrt{\{W^2/a^2 - 1\}}} . \qquad\qquad (28)$$

Also $\phi = \int w\, dz = Wz + \beta^{-1} H \sin \beta z . \cos\{\beta y \sqrt{(W^2/a^2 - 1)}\},$

$$\left(\frac{d\phi}{dy}\right)_{\frac{1}{2}b} = -H \sqrt{\{W^2/a^2 - 1\}} . \sin \beta z.$$

Thus $\frac{1}{2}\eta = \frac{1}{W} \int \left(\frac{d\phi}{dy}\right)_{\frac{1}{2}b} dz = \frac{H \cos \beta z . \sqrt{\{W^2/a^2 - 1\}}}{\beta W}.$

Accordingly

$$\Delta = C - \frac{H \cos \beta z}{\beta W} \left[\rho_1 \sqrt{\{W^2/a^2 - 1\}} + \frac{\rho}{\sqrt{\{W^2/a^2 - 1\}}} \right] ; \qquad (29)$$

so that the retardation is greatest at the places where η is least, that is where the jet is narrowest. This is in agreement with observation, since the places of maximum retardation act after the manner of a convex lens. Although a complete theory of the optical effects in the case of a symmetrical jet is lacking, there seems no reason to question Emden's opinion that they are natural consequences of the constitution of the jet.

But although many features are more or less perfectly explained, we are far from anything like a complete mathematical theory of the jet escaping from high pressure, even in the simplest case. A preliminary question is— are we justified at all in assuming the adiabatic law as approximately governing the expansions throughout? Is there anything like the "bore" which forms in front of a bullet advancing with a velocity exceeding that of sound?* It seems that the latter question may be answered in the negative, since here the passage of air is always from a greater to a less pressure, so that the application of the adiabatic law is justified. The conditions appear to be simplest if we suppose the nozzle to end in a parallel part within which the motion may be uniform and the velocity that of sound. But even then there seems to be no reason to suppose that this state of things terminates exactly at the plane of the mouth. As the issuing gas becomes free from the constraining influence of the nozzle walls, it must begin to expand, the pressure at the boundary suddenly falling to that of the environment. Subsequently vibrations must set in; but the circumstances are not precisely those of Prandtl's calculation, inasmuch as the variable part of the velocity is not small in comparison with the difference between the mean velocity and that of sound. It is scarcely necessary to call attention to the violence of the assumption that viscosity may be neglected when a jet moves with high velocity through quiescent air.

* *Proc. Roy. Soc.* A, Vol. LXXXIV. p. 247 (1910); *Scientific Papers*, Vol. v. Art. 346, p. 608.

On the experimental side it would be of importance to examine, with more accuracy than has hitherto been attained, whether the asserted independence of the discharge of the pressure in the receiving vessel (supposed to be less than a certain fraction of that in the discharging vessel) is absolute, and if not to ascertain the precise law of departure. To this end it would seem necessary to abandon the method followed by more recent workers in which compressed gas discharges into the open, and to fall back upon the method of Saint-Venant and Wantzel where the discharge is from atmospheric pressure to a lower pressure. The question is whether any alteration of discharge is caused by a reduction of this lower pressure beyond a certain point. To carry out the investigation on a sufficient scale would need a powerful air-pump capable of absorbing the discharge, but otherwise the necessary apparatus is simple. In order to measure the discharge, or at any rate to determine whether it varies or not, the passage of atmospheric air to the nozzle might be somewhat choked. The accompanying diagram will explain the idea. *A* is the nozzle, which would be varied in different series of experiments; *B* the recipient, partially exhausted, vessel; *C* the passage to the air-pump. Above the nozzle is provided a closed chamber *E*

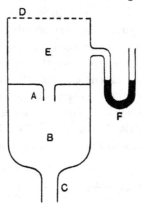

into which the external air has access through a metal gauze *D*, and where consequently the pressure is a little below atmospheric. *F* represents (diagrammatically) a pressure-gauge, or micromanometer, whose reading would be constant as long as the discharge remains so. Possibly an aneroid barometer would suffice; in any case there is no difficulty in securing the necessary delicacy*. Another manometer of longer range, but only ordinary sensitiveness, would register the low pressure in *B*. In this way there should be no difficulty in attaining satisfactory results. If *F* remains unaffected, notwithstanding large alterations of pressure in *B*, there are no complications to confuse the interpretation.

* See for example *Phil. Trans.* cxcvi. A, p. 205 (1901); *Scientific Papers*, Vol. iv. p. 510.

[1918. The experiments here proposed have been skilfully carried into effect by Hartshorn, working in my son's laboratory, *Proc. Roy. Soc.* A, Vol. xciv. p. 155, 1917.]

409.

ON THE ENERGY ACQUIRED BY SMALL RESONATORS FROM INCIDENT WAVES OF LIKE PERIOD.

[*Philosophical Magazine*, Vol. XXXII. pp. 188–190, 1916.]

IN discussions on photo-electricity it is often assumed that a resonator can operate only upon so much of the radiation incident upon it as corresponds to its own cross-section*. As a general proposition this is certainly not true and may indeed differ from the truth very widely. Since 1878† it has been known that an ideal *infinitely small* acoustical resonator may disperse energy corresponding to an area of wave-front of the primary waves equal to λ^2/π, an efficiency exceeding to any extent the limit fixed by the above mentioned rule. The questions of how much energy can be absorbed into the resonator itself and how long the absorption may take are a little different, but they can be treated without difficulty by the method explained in a recent paper‡. The equation (49) there found for the free vibration of a small symmetrical resonator was

$$M \frac{d^2\rho}{dt^2} + \mu\rho + 4\pi\sigma r^3 (1 - ikr) \frac{d^2\rho}{dt^2} = 0, \quad\ldots\ldots\ldots\ldots\ldots(1)$$

in which ρ denotes the radial displacement of the spherical surface from its equilibrium value r, M the mass, μ the coefficient of restitution, σ the density of the surrounding gas, and $k = 2\pi \div$ wave-length (λ) of vibrations in the gas. The first of the two terms containing σ operates merely as an addition to M. If we write

$$M' = M + 4\pi\sigma r^3, \quad\ldots\ldots\ldots\ldots\ldots\ldots\ldots\ldots(2)$$

(1) becomes

$$M' \frac{d^2\rho}{dt^2} + \mu\rho - i \cdot 4\pi\sigma k r^4 \frac{d^2\rho}{dt^2} = 0. \quad\ldots\ldots\ldots\ldots\ldots(3)$$

---

* See for example Millikan's important paper on a direct determination of Planck's constant "*h*"; *Physical Review*, Vol. VII. March 1916, p. 385.

  † *Theory of Sound*, § 319 ; λ = wave-length.

  ‡ *Phil. Mag.* Vol. XXIX. Feb. 1915, p. 210. [This volume, p. 289.]

Thus, if in free vibration ρ is proportional to e^{int}, where n is complex, the equation for n is

$$n^2\left(-M' + i\,.\,4\pi\sigma kr^4\right) + \mu = 0. \qquad \qquad (4)$$

The free vibrations are assumed to have considerable persistence, and the co-efficient of decay is e^{-qt}, where

$$q = 2\pi\sigma kr^4\,\sqrt{(\mu/M'^3)} = 2\pi\sigma pkr^4/M', \qquad \qquad (5)$$

if $p^2 = \mu/M'$.

We now suppose that the resonator is exposed to primary waves whose velocity-potential is there

$$\phi = \alpha e^{ipt}. \qquad \qquad (6)$$

The effect is to introduce on the right hand of (3) the term $4\pi r^2\sigma\alpha\,.\,ipe^{ipt}$; and since the resonance is supposed to be accurately adjusted, $p^2 = \mu/M'$. Under the same conditions $id^2\rho/dt^2$ in the third term on the left of (3) may be replaced by $-pd\rho/dt$, whether we are dealing with the permanent forced vibration or with free vibrations of nearly the same period which gradually die away. Thus our equation becomes on rejection of the imaginary part

$$M'\frac{d^2\rho}{dt^2} + 4\pi\sigma pkr^4\frac{d\rho}{dt} + \mu\rho = -4\pi r^2\sigma\alpha p\sin pt, \qquad \qquad (7)$$

which is of the usual form for vibrations of systems of one degree of freedom. For the permanent forced vibration $M'd^2\rho/dt^2 + \mu\rho = 0$ absolutely, and

$$\frac{d\rho}{dt} = -\frac{\alpha\sin pt}{kr^2}. \qquad \qquad (8)$$

The energy located in the resonator is then

$$\frac{M\alpha^2}{2k^2r^4}, \qquad \qquad (9)$$

and it may become very great when M is large and r small.

But when M is large, it may take a considerable *time* to establish the permanent regime after the resonator starts from rest. The approximate solution of (7), applicable in that case, is

$$\rho = \frac{\alpha\cos pt}{pkr^2}(1 - e^{-qt}), \qquad \qquad (10)$$

q being regarded as small in comparison with p; and the energy located in the resonator at time t

$$= \tfrac{1}{2}M\left(\frac{d\rho}{dt}\right)^2_{\text{max.}} = \frac{M\alpha^2}{2k^2r^4}(1 - e^{-qt})^2. \qquad \qquad (11)$$

We may now inquire what time is required for the accumulation of energy equal (say) to one quarter of the limiting value. This occurs when $e^{-qt} = \tfrac{1}{2}$, or by (5) when

$$t = \frac{\log 2}{q} = \frac{\log 2\,.\,M'}{p\,.\,kr\,.\,2\pi\sigma r^3}. \qquad \qquad (12)$$

The energy propagated in time t across the area S of primary wave-front is (*Theory of Sound*, § 245)

$$\tfrac{1}{2} S \sigma a k^2 \alpha^2 t, \dots\dots\dots\dots\dots\dots\dots\dots\dots(13)$$

where a is the velocity of propagation, so that $p = ak$. If we equate (13) to one quarter of (9) and identify t with the value given by (12), neglecting the distinction between M and M', we get

$$S = \frac{\pi}{2 \log 2 . k^2} = \frac{\lambda^2}{8\pi \log 2} . \dots\dots\dots\dots\dots(14)^*$$

The resonator is thus able to capture an amount of energy equal to that passing in the same time through an area of primary wave-front comparable with λ^2/π, an area which may exceed any number of times the cross-section of the resonator itself.

* $\log 2 = 0\cdot 693$.

410.

ON THE ATTENUATION OF SOUND IN THE ATMOSPHERE.

[*Advisory Committee for Aeronautics. August*, 1916.]

In T. 749, Major Taylor presents some calculations which "shew that the chief cause of the dissipation of sound during its transmission through the lower atmosphere must be sought for in the eddying motion which is known to exist there. The amount of dissipation which these calculations would lead us to expect from our knowledge of the structure of the lower atmosphere agrees, as well as the rough nature of the observations permit, with the amount of dissipation given by Mr Lindemann."

The problem discussed is one of importance and it is attended with considerable difficulties. There can be no doubt that on many occasions, perhaps one might say normally, the attenuation is much more rapid than according to the law of inverse squares. Some 20 years ago (*Scientific Papers*, Vol. IV. p. 298) I calculated that according to this law the sound of a Trinity House syren, absorbing 60 horse-power, should be audible to 2700 kilometres!

A failure to propagate, so far as it is uniform on all occasions, would naturally be attributed to dissipative action. I am here using the word in the usual and narrower technical sense, implying a degradation of energy from the mechanical form into heat, or a passage of heat from a higher to a lower temperature. Although there must certainly be dissipation consequent upon radiation and conduction of heat, it does not appear that these causes are adequate to explain the attenuation of sound sometimes observed, even at moderate distances. This question is discussed in *Phil. Mag.* XLVII. p. 308, 1899 (*Scientific Papers*, Vol. IV. p. 376) in connexion with some observations of Wilmer Duff.

If we put dissipation out of account, the energy of a sound wave, advancing on a broad front, remains mechanical, and we have to consider what becomes of it. Part of the sound may be reflected, and there is no doubt at all that, whatever may be the mechanism, reflection does really occur, even when no obstacles are visible. At St Catherine's Point in 1901, I heard strong echoes from over the sea for at least 12 seconds after the syren had ceased sounding. The sky was clear and there were no waves to speak of. Reflection in the narrower sense (which does not include so called total reflection!) requires irregularities in the medium whose outlines are somewhat sharply defined, the linear standard being the wave-length of the vibration; but this requirement is probably satisfied by ascending streams of heated air.

In considering the effect of eddies on maintained sounds of given pitch, Major Taylor does not include either dissipation (in the narrower sense) or reflection. I do not understand how, under such conditions, there can be any general attenuation of plane waves. What is lost in one position in front of the phase-disturbing obstacles, must be gained at another. The circumstances are perhaps more familiar in Optics. Consider the passage of light of given wave-length through a grating devoid of absorbing and reflecting power. The whole of the incident light is then to be found distributed between the central image and the lateral spectra. At a sufficient distance behind the grating, supposed to be of limited width, the spectra are separated, and as I understand it the calculation refers to what would be found in the beam going to form the central image. But close behind the grating, or at any distance behind if the width be unlimited, there is no separation, and the average intensity is the same as before incidence. The latter appears to be the case with which we are now concerned. The problem of the grating is treated in *Theory of Sound*, 2nd edition, § 272 a.

Of course, the more important anomalies, such as the usual failure of sound up wind, are to be explained after Stokes and Reynolds by a refraction which is approximately regular.

In connexion with eddies it may be worth while to mention the simple case afforded by a vortex in two dimensions whose axis is parallel to the plane of the sound waves. The circumferential velocity at any point is proportional to $1/r$, where r is the distance from the axis. By integration, or more immediately by considering what Kelvin called the "circulation," it is easy to prove that the whole of the wave which passes on one side of the axis is uniformly advanced by a certain amount and the whole on the other side retarded by an equal amount. A *fault* is thus introduced into the otherwise plane character of the wave.

[1918. Major Taylor sends me the following observations:

NOTE ON THE DISPERSION OF SOUND.

Observations have shown that sound is apparently dissipated at a much greater rate than the inverse square law both up and down wind. The effect of turbulence on a plane wave front is to cause it to deviate locally from its plane form. The wave train cannot then be propagated forward without further change, but it may be regarded as being composed of a plane wave train of smaller amplitude, together with waves which are dispersed in all directions, and are due to the effect of the turbulence of the original train. If d is the diameter of an eddy, λ is the wave length of the sound, U is the velocity of the air due to the eddy, and V is the velocity of sound, the amount of sound energy dispersed from unit volumes of the main wave is

$$E \cdot 4\pi^2 \cdot \frac{U^2 d}{\lambda^2 V^2},$$

where E is the energy of the sound per unit volume. If the turbulence is uniformly distributed round the source of sound then, as Lord Rayleigh points out, the sound energy will be uniformly distributed because the energy dispersed from one part of the wave front will be replaced by energy dispersed from other parts; but if the turbulence is a maximum in any particular direction then more sound energy will be dispersed from the wave fronts as they proceed in that direction than will be received from the less turbulent regions. Regions of maximum turbulence should, therefore, be regions of minimum sound. The turbulence is usually a maximum near the ground. The intensity of sound should, therefore, fall off near the ground at a greater rate than the inverse square law, even although there is no solid obstacle between the source of sound and the listener.]

411.

ON VIBRATIONS AND DEFLEXIONS OF MEMBRANES, BARS, AND PLATES.

[*Philosophical Magazine*, Vol. XXXII. pp. 353—364, 1916.]

IN *Theory of Sound*, § 211, it was shown that "any contraction of the fixed boundary of a vibrating membrane must cause an elevation of pitch, because the new state of things may be conceived to differ from the old merely by the introduction of an additional constraint. Springs, without inertia, are supposed to urge the line of the proposed boundary towards its equilibrium position, and gradually to become stiffer. At each step the vibrations become more rapid, until they approach a limit corresponding to infinite stiffness of the springs and absolute fixity of their points of application. It is not necessary that the part cut off should have the same density as the rest, or even any density at all."

From this principle we may infer that the gravest mode of vibration for a membrane of any shape and of any variable density is devoid of internal nodal lines. For suppose that $ACDB$ (fig. 1) vibrating in its longest period

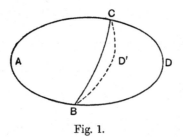

Fig. 1.

(τ) has an internal nodal line CB. This requires that a membrane with the fixed boundary ACB shall also be capable of vibration in period τ. The impossibility is easily seen. As $ACDB$ gradually contracts through $ACD'B$ to ACB, the longest period diminishes, so that the longest period of ACB is less than τ. No period possible to ACB can be equal to τ.

If we replace the reactions against acceleration by external forces, we may obtain the solution of a statical problem. When a membrane of any shape is submitted to transverse forces, all in one direction, the displacement is *everywhere* in the direction of the forces.

Similar conclusions may be formulated for the conduction of heat in two dimensions, which depends upon the same fundamental differential equation. Here the boundary is maintained at a constant temperature taken as zero, and "persistences" replace the periods of vibration. Any closing in of the boundary reduces the principal persistence. In this mode there can be no internal place of zero temperature. In the steady state under positive sources of heat, however distributed, the temperature is above zero everywhere. In the application to the theory of heat, extension may evidently be made to three dimensions.

Arguments of a like nature may be used when we consider a bar vibrating transversely in virtue of rigidity, instead of a stretched membrane. In *Theory of Sound*, §184, it is shown that whatever may be the constitution of the bar in respect of stiffness and mass, a curtailment at either end is associated with a rise of pitch, and this whether the end in question be free, clamped, or merely " supported."

In the statical problem of the deflexion of a bar by a transverse force locally applied, the question may be raised whether the linear deflexion must everywhere be in the same direction as the force. It can be shown that the answer is in the affirmative. The equation governing the deflexion (w) is

$$\frac{d^2}{dx^2}\left(B\frac{d^2w}{dx^2}\right) = Z, \quad\ldots\ldots\ldots\ldots\ldots\ldots\ldots\ldots(1)$$

where $Z\,dx$ is the transverse force applied at dx, and B is a coefficient of stiffness. In the case of a uniform bar B is constant and w may be found by simple integration. It suffices to suppose that Z is localized at one point, say at $x=b$; and the solution shows that whether the ends be clamped or supported, or if one end be clamped and the other free or supported, w is everywhere of the same sign as Z. The conclusion may evidently be extended to a force variable in any manner along the length of the bar, provided that it be of the same sign throughout.

But there is no need to lay stress upon the case of a uniform bar, since the proposition is of more general application. The first integration of (1) gives

$$\frac{d}{dx}\left(B\frac{d^2w}{dx^2}\right) = \int_0^x Z\,dx + C, \quad\ldots\ldots\ldots\ldots\ldots\ldots(2)$$

and $\int Z\,dx = 0$ from $x=0$ at one end to $x=b$, and takes another constant value (Z_1) from $x=b$ to the other end at $x=l$. A second integration now shows that $B d^2w/dx^2$ is a linear function of x between 0 and b, and again a linear function between b and l, the two linear functions assuming the same value at $x = b$. Since B is everywhere positive, it follows that the curvature cannot vanish more than twice in the whole range from 0 to l, ends included, unless indeed it vanish everywhere over one of the parts. If one end be supported, the curvature vanishes there. If the other end also be supported, the curvature is of one sign throughout, and the curve of deflexion can nowhere cross the axis. If the second end be clamped, there is but one internal point of inflexion, and again the axis cannot be crossed. If both ends are clamped, the two points of inflexion are internal, but the axis cannot be crossed, since a crossing would involve three points of inflexion. If one end be free, the curvature vanishes there, and not only the curvature but also the rate of change of curvature. The part of the rod from this end up to the point of application of the force remains unbent and one of the linear functions spoken of is zero throughout. Thus the curvature never changes sign, and the axis cannot be crossed. In this case equilibrium requires that the other end be clamped. We conclude that in no case can there be a deflexion anywhere of opposite sign to that of the force applied at $x = b$, and the conclusion may be extended to a force, however distributed, provided that it be one-signed throughout.

Leaving the problems presented by the membrane and the bar, we may pass on to consider whether similar propositions are applicable in the case of a flat plate, whose stiffness and density may be variable from point to point. An argument similar to that employed for the membrane shows that when the boundary is clamped any contraction of it is attended by a rise of pitch. But (*Theory of Sound*, § 230) the statement does *not* hold good when the boundary is free.

When a localized transverse force acts upon the plate, we may inquire whether the displacement is at all points in the same direction as the force. This question was considered in a former paper* in connexion with a hydrodynamical analogue, and it may be convenient to repeat the argument. Suppose that the plate (fig. 2), clamped at a distant boundary, is almost divided into two independent parts by a straight partition CD extending across, but perforated by a narrow aperture AB; and that the force is applied at a distance from CD on the left. If the partition were complete, w and dw/dn would be zero over the whole (in virtue of the clamping), and the displacement in the neighbourhood on the left would be simple one-dimensional bending, with w positive throughout. On the right w would vanish. In order to maintain this condition of things a certain couple acts upon the plate in virtue of the supposed constraints along CD.

Fig. 2.

* *Phil. Mag.* Vol. XXXVI. p. 354 (1893); *Scientific Papers*, Vol. IV. p. 88.

Along the perforated portion AB the couple required to produce the one-dimensional bending fails. The actual deformation accordingly differs from the one-dimensional bending by the deformation that would be produced by a couple over AB acting upon the plate, as clamped along CA, BD, but otherwise free from force. This deformation is evidently symmetrical with change of sign upon the two sides of CD, w being positive on the left, *negative* on the right, and vanishing on AB itself. Thus upon the whole a downward force acting on the left gives rise to an upward motion on the right, in opposition to the general rule proposed for examination.

If we suppose a load attached at the place where the force acts, but that otherwise the plate is devoid of mass, we see that a clamped plate vibrating freely in its gravest mode may have internal nodes in the sense that w is there evanescent, but of course not in the full sense of places which behave as if they were clamped.

In the case of a plate whose boundary is merely supported, *i.e.* acted upon by a force (without couple) constraining w to remain zero *, it is still easier to recognize that a part of the plate may move in the direction opposite to that of an applied force. We may contemplate the arrangement of fig. 2, where, however, the partition CD is now merely supported and not clamped. Along the unperforated parts CA, BD the plate must be supposed cut through so that no couple is transmitted. And in the same way we infer that internal nodes are possible when a supported plate vibrates freely in its gravest mode.

But although a movement opposite to that of the impressed force may be possible in a plate whose boundary is clamped or supported, it would seem that this occurs only in rather extreme cases when the boundary is strongly re-entrant. One may suspect that such a contrary movement is excluded when the boundary forms an oval curve, *i.e.* a curve whose curvature never changes sign. A rectangular plate comes under this description; but according to M. Mesnager[†], "M. J. Résal a montré qu'en applicant une charge au centre d'une plaque rectangulaire de proportions convenables, on produit très probablement le soulèvement de certaines régions de la plaque." I understand that the boundary is supposed to be "supported" and that suitable proportions are attained when one side of the rectangle is relatively long. It seems therefore desirable to inquire more closely into this question.

The general differential equation for the equilibrium of a uniform elastic plate under an impressed transverse force proportional to Z is[‡]

$$\nabla^4 w = (d^2/dx^2 + d^2/dy^2)^2\, w = Z. \quad\quad\quad\quad (3)$$

---

* It may be remarked that the substitution of a supported for a clamped boundary is equivalent to the abolition of a constraint, and is in consequence attended by a fall in the frequency of free vibrations.

† *C. R. t.* CLXII. p. 826 (1916).

‡ *Theory of Sound*, §§ 215, 225 ; Love's *Mathematical Theory of Elasticity*, Chapter XXII.

We will apply this equation to the plate bounded by the lines $y = 0$, $y = \pi$, and extending to infinity in both directions along x, and we suppose that external transverse forces act only along the line $x = 0$. Under the operation of these forces the plate deflects symmetrically, so that w is the same on both sides of $x = 0$ and along this line $dw/dx = 0$. Having formulated this condition, we may now confine our attention to the positive side, regarding the plate as bounded at $x = 0$.

The conditions for a supported edge parallel to x are

$$w = 0, \quad d^2w/dy^2 = 0 ; \qquad \dots\dots\dots\dots\dots\dots(4)$$

and they are satisfied at $y = 0$ and $y = \pi$ if we assume that w as a function of y is proportional to $\sin ny$, n being an integer. The same assumption introduced into (3) with $Z = 0$ gives

$$(d^2/dx^2 - n^2)^2 \, w = 0, \qquad \dots\dots\dots\dots\dots\dots(5)$$

of which the general solution is

$$w = \{(A + Bx)\, e^{-nx} + (C + Dx)\, e^{nx}\} \sin ny, \quad \dots\dots\dots\dots(6)$$

where A, B, C, D, are constants. Since $w = 0$ when $x = +\infty$, C and D must here vanish; and by the condition to be satisfied when $x = 0$, $B = nA$. The solution applicable for the present purpose is thus

$$w = A \sin ny \, . \, (1 + nx)\, e^{-nx} \dots\dots\dots\dots\dots\dots(7)$$

The force acting at the edge $x = 0$ necessary to maintain this displacement is proportional to

$$\frac{d\nabla^2 w}{dx} + (1 - \mu)\, \frac{d^2}{dy^2} \frac{dw}{dx}, \text{ that is } \frac{d^3 w}{dx^3} \text{ simply, } \dots\dots\dots\dots(8)$$

in virtue of the condition there imposed. Introducing the value of w from (7), we find that

$$d^3 w / dx^3 = 2n^3 A \sin ny, \qquad \dots\dots\dots\dots\dots\dots(9)$$

which represents the force in question. When $n = 1$,

$$w = A \sin y \, . \, (1 + x)\, e^{-x} ; \qquad \dots\dots\dots\dots\dots\dots(10)$$

and it is evident that w retains the same sign over the whole plate from $x = 0$ to $x = \infty$. On the negative side (10) is not applicable as it stands, but we know that w has identical values at $\pm x$.

The solution expressed in (10) suggests strongly that Résal's expectation is not fulfilled, but two objections may perhaps be taken. In the first place the force expressed in (9) with $n = 1$, though preponderant at the centre $y = \frac{1}{2}\pi$, is not entirely concentrated there. And secondly, it may be noticed that we have introduced no special boundary condition at $x = \infty$. It might be argued that although w tends to vanish when x is very great, the manner of its evanescence may not exclude a reversal of sign.

We proceed then to examine the solution for a plate definitely terminated at distances l, and there *supported*. For this purpose we resume the general solution (6),

$$w = \sin ny \{(A + Bx)\, e^{-nx} + (C + Dx)\, e^{nx}\},\ldots\ldots\ldots\ldots(11)$$

which already satisfies the conditions of a supported edge at $y = 0$, $y = \pi$. At $x = 0$, the condition is as before $dw/dx = 0$. At $x = l$ the conditions for a supported edge give first $w = 0$, and therefore $d^2w/dy^2 = 0$. The second condition then reduces to $d^2w/dx^2 = 0$. Applying these conditions to (11) we find

$$D = Be^{-2nl}, \quad C = -e^{-2nl}\,(A + 2lB).\ \ldots\ldots\ldots\ldots(12)$$

It remains to introduce the condition to be satisfied at $x = 0$. In general

$$\frac{dw}{dx} = \sin ny\,[e^{-nx}\,\{-n\,(A + Bx) + B\} + e^{nx}\,\{n\,(C + Dx) + D\}];\ \ldots(13)$$

and since this is to vanish when $x = 0$,

$$-nA + B + nC + D = 0.\ \ldots\ldots\ldots\ldots\ldots\ldots(14)$$

By means of (12), (14) A, C, D may be expressed in terms of B, and we find

$$\frac{dw}{dx} = \frac{nB \sin ny}{1 + e^{-2nl}}\,[e^{-nx}\,\{-x + (2l - x)\,e^{-2nl}\} + e^{-n(2l-x)}\,\{-(2l - x) + xe^{-2nl}\}].\ (15)$$

In (15) the square bracket is negative for any value of x between 0 and l, for it may be written in the form

$$-xe^{-nx}\,\{1 - e^{-2n(2l-x)}\} - (2l - x)e^{-2nl}\,\{e^{nx} - e^{-nx}\}.\ \ldots\ldots\ldots(16)$$

When $x = 0$ it vanishes, and when $x = l$ it becomes

$$-2le^{-2nl}\,(e^{nl} - e^{-nl}).$$

It appears then that for any fixed value of y there is no change in the sign of dw/dx over the whole range from $x = 0$ to $x = l$. And when $n = 1$, this sign does not alter with y. As to the sign of w when $x = 0$, we have then from (11)

$$w = \sin ny\,(A + C) = B \sin ny\,\frac{e^{2nl} - e^{-2nl} - 4nl}{n\,(e^{nl} + e^{-nl})}\ *,$$

so that dw/dx in (15) has throughout the opposite sign to that of the initial value of w. And since $w = 0$ when $x = l$, it follows that for every value of y the sign of w remains unchanged from $x = 0$ to $x = l$. Further, if $n = 1$, this sign is the same whatever be the value of y. Every point in the plate is deflected in the same direction.

Let us now suppose that the plate is clamped at $x = \pm l$, instead of merely supported. The conditions are of course $w = 0$, $dw/dx = 0$. They give

$$D = e^{-2nl}\,\{2nA + B\,(2nl - 1)\},\ \ldots\ldots\ldots\ldots\ldots(17)$$

$$C = -e^{-2nl}\,\{A\,(1 + 2nl) + 2nl^2B\}.\ \ldots\ldots\ldots\ldots(18)$$

The condition at $x = 0$ is that already expressed in (14).

* [The factor e^{nl} has been omitted from the denominator; with $l = \infty$ the corrected result agrees with (7) when $x = 0$, if $B = nA$. W. F. S.]

As before, A, C, D may be expressed in terms of B. For shortness we may set $B = 1$, and write

$$H = 1 + e^{-2nl}(2nl - 1). \qquad\qquad\qquad\ldots\ldots\ldots\ldots\ldots(19)$$

We find

$$-nA + 1 = 2n^2l^2 e^{-2nl}/H,$$

$$D = (2nl + 1 - e^{-2nl})\,e^{-2nl}/H,$$

$$nC + D = -e^{-2nl}\,.\,2n^2l^2/H.$$

Thus

$$\frac{dw}{dx} = \sin ny\,[e^{-nx}(-nA + 1 - nx) + e^{nx}(nC + D + nDx)]$$

$$= H^{-1}\sin ny\,.\,e^{-nx}\,[2n^2l^2 e^{-2nl} - nx\{1 + e^{-2nl}(2nl - 1)\}]$$

$$+\,H^{-1}\sin ny\,.\,e^{n(x-2l)}\,[-2n^2l^2 + nx\{2nl + 1 - e^{-2nl}\}],$$

vanishing when $x = 0$, and when $x = l$.

This may be put into the form

$$\frac{dw}{dx} = -H^{-1}\sin ny\,[2n^2l\,(l - x)\,e^{-2nl}(e^{nx} - e^{-nx})$$

$$+\,nx\,e^{-nl}(1 - e^{-2nl})(e^{n(l-x)} - e^{-n(l-x)})], \qquad\ldots\ldots\ldots\ldots(20)$$

in which the square bracket is positive from $x = 0$ to $x = l$.

It is easy to see that H also is positive.   When nl is small, (19) is positive, and it cannot vanish, since

$$e^{2nl} > 1 > 1 - 2nl.$$

It remains to show that the sign of w follows that of $\sin ny$ when $x = 0$. In this case

$$w = (A + C)\sin ny\,; \qquad\qquad\qquad\ldots\ldots\ldots\ldots\ldots(21)$$

and

$$n\,(A + C)\,H = 1 - e^{-2nl}(2 + 4n^2l^2) + e^{-4nl}$$

$$= e^{-2nl}(e^{2nl} + e^{-2nl} - 2 - 4n^2l^2). \qquad\ldots\ldots\ldots\ldots(22)*$$

The bracket on the right of (22) is positive, since

$$e^{2nl} + e^{-2nl} = 2\left(1 + \frac{4n^2l^2}{2} + \frac{16n^4l^4}{4!} + \ldots\right).$$

We see then that for any value of y, the sign of dw/dx over the whole range from $x = 0$ to $x = l$ is the opposite of the sign of w when $x = 0$†; and since $w = 0$ when $x = l$, it follows that it cannot vanish anywhere between. When $n = 1$, w retains the same sign at $x = 0$ whatever be the value of y, and therefore also at every point of the whole plate.   No more in this case than when the edges at $x = \pm l$ are merely supported, can there be anywhere a deflexion in the reverse direction.

In both the cases just discussed the force operative at $x = 0$ to which the deflexion is due is, as in (8), proportional simply to d^3w/dx^3, and therefore to

---

* [Some corrections have been made in this equation. W. F. S.]

† This follows at once if we start from $x = l$ where $w = 0$.

sin ny, and is of course in the same direction as the displacement along the same line. When $n = 1$, both forces and displacements are in a fixed direction. It will be of interest to examine what happens when the force is concentrated at a single point on the line $x = 0$, instead of being distributed over the whole of it between $y = 0$ and $y = \pi$. But for this purpose it may be well to simplify the problem by supposing l infinite.

On the analogy of (7) we take

$$w = \Sigma A_n (1 + nx) e^{-nx} \sin ny, \quad \dots \dots \dots \dots \dots (23)$$

making, when $x = 0$,

$$d^3 w / dx^3 = 2\Sigma n^3 A_n \sin ny. \quad \dots \dots \dots \dots \dots (24)$$

If, then, Z represent the force operative upon dy, analysable by Fourier's theorem into

$$Z = Z_1 \sin y + Z_2 \sin 2y + Z_3 \sin 3y + \dots, \quad \dots \dots \dots \dots (25)$$

we have

$$Z_n = \frac{2}{\pi} \int_0^\pi Z \sin ny \, dy = \frac{2}{\pi} Z_\eta \sin n\eta, \quad \dots \dots \dots \dots (26)$$

if the force is concentrated at $y = \eta$. Hence by (24)

$$A_n = \frac{Z_\eta}{\pi} \frac{\sin n\eta}{n^3}, \quad \dots \dots \dots \dots \dots \dots (27)$$

so that

$$w = \frac{Z_\eta}{2\pi} \Sigma \frac{\cos n(y-\eta) - \cos n(y+\eta)}{n^3} e^{-nx}(1 + nx), \quad \dots \dots (28)$$

where $n = 1, 2, 3$, etc. It will be understood that a constant factor, depending upon the elastic constants and the thickness of the plate, but not upon n, has been omitted.

The series in (28) becomes more tractable when differentiated. We have

$$\frac{dw}{dx} = -\frac{x Z_\eta}{2\pi} \Sigma \frac{\cos n(y - \eta) - \cos n(y + \eta)}{n} e^{-nx}; \quad \dots \dots (29)$$

and the summations to be considered are of the form

$$\Sigma n^{-1} \cos n\beta \, e^{-nx}. \quad \dots \dots \dots \dots \dots (30)$$

This may be considered as the real part of

$$\Sigma n^{-1} e^{-n(x-i\beta)}, \quad \dots \dots \dots \dots \dots (31)$$

that is, of

$$- \log (1 - e^{-(x-i\beta)}). \quad \dots \dots \dots \dots \dots (32)$$

Thus, if we take

$$\Sigma n^{-1} e^{-n(x-i\beta)} = X + iY, \quad \dots \dots \dots \dots (33)$$

$$e^{-X-iY} = 1 - e^{-(x-i\beta)}, \text{ and } e^{-X+iY} = 1 - e^{-(x+i\beta)},$$

so that

$$e^{-2X} = 1 + e^{-2x} - 2e^{-x} \cos \beta. \quad \dots \dots \dots \dots (34)$$

Accordingly

$$\Sigma n^{-1} \cos n\beta \, e^{-nx} = -\tfrac{1}{2} \log (1 + e^{-2x} - 2e^{-x} \cos \beta); \quad \dots\dots\dots (35)$$

and

$$\frac{dw}{dx} = \frac{x \, Z_\eta}{4\pi} \log \frac{1 + e^{-2x} - 2e^{-x} \cos (y - \eta)}{1 + e^{-2x} - 2e^{-x} \cos (y + \eta)}. \quad \dots\dots\dots (36)$$

From the above it appears that

$$W = x \log \{1 + e^{-2x} - 2e^{-x} \cos (y + \eta)\} = x \log h$$

must satisfy $\nabla^4 W = 0$. This may readily be verified by means of

$$\nabla^2 \log h = 0, \quad \text{and} \quad \nabla^2 W = x \, \nabla^2 \log h + 2d \log h/dx.$$

We have now to consider the sign of the logarithm in (36), or, as it may be written,

$$\log \frac{e^x + e^{-x} - 2 \cos (y - \eta)}{e^x + e^{-x} - 2 \cos (y + \eta)}. \quad \dots\dots\dots\dots (37)$$

Since the cosines are less than unity, both numerator and denominator are positive. Also the numerator is less than the denominator, for

$$\cos (y - \eta) - \cos (y + \eta) = 2 \sin y \sin \eta = +,$$

so that $\cos (y - \eta) > \cos (y + \eta)$. The logarithm is therefore negative, and dw/dx has everywhere the opposite sign to that of Z_η. If this be supposed positive, w on every line $y = $ const. increases as we pass inwards from $x = \infty$ where $w = 0$ to $x = 0$. Over the whole plate the displacement is positive, and this whatever the point of application (η) of the force. Obviously extension may be made to any distributed one-signed force.

It may be remarked that since the logarithm in (37) is unaltered by a reversal of x, (36) is applicable on the negative as well as on the positive side of $x = 0$. If $y = \eta$, $x = 0$, the logarithm becomes infinite, but dw/dx is still zero in virtue of the factor x.

I suppose that w cannot be expressed in finite terms by integration of (36), but there would be no difficulty in dealing arithmetically with particular cases by direct use of the series (28). If, for example, $\eta = \tfrac{1}{2}\pi$, so that the force is applied at the centre, we have to consider

$$\Sigma n^{-3} \sin \tfrac{1}{2} n\pi \, . \, \sin ny \, . \, e^{-nx} (1 + nx), \quad \dots\dots\dots (38)$$

and only odd values of n enter. Further, (38) is symmetrical on the two sides of $y = \tfrac{1}{2}\pi$. Two special cases present themselves when $x = 0$ and when $y = \tfrac{1}{2}\pi$. In the former w is proportional to

$$\sin y - \frac{1}{3^3} \sin 3y + \frac{1}{5^3} \sin 5y - \dots, \quad \dots\dots\dots (39)$$

and in the latter to

$$e^{-x} (1 + x) + \frac{1}{3^3} e^{-3x} (1 + 3x) + \frac{1}{5^3} e^{-5x} (1 + 5x) + \dots . \quad \dots\dots (40)$$

August 2, 1916.

Added August 21.

The accompanying tables show the form of the curves of deflexion defined by (39), (40).

y	(39)	y	(39)
0°	·0000	50	·7416
10	·1594	60	·8574
20	·3162	70	·9530
30	·4675	80	1·0217
40	·6104	90	1·0518

x	(40)	x	(40)
0·0	1·0518	3·0	·1992
0·5	·9333	4·0	·0916
1·0	·7435	5·0	·0404
2·0	·4066	10·0	·0005

In a second communication* Mesnager returns to the question and shows by very simple reasoning that all points of a rectangular plate supported at the boundary move in the direction of the applied transverse forces.

If z denote $\nabla^2 w$, then $\nabla^2 z$, $= \nabla^4 w$, is positive over the plate if the applied forces are everywhere positive. At a straight portion of the boundary of a supported plate $z = 0$, and this is regarded as applicable to the whole boundary of the rectangular plate, though perhaps the corners may require further consideration. But if $\nabla^2 z$·is everywhere positive within a coutour and z vanish on the contour itself, z must be negative over the interior, as is physically obvious in the theory of the conduction of heat. Again, since $\nabla^2 w$ is negative throughout the interior, and w vanishes at the boundary, it follows in like manner that w is positive throughout the interior.

It does not appear that an argument on these lines can be applied to a rectangular plate whose boundary is clamped, or to a supported plate whose boundary is in part curved.

P.S. In connexion with a recent paper on the "Flow of Compressible Fluid past an Obstacle" (*Phil. Mag.* July 1916)†, I have become aware that the subject had been treated with considerable generality by Prof. Cisotti of Milan, under the title "Sul Paradosso di D'Alembert" (*Atti R. Istituto Veneto*, t. lxv. 1906). There was, however, no reference to the limitation necessary when the velocity exceeds that of sound in the medium. I understand that this matter is now engaging Prof. Cisotti's attention.

* *C. R.* July 24, 1916, p. 84.   † [This volume, p. 402.]

412.

ON CONVECTION CURRENTS IN A HORIZONTAL LAYER OF FLUID, WHEN THE HIGHER TEMPERATURE IS ON THE UNDER SIDE.

[*Philosophical Magazine*, Vol. XXXII. pp. 529—546, 1916.]

THE present is an attempt to examine how far the interesting results obtained by Bénard'* in his careful and skilful experiments can be explained theoretically. Bénard worked with very thin layers, only about 1 mm. deep, standing on a levelled metallic plate which was maintained at a uniform temperature. The upper surface was usually free, and being in contact with the air was at a lower temperature. Various liquids were employed—some, indeed, which would be solids under ordinary conditions.

The layer rapidly resolves itself into a number of *cells*, the motion being an ascension in the middle of a cell and a descension at the common boundary between a cell and its neighbours. Two phases are distinguished, of unequal duration, the first being relatively very short. The limit of the first phase is described as the "semi-regular cellular regime"; in this state all the cells have already acquired surfaces *nearly* identical, their forms being nearly regular convex polygons of, in general, 4 to 7 sides. The boundaries are vertical, and the circulation in each cell approximates to that already indicated. This phase is brief (1 or 2 seconds) for the less viscous liquids (alcohol, benzine, etc.) at ordinary temperatures. Even for paraffin or spermaceti, melted at 100° C., 10 seconds suffice; but in the case of very viscous liquids (oils, etc.), if the flux of heat is small, the deformations are extremely slow and the first phase may last several minutes or more.

The second phase has for its limit a permanent regime of regular hexagons. During this period the cells become equal and regular and align

---

* *Revue générale des Sciences*, Vol. XII. pp. 1261, 1309 (1900); *Ann. d. Chimie et de Physique*, t. XXIII. p. 62 (1901). M. Bénard does not appear to be acquainted with James Thomson's paper "On a Changing Tesselated Structure in certain Liquids" (*Proc. Glasgow Phil. Soc.* 1881—2), where is described a like structure in much thicker layers of soapy water cooling from the surface.

themselves. It is extremely protracted, if the limit is regarded as the complete attainment of regular hexagons. And, indeed, such perfection is barely attainable even with the most careful arrangements. The tendency, however, seems sufficiently established.

The theoretical consideration of the problem here arising is of interest for more than one reason. In general, when a system falls away from unstable equilibrium it may do so in several principal modes, in each of which the departure at time t is proportional to the small displacement or velocity supposed to be present initially, and to an exponential factor e^{qt}, where q is positive. If the initial disturbances are small enough, that mode (or modes) of falling away will become predominant for which q is a maximum. The simplest example for which the number of degrees of freedom is infinite is presented by a cylindrical rod of elastic material under a longitudinal compression sufficient to overbalance its stiffness. But perhaps the most interesting hitherto treated is that of a cylinder of fluid disintegrating under the operation of capillary force as in the beautiful experiments of Savart and Plateau upon jets. In this case the surface remains one of revolution about the original axis, but it becomes *varicose*, and the question is to compare the effects of different wave-lengths of varicosity, for upon this depends the number of detached masses into which the column is eventually resolved. It was proved by Plateau that there is no instability if the wave-length be less than the circumference of the column. For all wave-lengths greater than this there is instability, and the corresponding modes of disintegration may establish themselves if the initial disturbances are suitable. But if the general disturbance is very small, those components only will have opportunity to develop themselves for which the wave-length lies near to that of maximum instability.

It has been shown* that the wave-length of maximum instability is 4·508 times the diameter of the jet, exceeding the wave-length at which instability first enters in the ratio of about 3 : 2. Accordingly this is the sort of disintegration to be expected when the jet is shielded as far as possible from external disturbance.

It will be observed that there is nothing in this theory which could fix the *phase* of the predominant disturbance, or the particular particles of the fluid which will ultimately form the centres of the detached drops. There remains a certain indeterminateness, and this is connected with the circumstance that absolute regularity is not to be expected. In addition to the wave-length of maximum instability we must include all those which lie sufficiently near to it, and the superposition of the corresponding modes will allow of a slow variation of phase as we pass along the column. The phase

---

* *Proc. Lond. Math. Soc.* Vol. x. p. 4 (1879); *Scientific Papers*, Vol. I. p. 361. Also *Theory of Sound*, 2nd ed. §§ 357, &c.

in any particular region depends upon the initial circumstances in and near that region, and these are supposed to be matters of chance*. The superposition of infinite trains of waves whose wave-lengths cluster round a given value raises the same questions as we are concerned with in considering the character of approximately homogeneous light.

In the present problem the case is much more complicated, unless we arbitrarily limit it to two dimensions. The cells of Bénard are then reduced to infinitely long strips, and when there is instability we may ask for what wave-length (width of strip) the instability is greatest. The answer can be given under certain restrictions, and the manner in which equilibrium breaks down is then approximately determined. So long as the two-dimensional character is retained, there seems to be no reason to expect the wave-length to alter afterwards. But even if we assume a natural disposition to a two-dimensional motion, the direction of the length of the cells as well as the phase could only be determined by initial circumstances, and could not be expected to be uniform over the whole of the infinite plane.

According to the observations of Bénard, something of this sort actually occurs when the layer of liquid has a general motion in its own plane at the moment when instability commences, the length of the cellular strips being parallel to the general velocity. But a little later, when the general motion has decayed, division-lines running in the perpendicular direction present themselves.

In general, it is easy to recognize that the question is much more complex. By Fourier's theorem the motion in its earlier stages may be analysed into components, each of which corresponds to rectangular cells whose sides are parallel to fixed axes arbitrarily chosen. The solution for maximum instability yields one relation between the sides of the rectangle, but no indication of their ratio. It covers the two-dimensional case of infinitely long rectangles already referred to, and the contrasted case of squares for which the length of the side is thus determined. I do not see that any plausible hypothesis as to the origin of the initial disturbances leads us to expect one particular ratio of sides in preference to another.

On a more general view it appears that the function expressing the disturbance which develops most rapidly may be assimilated to that which represents the free vibration of an infinite stretched membrane vibrating with given frequency.

The calculations which follow are based upon equations given by Boussinesq, who has applied them to one or two particular problems. The special limitation which characterizes them is the neglect of variations of density,

---

* When a jet of liquid is acted on by an external vibrator, the resolution into drops may be regularized in a much higher degree.

except in so far as they modify the action of gravity. Of course, such neglect can be justified only under certain conditions, which Boussinesq has discussed. They are not so restrictive as to exclude the approximate treatment of many problems of interest.

When the fluid is inviscid and the higher temperature is below, all modes of disturbance are instable, even when we include the conduction of heat during the disturbance. But there is one class of disturbances for which the instability is a maximum.

When viscosity is included as well as conduction, the problem is more complicated, and we have to consider boundary conditions. Those have been chosen which are simplest from the mathematical point of view, and they deviate from those obtaining in Bénard's experiments, where, indeed, the conditions are different at the two boundaries. It appears, a little unexpectedly, that the equilibrium may be thoroughly stable (with higher temperature below), if the coefficients of conductivity and viscosity are not too small. As the temperature gradient increases, instability enters, and at first only for a particular kind of disturbance.

The second phase of Bénard, where a tendency reveals itself for a slow transformation into regular hexagons, is not touched. It would seem to demand the inclusion of the squares of quantities here treated as small. But the size of the hexagons (under the boundary conditions postulated) is determinate, at any rate when they assert themselves early enough.

An appendix deals with a related analytical problem having various physical interpretations, such as the symmetrical vibration in two dimensions of a layer of air enclosed by a nearly circular wall.

———————

The general Eulerian equations of fluid motion are in the usual notation:—

$$\frac{Du}{Dt} = X - \frac{1}{\rho}\frac{dp}{dx}, \quad \frac{Dv}{Dt} = Y - \frac{1}{\rho}\frac{dp}{dy}, \quad \frac{Dw}{Dt} = Z - \frac{1}{\rho}\frac{dp}{dz}, \quad \ldots\ldots(1)$$

where

$$\frac{D}{Dt} = \frac{d}{dt} + u\frac{d}{dx} + v\frac{d}{dy} + w\frac{d}{dz}, \quad \ldots\ldots\ldots\ldots\ldots(2)$$

and X, Y, Z are the components of extraneous force reckoned per unit of mass. If, neglecting viscosity, we suppose that gravity is the only impressed force,

$$X = 0, \qquad Y = 0, \qquad Z = -g, \quad \ldots\ldots\ldots\ldots\ldots(3)$$

z being measured upwards. In equations (1) ρ is variable in consequence of variable temperature and variable pressure. But, as Boussinesq* has shown, in the class of problems under consideration the influence of pressure is

* *Théorie Analytique de la Chaleur*, t. II. p. 172 (1903).

unimportant and even the variation with temperature may be disregarded except in so far as it modifies the operation of *gravity*. If we write $\rho = \rho_0 + \delta\rho$, we have

$$g\rho = g\rho_0 \left(1 + \delta\rho/\rho_0\right) = g\rho_0 - g\rho_0 \alpha\theta,$$

where θ is the temperature reckoned from the point where $\rho = \rho_0$ and α is the coefficient of expansion. We may now identify ρ in (1) with ρ_0, and our equations become

$$\frac{Du}{Dt} = -\frac{1}{\rho}\frac{dP}{dx}, \quad \frac{Dv}{Dt} = -\frac{1}{\rho}\frac{dP}{dy}, \quad \frac{Dw}{Dt} = -\frac{1}{\rho}\frac{dP}{dz} + \gamma\theta, \quad ......(4)$$

where ρ is a constant, γ is written for $g\alpha$, and P for $p + g\rho z$. Also, since the fluid is now treated as incompressible,

$$\frac{du}{dx} + \frac{dv}{dy} + \frac{dw}{dz} = 0. ....(5)$$

The equation for the conduction of heat is

$$\frac{D\theta}{Dt} = \kappa \left(\frac{d^2\theta}{dx^2} + \frac{d^2\theta}{dy^2} + \frac{d^2\theta}{dz^2}\right), \quad ....(6)$$

in which κ is the diffusibility for temperature. These are the equations employed by Boussinesq.

In the particular problems to which we proceed the fluid is supposed to be bounded by two infinite fixed planes at $z = 0$ and $z = \zeta$, where also the temperatures are maintained constant. In the equilibrium condition u, v, w vanish and θ being a function of z only is subject to $d^2\theta/dz^2 = 0$, or $d\theta/dz = \beta$, where β is a constant representing the temperature gradient. If the equilibrium is stable, β is positive; and if unstable with the higher temperature below, β is negative. It will be convenient, however, to reckon θ as the departure from the equilibrium temperature Θ. The only change required in equations (4) is to write ϖ for P, where

$$\varpi = P - \rho\gamma \int \Theta dz. ....(7)$$

In equation (6) $D\theta/Dt$ is to be replaced by $D\theta/Dt + w\beta$.

The question with which we are principally concerned is the effect of a small departure from the condition of equilibrium, whether stable or unstable. For this purpose it suffices to suppose u, v, w, and θ to be small. When we neglect the squares of the small quantities, D/Dt identifies itself with d/dt and we get

$$\frac{du}{dt} = -\frac{1}{\rho}\frac{d\varpi}{dx}, \quad \frac{dv}{dt} = -\frac{1}{\rho}\frac{d\varpi}{dy}, \quad \frac{dw}{dt} = -\frac{1}{\rho}\frac{d\varpi}{dz} + \gamma\theta, \quad ....(8)$$

$$\frac{d\theta}{dt} + \beta w = \kappa \left(\frac{d^2\theta}{dx^2} + \frac{d^2\theta}{dy^2} + \frac{d^2\theta}{dz^2}\right), \quad ....(9)$$

which with (5) and the initial and boundary conditions suffice for the solution of the problem. The boundary conditions are that $w = 0$, $\theta = 0$, when $z = 0$ or ζ.

We now assume in the usual manner that the small quantities are proportional to

$$e^{ilx} e^{imy} e^{nt}, \quad \dots\dots\dots\dots\dots\dots\dots\dots\dots(10)$$

so that (8), (5), (9) become

$$nu = -\frac{il\varpi}{\rho}, \quad nv = -\frac{im\varpi}{\rho}, \quad nw = -\frac{1}{\rho}\frac{d\varpi}{dz} + \gamma\theta, \quad \dots\dots(11)$$

$$ilu + imv + dw/dz = 0, \quad \dots\dots\dots\dots\dots(12)$$

$$n\theta + \beta w = \kappa (d^2/dz^2 - l^2 - m^2)\,\theta, \quad \dots\dots\dots\dots(13)$$

from which by elimination of u, v, ϖ, we derive

$$\frac{n}{l^2 + m^2}\frac{d^2w}{dz^2} = nw - \gamma\theta. \quad \dots\dots\dots\dots\dots(14)$$

Having regard to the boundary conditions to be satisfied by w and θ, we now assume that these quantities are proportional to $\sin sz$, where $s = q\pi/\zeta$, and q is an integer. Hence

$$\beta w + \{n + \kappa (l^2 + m^2 + s^2)\}\,\theta = 0, \quad \dots\dots\dots\dots(15)$$

$$n(l^2 + m^2 + s^2)\,w - \gamma (l^2 + m^2)\,\theta = 0, \quad \dots\dots\dots\dots(16)$$

and the equation determining n is the quadratic

$$n^2 (l^2 + m^2 + s^2) + n\kappa (l^2 + m^2 + s^2)^2 + \beta\gamma (l^2 + m^2) = 0. \quad \dots\dots(17)$$

When $\kappa = 0$, there is no conduction, so that each element of the fluid retains its temperature and density. If β be positive, the equilibrium is stable, and

$$n = \frac{\pm i \sqrt{\{\beta\gamma (l^2 + m^2)\}}}{\sqrt{\{l^2 + m^2 + s^2\}}}, \quad \dots\dots\dots\dots\dots(18)$$

indicating vibrations about the condition of equilibrium. If, on the other hand, β be negative, say $-\beta'$,

$$n = \frac{\pm \sqrt{\{\beta'\gamma (l^2 + m^2)\}}}{\sqrt{\{l^2 + m^2 + s^2\}}}. \quad \dots\dots\dots\dots\dots(19)$$

When n has the positive value, the corresponding disturbance increases exponentially with the time.

For a given value of $l^2 + m^2$, the numerical values of n diminish without limit as s increases—that is, the more subdivisions there are along z. The greatest value corresponds with $q = 1$ or $s = \pi/\zeta$. On the other hand, if s be given, $|n|$ increases from zero as $l^2 + m^2$ increases from zero (great wavelengths along x and y) up to a finite limit when $l^2 + m^2$ is large (small wavelengths along x and y). This case of no conductivity falls within the scope of a former investigation where the fluid was supposed from the beginning to be incompressible but of variable density *.

Returning to the consideration of a finite conductivity, we have again to distinguish the cases where β is positive and negative. When β is negative (higher temperature below) both values of n in (17) are real and one is positive. The equilibrium is unstable for all values of $l^2 + m^2$ and of s. If β be positive, n may be real or complex. In either case the real part of n is negative, so that the equilibrium is stable whatever $l^2 + m^2$ and s may be.

When β is negative $(-\beta')$, it is important to inquire for what values of $l^2 + m^2$ the instability is greatest, for these are the modes which more and more assert themselves as time elapses, even though initially they may be quite subordinate. That the positive value of n must have a maximum appears when we observe it tends to vanish both when $l^2 + m^2$ is small and also when $l^2 + m^2$ is large. Setting for shortness $l^2 + m^2 + s^2 = \sigma$, we may write (17)

$$n^2\sigma + n\kappa\sigma^2 - \beta'\gamma(\sigma - s^2) = 0, \quad \ldots\ldots\ldots\ldots\ldots(20)$$

and the question is to find the value of σ for which n is greatest, s being supposed given. Making $dn/d\sigma = 0$, we get on differentiation

$$n^2 + 2n\kappa\sigma - \beta'\gamma = 0; \quad \ldots\ldots\ldots\ldots\ldots(21)$$

and on elimination of n^2 between (20), (21)

$$n = \frac{\beta'\gamma s^2}{\kappa\sigma^2}. \quad \ldots\ldots\ldots\ldots\ldots(22)$$

Using this value of n in (21), we find as the equation for σ

$$\frac{2s^2}{\sigma} = 1 - \frac{\beta'\gamma s^4}{\kappa^2\sigma^4}. \quad \ldots\ldots\ldots\ldots\ldots(23)$$

When κ is relatively great, $\sigma = 2s^2$, or

$$l^2 + m^2 = s^2. \quad \ldots\ldots\ldots\ldots\ldots(24)$$

A second approximation gives

$$l^2 + m^2 = s^2 + \frac{\beta'\gamma}{8\kappa^2 s^2}. \quad \ldots\ldots\ldots\ldots\ldots(25)$$

The corresponding value of n is

$$n = \frac{\beta'\gamma}{4\kappa s^2}\left\{1 - \frac{\beta'\gamma}{8\kappa^2 s^4}\right\}. \quad \ldots\ldots\ldots\ldots\ldots(26)$$

The modes of greatest instability are those for which s is smallest, that is equal to π/ζ, and

$$l^2 + m^2 = \frac{\pi^2}{\zeta^2} + \frac{\beta'\gamma}{8\kappa^2\pi^2/\zeta^2}. \quad \ldots\ldots\ldots\ldots\ldots(27)$$

* *Proc. Lond. Math. Soc.* Vol. xiv. p. 170 (1883); *Scientific Papers*, Vol. ii. p. 200.

For a two-dimensional disturbance we may make $m = 0$ and $l = 2\pi/\lambda$, where λ is the wave-length along x. The λ of maximum instability is thus approximately

$$\lambda = 2\zeta. \quad\dots\dots\dots\dots\dots\dots\dots\dots(28)$$

Again, if $l = m = 2\pi/\lambda$, as for square cells,

$$\lambda = 2\sqrt{2} \cdot \zeta, \quad\dots\dots\dots\dots\dots\dots\dots(29)$$

greater than before in the ratio $\sqrt{2} : 1$.

We have considered especially the cases where κ is relatively small and relatively large. Intermediate·cases would need to be dealt with by a numerical solution of (23).

When w is known in the form

$$w = We^{ilx}e^{imy}\sin sz \cdot e^{nt}, \quad\dots\dots\dots\dots\dots\dots(30)$$

n being now a known function of l, m, s, u and v are at once derived by means of (11) and (12). Thus -

$$u = \frac{il}{l^2 + m^2}\frac{dw}{dz}, \quad v = \frac{im}{l^2 + m^2}\frac{dw}{dz} . \quad\dots\dots\dots(31)$$

The connexion between w and θ is given by (15) or (16). When β is negative and n positive, θ and w are of the same sign.

As an example in two dimensions of (30), (31), we might have in real form

$$u = W\cos x \cdot \sin z \cdot e^{nt}, \quad\dots\dots\dots\dots\dots(32)$$

$$u = -W\sin x \cdot \cos z \cdot e^{nt}, \quad v = 0. \quad\dots\dots\dots(33)$$

Hitherto we have supposed the fluid to be destitute of viscosity. When we include viscosity, we must add $\nu(\nabla^2 u, \nabla^2 v, \nabla^2 w)$ on the right of equations (1), (8), and (11), ν being the kinematic coefficient. Equations (12) and (13) remain unaffected. And in (11)

$$\nabla^2 = d^2/dz^2 - l^2 - m^2. \quad\dots\dots\dots\dots\dots(34)$$

We have also to reconsider the boundary conditions at $z = 0$ and $z = \zeta$. We may still suppose $\theta = 0$ and $w = 0$; but for a further condition we should probably prefer $dw/dz = 0$, corresponding to a fixed solid wall*. But this entails much complication, and we may content ourselves with the supposition $d^2w/dz^2 = 0$, which (with $w = 0$) is satisfied by taking as before w proportional to $\sin sz$ with $s = q\pi/\zeta$. This is equivalent to the annulment of lateral *forces* at the wall. For (Lamb's *Hydrodynamics*, §§ 323, 326) these forces are expressed in general by

$$p_{xz} = \frac{dw}{dx} + \frac{du}{dz}, \quad p_{yz} = \frac{dw}{dy} + \frac{dv}{dz}, \quad\dots\dots\dots(35)$$

* [It would appear that the immobility and solidity of the walls are sufficiently provided for by the condition $w = 0$, and that for "a fixed solid wall" there should be substituted "no slipping at the walls." W. F. S.]

while here $w = 0$ at the boundaries requires also $dw/dx = 0$, $dw/dy = 0$. Hence, at the boundaries, $d^2u/dx\,dz$, $d^2v/dy\,dz$ vanish, and therefore by (5), d^2w/dz^2.

Equation (15) remains unaltered :—

$$\beta w + \{n + \kappa\,(l^2 + m^2 + s^2)\}\,\theta = 0, \quad\dots\dots\dots\dots(15)$$

and (16) becomes

$$\{n + \nu\,(l^2 + m^2 + s^2)\}\,(l^2 + m^2 + s^2)\,w - \gamma\,(l^2 + m^2)\,\theta = 0. \quad\dots\dots(36)$$

Writing as before $\sigma = l^2 + m^2 + s^2$, we get the equation in n

$$(n + \kappa\sigma)\,(n + \nu\sigma)\,\sigma + \beta\gamma\,(l^2 + m^2) = 0, \quad\dots\dots\dots\dots(37)$$

which takes the place of (17).

If $\gamma = 0$ (no expansion with heat), the equations degrade and we have two simple alternatives. In the first $n + \kappa\sigma = 0$ with $w = 0$, signifying conduction of heat with no motion. In the second $n + \nu\sigma = 0$, when the relation between w and θ becomes

$$\beta w + \sigma\,(\kappa - \nu)\,\theta = 0. \quad\dots\dots\dots\dots\dots(38)$$

In both cases, since n is real and negative, the disturbance is stable.

If we neglect κ in (37), the equation takes the same form (20) as that already considered when $\nu = 0$. Hence the results expressed in (22), (23), (24), (25), (26), (27) are applicable with simple substitution of ν for κ.

In the general equation (37) if β be positive, as γ is supposed always to be, the values of n may be real or complex. If real they are both negative, and if complex the real part is negative. In either case the disturbance dies down. As was to be expected, when the temperature is higher above, the equilibrium is stable.

In the contrary case when β is negative $(-\beta')$ the roots of the quadratic are always real, and one at least is negative. There is a positive root only when

$$\beta'\gamma\,(l^2 + m^2) > \kappa\nu\sigma^3. \quad\dots\dots\dots\dots\dots\dots(39)$$

If κ, or ν, vanish there is instability ; but if κ and ν are finite and large enough, the equilibrium for this disturbance is stable, although the higher temperature is underneath.

Inequality (39) gives the condition of instability for the particular disturbance (l, m, s). It is of interest to inquire at what point the equilibrium becomes unstable when there is no restriction upon the value of $l^2 + m^2$. In the equation

$$\beta'\gamma\,(l^2 + m^2) - \kappa\nu\sigma^3 = \beta'\gamma\,(\sigma - s^2) - \kappa\nu\sigma^3 = 0, \quad\dots\dots\dots(40)$$

we see that the left-hand member is negative when $l^2 + m^2$ is small and also when it is large. When the conditions are such that the equation can only just be satisfied with some value of $l^2 + m^2$, or σ, the derived equation

$$\beta'\gamma - 3\kappa\nu\sigma^2 = 0 \quad\dots\dots\dots\dots\dots\dots\dots(41)$$

must also hold good, so that

$$\sigma = 3s^2/2, \qquad l^2 + m^2 = \tfrac{1}{2}s^2, \qquad \text{.................(42)}$$

and
$$\beta'\gamma = 27\kappa\nu s^4/4. \qquad \text{...........................(43)}$$

Unless $\beta'\gamma$ exceeds the value given in (43) there is no instability, however l and m are chosen. But the equation still contains s, which may be as large as we please. The smallest value of s is π/ζ. The condition of instability when l, m, and s are all unrestricted is accordingly

$$\beta'\gamma > \frac{27\pi^4\kappa\nu}{4\zeta^4}. \qquad \text{...........................(44)}$$

If $\beta'\gamma$ falls below this amount, the equilibrium is altogether stable. I am not aware that the possibility of complete stability under such circumstances has been contemplated.

To interpret (44) more conveniently, we may replace β' by $(\Theta_2 - \Theta_1)/\zeta$ and γ by $g(\rho_2 - \rho_1)/\rho_1(\Theta_2 - \Theta_1)$*, so that

$$\beta'\gamma = \frac{g}{\zeta}\frac{\rho_2 - \rho_1}{\rho_1}, \qquad \text{.........................(45)}$$

where Θ_2, Θ_1, ρ_2, and ρ_1 are the extreme temperatures and densities in equilibrium. Thus (44) becomes

$$\frac{\rho_2 - \rho_1}{\rho_1} > \frac{27\pi^4\kappa\nu}{4g\zeta^3}. \qquad \text{.........................(46)}$$

In the case of air at atmospheric conditions we may take in C.G.S. measure

$$\nu = \cdot 14, \qquad \text{and} \qquad \kappa = \tfrac{5}{2}\nu \text{ (Maxwell's Theory)}.$$

Also $g = 980$, and thus

$$\frac{\rho_2 - \rho_1}{\rho_1} > \frac{\cdot 033}{\zeta^3}. \qquad \text{...........................(47)}$$

For example, if $\zeta = 1$ cm., instability requires that the density at the top exceed that at the bottom by one-thirtieth part, corresponding to about $9°$ C. of temperature. We should not forget that our method postulates a small value of $(\rho_2 - \rho_1)/\rho_1$. Thus if $\kappa\nu$ be given, the application of (46) may cease to be legitimate unless ζ be large enough.

It may be remarked that the influence of viscosity would be increased were we to suppose the horizontal velocities (instead of the horizontal forces) to be annulled at the boundaries.

The problem of determining for what value of $l^2 + m^2$, or σ, the instability, when finite, is a maximum is more complicated. The differentiation of (37) with respect to σ gives

$$n^2 + 2n\sigma(\kappa + \nu) + 3\kappa\nu\sigma^2 - \beta'\gamma = 0, \qquad \text{.................(48)}$$

whence
$$n = \frac{\beta'\gamma s^2 - 2\kappa\nu\sigma^3}{\sigma^2(\kappa + \nu)}, \qquad \text{..........................(49)}$$

* [If ρ_1 is taken to correspond to Θ_1, and ρ_2 to Θ_2, "$\rho_1 - \rho_2$" must be substituted for "$\rho_2 - \rho_1$" throughout this page. W. F. S.]

expressing n in terms of σ. To find σ we have to eliminate n between (48) and (49). The result is

$$\sigma^6 \kappa \nu (\kappa - \nu)^2 + \sigma^4 \beta' \gamma (\kappa + \nu)^2 - \sigma^3 . 2\beta' \gamma s^2 (\kappa^2 + \nu^2) - \beta'^2 \gamma^2 s^4 = 0, \quad ...(50)$$

from which, in particular cases, σ could be found by numerical computation. From (50) we fall back on (23) by supposing $\nu = 0$, and again on a similar equation if we suppose $\kappa = 0$.

But the case of a nearly evanescent n is probably the more practical. In an experiment the temperature gradient could not be established all at once and we may suppose the progress to be very slow. In the earlier stages the equilibrium would be stable, so that no disturbance of importance would occur until n passed through zero to the positive side, corresponding to (44) or (46). The breakdown thus occurs for $s = \pi/\zeta$, and by (42) $l^2 + m^2 = \pi^2/2\zeta^2$. And since the evanescence of n is equivalent to the omission of d/dt in the original equations, the motion thus determined has the character of a *steady* motion. The constant multiplier is, however, arbitrary; and there is nothing to determine it so long as the squares of u, v, w, θ are neglected.

In a particular solution where w as a function of x and y has the simplest form, say

$$w = 2 \cos x . \cos y, \quad ....(51)$$

the particular coefficients of x and y which enter have relation to the particular axes of reference employed. If we rotate these axes through an angle ϕ, we have

$$\begin{aligned}
w &= 2 \cos \{x' \cos \phi - y' \sin \phi\} . \cos \{x' \sin \phi + y' \cos \phi\} \\
&= \cos \{x' (\cos \phi - \sin \phi)\} . \cos \{y' (\cos \phi + \sin \phi)\} \\
&+ \sin \{x' (\cos \phi - \sin \phi)\} . \sin \{y' (\cos \phi + \sin \phi)\} \\
&+ \cos \{x' (\cos \phi + \sin \phi)\} . \cos \{y' (\cos \phi - \sin \phi)\} \\
&- \sin \{x' (\cos \phi + \sin \phi)\} . \sin \{y' (\cos \phi - \sin \phi)\}. \quad ....(52)
\end{aligned}$$

For example, if $\phi = \frac{1}{4}\pi$, (52) becomes

$$w = \cos (y' \sqrt{2}) + \cos (x' \sqrt{2}). \quad ....(53)$$

It is to be observed that with the general value of ϕ, if we call the coefficients of x', y', l and m respectively, we have in every part $l^2 + m^2 = 2$, unaltered from the original value in (51).

The character of w, under the condition that all the elementary terms of which it is composed are subject to $l^2 + m^2 = $ constant (k^2), is the same as for the transverse displacement of an infinite stretched membrane, vibrating with one definite frequency. The limitation upon w is, in fact, merely that it satisfies

$$(d^2/dx^2 + d^2/dy^2 + k^2) w = 0. \quad ....(54)$$

The character of w in particular solutions of the membrane problem is naturally associated with the nodal system ($w = 0$), where the membrane may be regarded as held fast; and we may suppose the nodal system to divide the plane into similar parts or cells, such as squares, equilateral triangles, or regular hexagons. But in the present problem it is perhaps more appropriate to consider divisions of the plane with respect to which w is symmetrical, so that dw/dn is zero on the straight lines forming the divisions of the cells. The more natural analogy is then with the two-dimensional vibration of air, where w represents velocity-potential and the divisions may be regarded as fixed walls.

The simplest case is, of course, that in which the cells are squares. If the sides of the squares be 2π, we may take with axes parallel to the sides and origin at centre

$$w = \cos x + \cos y, \quad\dots\dots\dots\dots\dots\dots\dots\dots\dots(55)$$

being thus composed by superposition of two parts for each of which $k^2 = 1$. This makes $dw/dx = -\sin x$, vanishing when $x = \pm\pi$. Similarly, dw/dy vanishes when $y = \pm\pi$, so that the sides of the square behave as fixed walls. To find the places where w changes sign, we write it in the form

$$w = 2\cos\frac{x+y}{2}\,.\,\cos\frac{x-y}{2}, \quad\dots\dots\dots\dots\dots(56)$$

giving $x + y = \pm\pi$, $x - y = \pm\pi$, lines which constitute the inscribed square (fig. 1). Within this square w has one sign (say +) and in the four right-angled triangles left over the − sign. When the whole plane is considered, there is no want of symmetry between the + and the − regions.

The principle is the same when the elementary cells are equilateral triangles or hexagons; but I am not aware that an analytical solution has been obtained for these cases. An experimental determination of k^2 might be made by observing the time of vibration under gravity of water contained in a trough with vertical sides and of corresponding section, which depends upon

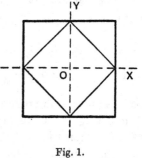

Fig. 1.

the same differential equation and boundary conditions*. The particular vibration in question is not the slowest possible, but that where there is a simultaneous rise at the centre and fall at the walls all round, with but one curve of zero elevation between.

In the case of the hexagon, we may regard it as deviating comparatively little from the circular form and employ the approximate methods then applicable. By an argument analogous to that formerly developed† for the boundary condition $w = 0$, we may convince ourselves that the value of k^2 for the hexagon cannot differ much from that appropriate to a circle of the same area. Thus if a be the radius of this circle, k is given by $J_0'(ka) = 0$,

---

* See *Phil. Mag.* Vol. I. p. 257 (1876); *Scientific Papers*, Vol. I. pp. 265, 271.

† *Theory of Sound*, § 209; compare also § 317. See Appendix.

J_0 being the Bessel's function of zero order, or $ka = 3\cdot832$. If b be the side of the hexagon, $a^2 = 3\sqrt{3} \cdot b^2/2\pi$.

APPENDIX.

On the nearly symmetrical solution for a nearly circular area, when w satisfies $(d^2/dx^2 + d^2/dy^2 + k^2)\, w = 0$ *and makes $dw/dn = 0$ on the boundary.*

Starting with the true circle of radius a, we have w a function of r (the radius vector) only, and the solution is $w = J_0(kr)$ with the condition $J_0'(ka) = 0$, yielding $ka = 3\cdot832$, which determines k if a be given, or a if k be given. In the problem proposed the boundary is only approximately circular, so that we write $r = a + \rho$, where a is the mean value and

$$\rho = \alpha_1 \cos\theta + \beta_1 \sin\theta + \ldots + \alpha_n \cos n\theta + \beta_n \sin n\theta. \quad\ldots\ldots\ldots(57)$$

In (57) θ is the vectorial angle and α_1 etc. are quantities small relatively to a. The general solution of the differential equation being

$$w = A_0 J_0(kr) + J_1(kr)\{A_1 \cos\theta + B_1 \sin\theta\}$$
$$+ \ldots + J_n(kr)\{A_n \cos n\theta + B_n \sin n\theta\}, \ldots(58)$$

we are to suppose now that A_1, etc., are small relatively to A_0. It remains to consider the boundary condition.

If ϕ denote the small angle between r and the normal dn measured outwards,

$$\frac{dw}{dn} = \frac{dw}{dr}\cos\phi - \frac{dw}{rd\theta}\sin\phi, \quad\ldots\ldots\ldots\ldots\ldots(59)$$

and $\qquad\qquad \tan\phi = \dfrac{dr}{rd\theta} = \dfrac{d\rho}{ad\theta} = \dfrac{n}{a}(-\alpha_n \sin n\theta + \beta_n \cos n\theta) \quad\ldots\ldots(60)$

with sufficient approximation, only the general term being written. In formulating the boundary condition $dw/dn = 0$ correct to the second order of small quantities, we require dw/dr to the second order, but $dw/d\theta$ to the first order only. We have

$$\frac{1}{k}\frac{dw}{dr} = A_0 \{J_0'(ka) + k\rho J_0''(ka) + \tfrac{1}{2}k^2\rho^2 J_0'''(ka)\}$$
$$+ \{J_n'(ka) + k\rho J_n''(ka)\}\{A_n \cos n\theta + B_n \sin n\theta\},$$
$$\frac{dw}{ad\theta} = \frac{n}{a} J_n(ka)\{-A_n \sin n\theta + B_n \cos n\theta\}$$

and for the boundary condition, setting $ka = z$ and omitting the argument in the Bessel's functions,

$$A_0 \{J_0'\cdot\cos\phi + k\rho J_0'' + \tfrac{1}{2}k^2\rho^2 J_0'''\}$$
$$+ \{J_n' + k\rho J_n''\}\{A_n \cos n\theta + B_n \sin n\theta\}$$
$$- \frac{n^2}{az} J_n\{-A_n \sin n\theta + B_n \cos n\theta\}\{-\alpha_n \sin n\theta + \beta_n \cos n\theta\} = 0. \quad (61)$$

If for the moment we omit the terms of the second order, we have

$$A_0 J_0' + k A_0 J_0'' \{\alpha_n \cos n\theta + \beta_n \sin n\theta\} + J_n' \{A_n \cos n\theta + B_n \sin n\theta\} = 0 \; ; \; (62)$$

so that
$$J_0'(z) = 0,$$

and
$$k A_0 J_0'' . \alpha_n + J_n' . A_n = 0, \qquad k A_0 J_0'' . \beta_n + J_n' . B_n = 0. \quad \ldots\ldots(63)$$

To this order of approximation z, $= ka$, has the same value as when $\rho = 0$; that is to say, the equivalent radius is equal to the mean radius, or (as we may also express it) k may be regarded as dependent upon the *area* only. Equations (63) determine A_n, B_n in terms of the known quantities α_n, β_n.

Since J_0' is a small quantity, $\cos \phi$ in (61) may now be omitted. To obtain a corrected evaluation of z, it suffices to take the mean of (61) for all values of θ. Thus

$$A_0 \{2 J_0' + \tfrac{1}{2} k^2 J_0''' (\alpha_n^2 + \beta_n^2)\} + \{k J_n'' - n^2 J_n / az\} \{\alpha_n A_n + \beta_n B_n\} = 0,$$

or on substitution of the approximate values of A_n, B_n from (63),

$$J_0' = \tfrac{1}{2} k^2 (\alpha_n^2 + \beta_n^2) \left\{ \frac{J_0''}{J_n'} \left(J_n'' - \frac{n^2 J_n}{z^2} \right) - \frac{J_0'''}{2} \right\}. \ldots\ldots\ldots(64)$$

This expression may, however, be much simplified. In virtue of the general equation for J_n,

$$J_n'' - \frac{n^2}{z^2} J_n = - \frac{J_n'}{z} - J_n \; ;$$

and since here $J_0' = 0$ approximately,

$$J_0'' = - J_0, \qquad J_0''' = - z^{-1} J_0'' = z^{-1} J_0.$$

Thus
$$J_0'(z) = \tfrac{1}{2} k^2 J_0 . \Sigma (\alpha_n^2 + \beta_n^2) \left\{ \frac{J_n}{J_n'} + \frac{1}{2z} \right\}, \ldots\ldots\ldots\ldots(65)$$

the sign of summation with respect to n being introduced.

Let us now suppose that $a + da$ is the equivalent radius, so that $J_0'(ka + kda) = 0$, that is the radius of the exact circle which corresponds to the value of k appropriate to the approximate circle. Then

$$J_0'(z) + kda J_0''(z) = 0,$$

and
$$da = - \frac{J_0'}{k J_0''} = k \Sigma (\alpha_n^2 + \beta_n^2) \left\{ \frac{J_n}{2 J_n'} + \frac{1}{4z} \right\}. \ldots\ldots\ldots(66)$$

Again, if $a + da'$ be the radius of the true circle which has the same *area* as the approximate circle

$$da' = \frac{1}{4a} \Sigma (\alpha_n^2 + \beta_n^2), \ldots\ldots\ldots\ldots\ldots(67)$$

and
$$da' - da = - \Sigma \frac{\alpha_n^2 + \beta_n^2}{2a} \frac{z J_n(z)}{J_n'(z)}, \ldots\ldots\ldots\ldots(68)$$

where z is the first root (after zero) of $J_0'(z) = 0$, viz. 3·832.

The question with which we are mainly concerned is the sign of $da' - da$ for the various values of n. When $n = 1$, $J_1(z) = - J_0'(z) = 0$, so that $da = da'$, a result which was to be expected, since the terms in α_1, β_1 represent approximately a displacement merely of the circle. without alteration of size or shape. We will now examine the sign of J_n/J_n' when $n = 2$, and 3.

For this purpose we may employ the sequence equations

$$J_{n+1} = \frac{2n}{z} J_n - J_{n-1}, \qquad J_n' = \tfrac{1}{2}J_{n-1} - \tfrac{1}{2}J_{n+1},$$

which allow J_n and J_n' to be expressed in terms of J_1 and J_0, of which the former is here zero. We find

$$J_2 = - J_0, \qquad J_3 = - 4z^{-1}J_0, \qquad J_4 = (1 - 24z^{-2}) J_0;$$

$$J_1' = J_0, \qquad J_2' = 2z^{-1} J_0, \qquad J_3' = (12z^{-2} - 1) J_0.$$

Thus
$$\frac{J_1}{J_1'} = 0, \qquad \frac{J_2}{J_2'} = -\frac{z}{2}, \qquad \frac{J_3}{J_3'} = \frac{4z}{z^2 - 12};$$

whence on introduction of the actual value of z, viz. 3·832, we see that J_2/J_2' is negative, and that J_3/J_3' is positive.

When $n > z$, it is a general proposition that $J_n(z)$ and $J_n'(z)$ are both positive*. Hence for $n = 4$ and onwards, J_n/J_n' is positive when $z = 3·832$. We thus arrive at the curious conclusion that when $n = 2$, $da' > da$, as happens for all values of n (exceeding unity) when the boundary condition is $w = 0$, but that when $n > 2$, $da' < da$. The existence of the exceptional case $n = 2$ precludes a completely general statement of the effect of a departure from the truly circular form; but if the terms for which $n = 2$ are absent, as they would be in the case of any regular polygon with an even number of sides, regarded as a deformed circle, we may say that $da' < da$. In the physical problems the effect of a departure from the circular form is then to depress the pitch when the area is maintained constant $(da' = 0)$. But for an elliptic deformation the reverse is the case.

At first sight it may appear strange that an elliptic deformation should be capable of raising the pitch. But we must remember that we are here dealing with a vibration such that the phase at both ends of the minor axis is the opposite of that at the centre. A parallel case which admits of complete calculation is that of the rectangle regarded as a deformed square, and vibrating in the gravest *symmetrical* mode†. It is easily shown that a departure from the square form raises the pitch. Of course, the one-dimensional vibration parallel to the longer side has its pitch depressed.

[1918. This problem had already been treated by Aichi (*Proc. Tokio Math.-Phys. Soc.* 1907).]

---

* See, for example, *Theory of Sound*, § 210.
† *Theory of Sound*, § 267 ($p=q=2$).

413.

ON THE DYNAMICS OF REVOLVING FLUIDS.

[*Proceedings of the Royal Society*, A, Vol. XCIII. pp. 148—154, 1916.]

So much of meteorology depends ultimately upon the dynamics of revolving fluid that it is desirable to formulate as clearly as possible such simple conclusions as are within our reach, in the hope that they may assist our judgment when an exact analysis seems impracticable. An important contribution to this subject is that recently published by Dr Aitken*. It formed the starting point of part of the investigation which follows, but I ought perhaps to add that I do not share Dr Aitken's views in all respects. His paper should be studied by all interested in these questions.

As regards the present contribution to the theory it may be well to premise that the limitation to symmetry round an axis is imposed throughout.

The motion of an inviscid fluid is governed by equations of which the first expressed by rectangular coordinates may be written

$$\frac{du'}{dt} + u'\frac{du'}{dx} + v'\frac{du'}{dy} + w'\frac{du'}{dz} = -\frac{dP}{dx}, \quad \dots \dots \dots \dots \dots (1)$$

where
$$P = \int dp/\rho + V, \quad \dots \dots \dots \dots \dots \dots \dots \dots (2)$$

and V is the potential of extraneous forces. In (2) the density ρ is either a constant, as for an incompressible fluid, or at any rate a known function of the pressure p. Referred to cylindrical coordinates r, θ, z, with velocities u, v, w, reckoned respectively in the directions of r, θ, z increasing, these equations become†

$$\frac{du}{dt} + u\frac{du}{dr} + v\left(\frac{du}{rd\theta} - \frac{v}{r}\right) + w\frac{du}{dz} = -\frac{dP}{dr}, \quad \dots \dots \dots \dots (3)$$

$$\frac{dv}{dt} + u\frac{dv}{dr} + v\left(\frac{dv}{rd\theta} + \frac{u}{r}\right) + w\frac{dv}{dz} = -\frac{dP}{rd\theta}, \quad \dots \dots \dots \dots (4)$$

$$\frac{dw}{dt} + u\frac{dw}{dr} + v\frac{dw}{rd\theta} + w\frac{dw}{dz} = -\frac{dP}{dz}. \quad \dots \dots \dots \dots (5)$$

---

* "The Dynamics of Cyclones and Anticyclones.—Part 3," *Roy. Soc. Edin. Proc.* Vol. XXXVI. p. 174 (1916).

† Compare Basset's *Hydrodynamics*, § 19.

For the present purpose we assume symmetry with respect to the axis of z, so that u, v, w, and P (assumed to be single-valued) are independent of θ. So simplified, the equations become

$$\frac{du}{dt} + u\frac{du}{dr} - \frac{v^2}{r} + w\frac{du}{dz} = -\frac{dP}{dr}, \quad \dots\dots\dots\dots\dots(6)$$

$$\frac{dv}{dt} + u\frac{dv}{dr} + \frac{uv}{r} + w\frac{dv}{dz} = 0, \quad \dots\dots\dots\dots\dots(7)$$

$$\frac{dw}{dt} + u\frac{dw}{dr} + w\frac{dw}{dz} = -\frac{dP}{dz}, \quad \dots\dots\dots\dots\dots(8)$$

of which the second may be written

$$\left(\frac{d}{dt} + u\frac{d}{dr} + w\frac{d}{dz}\right)(rv) = 0, \quad \dots\dots\dots\dots\dots(9)$$

signifying that (rv) may be considered to move with the fluid, in accordance with Kelvin's general theorem respecting "circulation." If r_0, v_0, be the initial values of r, v, for any particle of the fluid, the value of v at any future time when the particle is at a distance r from the axis is given by $rv = r_0 v_0$.

Respecting the motion expressed by u, w, we see that it is the same as might take place with $v = 0$, that is when the whole motion is in planes passing through the axis, provided that we introduce a force along r equal to v^2/r. We have here the familiar idea of "centrifugal force," and the conclusion might have been arrived at immediately, at any rate in the case where there is no (u, w) motion.

It will be well to consider this case $(u = 0, w = 0)$ more in detail. The third equation (8) shows that P is then independent of z, that is a function of r (and t) only. It follows from the first equation (6) that v also is a function of r only, and $P = \int v^2 dr/r$. Accordingly by (2)

$$\int dp/\rho = -V + \int v^2 r^{-1} dr. \quad \dots\dots\dots\dots\dots(10)$$

If V, the potential of impressed forces, is independent of z, so also will be p and ρ, but not otherwise. For example, if gravity (g) act parallel to z (measured downwards),

$$\int dp/\rho = C + gz + \int v^2 dr/r, \quad \dots\dots\dots\dots\dots(11)$$

gravity and centrifugal force contributing independently. In (11) ρ will be constant if the fluid is incompressible. For gases following Boyle's law $(p = a^2\rho)$,

$$a^2(\log\rho, \text{ or } \log p) = C + gz + \int v^2 dr/r. \quad \dots\dots\dots(12)$$

At a constant level the pressure diminishes as we pass inwards. But the corresponding rarefaction experienced by a compressible fluid does not cause such fluid to ascend. The heavier part outside is prevented from coming in below to take its place by the centrifugal force*.

The condition for equilibrium, taken by itself, still leaves v an arbitrary function of r, but it does not follow that the equilibrium is stable. In like manner an incompressible liquid of variable density is in equilibrium under gravity when arranged in horizontal strata of constant density, but stability requires that the density of the strata everywhere increase as we pass downwards. This analogy is, indeed, very helpful for our present purpose. As the fluid moves (u and w finite) in accordance with equations (6), (7), (8), (vr) remains constant (k) for a ring consisting always of the same matter, and $v^2/r = k^2/r^3$, so that the centrifugal force acting upon a *given portion* of the fluid is inversely as r^3, and thus a known function of position. The only difference between this case and that of an incompressible fluid of variable density, moving under extraneous forces derived from a potential, is that here the inertia concerned in the (u, w) motion is uniform, whereas in a variably dense fluid moving under gravity, or similar forces, the inertia and the weight are proportional. As regards the question of stability, the difference is immaterial, and we may conclude that the equilibrium of fluid revolving one way round in cylindrical layers and included between coaxial cylindrical walls is stable only under the condition that the circulation (k) always increases with r. In any portion where k is constant, so that the motion is there "irrotational," the equilibrium is neutral.

An important particular case is that of fluid moving between an inner cylinder ($r = a$) revolving with angular velocity ω and an outer fixed cylinder ($r = b$). In the absence of viscosity the rotation of the cylinder is without effect. But if the fluid were viscous, equilibrium would require†

$$k = vr = a^2\omega\,(b^2 - r^2)/(b^2 - a^2),$$

expressing that the circulation diminishes outwards. Accordingly a fluid without viscosity cannot stably move in this manner. On the other hand, if it be the outer cylinder that rotates while the inner is at rest,

$$k = vr = b^2\omega\,(r^2 - a^2)/(b^2 - a^2),$$

and the motion of an inviscid fluid according to this law would be stable.

We may also found our argument upon a direct consideration of the kinetic energy (T) of the motion. For T is proportional to $\int v^2 r\,dr$, or $\int k^2 dr^2/r^2$.

---

* When the fluid is viscous, the loss of circulation near the bottom of the containing vessel modifies this conclusion, as explained by James Thomson.

† Lamb's *Hydrodynamics*, § 333.

Suppose now that two rings of fluid, one with $k = k_1$ and $r = r_1$ and the other with $k = k_2$ and $r = r_2$, where $r_2 > r_1$, and of equal areas dr_1^2 or dr_2^2, are interchanged. The corresponding increment in T is represented by

$$(dr_1^2 = dr_2^2)\,\{k_2^2/r_1^2 + k_1^2/r_2^2 - k_1^2/r_1^2 - k_2^2/r_2^2\} = dr^2\,(k_2^2 - k_1^2)\,(r_1^{-2} - r_2^{-2}),$$

and is positive if $k_2^2 > k_1^2$; so that a circulation always increasing outwards makes T a minimum and thus ensures stability.

The conclusion above arrived at may appear to conflict with that of Kelvin[*], who finds as the condition of minimum energy that the *vorticity*, proportional to $r^{-1}dk/dr$, must increase outwards. Suppose, for instance, that $k = r^{\frac{1}{2}}$, increasing outwards, while $r^{-1}dk/dr$ decreases. But it would seem that the variations contemplated differ. As an example, Kelvin gives for maximum energy

$$v = r \text{ from } r = 0 \text{ to } r = b,$$

$$v = b^2/r \text{ from } r = b \text{ to } r = a;$$

and for minimum energy

$$v = 0 \text{ from } r = 0 \text{ to } r = \sqrt{(a^2 - b^2)},$$

$$v = r - (a^2 - b^2)/r \text{ from } r = \sqrt{(a^2 - b^2)} \text{ to } r = a.$$

In the first case
$$\int_0^a vr^2 dr = \tfrac{1}{4}b^2\,(2a^2 - b^2),$$

and in the second case
$$\int_0^a vr^2 dr = \tfrac{1}{4}b^4;$$

so that the moment of momentum differs in the two cases. In fact Kelvin supposes operations upon the boundary which alter the moment of momentum. On the other hand, he maintains the strictly two-dimensional character of the admissible variations. In the problem that I have considered, symmetry round the axis is maintained and there can be no alteration in the moment of momentum, since the cylindrical walls are fixed. But the variations by which the passage from one two-dimensional condition to another may be effected are not themselves two-dimensional.

The above reasoning suffices to fix the criterion for stable equilibrium; but, of course, there can be no actual transition from a configuration of unstable equilibrium to that of permanent stable equilibrium without dissipative forces, any more than there could be in the case of a heterogeneous liquid under gravity. The difference is that in the latter case dissipative forces exist in any real fluid, so that the fluid ultimately settles down into stable equilibrium, it may be after many oscillations. In the present problem ordinary viscosity does not meet the requirements, as it would interfere with the constancy of the circulation of given rings of fluid on which our reasoning depends. But

---

[*] *Nature*, Vol. XXIII. October, 1880 ; *Collected Papers*, Vol. IV. p. 175.

for purely theoretical purposes there is no inconsistency in supposing the (u, w) motion resisted while the v motion is unresisted.

The next supposition to $u = 0$, $w = 0$ in order of simplicity is that u is a function of r and t only, and that $w = 0$, or at most a finite constant. It follows from (8) that P is independent of z, while (6) becomes

$$\frac{du}{dt} + u\frac{du}{dr} - \frac{v^2}{r} = -\frac{dP}{dr}, \quad\ldots\ldots\ldots\ldots\ldots\ldots(13)$$

determining the pressure. In the case of an incompressible fluid u as a function of r is determined by the equation of continuity $ur = C$, where C is a function of t only; and when u and the initial circumstances are known, v follows. As the motion is now two-dimensional, it may conveniently be expressed by means of the vorticity ζ, which moves with the fluid, and the stream-function ψ, connected with ζ by the equation

$$\frac{1}{r}\frac{d}{dr}\left(r\frac{d\psi}{dr}\right) + \frac{1}{r^2}\frac{d^2\psi}{d\theta^2} = 2\zeta. \quad\ldots\ldots\ldots\ldots\ldots(14)$$

The solution, appropriate to our purpose, is

$$\psi = 2\int dr\, r^{-1}\int \zeta r\, dr + A\log r - B\theta, \ldots\ldots\ldots\ldots\ldots(15)$$

where A and B are arbitrary constants of integration. Accordingly

$$u = -\frac{d\psi}{r\, d\theta} = \frac{B}{r}, \qquad v = \frac{d\psi}{dr} = \frac{2}{r}\int \zeta r\, dr + \frac{A}{r}. \quad\ldots\ldots\ldots(16)$$

In general, A and B are functions of the time, and ζ is a function of the time as well as of r.

A simple particular case is when ζ is initially, and therefore permanently, uniform throughout the fluid. Then

$$v = \zeta r + A r^{-1}. \quad\ldots\ldots\ldots\ldots\ldots\ldots(17)^*$$

Let us further suppose that initially the motion is one of pure rotation, as of a solid body, so that initially $A = 0$, and that then the outer wall closes in. If the outer radius be initially R_0 and at time t equal to R, then at time t

$$A = \zeta(R_0^2 - R^2), \quad\ldots\ldots\ldots\ldots\ldots\ldots(18)$$

since vr remains unchanged for a given ring of the fluid; and correspondingly,

$$v = \zeta\{r + (R_0^2 - R^2)\, r^{-1}\}. \quad\ldots\ldots\ldots\ldots\ldots(19)$$

Thus, in addition to the motion as of a solid body, the fluid acquires that of a simple vortex of intensity increasing as R diminishes.

---

* It may be remarked that (17) is still applicable under appropriate boundary conditions even when the fluid is viscous.

If at any stage the u motion ceases, (6) gives

$$dp/dr = \rho v^2/r, \quad\ldots\ldots\ldots\ldots\ldots\ldots\ldots\ldots(20)$$

and thus

$$p/\rho = \zeta^2 \{\tfrac{1}{2} r^2 + 2 (R_0{}^2 - R^2) \log r - \tfrac{1}{2} (R_0{}^2 - R^2)^2 r^{-2}\} + \text{const.} \ldots(21)$$

Since, as a function of r, v^2 continually increases as R diminishes, the same is true for the difference of pressures at two given values of r, say r_1 and r_2, where $r_2 > r_1$. Hence, if the pressure be supposed constant at r_1, it must continually increase at r_2.

If the fluid be supposed to be contained between two coaxial cylindrical walls, both walls must move inwards together, and the process comes to an end when the inner wall reaches the axis. But we are not obliged to imagine an inner wall, or, indeed, any wall. The fluid passing inwards at $r = r_1$ may be supposed to be removed. And it remains true that, if it there pass at a constant pressure, the pressure at $r = r_2$ must continually increase. If this pressure has a limit, the inwards flow must cease.

It would be of interest to calculate some case in which the (u, w) motion is less simple, for instance, when fluid is removed at a point instead of uniformly along an axis, or inner cylindrical boundary. But this seems hardly practicable. The condition by which v is determined requires the expression of the motion of individual particles, as in the so-called Lagrangian method, and this usually presents great difficulties. We may, however, formulate certain conclusions of a general character.

When the (u, w) motion is slow relatively to the v motion, a kind of "equilibrium theory" approximately meets the case, much as when the slow motion under gravity of a variably dense liquid retains as far as possible the horizontal stratification. Thus oil standing over water is drawn off by a syphon without much disturbing the water underneath. When the density varies continuously the situation is more delicate, but the tendency is for the syphon to draw from the horizontal stratum at which it opens. Or if the liquid escapes slowly through an aperture in the bottom of the containing vessel, only the lower strata are disturbed. In like manner when revolving fluid is drawn off in the neighbourhood of a point situated on the axis of rotation, there is a tendency for the surfaces of constant circulation to remain cylindrical and the tendency is the more decided the greater the rapidity of rotation. The escaping liquid is drawn always from along the axis and not symmetrically in all directions, as when there is no rotation. The above is, in substance, the reasoning of Dr Aitken, who has also described a simple experiment in illustration.

P.S.—It may have been observed that according to what has been said above the stability of fluid motion in cylindrical strata requires only that the *square* of the circulation increase outwards. If the circulation be in both directions, this disposition involves discontinuities, and the stability exists only under the condition that symmetry with respect to the axis is rigorously maintained. If this limitation be dispensed with, the motion is certainly unstable, and thus the stability of motion in cylindrical layers really requires that the circulation be one-signed. On the general question of the *two-dimensional* motion of liquids between fixed coaxial cylindrical walls reference may be made to a former paper*. The motion in cylindrical strata is stable provided that the "rotation either continually increase or continually decrease in passing outwards from the axis." The demonstration is on the same lines as there set out for plane strata.

* *Proc. Lond. Math. Soc.* Vol. XI. p. 57 (1880); *Scientific Papers*, Vol. I. p. 487. See last paragraph.

414.

PROPAGATION OF SOUND IN WATER.

[Not hitherto published.]

FROM the theoretical point of view there is little to distinguish propagation of sound in an unlimited mass of water from the corresponding case of air; of course the velocity is greater (about four times). It is probable that at a great depth the velocity increases, the effect of diminishing compressibility out-weighing increased density.

As regards absorption, it would appear that it is likely to be less in water than in air. The viscosity (measured kinematically) is less in water.

But the practical questions are largely influenced by the presence of a free surface, which must act as a nearly perfect reflector. So far the case is analogous to that of a fixed wall reflecting sound waves in air; but there is an important difference. In order to imitate the wall in air, we must suppose the image of the source of sound to be exactly similar to the original; but the image of the source of sound reflected from the free surface of water must be taken negatively, viz., in the case of a pure tone with phase altered by 180°. In practice the case of interest is when both source and place of observation are somewhat near the reflecting surface. We must expect phenomena of interference varying with the precise depth below the surface. The analogy is with Lloyd's interference bands in Optics. If we suppose the distance to be travelled very great, the paths of the direct and reflected sounds will be nearly equal. Here the distinction of the two problems comes in.

For air and wall the phases of the direct and reflected waves on arrival would be the same, and the effect a maximum. But for the free surface of water the phases would be opposite and the effect approximately zero. This is what happens close to the surface. By going lower down the sound would be recovered. It is impossible to arrive at quantitative results unless all the circumstances are specified—distance, depths, and wave-length. If there are waves upon the surface of the water there is further complication; but in any case the surface acts as a nearly perfect reflector. The analogy is with a rough wall in air.

There is also the bottom to be considered. This, too, must act as a reflector in greater or less degree. With a rocky bottom and nearly grazing incidence, the reflection would be nearly perfect. Presumably a muddy or sandy bottom would reflect less. But I imagine that at grazing incidence—as when the distance between source and place of observation is a large multiple of the depth—the reflection would be good. This makes another complication.

415.

ON METHODS FOR DETECTING SMALL OPTICAL RETARDA-TIONS, AND ON THE THEORY OF FOUCAULT'S TEST.

[*Philosophical Magazine*, Vol. XXXIII. pp. 161—178, 1917.]

As was, I think, first emphasized by Foucault, the standard of accuracy necessary in optical surfaces is a certain fraction of the wave-length (λ) of the light employed. For glass surfaces refracting at nearly perpendicular incidence the error of linear retardation is about the half of that of the surface; but in the case of perpendicular reflexion the error of retardation is the double of that of the surface. The admissible error of retardation varies according to circumstances. In the case of lenses and mirrors affected with "spherical aberration," an error of $\frac{1}{4}\lambda$ begins to influence the illumination at the geometrical focus, and so to deteriorate the image. For many purposes an error less than this is without importance. The subject is discussed in former papers*.

But for other purposes, especially when measurements are in question, a higher standard must be insisted on. It is well known that the parts of the surfaces actually utilized in interferometers, such as those of Michelson and of Fabry and Perot, should be accurate to $\frac{1}{10}\lambda$ to $\frac{1}{20}\lambda$, and that a still higher degree of accuracy would be advantageous. Even under difficult conditions interference-bands may be displayed in which a local departure from ideal straightness amounting to $\frac{1}{20}$ of the band period can be detected on simple inspection. I may instance some recent observations in which the rays passing a fine vertical slit backed by a common paraffin-flame fell upon the object-glass of a 3-inch telescope placed some 20 feet away at the further end of a dark room. No collimator was needed. The object-glass was provided with a cardboard cap, pierced by two vertical slits, each $\frac{1}{10}$ inch wide, and so placed that the distance between the inner edges was $\frac{8}{10}$ inch. The parallelism of the three slits could be tested with a plumb-line. To observe the bands formed at the focus of the object-glass, a high magnifying-power

---

* *Phil. Mag.* Vol. VIII. pp. 403, 477 (1879); *Scientific Papers*, Vol. I. p. 415, §§ 3, 4.

is required.   This was afforded by a small cylinder lens, acting as sole eye-piece, whose axis is best adjusted by trial to the required parallelism with the slits.   Fairly good results were obtained with a glass tube of external diameter equal to about 3 mm., charged with water or preferably nitro-benzol.   Latterly, I have used with advantage a solid cylinder lens of about the same diameter kindly placed at my disposal by Messrs Hilger.   With this arrangement a wire stretched horizontally across the object-glass in front of the slits is seen in fair focus.   When the adjustment is good, the bands are wide and the blacknesses well developed, so that a local retardation of $\frac{1}{20}\lambda$ or less is evident if suitably presented.   The bands are much disturbed by heated air rising from the hand held below the path of the light.

The necessity for a high magnifying-power is connected with the rather wide separation of the interfering pencils as they fall upon the object-glass. The conditions are most favourable for the observation of very small retar-dations when the interfering pencils travel along precisely the same path, as may happen in the interference of polarized light, whether the polarization be rectilinear, as in ordinary double refraction, or circular, as along the axis of quartz.   In some experiments directed to test whether " motion through the æther causes double refraction*," it appeared that a relative retardation of the two polarized components could be detected when it amounted to only $\lambda/12000$, and, if I remember rightly, Brace was able to achieve a still higher sensibility.   The sensibility would increase with the intensity of the light employed and with the transparency of the optical parts (nicols, &c.), and it can scarcely be said that there is any theoretical limit.

Another method by which moderately small retardations can be made evident is that introduced by Foucault† for the figuring of optical surfaces. According to geometrical optics rays issuing from a point can be focussed at another point, if the optical appliances are perfect.   An eye situated just behind the focus observes an even field of illumination ; but if a screen with a sharp edge is gradually advanced in the focal plane, all light is gradually cut off, and the entire field becomes dark simultaneously.   At this moment any irregularity in the optical surfaces, by which rays are diverted from their proper course so as to escape the screening, becomes luminous ; and Foucault explained how the appearances are to be interpreted and information gained as to the kind of correction necessary.   He does not appear to have employed the method to observe irregularities arising otherwise than in optical surfaces, but H. Draper, in his memoir of 1864 on the Construction of a Spherical Glass Telescope‡, gives a picture of the disturbances due to the heating action of the hand held near the telescope mirror.   Töpler's work dates from

---

* *Phil. Mag.* Vol. IV. p. 678 (1902); *Scientific Papers*, Vol. v. p. 66.
† *Ann. de l'Observ. de Paris*, t. v. ; *Collected Memoirs*, Paris, 1878.
‡ *Smithsonian Contribution to Knowledge*, Jan. 1864.

the same year, and in subsequent publications* he made many interesting applications, such as to sonorous waves in air originating in electric sparks, and further developed the technique. His most important improvements were perhaps the introduction of a larger source of light bounded by a straight edge parallel to that of the screen at the observing end, and of a small telescope to assist the eye. Worthy of notice is a recent application by R. Cheshire† to determine with considerable precision for practical purposes the refractive index of irregular glass fragments. When the fragment is surrounded by liquid‡ of slightly different index contained in a suitable tank, it appears luminous as an irregularity, but by adjusting the composition of the liquid it may be made to disappear. The indices are then equal, and that of the liquid may be determined by more usual methods.

We have seen that according to geometrical optics ($\lambda = 0$) the regular light from an infinitely fine slit may be cut off suddenly, and that an irregularity will become apparent in full brightness however little (in the right direction) it may deflect the proper course of the rays. In considering the limits of sensibility we must remember that with a finite λ the image of the slit cannot be infinitely narrow, but constitutes a diffraction pattern of finite size. If we suppose the aperture bounding the field of view to be rectangular, we may take the problem to be in two dimensions, and the image consists of a central band of varying brightness bounded by dark edges and accompanied laterally by successions of bands of diminishing brightness. A screen whose edge is at the geometrical focus can cut off only half the light and, even if the lateral bands could be neglected altogether, it must be further advanced through half the width of the central band before the field can become dark. The width of the central band depends upon the horizontal aperture a (measured perpendicularly to the slit supposed vertical), the distance f between the lens and the screen, and the wave-length λ. By elementary diffraction theory the first darkness occurs when the difference of retardations of the various secondary rays issuing from the aperture ranges over one complete wave-length, i.e. when the projection of the aperture on the central secondary ray is equal to λ. The half-width (ξ) of the central band is therefore expressed by $\xi = f\lambda/a$.

If a prism of relative index μ, and of small angle i, be interposed near the lens, the geometrical focus of rays passing through the prism will be displaced through a distance $(\mu - 1)\,if$. If we identify this with ξ as expressed above, we have

$$(\mu - 1)\,i = \lambda/a, \quad\quad\quad\quad\quad\quad\quad\quad\quad\quad (1)$$

* Pogg. *Ann.* Bd. cxxviii. p. 126 (1866); Bd. cxxxi. pp. 33, 180 (1867).

† *Phil. Mag.* Vol. xxxii. p. 409 (1916).

‡ The liquid employed was a solution of mercuric iodide, and is spoken of as Thoulet's solution. Liveing (*Camb. Phil. Proc.* Vol. iii. p. 258, 1879), who made determinations of the dispersive power, refers to Sonstadt (*Chem. News*, Vol. xxix. p. 128, 1874). I do not know the date of Thoulet's use of the solution, but suspect that it was subsequent to Sonstadt's.

as the condition that the half maximum brightness of the prism shall coincide with approximate extinction of the remainder of the field of view. If the linear aperture of the prism be b, supposed to be small in comparison with a, the maximum retardation due to it is

$$(\mu - 1)\, ib = \lambda \,.\, b/a \,; \dots\dots\dots\dots\dots\dots\dots\dots(2)$$

and we recognize that easy visibility of the prism on the darkened field is consistent with a maximum retardation which is a small fraction of λ.

In Cheshire's application of Foucault's method (for I think it should be named after him) the prism had an angle i of $10°$, and the aperture a was 8 cms., although it would appear from the sketch that the whole of it was not used. Thus in (1) λ/ia would be about 5×10^{-5}; and the accuracy with which μ was determined (about $\pm \cdot00002$) is of the order that might be expected.

It is of interest to trace further and more generally what the wave theory has to tell us, still supposing that the source of light is from an infinitely narrow slit (or, what comes to the same, a slit of finite width at an infinite distance), and that the apertures are rectangular. The problem may then be supposed to be in two dimensions*, although in strictness this requires that the elementary sources distributed uniformly along the length of the slit should be all in one phase. The calculation makes the usual assumption, which cannot be strictly true, that the effect of a screen is merely to stop those parts of the wave which impinge upon it, without influencing the neighbouring parts. In fig. 1, A represents the lens with its rectangular

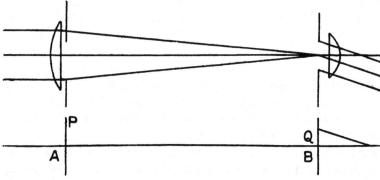

Fig. 1.

aperture, which brings parallel rays to a focus. In the focal plane B are two adjustable screens with vertical edges, and immediately behind is the eye or objective of a small telescope. The rays from the various points Q of the second aperture, which unite at a point in the focal plane of the telescope, or of the retina, may be regarded as a parallel pencil inclined to the axis at

* Compare " Wave Theory," *Encyc. Brit.* 1888 ; *Scientific Papers*, Vol. III. p. 84.

a small angle ϕ. P is a point in the first aperture, $AP = x$, $BQ = \xi$, $AB = f$. Any additional linear retardation operative at P may be denoted by R, a function of x. Thus if V be the velocity of propagation and $\kappa = 2\pi/\lambda$, the vibration at the point ξ of the second aperture will be represented by

$$\int dx \sin \kappa \left(Vt - f - R + \frac{x\xi}{f} \right),$$

or, if $x/f = \theta$, by

$$\int d\theta \sin \kappa (Vt - f - R + \theta\xi), \quad \ldots\ldots\ldots\ldots\ldots\ldots(3)$$

the limits for θ corresponding to the angular aperture of the lens A. For shortness we shall omit κ*, which can always be restored on considering "dimensions," and shall further suppose that R is at most a linear function of θ, say $\rho + \sigma\theta$, or, at any rate, that the whole aperture can be divided into parts for each of which R is a linear function. In the former case the constant part ρ may be associated with $Vt - f$, and if T be written for $Vt - f - \rho$, (3) becomes

$$\sin T \int d\theta \cos (\xi - \sigma)\, \theta + \cos T \int d\theta \sin (\xi - \sigma)\, \theta. \quad \ldots\ldots\ldots\ldots(4)$$

Since the same values of ρ, σ apply over the whole aperture, the range of integration is between $\pm\, \theta$, where θ denotes the angular semi-aperture, and then the second term, involving $\cos T$, disappears, while the effect of σ is represented by a shift in the origin of ξ, as was to be expected. There is now no real loss of generality in omitting R altogether, so that (4) becomes simply

$$2 \sin T\, \frac{\sin \xi\theta}{\xi}, \quad \ldots\ldots\ldots\ldots\ldots\ldots\ldots\ldots\ldots(5)$$

as in the usual theory. The borders of the central band correspond to $\xi\theta$, or rather $\kappa\xi\theta, = \pm\, \pi$, or $\xi\theta = \pm\, \tfrac{1}{2}\lambda$, which agrees with the formula used above, since $2\theta = a/f$.

When we proceed to inquire what is to be observed at angle ϕ we have to consider the integral

$$2 \int d\xi \sin (T + \phi\xi) \frac{\sin \theta\xi}{\xi} = \sin T \int \frac{\sin (\theta + \phi)\, \xi + \sin (\theta - \phi)\, \xi}{\xi}\, d\xi$$

$$+ \cos T \int \frac{\cos (\theta - \phi)\, \xi - \cos (\theta + \phi)\, \xi}{\xi}\, d\xi. \quad \ldots\ldots\ldots(6)$$

It will be observed that, whatever may be the limits for ξ, the first integral is an even and the second an odd function of ϕ, so that the intensity (I), represented by the sum of the squares of the integrals, is an even function. The field of view is thus symmetrical with respect to the axis.

---

* Equivalent to supposing $\lambda = 2\pi$.

The integrals in (6) may be at once expressed in terms of the so-called sine-integral and cosine-integral defined by

$$\text{Si}\,(x) = \int_0^x \frac{\sin x}{x}\,dx, \quad \text{Ci}\,(x) = \int_\infty^x \frac{\cos x}{x}\,dx.$$

If the limits of ξ be ξ_1 and ξ_2 we get

$$\sin T[\text{Si}\,\{(\theta+\phi)\,\xi_2\} - \text{Si}\,\{(\theta+\phi)\,\xi_1\} + \text{Si}\,\{(\theta-\phi)\,\xi_2\} - \text{Si}\,\{(\theta-\phi)\,\xi_1\}]$$
$$+ \cos T[\text{Ci}\,\{(\theta-\phi)\,\xi_2\} - \text{Ci}\,\{(\theta-\phi)\,\xi_1\} - \text{Ci}\,\{(\theta+\phi)\,\xi_2\} + \text{Ci}\,\{(\theta+\phi)\,\xi_1\}].$$
$$\dots\dots(7)$$

If $\xi_1 = -\xi_2 = -\xi$, so that the second aperture is symmetrical with respect to the axis, the Ci's, being even functions, disappear, and we have simply

$$2\sin T[\text{Si}\,\{(\theta+\phi)\,\xi\} + \text{Si}\,\{(\theta-\phi)\,\xi\}]. \quad \dots\dots\dots(8)$$

If the aperture of the telescope be not purposely limited, the value of ξ, or rather of $\kappa\xi$, is very great, and for most purposes the error will be small in supposing it infinite. Now $\text{Si}\,(\pm\infty) = \pm\frac{1}{2}\pi$, so that if ϕ is numerically less than θ, $I = 4\pi^2$, but if ϕ is numerically greater than θ, $I = 0$. The angular field of view 2θ is thus uniformly illuminated and the transition to darkness at angles $\pm\theta$ is sudden—that is, the edges are seen with infinite sharpness. Of course, ξ cannot really be infinite, nor consequently the resolving power of the telescope; but we may say that the edges are defined with full sharpness. The question here is the same as that formerly raised under the title "An Optical Paradox*," the paradox consisting in the full definition of the edges of the first aperture, although nearly the whole of the light at the second aperture is concentrated in a very narrow band, which might appear to preclude more than a very feeble resolving power.

It may be well at this stage to examine more closely what is actually the distribution of light between the central and lateral bands in the diffraction pattern formed at the plane of the second aperture. By (5) the intensity of light at ξ is proportional to $\xi^{-2}\sin^2\theta\xi$ or, if we write η for $\theta\xi$, to $\eta^{-2}\sin^2\eta$. The whole light between 0 and η is thus represented by

$$J = \int_0^\eta \frac{\sin^2\eta}{\eta^2}\,d\eta. \quad \dots\dots\dots\dots(9)$$

J can be expressed by means of the Si-function. As may be verified by differentiation,

$$J = \text{Si}\,(2\eta) - \eta^{-1}\sin^2\eta, \quad \dots\dots\dots\dots(10)$$

vanishing when $\eta = 0$. The places of zero illumination are defined by $\eta = n\pi$, when $n = 1, 2, 3$, &c.; and, if η assume one of these values, we have simply

$$J = \text{Si}\,(2\eta) = \text{Si}\,(2n\pi). \quad \dots\dots\dots\dots(11)$$

---

* *Phil. Mag.* Vol. IX. p. 779 (1905); *Scientific Papers*, Vol. v. p. 254.

Thus, setting $n = 1$, we find for half the light in the central band

$$J = \mathrm{Si}\,(2\pi) = \tfrac{1}{2}\pi - \cdot 15264.$$

On the same scale half the whole light is $\mathrm{Si}\,(\infty)$, or $\tfrac{1}{2}\pi$, so that the fraction of the whole light to be found in the central band is

$$1 - \frac{2 \times \cdot 15264}{\pi} = 1 - \cdot 097174, \quad\dots\dots\dots\dots\dots(12)$$

or more than nine-tenths. About half the remainder is accounted for by the light in the two lateral bands immediately adjacent (on the two sides) to the central band.

We are now in a position to calculate the appearance of the field when the second aperture is actually limited by screens, so as to allow only the passage of the central band of the diffraction pattern. For this purpose we have merely to suppose in (8) that $\theta\xi = \pi$. The intensity at angle ϕ is thus

$$4\left[\mathrm{Si}\left(\frac{\theta + \phi}{\theta}\,\pi\right) + \mathrm{Si}\left(\frac{\theta - \phi}{\theta}\,\pi\right)\right]^2. \quad\dots\dots\dots\dots(13)$$

The further calculation requires a knowledge of the function Si, and a little later we shall need the second function Ci. In ascending series

$$\mathrm{Si}\,(x) = x - \frac{1}{3}\frac{x^3}{1\,.\,2\,.\,3} + \frac{1}{5}\frac{x^5}{1\,.\,2\,.\,3\,.\,4\,.\,5} - \frac{1}{7}\frac{x^7}{1\,.\,2\,\dots\,7} + \dots\dots(14)$$

$$\mathrm{Ci}\,(x) = \gamma + \frac{1}{2}\log\,(x^2) - \frac{1}{2}\frac{x^2}{1\,.\,2} + \frac{1}{4}\frac{x^4}{1\,.\,2\,.\,3\,.\,4} - \dots; \quad\dots\dots(15)$$

γ is Euler's constant $\cdot 5772157$, and the logarithm is to base e

These series are always convergent and are practically available when x is moderate. When x is great, we may use the semi-convergent series

$$\mathrm{Si}\,(x) = \frac{\pi}{2} - \cos x\left\{\frac{1}{x} - \frac{1\,.\,2}{x^3} + \frac{1\,.\,2\,.\,3\,.\,4}{x^5} - \frac{1\,.\,2\,\dots\,6}{x^7} + \dots\right\}$$

$$- \sin x\left\{\frac{1}{x^2} - \frac{1\,.\,2\,.\,3}{x^4} + \frac{1\,.\,2\,.\,3\,.\,4\,.\,5}{x^6} - \dots\right\}, \quad\dots\dots\dots(16)$$

$$\mathrm{Ci}\,(x) = \sin x\left\{\frac{1}{x} - \frac{1\,.\,2}{x^3} + \frac{1\,.\,2\,.\,3\,.\,4}{x^5} - \dots\right\}$$

$$- \cos x\left\{\frac{1}{x^2} - \frac{1\,.\,2\,.\,3}{x^4} + \frac{1\,.\,2\,.\,3\,.\,4\,.\,5}{x^6} - \dots\right\}. \quad\dots\dots\dots(17)$$

Tables of the functions have been calculated by Glaisher*. For our present purpose it would have been more convenient had the argument been πx, rather than x. Between $x = 5$ and $x = 15$, the values of $\mathrm{Si}\,(x)$ are given for integers only, and interpolation is not effective. For this reason some

* Phil. Trans. Vol. CLX. p. 367 (1870).

values of ϕ/θ are chosen which make $(1 + \phi/\theta)\,\pi$ integral. The calculations recorded in Table I refer in the first instance to the values of

$$\mathrm{Si}\,(1 + \phi/\theta)\,\pi + \mathrm{Si}\,(1 - \phi/\theta)\,\pi. \quad\dots\dots\dots\dots(18)$$

TABLE I.

$$\kappa\theta\xi_1 = -\pi, \quad \kappa\theta\xi_2 = +\pi.$$

ϕ/θ	(18)	$(18)^2$
0·0000	3·704	13·72
0·2732	3·475	12·08
0·5000	2·979	8·87
0·5915	2·721	7·40
0·9099	1·707	2·91
1·0000	1·418	2·01
1·2282	0·758	0·57
1·5465	0·115	0·01
2·0000	−0·177	0·03

It will be seen that, in spite of the fact that nine-tenths of the whole light passes, the definition of what should be the edge of the field at $\phi = \theta$ is very bad. Also that the illumination at $\phi = 0$ is *greater* than what it would be (π^2) if the second screening were abolished altogether ($\pm\,\xi = \infty$).

So far we have dealt only with cases where the second aperture is symmetrically situated with respect to the geometrical focus. This restriction we will now dispense with, considering first the case where $\xi_1 = 0$ and $\xi_2 (= \xi)$ is positive and of arbitrary value. The coefficient of $\sin T$ in (7) becomes simply

$$\mathrm{Si}\,\{(\theta + \phi)\,\xi\} + \mathrm{Si}\,\{(\theta - \phi)\,\xi\}. \quad\dots\dots\dots\dots(19)$$

In the coefficient of $\cos T$, $\mathrm{Ci}\,\{(\theta + \phi)\,\xi_1\}$, $\mathrm{Ci}\,\{(\theta - \phi)\,\xi_1\}$ assume infinite values, but by (15) we see that

$$\mathrm{Ci}\,\{(\theta + \phi)\,\xi_1\} - \mathrm{Ci}\,\{(\theta - \phi)\,\xi_1\} = \log\left|\frac{\theta + \phi}{\theta - \phi}\right|, \quad\dots\dots(20)$$

so that the coefficient of $\cos T$ is

$$\mathrm{Ci}\,\{(\theta - \phi)\,\xi\} - \mathrm{Ci}\,\{(\theta + \phi)\,\xi\} + \log\left|\frac{\theta + \phi}{\theta - \phi}\right|. \quad\dots\dots(21)$$

The intensity I at angle ϕ is represented by the sum of the squares of (19) and (21). When $\phi = 0$ at the centre of the field of view, $I = 4\,(\mathrm{Si}\,\theta\xi)^2$, but at the edges for which it suffices to suppose $\phi = +\,\theta$, a modification is called for, since $\mathrm{Ci}\,\{(\theta - \phi)\,\xi\}$ must then be replaced by $\gamma + \log\,|(\theta - \phi)\,\xi|$. Under these circumstances the coefficient of $\cos T$ becomes

$$\gamma + \log\,(2\theta\xi) - \mathrm{Ci}\,(2\theta\xi),$$

and $$I = \{\mathrm{Si}\,(2\theta\xi)\}^2 + \{\gamma + \log\,(2\theta\xi) - \mathrm{Ci}\,(2\theta\xi)\}^2. \quad\dots\dots\dots(22)$$

If in (22) ξ be supposed to increase without limit, we find

$$I = \tfrac{1}{4}\pi^2 + \{\log \theta\xi\}^2, \quad\dots\dots\dots\dots\dots\dots\dots(23)$$

becoming logarithmically infinite.

Since in practice ξ, or rather $\kappa\xi$, is large, the edges of the field may be expected to appear very bright.

As may be anticipated, this conclusion does not depend upon our supposition that $\xi_1 = 0$. Reverting to (7) and supposing $\phi = \theta$, we have

$$\sin T\,[\text{Si}\,(2\theta\xi_2) - \text{Si}\,(2\theta\xi_1)] + \cos T\,[\text{Ci}\,(2\theta\xi_1) - \text{Ci}\,(2\theta\xi_2) + \log(\xi_2/\xi_1)], \quad(24)$$

and $I = \infty$, when $\xi_2 = \infty$. If ξ_1 vanishes in (24), we have only to replace $\text{Ci}\,(2\theta\xi_1)$ by $\gamma + \log(2\theta\xi_1)$ in order to recover (22).

We may perhaps better understand the abnormal increase of illumination at the edges of the field by a comparison with the familiar action of a grating in forming diffraction spectra. Referring to (5) we see that if positive values of ξ be alone regarded, the vibration in the plane of the second aperture, represented by $\xi^{-1} \sin(\theta\xi)$, is the same in respect of phase as would be due to a theoretically simple grating receiving a parallel beam perpendicularly, and the directions $\phi = \pm\theta$ are those of the resulting lateral spectra of the first order. On account, however, of the factor ξ^{-1}, the case differs somewhat from that of the simple grating, but not enough to prevent the illumination becoming logarithmically infinite with infinite aperture. But the approximate resemblance to a simple grating fails when we include negative as well as positive values of ξ, since there is then a reversal of phase in passing zero. Compare fig. 2, where positive values are represented by full lines and

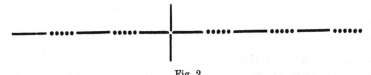

Fig. 2.

negative by dotted lines. If the aperture is symmetrically bounded, the parts at a distance from the centre tend to compensate one another, and the intensity at $\phi = \pm\theta$ does not become infinite with the aperture.

We now proceed to consider the actual calculation of $I = (19)^2 + (21)^2$ for various values of ϕ/θ, which we may suppose to be always positive, since I is independent of the sign of ϕ. When $\xi\theta$ is very great and ϕ/θ is not nearly equal to unity, $\text{Si}\,\{(\theta + \phi)\,\xi\}$ in (19) may be replaced by $\tfrac{1}{2}\pi$ and $\text{Si}\,\{(\theta - \phi)\,\xi\}$ by $\pm\tfrac{1}{2}\pi$, according as ϕ/θ is less or greater than unity. Under the same conditions the Ci's in (21) may be omitted, so that

$$I = \pi^2\,(1, \text{ or } 0) + \left\{\log\left|\frac{\theta + \phi}{\theta - \phi}\right|\right\}^2. \quad\dots\dots\dots\dots(25)$$

But if we wish to avoid the infinity when $\phi = \theta$, we must make some supposition as to the actual value of $\theta\xi$, or rather of $2\pi\theta\xi/\lambda$. In some observations to be described later $a = 1$ inch, $\xi = \frac{1}{2}$ inch, $1/\lambda = 40,000$, and $\theta = \frac{1}{2}a/f$. Also f was about 10 feet $= 120$ inches. For simplicity we may suppose $f = 40\pi$, so that $2\pi\theta\xi/\lambda = 500$, or in our usual notation $\theta\xi = 500$. Thus

$$(19) = \mathrm{Si}\,\{500\,(1 + \phi/\theta)\} + \mathrm{Si}\,\{500\,(1 - \phi/\theta)\}, \quad \ldots\ldots\ldots(26)$$

and
$$(21) = \mathrm{Ci}\,\{500\,(1 - \phi/\theta)\} - \mathrm{Ci}\,\{500\,(1 + \phi/\theta)\}$$

$$+ \log\,(1 + \phi/\theta) - \log\,|\,1 - \phi/\theta\,|. \quad \ldots\ldots\ldots\ldots\ldots(27)$$

For the purposes of a somewhat rough estimate we may neglect the second Ci in (27) and identify the first Si in (26) with $\frac{1}{2}\pi$ for all (positive) values of ϕ/θ. Thus when $\phi = 0$, $I = \pi^2$; and when $\phi = \infty$, $I = 0$.

When $\phi/\theta = 1$, we take

$$(26) = \tfrac{1}{2}\pi = 1\cdot571, \quad (26)^2 = 2\cdot467.$$

In (27) $\mathrm{Ci}\,\{500\,(1 - \phi/\theta)\} = \gamma + \log 500 + \log\,(1 - \phi/\theta),$

so that
$$(27) = \gamma + \log 1000 = 7\cdot485, \quad (27)^2 = 56\cdot03\,;$$

and
$$I = 58\cdot50.$$

For the values of ϕ/θ in the neighbourhood of unity we may make similar calculations with the aid of Glaisher's Tables. For example, if $\phi/\theta = 1 \mp \cdot02$, we have
$$500\,(1 - \phi/\theta) = \pm\,10.$$

From the Tables
$$\mathrm{Si}\,(\pm\,10) = \pm\,1\cdot6583, \quad \mathrm{Ci}\,(\pm\,10) = -\,\cdot0455,$$

and thence
$$I\,(\cdot98) = 31\cdot13, \quad I\,(1\cdot02) = 20\cdot89.$$

As regards values of the argument outside these units, we may remark that when x exceeds 10, $\mathrm{Si}\,(x) - \frac{1}{2}\pi$ and $\mathrm{Ci}\,(x)$ are approximately periodic in period 2π and of order x^{-1}. It is hardly worth while to include these fluctuations, which would manifest themselves as rather feeble and narrow bands, superposed upon the general ground, and we may thus content ourselves with (25). If we apply this to $\pm\,10$, we get

$$I\,(\cdot98) = 30\cdot98, \quad I\,(1\cdot02) = 21\cdot30\,;$$

and the smoothed values differ but little from those calculated for $\pm\,10$ more precisely. The Table (II) annexed shows the values of I for various values of ϕ/θ. Those in the 2nd and 8th columns are smoothed values as explained, and they would remain undisturbed if the value of $\theta\xi$ were increased. It will be seen that the maximum illumination near the edges is some 6 times that at the centre.

TABLE II.

$$\kappa\theta\xi_1 = 0, \quad \kappa\theta\xi_2 = 500.$$

ϕ/θ	I	ϕ/θ	I	ϕ/θ	I	ϕ/θ	I
0·000	9·87	0·980	31·13	1·001	56·28	1·05	13·76
0·250	10·13	0·990	35·78	1·002	52·89	1·10	9·24
0·500	11·08	0·992	39·98	1·004	44·09	1·20	5·76
0·800	14·71	0·994	46·81	1·006	35·27	1·50	2·59
0·900	18·51	0·996	54·13	1·008	29·03	2·00	1·21
0·950	23·27	0·998	58·81	1·010	26·14	∞	0
		0·999	59·36	1·020	20·89		
		1·000	58·50				

TABLE III.

$$\kappa\theta\xi_1 = \pi, \quad \kappa\theta\xi_2 = 500.$$

ϕ/θ	I	ϕ/θ	I
0·00	0·32	1·01	8·98
0·50	0·48	1·02	6·57
0·91	2·46	1·23	0·58
0·98	7·55	1·55	0·13
0·99	9·90	1·86	0·05
1·00	25·51	∞	0·00

In the practical use of Foucault's method the general field would be darkened much more than has been supposed above where half the whole light passes. We may suppose that the screening just cuts off the central band, as well as all on one side of it, so that $\theta\xi_1 = \pi$. In this case (7) becomes

$$\sin T\,[\mathrm{Si}\,(\theta + \phi)\,\xi + \mathrm{Si}\,(\theta - \phi)\,\xi - \mathrm{Si}\,(1 + \phi/\theta)\,\pi - \mathrm{Si}\,(1 - \phi/\theta)\,\pi]$$
$$+ \cos T\,[\mathrm{Ci}\,(\theta - \phi)\,\xi - \mathrm{Ci}\,(\theta + \phi)\,\xi + \mathrm{Ci}\,(1 + \phi/\theta)\,\pi - \mathrm{Ci}\,(1 - \phi/\theta)\,\pi].$$
$$\dots\dots(28)$$

We will apply it to the case already considered, where $\theta\xi = 500$, as before omitting $\mathrm{Ci}\,(\theta + \phi)\,\xi$ and equating $\mathrm{Si}\,(\theta + \phi)\,\xi$ to $\tfrac{1}{2}\pi$. Thus

$$I = [\tfrac{1}{2}\pi + \mathrm{Si}\,500\,(1 - \phi/\theta) - \mathrm{Si}\,(1 + \phi/\theta)\,\pi - \mathrm{Si}\,(1 - \phi/\theta)\,\pi]^2$$
$$+ [\mathrm{Ci}\,500\,(1 - \phi/\theta) + \mathrm{Ci}\,(1 + \phi/\theta)\,\pi - \mathrm{Ci}\,(1 - \phi/\theta)\,\pi]^2.$$
$$\dots\dots(29)$$

When $\phi = \infty$, $I = 0$. When $\phi = 0$,

$$I = [\pi - 2\,\mathrm{Si}\,\pi]^2 = ·3162.$$

When $\phi = \theta$,

$$I = [\tfrac{1}{2}\pi - \mathrm{Si}\,(2\pi)]^2 + [\log\,(500/\pi) + \mathrm{Ci}\,(2\pi)]^2 = 25·51;$$

so that the brightness of the edges is now about 80 times that at the centre of the field. The remaining values of I in Table III have been calculated as before with omission of the terms representing minor periodic fluctuations.

Hitherto we have treated various kinds of screening, but without additional retardation at the plane of the first aperture. The introduction of such retardation is, of course, a complication, but in principle it gives rise to no difficulty, provided the retardation be linear in θ over the various parts of the aperture. The final illumination as a function of ϕ can always be expressed by means of the Si- and Ci-functions.

As the simplest case which presents something essentially novel, we may suppose that an otherwise constant retardation (R) changes sign when $\theta = 0$, is equal (say) to $+\rho$ when θ is positive and to $-\rho$ when θ is negative. Then (3) becomes

$$
\int_{-\theta}^{0} \sin (T + \rho + \theta\xi)\, d\theta + \int_{0}^{\theta} \sin (T - \rho + \theta\xi)\, d\theta
$$

$$
= 2 \sin T \left[\cos \rho\, \frac{\sin \theta\xi}{\xi} + \sin \rho\, \frac{1 - \cos \theta\xi}{\xi} \right], \quad \ldots(30)
$$

reducing to (5) when $\rho = 0$. This gives the vibration at the point ξ of the second aperture. If $\xi = 0$, (30) becomes $2\theta \cos \rho \sin T$, and vanishes when $\cos \rho = 0$; for instance, when the whole difference of retardation $2\rho = \pi$, or (reckoned in wave-lengths) $\tfrac{1}{2}\lambda$.

The vibration in direction ϕ behind the second aperture is to be obtained by writing $T + \phi\xi$ for T in (30) and integrating with respect to ξ. This gives

$$
2 \sin T \int d\xi \cos \phi\xi \left\{ \cos \rho\, \frac{\sin \theta\xi}{\xi} + \sin \rho\, \frac{1 - \cos \theta\xi}{\xi} \right\}
$$

$$
+ 2 \cos T \int d\xi \sin \phi\xi \left\{ \cos \rho\, \frac{\sin \theta\xi}{\xi} + \sin \rho\, \frac{1 - \cos \theta\xi}{\xi} \right\}, \quad \ldots(31)
$$

and the illumination (I) is independent of the sign of ϕ, which we may henceforward suppose to be positive.

If the second aperture be symmetrically placed, we may take the limits to be expressed as $\pm \xi$, and (31) becomes

$$
2 \sin T \cos \rho \int_{0}^{\xi} \frac{\sin (\theta + \phi)\, \xi + \sin (\theta - \phi)\, \xi}{\xi}\, d\xi
$$

$$
+ 2 \cos T \sin \rho \int_{0}^{\xi} \frac{2 \sin \phi\xi - \sin (\theta + \phi)\, \xi + \sin (\theta - \phi)\, \xi}{\xi}\, d\xi. \quad \ldots(32)
$$

If we apply this to $\xi = \infty$ to find what occurs when there is no screening, we fall upon ambiguities, for (32) becomes

$$
2 \sin T \cos \rho \left\{ \tfrac{1}{2}\pi \pm \tfrac{1}{2}\pi \right\} + 2 \cos T \sin \rho \left\{ 2\, \mathrm{Si}\, (\phi\xi) - \tfrac{1}{2}\pi \pm \tfrac{1}{2}\pi \right\},
$$

the alternatives following the sign of $\theta - \phi$, with exclusion of the case $\phi = \theta$. If ϕ is finite, $2\operatorname{Si}(\phi\xi)$ may be equated to π, and we get

$$I = 4\pi^2 (1 \text{ or } 0),$$

according as $\theta - \phi$ is positive or negative. But if $\phi = 0$ absolutely, $\operatorname{Si}(\phi\xi)$ disappears, however great ξ may be; and when ϕ is small,

$$I = 4\pi^2 \cos^2 \rho + 4 \sin^2 \rho \{2\operatorname{Si}(\phi\xi)\}^2,$$

in which the value of the second term is uncertain, unless indeed $\sin \rho = 0$.

It would seem that the difficulty depends upon the assumed discontinuity of R when $\theta = 0$. If the limits for θ be $\pm \alpha$ (up to the present written as $\pm \theta$), what we have to consider is

$$\int_{-\infty}^{+\infty} d\xi \left[\int_{-\alpha}^{+\alpha} d\theta \sin \{T - R + (\theta + \phi)\xi\} \right],$$

in which hitherto we have taken first the integration with respect to θ. We propose now to take first the integration with respect to ξ, introducing the factor $e^{\pm\mu\xi}$ to ensure convergency. We get

$$2 \sin (T - R) \int_0^\infty e^{-\mu\xi} \cos (\theta + \phi) \, \xi \, . \, d\xi = \frac{2\mu \sin (T - R)}{\mu^2 + (\theta + \phi)^2}. \quad \ldots(33)$$

There remains the integration with respect to θ, of which R is supposed to be a continuous function. As μ tends to vanish, the only values of θ which contribute are confined more and more to the neighbourhood of $-\phi$, so that ultimately we may suppose θ to have this value in R. And

$$\int_{-\alpha}^{+\alpha} \frac{\mu \, d\theta}{\mu^2 + (\theta + \phi)^2} = \tan^{-1} \frac{\phi + \alpha}{\mu} - \tan^{-1} \frac{\phi - \alpha}{\mu},$$

which is π, if ϕ lies between $\pm \alpha$, and 0 if ϕ lies outside these limits, when μ is made vanishing small. The intensity in any direction ϕ is thus independent of R altogether. This procedure would fail if R were discontinuous for any values of θ.

Resuming the suppositions of equation (31), let us now further suppose that the aperture extends from ξ_1 to ξ_2, where both ξ_1 and ξ_2 are positive and $\xi_2 > \xi_1$. Our expression for the vibration in direction ϕ becomes

$$\sin T \left[\cos \rho \left\{\operatorname{Si}(\theta + \phi)\xi + \operatorname{Si}(\theta - \phi)\xi\right\}\right.$$
$$\left. + \sin \rho \left\{2\operatorname{Ci}(\phi\xi) - \operatorname{Ci}(\theta + \phi)\xi - \operatorname{Ci}(\theta - \phi)\xi\right\}\right]_{\xi_1}^{\xi_2}$$
$$+ \cos T \left[\cos \rho \left\{\operatorname{Ci}(\theta - \phi)\xi - \operatorname{Ci}(\theta + \phi)\xi\right\}\right.$$
$$\left. + \sin \rho \left\{2\operatorname{Si}(\phi\xi) - \operatorname{Si}(\theta + \phi)\xi + \operatorname{Si}(\theta - \phi)\xi\right\}\right]_{\xi_1}^{\xi_2}.$$
$$\ldots\ldots\ldots(34)$$

We will apply this to the case already considered where $\xi_2\theta = 500$, $\xi_1\theta = \pi$; and since we are now concerned mainly with what occurs in the neighbourhood of $\phi = 0$, we may confine ϕ to lie between the limits 0 and $\frac{1}{2}\theta$. Under these circumstances, and putting minor rapid fluctuations out of account, we may neglect $\mathrm{Ci}\,(\theta \pm \phi)\,\xi_2$ and equate $\mathrm{Si}\,(\theta \pm \phi)\,\xi_2$ to $\tfrac{1}{2}\pi$. A similar simplification is admissible for $\mathrm{Si}\,(\phi\xi_2)$, $\mathrm{Ci}\,(\phi\xi_2)$, unless ϕ/θ is very small.

When $\phi = 0$, (34) gives

$$\sin T\,[\cos \rho\,\{\pi - 2\,\mathrm{Si}\,(\pi)\} + \sin \rho\,\{2\log (500/\pi) + 2\,\mathrm{Ci}\,(\pi)\}],$$

in which

$$\pi - 2\,\mathrm{Si}\,(\pi) = -\,\cdot5623, \quad \mathrm{Ci}\,(\pi) = \cdot0738, \quad \log (500/\pi) = 5\cdot0699.$$

Thus for the intensity

$$I\,(0) = [-\,\cdot5623 \cos \rho + 10\cdot2874 \sin \rho]^2. \quad \ldots\ldots\ldots\ldots(35)$$

If $\rho = 0$, we fall back upon a former result ($\cdot3162$). If $\rho = \tfrac{1}{4}\pi$, $I\,(0) = 47\cdot3$.

Interest attaches mainly to small values of ρ, and we see that the effect depends upon the sign of ρ. A positive ρ means that the retardation at the first aperture takes place on the side opposite to that covered by the screen at the second aperture. As regards magnitude, we must remember that ρ stands for an angular retardation $\kappa\rho$, or $2\pi\rho/\lambda$; so that, for example, $\rho = \tfrac{1}{4}\pi$ above represents a linear retardation $\lambda/8$, and a total relative retardation between the two halves of the first aperture equal to $\lambda/4$.

The second column of Table IV gives the general expression for the vibration in terms of ρ for various values of ϕ/θ, followed by the values of the intensity (I) for $\sin \rho = \pm 1/10$ and $\sin \rho = \pm 1/\sqrt{2}$.

TABLE IV.

$$\kappa\theta\xi_1 = \pi, \quad \kappa\theta\xi_2 = 500.$$

$\dfrac{\phi}{\theta}$	Formula for Vibration	I		I	
		$\sin \rho$		$\sin \rho$	
		$+\cdot1$	$-\cdot1$	$+1/\sqrt{2}$	$-1/\sqrt{2}$
0	$\sin T\{-\cdot56 \cos \rho + 10\cdot29 \sin \rho\}$	$\cdot22$	$2\cdot53$	$47\cdot3$	$58\cdot9$
$\cdot001$	$\sin T\{-\cdot56 \cos \rho + 10\cdot16 \sin \rho\}$ $+\cos T \times \cdot99 \sin \rho$	$\cdot22$	$2\cdot50$	$46\cdot6$	$58\cdot0$
$\cdot010$	$\sin T\{-\cdot56 \cos \rho + 5\cdot53 \sin \rho\}$ $+\cos T \times 3\cdot10 \sin \rho$	$\cdot10$	$1\cdot34$	$17\cdot2$	$23\cdot4$
$\cdot050$	$\sin T\{-\cdot55 \cos \rho + 2\cdot71 \sin \rho\}$ $+\cos T\{-\cdot10 \cos \rho + 2\cdot83 \sin \rho\}$	$\cdot11$	$\cdot83$	$6\cdot0$	$9\cdot6$
$\cdot100$	$\sin T\{-\cdot53 \cos \rho + 1\cdot37 \sin \rho\}$ $+\cos T\{-\cdot20 \cos \rho + 2\cdot52 \sin \rho\}$	$\cdot16$	$\cdot66$	$3\cdot0$	$5\cdot5$
$\cdot250$	$\sin T\{-\cdot37 \cos \rho - \cdot17 \sin \rho\}$ $+\cos T\{-\cdot46 \cos \rho + 1\cdot66 \sin \rho\}$	$\cdot23$	$\cdot52$	$\cdot86$	$2\cdot3$
$\cdot500$	$\sin T\{+\cdot16 \cos \rho - \cdot67 \sin \rho\}$ $+\cos T\{-\cdot67 \cos \rho + \cdot64 \sin \rho\}$	$\cdot38$	$\cdot59$	$\cdot13$	$1\cdot2$

It will be seen that the direction of the discontinuity ($\phi = 0$) is strongly marked by excess of brightness, and that especially when ρ is small there is a large variation with the sign of ρ.

Perhaps the next case in order of simplicity of a variable R is to suppose $R = 0$ from $\theta = -\theta$ to $\theta = 0$, and $R = \sigma\theta$ from $\theta = 0$ to $\theta = +\theta$, corresponding to the introduction of a prism of small angle, whose edge divides equally the field of view. For the vibration in the focal plane we get

$$\sin T \left[\frac{\sin \theta \xi}{\xi} + \frac{\sin (\xi - \sigma)\,\theta}{\xi - \sigma} \right] + \cos T \left[\frac{1 - \cos (\xi - \sigma)\,\theta}{\xi - \sigma} - \frac{1 - \cos \xi \theta}{\xi} \right].$$
$$\dots\dots\dots(36)$$

In order to find what would be seen in direction ϕ, we should have next to write $(T + \phi\xi)$ for T and integrate again with respect to ξ between the appropriate limits. As to this there is no difficulty, but the expressions are rather long. It may suffice to notice that whatever the limits may be, no infinity enters at $\phi = 0$, in which case we have merely to integrate (36) as it stands. For although the denominators become zero when $\xi = 0$ or $\xi = \sigma$, the four fractions themselves always remain finite. The line of transition between the two halves of the field is not so marked as when there was an actual discontinuity in the retardation itself.

In connection with these calculations I have made for my own satisfaction a few observations, mainly to examine the enhanced brightness at the edges of the field of view. The luminous border is shown in Draper's drawing, and is described by Töpler as due to diffraction. The slit and focussing lens were those of an ordinary spectroscope, the slit being drawn back from the "collimating" lens. The telescope was from the same instrument, now mounted independently at a distance so as to receive an image of the slit, and itself focussed upon the first lens. The rectangular aperture at the first lens was originally cut out of the black card. The principal dimensions have already been given. A flat paraffin-flame afforded sufficient illumination. The screens used in front of the telescope were razor-blades (Gillettes), and were adjusted in position with the aid of an eyepiece, the telescope being temporarily removed. It is not pretended that the arrangements used corresponded fully to the suppositions of theory.

The brightness of the vertical edge of the field of view is very conspicuous when the light is partly cut off by the advancing screen. A question may arise as to how much of it may be due to light ordinarily reflected at the edges of the first aperture. With the aperture cut in cardboard, I think this part was appreciable, but the substitution of a razor-edge at the first aperture made no important difference. The strongly illuminated border must often have been seen in repetitions of Foucault's experiment, but I am not aware that it has been explained.

To examine the sudden transition from one uniform retardation to another, I used a piece of plate glass which had been etched in alternate strips with hydrofluoric acid to a depth of about $\frac{1}{4}\lambda$*. When this was set up in front of the first aperture with strips vertical, the division-lines shone out brightly, when the intervening areas were uniformly dark or nearly so. No marked difference was seen between the alternate division-lines corresponding to opposite signs of ρ. Perhaps this could hardly be expected. The whole relative retardation, reckoned as a distance, is $\frac{1}{2}\lambda$, and is thus intermediate between the values specified in Table IV. It would be of interest to make a similar experiment with a shallower etching.

[1919. For further developments reference may be made to Banerji, *Phil. Mag.* Vol. XXXVII. p. 112, 1919.]

* Compare *Nature*, Vol. LXIV. p. 385 (1901); *Scientific Papers*, Vol. IV. pp. 546, 547.

416.

TALBOT'S OBSERVATIONS ON FUSED NITRE.

[*Nature*, Vol. xcviii. p. 428, 1917.]

AMONG the little remembered writings of that remarkable man H. F. Talbot, there is an optical note in which he describes the behaviour of fused nitre (nitrate of potash) as observed under the polarizing microscope. The experiments are interesting and easily repeated by any one who has access to a suitable instrument, by preference one in which the nicols can be made to revolve together so as to maintain a dark field in the absence of any interposed crystal.

"Put a drop of a solution of nitre on a small plate of glass, and evaporate it to dryness over a spirit-lamp; then invert the glass, and hold it with the salt downwards and in contact with the flame. By this means the nitre may be brought into a state of fusion, and it will spread itself in a thin transparent film over the surface of the glass.

"Removed from the lamp it immediately solidifies, and the film in cooling cracks irregularly. As soon as the glass is cool enough, let it be placed beneath the microscope (the polariness being crossed, and the field of view consequently dark)."

I have found it better to use several drops spread over a part of the glass. And instead of inverting the plate in order to melt the nitre, I prefer to employ the flame from a mouth blow-pipe, caused to play upon the already heated salt. The blow-pipe may also be used to clean the glass in the first instance, after a preliminary heating over the flame to diminish the risk of fracture. Further security is afforded by keeping down the width of the strip, for which half an inch suffices.

Talbot describes how under the microscope there appeared crystalline plates of irregular shape, often fitted together like a tesselated pavement, each plate forming a single crystal. If one plate is darkened by rotation of the nicols, the others remain visible in varying degrees of brightness. If the plates are thin, the light is white; but with more salt they display colour, and the colour is not always uniform over the whole plate, indicating a variable thickness. But this condition of things is not permanent. After perhaps a quarter of an hour, the plates break up in a surprising fashion and the general appearance is totally changed.

Moreover the transformation may be accelerated. "Let a film of fused nitre be obtained in the manner already mentioned, and let it be allowed to cool during three or four minutes. The plate of glass should be turned round upon the stage of the microscope until the crystalline film is darkened as accurately as possible. Things being thus adjusted, let the observer touch the film with the point of a needle, while he is observing it in the microscope. He will perceive that the touch immediately produces a luminous spot on the dark surface, and this spot will slowly expand itself in all directions like a luminous wave. This is a very curious object, but difficult to describe." And further on "If however we touch it prematurely, as, for instance, during the first minute after it has become solid, this change does not take place."

I have made a few trials to ascertain whether the life-of the plates can be prolonged. Protection from atmospheric moisture did little good. Another plate kept for five hours at a temperature not much short of that of boiling water was found to have undergone transformation. But, as might be expected, a higher temperature over a diminutive gas flame acted as a safeguard, and the plate after removal behaved like one newly formed.

I have found that nitre may be replaced by *chlorate* of potash, with the advantage that the plates will keep (at any rate in an artificially warmed room) for weeks and perhaps indefinitely. The appearances are similar but less beautiful, as colour is not so often developed. The chlorate is more fusible than nitre, and the heat should not be pushed beyond what is needed for fusion.

Other salts, for example silver nitrate, which fuse in the anhydrous state without decomposition may also be employed, as is probably known to those who prepare objects for the microscope. But Talbot's early observations on nitre are rather special and deserve recall as they seem to be but little known.

417.

CUTTING AND CHIPPING OF GLASS.

[*Engineering*, Feb. 2, 1917, p. 111.]

WITH all its advantages, the division of labour, so much accentuated in modern times, tends to carry with it a regrettable division of information. Much that is familiar to theorists and experimenters in laboratories percolates slowly into the workshop, and, what is more to my present purpose, much practical knowledge gained in the workshop fails to find its way into print. At the moment I am desirous of further information on two matters relating to the working of glass in which I happen to be interested, and I am writing in the hope that some of your readers may be able to assist.

Almost the only discussion that I have seen of the cutting of glass by the diamond is a century old, by the celebrated W. H. Wollaston (*Phil. Trans.* 1816, p. 265). Wollaston's description is brief and so much to the point that it may be of service to reproduce it from the " Abstracts," p. 43 :—

"The author, having never met with a satisfactory explanation of the property which the diamond possesses of cutting glass, has endeavoured, by experiment, to determine the conditions necessary for this effect, and the mode in which it is produced. The diamonds chosen for this purpose are naturally crystallised, with curved surfaces, so that the edges are also curvilinear. In order to cut glass, a diamond of this form requires to be so placed that the surface of the glass is a tangent to a curvilinear edge, and equally inclined laterally to the two adjacent surfaces of the diamond. Under these circumstances the parts of the glass to which the diamond is applied are forced asunder, as by an obtuse wedge, to a most minute distance, without being removed ; so that a superficial and continuous crack is made from one end of the intended cut to the other. After this, any small force applied to one extremity is sufficient to extend this crack through the whole substance, and successively across the whole breadth of the glass. For since the strain at each instant in the progress of the crack is confined nearly to a mathematical point at the bottom of the fissure, the effort necessary for carrying it through is proportionately small.

"The author found by trial that the cut caused by the mere passage of the diamond need not penetrate so much as $\frac{1}{200}$ of an inch.

"He found also that other mineral bodies recently ground into the same form are also capable of cutting glass, although they cannot long retain the power, from want of the requisite hardness."

I suppose that no objection will be taken to Wollaston's general description of the action, but it may be worth while to consider it rather more closely in the light of mathematical solutions of related elastic problems by Boussinesq and Hertz; at the same time we may do well to remember Mr W. Taylor's saying that everything calculated by theorists is concerned with what happens within the elastic limit of the material, and everything done in the workshop lies beyond that limit. A good account of these theoretical investigations will be found in Love's *Elasticity*, Chap. VIII. It appears that when a pressure acts locally near a point on the plane surface of an elastic solid, the material situated along the axis is in a state of strain, which varies rapidly with the distance from the surface. The force transmitted across internal surfaces parallel to the external surface is a *pressure* all along, but the force transmitted in a perpendicular direction, although at first a pressure, at a very small distance below changes to a *tension*, which soon reaches a maximum and afterwards gradually diminishes. I suppose it is this tension which determines the crack, an action favoured by the longitudinal character of the pressure on the surface, and, once started, easily propagated as the diamond travels. Doubtless cutters of hardened steel discs, sharpened on the edge, act in a similar manner. It is possible that examination under the microscope by a skilled observer would throw light upon the matter. Among the questions which suggest themselves, one may ask whether the diamond cut necessarily involves a crushing at the surface, and what materials, besides glass, can be dealt with in this way. Would a bending force, such as is afterwards applied to separate the parts, facilitate the original formation of the crack?

The other matter in which I have been interested is the preparation of what I believe is called "chipped" glass. The only mention of it that I know is a casual one in Threlfall's *Laboratory Arts*. In an experiment tried some years ago, a glass plate was coated thickly with a warm solution of gelatine and allowed to dry on a levelling stand. Nothing particular happened afterwards for days or weeks; but eventually parts of the gelatine film lifted, carrying up with them material torn away from the glass. The plate is still in my possession, and there is now but little of the original glass surface left. If the process is in regular use, I should much like to know the precise procedure. It seems rather mysterious that a film of gelatine, scarcely thicker than thick paper, should be able to tear out fragments of solid glass, but there is no doubt of the fact.

[1919. Interesting information in response to the above will be found in *Engineering* for March 11 and 16, and April 27, 1917.]

418.

THE LE CHATELIER-BRAUN PRINCIPLE.

[*Transactions of the Chemical Society*, Vol. CXI. pp. 250—252, 1917.]

IN a paper with the above title, Ehrenfest (*Zeitsch. physikal. Chem.* 1911, **77**, 2) has shown that, as usually formulated, the principle is entirely ambiguous, and that nothing definite can be stated without a discrimination among the parameters by which the condition of a system may be defined. The typical example is that of a gas, the expansions and contractions of which may be either (α) isothermal or (β) adiabatic, and the question is a comparison of the contractions in the two cases due to an increment of pressure δp. It is known, of course, that if δp be given, the contraction $|\delta v|$ is less in case (β) than in case (α). The response of the system is said to be less in case (β), where the temperature changes spontaneously. But we need not go far to encounter an ambiguity. For if we regard δv as given instead of δp, the effect δp is now *greater* in (β) than in (α). Why are we to choose the one rather than the other as the independent variable?

When we attempt to answer this question, we are led to recognise that the treatment should commence with purely mechanical systems. The equilibrium of such a system depends on the potential energy function, and the investigation of its character presents no difficulty. Afterwards we may endeavour to extend our results to systems dependent on other, for example, thermodynamic, potentials.

As regards mechanical systems, the question may be defined as relating to the operation of *constraints*. A general treatment (*Phil. Mag.* 1875, [iv], Vol. XLIX. p. 218; *Scientific Papers*, Vol. I. p. 235: also *Theory of Sound*, §75) shows that "the introduction of a constraint has the effect of diminishing the potential energy of deformation of a system acted on by given forces; and the amount of the diminution is the potential energy of the difference of the deformations.

"For an example take the case of a horizontal rod clamped at one end and free at the other, from which a weight may be suspended at the point Q. If a constraint is applied holding a point P of the rod in its place (for example, by a support situated under it), the potential energy of the bending due to the weight at Q is less than it would be without the constraint by the potential energy of the difference of the deformations. And since the potential energy in either case is proportional to the descent of the point Q, we see that the effect of the constraint is to diminish this descent."

It may suffice here to sketch the demonstration for the case of two degrees of freedom, the results of which may, indeed, be interpreted so as to cover most of the ground. The potential energy of the system, slightly displaced from stable equilibrium at $x = 0$, $y = 0$, may be expressed

$$V = \tfrac{1}{2}ax^2 + bxy + \tfrac{1}{2}cy^2,$$

where, in virtue of the stability, a, c, and $ac - b^2$ are positive. The forces X, Y, corresponding with the displacements x, y, and necessary to maintain these displacements, are:

$$X = dV/dx = ax + by, \qquad Y = dV/dy = bx + cy.$$

If only X act, that is, if $Y = 0$, $y = - bx/c$, and

$$x = \frac{X}{a - b^2/c} \, .$$

This is the case of no constraint. On the other hand, if y is constrained to remain zero by the application of a suitable force Y, the relation between the new x (say x') and X is simply

$$x' = \frac{X}{a} \, .$$

Thus
$$\frac{x'}{x} = 1 - \frac{b^2}{ac} \, ;$$

so that x', having the same sign as x, is numerically less, or the effect of the constraint is to diminish the displacement x due to the force X. An exception occurs if $b = 0$, when $x = X/a$, whatever y and Y may be, so that the constraint has no effect.

An example, mentioned by Ehrenfest, may be taken from a cylindrical rod of elastic material subject to a longitudinal pressure, X, by which the length is shortened (x). In the first case the curved wall is free, and in the second the radius is prevented from changing by the application of a suitable pressure. The theorem asserts that in the second case the shortening due to the longitudinal pressure X is less, in virtue of the constraint applied to the walls.

Returning to the compressed gas, we now recognise that it is the pressure δp which is the *force* and $- \delta v$ the *effect*, corresponding respectively with X and x of the general theorem. But we may still feel a doubt as to which is the constrained condition, the isothermal or the adiabatic, and without a decision on this point no statement can be made. It is, however, evident that if the general theorem is applicable at all, the adiabatic condition must be regarded as the constrained one, since the response is to be *diminished* by a constraint.

The justification of this view does not seem difficult. The gas may be supposed to be confined in a cylinder under a piston, and the walls of the cylinder may be taken to be so massive as to control the temperature of the gas when undergoing slow alterations of volume. The necessary interchanges of heat take place of themselves, and the condition is one of freedom from constraint. We pass to the adiabatic condition by preventing this accommodation. The flow of heat may be stopped by the introduction of a non-conducting layer or in any other way, and the operation has the character of a constraint. Since the motion of heat in a conductor is due to differences of temperature, the former is assimilated to the displacement and the latter to the force of the purely mechanical problem. The same conclusion follows from a consideration of the thermodynamic potential.

Instead of a gas we may take a vapour in contact with liquid, say steam in contact with water. The pressure is now a function of temperature only, so that if the pressure is increased while the temperature remains unchanged, the whole of the steam is condensed, and the volume is greatly reduced. If by a constraint the outward passage of heat is prevented, the temperature rises and the reduction of volume soon ceases, in accordance with the principle.

Or again, we may suppose that the temperature is raised by a given amount, in the first case under constant pressure (no constraint), or, secondly, under constant volume, namely, with constraint. The passage of heat is less in the second case.

Electrostatic problems, governed by a potential energy function, are seen to be included under those of ordinary mechanics. Imagine two conductors, near enough to influence one another, of which the first can be connected with a battery (the other pole of which is earthed), whilst the second can be connected directly to earth. In the first case (of no constraint), the second conductor is earthed, and a certain charge enters the first conductor as the result of the battery contact. In the second case, the earth connexion of the second conductor is broken before battery contact is made. The breaking of this contact introduces a constraint, and the charge on the first conductor is reduced. In all such problems potential corresponds with force and charge corresponds with displacement.

In problems relating to steady electric currents maintained against resistance, the dissipation function takes the place of the energy function. If an electromotive force act on any branch of a network of conductors, it generates less current, and accordingly does less work, when an interruption occurs, as by breaking a contact in any part of the system.

419.

ON PERIODIC IRROTATIONAL WAVES AT THE SURFACE OF DEEP WATER.

[*Philosophical Magazine*, Vol. XXXIII. pp. 381—389, 1917.]

THE treatment of this question by Stokes, using series proceeding by ascending powers of the height of the waves, is well known. In a paper with the above title* it has been criticised rather severely by Burnside, who concludes that "these successive approximations can not be used for purposes of numerical calculation...." Further, Burnside considers that a numerical discrepancy which he encountered may be regarded as suggesting the non-existence of permanent irrotational waves. It so happens that on this point I myself expressed scepticism in an early paper†, but afterwards I accepted the existence of such waves on the later arguments of Stokes, McCowan‡, and of Korteweg and De Vries§. In 1911‖ I showed that the method of the early paper could be extended so as to obtain all the later results of Stokes.

The discrepancy that weighed with Burnside lies in the fact that the value of β (see equation (1) below) found best to satisfy the conditions in the case of $\alpha = \frac{1}{4}$ differs by about 50 per cent. from that given by Stokes' formula, viz. $\beta = -\frac{1}{2}\alpha^4$. It seems to me that too much was expected. A series proceeding by powers of $\frac{1}{4}$ need not be very convergent. One is reminded of a parallel instance in the lunar theory where the motion of the moon's apse, calculated from the first approximation, is doubled at the next step. Similarly here the next approximation largely increases the numerical value of β. When a smaller α is chosen ($\frac{1}{10}$), series developed on Stokes' plan give satisfactory results, even though they may not converge so rapidly as might be wished.

The question of the convergency of these series is distinct from that of the existence of permanent waves. Of course a strict mathematical proof of their existence is a desideratum; but I think that the reader who follows the results of the calculations here put forward is likely to be convinced that

* *Proc. Lond. Math. Soc.* Vol. xv. p. 26 (1915).
† *Phil. Mag.* Vol. I. p. 257 (1876); *Scientific Papers*, Vol. I. p. 261.
‡ *Phil. Mag.* Vol. XXXII. pp. 45, 553 (1891).
§ *Phil. Mag.* Vol. XXXIX. p. 422 (1895).
‖ *Phil. Mag.* Vol. XXI. p. 183 (1911). [This volume, p. 11.]

permanent waves of moderate height do exist. If this is so, and if Stokes' series are convergent in the mathematical sense for such heights, it appears very unlikely that the case will be altered until the wave attains the greatest admissible elevation, when, as Stokes showed, the crest comes to an edge at an angle of 120°.

It may be remarked that most of the authorities mentioned above express belief in the existence of permanent waves, even though the water be not deep, provided of course that the bottom be flat. A further question may be raised as to whether it is necessary that gravity be constant at different levels. In the paper first cited I showed that, under a gravity inversely as the cube of the distance from the bottom, *very long* waves are permanent. It may be that under a wide range of laws of gravity permanent waves exist.

Following the method of my paper of 1911, we suppose for brevity that the wave-length is 2π, the velocity of propagation unity*, and we take as the expression for the stream-function of the waves, reduced to rest,

$$\psi = y - \alpha e^{-y} \cos x - \beta e^{-2y} \cos 2x - \gamma e^{-3y} \cos 3x$$
$$- \delta e^{-4y} \cos 4x - \epsilon e^{-5y} \cos 5x, \quad \ldots\ldots(1)$$

in which x is measured horizontally and y vertically downwards. This expression evidently satisfies the differential equation to which ψ is subject, whatever may be the values of the constants α, β, &c. And, much as before, we shall find that the surface condition can be satisfied to the order of α^7 inclusive; β, γ, δ, ϵ being respectively of orders α^4, α^5, α^6, α^7.

We suppose that the free surface is the stream-line $\psi = 0$, and the constancy of pressure there imposed requires the constancy of $U^2 - 2gy$, where U, representing the resultant velocity, is equal to $\sqrt{\{(d\psi/dx)^2 + (d\psi/dy)^2\}}$, and g is the constant acceleration of gravity now to be determined. Thus when $\psi = 0$,

$$U^2 - 2gy = 1 + 2(1-g)y + \alpha^2 e^{-2y} + 2\beta e^{-2y} \cos 2x$$
$$+ 4\gamma e^{-3y} \cos 3x + 6\delta e^{-4y} \cos 4x + 8\epsilon e^{-5y} \cos 5x$$
$$+ 4\alpha\beta e^{-3y} \cos x + 6\alpha\gamma e^{-4y} \cos 2x + 8\alpha\delta e^{-5y} \cos 3x \quad \ldots\ldots(2)$$

correct to α^7 inclusive. On the right of (2) we have to expand the exponentials and substitute for the various powers of y expressions in terms of x.

It may be well to reproduce the process as formerly given, omitting δ and ϵ, and carrying (2) only to the order α^5. We have from (1) as successive approximations to y:—

$$y = \alpha e^{-y} \cos x = \alpha \cos x; \quad \ldots\ldots\ldots\ldots\ldots\ldots(3)$$

---

* The extension to arbitrary wave-lengths and velocities may be effected at any time by attention to dimensions.

$$y = \alpha(1-y)\cos x = -\tfrac{1}{2}\alpha^2 + \alpha\cos x - \tfrac{1}{2}\alpha^2\cos 2x; \quad \ldots\ldots\ldots\ldots(4)$$

$$y = \alpha(1 - y + \tfrac{1}{2}y^2)\cos x$$
$$= -\tfrac{1}{2}\alpha^2 + \alpha(1 + \tfrac{9}{8}\alpha^2)\cos x - \tfrac{1}{2}\alpha^2\cos 2x + \tfrac{3}{8}\alpha^3\cos 3x, \quad \ldots\ldots(5)$$

which is correct to α^3 inclusive, β being of order α^4. In calculating (2) to the approximation now intended we omit the term in $\alpha\gamma$. In association with $\alpha\beta$ and γ we take $e^{-3y} = 1$; in association with β, $e^{-2y} = 1 - 2y$; while

$$\alpha^2 e^{-2y} = \alpha^2(1 - 2y + 2y^2 - \tfrac{4}{3}y^3).$$

Thus on substitution for y^2 and y^3 from (5)

$$\alpha^2 e^{-2y} = \alpha^2\{1 - 2y + \alpha^2 - 4\alpha^3\cos x + \alpha^2\cos 2x - \tfrac{4}{3}\alpha^3\cos 3x\}.$$

In like manner

$$2\beta e^{-2y}\cos 2x = 2\beta\cos 2x - 2\alpha\beta(\cos x + \cos 3x).$$

Since the terms in $\cos x$ are of the fifth order, we may replace $\alpha\cos x$ by y, and we get

$$U^2 - 2gy = 1 + \alpha^2 + \alpha^4 + 2y(1 - g - \alpha^2 - 2\alpha^4 + \beta)$$
$$+ (\alpha^4 + 2\beta)\cos 2x + (-\tfrac{4}{3}\alpha^5 + 4\gamma - 2\alpha\beta)\cos 3x. \quad \ldots\ldots(6)$$

The constancy of (6) requires the annulment of the coefficients of y and of $\cos 2x$ and $\cos 3x$, so that

$$\beta = -\tfrac{1}{2}\alpha^4, \qquad \gamma = \tfrac{1}{12}\alpha^5, \quad \ldots\ldots\ldots\ldots\ldots\ldots\ldots(7)$$
and
$$g = 1 - \alpha^2 - \tfrac{5}{2}\alpha^4. \quad \ldots\ldots\ldots\ldots\ldots\ldots\ldots\ldots\ldots(8)$$

The value of g in (8) differs from that expressed in equation (11) of my former paper. The cause is to be found in the difference of suppositions with respect to ψ. Here we have taken $\psi = 0$ at the free surface, which leads to a constant term in the expression for y, as seen in (5), while formerly the constant term was made to disappear by a different choice of ψ.

There is no essential difficulty in carrying the approximation to y two stages further than is attained in (5). If δ, ϵ are of the 6th and 7th order, they do not appear. The longest part of the work is the expression of e^{-y} as a function of x. We get

$$e^{-y} = 1 + \frac{3\alpha^2}{4} + \frac{125\alpha^4}{64} - \cos x\{\alpha + 2\alpha^3\}$$
$$+ \cos 2x\left\{\frac{3\alpha^2}{4} + \frac{125\alpha^4}{48} - \beta\right\} - \frac{2\alpha^3}{3}\cos 3x + \frac{125\alpha^4}{192}\cos 4x, \ldots(9)$$

and thence from (1)

$$y = -\tfrac{1}{2}\alpha^2 - \alpha^4 + \cos x\left\{\alpha + \frac{9\alpha^3}{8} + \frac{625\alpha^5}{192} - \frac{3\alpha\beta}{2}\right\}$$
$$- \cos 2x\left\{\tfrac{1}{2}\alpha^2 + \frac{4\alpha^4}{3} - \beta\right\} + \cos 3x\left\{\frac{3\alpha^3}{8} + \frac{625\alpha^5}{384} - \frac{3\alpha\beta}{2} + \gamma\right\}$$
$$- \frac{\alpha^4}{3}\cos 4x + \frac{125\alpha^5}{384}\cos 5x. \quad \ldots\ldots\ldots\ldots\ldots\ldots\ldots\ldots(10)$$

When we introduce the values of β and γ, already determined in (7) with sufficient approximation, we have

$$y = -\tfrac{1}{2}\alpha^2 - \alpha^4 + \cos x \left\{ \alpha + \frac{9\alpha^3}{8} + \frac{769\alpha^5}{192} \right\}$$

$$- \cos 2x \left\{ \frac{\alpha^2}{2} + \frac{11\alpha^4}{6} \right\} + \cos 3x \left\{ \frac{3\alpha^3}{8} + \frac{315\alpha^5}{128} \right\} - \frac{\alpha^4}{3}\cos 4x + \frac{125\alpha^5}{384}\cos 5x, \ldots (11)$$

in agreement with equations (13), (18) of my former paper when allowance is made for the different suppositions with respect to ψ, as may be effected by expressing both results in terms of a, the coefficient of $\cos x$, instead of α.

The next step is the further development of the pressure equation (2), so as to include terms of the order α^7. Where β, γ, etc. occur as factors, the expression for y to the third order, as in (5), suffices; but a more accurate value is required in $\alpha^2 e^{-2y}$. Expanding the exponentials and replacing products of cosines by cosines of sums and differences, we find in the first place

$$U^2 - 2gy = 2(1 - g - \alpha^2)\,y + 1 + \alpha^2 + \alpha^4 + \frac{19\alpha^6}{4} - 4\alpha^2\beta$$

$$+ \cos x \left\{ -4\alpha^5 + 2\alpha\beta - \frac{37\alpha^7}{2} + \frac{97\alpha^3\beta}{6} - \frac{9\alpha^2\gamma}{2} \right\}$$

$$+ \cos 2x \left\{ \alpha^4 + 2\beta + \frac{19\alpha^6}{3} - 2\alpha^2\beta \right\}$$

$$+ \cos 3x \left\{ -\frac{4\alpha^5}{3} - 2\alpha\beta + 4\gamma - \frac{37\alpha^7}{4} + \frac{13\alpha^3\beta}{4} + 3\alpha^2\gamma - 4\alpha\delta \right\}$$

$$+ \cos 4x \left\{ \frac{19\alpha^6}{12} + 2\alpha^2\beta - 6\alpha\gamma + 6\delta \right\}$$

$$+ \cos 5x \left\{ -\frac{37\alpha^7}{20} - \frac{25\alpha^3\beta}{12} + \frac{15\alpha^2\gamma}{2} - 12\alpha\delta + 8\epsilon \right\}. \ldots\ldots\ldots\ldots (12)*$$

From the terms in $\cos x$ we now eliminate $\cos x$ by means of

$$\alpha \cos x = y(1 - \tfrac{9}{8}\alpha^2) + \tfrac{1}{2}\alpha^2 + \tfrac{1}{2}\alpha^2 \cos 2x\dagger,$$

thus altering those terms of (12) which are constant, and which contain y and $\cos 2x$. Thus modified, (12) becomes

$$U^2 - 2gy = 1 + \alpha^2 + \alpha^4 + \frac{11\alpha^6}{4} - 3\alpha^2\beta$$

$$+ 2y \left\{ 1 - g - \alpha^2 - 2\alpha^4 + \beta - 7\alpha^6 + \frac{167\alpha^2\beta}{24} - \frac{9\alpha\gamma}{4} \right\}$$

$$+ \cos 2x \left\{ \alpha^4 + 2\beta + \frac{13\alpha^6}{3} - \alpha^2\beta \right\}$$

$$+ \cos 3x \left\{ -\frac{4\alpha^5}{3} - 2\alpha\beta + 4\gamma - \frac{37\alpha^7}{4} + \frac{13\alpha^3\beta}{4} + 3\alpha^2\gamma - 4\alpha\delta \right\}$$

[* The terms in $\alpha^3\beta\,(\cos x,\ \cos 3x)$ should read $+\frac{85}{6}\alpha^3\beta\cos x$, $+\frac{5}{4}\alpha^3\beta\cos 3x$; apparently the term $-4\alpha^3\beta\cos x\cos 2x$ had been omitted from the development of $2\beta e^{-2y}\cos 2x$.

† Since terms of order α^7 are retained, the term $-\frac{3}{8}\alpha^3\cos 3x$ should be added to the expression for $\alpha\cos x$. W. F. S.]

$$+ \cos 4x \left\{ \frac{19\alpha^6}{12} + 2\alpha^2\beta - 6\alpha\gamma + 6\delta \right\}$$

$$+ \cos 5x \left\{ -\frac{37\alpha^7}{20} - \frac{25\alpha^3\beta}{12} + \frac{15\alpha^2\gamma}{2} - 12\alpha\delta + 8\epsilon \right\}. \quad\ldots\ldots\ldots\ldots(13)^*$$

The constant part has no significance for our purpose, and the term in y can be made to vanish by a proper choice of g.

If we use only α, none of the cosines can be made to disappear, and the value of g is

$$g = 1 - \alpha^2 - 2\alpha^4 - 7\alpha^6. \quad\ldots\ldots\ldots\ldots\ldots\ldots(14)$$

When we include also β, we can annul the term in $\cos 2x$ by making

$$\beta = -\frac{\alpha^4}{2}\left(1 + \frac{29\alpha^2}{6}\right), \quad\ldots\ldots\ldots\ldots\ldots\ldots(15)$$

and with this value of β

$$g = 1 - \alpha^2 - \frac{5\alpha^4}{2} - \frac{619\alpha^6}{48}. \quad\ldots\ldots\ldots\ldots\ldots(16)^*$$

But unless α is very small, regard to the term in $\cos 3x$ suggests a higher value of β as the more favourable on the whole.

With the further aid of γ we can annul the terms both in $\cos 2x$ and in $\cos 3x$. The value of β is as before. That of γ is given by

$$\gamma = \frac{\alpha^5}{12}\left(1 + \frac{139\alpha^2}{8}\right), \quad\ldots\ldots\ldots\ldots\ldots\ldots(17)^*$$

and with this is associated

$$g = 1 - \alpha^2 - \frac{5\alpha^4}{2} - \frac{157\alpha^6}{12}. \quad\ldots\ldots\ldots\ldots\ldots(18)^*$$

The inclusion of δ and ϵ does not alter the value of g in this order of approximation, but it allows us to annul the terms in $\cos 4x$ and $\cos 5x$. The appropriate values are

$$\delta = -\frac{\alpha^6}{72}, \qquad \epsilon = \frac{\alpha^7}{480}, \quad\ldots\ldots\ldots\ldots\ldots\ldots(19)$$

and the accompanying value of γ is given by

$$\gamma = \frac{\alpha^5}{12}\left(1 + \frac{413\alpha^2}{24}\right), \quad\ldots\ldots\ldots\ldots\ldots(20)^*$$

while β remains as in (15).

We now proceed to consider how far these approximations are successful, for which purpose we must choose a value for α. Prof. Burnside took $\alpha = \frac{1}{4}$. With this value the second term of β in (15) is nearly one-third of the first (Stokes') term, and the second term of γ in (20) is actually larger* than the

[* With the alterations specified in the footnotes on p. 481, the terms in (13) involving $\alpha^2\beta y$, and $(\alpha^7, \alpha^3\beta) \cos 3x$, become $2y \cdot \frac{143}{24}\alpha^2\beta$, and $\cos 3x \left(-\frac{31}{4}\alpha^7 + \frac{1}{2}\alpha^3\beta\right)$. Then the highest terms in (16), (17), (18), and (20) become respectively $-\frac{595}{48}\alpha^6$, $\frac{\alpha^5}{12}\left(+\frac{35}{4}\alpha^2\right)$, $-\frac{151}{12}\alpha^6$, and $\frac{\alpha^5}{12}\left(+\frac{103}{12}\alpha^2\right)$; the second term in (20) being now little more than half the first when $\alpha = \frac{1}{4}$. W. F. S.]

first. If the series are to be depended upon, we must clearly take a smaller value. I have chosen $\alpha = \frac{1}{10}$, and this makes by (15), (18), (20)

$$\beta = - \cdot000,052,42, \quad \gamma = \cdot000,000,976, \quad g = \cdot989,736,92. \ldots\ldots(21)*$$

The next step is the calculation of approximate values of y from (11), which now takes the form

$$y = - \cdot0051 + \cdot101,165,0 \cos x$$
$$- \cdot005,183,3 \cos 2x + \cdot000,399,6 \cos 3x$$
$$- \cdot000,033,3 \cos 4x + \cdot000,003,3 \cos 5x. \ldots\ldots\ldots(22)$$

For example, when $x = 0$, $y = \cdot091,251,3$. The values of y calculated from (22) at steps of $22\frac{1}{2}°$ (as in Burnside's work) are shown in column 2 of Table I.

We have next to examine how nearly the value of y afforded by (22) really makes ψ vanish, and if necessary to calculate corrections. To this δ and ϵ in (1) do not contribute sensibly and we find $\psi = + \cdot000,015,4$ for $x = 0$. In order to reduce ψ to zero, we must correct the value of y. With sufficient approximation we have in general

$$\delta\psi = \delta y \left(1 + \tfrac{1}{10} e^{-y} \cos x\right)\dagger,$$

or in the present case

$$\delta y = - \frac{\cdot000,015,4}{1\cdot091} = - \cdot000,014,1,$$

so that the corrected value of y for $x = 0$ is $\cdot091,237,2$. If we repeat the calculation, using the new value of y, we find $\psi = 0$.

TABLE I.‡

x	y from (22)	y corrected	$U^2 - 2gy - 1$	Corrected by $\delta\beta$
0	+ ·091,251,3	+ ·091,237,2	·010,104,9	45
22½	+ ·084,839,7	+ ·084,841,9	 4,7	44
45	+ ·066,182,8	+ ·066,181,8	 4,3	43
67½	+ ·036,913,1	+ ·036,915,1	 4,1	44
90	+ ·000,050,0	+ ·000,052,4	 4,2	46
112½	− ·039,782,7	− ·039,780,2	 4,4	47
135	− ·076,316,2	− ·076,317,5	 4,3	43
157½	− ·102,381,1	− ·102,395,1	 4,7	44
180	− ·111,884,7	− ·111,907,9	·010,105,1	47

[* With the corrections specified in the footnote on p. 482 we have $\gamma = \cdot000,000,905$, $g = \cdot989,737,42$. W. F. S.]

† The double use of δ will hardly cause confusion.

[‡ With the corrections specified in the footnotes on pp. 481, 482, and calculating direct from (2), with the inclusion of the term $6\delta e^{-4y} \cos 4x$, I find that the first 5 figures in the value of $U^2 - 2gy - 1$ are as in the table, whilst the last 2 figures, proceeding in order from $x = 0$ to $x = 180$, become 45, 45, 44, 43, 42, 42, 45, 51, 53; after making 6 modifications in "y corrected" (third column), the first 6 figures of which remain as printed, whilst the last becomes, taken in the same order, 1, 9, 9, 1, 4, 3, 6, 3, 8, these modified values of y in every case reducing ψ to zero to 7 places of decimals. W. F. S.]

In the fourth column are recorded the values of $U^2 - 2gy - 1$, calculated from (1) with omission of δ and ϵ, and with the corrected values of y. $d\psi/dx$, $d\psi/dy$ were first found separately, and then U^2 as the sum of the two squares. The values of β, γ, g employed are those given in (15), (18), (20). The form of ψ in (1) with these values of the constants vanishes when y takes the values of the third column, and the pressure at the surface is also constant to a high degree of approximation. The greatest difference is ($\cdot$000,001,0), which may be compared with $\cdot$4, the latter amount representing the corresponding statical difference at the crest and trough of the wave. According to this standard the pressure at the surface is constant to $2\frac{1}{2}$ parts in a million*.

The advantage gained by the introduction of β and γ will be better estimated by comparison with a similar calculation where only α (still equal to $\frac{1}{10}$) and g are retained. By (2) in this case

$$U^2 - 2gy - 1 = \alpha^2 e^{-2y} + 2(1 - g)y. \quad\ldots\ldots\ldots\ldots(23)$$

Table II shows the values of y and of $\alpha^2 e^{-2y}$ corresponding to the same values of x as before. The fourth column gives (23) when g is so determined as to make the values equal at $0°$ and $180°$. It appears that the discrepancy in the values of $U^2 - 2gy$ is reduced 200 times by the introduction of β and γ, even when we tie ourselves to the values of β, γ, g prescribed by approximations on the lines of Stokes.

TABLE II.

x	y	$\alpha^2 e^{-2y}$	$U^2 - 2gy - 1$
0	$+ \cdot$091,276,5	$\cdot$008,331,4	$\cdot$010,207,7
$22\frac{1}{2}$	$\cdot$084,870,5	$\cdot$008,438,8	$\ldots$183,4
45	$\cdot$066,182,4	$\cdot$008,760,2	$\ldots$120,7
$67\frac{1}{2}$	$\cdot$036,882,6	$\cdot$009,288,9	$\ldots$047,1
90	0	$\cdot$010,000,0	$\ldots$000,0
$112\frac{1}{2}$	$- \cdot$039,823,1	$\cdot$010,829,0	$\ldots$010,4
135	$- \cdot$076,318,5	$\cdot$011,649,0	$\ldots$080,2
$157\frac{1}{2}$	$- \cdot$102,344,1	$\cdot$012,271,4	$\ldots$167,6
180	$- \cdot$111,832,6	$\cdot$012,506,5	$\cdot$010,207,7

A cursory inspection of the numbers in column 4 of Table I suffices to show that an improvement can be effected by a slight alteration in the value of β. For small corrections of this kind it is convenient to use a formula which may be derived from (2). We suppose that while α and ψ are maintained constant, small alterations $\delta\beta$, $\delta\gamma$, δg are incurred. Neglecting the small variations of β, γ, g when multiplied by α^2 and higher powers of α, we get

$$\delta y = \delta\beta \left\{\cos 2x - \tfrac{3}{2}\alpha \cos x - \tfrac{3}{2}\alpha \cos 3x\right\}$$
$$+ \delta\gamma \left\{\cos 3x - 2\alpha \cos 2x - 2\alpha \cos 4x\right\}, \quad\ldots\ldots\ldots\ldots(24)$$

[* With the alterations specified in footnote $\ddagger$ on p. 483, the greatest difference becomes $\cdot$000,001,1, so that the surface pressure is constant to $2\frac{3}{4}$ parts in a million. W. F. S.]

and $\delta \left(U^{2} - 2gy \right) = 2\alpha \left(\delta\beta - \delta g \right) \cos x + 2\delta\beta \cos 2x$

$$+ 2 \left(2\delta\gamma - \alpha\delta\beta \right) \cos 3x - 6\alpha\delta\gamma \cos 4x. \quad\ldots\ldots\ldots\ldots(25)$$

For the present purpose we need only to introduce $\delta\beta$, and with sufficient accuracy we may take

$$\delta \left(U^{2} - 2gy \right) = 2\delta\beta \cos 2x. \quad\ldots\ldots\ldots\ldots\ldots(26)$$

We suppose $\delta\beta = -\ \cdot000,000,2$, so that the new value of β is $-\ \cdot000,052,6$. Introducing corrections according to (26) and writing only the last two figures, we obtain column 5 of Table I, in which the greatest discrepancy is reduced from 10 to 4—almost as far as the arithmetic allows—and becomes but one-millionth of the statical difference between crest and trough. This is the degree of accuracy attained when we take simply

$$\psi = y - \alpha e^{-y} \cos x - \beta e^{-2y} \cos 2x - \gamma e^{-3y} \cos 3x, \ldots\ldots\ldots\ldots(27)$$

with $\alpha = \frac{1}{10}$, g and γ determined by Stokes' method, and β determined so as to give the best agreement*.

[1919.   Reference may be made to Wilton, *Phil. Mag.* Vol. 27, p. 385, 1914; also to Havelock, *Roy. Soc. Proc.*, Vol. A 95, p. 38, 1918.]

[* If we include the first 3 terms of (25), and write

$$\delta \left(U^{2} - 2gy \right) = \cdot000,000,2 \cos x - \cdot000,000,4 \cos 2x + \cdot000,000,2 \cos 3x,$$

corresponding to $\delta\beta = -\ \cdot000,000,2$, $\delta\gamma = +\ \cdot000,000,04$, $\delta g = -\ \cdot000,001,2$, we find that the corrected values of the last two figures of $U^{2} - 2gy - 1$, given in footnote ‡ on p. 483, become 45, 45, 44, 45, 46, 46, 45, 46, 45, taken in the same order; these results would not be affected by including the term in (25) involving $\cos 4x$. Thus the greatest discrepancy is reduced from 11 to 2, becoming only half one-millionth of the statical difference. The new values of β, γ, and g, thus determined so as to give the best agreement, are $\beta = -\ \cdot000,052,6$, $\gamma = \cdot000,000,94$, $g = \cdot989,736,2$.   W. F. S.]

420.

ON THE SUGGESTED ANALOGY BETWEEN THE CONDUCTION OF HEAT AND MOMENTUM DURING THE TURBULENT MOTION OF A FLUID.

[*Advisory Committee for Aeronautics*, T. 941, 1917.]

THE idea that the passage of heat from solids to liquids moving past them is governed by the same principles as apply in virtue of viscosity to the passage of *momentum*, originated with Reynolds (*Manchester Proc.*, 1874); and it has been further developed by Stanton (*Phil. Trans.*, Vol. CXC. p. 67, 1897; *Tech. Rep. Adv. Committee*, 1912–13, p. 45) and Lanchester (same *Report*, p. 40). Both these writers express some doubt as to the exactitude of the analogy, or at any rate of the proofs which have been given of it. The object of the present note is to show definitely that the analogy is not complete.

The problem which is the simplest, and presumably the most favourable to the analogy, is that of fluid enclosed between two parallel plane solid surfaces. One of these surfaces at $y = 0$ is supposed to be fixed, while the other at $y = 1$ moves in the direction of x in its own plane with unit velocity. If the motion of the fluid is in plane strata, as would happen if the viscosity were high enough, the velocity u in permanent régime of any stratum y is represented by y simply. And by definition, if the viscosity be unity, the tangential traction per unit area on the bounding planes is also unity.

Let us now suppose that the fixed surface is maintained at temperature 0, and the moving surface at temperature 1. So long as the motion is stratified, the flow of heat is the same as if the fluid were at rest, and the temperature (θ) at any stratum y has the same value y as has u. If the conductivity is unity, the passage of heat per unit area and unit time is also unity. In this case, the analogy under examination is seen to be complete. The question is—will it still hold when the motion becomes turbulent? It appears that the identity in the values of θ and u then fails.

The equations for the motion of the fluid when there are no impressed forces are

$$\frac{Du}{Dt} = -\frac{1}{\rho}\frac{dp}{dx} + \nu\nabla^2 u,$$

with two similar equations, where

$$\frac{D}{Dt} = \frac{d}{dt} + u\frac{d}{dx} + v\frac{d}{dy} + w\frac{d}{dz},$$

representing differentiation with respect to time when a particle of the fluid is followed.

In like manner, the equation for the conduction of heat is

$$\frac{D\theta}{Dt} = k\nabla^2\theta.$$

Although we identify the values of k and ν, and impose the same boundary conditions upon u and θ, we see that the same values will not serve for both u and θ in the interior of the fluid on account of the term in dp/dx, which is not everywhere zero.

It is to be observed that turbulent motion is not *steady* in the hydrodynamical sense, and that a uniform régime can be spoken of only when we contemplate averages of u and θ for all values of x or for all values of t. It is conceivable that, although there is no equality between the passage of heat and the tangential traction at a particular time and place, yet that the *average* values of these quantities might still be equal. This question must for the present remain open, but the suggested equality does not seem probable.

The principle of similitude may be applied in the present problem to find a general form for H, the heat transmitted per unit area and per unit time (compare *Nature*, Vol. xcv. p. 67, 1915)*. In the same notation as there used, let a be the distance between the planes, v the mean velocity of the stream, θ the temperature difference between the planes, κ the conductivity of the fluid, c the capacity for heat per unit volume, ν the kinematic viscosity. Then

$$H = \frac{\kappa\theta}{a} \cdot F\left(\frac{avc}{\kappa}, \frac{c\nu}{\kappa}\right),$$

or, which comes to the same,

$$H = \frac{\kappa\theta}{a} \cdot F_1\left(\frac{av}{\nu}, \frac{c\nu}{\kappa}\right),$$

where F, F_1 denote arbitrary functions of two variables. When $v = 0$, $F(0, c\nu/\kappa) = 1$.

For a given fluid $c\nu/\kappa$ is constant and may be omitted. Dynamical similarity is attained when av is constant, so that a complete determination of F (experimentally or otherwise) does not require the variation of *both* a and v. There is advantage in keeping a constant; for if a be varied, geometrical similarity demands that any roughnesses shall be in proportion.

The objection that κ, c, ν are not constants, but functions of the temperature, may be obviated by supposing that θ is small.

[* This volume, p. 300.]

421.

THE THEORY OF ANOMALOUS DISPERSION.

[*Philosophical Magazine*, Vol. XXXIII. pp. 496—499, 1917.]

IN a short note* with the above title I pointed out that Maxwell as early as 1869 in a published examination paper had given the appropriate formulæ, thus anticipating the work of Sellmeier† and Helmholtz‡. It will easily be understood that the German writers were unacquainted with Maxwell's formulæ, which indeed seem to have been little known even in England. I have thought that it would be of more than historical interest to examine the relation between Maxwell's and Helmholtz's work. It appears that the generalization attempted by the latter is nugatory, unless we are prepared to accept a refractive index in the dispersive medium becoming infinite with the wave-length in vacuo.

In the æther the equation of plane waves propagated in the direction of x is in Maxwell's notation

$$\rho\, d^2\eta/dt^2 = E\, d^2\eta/dx^2, \quad\quad\quad\quad\quad\quad (1)$$

where η is the transverse displacement at any point x and time t, ρ is the density and E the coefficient of elasticity. Maxwell supposes "that every part of this medium is connected with an atom of other matter by an attractive force varying as distance, and that there is also a force of resistance between the medium and the atoms varying as their relative velocity, the atoms being independent of each other"; and he shows that the equations of propagation in this compound medium are

$$\rho\frac{d^2\eta}{dt^2} - E\frac{d^2\eta}{dx^2} = \sigma\left(p^2\zeta + R\frac{d\zeta}{dt}\right) = -\sigma\left(\frac{d^2\eta}{dt^2} + \frac{d^2\zeta}{dt^2}\right), \quad\quad (2)$$

where ρ and σ are the quantities of the medium and of the atoms respectively in unit of volume, η is the displacement of the medium, and $\eta + \zeta$ that of the atoms, $\sigma p^2\zeta$ is the attraction, and $\sigma R\, d\zeta/dt$ is the resistance to the relative motion per unit of volume.

---

* *Phil. Mag.* Vol. XLVIII. p. 151 (1899); *Scientific Papers*, Vol. IV. p. 413. A misprint is now corrected, see (4) below.

† *Pogg. Ann.* CXLIII. p. 272 (1871).

‡ *Pogg. Ann.* CLIV. p. 582 (1874); *Wissenschaftliche Abhandlungen*, Band II. p. 213.

On the assumption that

$$\eta, \zeta = (C, D)\, e^{int} e^{-(1/l + in/v)x}, \quad \dots\dots\dots\dots\dots(3)$$

we get Maxwell's results*

$$\frac{1}{v^2} - \frac{1}{l^2 n^2} = \frac{\rho + \sigma}{E} + \frac{\sigma n^2}{E}\, \frac{p^2 - n^2}{(p^2 - n^2)^2 + R^2 n^2}, \quad \dots\dots\dots(4)$$

$$\frac{2}{vln} = \frac{\sigma n^2}{E}\, \frac{Rn}{(p^2 - n^2)^2 + R^2 n^2}. \quad \dots\dots\dots\dots(5)$$

Here v is the velocity of propagation of phase, and l is the distance the waves must run in order that the amplitude of vibration may be reduced in the ratio $e : 1$.

When we suppose that $R = 0$, and consequently that $l = \infty$, (4) simplifies. If v_0 be the velocity in æther ($\sigma = 0$), and ν be the refractive index,

$$\nu^2 = \frac{v_0^2}{v^2} = 1 + \frac{\sigma}{\rho}\, \frac{p^2}{p^2 - n^2}. \quad \dots\dots\dots\dots\dots(6)$$

For comparison with experiment, results are often conveniently expressed in terms of the wave-lengths in free æther corresponding with the frequencies in question. Thus, if λ correspond with n and Λ with p, (6) may be written

$$\nu^2 = 1 + \frac{\sigma}{\rho}\, \frac{\lambda^2}{\lambda^2 - \Lambda^2}, \quad \dots\dots\dots\dots\dots(7)$$

—the dispersion formula commonly named after Sellmeier. It will be observed that p, Λ refer to the vibrations which the atoms might freely execute when the æther is maintained at rest ($\eta = 0$).

If we suppose that n is infinitely small, or λ infinitely great,

$$\nu_\infty^2 = 1 + \sigma/\rho, \quad \dots\dots\dots\dots\dots\dots(8)$$

thus remaining finite.

Helmholtz in his investigation also introduces a dissipative force, as is necessary to avoid infinities when $n = p$, but one differing from Maxwell's, in that it is dependent upon the absolute velocity of the atoms instead of upon the *relative* velocity of æther and matter. A more important difference is the introduction of an additional force of restitution ($a^2 x$), proportional to the absolute displacement of the atoms. His equations are

$$\mu \frac{d^2 \xi}{dt^2} = a^2 \frac{d^2 \xi}{dy^2} + \beta^2 (x - \xi), \quad \dots\dots\dots\dots(9)\dagger$$

$$m \frac{d^2 x}{dt^2} = \beta^2 (\xi - x) - a^2 x - \gamma^2 \frac{dx}{dt}. \quad \dots\dots\dots(10)$$

---

* Thus in Maxwell's original statement. In my quotation of 1899 the sign of the second term in (4) was erroneously given as *plus*.

† What was doubtless meant to be $d^2\xi/dy^2$ appears as $d^2\xi/dx^2$, bringing in x in two senses.

This notation is so different from Maxwell's, that it may be well to exhibit explicitly the correspondence of symbols.

Helmholtz...	ξ	μ	a^2	y	$x-\xi$	β^2	m	a^2	c	k
Maxwell......	η	ρ	E	x	ζ	σp^2	σ	0	v	$1/l$

When there is no dissipation ($R = 0$, $\gamma^2 = 0$), these interchanges harmonize the two pairs of equations. The terms involving respectively R and γ^2 follow different laws.

Similarly Helmholtz's results

$$\frac{1}{c^2} - \frac{k^2}{n^2} = \frac{\mu}{a^2} - \frac{\beta^2}{a^2 n^2} - \frac{\beta^4}{a^2 n^2} \frac{mn^2 - a^2 - \beta^2}{(mn^2 - a^2 - \beta^2)^2 + \gamma^4 n^2}, \quad \dots\dots\dots(11)$$

$$\frac{2k}{cn} = -\frac{\beta^4 \gamma^2}{a^2 n} \frac{1}{(mn^2 - a^2 - \beta^2)^2 + \gamma^4 n^2}, \quad \dots\dots\dots\dots(12)*$$

identify themselves with Maxwell's, when we omit R and γ^2 and make $a^2 = 0$.

In order to examine the effect of a^2, we see that when $\gamma = 0$, (11) becomes

$$\frac{1}{c^2} = \frac{\mu}{a^2} - \frac{\beta^2}{a^2 n^2} \frac{mn^2 - a^2}{mn^2 - a^2 - \beta^2}, \quad \dots\dots\dots\dots(13)$$

or in terms of $\nu^2 (= c_0^2/c^2)$,

$$\nu^2 = 1 - \frac{\beta^2}{\mu} \frac{m - a^2/n^2}{mn^2 - a^2 - \beta^2}. \quad \dots\dots\dots\dots(14)$$

If now in (14) we suppose $n = 0$, or $\lambda = \infty$, we find that $\nu = \infty$, unless $a^2 = 0$. If $a^2 = 0$, we get, in harmony with (6),

$$\nu^2 = 1 - \frac{m\beta^2}{\mu (mn^2 - \beta^2)}, \quad \dots\dots\dots\dots\dots(15)$$

which is finite, unless $mn^2 = \beta^2$. It is singular that Helmholtz makes precisely opposite statements† :—" Wenn $a = 0$, wird $k = 0$ und $1/c = \infty$; sonst werden beide Werthe endlich sein."

The same conclusion may be deduced immediately from the original equations (9), (10). For if the frequency be zero and the velocity of propagation in the medium finite, all the differential coefficients may be omitted; so that (9) requires $x - \xi = 0$ and (10) then gives $a^2 = 0$.

Wüllner‡, retaining a^2 in Helmholtz's equation, writes (14) in the form

$$\nu^2 = 1 - P\lambda^2 + \frac{Q\lambda^4}{\lambda^2 - \Lambda^2}, \quad \dots\dots\dots\dots\dots(16)$$

[* The result (12) is so given by Helmholtz; but the first " $-$ " should be " $+$ ", involving some further corrections in Helmholtz's paper.

† Helmholtz, however, supposes $\gamma \neq 0$, and on that supposition his statements appear to be correct. They cannot, however, legitimately be deduced, as appears to be assumed by Helmholtz, from the equations which in his paper immediately precede those statements, since those equations are obtained on the understanding that the ratio of the right-hand side of (12) to that of (11) is zero when $n = 0$, which is not the case when a absolutely $= 0$. W. F. S.]

‡ Wied. Ann. xvii. p. 580; xxiii. p. 306.

applicable when there is no absorption. And he finds that in many cases the facts of observation require us to suppose $P = Q$. This is obviously the condition that v^2 shall remain finite when $\lambda = \infty$, and it requires that a^2 in Helmholtz's equation be zero. It is true that in some cases a better agreement with observation may be obtained by allowing Q to differ slightly from P, but this circumstance is of little significance. The introduction of a new arbitrary constant into an empirical formula will naturally effect some improvement over a limited range.

It remains to consider whether *a priori* we have grounds for the assumption that v is finite when $\lambda = \infty$. On the electromagnetic theory this should certainly be the case. Moreover, an infinite refractive index must entail *complete* reflexion when radiation falls upon the substance, even at perpendicular incidence. So far as observation goes, there is no reason for thinking that dark heat is so reflected. It would seem then that the introduction of a^2 is a step in the wrong direction and that Helmholtz's formulæ are no improvement upon Maxwell's *.

It is scarcely necessary to add that the full development of these ideas requires the recognition of more than one resonance as admissible (Sellmeier).

[* Similarly, the substitution of a dissipative force " dependent upon the absolute velocity of the atoms instead of upon the *relative* velocity of æther and matter " (p. 489 above) appears to be the reverse of an improvement, since Maxwell's results (4) and (5) above lead to a finite v when $n = 0$, but $R \neq 0$ (cf. p. 490 and footnote †).   W. F. S.]

422.

ON THE REFLECTION OF LIGHT FROM A REGULARLY STRATIFIED MEDIUM.

[*Proceedings of the Royal Society*, A, Vol. XCIII. pp. 565—577, 1917.]

THE remarkable coloured reflection from certain crystals of chlorate of potash described by Stokes[*], the colours of old decomposed glass, and probably those of some beetles and butterflies, lend interest to the calculation of reflection from a regular stratification, in which the alternate strata, each uniform and of constant thickness, differ in refractivity. The higher the number of strata, supposed perfectly regular, the nearer is the approach to homogeneity in the light of the favoured wave-lengths. In a crystal of chlorate described by R. W. Wood, the purity observed would require some 700 alternations combined with a very high degree of regularity. A general idea of what is to be expected may be arrived at by considering the case where a single reflection is very feeble, but when the component reflections are more vigorous, or when the number of alternations is very great, a more detailed examination is required. Such is the aim of the present communication.

The calculation of the aggregate reflection and transmission by a single parallel plate of transparent material has long been known, but it may be convenient to recapitulate it. At each reflection or refraction the amplitude of the incident wave is supposed to be altered by a certain factor. When the light proceeds at A from the surrounding medium to the plate, the factor for reflection will be supposed to be b', and for refraction c'; the corresponding quantities when the progress at B is from the plate to the surrounding medium may be denoted by e', f'. Denoting the incident vibration by unity, we have then for the first component of the reflected wave b', for the second $c'e' f'e^{-ik\delta}$, for the third $c'e'^3 f'e^{-2ik\delta}$, and so on, all reckoned as at the first surface A. Here δ denotes the linear retardation of the second reflection as compared with the first, due to the thickness of the plate, and it is given by

$$\delta = 2\mu T \cos \alpha', \quad \dots\dots\dots\dots\dots\dots\dots\dots(1)$$

* *Roy. Soc. Proc.*, February, 1885. See also Rayleigh, *Phil. Mag.* Vol. XXIV. p. 145 (1887), Vol. XXVI. pp. 241, 256 (1888); *Scientific Papers*, Vol. III. pp. 1, 190, 204, 264.

where μ is the refractive index, T the thickness, and α' the angle of refraction within the plate. Also $k = 2\pi/\lambda$, λ being the wave-length. Adding together the various reflections and summing the infinite geometric series, we find

$$b' + \frac{c'e'f'e^{-ik\delta}}{1 - e'^2e^{-ik\delta}}. \qquad\qquad (2)$$

In like manner for the wave transmitted through the plate we get

$$c'f' + c'f'e'^2e^{-ik\delta} + \ldots = \frac{c'f'}{1 - e'^2e^{-ik\delta}}, \qquad (3)$$

the incident and transmitted waves being reckoned as at A.

The quantities b', c', e', f' are not independent. The simplest way to find the relations between them is to trace the consequences of supposing $\delta = 0$ in (2) and (3). For it is evident *a priori* that, with a plate of vanishing thickness, there must be a vanishing reflection and an undisturbed total transmission*. Accordingly,

$$b' + e' = 0, \quad c'f' = 1 - e'^2, \qquad\qquad (4)$$

the first of which embodies Arago's law of the equality of reflections, as well as the famous "loss of half an undulation." Using these, and substituting η for e', we find for the reflected vibration,

$$-\frac{\eta(1 - e^{-ik\delta})}{1 - \eta^2e^{-ik\delta}}, \qquad\qquad (5)$$

and for the transmitted vibration

$$\frac{1 - \eta^2}{1 - \eta^2e^{-ik\delta}}. \qquad\qquad (6)$$

In dealing with a single plate, we are usually concerned only with intensities, represented by the squares of the moduli of these expressions. Thus,

$$\text{Intensity of reflected light} = \eta^2\frac{(1 - \cos k\delta)^2 + \sin^2 k\delta}{(1 - \eta^2\cos k\delta)^2 + \eta^4\sin^2 k\delta}$$

$$= \frac{4\eta^2\sin^2(\frac{1}{2}k\delta)}{1 - 2\eta^2\cos k\delta + \eta^4}; \qquad (7)$$

$$\text{Intensity of transmitted light} = \frac{(1 - \eta^2)^2}{1 - 2\eta^2\cos k\delta + \eta^4}, \qquad (8)$$

the sum of the two expressions being unity, as was to be expected.

According to (7), not only does the reflected light vanish completely when $\delta = 0$, but also whenever $\frac{1}{2}k\delta = s\pi$, s being an integer; that is, whenever $\delta = s\lambda$.

Returning to (5) and (6), we may remark that, in supposing k real, we are postulating a transparent plate. The effect of absorption might be included by allowing k to be complex.

---

* "Wave Theory of Light," *Ency. Brit.* Vol. xxiv. 1888; *Scientific Papers*, Vol. iii. p. 64.

When we pass from a single plate to consider the operation of a number of plates of equal thicknesses and separated by equal intervals, the question of phase assumes importance. It is convenient to refer the vibrations to points such as O, O', bisecting the intervals between the plates; see figure, where for simplicity the incidence is regarded as perpendicular. When we

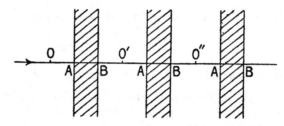

reckon the incident and reflected waves from O instead of A, we must introduce the additional factor $e^{-\frac{1}{2}ik\delta'}$, δ' for the interval corresponding to δ for the plate. Thus (5) becomes

$$-\frac{\eta\,(1-e^{-ik\delta})\,e^{-\frac{1}{2}ik\delta'}}{1-\eta^2e^{-ik\delta}}=r. \qquad\qquad\ldots\ldots\ldots\ldots\ldots(9)$$

So also if we reckon the transmitted wave at O', instead of A, we must introduce the factor $e^{-\frac{1}{2}ik\,(\delta+\delta')}$, and (6) becomes

$$\frac{(1-\eta^2)\,e^{-\frac{1}{2}ik\,(\delta+\delta')}}{1-\eta^2e^{-ik\delta}}=t. \qquad\qquad\ldots\ldots\ldots\ldots\ldots(10)$$

The introduction of the new exponential factors does not interfere with the moduli, so that still

$$|\,r^2\,|+|\,t^2\,|=1. \qquad\qquad\ldots\ldots\ldots\ldots\ldots\ldots(11)$$

Further, we see that

$$\frac{r}{t}=-\frac{\eta\,(1-e^{-ik\delta})}{(1-\eta^2)\,e^{-\frac{1}{2}ik\delta}}=-\frac{2i\eta\sin\frac{1}{2}k\delta}{1-\eta^2}, \qquad\ldots\ldots\ldots(12)$$

and thus (in the case of transparency) r/t is a pure imaginary. In accordance with (11) and (12) it is permitted to write

$$r=\sin\theta\,.\,e^{i\rho}, \qquad t=i\cos\theta\,.\,e^{i\rho}, \qquad\ldots\ldots\ldots\ldots(13)$$

in which θ and ρ are real and

$$\tan\theta=\frac{2\eta\sin\frac{1}{2}k\delta}{1-\eta^2}. \qquad\qquad\ldots\ldots\ldots\ldots\ldots(14)$$

Also from (9), (13)

$$\rho=s\pi-\chi-\tfrac{1}{2}k\,(\delta+\delta')-\tfrac{1}{2}\pi, \qquad\ldots\ldots\ldots\ldots(15)$$

where s is an integer and

$$\tan\chi=\frac{\eta^2\sin k\delta}{1-\eta^2\cos k\delta}. \qquad\qquad\ldots\ldots\ldots\ldots(16)$$

The calculation for a set of equal and equidistant plates may follow the lines of Stokes' work for a pile of plates, where intensities were alone regarded*.

* Roy. Soc. Proc. 1862; Math. and Phys. Papers, Vol. IV. p. 145.

In that case there was no need to refer the vibrations to particular points, but for our purpose we refer the vibrations always to the points O, O', etc., bisecting the intervals between the plates. On this understanding the formal expressions are the same. ϕ_m denotes the reflection from m plates, referred to the point O in front of the plates; ψ_m the transmission referred to a point O_m behind the last plate. " Consider a system of $m + n$ plates, and imagine these grouped into two systems, of m and n plates respectively. The incident light being represented by unity, the light ϕ_m will be reflected from the first group, and ψ_m will be transmitted. Of the latter the fraction ψ_n will be transmitted by the second group, and ϕ_n reflected. Of the latter the fraction ψ_m will be transmitted by the first group, and ϕ_m reflected, and so on. Hence we get for the light reflected by the whole system,

$$\phi_m + \psi_m{}^2 \phi_n + \psi_m{}^2 \phi_m \phi_n{}^2 + \cdots,$$

and for the light transmitted

$$\psi_m \psi_n + \psi_m \phi_n \phi_m \psi_n + \psi_m \phi_n{}^2 \phi_m{}^2 \psi_n + \cdots,$$

which gives, by summing the two geometric series,

$$\phi_{m+n} = \phi_m + \frac{\psi_m{}^2 \phi_n}{1 - \phi_m \phi_n} ; \quad \dots \dots \dots \dots \dots \dots (17)$$

$$\psi_{m+n} = \frac{\psi_m \psi_n}{1 - \phi_m \phi_n}. " \quad \dots \dots \dots \dots \dots \dots \dots \dots (18)$$

The argument applies equally in our case, only ϕ_m, etc., now denote complex quantities by which the amplitudes of vibration are multiplied, instead of real positive quantities, less than unity, relating to intensities. By definition $\phi_1 = r$, $\psi_1 = t$.

Before proceeding further, we may consider the comparatively simple cases of two or three plates. Putting $m = n = 1$, we get from (17), (18)

$$\phi_2 = \frac{r(1 - r^2 + t^2)}{1 - r^2}, \qquad \psi_2 = \frac{t^2}{1 - r^2} \dots \dots \dots \dots \dots (19)$$

By (13), $1 - r^2 + t^2 = 1 - e^{2i\rho}$, and thus

$$\phi_2 = \frac{r(1 - e^{2i\rho})}{1 - r^2}. \quad \dots \dots \dots \dots \dots \dots (20)$$

It appears that ϕ_2 vanishes not only when $r = 0$, but also independently of r when $\cos 2\rho = 1$. In this case $\psi_2 = -1$.

When $\cos 2\rho = 1$, $r = \pm \sin \theta$, $t = \pm i \cos \theta$, so that r is real and t is a pure imaginary. From (9) we find that a real r requires that

$$\cos \tfrac{1}{2} k (\delta + \delta') = \eta^2 \cos \tfrac{1}{2} k (\delta' - \delta), \quad \dots \dots \dots \dots (21)$$

or, as it may also be written,

$$\tan \tfrac{1}{2} k \delta . \tan \tfrac{1}{2} k \delta' = \frac{1 - \eta^2}{1 + \eta^2}. \quad \dots \dots \dots \dots (22)$$

When η is small we see that

$$k(\delta + \delta') = (2s + 1)\pi, \qquad \text{or} \qquad \delta + \delta' = (2s + 1)\lambda/2.$$

In this case only the first and second components of the aggregate reflection are sensible.

If there are three plates we may suppose in (17) $m = 2$, $n = 1$.

Thus
$$\phi_3 = \phi_2 + \frac{\psi_2^2 r}{1 - \phi_2 r}, \quad \dots\dots\dots\dots\dots\dots(23)$$

ϕ_2 and ψ_2 being given by (19). If $\phi_3 = 0$,

$$\phi_2(1 - r\phi_2) + r\psi_2^2 = 0. \quad \dots\dots\dots\dots\dots(24)$$

In terms of ρ and θ

$$\phi_2 = \frac{\sin\theta\,(1 - e^{2i\rho})\,e^{i\rho}}{1 - \sin^2\theta\,e^{2i\rho}}, \qquad \psi_2 = -\frac{\cos^2\theta\,e^{2i\rho}}{1 - \sin^2\theta\,e^{2i\rho}}. \quad \dots\dots\dots(25)$$

Using these in (24), we find that either $\sin\theta$, and therefore r, is equal to zero, or else that

$$E^2\cos^4\theta + E(2 - E)(1 - E)\cos^2\theta + (1 - E)^3 = 0, \quad \dots\dots\dots(26)$$

E being written for $e^{2i\rho}$. By solution of the quadratic

$$\cos^2\theta = -(1 - E)^2/E \qquad \text{or} \qquad 1 - E^{-1}.$$

The second alternative is inadmissible, since it makes the denominators zero in (25). The first alternative gives

$$E = \cos 2\rho + i\sin 2\rho = 1 - \tfrac{1}{2}\cos^2\theta \pm i\cos\theta\,\sqrt{(1 - \tfrac{1}{4}\cos^2\theta)},$$

whence
$$\cos\theta = \pm 2\sin\rho. \quad \dots\dots\dots\dots\dots\dots\dots(27)$$

When η, and therefore r, is small, $\cos\theta = 1$ nearly, and χ in (15) may be omitted. Hence

$$\delta + \delta' = \lambda\,(\tfrac{1}{3} \text{ or } \tfrac{2}{3}) + s\lambda, \quad \dots\dots\dots\dots\dots(28)$$

as might have been expected.

If we suppose $e^{2i\rho} = 1$, $\phi_2 = 0$, $\psi_2 = -1$, and (23) gives $\phi_3 = r$. It is easy to recognize that for every odd number $\phi_m = r$, and for every even number $\phi_m = 0$.

In his solution of the functional equations (17), (18)*, Stokes regards ϕ and ψ as functions of continuous variables m and n, and he obtains it with the aid of a differential equation. The following process seems simpler, and has the advantage of not introducing other than integral values of m and n. If we make $m = 1$ in (17),

$$\phi_{n+1} = r + \frac{t^2\phi_n}{1 - r\phi_n}, \quad \dots\dots\dots\dots\dots\dots(29)$$

or if we write $u_n = r\phi_n - 1$,

$$u_{n+1}u_n + (1 - r^2 + t^2)\,u_n + t^2 = 0. \quad \dots\dots\dots\dots(30)$$

---

* Stirling has shown, *Roy. Soc. Proc.* A, Vol. xc. p. 237 (1914), that the two equations are not independent, (18) being derivable from (17).

In this we assume $u_n = v_{n+1}/v_n$, so that

$$v_{n+2} + (1 - r^2 + t^2)\, v_{n+1} + t^2 v_n = 0. \quad \ldots\ldots\ldots\ldots\ldots(31)$$

The solution of (31) is

$$v_n = H p^n + K q^n,$$

where
$$p + q = r^2 - t^2 - 1, \qquad pq = t^2, \quad \ldots\ldots\ldots\ldots\ldots(32)$$

and H, K are arbitrary constants. Accordingly

$$u_n = \frac{H p^{n+1} + K q^{n+1}}{H p^n + K q^n}, \quad \ldots\ldots\ldots\ldots\ldots(33)$$

in which there is but one constant of integration effectively.

This constant may be determined from the case of $n = 1$, for which $u_1 = r^2 - 1$. By means of (32) we get $(p+1)\,H + (q+1)\,K = 0$,

so that
$$u_n = \frac{(q+1)\, p^{n+1} - (p+1)\, q^{n+1}}{(q+1)\, p^n - (p+1)\, q^n}, \quad \ldots\ldots\ldots\ldots(34)$$

and
$$\phi_n = \frac{(p+1)(q+1)\,\{p^n - \bar{q}^n\}}{r\,\{(q+1)\, p^n - (p+1)\, q^n\}},$$

or since by (32) $r^2 = (p+1)(q+1)$,

$$\phi_n = \frac{b^n - b^{-n}}{a b^n - a^{-1} b^{-n}}, \quad \ldots\ldots\ldots\ldots\ldots(35)$$

where
$$\sqrt{\left(\frac{p}{q}\right)} = b, \qquad \sqrt{\left(\frac{q+1}{p+1}\right)} = a. \quad \ldots\ldots\ldots\ldots(36)$$

In order to find ψ_m we may put $n = 1$ in (17); and by use of (29), with m substituted for n, we get

$$\psi_m^2 = 1 - \phi_m \frac{r^2 + 1 - t^2}{r} + \phi_m^2,$$

and on reduction with use of (35), (32),

$$\psi_m = \pm \frac{a - a^{-1}}{a b^m - a^{-1} b^{-m}}. \quad \ldots\ldots\ldots\ldots(37)$$

By putting $m = 0$, we see that the upper sign is to be taken.

The expressions thus obtained are those of Stokes:

$$\frac{\phi_m}{b^m - b^{-m}} = \frac{\psi_m}{a - a^{-1}} = \frac{1}{a b^m - a^{-1} b^{-m}} \quad \ldots\ldots\ldots\ldots(38)$$

The connexion between a, b and r, t is established by setting $m = 1$. Thus

$$\frac{r}{b - b^{-1}} = \frac{t}{a - a^{-1}} = \frac{1}{ab - a^{-1} b^{-1}}. \quad \ldots\ldots\ldots\ldots(39)$$

In Stokes' problem, where r, t, ϕ, ψ represent intensities, a and b are real. If there is no absorption, $r + t = 1$, so that $a - 1, b - 1$ are vanishing quantities. In this case

$$\frac{r}{b - 1} = \frac{t}{a - 1} = \frac{1}{a - 1 + b - 1},$$

and
$$\frac{\phi_m}{mr} = \frac{\psi_m}{1-r} = \frac{1}{1+(m-1)r}. \quad\text{...................(40)}$$

When m tends to infinity, ϕ_m approaches unity, and ψ_m approaches zero.

For many purposes, equations (38), (39) may conveniently be written in another form, by making $b = e^\beta$, $a = e^\alpha$. Thus

$$\frac{\phi_m}{\sinh m\beta} = \frac{\psi_m}{\sinh \alpha} = \frac{1}{\sinh(\alpha+m\beta)}, \quad\text{.................(41)}$$

$$\frac{r}{\sinh \beta} = \frac{t}{\sinh \alpha} = \frac{1}{\sinh(\alpha+\beta)}, \quad\text{....................(42)}$$

where in Stokes' problem α and β are real, and are uniquely determined in terms of r and t by (44), (46) below*.

If we form the expression for $(1 + r^2 - t^2)/2r$ by means of (42), we find that it is equal to $\cosh \alpha$. Also

$$\sinh^2 \alpha = \frac{\{(1+r)^2 - t^2\}\,\{(1-r)^2 - t^2\}}{4r^2}, \quad\text{...............(43)}$$

from which we see that, if r and t are real positive quantities, such that $r + t < 1$, $\sinh \alpha$ is real. Similarly, $\sinh \beta$, $\sinh(\alpha + \beta)$ are real.

Passing now to my proper problem, where r and t are complex factors, represented (when there is no absorption) by (13), we have

$$\cosh \alpha = \frac{1 + r^2 - t^2}{2r} = \frac{\cos \rho}{\sin \theta}, \quad\text{....................(44)}$$

so that $\cosh \alpha$ is real. Also

$$\sinh^2 \alpha = \frac{\cos^2 \rho}{\sin^2 \theta} - 1. \quad\text{.........................(45)}$$

If we write $\alpha = \alpha_1 + i\alpha_2$, $\beta = \beta_1 + i\beta_2$, where α_1, α_2, β_1, β_2 are real,

$$\sinh \alpha = \sinh \alpha_1 \cos \alpha_2 + i \cosh \alpha_1 \sin \alpha_2,$$

$$\cosh \alpha = \cosh \alpha_1 \cos \alpha_2 + i \sinh \alpha_1 \sin \alpha_2.$$

Since $\cosh \alpha$ is real, either α_1 or $\sin \alpha_2$ must vanish. In the first case, $\sinh \alpha = i \sin \alpha_2$, and (45) shows that this can occur only when $\sin^2 \theta > \cos^2 \rho$. In the second case ($\sin \alpha_2 = 0$), $\sinh^2 \alpha = \sinh^2 \alpha_1$, which requires that $\sin^2 \theta < \cos^2 \rho$.

Similarly if we interchange r and t,

$$\cosh \beta = \frac{1 + t^2 - r^2}{2t} = -\frac{\sin \rho}{\cos \theta}, \quad\text{...............(46)}$$

so that $\cosh \beta$ is real, requiring either $\beta_1 = 0$, or $\sin \beta_2 = 0$. Also

$$\sinh^2 \beta = \frac{\sin^2 \rho}{\cos^2 \theta} - 1. \quad\text{.........................(47)}$$

---

* Except as to sign, which is a matter of indifference. It may be remarked that his equation (13) can at once be put into this form by making his α and β pure imaginaries.

If $\beta_1 = 0$, $\sinh \beta = i \sin \beta_2$, which can occur only when $\sin^2 \rho < \cos^2 \theta$, or, which is the same, $\sin^2 \theta < \cos^2 \rho$. Again, if $\sin \beta_2 = 0$, $\sinh^2 \beta = \sinh^2 \beta_1$, occurring when $\sin^2 \theta > \cos^2 \rho$.

It thus appears that, of the four cases at first apparently possible, $\alpha_1 = \beta_1 = 0$, $\sin \alpha_2 = \sin \beta_2 = 0$, are excluded. There are two remaining alternatives:

(i) $\sinh^2 \alpha = -$; $\sin^2 \theta > \cos^2 \rho$; $\alpha_1 = 0$, $\sin \beta_2 = 0$;

(ii) $\sinh^2 \alpha = +$; $\sin^2 \theta < \cos^2 \rho$; $\beta_1 = 0$, $\sin \alpha_2 = 0$.

Between these there is an important distinction in respect of what happens when m is increased. For

$$\phi_m = \sinh m\beta / \sinh (\alpha + m\beta).$$

In case (i) this becomes

$$1/\phi_m = \cos \alpha_2 + i \coth m\beta_1 \sin \alpha_2, \quad \dots \dots \dots \dots (48)$$

and

$$1/|\phi_m|^2 = 1 + \sin^2 \alpha_2 / \sinh^2 m\beta_1. \quad \dots \dots \dots \dots (48 \text{ bis})$$

If β_1 be finite, $\sinh^2 m\beta_1$ tends to ∞ when m increases, so that $|\phi_m|^2$ tends to unity, that is, the reflection tends to become complete. We see also that, whatever m may be, ϕ_m cannot vanish, unless $\beta_1 = 0$, when also $r = 0$.

In case (ii)

$$\pm 1/\phi_m = \cosh \alpha_1 - i \cot m\beta_2 \sinh \alpha_1, \quad \dots \dots \dots (49)$$

and

$$1/|\phi_m|^2 = 1 + \sinh^2 \alpha_1 / \sin^2 m\beta_2, \quad \dots \dots \dots (49 \text{ bis})$$

so that ϕ_m continues to fluctuate, however great m may be. Here ϕ_m may vanish, since there is nothing to forbid $m\beta_2 = s\pi$. Of this behaviour we have already seen an example, where $\cos^2 \rho = 1$.

In order to discriminate the two cases more clearly, we may calculate the value of $\sinh^2 \alpha$ from (43), writing temporarily for brevity

$$e^{\frac{1}{2}ik\delta} = \Delta, \qquad e^{\frac{1}{2}ik\delta'} = \Delta'. \quad \dots \dots \dots \dots \dots (50)$$

Thus by (9) and (10)

$$r = -\frac{\eta(\Delta^2 - 1)}{(\Delta^2 - \eta^2)\Delta'}, \qquad t = \frac{(1 - \eta^2)\Delta}{(\Delta^2 - \eta^2)\Delta'}, \quad \dots \dots \dots (51)$$

so that

$$r \pm t = \frac{1 - \eta\Delta}{(\Delta - \eta)\Delta'}, \quad \text{or} \quad -\frac{1 + \eta\Delta}{(\Delta + \eta)\Delta'}; \quad \dots \dots \dots (52)$$

whence

$$\sinh^2 \alpha = \frac{\{(\Delta - \eta)^2 \Delta'^2 - (1 - \eta\Delta)^2\}\{(\Delta + \eta)^2 \Delta'^2 - (1 + \eta\Delta)^2\}}{4\eta^2 \Delta'^2 (\Delta^2 - 1)^2}. \quad \dots (53)$$

The two factors in the numerator of the fraction differ only by the sign of η, so that the fraction itself is an even function of η. The first factor may be written

$$\{(\Delta - \eta)\Delta' + 1 - \eta\Delta\}\{(\Delta - \eta)\Delta' - (1 - \eta\Delta)\}$$

$$= -\{1 + \Delta\Delta' - \eta(\Delta + \Delta')\}\{1 - \Delta\Delta' + \eta(\Delta' - \Delta)\};$$

32—2

and similarly the second factor may be written with change of sign of η

$$- \{1 + \Delta\Delta' + \eta(\Delta + \Delta')\}\{1 - \Delta\Delta' - \eta(\Delta' - \Delta)\}.$$

Accordingly

$$\sinh^2 \alpha = \frac{\{(1+\Delta\Delta')^2 - \eta^2(\Delta+\Delta')^2\}\{(1-\Delta\Delta')^2 - \eta^2(\Delta-\Delta')^2\}}{4\eta^2\Delta'^2(\Delta^2-1)^2} \quad \ldots(54)$$

In this, on restoring the values of Δ, Δ',

$$1 + \Delta\Delta' \pm \eta(\Delta + \Delta') = 2e^{\frac{1}{2}ik(\delta+\delta')}\{\cos \tfrac{1}{4}k(\delta+\delta') \pm \eta \cos \tfrac{1}{4}k(\delta-\delta')\},$$

and

$$1 - \Delta\Delta' \pm \eta(\Delta - \Delta') = -2ie^{\frac{1}{2}ik(\delta+\delta')}\{\sin \tfrac{1}{4}k(\delta+\delta') \mp \eta \sin \tfrac{1}{4}k(\delta-\delta')\}.$$

Also

$$4\Delta'^2(\Delta^2-1)^2 = -16e^{ik(\delta+\delta')}\sin^2\tfrac{1}{2}k\delta,$$

and thus

$$\sinh^2\alpha = \frac{\{\cos^2 \tfrac{1}{4}k(\delta+\delta') - \eta^2 \cos^2 \tfrac{1}{4}k(\delta-\delta')\}}{\eta^2\sin^2\tfrac{1}{2}k\delta}$$
$$\times \{\sin^2 \tfrac{1}{4}k(\delta+\delta') - \eta^2 \sin^2 \tfrac{1}{4}k(\delta-\delta')\}. \quad \ldots\ldots(55)$$

The transition between the two cases (of opposite behaviour when $m = \infty$) occurs when $\sinh\alpha = 0$. In general, this requires either

$$\eta = \pm \frac{\cos \tfrac{1}{4}k(\delta+\delta')}{\cos \tfrac{1}{4}k(\delta-\delta')}, \quad \text{or} \quad \eta = \pm \frac{\sin \tfrac{1}{4}k(\delta+\delta')}{\sin \tfrac{1}{4}k(\delta-\delta')}, \quad \ldots\ldots(56)$$

conditions which are symmetrical with respect to δ and δ', as clearly they ought to be*. In (55), (56), η^2 is limited to values less than unity.

Reverting to (43), we see that the evanescence of $\sinh^2\alpha$ requires that $r = \pm 1 \mp t$, or, if we separate the real and imaginary parts of r and t, $r = \pm 1 \mp t_1 \mp it_2$.

If, for example, we take $r = -1 - t$, we have

$$|r|^2 = (1+t_1)^2 + t_2^2 = 1 + |t|^2 + 2t_1.$$

Also

$$|r|^2 = 1 - |t|^2;$$

so that

$$|r|^2 = 1 + t_1, \qquad |t|^2 = -t_1.$$

In like manner by interchange of r and t,

$$|t|^2 = 1 + r_1, \qquad |r|^2 = -r_1,$$

showing that in this case r_1, t_1 are both negative.

The general equation (55) shows that $\sinh^2\alpha$ is negative, when η^2 lies between

$$\frac{\cos^2 \tfrac{1}{4}k(\delta+\delta')}{\cos^2 \tfrac{1}{4}k(\delta-\delta')} \quad \text{and} \quad \frac{\sin^2 \tfrac{1}{4}k(\delta+\delta')}{\sin^2 \tfrac{1}{4}k(\delta-\delta')}.$$

This is the case (i) above defined where an increase in m leads to complete reflection. On the other hand, $\sinh^2\alpha$ is positive when η^2 lies outside the

---

* That is with reversal of the sign of η, which makes no difference here.

above limits, and then (ii) the reflection (and transmission) remain fluctuating however great m may be. When η^2 is small, case (ii) usually obtains, though there are exceptions for specially related values of δ and δ'.

Particular cases, worthy of notice, occur when $\delta' \pm \delta = s\lambda$, where s is an integer. If $\delta' + \delta = s\lambda$,

$$\sinh^2 \alpha = \eta^2 \cos^2 \tfrac{1}{2} k\delta - 1, \qquad \ldots\ldots\ldots\ldots\ldots\ldots(57)$$

and is negative for all admissible values of η, case (i). If $\delta' - \delta = s\lambda$,

$$\sinh^2 \alpha = \cos^2 \tfrac{1}{2} k\delta / \eta^2 - 1, \qquad \ldots\ldots\ldots\ldots\ldots\ldots(58)$$

and we have case (i) or case (ii), according as η^2 is greater or less than $\cos^2 \tfrac{1}{2} k\delta$.

When η is given, as would usually happen in calculations with an optical purpose, it may be convenient to express the limiting values of (56) in another form. We have

$$\frac{1 \mp \eta}{1 \pm \eta} = \tan \tfrac{1}{4} k\delta \cdot \tan \tfrac{1}{4} k\delta', \qquad \frac{1 \mp \eta}{1 \pm \eta} = - \cot \tfrac{1}{4} k\delta \cdot \tan \tfrac{1}{4} k\delta'. \quad \ldots(59)$$

When the passage is perpendicular, Young's formula, viz. $\eta = (\mu - 1)/(\mu + 1)$, gives

$$(1 \mp \eta)/(1 \pm \eta) = \mu^{\mp 1}, \qquad \ldots\ldots\ldots\ldots\ldots\ldots\ldots(60)$$

μ being the relative refractive index.

We will now consider more in detail some special cases of optical interest. We choose a value of δ such as will give the maximum reflection from a single plate. From (5) or (9)

$$\frac{1}{|r|^2} = 1 + \frac{(1 - \eta^2)^2}{2\eta^2 (1 - \cos k\delta)}, \qquad \ldots\ldots\ldots\ldots\ldots(61)$$

so that $|r|$ is greatest for a given η when $\cos k\delta = -1$. And then

$$r = -\frac{2\eta e^{-\frac{1}{2}ik\delta'}}{1 + \eta^2}, \qquad t = \pm \frac{i(1 - \eta^2) e^{-\frac{1}{2}ik\delta'}}{1 + \eta^2}. \qquad \ldots\ldots\ldots\ldots(62)$$

We may expect the greatest aggregate reflection when the components from the various plates co-operate. This occurs when $e^{-ik(\delta + \delta')} = 1$, so that in the notation of (50), $\Delta^2 = \Delta'^2 = -1$. The introduction of these values into (54) yields

$$\sinh^2 \alpha = -1, \qquad \ldots\ldots\ldots\ldots\ldots\ldots\ldots(63)$$

coming under (i). The same result may be derived from (57), since here $\cos \tfrac{1}{2} k\delta = 0$. In addition to $\alpha_1 = 0$, $\sin \beta_2 = 0$, we now have by (63) $\sin \alpha_2 = \pm 1$, $\cos \alpha_2 = 0$, and (48) gives

$$|\phi_m|^2 = \tanh^2 m\beta_1, \qquad |r|^2 = \tanh^2 \beta_1. \qquad \ldots\ldots\ldots\ldots(64)$$

We are now in a position to calculate the reflection for various values of m, since by (62)

$$\tanh \beta_1 = \pm \frac{2\eta}{1 + \eta^2} = \pm \tanh 2\zeta,$$

if $\eta = \tanh \zeta$, so that

$$\beta_1 = \pm 2 \tanh^{-1} \eta. \quad \dots\dots\dots\dots\dots\dots\dots(65)$$

Let us suppose that, as for glass and air, $\mu = 1\cdot5$, $\eta = \frac{1}{5}$, making $\beta_1 = 0\cdot40546$. The following were calculated with the aid of the Smithsonian Tables of Hyperbolic Functions. It appears that under these favourable conditions as regards δ and δ', the intensity of the reflected light $|\phi_m|^2$ approaches its limit (unity) when m reaches 4 or 5.

TABLE I.

m	$m\beta_1$	$\tanh m\beta_1$	$|\phi m|^2 = \tanh^2 m\beta_1$
1	0·4055	0·3846	0·1479
2	0·8109	0·6701	0·4490
3	1·2164	0·8386	0·7032
4	1·6218	0·9249	0·8554
5	2·0273	0·9659	0·9330
6	2·4328	0·9847	0·9696
7	2·8382	0·9932	0·9864
10	4·055	0·9994	0·9988
∞	∞	1·0000	1·0000

In the case of chlorate of potash crystals with periodic twinning η is very small at moderate incidences. As an example of the sort of thing to be expected, we may take $\beta_1 = 0\cdot04$, corresponding to $\eta = 0\cdot02$.

TABLE II.

m	$\tanh m\beta_1$	$|\phi m|^2$
1	0·0400	0·00160
2	0·0798	0·00637
4	0·1586	0·02517
8	0·3095	0·09579
16	0·5649	0·3191
32	0·8565	0·7336
64	0·9881	0·9763

According to (58), if $\delta' - \delta = s\lambda$, the same value of $\sinh^2 \alpha$ obtains as in (63), since we are supposing $\cos \frac{1}{2}k\delta = 0$, and the same consequences follow*.

Retaining the same values of δ, that is those included under $\delta = (s + \frac{1}{2})\lambda$, we will now suppose $\delta' = s'\lambda$, where s' also is an integer. From (55)

$$\sinh^2 \alpha = \frac{(1 - \eta^2)^2}{4\eta^2} = \sinh^2 \alpha_1, \quad \dots\dots\dots\dots\dots(66)$$

* But when η is small, a slight departure from $\cos \frac{1}{2}k\delta = 0$ produces very different effects in the two cases.

since $\sin \alpha_2 = 0$ in this case (ii). By (49 *bis*) we have now, setting $m = 1$,

$$\frac{1}{|r|^2} = 1 + \frac{\sinh^2 \alpha_1}{\sin^2 \beta_2} = \frac{(1+\eta^2)^2}{4\eta^2},$$

as we see from (62). Comparing with (66), we find $\sin^2 \beta_2 = 1$, $\beta_2 = (s + \tfrac{1}{2})\pi$. Thus $\sin^2 m\beta_2$ is equal to 1 or 0, according as m is odd or even; and (49 *bis*) shows that when m is odd

$$|\phi_m|^2 = r^2 = 4\eta^2/(1+\eta^2)^2, \quad\ldots\ldots\ldots\ldots\ldots\ldots(67)$$

and that when m is even, $|\phi_m|^2 = 0$. The second plate neutralizes the reflection from the first plate, the fourth plate that from the third, and so on. The simplest case under this head is when $\delta = \tfrac{1}{2}\lambda$, $\delta' = \lambda$.

A variation of the latter supposition leads to a verification of the general formulæ worth a moment's notice. We assume, as above, $\delta' = s'\lambda$, but leave δ open. Since $e^{\frac{1}{2}ik\delta'} = \pm 1$, (9) and (10) become

$$r = \mp \frac{e^{\frac{1}{2}ik\delta} - e^{-\frac{1}{2}ik\delta}}{\eta^{-1}e^{\frac{1}{2}ik\delta} - \eta e^{-\frac{1}{2}ik\delta}}, \qquad t = \pm \frac{\eta^{-1} - \eta}{\eta^{-1}e^{\frac{1}{2}ik\delta} - \eta e^{-\frac{1}{2}ik\delta}}; \quad \ldots\ldots(68)$$

and these are of the form (39), if we suppose $a = \eta^{-1}$, $b = e^{\frac{1}{2}ik\delta}$. The reflection ϕ_m from m plates is derived from r by merely writing b^m for b, that is, $e^{\frac{1}{2}imk\delta}$ for $e^{\frac{1}{2}ik\delta}$, leaving $|\phi_m|$ equal to $|r|$*, as should evidently be the case, at least when $\delta' = 0$.

[* This statement does not hold in general, when $\delta' = s'\lambda$, where s' is an integer and may be zero. We have

$$r = \mp \frac{2i \sin \tfrac{1}{2}k\delta}{(\eta^{-1} - \eta)\cos \tfrac{1}{2}k\delta + i(\eta^{-1} + \eta)\sin \tfrac{1}{2}k\delta},$$

so that

$$\frac{1}{|r|^2} = \frac{(\eta^{-1} - \eta)^2}{4\sin^2 \tfrac{1}{2}k\delta} + 1.$$

Hence

$$\frac{1}{|\phi_m|^2} = \frac{(\eta^{-1} - \eta)^2}{4\sin^2 \tfrac{1}{2}mk\delta} + 1;$$

consequently, if $|\phi_m| = |r|$, we must have

$$\sin^2 \tfrac{1}{2}mk\delta = \sin^2 \tfrac{1}{2}k\delta,$$

or

$$\tfrac{1}{2}mk\delta = n\pi \pm \tfrac{1}{2}k\delta,$$

where n is an integer, so that

$$\delta = \frac{n\lambda}{m \pm 1}.$$

This result may be verified for $m=2$ or 3 from (19), (23), and (68). It includes as a special case that dealt with in the preceding paragraph, if, when m is odd, we write $n = (s + \tfrac{1}{2})(m \pm 1)$, where s is an integer. When $\delta' = 0$ the strata intervening between the plates disappear, but the theory is only applicable on the supposition that reflection and refraction continue to take place as before at each of the contiguous surfaces of the plates. W. F. S.]

423.

ON THE PRESSURE DEVELOPED IN A LIQUID DURING THE COLLAPSE OF A SPHERICAL CAVITY.

[*Philosophical Magazine*, Vol. XXXIV. pp. 94—98, 1917.]

WHEN reading O. Reynold's description of the sounds emitted by water in a kettle as it comes to the boil, and their explanation as due to the partial or complete collapse of bubbles as they rise through cooler water, I proposed to myself a further consideration of the problem thus presented; but I had not gone far when I learned from Sir C. Parsons that he also was interested in the same question in connexion with cavitation behind screw-propellers, and that at his instigation Mr S. Cook, on the basis of an investigation by Besant, had calculated the pressure developed when the collapse is suddenly arrested by impact against a rigid concentric obstacle. During the collapse the fluid is regarded as incompressible.

In the present note I have given a simpler derivation of Besant's results, and have extended the calculation to find the pressure in the interior of the fluid during the collapse. It appears that before the cavity is closed these pressures may rise very high in the fluid near the inner boundary.

As formulated by Besant*, the problem is—

"An infinite mass of homogeneous incompressible fluid acted upon by no forces is at rest, and a spherical portion of the fluid is suddenly annihilated; it is required to find the instantaneous alteration of pressure at any point of the mass, and the time in which the cavity will be filled up, the pressure at an infinite distance being supposed to remain constant."

Since the fluid is incompressible, the whole motion is determined by that of the inner boundary. If U be the velocity and R the radius of the boundary at time t, and u the simultaneous velocity at any distance r (greater than R) from the centre, then

$$u/U = R^2/r^2 ; \qquad \dots\dots\dots\dots\dots\dots\dots\dots\dots(1)$$

* Besant's *Hydrostatics and Hydrodynamics*, 1859, § 158.

and if ρ be the density, the whole kinetic energy of the motion is

$$\tfrac{1}{2}\rho \int_R^\infty u^2 . 4\pi r^2\, dr = 2\pi\rho U^2 R^3. \quad\ldots\ldots\ldots\ldots\ldots(2)$$

Again, if P be the pressure at infinity and R_0 the initial value of R, the work done is

$$\frac{4\pi P}{3}(R_0{}^3 - R^3). \quad\ldots\ldots\ldots\ldots\ldots\ldots(3)$$

When we equate (2) and (3) we get

$$U^2 = \frac{2P}{3\rho}\left(\frac{R_0{}^3}{R^3} - 1\right), \quad\ldots\ldots\ldots\ldots\ldots(4)$$

expressing the velocity of the boundary in terms of the radius. Also, since $U = dR/dt$,

$$t = \sqrt{\left(\frac{3\rho}{2P}\right)} . \int_R^{R_0} \frac{(R^{3/2}\,dR)}{(R_0{}^3 - R^3)^{\frac{1}{2}}} = R_0\sqrt{\left(\frac{3\rho}{2P}\right)} . \int_\beta^1 \frac{\beta^{3/2}\,d\beta}{(1 - \beta^3)^{\frac{1}{2}}} \quad\ldots\ldots(5)$$

if $\beta = R/R_0$. The time of collapse to a given fraction of the original radius is thus proportional to $R_0\rho^{\frac{1}{2}}P^{-\frac{1}{2}}$, a result which might have been anticipated by a consideration of "dimensions." The time τ of *complete* collapse is obtained by making $\beta = 0$ in (5). An equivalent expression is given by Besant, who refers to Cambridge Senate House Problems of 1847.

Writing $\beta^3 = z$, we have

$$\int_0^1 \frac{\beta^{3/2}\,d\beta}{(1 - \beta^3)^{\frac{1}{2}}} = \tfrac{1}{3}\int_0^1 z^{-\frac{1}{6}}(1 - z)^{-\frac{1}{2}}\,dz,$$

which may be expressed by means of Γ functions. Thus

$$\tau = R_0\sqrt{\left(\frac{\rho}{6P}\right)} . \frac{\Gamma\left(\tfrac{1}{6}\right).\Gamma\left(\tfrac{1}{2}\right)}{\Gamma\left(\tfrac{4}{3}\right)} = \cdot 91468\, R_0\, \sqrt{(\rho/P)}. \quad\ldots\ldots(6)$$

According to (4) U increases without limit as R diminishes. This indefinite increase may be obviated if we introduce, instead of an internal pressure zero or constant, one which increases with sufficient rapidity. We may suppose such a pressure due to a permanent gas obedient to Boyle's law. Then, if the initial pressure be Q, the work of compression is $4\pi Q R_0{}^3 \log (R_0/R)$, which is to be subtracted from (3). Hence

$$U^2 = \frac{2P}{3\rho}\left(\frac{R_0{}^3}{R^3} - 1\right) - \frac{2Q}{\rho}\frac{R_0{}^3}{R^3}\log\frac{R_0}{R}; \quad\ldots\ldots\ldots\ldots(7)$$

and $U = 0$ when

$$P(1 - z) + Q\log z = 0, \quad\ldots\ldots\ldots\ldots\ldots(8)$$

z denoting (as before) the ratio of volumes $R^3/R_0{}^3$. Whatever be the (positive) value of Q, U comes again to zero before complete collapse, and if $Q > P$ the first movement of the boundary is outwards. The boundary oscillates between two positions, of which one is the initial.

The following values of P/Q are calculated from (8):

z	P/Q	z	P/Q
$\frac{1}{1000}$	6·9147	1	arbitrary
$\frac{1}{100}$	4·6517	2	0·6931
$\frac{1}{10}$	2·5584	4	0·4621
$\frac{1}{4}$	1·8484	10	0·2558
$\frac{1}{2}$	1·3863	100	0·0465
1	arbitrary	1000	0·0069

Reverting to the case where the pressure inside the cavity is zero, or at any rate constant, we may proceed to calculate the pressure at any internal point. The general equation of pressure is

$$\frac{1}{\rho}\frac{dp}{dr} = -\frac{Du}{Dt} = -\frac{du}{dt} - u\frac{du}{dr}, \quad \ldots\ldots\ldots\ldots\ldots(9)$$

u being a function of r and t, reckoned positive in the direction of increasing r. As in (1), $u = UR^2/r^2$, and

$$\frac{du}{dt} = \frac{1}{r^2}\frac{d}{dt}(UR^2).$$

Also

$$\frac{d(UR^2)}{dt} = 2R\frac{dR}{dt}U + R^2\frac{dU}{dt} = 2RU^2 + R^2\frac{dU}{dt},$$

and by (4)

$$\frac{dU}{dt} = -\frac{P}{\rho}\frac{R_0^3}{R^4},$$

so that

$$\frac{d(UR^2)}{dt} = 2RU^2 - \frac{P}{\rho}\frac{R_0^3}{R^2}.$$

Thus, suitably determining the constant of integration, we get

$$\frac{p}{P} - 1 = \frac{R}{3r}\left\{\frac{R_0^3}{R^3} - 4\right\} - \frac{R^4}{3r^4}\left\{\frac{R_0^3}{R^3} - 1\right\}. \quad \ldots\ldots\ldots\ldots(10)$$

At the first moment after release, when $R = R_0$, we have

$$p = P(1 - R_0/r). \quad \ldots\ldots\ldots\ldots\ldots(11)$$

When $r = R$, that is on the boundary, $p = 0$, whatever R may be, in accordance with assumptions already made.

Initially the maximum p is at infinity, but as the contraction proceeds, this ceases to be true. If we introduce z to represent R_0^3/R^3, (10) may be written

$$\frac{p}{P} - 1 = \frac{R}{3r}(z - 4) - \frac{R^4}{3r^4}(z - 1), \quad \ldots\ldots\ldots\ldots(12)$$

and

$$\frac{dp/P}{dr} = \frac{R}{3r^2}\left\{\frac{(4z - 4)R^3}{r^3} - (z - 4)\right\}. \quad \ldots\ldots\ldots(13)$$

The maximum value of p occurs when

$$\frac{r^3}{R^3} = \frac{4z - 4}{z - 4} ; \quad \dots\dots\dots\dots\dots\dots(14)$$

and then

$$\frac{p}{P} = 1 + \frac{(z - 4) R}{4r} = 1 + \frac{(z - 4)^{\frac{4}{3}}}{4^{\frac{4}{3}} (z - 1)^{\frac{1}{3}}}. \quad \dots\dots\dots\dots(15)$$

So long as z, which always exceeds 1, is less than 4, the greatest value of p, viz. P, occurs at infinity; but when z exceeds 4, the maximum p occurs at a finite distance given by (14) and is greater than P. As the cavity fills up, z becomes great, and (15) approximates to

$$\frac{p}{P} = \frac{z}{4^{\frac{4}{3}}} = \frac{R_0^3}{4^{\frac{4}{3}} R^3}, \quad \dots\dots\dots\dots\dots(16)$$

corresponding to

$$r = 4^{\frac{1}{3}} R = 1\cdot587 R. \quad \dots\dots\dots\dots\dots(17)$$

It appears from (16) that before complete collapse the pressure near the boundary becomes very great. For example, if $R = \frac{1}{20} R_0$, $p = 1260P$.

This pressure occurs at a relatively moderate distance outside the boundary. At the boundary itself the pressure is zero, so long as the motion is free. Mr Cook considers the pressure here developed when the fluid strikes an absolutely rigid sphere of radius R. If the supposition of incompressibility is still maintained, an infinite pressure momentarily results; but if at this stage we admit compressibility, the instantaneous pressure P' is finite, and is given by the equation

$$\frac{P'^2}{2\beta'} = \tfrac{1}{2} \rho U^2 = \frac{P}{3} \left(\frac{R_0^3}{R^3} - 1 \right), \quad \dots\dots\dots\dots(18)$$

β' being the coefficient of compressibility. P, P', β' may all be expressed in atmospheres. Taking (as for water) $\beta' = 20,000$, $P = 1$, and $R = \frac{1}{20} R_0$, Cook finds

$$P' = 10,300 \text{ atmospheres} = 68 \text{ tons per sq. inch,}$$

and it would seem that this conclusion is not greatly affected by the neglect of compressibility before impact.

The subsequent course of events might be traced as in *Theory of Sound*, § 279, but it would seem that for a satisfactory theory compressibility would have to be taken into account at an earlier stage.

424.

ON THE COLOURS DIFFUSELY REFLECTED FROM SOME COLLODION FILMS SPREAD ON METAL SURFACES.

[*Philosophical Magazine*, Vol. XXXIV. pp. 423—428, 1917.]

It is known that "when a thin transparent film is backed by a perfect reflector, no colours should be visible, all the light being ultimately reflected, whatever the wave-length may be. The experiment may be tried with a thin layer of gelatine on a polished silver plate*." An apparent exception has been described by R. W. Wood†: "A thin film of collodion deposited on a bright surface of silver shows brilliant colours in reflected light. It, moreover, *scatters* light of a colour complementary to the colour of the directly reflected light. This is apparently due to the fact that the collodion film "frills," the mesh, however, being so small that it can be detected only with the highest powers of the microscope. Commercial ether and collodion should be used. If chemically pure ether obtained by distillation is used, the film does not frill, and no trace of colour is exhibited. Still more remarkable is the fact that if sunlight be thrown down upon the plate at normal incidence, brilliant colours are seen at grazing emergence, if a Nicol prism is held before the eye. These colours change to the complementary tints if the Nicol is rotated through 90°, *i.e.* in the scattered light, one half of the spectrum is polarized in one plane, and the remainder in a plane perpendicular to it."

I have lately come across an entirely forgotten letter from Rowland in which he describes a similar observation. Writing to me in March 1893, he says:—"While one of my students was working with light reflected from a metal, it occurred to me to try a thin collodion film on the metal. This not only had a remarkable effect on the polarization and the phase but I was astonished to find that it gave remarkably bright colours, both by direct reflexion and by diffused light, the two being complementary to each other. I have not gone into the theory but it looks like the phenomenon of thick plates as described by Newton in a different form. The curious point is

---

* " Wave Theory of Light," *Enc. Brit.* 1888; *Scientific Papers*, Vol. III. p. 67.

† *Physical Optics*, Macmillan, 1914, p. 172.

that I cannot get the effect by making the film on glass and then pressing it down *hard* upon speculum metal or mercury although I think the contact is very good in the case of the speculum metal. Possibly, however, it is not. Gelatine films on metal give good colours by direct reflexion but *not by diffused light*: only faint ones. It would seem that the collodion film must be of variable density or full of fine particles. However, I leave it to you. I send by express two of the plates used." Probably it was preoccupation with other work (weighing of gases) that prevented my giving attention to the matter at the time.

Wishing to repeat the observation of the diffusely scattered colours, I made some trials, but at first without success. On application to Prof. Wood, I was kindly supplied with further advice and with a specimen of a suitably coated plate of speculum metal. Acting on this advice, I have since obtained good results, using very dilute collodion poured upon a slightly warmed silvered plate (plated copper) warmed again as soon as the collodion was set. That the film is no longer a thin homogeneous plate seems certain. Wood speaks of "frilling," a word which rather suggests a wrinkling in parallel lines, but the suggestion seems negatived by the subsequent use of "mesh." I should suppose the disintegration to be like that sometimes seen on varnished paint, where under exposure to sunshine the varnish gathers itself into small detached heaps. At any rate there is no apparent change when the plate is turned round in its own plane, showing that the structure is effectively symmetrical with respect to the normal of the plate.

As regards Rowland's suggestion as to the origin of the colours, it does not seem that they can be assimilated to those of "thick plates." The latter require a highly localized source of light and are situated near the light or its image, whereas the colours now under consideration are seen when the plate is held near a large window backed by an overcast sky, and are localized on the plate itself, the passage from one colour to another depending presumably upon an altered scale in the structure of the film. The formation of well-developed colour at the various parts of the plate requires that the structure be, in a certain sense, uniform locally. The case is similar to that of coronas, as in experiments with lycopodium, only that here the grains must be very much smaller.

When examined by polarized light the behaviour of different plates is found to vary a good deal. We may take the case where sunlight is incident normally and the diffuse reflexion observed is nearly grazing. In the case of the specimen (on speculum metal) sent me by Prof. Wood, the light is practically extinguished in one position (α) of the nicol, that namely required to darken the reflexion from glass. In the perpendicular position (β) of the nicol good colours are seen, and also of course when the nicol is removed from the eye. At angles of scattering less nearly grazing there is some light in-both positions of the nicol, the fainter light in (α) showing much the same colour as in (β).

It will be noticed that this behaviour differs from that observed by Wood (on another plate) and already quoted. On the other hand, one of the (silvered) plates prepared by me shows a better agreement, more light than before being scattered at a grazing angle when the nicol is in the (α) position, while the colours in the (α) and (β) positions of the nicol are roughly complementary.

No more than Rowland have I succeeded in getting diffusely reflected colours from collodion films on glass or, I may add, quartz, either with or without the treatment with the breath suggested by Wood. The latter observer describes an experiment (p. 174) in which a film, deposited on the face of a prism, frilled under the action of the breath and then afforded a nearly three-fold reflexion. But, as I understand it, this augmented reflexion was specular. The only thing that I have seen at all resembling this was when I treated a coated glass with dilute hydrofluoric acid with the intention of loosening the film. Even when dry, the film remained out of optical contact with the glass, except I suppose at detached points, and gave an augmented specular reflexion, as was to be expected, inasmuch as three surfaces were operative.

Two views are possible with regard to the different behaviour of films on metal and on glass. One is to suppose that the actual structure is different in the two cases; the other, apparently favoured by Wood, refers the difference to the copious reflexion of light from metallic surfaces. The first view would seem the more probable *a priori* and is to a certain extent supported by Rowland's experiment. I have not succeeded in carrying out any decisive test. On either view we may expect the result to be modified by the metallic reflexion.

As to the explanation of the colours, anything more than a rough outline can hardly be expected. We do not know with any precision the constitution of the film as modified by frilling. And, even if we did, a rigorous calculation of the consequences would probably be impracticable. But some idea may be gained from considering the action of an obstacle, *e.g.* a sphere, of material slightly differing optically from its environment and situated in the neighbourhood of a perfectly reflecting plane surface upon which the light is incident perpendicularly. Under this condition the reflected light may still be supposed to consist of plane waves undisturbed by the previous passage through and past the obstacle.

The calculation, applying in the absence of a reflector but without limitation to the spherical form of obstacle, was given in an early paper*. In Maxwell's notation, f, g, h are the electric displacements. The magnetic

---

* "On the Electro-magnetic Theory of Light," *Phil. Mag.* Vol. xii. p. 81 (1881); *Scientific Papers*, Vol. i. p. 518.

susceptibility is supposed to be uniform throughout; the specific inductive capacity to be K, altered within the obstacle to $K + \Delta K$. The suffixes 0 and 1 refer respectively to the primary and scattered waves. The direction of propagation being supposed parallel to x and that of vibration parallel to z, we have $f_0 = g_0 = 0$, and

$$h_0 = e^{int} e^{ikx}, \quad \dots\dots\dots\dots\dots\dots\dots\dots\dots(1)$$

e^{int} being the time factor for simple progressive waves. For the scattered vibration at the point (α, β, γ) distant r from the element of volume $(dx\,dy\,dz)$ of the obstacle, we have

$$f_1, g_1, h_1 = \frac{k^2 P}{4\pi r^3} \{\alpha\gamma, \beta\gamma, -(\alpha^2 + \beta^2)\}, \quad \dots\dots\dots\dots(2)$$

where

$$P = -\frac{\Delta K}{K} \iiint h_0 e^{-ikr} dx\,dy\,dz, \quad \dots\dots\dots\dots(3)$$

and the integration is over the volume of the obstacle. If the obstacle is very small in comparison with the wave-length (λ) of the vibrations, $h_0 e^{-ikr}$ may be removed from under the integral sign and

$$P = -\frac{T \cdot \Delta K \cdot h_0 e^{-ikr}}{K}, \quad \dots\dots\dots\dots\dots\dots(4)$$

T denoting the volume of the obstacle. In the direction of primary vibration $\alpha = \beta = 0$, so that in this direction there is *no scattered vibration*. It will be understood that our suppositions correspond to primary light already polarized. If, as usually in experiment, the primary light is unpolarized, the light scattered perpendicularly to the incident rays is plane polarized and can be extinguished with a nicol.

The formation of colour depends upon other factors. When the obstacle is very small, P is constant, and the secondary vibration varies as k^2, so that the intensity is as the inverse fourth power of the wave-length, as in the theory of the blue of the sky   In this case it is immaterial whether the obstacles are of the same size or not, but for larger sizes when the colour depends mainly upon the variation of P, strongly marked effects require an approximate uniformity. If the distribution be at random, the colours due to a large number may then be inferred from the calculation relating to a single obstacle; but if the distribution were in regular patterns, complications would ensue from the necessity for taking phases into account, as in the theory of gratings. For the present purpose it suffices to consider a random distribution, although we may suppose that the centres, or more generally corresponding points, of the obstacles lie in a plane perpendicular to the direction of the primary light.

When the obstacle is a sphere, the integral in (3) can be evaluated*. The centre of the sphere, of radius R, is taken as the origin of coordinates. It is

---

* *Proc. Roy. Soc.* A, Vol. xc. p. 219 (1914). [This volume, p. 220.]

evident that, so far as the secondary ray is concerned, P depends only on the angle (χ) which this ray makes with the primary ray. We suppose that $\chi = 0$ in the direction backwards along the primary ray, and that $\chi = \pi$ along the primary ray continued. Then with introduction of the value of h_0 from (1), we find

$$P = -\frac{\Delta K \cdot 4\pi R^3 \cdot e^{i(nt-kr)}}{K}\left(\frac{\sin m}{m^3} - \frac{\cos m}{m^2}\right), \quad \ldots\ldots\ldots\ldots(5)$$

where

$$m = 2kR \cos \tfrac{1}{2}\chi. \quad \ldots\ldots\ldots\ldots\ldots\ldots\ldots\ldots\ldots(6)$$

The secondary disturbance vanishes with P, viz. when $\tan m = m$, and on these lines the formation of colour may be understood. Some further particulars are given in the paper just referred to.

The solution here expressed may be applied to illustrate the scattering of light by a series of equal spheres distributed at random over a plane perpendicular to the direction of primary propagation. The effect of a reflector will be represented by taking, instead of (1),

$$h_0 = e^{int}(e^{ikx} + e^{-ik(x+2x_0)}), \quad \ldots\ldots\ldots\ldots\ldots\ldots(7)*$$

x_0 expressing the distance between the plane of the reflector and that containing the centres of the spheres. The only difference is that

$$m^{-3}\sin m - m^{-2}\cos m$$

is now replaced by

$$\frac{\sin m}{m^3} - \frac{\cos m}{m^2} + e^{-2ikx_0}\left(\frac{\sin m'}{m'^3} - \frac{\cos m'}{m'^2}\right), \quad \ldots\ldots\ldots\ldots(8)*$$

where m is as before, and $m' = 2kR \sin \tfrac{1}{2}\chi$. In the special case where, while the incidence is perpendicular, the scattered light is nearly grazing, $\chi = \tfrac{1}{2}\pi$, $\sin \tfrac{1}{2}\chi = \cos \tfrac{1}{2}\chi = 1/\sqrt{2}$, and $m = m' = \sqrt{2} \cdot kR$; so that (8) becomes

$$(1 + e^{-2ikx_0})\left(\frac{\sin m}{m^3} - \frac{\cos m}{m^2}\right). \quad \ldots\ldots\ldots\ldots\ldots(9)*$$

This vanishes if $\cos 2kx_0* = -1$; otherwise the reflector merely introduces a constant factor, not affecting the character of the scattering. At other angles the reflector causes more complication on account of the different values of m and m'.

[* The results given in the original text have been corrected by the substitution of $-2x_0$ for x_0. It is assumed, as apparently in the original, that no change of phase occurs at the reflector. W. F. S.]

425.

MEMORANDUM ON SYNCHRONOUS SIGNALLING.

[Report to Trinity House, 1917.]

I HAVE been impressed for some time with the unsatisfactory character of the present fog signals. We must recognize that powerful siren signals are sometimes inaudible at distances but little exceeding a mile. It is true that these worst cases of inaudibility may not recur during fogs—as to this there seems to be insufficient evidence. But even when a sound-in-air signal is audible, the information conveyed is far from precise. The bearing of the source cannot be told with much accuracy, indeed some say that it cannot be told at all. The distance is still more uncertain. I should say that no system is satisfactory which does not give either the one or the other element, bearing or distance.

The system of synchronous signalling explained by Prof. Joly claims to give the *distance* with sufficient precision, and the American and Russian trials show that the claim is justified, as might indeed have been expected with some confidence, provided both signals themselves are well defined in time. The wireless electric signals are easily made sharp. Submarine signals from a bell, or explosive, would also be sharp enough. So probably would be explosive signals in air. The case of siren signals is more doubtful. Possibly the *end* might be sharp enough. Even so, the objection of the uncertain carrying of air signals remains.

I do not know whether there is already sufficient experience of submarine signals. If it be true that they can be depended upon up to distances of at least 4 or 5 miles, the case is strong for a combination of them with electric signals.

In some respects the system described in my former memorandum of 1916* has its advantages. It would give the *bearing* with electric signals *only*, but requires further experimenting, which if desired could be arranged for at the National Physical Laboratory but perhaps not during the war.

I am strongly of opinion that whatever is possible at the present time should be done to prepare the way for a better system.

* This volume, p. 398.

426.

A SIMPLE PROBLEM IN FORCED LUBRICATION.

[*Engineering*, Dec. 14, 28, 1917*.]

THE important case of a shaft or journal running in bearings has been successfully treated by Reynolds, Sommerfeld and others. As Tower showed, the combination acts as a *pump*, and of itself maintains the layer of lubricant between the opposed solid surfaces†. There are other cases, and some of them are of practical importance, where the layer can be maintained only with the aid of special devices, such as Michell bearings, or by the forcible introduction of fluid from outside, in order to compensate inevitable escapes. Thus, Fig. 1,

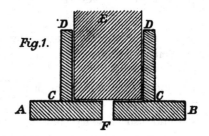

Fig.1.

when a shaft E with a flat end bears against a flat surface AB, the included oil tends to escape from the pressure, whether at C when the flat surface is continued, or at D when it is surmounted by a cylindrical cup. The permanence of the layer requires a continuous forcible feed, which may be through an axial perforation at F; for here, in contradistinction to the case of the journal, the rotation of the shaft does not avail. It is proposed to consider the problem thus presented, supposing in the first instance, that there is no cup. The small distance between the flat surfaces, *i.e.*, the thickness of the oil layer, is denoted by h, and the angular velocity of the shaft by ω. The motion is referred to cylindrical coordinates r, θ, z, where z is measured parallel

---

* In the original statement there was an error, pointed out by Mr W. Pettingill.

† As was noticed at an early date by the present writer (President's address to British Association in 1884), this requires that the layer be thicker on the ingoing than on the outgoing side. [This collection, Vol. II. p. 344.]

to the axis of symmetry, and r is the distance of any point from that axis, Fig. 2. The velocities in the three directions are respectively u, v, w, and in virtue of the symmetry, they are all independent of θ. The motion is supposed to be *steady*, that is, the same at all times, and the inertia of the fluid is neglected. Under these conditions it is easy to recognize that w may be supposed to vanish throughout, and that v is given by

$$v = \omega z r / h, \quad \dots\dots\dots\dots\dots\dots\dots\dots\dots(1)$$

where z is measured from the fixed surface, so that v there vanishes.

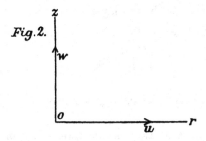

Fig. 2.

In like manner the boundary conditions at $z = 0$ and $z = h$, as well as the equation of continuity, are satisfied by

$$u = C \frac{hz - z^2}{r}, \quad \dots\dots\dots\dots\dots\dots\dots\dots(2)$$

where C is a constant. The total flow U, representing the volume of lubricant fed in unit time, which flows past every cylindrical surface of radius r, is

$$U = 2\pi r \int_0^h u\, dz = \frac{\pi h^3 C}{3}. \quad \dots\dots\dots\dots\dots\dots(3)$$

When the inertia terms are neglected, and attention is paid to the symmetry, the formal equations in cylindrical coordinates* are

$$\frac{dp}{dr} = \mu \left(\nabla^2 u - \frac{u}{r^2} \right), \quad \dots\dots\dots\dots\dots\dots(4)$$

$$0 = \nabla^2 v - \frac{v}{r^2}, \quad \dots\dots\dots\dots\dots\dots\dots\dots(5)$$

$$\frac{dp}{dz} = \mu \nabla^2 w, \quad \dots\dots\dots\dots\dots\dots\dots\dots(6)$$

p denoting the pressure and μ the viscosity, where

$$\nabla^2 = \frac{d^2}{dr^2} + \frac{1}{r}\frac{d}{dr} + \frac{d^2}{dz^2}. \quad \dots\dots\dots\dots\dots(7)$$

---

* Basset's *Hydrodynamics*, Vol. II. p. 244, 1888.

Of these (5) is satisfied by v in (1), and (6) is satisfied when $w = 0$ and p is independent of z. Also, with use of (2), (4) becomes

$$\frac{dp}{dr} = -\frac{2\mu C}{r} = -\frac{6\mu U}{\pi h^3 r}, \qquad \dots\dots\dots\dots\dots\dots(8)$$

so that

$$p - p_1 = \frac{6\mu U}{\pi h^3} \log \frac{r_1}{r}, \qquad \dots\dots\dots\dots\dots\dots(9)$$

where p_1 is the pressure at the outer radius r_1. If the layer is open at r_1, and we reckon only pressures above atmosphere, p_1 may be omitted.

The whole force sustained by the layer of fluid between the radii r_0 and r_1 is independent of ω, being given by

$$P = \frac{12\mu U}{h^3} \int_{r_0}^{r_1} r \log \frac{r_1}{r} \, dr$$

$$= \frac{3\mu U}{h^3} \left\{ r_1^2 - r_0^2 - 2r_0^2 \log \frac{r_1}{r_0} \right\}. \qquad \dots\dots\dots\dots(10)$$

If we suppose $r_0 = 0$, so that the supply takes place on the axis itself, this becomes simply

$$P = 3\mu U r_1^2 / h^3, \qquad \dots\dots\dots\dots\dots\dots\dots(11)$$

but we have then to face an infinite pressure at the axis. In practice r_0 would have to be finite though small, and would correspond to the radius of the perforation in the lower fixed plate, not much disturbing (11). In fact, if p_0 be the pressure of the feed corresponding to r_0,

$$p_0 = \frac{6\mu U}{\pi h^3} \log \frac{r_1}{r_0} = \frac{2P}{\pi r_1^2} \log \frac{r_1}{r_0}. \qquad \dots\dots\dots\dots\dots(12)$$

The moment of the forces due to viscosity, by which the rotation is resisted, has the expression

$$M = 2\pi\mu \int r^2 dr \left(\frac{dv}{dz}\right)_0 = \frac{\pi\mu\omega}{2h} (r_1^4 - r_0^4). \qquad \dots\dots\dots\dots(13)$$

It may be worth remarking that if geometric similarity is preserved, so that r_1, r_0, h are in constant ratios, a consideration of "dimensions" suffices to show that P is proportional to $\mu U r_1^{-1}$, at least when we assume independence of the rotation (ω) which does not influence u. A deficiency of viscosity may thus always be compensated by an increase of supply.

The work which must be done in unit time to maintain the rotation is $M\omega$. In addition to this, there is the work required to introduce the feed of lubricant, represented by $p_0 U$. Thus, altogether, for the work required

$$W = \frac{\pi\mu\omega^2 r_1^4}{2h} + \frac{2h^3 P^2}{3\pi\mu r_1^4} \log \frac{r_1}{r_0}. \qquad \dots\dots\dots\dots(14)$$

In practice the diminution of h calls for the utmost accuracy in fitting together the two opposed surfaces, which, however, need not be accurately plane, as well as the removal of all suspended solid matter from the lubricant.

When this is attended to, there should be *no wear* of the solid surfaces, which should never come into contact. To attain this ideal it is evidently necessary that the feed of lubricant should be established *before* the rotation commences.

It should be observed that no property of oil beyond viscosity is involved, and that the investigation may be expected to remain valid until the thickness (h) of the layer is approaching molecular limits.

P.S.—I may perhaps mention that I have made a small model, in which the opposed surfaces are those of two pennies ground to a fit, and the "lubricant" is water supplied from a tap.

427.

ON THE SCATTERING OF LIGHT BY SPHERICAL SHELLS, AND BY COMPLETE SPHERES OF PERIODIC STRUCTURE, WHEN THE REFRACTIVITY IS SMALL.

[Proceedings of the Royal Society, A, Vol. XCIV. pp. 296—300, 1918.]

THE problem of a small sphere of uniform optical quality has been treated in several papers*. In general, the calculations can be carried to an arithmetical conclusion only when the circumference of the sphere does not exceed a few wave-lengths. But when the relative refractivity is small enough, this restriction can be dispensed with, and a general result formulated.

In the present paper some former results are quoted, but the investigation is now by an improved method. It commences with the case of an infinitely thin spherical *shell*, from which the result for the complete uniform sphere is derived by integration. Afterwards application is made to a complete sphere, of which the structure is symmetrical but periodically variable along the radius, a problem of interest in connexion with the colours, changing with the angle, often met with in the organic world.

The specific inductive capacity of the general medium being unity, that of the sphere of radius R is supposed to be K, where $K-1$ is very small. Electric displacements being denoted by f, g, h, the primary wave is taken to be

$$h_0 = e^{int} e^{ikx}, \dots\dots\dots\dots\dots\dots\dots\dots\dots(1)$$

so that the direction of propagation is along x (negatively), and that of vibration parallel to z. The electric displacements in the scattered wave, so far as they depend upon the *first power* of $(K-1)$, have at a great distance the values

$$f_1, g_1, h_1 = \frac{k^2 P}{4\pi r} \left(\frac{\alpha\gamma}{r^2}, \frac{\beta\gamma}{r^2}, -\frac{\alpha^2+\beta^2}{r^2} \right), \dots\dots\dots\dots(2)$$

in which
$$P = -(K-1) \cdot e^{int} \iiint e^{ik(x-r)} \, dx\, dy\, dz. \dots\dots\dots(3)$$

---

* *Phil. Mag.* Vol. XLI. pp. 107, 274, 447 (1871); Vol. XII. p. 81 (1881); Vol. XLVII. p. 375 (1889); *Roy. Soc. Proc.* A, Vol. LXXXIV. p. 25 (1910); Vol. XC. p. 219 (1914); *Scientific Papers*, Vol. I. pp. 87, 104, 518; Vol. IV. p. 397; Vol. V. p. 547; Vol. VI. p. 220.

In these equations r denotes the distance between the point (α, β, γ), where the disturbance is to be estimated, and the element of volume $(dx\,dy\,dz)$ of the obstacle. The centre of the sphere R will be taken as the origin of coordinates. It is evident that, so far as the secondary ray is concerned, P depends only on the angle (χ) which this ray makes with the primary ray. We will suppose that $\chi = 0$ in the direction backwards along the primary ray, and that $\chi = \pi$ along the primary ray continued. The integral in (3) may then be found in the form

$$\frac{2\pi R^2 e^{-ik\rho}}{k \cos \tfrac{1}{2}\chi} \int_0^{\tfrac{1}{2}\pi} J_1 \left(2kR \cos \tfrac{1}{2}\chi \cdot \cos \phi\right) \cos^2 \phi\, d\phi, \quad \ldots\ldots\ldots\ldots(4)^*$$

ρ denoting the distance of the point of observation from the *centre* of the sphere. In the paper of 1914 I showed that the integral in (4) can be simply expressed by circular functions in virtue of a theorem given by Hobson, so that

$$P = -(K-1) \cdot 4\pi R^3 \cdot e^{i\,(nt-k\rho)} \left(\frac{\sin m}{m^3} - \frac{\cos m}{m^2}\right), \quad \ldots\ldots\ldots\ldots(5)$$

where

$$m = 2kR \cos \tfrac{1}{2}\chi. \quad \ldots\ldots\ldots\ldots\ldots\ldots\ldots\ldots(6)$$

In (5) the optical quality of the sphere, expressed by $(K-1)$, is supposed to be uniform throughout. In view of an application presently to be considered, it was desired to obtain the expression for a spherical *shell* of infinitesimal thickness dR, from which could be derived the value of P for a complete symmetrical sphere whose optical quality varies along the radius. The required result is obtained at once from (5) and (6) by differentiation. We find

$$dP = -(K-1) \cdot 4\pi R^2 dR \cdot e^{i\,(nt-k\rho)} \cdot \sin m/m, \quad \ldots\ldots\ldots\ldots(7)$$

expressing the value of P for a spherical shell of volume $4\pi R^2 dR$. The simplicity of (7) suggested that the reasoning by which it had been arrived at is needlessly indirect, and that a better procedure would be an inverse one, in which (7) was established first, and the result for the complete sphere derived from it by integration. And this anticipation was easily confirmed.

Commencing then with a spherical shell of centre O and radius OA equal to R, let xO be the direction of the primary and $O\rho$ that of the secondary ray (Fig. 1). Draw $O\zeta$ in the plane of Ox, $O\rho$, and bisecting the angle between these lines, and let ζ be a coordinate measured from O in the direction $O\zeta$, so that the plane AOA, perpendicular to $O\zeta$, is represented by $\zeta = 0$. The angle $xO\zeta$ is $\tfrac{1}{2}\chi$, as in our former notation. We have now to consider the phases represented by the factor $e^{ik\,(x-r)}$ in P. For the point O, $x = 0$, $r = \rho$, and the exponential factor is $e^{-ik\rho}$. As in the ordinary theory of specular reflection, the same is true for every point in the plane AOA and therefore for the element of surface at AA whose volume is $2\pi R\,dR\,d\zeta$. For points in a plane

* Given in the 1881 paper.

BB parallel to AA at a distance ζ the linear retardation is $-2\zeta\cos\frac{1}{2}\chi$, as in the theory of thin plates; and the exponential factor is $e^{-ik\rho}e^{2ik\zeta\cos\frac{1}{2}\chi}$. The

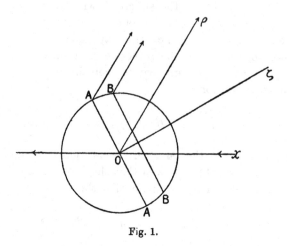

Fig. 1.

elementary volume at BB is still expressed by $2\pi R\,dR\,d\zeta$, and accordingly by (3)

$$dP = -(K-1).2\pi R\,dR.\,e^{i\,(nt-k\rho)}\int_{-R}^{+R}d\zeta e^{2ik\zeta\cos\frac{1}{2}\chi}. \quad\ldots\ldots\ldots(8)$$

The integral in (8) is $2R\sin m/m$, m being given by (6), and we recover (7) as expressing the value of dP for a spherical shell of volume $4\pi R^2 dR$.

The value of dP for a spherical shell having been now obtained independently, we can pass at once by integration to the corresponding expression for a complete sphere of uniform optical quality, thus recovering (5) by a simpler method not involving Bessel's functions at all. And a comparison of the two processes affords a demonstration of Hobson's theorem formerly employed as a stepping stone.

When P is known, the secondary vibration is given by (2), in which we may replace r by ρ. So far as it depends upon P, the angular distribution, being a function of χ, is symmetrical round Ox, the direction of primary propagation. So far as it depends on the other factors $\alpha\gamma/\rho^2$, etc., it is the same as for an infinitely small sphere; in particular no ray is emitted in the direction defined by $\alpha = \beta = 0$, that is in the direction of primary vibration. There is no limitation upon the value of R if $(K-1)$ be small *enough*; but the reservation is important, since it is necessary that at every point of the obstacle the retardation of the primary waves *due to the obstacle* be negligible.

When R is great compared with $\lambda\,(=2\pi/k)$, m usually varies rapidly with R or k, and so does P, as given for the complete uniform sphere in (5). An exception occurs when χ is nearly equal to π, that is when the secondary ray is nearly in the direction of the primary ray continued ($\beta = \gamma = 0$). In this case m is very small,

$$\frac{\sin m}{m^3} - \frac{\cos m}{m^2} = \frac{1}{3},$$

and $|P|$ is independent of k, and is proportional to R^3. The *intensity* is then

$$|P^2| = \frac{16 (K - 1)^2 \pi^2 R^6}{9}. \quad\dots\dots\dots\dots(9)$$

The haze immediately surrounding a small source of light seen through a foggy medium is of relatively great intensity. And the cause is simply that the contributions from the various parts of a small obstacle agree in phase.

But in general when R is great, so also is m, and $|P|$ varies rapidly and periodically with k along the spectrum. We might then be concerned mainly with the mean value of $|P^2|$. Now

$$|P^2| = (K - 1)^2 . 4^2 \pi^2 R^6 (\sin m - m \cos m)^2 m^{-6},$$

of which the mean value is

$$(K - 1)^2 . 8\pi^2 R^6 (1 + m^2) m^{-6},$$

or approximately, since m is great,

$$(K - 1)^2 . 8\pi^2 R^6 m^{-4}.$$

When we introduce the value of m from (6), this becomes

$$\text{Mean } |P^2| = \frac{(K - 1)^2 \pi^2 R^2}{2k^4 \cos^4 \frac{1}{2}\chi} = \frac{(K - 1)^2 R^2 \lambda^4}{32\pi^2 \cos^4 \frac{1}{2}\chi}. \quad\dots\dots\dots(10)$$

The occurrence of λ^4 shows that this is in general very small in comparison with (9).

If, instead of a sphere of uniform quality, we have to deal with one where $(K - 1)$ is variable, we must employ (7). The case of greatest interest is when $(K - 1)$, besides a constant, includes also a periodic part. For the constant part the integration proceeds as before, and for the periodic part, where $(K - 1)$ varies as a circular function of R, it presents no difficulty. It may suffice to consider the particular case where $(K - 1)$ is proportional to $\sin m$, m as before being given by (6); for this supposition evidently leads to a large augmentation of P, analogous to what occurs in crystals of chlorate of potash, to which a plane periodic structure is attributed[*]. It will be observed that the wave-length of the structure now supposed varies with χ, as well as with k or λ. Thus, if $K - 1 = \beta \sin m$,

$$P = -\pi\beta e^{i (nt - kp)} R^3 \frac{m^2 - m \sin 2m + \frac{1}{2}(1 - \cos 2m)}{m^3} \quad\dots\dots(11)$$

[*] *Phil. Mag.* Vol. xxvi. p. 256 (1888) ; *Scientific Papers*, Vol. iii. p. 204.

when the integration is taken for a complete sphere of radius R. If m is moderately great, that is, if R be a large multiple of λ, the first term on the right of (11) preponderates, and we may use approximately

$$P = -\frac{\pi \beta R^2 e^{i\,(nt-k\rho)}}{2k \cos \frac{1}{2}\chi}. \qquad \dots\dots\dots\dots\dots(12)$$

Thus, if $(K-1)$ has no constant part,

$$|P| = \frac{\pi \beta R^2}{2k \cos \frac{1}{2}\chi} = \frac{\beta R^2 \lambda}{4 \cos \frac{1}{2}\chi}. \qquad \dots\dots\dots\dots\dots(13)$$

The relation between the wave-length of the structure (Λ) and that of the light is expressed by

$$\Lambda = \tfrac{1}{2}\lambda / \cos \tfrac{1}{2}\chi. \qquad \dots\dots\dots\dots\dots\dots(14)$$

It seems probable that a structure of this sort is the cause of the remarkable colours, variable with the angle of observation, which are so frequent in beetles, butterflies, and feathers.

428.

NOTES ON THE THEORY OF LUBRICATION.

[*Philosophical Magazine*, Vol. XXXV. pp. 1—12, 1918.]

MODERN views respecting mechanical lubrication are founded mainly on the experiments of B. Tower[*], conducted upon journal bearings. He insisted upon the importance of a complete film of oil between the opposed solid surfaces, and he showed how in this case the maintenance of the film may be attained by the dragging action of the surfaces themselves, playing the part of a pump. To this end it is "necessary that the layer should be thicker on the ingoing than on the outgoing side[†]," which involves a slight displacement of the centre of the journal from that of the bearing. The theory was afterwards developed by O. Reynolds, whose important memoir[‡] includes most of what is now known upon the subject. In a later paper Sommerfeld has improved considerably upon the mathematics, especially in the case where the bearing completely envelops the journal, and his exposition[§] is much to be recommended to those who wish to follow the details of the investigation. Reference may also be made to Harrison[||], who includes the consideration of compressible lubricants (air).

In all these investigations the question is treated as two-dimensional. For instance, in the case of the journal the width—axial dimension—of the bearing must be large in comparison with the arc of contact, a condition not usually fulfilled in practice. But Michell[¶] has succeeded in solving the problem for a plane rectangular block, moving at a slight inclination over another plane surface, free from this limitation, and he has developed a system of pivoted bearings with valuable practical results.

It is of interest to consider more generally than hitherto the case of two dimensions. In the present paper attention is given more especially to the case where one of the opposed surfaces is plane, but the second not necessarily

[*] *Proc. Inst. Mech. Eng.*, 1883, 1884.

[†] British Association Address at Montreal, 1884; *Scientific Papers*, Vol. II. p. 344.

[‡] *Phil. Trans.* Vol. 177, p. 157 (1886).

[§] *Zeitschr. f. Math.* t. 50, p. 97 (1904).

[||] *Camb. Trans.* Vol. XXII. p. 39 (1913).

[¶] *Zeitschr. f. Math.* t. 52, p. 123 (1905).

so. As an alternative to an inclined plane surface, consideration is given to a broken surface consisting of two parts, each of which is parallel to the first plane surface but at a different distance from it. It appears that this is the form which must be approached if we wish the total pressure supported to be a maximum, when the length of the bearing and the closest approach are prescribed. In these questions we may anticipate that our calculations correspond pretty closely with what actually happens,—more than can be said of some branches of hydrodynamics.

In forming the necessary equation it is best, following Sommerfeld, to begin with the simplest possible case. The layer of fluid is contained between two parallel planes at $y = 0$ and at $y = h$. The motion is everywhere parallel to x, so that the velocity-component u alone occurs, v and w being everywhere zero. Moreover u is a function of y only. The tangential traction acting across an element of area represented by dx is $\mu \, (du/dy) \, dx$, where μ is the viscosity, so that the element of volume $(dx \, dy)$ is subject to the force $\mu \, (d^2u/dy^2) \, dx \, dy$. Since there is no acceleration, this force is balanced by that due to the pressure, viz. $-(dp/dx) \, dx \, dy$, and thus

$$\frac{dp}{dx} = \mu \, \frac{d^2u}{dy^2}. \quad \dots\dots\dots\dots\dots\dots\dots\dots(1)$$

In this equation p is independent of y, since there are in this direction neither motion nor components of traction, and (1), which may also be derived directly from the general hydrodynamical equations, is immediately integrable. We have

$$u = \frac{1}{2\mu} \frac{dp}{dx} y^2 + A + By, \quad \dots\dots\dots\dots\dots\dots(2)$$

where A and B are constants of integration. We now suppose that when $y = 0$, $u = -U$, and that when $y = h$, $u = 0$. Thus

$$u = \frac{y^2 - hy}{2\mu} \frac{dp}{dx} - \left(1 - \frac{y}{h}\right) U. \quad \dots\dots\dots\dots(3)$$

The whole flow of liquid, regarded as incompressible, between 0 and h is

$$\int_0^h u \, dy = -\frac{h^3}{12\mu} \frac{dp}{dx} - \frac{hU}{2} = -Q,$$

where Q is a constant, so that

$$\frac{dp}{dx} = -\frac{6\mu U}{h^3} \left(h - \frac{2Q}{U}\right). \quad \dots\dots\dots\dots\dots\dots(4)$$

If we suppose the passage to be absolutely blocked at a place where x is negatively great, we are to make $Q = 0$ and (4) gives the rise of pressure as x decreases algebraically. But for the present purpose Q is to be taken finite. Denoting $2Q/U$ by H, we write (4)

$$\frac{dp}{dx} = -\frac{6\mu U}{h^3} (h - H). \quad \dots\dots\dots\dots\dots\dots(5)$$

When $y = 0$, we get from (3) and (5)

$$\mu \frac{du}{dy} = \mu U \frac{4h - 3H}{h^2}, \quad \dots\dots\dots\dots\dots\dots(6)$$

which represents the tangential traction exercised by the liquid upon the moving plane.

It may be remarked that in the case of a simple shearing motion $Q = \frac{1}{2}hU$, making $H = h$, and accordingly

$$dp/dx = 0, \quad du/dy = U/h.$$

Our equations allow for a different value of Q and a pressure variable with x.

So far we have regarded h as absolutely constant. But it is evident that Reynolds' equation (5) remains approximately applicable to the lubrication problem in two dimensions even when h is variable, though always very small, provided that the changes are not too sudden, x being measured circumferentially and y normally to the opposed surfaces. If the whole changes of direction are large, as in the journal bearing with a large arc of contact, complication arises in the reckoning of the resultant forces operative upon the solid parts concerned; but this does not interfere with the applicability of (5) when h is suitably expressed as a function of x. In the present paper we confine ourselves to the case where one surface (at $y = 0$) may be treated as absolutely plane. The second surface is supposed to be limited at $x = a$ and at $x = b$, where h is equal to h_1 and h_2 respectively, and the pressure at both these places is taken to be zero.

For the total pressure, or load, (P) we have

$$P = \int_a^b p\,dx = -\int_a^b x \frac{dp}{dx}\,dx,$$

on integration by parts with regard to the evanescence of p at both limits. Hence by (5)

$$\frac{P}{6\mu U} = \int_a^b \frac{x\,dx}{h^2} - H \int_a^b \frac{x\,dx}{h^3}. \quad \dots\dots\dots\dots\dots(7)$$

Again, by direct integration of (5),

$$0 = \int_a^b \frac{dx}{h^2} - H \int_a^b \frac{dx}{h^3}, \quad \dots\dots\dots\dots\dots\dots(8)$$

by which H is determined. It is the thickness of the layer at the place, or places, where p is a maximum or a minimum. A change in the sign of U reverses also that of P.

Again, if $\bar{x}$ be the value of x which gives the point of application of the resultant force,

$$\bar{x} \cdot P = \int_a^b px\,dx = -\frac{1}{2}\int_a^b x^2 \frac{dp}{dx}\,dx,$$

so that
$$\frac{\bar{x} \cdot P}{3\mu U} = \int_a^b \frac{x^2 dx}{h^2} - H \int_a^b \frac{x^2 dx}{h^3}. \qquad \text{......................(9)}$$

By (7), (8), (9) $\bar{x}$ is determined.

As regards the total friction (F), we have by (6)
$$\frac{F}{\mu U} = 4 \int_a^b \frac{dx}{h} - 3H \int_a^b \frac{dx}{h^2}. \qquad \text{......................(10)}$$

Comparing (7) and (10), we see that the ratio of the total friction to the total load is *independent of μ and of U*. And, since the right-hand members of (7) and (10) are dimensionless, the ratio is also independent of the linear scale. But if the scale of h only be altered, F/P varies as h.

We may now consider particular cases, of which the simplest and the most important is when the second surface also is flat, but inclined at a very small angle to the first surface. We take
$$h = mx, \qquad \text{..............................(11)}$$
and we write for convenience
$$b - a = c, \quad h_2/h_1 = b/a = k, \qquad \text{....................(12)}$$
so that
$$m = (k - 1) h_1/c. \qquad \text{..........................(13)}$$

We find in terms of c, k, and h_1
$$H = \frac{2kh_1}{k + 1}, \qquad \text{..........................(14)}$$

$$\frac{P}{6\mu U} = \frac{c^2}{(k-1)^2 h_1^2} \left\{ \log_e k - \frac{2(k-1)}{k+1} \right\}, \qquad \text{................(15)}$$

$$\frac{\bar{x}}{\frac{1}{2}c} = \frac{k^2 - 1 - 2k \log k}{(k^2 - 1) \log k - 2(k-1)^2}, \qquad \text{................(16)}$$

$$\frac{F}{P} = \frac{h_1}{c} \frac{2(k^2-1)\log k - 3(k-1)^2}{3(k+1)\log k - 6(k-1)} \qquad \text{................(17)}$$

U being positive, the sign of P is that of
$$\log k - \frac{2(k-1)}{k+1}.$$

If $k > 1$, that is when $h_2 > h_1$, this quantity is positive. For its derivative is positive, as is also the initial value when k exceeds unity but slightly. In order that a load may be sustained, the layer must be thicker where the liquid enters.

In the above formulæ we have taken as data the length of the bearing c and the minimum distance h_1 between the surfaces. So far k, giving the maximum distance, is open. It may be determined by various considerations. Reynolds examines for what value P, as expressed in (15), is a maximum, and he gives (in a different notation) $k = 2\cdot2$.   For values of k equal to $2\cdot0$, $2\cdot1$, $2\cdot2$, $2\cdot3$ I find for the coefficient of c^2/h_1^2 on the right of (15) respectively

$$\cdot02648, \quad \cdot02665, \quad \cdot02670, \quad \cdot02663.$$

In agreement with Reynolds the maximum occurs when $k = 2\cdot2$ nearly, and the maximum value is

$$P = 0\cdot1602 \frac{\mu U c^2}{h_1^2}. \qquad \qquad (18)^*$$

It should be observed—and it is true whatever value be taken for k—that P varies as the *square* of c/h_1.

With the above value of k, viz. $2\cdot2$,

$$H = 1\cdot37\, h_1, \qquad \qquad (19)$$

fixing the place of maximum pressure.

Again, from (16) with the same value of k,

$$\bar{x} - a = 0\cdot4221\, c, \qquad \qquad (20)$$

which gives the distance of the centre of pressure from the trailing edge.

And, again with the same value of k, by (17)

$$F/P = 4\cdot70\, h_1/c. \qquad \qquad (21)$$

Since h_1 may be very small, it would seem that F may be reduced to insignificance†.

In (18)—(21) the choice of k has been such as to make P a maximum. An alternative would be to make F/P a minimum.   But it does not appear that this would make much practical difference.   In Michell's bearings it is the position of the centre of pressure which determines the value of k by (16). If we use (20), k will be $2\cdot2$, or thereabouts, as above.

When in (16) k is very large, the right-hand member tends to zero, as also does a/c, so that $\bar{x} - a$ tends to vanish, c being given.   As might be expected, the centre of pressure is then close to the trailing edge.   On the other hand, when k exceeds unity but little, the right-hand member of (16) assumes an indeterminate form.   When we evaluate it, we find

$$\bar{x} - a = \tfrac{1}{2}c.$$

For all values of $k\,(>1)$ the centre of pressure lies nearer the narrower end of the layer of fluid.

The above calculations suppose that the second surface is *plane*.   The question suggests itself whether any advantage would arise from another choice of form.   The integrations are scarcely more complicated if we take

$$h = mx^n. \qquad \qquad (22)$$

[* It may be proved that P has only one maximum when $k > 1$.

† Although the ratio F/P diminishes with h_1, F itself increases as $1/h_1$.   By (18) and (21) we have $F = \cdot75\, \dfrac{\mu U c}{h_1}$, when $k = 2\cdot2$.   W. F. S.]

We denote, as before, the ratio of the extreme thicknesses (h_2/h_1) by k, and c still denotes $b - a$. For the total pressure we get from (7) and (8)

$$\frac{P}{6\mu U} = \frac{c^2}{(k^{1/n} - 1)^2 h_1{}^2} \left\{ \frac{3n-1}{(2n-1)(3n-2)} \frac{(k^{-2+1/n} - 1)(k^{-3+2/n} - 1)}{k^{-3+1/n} - 1} - \frac{k^{-2+2/n} - 1}{2n-2} \right\},$$
$$\dots\dots(23)$$

from which we may fall back on (15) by making $n = 1$.

For example, if $n = 2$, so that the curve of the second surface is part of a common parabola, P is a maximum at

$$P = 0\cdot163 \frac{\mu U c^2}{h_1{}^2}, \quad \dots\dots\dots\dots\dots\dots\dots(24)^*$$

when $k = 2\cdot3$. The departure from (18) with $k = 2\cdot2$ is but small. In order to estimate the curvature involved we may compare $\frac{1}{2}(h_1 + h_2)$ with the middle ordinate of the curve, viz.

$$\tfrac{1}{4} m (a + b)^2 = \tfrac{1}{4} \{\sqrt{h_1} + \sqrt{(2\cdot3 \, h_1)}\}^2 = 1\cdot58 \, h_1,$$

which is but little less than

$$\tfrac{1}{2}(h_1 + h_2) = \tfrac{1}{2} h_1 (1 + 2\cdot3) = 1\cdot65 \, h_1.$$

It appears that curvature following the parabolic law is of small advantage.

I have also examined the case of $n = \infty$. It is perhaps simpler and comes to the same to assume

$$h = e^{\beta x}. \quad \dots\dots\dots\dots\dots\dots\dots\dots(25)$$

The integrals required in (7), (8) are easily evaluated. Thus

$$\int \frac{dx}{h^2} = \frac{e^{-2\beta a} - e^{-2\beta b}}{2\beta} = \frac{k^2 - 1}{2\beta k^2 h_1{}^2},$$

$$\int \frac{dx}{h^3} = \frac{e^{-3\beta a} - e^{-3\beta b}}{3\beta} = \frac{k^3 - 1}{3\beta k^3 h_1{}^3},$$

making

$$H = \frac{3 k h_1 (k^2 - 1)}{2 (k^3 - 1)}. \quad \dots\dots\dots\dots\dots\dots(26)$$

In like manner

$$\int \frac{x\,dx}{h^2} = \frac{k^2 (1 + 2\beta a) - 1 - 2\beta b}{4\beta^2 k^2 h_1{}^2},$$

$$\int \frac{x\,dx}{h^3} = \frac{k^3 (1 + 3\beta a) - 1 - 3\beta b}{9\beta^2 k^3 h_1{}^3}.$$

Using these in (7), we get on reduction

$$P = \frac{3\mu U}{\beta^2 k^2 h_1{}^2} \left\{ \frac{k^2 - 1}{6} + \frac{\beta (k^2 - k^3)(b - a)}{k^3 - 1} \right\},$$

or, since $\beta c = \log k$,

$$P = \frac{3\mu U \cdot c^2}{k^2 (\log k)^2 h_1{}^2} \left\{ \frac{k^2 - 1}{6} - \frac{k^2 (k - 1) \log k}{k^3 - 1} \right\}. \quad \dots\dots\dots(27)$$

[* It may be proved that P has only one maximum when $n = 2$, $k > 1$.   W. F. S.]

If we introduce the value of β, the equation of the curve may be written

$$h = k^{x/c}. \qquad \dots\dots\dots\dots\dots\dots\dots\dots\dots\dots(28)$$

When we determine k so as to make P a maximum, we get $k = 2\cdot3$, and

$$P = 0\cdot165\,\frac{\mu U c^2}{h_1^{\,2}}, \qquad \dots\dots\dots\dots\dots\dots\dots(29)*$$

again with an advantage which is but small.

In all the cases so far considered the thickness h increases all the way along the length, and the resultant pressure is proportional to the square of this length (c). In view of some suggestions which have been made, it is of interest to inquire what is the effect of (say) r repetitions of the same curve, as, for instance, a succession of inclined lines $ABCDEF$ (Fig. 1). It

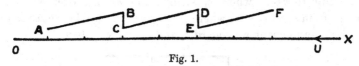

Fig. 1.

appears from (8) that H has the same value for the aggregate as for each member singly, and from (5) that the increment of p in passing along the series is r times the increment due to one member. Since the former increment is zero, it follows that the pressure is zero at the beginning and end of each member. The circumstances are thus precisely the same for each member, and the total pressure is r times that due to the first, supposed to be isolated. But if we imagine the curve spread once over the entire length by merely increasing the scale of x, we see that the resultant pressure would be increased r^2 times, instead of merely r times. Accordingly a repetition of a curve is very unfavourable. But at this point it is well to recall that we are limiting ourselves to the case of two dimensions. An extension in the third dimension, which would suffice for a particular length, might be inadequate when this length is multiplied r times.

The forms of curve hitherto examined have been chosen with regard to practical or mathematical convenience, and it remains open to find the form which according to (5) makes P a maximum, subject to the conditions of a given length and a given minimum thickness (h_1) of the layer of liquid. If we suppose that h becomes $h + \delta h$, where δ is the symbol of the calculus of variations, (8) gives

$$2\int \frac{\delta h}{h^3}\,dx - 3H \int \frac{\delta h}{h^4}\,dx + \delta H \int \frac{dx}{h^3} = 0, \qquad \dots\dots\dots\dots(30)$$

and from (7)

$$\frac{\delta P}{6\mu U} = \int \frac{\delta h\,(-2h + 3H)\,x\,dx}{h^4} - \delta H \int \frac{x\,dx}{h^3}, \qquad \dots\dots\dots\dots(31)$$

[* P increases rapidly from zero when $k=1$ to the maximum given by (29), and then decreases slowly ($P = 0\cdot141\mu U c^2/h_1^{\,2}$, when $k=4$). W. F. S.]

the integrations being always over the length. Eliminating δH, we get

$$\frac{\delta P}{12\mu U} = -\int \frac{\delta h}{h^4} \left\{ x - \frac{\int h^{-3} x\, dx}{\int h^{-3}\, dx} \right\} \left\{ h - \tfrac{3}{2} H \right\} dx. \quad \ldots\ldots\ldots(32)$$

The evanescence of δP for all possible variations δh would demand that over the whole range either

$$x = \frac{\int h^{-3} x\, dx}{\int h^{-3}\, dx}, \quad \text{or} \quad h = \tfrac{3}{2} H. \quad \ldots\ldots\ldots\ldots(33)$$

But this is not the requirement postulated. It suffices that the coefficient of δh on the right of (32) vanish over that part of the range where $h > h_1$, and that it be negative when $h = h_1$, so that a *positive* δh in this region involves a decrease in P, a negative δh here being excluded *a priori*. These conditions may be satisfied if we make $h = h_1$ from $x = 0$ at the edge where the layer is thin to $x = c_1$, where c_1 is finite, and $h = \tfrac{3}{2} H$ over the remainder of the range from c_1 to $c_1 + c_2$, where $c_1 + c_2 = c$, the whole length concerned (Fig. 2). For the moment we regard c_1 and c_2 as prescribed.

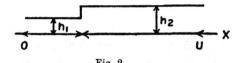

Fig. 2.

For the first condition we have by (8)

$$\tfrac{2}{3} h_2 = H = \frac{c_1/h_1{}^2 + c_2/h_2{}^2}{c_1/h_1{}^3 + c_2/h_2{}^3},$$

so that

$$c_2/c_1 = k^2 (2k - 3), \quad \ldots\ldots\ldots\ldots\ldots\ldots(34)$$

determining k, where as before $k = h_2/h_1$. The fulfilment of (34) secures that $h = \tfrac{3}{2} H$ over that part of the range where $h = h_2$. When $h = h_1$, $h - \tfrac{3}{2} H$ is negative; and the second condition requires that over the range from 0 to c_1

$$\frac{\int h^{-3} x\, dx}{\int h^{-3}\, dx} - x$$

be positive, or since c_1 is the greatest value of x involved, that

$$\int h^{-3} x\, dx - c_1 \int h^{-3}\, dx = +. \quad \ldots\ldots\ldots\ldots(35)$$

The integrals can be written down at once, and the condition becomes

$$k^3 < c_2{}^2/c_1{}^2, \quad \ldots\ldots\ldots\ldots\ldots\ldots(36)$$

whence on substitution of the value of c_2/c_1 from (34),

$$k (2k - 3)^2 > 1. \quad \ldots\ldots\ldots\ldots\ldots\ldots(37)^*$$

If k be such as to satisfy (37) and c_2/c_1 be then chosen in accordance with (34) and regarded as fixed, every admissible variation of h diminishes P. But the ratio c_2/c_1 is still at disposal within certain limits, while $c_1 + c_2 (= c)$ is prescribed.

[* This inequality may be written $(k-1)\{4(k-1)^2 - 3\} > 0$; showing that, since $k > 1$ and the conditions are satisfied when (37) becomes an equality, we must have $k \ne 1 + \dfrac{\sqrt{3}}{2} \ne 1\cdot 866$. W. F. S.]

In terms of k and c by (34)

$$c_1 = \frac{c}{1 + 2k^3 - 3k^2}, \quad c_2 = \frac{c(2k^3 - 3k^2)}{1 + 2k^3 - 3k^2}, \quad \dots\dots\dots(38)$$

and by (7)

$$\frac{P}{\mu U} = \frac{1}{h_1^2}\left\{c_1^2(3 - 2k) + \frac{2c_1c_2 + c_2^2}{k^2}\right\} = \frac{c^2}{h_1^2}\frac{2k - 3}{1 + 2k^3 - 3k^2} = \frac{c^2}{h_1^2}f(k). \quad (39)$$

The maximum* of $f(k)$ is 0·20626, and it occurs when $k = 1·87$. The following shows also the neighbouring values:

k	$f(k)$	$k(2k-3)^2$
1·86	0·20625	0·964
1·87	0·20626	1·024
1·88	0·20618	1·086

It will be seen that while $k = 1·86$ is inadmissible as not satisfying (37), $k = 1·87$ is admissible and makes

$$P = 0·20626 \frac{\mu U c^2}{h_1^2}, \quad \dots\dots\dots\dots(40)*$$

no great increase on (18). It may be repeated that k is the ratio of the two thicknesses of the layer (h_2/h_1), and that by (34)

$$c_2/c_1 = 2·588. \quad \dots\dots\dots\dots\dots(41)*$$

This defines the form of the upper surface which gives the maximum total pressure when the minimum thickness and the total length are given, and it is the solution of the problem as proposed. But it must not be overlooked that it violates the supposition upon which the original equation (5) was founded. The solution of an accurate equation would probably involve some rounding off of the sharp corners, not greatly affecting the numerical results.

The distance $\bar{x}$ of the centre of pressure from the narrow end is given by

$$\bar{x} = 0·4262\,c, \quad \dots\dots\dots\dots(42)*$$

differing very little from the value found in (20). From (10) with use of (38) we get

$$\frac{F}{\mu U} = \frac{4c(k-1)^2}{h_1(1 + 2k^3 - 3k^2)} = \frac{4c}{(2k+1)h_1}, \quad \dots\dots\dots(43)$$

and

$$\frac{F}{P} = \frac{4h_1(k-1)}{c(2k-3)}. \quad \dots\dots\dots\dots(44)$$

If $k = 1·87$,

$$F/P = 4·091\,h_1/c, \quad \dots\dots\dots\dots(45)*$$

a little less than was found in (21). The maximum total pressure and the

[* The maximum occurs when $4k^2 - 8k + 1 = 0$, precisely when (37) begins to hold (cf. footnote on p. 530); this explains the numerical coincidence. Taking $k = 1 + \frac{\sqrt{3}}{2}$ we have $P = 0·20627\mu Uc^2/h_1^2$, $c_2/c_1 = 2·549$, $\bar{x} = (c + c_1)/3 = ·4273c$, $F/P = 4·098h_1/c$. W. F. S.]

corresponding ratio F/P are both rather more advantageous in the arrangement now under discussion than for the simply inclined line. But the choice would doubtless depend upon other considerations.

The particular case treated above is that which makes P a maximum. We might inquire as to the form of the curve for which F/P is a minimum, for a given length and closest approach to the axis of x. In the expression corresponding with (32), instead of a product of two linear factors*, the coefficient of δh will involve a quadratic factor of the form

$$Bxh + Ch^2 + Dx + Eh + F, \quad\quad\quad\quad (46)$$

so that the curve is again hyperbolic in the general sense. But its precise determination would be troublesome and probably only to be effected by trial and error. It is unlikely that any great reduction in the value of F/P would ensue.

Fig. 3 is a sketch of a suggested arrangement for a footstep. The white parts are portions of an original plane surface. The four black radii represent grooves for the easy passage of lubricant. The shaded parts are slight depressions of uniform depth, such as might be obtained by etching with acid. It is understood that the opposed surface is plane throughout.

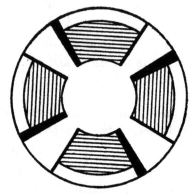

Fig. 3.

[* This statement appears to be due to an oversight. We have in fact

$$\delta\,(F/P)\,P^2/(6\mu^2 U^2) \int h^{-3} dx = \int h^{-4}\,(-2h + 3H)\,\delta h\,\{-3\int h^{-2} dx \int h^{-2} x\,dx + 4\int h^{-1} dx \int h^{-3} x\,dx$$
$$+ hP/(3\mu U)\int h^{-3} dx - xF/(\mu U)\int h^{-3} dx\}\,dx,$$

the integrations being over the length (c). Hence for a minimum of F/P the boundary may be taken as $h = h_1$ from $x = 0$ to $x = c_1$, as $h = h_3 = 3H/2$ from $c_1 + c_2$ to c, and from c_1 to $c_1 + c_2$ as an oblique line with an equation which must be made to coincide with the second factor equated to zero. This line must be continuous with the first at (c_1, h_1), in order that over the latter the second factor may be positive, and it is inclined to the axis of x at an angle $\tan^{-1} 3F/P$. If there is a discontinuity $h_3 - h_2$ at $x = c_1 + c_2$, and $h_3 = k'h_1$, $h_1 = lh_2$, where $k > 1$, $l < 1$, the condition $3H = 2h_3$ yields by (8)

$$c_1\,(2k' + 1)\,(k' - 1)^2 + c_2\,\{k'^3 l\,(1 + l) - 3k'^2 l + 1\} = c.$$

The remaining two conditions, to be derived from proportionating the second factor to $-h_1 + mc_1 + h - mx$, where $mc_2 = h_1\,(1 - l)/l$, provide two equations of the second and third degrees respectively in c_1/c_2, and lead to very complicated expressions. Without, however, including the oblique line, it may be shown that the two lines $h = h_1$ from $x = 0$ to c_1, and $h = h_2 = kh_1$ from c_1 to $c_1 + c_2$, with $2h_2 = 3H$, as on pp. 530, 531, make F/P a minimum when $h + h_1$, provided $4k^3 - 8k^2 + k - 3 \lessgtr 0$, leading to $k \lessgtr 2\cdot06$ approximately: since these lines have been shown on pp. 530, 531 to make P a maximum, they therefore also make F a minimum when $k \lessgtr 2\cdot06$. With $k = 2\cdot06$, (44) and (39) make $F/P = 4\cdot013 h_1/c$, $P = \cdot19469 \mu Uc^2/h_1^2$. But, with F/P positive, the minimum value of (44) occurs at $k = 2$, when $F/P = 4h_1/c$, $P = 0\cdot2\mu Uc^2/h_1^2$. Accordingly, as this value of F/P is not a minimum for all variations outside the straight line $h = h_1$, the actual minimum value of F/P must be less than $4h_1/c$. W. F. S.]

P.S.　Dec. 13.—In a small model the opposed pieces were two pennies ground with carborundum to a fit.　One of them—the stationary one—was afterwards grooved by the file and etched with dilute nitric acid according to Fig. 3, sealing-wax, applied· to the hot metal, being used as a " resist."　They were mounted in a small cell of tin plate, the upper one carrying an inertia bar.　With oil as a lubricant, the contrast between the two directions of rotation was very marked.

Opportunity has not yet been found for trying polished glass plates, such as are used in optical observations on "interference."　In this case the etching would be by hydrofluoric acid*, and air should suffice as a lubricant.

* Compare *Nature*, Vol. LXIV. p. 385 (1901) ; *Scientific Papers*, Vol. IV. p. 546.

429.

ON THE LUBRICATING AND OTHER PROPERTIES OF THIN OILY FILMS.

[*Philosophical Magazine*, Vol. XXXV. pp. 157—162, 1918.]

THE experiments about to be described were undertaken to examine more particularly a fact well known in most households. A cup of tea, standing in a dry saucer, is apt to slip about in an awkward manner, for which a remedy is found in the introduction of a few drops of water, or tea, wetting the parts in contact. The explanation is not obvious, and I remember discussing the question with Kelvin many years ago, with but little progress.

It is true that a drop of liquid between two curved surfaces draws them together and so may increase the friction. If d be the distance between the plates at the edge of the film, T the capillary tension, and α the angle of contact, the whole force is*

$$\frac{2AT\cos\alpha}{d} + BT\sin\alpha,$$

A being the area of the film between the plates and B its circumference. If the fluid wets the plate, $\alpha = 0$ and we have simply $2AT/d$. For example, if $d = 6 \times 10^{-5}$ cm., equal to a wave-length of ordinary light, and T (as for water) be 74 dynes per cm., the force per sq. cm. is 25×10^5 dynes, a suction of $2\frac{1}{2}$ atmospheres. For the present purpose we may express d in terms of the radius of curvature (ρ) of one of the surfaces, the other being supposed flat, and the distance (x) from the centre to the edge of the film. In two dimensions $d = x^2/2\rho$, and A (per unit of length in the third dimension) $= 2x$, so that the force per unit of length is $8\rho T/x$, inversely as x. On the other hand, in the more important case of symmetry round the common normal $A = \pi x^2$, and the whole force is $4\pi\rho T$, *independent of x*, but increasing with the radius of curvature. For example, if $T = 74$ dynes per cm., and $\rho = 100$ cm., the force is 925 dynes, or the weight of about 1 gram†. The radius of curvature (ρ) might of course be much greater. There are circumstances where this force is of importance; but, as we shall see presently, it does not avail to explain the effects now under consideration.

---

* See for example Maxwell on Capillarity. *Collected Papers*, Vol. II. p. 571.

[† This result does not correspond to the stated values of T and ρ, which imply a force of 93,000 dynes, or the weight of about 95 grammes. W. F. S.]

My first experiments were very simple ones, with a slab of thick plate glass and a small glass bottle weighing about 4 oz. The diameter of the bottle is $4\frac{1}{2}$ cm., and the bottom is concave, bounded by a rim which is not ground but makes a fairly good fit with the plate. The slab is placed upon a slope, and the subject of observation is the slipping of the bottle upon it. If we begin with surfaces washed and well rubbed with an ordinary cloth, or gone over with a recently wiped hand, we find that at a suitable inclination the conditions are uniform, the bottle starting slowly and moving freely from every position. If now we breathe upon the slab, maintained in a fixed position, or upon the bottle, or upon both, we find that the bottle *sticks* and requires very sensible forces to make it move down. A like result ensues when the contacts are thoroughly wetted with water instead of being merely damped. When, after damping with the breath, evaporation removes the moisture, almost complete recovery of the original slipperiness recurs.

In the slippery condition the surfaces, though apparently clean, are undoubtedly coated with an invisible greasy layer. If, after a thorough washing and rubbing under the tap, the surfaces are dried by evaporation after shaking off as much of the water as possible, they are found to be sticky as compared with the condition after wiping. A better experiment was made with substitution of a strip of thinner glass about 5 cm. wide for the thick slab. This was heated strongly by an alcohol flame, preferably with use of a blow-pipe. At a certain angle of inclination the bottle was held everywhere, but on going over the surface with the fingers, not purposely greased, free movement ensued. As might have been expected, the clean surface is sticky as compared with one slightly greased; the difficulty so far is to explain the effect of moisture upon a surface already slightly greased. It was not surprising that the effect of alcohol was similar to that of water.

At this stage it was important to make sure that the stickiness due to water was not connected with the minuteness of the quantity in operation. Accordingly a glass plate was mounted at a suitable angle in a dish filled with water. Upon this fully drowned surface the bottle stuck, the inclination being such that on the slightest greasing the motion became free. In another experiment the water in the dish was replaced by paraffin oil. There was decided stickiness as compared with surfaces slightly greased.

The better to guard against the ordinary operation of surface tension, the weight of the bottle was increased by inclusion of mercury until it reached 20 oz., but without material modification of the effects observed. The moisture of the breath, or drowning in water whether clean or soapy, developed the same stickiness as before.

The next series of experiments was a little more elaborate. In order to obtain measures more readily, and to facilitate drowning of the contacts, the slab was used in the horizontal position and the movable piece was pulled by a thread which started horizontally, and passing over a pulley carried a small pan into which weights could be placed. The pan itself weighed 1 oz. (28 grams). Another change was the substitution for the bottle of a small carriage standing on glass legs terminating in three feet of hemispherical form and 5 mm. in diameter. The whole weight of the carriage, as loaded, was $7\frac{3}{4}$ oz. The object of the substitution was to eliminate any effects which might arise from the comparatively large area of approximate contact presented by the rim of the bottle, although in that case also the actual contacts would doubtless be only three in number and of very small area.

With $\frac{1}{2}$ oz. in pan and surfaces treated with the hand, the carriage would move within a second or two after being placed in position, but after four or five seconds' contact would stick. After a few minutes' contact it may require $1\frac{1}{2}$ oz. in pan to start it. When the slab is breathed upon it requires, even at first, $3\frac{1}{2}$ oz. in the pan to start the motion. As soon as the breath has evaporated, $\frac{1}{2}$ oz. in pan again suffices. When the weight of the pan is included, the forces are seen to be as $1:3$. When the feet stand in a pool of water the stickiness is nearly the same as with the breath, and the substitution of soapy for clean water makes little difference.

In another day's experiment paraffin (lamp) oil was used. After handling, there was free motion with 1 oz. in pan. When the feet stood in the oil, from $2\frac{3}{4}$ to 3 oz. were needed in the pan. Most of the oil was next removed by rubbing with blotting-paper until the slab looked clean. At this stage $\frac{3}{4}$ oz. in pan sufficed to start the motion. On again wetting with oil 2 oz. sufficed instead of the $2\frac{3}{4}$ oz. required before. After another cleaning with blotting-paper $\frac{1}{2}$ oz. in pan sufficed. From these results it appears that the friction is greater with a large dose than with a minute quantity of the *same* oil, and this is what is hard to explain. When olive oil was substituted for the paraffin oil, the results were less strongly marked.

Similar experiments with a carriage standing on *brass* feet of about the same size and shape as the glass ones gave different results. It should, however, be noticed that the brass feet, though fairly polished, could not have been so smooth as the fine surfaces of the glass. The present carriage weighed (with its load) $6\frac{1}{2}$ oz., and on the well-handled glass slide moved with $\frac{1}{4}$ oz. in pan. When the slide was breathed upon, the motion was as free as, perhaps more free than, before. And when the feet stood in a pool of water, there was equal freedom. A repetition gave confirmatory results. On another day paraffin oil was tried. At the beginning $\frac{1}{4}$ oz. in pan sufficed on the handled slab. With a pool of oil the carriage still moved with $\frac{1}{4}$ oz. in pan, but perhaps not quite so certainly. As the oil was removed with blotting-paper the motion became freer, and when the oil-film had visibly disappeared the $\frac{1}{4}$ oz. in pan could about be dispensed with. Doubtless a trace of oil remained.

The blotting-paper was of course applied to the feet and legs of the carriage, as well as to the slab.

In attempting to interpret these results, it is desirable to know what sort of thickness to attribute to the greasy films on handled surfaces. But this is not so easy a matter as when films are spread upon water. In an experiment made some years ago * I found that the mean thickness of the layer on a glass plate, heavily greased with fingers which had touched the hair, was about $\frac{1}{8}$ of the wave-length of visible light, viz. about 10^{-4} mm. The thickness of the layer necessary to induce slipperiness must be a small fraction of this, possibly $\frac{1}{10}$, but perhaps much less. We may compare this with the thickness of olive oil required to stop the camphor-movements on water, which I found † to be about 2×10^{-6} mm. It may well be that there is little difference in the quantities required for the two effects.

In view of the above estimate and of the probability that the point at which surface-tension begins to fall corresponds to a thickness of a single layer of molecules ‡, we see that the phenomena here in question probably lie out-side the field of the usual theory of lubrication, where the layer of lubricant is assumed to be at least many molecules thick. We are rather in the region of incipient *seizing*, as is perhaps not surprising when we consider the small-ness of the surfaces actually in contact. And as regards seizing, there is difficulty in understanding why, when it actually occurs, rupture should ensue at another place rather than at the recently engaged surfaces.

It may perhaps be doubted whether the time is yet ripe for a full dis cussion of the behaviour of the thinnest films, but I will take this opportunity to put forward a few remarks. Two recent French writers, Devaux § and Marcelin ‖, who have made interesting contributions to the subject, accept my suggestion that the drop of tension in contaminated surfaces commences when the layer is one molecule thick; but Hardy ¶ points out a difficulty in the case of pure oleic acid, where it appears that the drop commences at a thick-ness of $1\cdot3 \times 10^{-6}$ mm., while the thickness of a molecule should be decidedly less. Many of Devaux' observations relate to the case where the quantity of oil exceeds that required for the formation of the mono-molecular layer, and he formulates a conclusion, not accepted by Marcelin, that the thickness of the layer depends upon the existence and dimensions of the globules into which most of the superfluous oil is collected, inasmuch as experiment proves that when a layer with fine globules exists beside a layer with large globules,

* *Phil. Mag.* Vol. xix. p. 96 (1910) ; *Scientific Papers*, Vol. v. p. 538.

† *Proc. Roy. Soc.* Vol. xlvii. p. 364 (1890) ; *Scientific Papers*, Vol. iii. p. 349.

‡ *Phil. Mag.* Vol. xlviii. p. 321 (1899) ; *Scientific Papers*, Vol. iv. p. 430.

§ A summary of Devaux' work, dating from 1903 onwards, will be found in the *Revue Gén. d. Sciences* for Feb. 28, 1913.

‖ *Annales d. Physique*, t. i. p. 19 (1914).

¶ *Proc. Roy. Soc.* A, Vol. lxxxviii. p. 319 (1913).

the former always contracts at the expense of the latter. As to this, it may be worth notice that the tension T of the contaminated surface could not be expressed as a function merely of the volume of the drop and of the two other tensions, viz. T_1 the tension of an air-oil surface and T_2 that of a water-oil surface. It would be necessary to introduce other quantities, such as gravity, or molecular dimensions. I am still of the opinion formerly expressed that these complications are the result of *impurity* in the oil. If the oil were really homogeneous, Devaux' views would lead one to regard the continued existence of two sizes of globules on the same surface as impossible. What would there be to hinder the rapid growth of the smaller at the expense of the greater until equality was established? On the other hand, an impurity, present only in small proportion, would naturally experience more difficulty in finding its way about.

The importance of impurities in influencing the transformations of oil-films was insisted on long ago by Tomlinson[*]; and as regards olive oil, Miss Pockels showed that the behaviour of purified oil is quite different from that of the common oil. She quotes Richter (*Nature*, Vol. XLIX. p. 488) as expressing the opinion that the tendency of oil to spread itself on water is only due to the free oleic acid contained in it, and that if it were possible to completely purify the oil from oleic acid, it would not spread at all[†]. Some confusion arises from the different meanings attached to the word "spreading." I suppose no one disputes the rapid spreading upon a clean surface which results in the formation of the invisible mono-molecular layer. Miss Pockels calls this a solution current—a rather misleading term, which had tended to obscure the meaning of her really valuable work. It is the second kind of spreading in a thicker layer, resulting in more or less rapid subsequent transformations, which is attributed to the presence of oleic acid. Miss Pockels says: "The Provence oil used in my experiment was shaken up twice with pure alcohol, and the rest (residue) of the latter being carefully removed, a drop of the oil was placed upon the freshly formed water-surface in a small dish by means of a brass wire previously cleaned by ignition. The oil did not really spread, but after a momentary centrifugal movement, during which several small drops were separated from it, it contracted itself in the middle of the surface, and a second drop deposited on the same vessel remained absolutely motionless." I have repeated this experiment, using oil which is believed to have come direct from Italy. A drop of this placed upon a clean water-surface at once drives dust to the boundary in forming the mono-molecular layer, and in addition flattens itself out into a disk of considerable size, which rapidly undergoes the transformations well described and figured by Devaux. The same oil, purified by means of alcohol on Miss Pockels' plan, behaves quite differently. The first spreading, driving dust to the boundary,

---

[*] *Phil. Mag.* Vol. XXVI. p. 187 (1863).

[†] *Nature*, Vol. L. p. 223 (1894).

takes place entirely as before. But the drop remains upon the water as a lens, and flattens itself out, if at all, only very slowly. Small admixtures of the original oil with the purified oil behave in an intermediate manner, flattening out slowly and allowing the beautiful transformations which follow to be observed at leisure.

Another point of importance does not appear to have been noticed. Water-surfaces on which purified olive oil stands in drops *still allow the camphor-movements.* Very small fragments spin merrily, while larger ones by their slower movements testify to the presence of the oil. Perhaps this was the reason why in my experiments of 1890 I found the approximate, rather than the absolute, stoppage of the movements to give the sharpest results. The absolute stoppage, dependent upon the presence of impurity, might well be less defined.

If, after the deposition of a drop of purified oil, the surface be again dusted over with sulphur or talc and then touched with a very small quantity of the original oil, the dust is driven away a second time and camphor-movements cease.

The manner in which impurity operates in these phenomena merits close attention. It seems pretty clear that from pure oil water will only take a layer one molecule thick. But when oleic acid is available, a further drop of tension ensues. The question arises how does this oleic acid distribute itself? Is it in substitution for the molecules of oil, or an addition to them constituting a second layer? The latter seems the more probable. Again, how does the impurity act when it leads the general mass into the unstable flattened-out form? In considering such questions Laplace's theory is of little service, its fundamental postulate of forces operating over distances large in comparison with molecular dimensions being plainly violated.

430.

ON THE SCATTERING OF LIGHT BY A CLOUD OF SIMILAR SMALL PARTICLES OF ANY SHAPE AND ORIENTED AT RANDOM.

[*Philosophical Magazine*, Vol. XXXV. pp. 373—381, 1918.]

FOR distinctness of conception the material of the particles may be supposed to be uniform and non-magnetic, but of dielectric capacity different from that of the surrounding medium; at the same time the results at which we shall arrive are doubtless more general. The smallness is, of course, to be understood as relative to the wave-length of the vibrations.

When the particles are spherical, the problem is simple, as their orientation does not then enter*. If the incident light be polarized, there is no scattered ray in the direction of primary electric vibration, or if the incident light be unpolarized there is complete polarization of the light scattered at right angles to the direction of primary propagation. The consideration of elongated particles shows at once that a want of symmetry must usually entail a departure from the above law of polarization and may be one of the causes, though probably not the most important, of the incomplete polarization of sky-light at 90° from the sun. My son's recent experiments upon light scattered by carefully filtered gases† reveal a decided deficiency of polarization in the light emitted perpendicularly, and seem to call for a calculation of what is to be expected from particles of arbitrary shape.

As a preliminary to a more complete treatment, it may be well to take first the case of particles symmetrical about an axis, or at any rate behaving as if they were such, for the calculation is then a good deal simpler. We may also limit ourselves to finding the *ratio* of intensities of the two polarized components in the light scattered at right angles, the principal component being that which vibrates parallel to the primary vibrations, and the subordinate component (vanishing for spherical particles) being that in which

---

* *Phil. Mag.* Vol. XLI. pp. 107, 274, 447 (1871), Vol. XII. p. 81 (1881), Vol. XLVII. p. 375 (1899); *Scientific Papers*, Vol. I. pp. 87, 104, 518, Vol. IV. p. 397.

† *Roy. Soc. Proc.* A, Vol. XCIV. p. 453 (1918); see also A, Vol. XCV. p. 155 (1918).

the vibrations are perpendicular to the primary vibrations. All that we are then concerned with are certain resolving factors, and the integration over angular space required to take account of the random orientations. In virtue of the postulated symmetry, a revolution of a particle about its own axis has no effect, so that in the integration we have to deal only with the direction of this axis. It is to be observed that the system of vibrations scattered by a particle depends upon the direction of primary *vibration* without regard to that of primary *propagation*. In the case of a spherical particle the system of scattered vibrations is symmetrical with respect to this direction and the amplitude of the scattered vibration is proportional to the cosine of the angle between the primary and secondary vibrations. When we pass to unsymmetrical particles, we have first to resolve the primary vibrations in directions corresponding to certain principal axes of the disturbing particle and to introduce separate coefficients of radiation for the different axes. Each of the three component radiations is symmetrical with respect to its own axis, and follows the same law as obtains for the sphere[*].

In Fig. 1 the various directions are represented by points on a spherical surface with centre O. Thus in the rectangular system XYZ, OZ is the direction of primary vibration, corresponding (we may suppose) to primary propagation parallel to OX. The rectangular system UVW represents in like manner the principal axes of a particle, so that UV, VW, WU are quadrants. Since symmetry of the particle round W has been postulated, there is no loss of generality in taking U upon the prolongation of ZW. As usual, we denote ZW by θ, and XZW by ϕ.

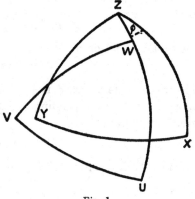

Fig. 1.

The first step is the resolving of the primary vibration Z in the directions U, V, W. We have

$$\cos ZU = -\sin \theta, \quad \cos ZV = 0, \quad \cos ZW = \cos \theta. \quad \ldots\ldots\ldots(1)$$

The coefficients, dependent upon the character of the particle, corresponding to U, V, W may be denoted by A, A, C; and we seek the effect along the scattered ray OY, perpendicular to both primary vibration and primary propagation. The ray scattered in this direction will not be completely polarized, and we consider separately vibrations parallel to Z and to X. As regards the former, we have the same set of factors over again, as in (1), so that the vibration is $A \sin^2 \theta + C \cos^2 \theta$, reducing to C simply, if $A = C$. This is the result for a single particle whose axis is at W. What we are aiming

---

[*] *Phil. Mag.* Vol. XLIV. p. 28 (1897); *Scientific Papers*, Vol. IV. p. 305.

at is the aggregate intensity due to a large number of particles with their positions and their axes distributed at random. The *mean* intensity is

$$\int_0^{\frac{1}{2}\pi} \{A + (C - A) \cos^2 \theta\}^2 \sin \theta d\theta \div \int_0^{\frac{1}{2}\pi} \sin \theta d\theta$$

$$= A^2 + \frac{2(C - A)A}{3} + \frac{(C - A)^2}{5} = \tfrac{1}{15}(8A^2 + 3C^2 + 4AC). \quad......(2)$$

This represents the intensity of that polarized component of the scattered light along OY whose vibrations are parallel to OZ.

For the vibrations parallel to OX the second set of resolving factors is $\cos UX$, $\cos VX$, $\cos WX$. Now from the spherical triangle UZX,

$$\cos UX = \sin (90° + \theta) \cos \phi = \cos \theta \cos \phi.$$

Also from the triangles VZX, WZX,

$$\cos VX = \cos VZX = \cos (90° + \phi) = - \sin \phi,$$

$$\cos WX = \sin \theta \cos \phi.$$

The first set of factors remains as before. Taking both sets into account, we get for the vibration parallel to X

$$- A \sin \theta \cos \theta \cos \phi + C \cos \theta \sin \theta \cos \phi,$$

the square of which is

$$(C - A)^2 \sin^2 \theta \cos^2 \theta \cos^2 \phi. \quad......................(3)$$

The mean value of $\cos^2 \phi$ is $\tfrac{1}{2}$. That of $\cos^2 \theta$ is $\tfrac{1}{3}$ and that of $\cos^4 \theta$ is $\tfrac{1}{5}$, as above, so that corresponding to (2) we have for the mean intensity of the vibrations parallel to X

$$\tfrac{1}{2}(C - A)^2(\tfrac{1}{3} - \tfrac{1}{5}) = \tfrac{1}{15}(C - A)^2. \quad...................(4)$$

The ratio of intensities of the two components is thus

$$\frac{(C - A)^2}{8A^2 + 3C^2 + 4AC}. \quad.............................(5)$$

Two particular cases are worthy of notice. If A can be neglected in comparison with C, (5) becomes simply *one-third*. On the other hand, if A is predominant, (5) reduces to *one-eighth*.

The above expressions apply when the primary light, propagated parallel to X, is completely polarized with vibrations parallel to Z, the direction of the secondary ray being along OY. If the primary light be unpolarized, we have further to include the effect of the primary vibrations parallel to Y. The two polarized components scattered along OY, resulting therefrom, both vibrate in directions perpendicular to OY, and accordingly are both represented by (4). In the case of unpolarized primary light we have therefore to double (4) for the secondary vibrations parallel to X, and to add together (2) and (4) for the vibrations parallel to Z. The latter becomes

$$\tfrac{1}{15}(9A^2 + 4C^2 + 2AC),$$

and for the ratio of intensities of the two components

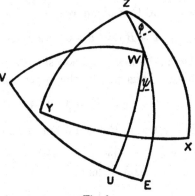

Fig. 2.

$$\frac{2(C-A)^2}{9A^2 + 4C^2 + 2AC} \cdot \quad \ldots\ldots(6)$$

When $A = 0$, this ratio is *one-half*.

For a more general treatment, which shall include all forms of particle, we must introduce another angle ψ to represent the inclination of WU to ZW produced, Fig. 2. The direction cosines of either set of axes with respect to the other are given by the formulæ*

$$\left. \begin{aligned} \cos XU &= -\sin\phi\sin\psi + \cos\phi\cos\psi\cos\theta \\ \cos YU &= \cos\phi\sin\psi + \sin\phi\cos\psi\cos\theta \\ \cos ZU &= -\sin\theta\cos\psi \end{aligned} \right\}, \quad \ldots\ldots\ldots\ldots(7)$$

$$\left. \begin{aligned} \cos XV &= -\sin\phi\cos\psi - \cos\phi\sin\psi\cos\theta \\ \cos YV &= \cos\phi\cos\psi - \sin\phi\sin\psi\cos\theta \\ \cos ZV &= \sin\theta\sin\psi \end{aligned} \right\}, \quad \ldots\ldots\ldots\ldots(8)$$

$$\left. \begin{aligned} \cos XW &= \sin\theta\cos\phi \\ \cos YW &= \sin\theta\sin\phi \\ \cos ZW &= \cos\theta \end{aligned} \right\}. \quad \ldots\ldots\ldots\ldots\ldots\ldots\ldots\ldots\ldots\ldots(9)$$

Supposing, as before, that the primary vibration is parallel to Z, we have as the first set of factors

$$\left. \begin{aligned} \cos ZU &= -\sin\theta\cos\psi \\ \cos ZV &= \sin\theta\sin\psi \\ \cos ZW &= \cos\theta \end{aligned} \right\}. \quad \ldots\ldots\ldots\ldots\ldots\ldots\ldots\ldots(10)$$

For the vibrations propagated along OY which are parallel to Z, we have the same factors over again with coefficients A, B, C as before, and the vibration is expressed by

$$A\sin^2\theta\cos^2\psi + B\sin^2\theta\sin^2\psi + C\cos^2\theta; \quad \ldots\ldots\ldots\ldots(11)$$

while for the intensity

$$\begin{aligned} I = {} & A^2\sin^4\theta\cos^4\psi + B^2\sin^4\theta\sin^4\psi + C^2\cos^4\theta \\ & + 2AB\sin^4\theta\cos^2\psi\sin^2\psi + 2BC\sin^2\theta\cos^2\theta\sin^2\psi \\ & + 2CA\sin^2\theta\cos^2\theta\cos^2\psi. \quad \ldots\ldots\ldots\ldots\ldots\ldots\ldots\ldots(12) \end{aligned}$$

---

* See, for example, Routh's *Rigid Dynamics*, Part I. § 258, 1897. ψ and ϕ are interchanged.

This is for a single particle, and we have now to take the mean for all orientations. The mean value of $\sin^4 \psi$, or $\cos^4 \psi$, is $\frac{3}{8}$; that of $\sin^2 \psi \cos^2 \psi$ is $\frac{1}{8}$; and that of $\sin^2 \psi$ is $\frac{1}{2}$. The averaging with respect to ψ thus yields

$$I = \tfrac{3}{8}(A^2 + B^2)\sin^4 \theta + C^2 \cos^4 \theta + \tfrac{1}{4}AB \sin^4 \theta + (A + B)C \sin^2 \theta \cos^2 \theta. \quad \ldots(13)$$

Again, the mean value of $\sin^4 \theta$ is $\frac{8}{15}$, that of $\cos^4 \theta$ is $\frac{1}{5}$, and that of $\sin^2 \theta \cos^2 \theta$ is $\frac{2}{15}$. Thus, finally, the mean value of I over the sphere is given by

$$\text{mean } I = \tfrac{1}{15}\{3(A^2 + B^2 + C^2) + 2(AB + BC + CA)\}. \quad \ldots\ldots(14)$$

This refers to the vibrations parallel to Z which are propagated along OY.

For the vibrations parallel to X, the second set of factors is $\cos XU$, $\cos XV$, $\cos XW$, as given above, and the vibration is expressed by

$$-A \sin \theta \cos \psi (-\sin \phi \sin \psi + \cos \phi \cos \psi \cos \theta)$$
$$+B \sin \theta \sin \psi (-\sin \phi \cos \psi - \cos \phi \sin \psi \cos \theta)$$
$$+C \cos \theta \sin \theta \cos \phi. \quad \ldots\ldots\ldots\ldots\ldots\ldots\ldots\ldots\ldots\ldots\ldots\ldots(15)$$

Accordingly for the intensity

$$I = A^2 \sin^2 \theta \cos^2 \psi (\sin^2 \phi \sin^2 \psi + \cos^2 \phi \cos^2 \psi \cos^2 \theta$$
$$\qquad\qquad\qquad - 2\sin \phi \cos \phi \sin \psi \cos \psi \cos \theta)$$
$$+ B^2 \sin^2 \theta \sin^2 \psi (\sin^2 \phi \cos^2 \psi + \cos^2 \phi \sin^2 \psi \cos^2 \theta$$
$$\qquad\qquad\qquad + 2\sin \phi \cos \phi \sin \psi \cos \psi \cos \theta)$$
$$+ C^2 \sin^2 \theta \cos^2 \theta \cos^2 \phi$$
$$- 2AB \sin^2 \theta \sin \psi \cos \psi (\sin^2 \phi \sin \psi \cos \psi - \cos^2 \phi \sin \psi \cos \psi \cos^2 \theta$$
$$\qquad\qquad + \sin \phi \cos \phi \sin^2 \psi \cos \theta - \sin \phi \cos \phi \cos^2 \psi \cos \theta)$$
$$+ 2BC \sin^2 \theta \cos \theta \sin \psi \cos \phi (-\sin \phi \cos \psi - \cos \phi \sin \psi \cos \theta)$$
$$- 2CA \sin^2 \theta \cos \theta \cos \psi \cos \phi (-\sin \phi \sin \psi + \cos \phi \cos \psi \cos \theta). \ \ldots(16)$$

In taking the mean with respect to ϕ, the terms which are odd in $\sin \phi$, or $\cos \phi$, disappear, while the mean value of $\sin^2 \phi$, or $\cos^2 \phi$, is $\frac{1}{2}$. We get for the mean

$$I = \tfrac{1}{2}A^2 \sin^2 \theta \cos^2 \psi (\sin^2 \psi + \cos^2 \psi \cos^2 \theta)$$
$$+ \tfrac{1}{2}B^2 \sin^2 \theta \sin^2 \psi (\cos^2 \psi + \sin^2 \psi \cos^2 \theta)$$
$$+ \tfrac{1}{2}C^2 \sin^2 \theta \cos^2 \theta$$
$$- AB \sin^2 \theta \sin \psi \cos \psi . \sin \psi \cos \psi \sin^2 \theta$$
$$- BC \sin^2 \theta \cos \theta \sin \psi . \sin \psi \cos \theta$$
$$- CA \sin^2 \theta \cos \theta \cos \psi . \cos \psi \cos \theta. \quad \ldots\ldots\ldots\ldots(17)$$

The averaging with respect to ψ now goes as before, and we obtain

$$\tfrac{1}{2}(A^2 + B^2)\sin^2 \theta (\tfrac{1}{8} + \tfrac{3}{8}\cos^2 \theta) + \tfrac{1}{2}C^2 \sin^2 \theta \cos^2 \theta$$
$$- \tfrac{1}{8}AB \sin^4 \theta - \tfrac{1}{2}(A + B)C \sin^2 \theta \cos^2 \theta; \quad \ldots(18)$$

and, finally, the averaging with respect to θ gives

$$\text{mean } I = \frac{A^2 + B^2}{16} (1 - \tfrac{1}{3} + \tfrac{6}{15}) + \frac{C^2}{15} - \frac{AB}{15} - \frac{(A+B)C}{15}$$

$$= \tfrac{1}{15} \{ A^2 + B^2 + C^2 - AB - BC - CA \}. \quad \dots\dots\dots(19)$$

This represents the intensity of the vibrations parallel to X dispersed along OY, due to primary vibrations parallel to Z. It vanishes, of course, if $A = B = C$; while, if $A = B$ merely, it reduces to (4).

The *ratio* of the two polarized components is

$$\frac{A^2 + B^2 + C^2 - AB - BC - CA}{3(A^2 + B^2 + C^2) + 2(AB + BC + CA)}, \quad \dots\dots\dots(20)$$

reducing to (5) when $B = A$.

If the primary light travelling in direction OX is unpolarized, we have also to include primary vibrations parallel to Y. The secondary vibrations scattered along OY are of the same intensity whether they are parallel to Z or to X. They are given by (19), where all that is essential is the perpendicularity of the primary and secondary vibrations. Thus, in order to obtain the effect along OY of unpolarized primary light travelling along OX, we have merely to add (19) to both components. The intensity of the component vibrating parallel to Z is thus

$$\tfrac{1}{15} \{ 3(A^2 + B^2 + C^2) + 2(AB + BC + CA) \}$$
$$+ \tfrac{1}{15} \{ A^2 + B^2 + C^2 - AB - BC - CA \}$$
$$= \tfrac{1}{15} \{ 4(A^2 + B^2 + C^2) + AB + BC + CA \}; \quad \dots\dots(21)$$

while that of the component vibrating parallel to X is simply

$$\tfrac{2}{15} \{ A^2 + B^2 + C^2 - AB - BC - CA \}. \quad \dots\dots\dots(22)$$

The ratio of the two intensities is

$$\frac{2(A^2 + B^2 + C^2 - AB - BC - CA)}{4(A^2 + B^2 + C^2) + AB + BC + CA}, \quad \dots\dots\dots(23)$$

reducing to (6) when $B = A$.

It may be observed that, since $(21) = (14) + (19)$, we obtain the same intensity whether we use a polarizer transmitting vibrations parallel to Z and no analyser, or whether we use an analyser transmitting vibrations parallel to Z and no polarizer.

If neither polarizing nor analysing apparatus is employed, we may add (21) and (22), thus obtaining

$$\tfrac{1}{15} [6(A^2 + B^2 + C^2) - AB - BC - CA]. \quad \dots\dots\dots(24)$$

When the particles are supposed to be of uniform quality, with a specific inductive capacity K' as compared with K for the undisturbed medium, and to be of *ellipsoidal* form with semi-axes a, b, c, we have

$$A^{-1} : B^{-1} : C^{-1} = 1 + \frac{K'-K}{4\pi K} L : 1 + \frac{K'-K}{4\pi K} M : 1 + \frac{K'-K}{4\pi K} N, \quad \dots\dots(25)$$

where
$$L = 2\pi abc \int_0^\infty \frac{d\lambda}{(a^2+\lambda)^{\frac{3}{2}}(b^2+\lambda)^{\frac{1}{2}}(c^2+\lambda)^{\frac{1}{2}}}, \dots\dots\dots(26)$$

with similar expressions for M and N.

If the ellipsoid be of revolution the case is simplified*. For example, if it be of the elongated or ovary form with eccentricity e,

$$a = b = c \sqrt{(1 - e^2)} ; \dots\dots\dots\dots\dots\dots\dots(27)$$

$$L = M = 2\pi \left\{ \frac{1}{e^2} - \frac{1-e^2}{2e^3} \log \frac{1+e}{1-e} \right\}, \dots\dots\dots(28)$$

$$N = 4\pi \left\{ \frac{1}{e^2} - 1 \right\} \left\{ \frac{1}{2e} \log \frac{1+e}{1-e} - 1 \right\}, \dots\dots\dots(29)$$

For the sphere $(e = 0)$ $L = M = N = \dfrac{4\pi}{3}. \dots\dots\dots\dots\dots\dots(30)$

In the case of a very elongated ovoid, L and M approximate to the value 2π, while N approximates to the form

$$N = 4\pi \frac{a^2}{c^2} \left(\log \frac{2c}{a} - 1 \right), \dots\dots\dots\dots\dots(31)$$

vanishing when $e = 1$. It appears that, when K'/K is finite, mere elongation does not suffice to render A and B negligible in comparison with C. The limiting value of $C : A$ is in fact $\frac{1}{2}(1 + K'/K)$. If, however, as for a perfectly conducting body, $K' = \infty$, then C becomes paramount, and the simplified values already given for this case acquire validity†.

Another question which naturally presents itself is whether a want of equality among the coefficients A, B, C interferes with the relation between attenuation and refractive index, explained in my paper of 1899‡. The answer appears to be in the affirmative, since the attenuation depends upon $A^2 + B^2 + C^2$, while the refractive index depends upon $A + B + C$, so that no simple relation obtains in general. But it may well be that in cases of interest the disturbance thus arising is not great.

The problem of an ellipsoidal particle of uniform dielectric quality can be no more than illustrative of what happens in the case of a molecule; but we may anticipate that the general form with suitable values of A, B, C still applies, except it may be under special circumstances where resonance occurs and where the effective values of the coefficients may vary greatly with the wave-length of the light.

* See the paper of 1897.

† But the particle must still be small relatively to the wave-length *within the medium* of which it is composed.

‡ An equivalent formula was given by Lorenz in 1890, *Œuvres Scientifiques*, t. i. p. 496, Copenhagen, 1898. See also Schuster's *Theory of Optics*, 2nd ed. p. 326 (1909).

431.

PROPAGATION OF SOUND AND LIGHT IN AN IRREGULAR ATMOSPHERE.

[*Nature*, Vol. CI. p. 284, 1918.]

I suppose that most of those who have listened to (single-engined) aeroplanes in flight must have noticed the highly uneven character of the sound, even at moderate distances. It would seem that the changes are to be attributed to atmospheric irregularities affecting the propagation rather than to variable emission. This may require confirmation; but, in any case, a comparison of what is to be expected in the analogous propagation of light and sound has a certain interest.

One point of difference should first be noticed. The velocity of propagation of sound through air varies indeed with temperature, but is independent of pressure (or density), while that of light depends upon pressure as well as upon temperature. In the atmosphere there is a variation of pressure with elevation, but this is scarcely material for our present purpose. And the kind of irregular local variations which can easily occur in temperature are excluded in respect of pressure by the mechanical conditions, at least in the absence of strong winds, not here regarded. The question is thus reduced to refractions consequent upon temperature variations.

The velocity of sound is as the square root of the absolute temperature. Accordingly for 1° C. difference of temperature the refractivity $(\mu - 1)$ is 0·00183. In the case of light the corresponding value of $(\mu - 1)$ is 0·000294 × 0·00366, the pressure being atmospheric. The effect of temperature upon sound is thus about 2000 times greater than upon light. If we suppose the system of temperature differences to be altered in this proportion, the course of *rays* of light and of sound will be the same.

When we consider mirage, and the twinkling of stars, and of terrestrial lights at no very great distances, we recognize how heterogeneous the atmosphere must often be for the propagation of sound, and we need no longer be surprised at the variations of intensity with which uniformly emitted sounds are received at moderate distances from their source.

It is true, of course, that the question is not exhausted by a consideration of rays, and that we must remember the immense disproportion of wave-lengths, greatly affecting all phenomena of diffraction. A twinkling star, as seen with the naked eye, may disappear momentarily, which means that then little or no light from it falls upon the eye. When a telescope is employed the twinkling is very much reduced, showing that the effects are entirely different at points so near together as the parts of an object-glass. In the case of sound, such sensitiveness to position is not to be expected, and the reproduction of similar phenomena would require the linear scale of the atmospheric irregularities to be very much enlarged.

432.

NOTE ON THE THEORY OF THE DOUBLE RESONATOR.

[*Philosophical Magazine*, Vol. XXXVI. pp. 231—234, 1918.]

IN my book on the *Theory of Sound** I have considered the case of a *double* resonator (Fig. 1), where two reservoirs of volumes S, S' communicate

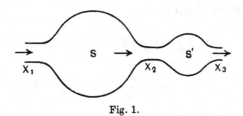

Fig. 1.

with each other and with the external atmosphere by narrow passages or necks. If we were to treat SS' as a single reservoir and apply the usual formula, we should be led to an erroneous result; for that formula is founded on the assumption that within the reservoir the inertia of the air may be left out of account, whereas it is evident that the energy of the motion through the connecting passage may be as great as through the two others. However, an investigation on the same general plan meets the case perfectly. Denoting by X_1, X_2, X_3 the total transfers of fluid through the three passages, we have for the kinetic energy the expression

$$T = \tfrac{1}{2}\rho \left\{ \frac{1}{c_1}\left(\frac{dX_1}{dt}\right)^2 + \frac{1}{c_2}\left(\frac{dX_2}{dt}\right)^2 + \frac{1}{c_3}\left(\frac{dX_3}{dt}\right)^2 \right\}, \quad\ldots\ldots\ldots\ldots(1)$$

and for the potential energy

$$V = \tfrac{1}{2}\rho a^2 \left\{ \frac{(X_2 - X_1)^2}{S} + \frac{(X_3 - X_2)^2}{S'} \right\}. \quad\ldots\ldots\ldots\ldots(2)$$

Here ρ denotes the density of the fluid, a the velocity of sound, while c_1, c_2, c_3 may be interpreted as the *electrical conductivities* of the passages. Thus for

* § 310, first edition 1878, second edition 1896, Macmillan. Also *Phil. Trans.* 1870; *Scientific Papers*, Vol. I. p. 41.

a long cylindrical neck of radius R and length L we should have $c = \pi R^2/L$. An application of Lagrange's method gives as the differential equations of motion,

$$\left.\begin{aligned}
\frac{1}{c_1}\frac{d^2 X_1}{dt^2} + a^2 \frac{X_1 - X_2}{S} &= 0, \\[1mm]
\frac{1}{c_2}\frac{d^2 X_2}{dt^2} + a^2 \left\{ \frac{X_2 - X_1}{S} + \frac{X_2 - X_3}{S'} \right\} &= 0, \\[1mm]
\frac{1}{c_3}\frac{d^2 X_3}{dt^2} + a^2 \frac{X_3 - X_2}{S'} &= 0.
\end{aligned}\right\} \quad \ldots\ldots\ldots\ldots(3)$$

By addition and integration

$$\frac{X_1}{c_1} + \frac{X_2}{c_2} + \frac{X_3}{c_3} = 0, \quad \ldots\ldots\ldots\ldots\ldots\ldots(4)$$

since in the case of free vibrations all the quantities X may be supposed proportional to e^{pt}, so that d/dt may be replaced by p.

From (3) and (4) by elimination of X_3,

$$\left(\frac{p^2}{a^2 c_1} + \frac{1}{S} \right) X_1 - \frac{X_2}{S} = 0, \quad \ldots\ldots\ldots\ldots\ldots(5)$$

$$\left(\frac{c_3}{c_1 S'} - \frac{1}{S} \right) X_1 + \left(\frac{p^2}{a^2 c_2} + \frac{1}{S} + \frac{c_2 + c_3}{c_2 S'} \right) X_2 = 0,$$

whence as the equation for p^2

$$\frac{p^4}{a^4} + \frac{p^2}{a^2} \left\{ \frac{c_1 + c_2}{S} + \frac{c_2 + c_3}{S'} \right\} + \frac{1}{SS'} \left\{ c_1(c_2 + c_3) + c_2 c_3 \right\} = 0. \quad \ldots\ldots\ldots(6)$$

In the use of double resonance to secure an exalted effect, as in the experiments of Boys and of Callendar, we may suppress the direct communication between the second resonator S' and the external air. Then $c_3 = 0$, and (6) becomes

$$\frac{p^4}{a^4} + \frac{p^2}{a^2} \left\{ \frac{c_1 + c_2}{S} + \frac{c_2}{S'} \right\} + \frac{c_1 c_2}{SS'} = 0. \quad \ldots\ldots\ldots\ldots\ldots(7)$$

To interpret the c's suppose first the passage between S and S' abolished, so that $c_2 = 0$. The first resonator then acts as a simple resonator, and if p_1 be the corresponding p, we have $p_1^2/a^2 = -c_1/S$, as usual. Again, if S be infinite, we have for the second resonator acting alone, $p_2^2/a^2 = -c_2/S'$; and (7) may be written

$$p^4 - p^2 \left(p_1^2 + p_2^2 + \frac{S'}{S} p_2^2 \right) + p_1^2 p_2^2 = 0. \quad \ldots\ldots\ldots\ldots(8)$$

In (8) if S'/S be very small, p^2 approximates to p_1^2 or to p_2^2, and this is the case of greatest importance in experiment.

If p_1^2 and p_2^2 differ sufficiently, we may pursue an approximation from (8) founded on the smallness of S'/S. But it is of more interest to suppose that p_1^2 and p_2^2 are absolutely equal, which nothing precludes. Then

$$p^4 - p^2\left(2p_1^2 + \frac{S'}{S}p_1^2\right) + p_1^4 = 0, \quad\ldots\ldots\ldots\ldots\ldots(9)$$

whence

$$\frac{p^2}{p_1^2} = 1 + \frac{S'}{2S} \pm \sqrt{\left(\frac{S'}{S} + \frac{S'^2}{4S^2}\right)}; \quad\ldots\ldots\ldots\ldots(10)$$

or, if S'/S be small enough,

$$\frac{p^2}{p_1^2} = 1 \pm \sqrt{\left(\frac{S'}{S}\right)}, \quad\ldots\ldots\ldots\ldots\ldots(11)$$

p^2 differing but little from p_1^2 or p_2^2.

Referring back to (5), we have

$$\frac{X_1 - X_2}{X_1} = -\frac{Sp^2}{a^2 c_1} = \frac{p^2}{p_1^2} = 1 \pm \sqrt{\left(\frac{S'}{S}\right)},$$

when we introduce the value of p^2 from (11). Thus

$$\frac{X_2}{X_1} = \mp \sqrt{\left(\frac{S'}{S}\right)}.\quad\ldots\ldots\ldots\ldots\ldots(12)$$

We may now compare effects in the two component resonators, and here a certain choice presents itself. The *condensations* in the interiors are $(X_1 - X_2)/S$ and X_2/S', and the *ratio* of condensations is

$$\frac{X_2/S'}{(X_1 - X_2)/S} = \frac{\sqrt{(S/S')}}{1 - \sqrt{(S'/S)}} = \sqrt{\left(\frac{S}{S'}\right)}\ldots\ldots\ldots\ldots(13)$$

approximately. It appears that the condensation in the second resonator may be made to exceed to any extent that in the first by making the second resonator small enough, which sufficiently explains the advantage found in experiment to attend the combination.

In some forms of the experiment we may have to do rather with the flow through the passages than with the condensations in the interiors. In (12) we have the ratio of the total flows already expressed. But we may be more concerned with a comparison of flows reckoned per unit of area of the passages. In the case of passages which are mere circular apertures of radii R and R' a simple result may be stated, for then $c_1 : c_2 = R : R'$; and, since $p_1^2 = p_2^2$, $c_1 : c_2 = S : S'$. Accordingly

$$\frac{X_2/R'^2}{X_1/R^2} = \sqrt{\left(\frac{S'}{S}\right)} \cdot \frac{S^2}{S'^2} = \left(\frac{S}{S'}\right)^{\frac{3}{2}}, \quad\ldots\ldots\ldots\ldots(14)$$

and the advantage of a small S' is even more pronounced than in (13).

433.

A PROPOSED HYDRAULIC EXPERIMENT.

[*Philosophical Magazine*, Vol. XXXVI. pp. 315—316, 1918.]

IN an early paper* Stokes showed "that in the case of a homogeneous incompressible fluid, whenever $udx + vdy + wdz$ is an exact differential, not only are the ordinary equations of fluid motion satisfied, but the equations obtained when friction is taken into account are satisfied likewise. It is only the equations of condition which belong to the boundaries of the fluid that are violated." In order to satisfy these also, it is only necessary to suppose that every part of the solid boundaries is made to move with the velocity which the fluid in irrotational motion would there assume. There is no difficulty in the supposition itself; but the only case in which it could readily be carried into effect with tolerable completeness is for the two-dimensional motion of fluid between coaxal cylinders, themselves made to rotate in the same direction with circumferential velocities which are inversely as the radii. Experiments upon these lines, but not I think quite satisfying the above conditions, have been made by Couette and Mallock. It would appear that, except at low velocities, the simple steady motion becomes unstable.

But the point of greatest interest is not touched in the above example. It arises when fluid passing along a uniform or contracting pipe, or channel, arrives at a place where the pipe expands. It is known that if the expansion be sufficiently gradual, the fluid generally speaking follows the walls, or, as it is often expressed, the pipe flows *full*; and the loss of velocity accompanying the increased section is represented by an augmentation of pressure, approximately according to Bernoulli's law. On the other hand, if in order to effect the conversion of velocity into pressure more rapidly, the expansion be made too violently, the fluid refuses to follow the walls, eddies result, and mechanical energy is lost by fluid friction. According to W. Froude's generally accepted view, the explanation is to be sought in the loss of velocity near the walls in consequence of fluid friction, which is such that the fluid

---

* *Camb. Trans.* Vol. IX. p. [8], 1850; *Math. and Phys. Papers*, Vol. III. p. 73.

in question is unable to penetrate into what should be the region of higher pressure beyond.

It would be a difficult matter to satisfy the necessary conditions for the walls of an expanding channel, even in two dimensions. The travelling bands of which the walls would be constituted should assume different velocities at different parts of their course. But it is quite possible that a very rough approximation to theoretical requirements would throw interesting light upon the subject, and I write in the hope of persuading some one with the necessary facilities, such as are to be found in some hydraulic laboratories, to undertake a comparatively simple experiment.

What I propose is the observation of the flow of liquid between two cylinders A, B (probably brass tubes), revolving about their axes in opposite directions. The diagram will sufficiently explain the idea. The circum-

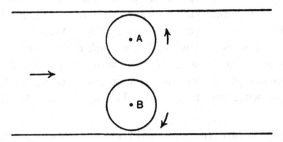

ferential velocity of the cylinders should not be less than that of irrotational fluid in contact with the walls at the narrowest place. The simple motion may be unstable; but, as I have had occasion to remark before*, the critical situation would be so quickly traversed that perhaps the instability may be of little consequence. If no marked difference in the character of the flow could be detected by colour streaks, whether the cylinders were turning or not, the inference would be that Froude's explanation is inadequate. In the contrary event the question would arise whether practical advantage could be taken by specially stimulating the motion of fluid near the walls of expanding channels, e.g. with the aid of steam jets.

* *Phil. Mag.* Vol. xxvi. p. 776 (1913). [This volume, Art. 376.]

434.

ON THE DISPERSAL OF LIGHT BY A DIELECTRIC CYLINDER.

[Philosophical Magazine, Vol. XXXVI. pp. 365—376, 1918.]

THE problem of the incidence of plane electric waves on an insulating dielectric cylinder was treated by me as long ago as 1881*. Further investigations upon the same subject have been published by Seitz† and by Ignatowski‡ who corrects some of Seitz's results. Neither of these authors appears to have been acquainted with my much earlier work. The purpose of the present paper is little more than numerical calculations from the expressions formerly given, but in order to make them intelligible it will be well to quote what was then said. The notation is for the most part Maxwell's.

"We will now return to the two-dimension problem with the view of determining the disturbance resulting from the impact of plane waves upon a cylindrical obstacle whose axis is parallel to the plane of the waves. There are, as in the problem of reflection from plane surfaces, two principal cases—(1) when the electric displacements are parallel to the axis of the cylinder taken as axis of z, (2) when the electric displacements are perpendicular to this direction.

"*Case* 1. [From the general equation with conductivity (C) zero and magnetic permeability (μ) constant],

$$\left(\frac{d^2}{dx^2} + \frac{d^2}{dy^2}\right)\frac{h}{K} + n^2\mu K \frac{h}{K} = 0; \quad \text{.................(1)§}$$

or if, as before, $k = 2\pi/\lambda$,

$$\left(\frac{d^2}{dx^2} + \frac{d^2}{dy^2} + k^2\right)\frac{h}{K} = 0, \quad \text{.........................(2)}$$

in which k is constant in each medium, but changes as we pass from one medium to another. From (2) we see that the problem now before us is

* *Phil. Mag.* Vol. XII. p. 81 (1881); *Scientific Papers*, Vol. I. p. 533.

† *Ann. d. Physik*, Vol. XVI. p. 746 (1905); Vol. XIX. p. 554 (1906).

‡ *Ann. d. Physik*, Vol. XVIII. p. 495 (1905).

§ The numbering of the equations is changed. h is the component of electric displacement parallel to z, K the specific inductive capacity, and λ the wave-length.

analytically identical with that treated in my book on Sound*, § 343, to which I must refer for more detailed explanations. The incident plane waves are represented by

$$e^{int} e^{ikx} = e^{int} e^{ikr \cos \theta}$$
$$= e^{int} \{ J_0(kr) + 2i J_1(kr) \cos \theta + \dots$$
$$+ 2i^m J_m(kr) \cos m\theta + \dots \} ; \dots \dots (3)$$

and we have to find for each value of m an internal motion finite at the centre, and an external motion representing a divergent wave, which shall in conjunction with (3) satisfy at the surface of the cylinder $(r = c)$ the condition that the function $[h/K]$ and its differential coefficient with respect to r shall be continuous. The divergent wave is expressed by

$$B_0 \psi_0 + B_1 \psi_1 \cos \theta + B_2 \psi_2 \cos 2\theta + \dots, \dots \dots (4)$$

where ψ_0, ψ_1, etc. are the functions of kr defined in § 341. The coefficients B are determined in accordance with

$$B_m \left\{ kc \frac{d\psi_m}{d . kc} J_m(k'c) - k'c \psi_m \frac{d}{d . k'c} J_m(k'c) \right\}$$
$$= 2i^m \{ k'c J_m(kc) J_m'(k'c) - kc J_m(k'c) J_m'(kc) \}, \dots \dots (5)$$

except in the case of $m = 0$, when $2i^m$ on the right-hand side is to be replaced by i^m†. In working out the result we suppose kc and $k'c$ to be small; and we find approximately for the secondary disturbance corresponding to (3)

$$\psi = \left(\frac{\pi}{2ikr} \right)^{\frac{1}{2}} e^{i(nt-kr)} \left[\frac{k'^2 c^2 - k^2 c^2}{2} - \frac{k^2 c^2 (k'^2 c^2 - k^2 c^2)}{8} \cos \theta \right] ; \dots \dots (6)$$

showing, as was to be expected, that the leading term is independent of θ.

"For case 2, which is of greater interest, we have [from the general equations]

$$\left(\frac{d}{dx} \frac{1}{k^2} \frac{d}{dx} + \frac{d}{dy} \frac{1}{k^2} \frac{d}{dy} + 1 \right) c = 0. \dots \dots (7)‡$$

This is of the same form as (2) within a uniform medium, but gives a different boundary condition at a surface of transition. In both cases the function itself is to be continuous; but in that with which we are now concerned the second condition requires the continuity of the differential coefficient *after division by* k^2. The equation for B_m [or B_m' as we may write it for distinctiveness] is therefore

$$B_m' \left\{ k'c \frac{d\psi_m}{d . kc} J_m(k'c) - kc \psi_m \frac{dJ_m(k'c)}{d . k'c} \right\}$$
$$= 2i^m \{ kc J_m(kc) J_m'(k'c) - k'c J_m(k'c) J_m'(kc) \}, \dots \dots (8)$$

---

* *Theory of Sound*, Vol. II. Macmillan, 1st ed. 1878, 2nd ed. 1896.

† Here k' relates to the cylindrical obstacle and k to the external medium.

‡ In (7) c is the magnetic component, and not the radius of the cylinder. So many letters are employed in the electromagnetic theory, that it is difficult to hit upon a satisfactory notation.

with the understanding that the 2 is to be omitted when $m = 0$. Corresponding to the primary wave $e^{i(nt+kx)}$, we find as the [approximate] expression of the secondary wave at a great distance from the cylinder,

$$\psi = \left(\frac{\pi}{2ikr}\right)^{\frac{1}{2}} e^{i(nt-kr)} \left[-\frac{k^2 c^2}{16} (k^2 c^2 - k'^2 c^2) \right.$$

$$\left. - k^2 c^2 \frac{k'^2 - k^2}{k'^2 + k^2} \cos\theta - \frac{1}{8} k^4 c^4 \frac{k^2 - k'^2}{k^2 + k'^2} \cos 2\theta \right]. \quad \ldots\ldots(9)$$

The term in $\cos\theta$ is now the leading term; so that the secondary disturbance approximately vanishes in the direction of the primary electrical displacements, agreeably with what has been proved before. It should be stated here that (9) is not complete to the order $k^4 c^4$ in the terms containing $\cos\theta$. The calculation of the part omitted is somewhat tedious in general; but if we introduce the supposition that the difference between k'^2 and k^2 is small, its effect is to bring in the factor $(1 - \frac{1}{4} k^2 c^2)$.

"Extracting the factor $(k'^2 - k^2)$, we may conveniently write (9)

$$\psi = - k^2 c^2 \frac{k'^2 - k^2}{k'^2 + k^2} \left(\frac{\pi}{2ikr}\right)^{\frac{1}{2}} e^{i(nt-kr)} \left[\cos\theta - \frac{k'^2 c^2 + k^2 c^2}{16} - \frac{k^2 c^2}{8} \cos 2\theta \right], \ldots(10)$$

in which

$$\cos\theta - \frac{k'^2 c^2 + k^2 c^2}{16} - \frac{k^2 c^2}{8} \cos 2\theta = \cos\theta - \frac{k'^2 c^2 - k^2 c^2}{16} - \frac{k^2 c^2}{4} \cos^2\theta. \ldots(11)$$

"In the directions $\cos\theta = 0$, the secondary light is thus not only of high order in kc, but is also of the second order in $(k' - k)$. For the direction in which the secondary light vanishes to the next approximation, we have

$$\tfrac{1}{2}\pi - \theta = \tfrac{1}{16} (k'^2 c^2 - k^2 c^2) = \frac{k^2 c^2}{16} \frac{K' - K}{K}. \quad \ldots\ldots\ldots\ldots(12)$$

This...is true if kc, $k'c$ be small enough, whatever may be the relation of k' and k. For the cylinder, as for the sphere, the direction is such that the primary light would be bent through an angle *greater* than a right angle...."

"If we suppose the cylinder to be extremely small, we may confine ourselves to the leading terms in (6) and (9). Let us compare the intensities of the secondary lights emitted in the two cases along $\theta = 0$, *i.e.* directly backwards. From (6)

$$\psi \propto \tfrac{1}{2} (k'^2 c^2 - k^2 c^2),$$

while from (9)

$$\psi \propto - k^2 c^2 (k'^2 - k^2)/(k'^2 + k^2).$$

The opposition of sign is apparent only, and relates to the different methods of measurement adopted in the two cases. In (6) the primary and secondary disturbances are represented by h/K, but in (9) by the magnetic function c...."

It may be remarked that Ignatowski's equation agrees with (5) for this case, and that his corresponding equation (11) for the second case also agrees with (8) after correction of some misprints. His function Q corresponds with my ψ, at least when we observe that the introduction of a constant multiplier, even if a function of m, does not influence the final result.

In proceeding to numerical calculations we must choose a refractive index. I take for this index 1·5, as in similar work for a transparent *sphere**, so that $k'/k = 1·5$. And before employing the more general formulæ, I commence with the approximations of (6) and (9), assuming $kc = ·10$, $k'c = ·15$. When we introduce these values into (6), we get

$$\psi = \frac{h}{K} = \left(\frac{\pi}{2ikr}\right)^{\frac{1}{2}} e^{i(nt-kr)} [·00625 - ·156 \times 10^{-4} \cos\theta], \quad ......(13)$$

in response to the incident wave $h/K = e^{i(nt+kx)}$. Again, from (9)

$$\psi = c = \left(\frac{\pi}{2ikr}\right)^{\frac{1}{2}} e^{i(nt-kr)} [10^{-4}(·0781 + ·0481\cos 2\theta) - ·00385\cos\theta],...(14)$$

corresponding with $c = e^{i(nt+kx)}$ for the incident wave.

In using the general formulæ the next step is to express ψ_m, representing a divergent wave, by means of functions already tabulated. I am indebted to Prof. Nicholson for valuable information under this head. It appears that we may take

$$\psi_m(z) = G_m(z) - \tfrac{1}{2} i\pi J_m(z), \quad ....(15)$$

where z is written for kr, and the real and imaginary parts are separated. When z is very great

$$\psi_m(z) = i^m \left(\frac{\pi}{2iz}\right)^{\frac{1}{2}} e^{-iz}. \quad .... (16)\dagger$$

$J_m(z)$ is the usual Bessel's function; the G-functions are tabulated in Brit. Assoc. Reports‡. The Bessel's functions satisfy the relations

$$J_{m+1} = \frac{2m}{z} J_m - J_{m-1}, \quad ....(17)$$

$$J_m' = J_{m-1} - \frac{m}{z} J_m; \quad ....(18)$$

and relations of the same form are satisfied by functions G. When $m = 0$, $J_0' = - J_1$, $G_0' = - G_1$.

Writing z for kc and z' for $k'c$ and with use of (18), we have for the coefficient D_m of $2i^m$ on the right-hand side of (5)

$$D_m = z' J_m(z) J_{m-1}(z') - z J_m(z') J_{m-1}(z); \quad ....(19)$$

and for the coefficient of B_m on the left

$$N_m + \tfrac{1}{2} i\pi D_m,$$

---

* *Proc. Roy. Soc.* A, Vol. LXXXIV. p. 25 (1910); *Scientific Papers*, Vol. v. p. 547.

[† A correction has been made in this formula. In order to yield (15), $\psi_m(z)$ must be deduced from $\psi_0(z)$ by (5) of § 341 of *Theory of Sound*, and not by (11) of that section, which gives a different law of signs; also $\psi_0(z)$ must be taken with the opposite sign from (18) of § 341. W. F. S.]

‡ *Reports for* 1913, p. 115; 1914, p. 75.

where
$$N_m = z\,G_m{}'(z)\,J_m(z') - z'\,G_m(z)\,J_m{}'(z')$$
$$= z\,G_{m-1}(z)\,J_m(z') - z'\,G_m(z)\,J_{m-1}(z'). \quad \ldots\ldots\ldots(20)$$

Thus
$$B_m = \frac{2i^m}{N_m/D_m + \tfrac12 i\pi}, \ldots\ldots\ldots\ldots\ldots\ldots(21)$$

where, however, the 2 is to be omitted when $m = 0$. Thus by (4) and (16) the divergent wave at a great distance r is expressed by

$$h/K = \left(\frac{\pi}{2ikr}\right)^{\frac12} e^{i(nt-kr)} \left[\frac{1}{N_0/D_0 + \tfrac12 i\pi} + \sum_1^\infty \frac{2(-1)^m \cos m\theta}{N_m/D_m + \tfrac12 i\pi}\right]. \quad \ldots(22)$$

Here N_m, D_m are given by (19), (20), and are real.

In like manner (8) may be put into the form

$$B_m{}' = \frac{2i^m}{N_m{}'/D_m{}' + \tfrac12 i\pi}, \quad \ldots\ldots\ldots\ldots\ldots(23)$$

where
$$N_m{}' = z'\,J_m(z)\,G_m{}'(z) - z\,J_m{}'(z')\,G_m(z)$$
$$= z'\,J_m(z')\,G_{m-1}(z) - z\,J_{m-1}(z')\,G_m(z)$$
$$- m\left(\frac{z'}{z} - \frac{z}{z'}\right)J_m(z')\,G_m(z), \quad \ldots\ldots\ldots\ldots\ldots(24)$$

$$D_m{}' = z\,J_m(z)\,J_{m-1}(z') - z'\,J_m(z')\,J_{m-1}(z)$$
$$+ m\left(\frac{z'}{z} - \frac{z}{z'}\right)J_m(z)\,J_m(z'). \quad \ldots\ldots\ldots\ldots\ldots(25)$$

And, as in (22), the expression for the diverging wave at a distance is

$$c = \left(\frac{\pi}{2ikr}\right)^{\frac12} e^{i(nt-kr)} \left[\frac{1}{N_0{}'/D_0{}' + \tfrac12 i\pi} + \sum_1^\infty \frac{2(-1)^m \cos m\theta}{N_m{}'/D_m{}' + \tfrac12 i\pi}\right]. \quad \ldots\ldots(26)$$

When we fix the refractive index at 1·5, the value of $z'/z - z/z'$ in (24), (25) is 5/6.

The values of N_1, $N_1{}'$, D_1, $D_1{}'$ may be deduced from the corresponding quantities with $m = 0$ by means of the relations

$$N_1 = N_0{}', \quad N_1{}' = N_0 - \tfrac56 J_1(z')\,G_1(z), \quad \ldots\ldots\ldots\ldots(27)$$
$$D_1 = D_0{}', \quad D_1{}' = D_0 + \tfrac56 J_1(z')\,J_1(z). \quad \ldots\ldots\ldots\ldots(28)$$

For numerical calculation we have also to specify the values of z, or kc. For this purpose we take $z = ·4, ·8, 1·2, 1·6, 2·0, 2·4$, where z denotes the ratio of the circumference of the cylinder to the wave-length in air; the corresponding values of $(N/D + \tfrac12 i\pi)^{-1}$ and of $(N'/D' + \tfrac12 i\pi)^{-1}$ may then be tabulated.

The next step is the calculation of the series included in the square brackets of (22) and (26) for various values of θ from $\theta = 0$ in the direction backwards along the primary ray to $\theta = 180°$ in the direction of the primary

TABLE I.

z	$[N_0/D_0 + \tfrac{1}{2}i\pi]^{-1}$	$[N_0'/D_0' + \tfrac{1}{2}i\pi]^{-1}$	
·4	·10624 − i × ·01825	·00202 − i × ·00001	
·8	·29104 − i × ·18940	·03397 − i × ·00182	
1·2	·31827 − i × ·32283	·17667 − i × ·05353	0
1·6	·31745 − i × ·34157	·31764 − i × ·33892	
2·0	·31565 − i × ·35939	·23337 − i × ·53480	
2·4	·26905 − i × ·48842	·19953 − i × ·56634	

z	$[N_1/D_1 + \tfrac{1}{2}i\pi]^{-1}$	$[N_1'/D_1' + \tfrac{1}{2}i\pi]^{-1}$	
·4	·00202 − i × ·00001	·03066 − i × ·00148	
·8	·03397 − i × ·00182	·10872 − i × ·01914	
1·2	·17667 − i × ·05353	·18711 − i × ·06080	1
1·6	·31764 − i × ·33892	·24560 − i × ·11581	
2·0	·23337 − i × ·53480	·30426 − i × ·22477	
2·4	·19953 − i × ·56634	·27720 − i × ·47478	

z	$[N_2/D_2 + \tfrac{1}{2}i\pi]^{-1}$	$[N_2'/D_2' + \tfrac{1}{2}i\pi]^{-1}$	
·4	·00001 − i × 0	·00061 − i × 0	
·8	·00084 − i × 0	·00931 − i × ·00014	
1·2	·00946 − i × ·00014	·04392 − i × ·00304	2
1·6	·05510 − i × ·00481	·12114 − i × ·02395	
2·0	·21352 − i × ·08223	·22506 − i × ·09321	
2·4	·28583 − i × ·45838	·30204 − i × ·21784	

z	$[N_3/D_3 + \tfrac{1}{2}i\pi]^{-1}$	$[N_3'/D_3' + \tfrac{1}{2}i\pi]^{-1}$	
·4	....	....	
·8	·00001 − i × 0	·00025 − i × 0	
1·2	·00027 − i × 0	·00262 − i × ·00001	3
1·6	·00259 − i × ·00001	·01346 − i × ·00028	
2·0	·01514 − i × ·00036	·04636 − i × ·00339	
2·4	·06724 − i × ·00718	·12088 − i × ·02384	

z	$[N_4/D_4 + \tfrac{1}{2}i\pi]^{-1}$	$[N_4'/D_4' + \tfrac{1}{2}i\pi]^{-1}$	
1·2	....	·00008 − i × 0	
1·6	·00008 − i × 0	·00072 − i × 0	4
2·0	·00071 − i × 0	·00388 − i × ·00002	
2·4	·00415 − i × ·00003	·01482 − i × ·00035	

z	$[N_5/D_5 + \tfrac{1}{2}i\pi]^{-1}$	$[N_5'/D_5' + \tfrac{1}{2}i\pi]^{-1}$	
1·6	....	·00002 − i × 0	
2·0	·00002 − i × 0	·00020 − i × 0	5
2·4	·00019 − i × 0	·00109 − i × 0	

z	$[N_6/D_6 + \tfrac{1}{2}i\pi]^{-1}$	$[N_6'/D_6' + \tfrac{1}{2}i\pi]^{-1}$	
2·0	....	·00001 − i × 0	6
2·4	·00001 − i × 0	·00005 − i × 0	

ray produced. If we add the terms due to even and odd values of m separately, we may include in one calculation the results for θ and for $180° - \theta$, since $(-1)^m \cos m (180° - \theta) = \cos m\theta$ simply.

In illustration we may take the numerically simple case where $\theta = 0$ and $\theta = 180°$, choosing as an example $z = 2\cdot4$ in (22). Thus

m		m	
0	$\cdot26905 - i \times \cdot48842$	1	$\cdot39906 - i \times 1\cdot13268$
2	$\cdot57166 - i \times \cdot91676$	3	$\cdot13448 - i \times \cdot01436$
4	$\cdot\ \ \cdot830 - i \times \ \cdot\ \ 6$	5	$\cdot\ \ 38 - i \qquad 0$
6	$\cdot\ \ \ 2 \qquad 0$		
$\Sigma_{\text{(even)}} = \cdot84903 - i \times 1\cdot40524$		$\Sigma_{\text{(odd)}} = \cdot53392 - i \times 1\cdot14704$	

Accordingly for $\theta = 0$, we have
$$\Sigma_{\text{even}} - \Sigma_{\text{odd}} = \cdot31511 - i \times \ \cdot25820,$$
and for $\theta = 180°$
$$\Sigma_{\text{even}} + \Sigma_{\text{odd}} = 1\cdot38295 - i \times 2\cdot55228.$$

These are the multipliers of
$$\left(\frac{\pi}{2ikr}\right)^{\frac{1}{2}} e^{i(nt-kr)}$$

in (22). For most purposes we need only the modulus. We find
$$(\cdot3151)^2 + (\cdot2582)^2 = (\cdot4074)^2,$$
and
$$(1\cdot383)^2 + (2\cdot552)^2 = (2\cdot903)^2.$$

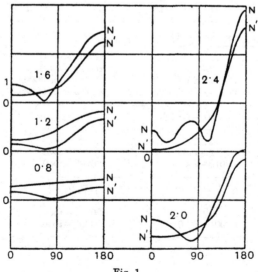

Fig. 1.

As might have been expected, the modulus, representing the amplitude of vibration, is greater in the second case, that is in the direction of the primary ray produced.

For other angles, except 90°, the calculation is longer on account of the factor $\cos m\theta$. The angles chosen as about sufficient are 0, 30°, 60°, 90° and their supplements. For 2 or 3 of the larger z's the angle 45° and its supplement were added. The results are embodied in Table II, and a plot of most of them is given in Fig. 1, where the abscissa is the angle θ and the ordinate the corresponding modulus from the table. The curve marked N corresponds to (22) and that marked N' to (26). A few points have been derived from values not tabulated. From the nature of the functions represented both curves are horizontal at the limits 0° and 180°.

When $z = \cdot 8$, the curves show the characteristics of a very thin cylinder. At 90° N' nearly vanishes, indicating that in this direction little light is scattered whose vibrations are perpendicular to the axis. When $z = 1\cdot2$, the maximum polarization is still pretty complete, but the direction in which it occurs is at a smaller angle θ. For $z = 1\cdot6$ the polarization is reversed over most of the range between 45° and 90°. By the time z has risen to $2\cdot4$ a good deal of complication enters, at any rate for the curve N.

TABLE II.

$z = \cdot 4.$

θ	[ ] in (22)	Modulus	[ ] in (26)	Modulus
0	$\cdot10222 - i \times \cdot01823$	$\cdot1038$	$- \cdot05808 + i \times \cdot00295$	$\cdot0582$
30	$\cdot10275 - i \times \cdot01824$	$\cdot1044$	$- \cdot05048 + i \times \cdot00255$	$\cdot0505$
60	$\cdot10421 - i \times \cdot01824$	$\cdot1058$	$- \cdot02925 + i \times \cdot00147$	$\cdot0293$
90	$\cdot10622 - i \times \cdot01825$	$\cdot1078$	$+ \cdot00080 - i \times \cdot00001$	$\cdot0008$
120	$\cdot10825 - i \times \cdot01826$	$\cdot1098$	$\cdot03207 - i \times \cdot00149$	$\cdot0321$
150	$\cdot10975 - i \times \cdot01826$	$\cdot1113$	$\cdot05574 - i \times \cdot00257$	$\cdot0558$
180	$\cdot11030 - i \times \cdot01827$	$\cdot1118$	$\cdot06456 - i \times \cdot00297$	$\cdot0646$

$z = \cdot 8.$

θ	[ ] in (22)	Modulus	[ ] in (26)	Modulus
0	$\cdot22476 - i \times \cdot18576$	$\cdot2916$	$- \cdot16535 + i \times \cdot03618$	$\cdot1693$
30	$\cdot23303 - i \times \cdot18625$	$\cdot2983$	$- \cdot14502 + i \times \cdot03119$	$\cdot1483$
60	$\cdot25625 - i \times \cdot18758$	$\cdot3176$	$- \cdot08356 + i \times \cdot01746$	$\cdot0854$
90	$\cdot28936 - i \times \cdot18940$	$\cdot3458$	$+ \cdot01535 - i \times \cdot00154$	$\cdot0154$
120	$\cdot32415 - i \times \cdot19122$	$\cdot3763$	$\cdot13288 - i \times \cdot02082$	$\cdot1345$
150	$\cdot35071 - i \times \cdot19255$	$\cdot4001$	$\cdot23158 - i \times \cdot03511$	$\cdot2342$
180	$\cdot36068 - i \times \cdot19304$	$\cdot4091$	$\cdot27053 - i \times \cdot04038$	$\cdot2735$

$$z = 1\cdot2.$$

θ	[ ] in (22)	Modulus	[ ] in (26)	Modulus
0	$-\cdot01669 - i \times \cdot21605$	$\cdot2167$	$-\cdot11469 + i \times \cdot06201$	$\cdot1305$
30	$+\cdot02173 - i \times \cdot23024$	$\cdot2313$	$-\cdot10357 + i \times \cdot04872$	$\cdot1145$
60	$\cdot13268 - i \times \cdot26916$	$\cdot3001$	$-\cdot04920 + i \times \cdot01029$	$\cdot0503$
90	$\cdot29935 - i \times \cdot32255$	$\cdot4401$	$+\cdot08899 - i \times \cdot04745$	$\cdot1009$
120	$\cdot48494 - i \times \cdot37622$	$\cdot6138$	$\cdot31454 - i \times \cdot11127$	$\cdot3337$
150	$\cdot63373 - i \times \cdot41570$	$\cdot7579$	$\cdot54459 - i \times \cdot16188$	$\cdot5681$
180	$\cdot69107 - i \times \cdot43017$	$\cdot8141$	$\cdot64413 - i \times \cdot18123$	$\cdot6691$

$$z = 1\cdot6.$$

θ	[ ] in (22)	Modulus	[ ] in (26)	Modulus
0	$-\cdot21265 + i \times \cdot32667$	$\cdot3898$	$\cdot04320 - i \times \cdot15464$	$\cdot1606$
30	$-\cdot17770 + i \times \cdot24065$	$\cdot2991$	$\cdot01272 - i \times \cdot16229$	$\cdot1628$
45	$-\cdot12826 + i \times \cdot13772$	$\cdot1882$	$-\cdot01207 - i \times \cdot17555$	$\cdot1760$
60	$-\cdot05019 + i \times \cdot00214$	$\cdot0502$	$-\cdot02292 - i \times \cdot19972$	$\cdot2010$
90	$+\cdot20741 - i \times \cdot33195$	$\cdot3914$	$+\cdot07680 - i \times \cdot29102$	$\cdot3010$
120	$\cdot57473 - i \times \cdot67566$	$\cdot8870$	$\cdot41448 - i \times \cdot43022$	$\cdot5974$
135	$\cdot76284 - i \times \cdot82086$	$1\cdot112$	$\cdot64445 - i \times \cdot50229$	$\cdot8171$
150	$\cdot92264 - i \times \cdot93341$	$1\cdot312$	$\cdot86340 - i \times \cdot56345$	$1\cdot031$
180	$1\cdot06827 - i \times 1\cdot02905$	$1\cdot483$	$1\cdot07952 - i \times \cdot61900$	$1\cdot244$

$$z = 2\cdot0.$$

θ	[ ] in (22)	Modulus	[ ] in (26)	Modulus
0	$\cdot24705 + i \times \cdot54647$	$\cdot5997$	$-\cdot01037 - i \times \cdot26494$	$\cdot2651$
30	$\cdot12429 + i \times \cdot48468$	$\cdot5004$	$-\cdot07211 - i \times \cdot23868$	$\cdot2493$
60	$-\cdot10169 + i \times \cdot25692$	$\cdot2763$	$-\cdot20729 - i \times \cdot22358$	$\cdot3049$
90	$-\cdot10997 - i \times \cdot19493$	$\cdot2238$	$-\cdot20901 - i \times \cdot34842$	$\cdot4063$
120	$+\cdot30453 - i \times \cdot81124$	$\cdot8665$	$+\cdot21619 - i \times \cdot65956$	$\cdot6941$
150	$\cdot93263 - i \times 1\cdot36792$	$1\cdot656$	$\cdot98119 - i \times 1\cdot01730$	$1\cdot413$
180	$1\cdot24117 - i \times 1\cdot59417$	$2\cdot020$	$1\cdot39291 - i \times 1\cdot17758$	$1\cdot824$

$$z = 2\cdot4.$$

θ	[ ] in (22)	Modulus	[ ] in (26)	Modulus
0	$\cdot31511 - i \times \cdot25820$	$\cdot4074$	$\cdot03501 - i \times \cdot00548$	$\cdot0354$
30	$\cdot20547 + i \times \cdot03416$	$\cdot2083$	$\cdot00846 + i \times \cdot03851$	$\cdot0394$
45	$\cdot07387 + i \times \cdot30240$	$\cdot3113$	$-\cdot04963 + i \times \cdot07206$	$\cdot0875$
60	$-\cdot08615 + i \times \cdot52197$	$\cdot5290$	$-\cdot15376 + i \times \cdot07895$	$\cdot1728$
90	$-\cdot29433 + i \times \cdot42828$	$\cdot5196$	$-\cdot37501 - i \times \cdot13136$	$\cdot3973$
120	$+\cdot04433 - i \times \cdot58199$	$\cdot5837$	$-\cdot08070 - i \times \cdot77525$	$\cdot7794$
135	$\cdot44763 - i \times 1\cdot27912$	$1\cdot355$	$+\cdot38943 - i \times 1\cdot20336$	$1\cdot265$
150	$\cdot89597 - i \times 1\cdot92770$	$2\cdot126$	$\cdot96494 - i \times 1\cdot60617$	$1\cdot874$
180	$1\cdot38295 - i \times 2\cdot55228$	$2\cdot903$	$1\cdot63169 - i \times 1\cdot99996$	$2\cdot581$

In Fig. 2 are plotted curves showing the variation with z at given angles of $\theta = 0°$, $60°$, and $90°$. At $0°$ the polarization is all in one direction over the whole range from 0 to $2\cdot4$. At $60°$ there are reversals of polarization at $z = 1\cdot5$ and $z = 2\cdot05$. At $90°$ these reversals occur when $z = 1\cdot7$ and $z = 2\cdot3$.

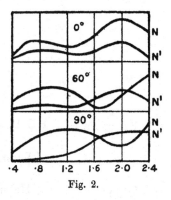

Fig. 2.

The curves stop at $z = 2\cdot4$. It would have been of interest to carry them further, but the calculations would soon become very laborious. As it is, they apply only to visible light dispersed by the very finest fibres, inasmuch as z is the ratio of the *circumference* of the cylinder to the wave-length of the light.

When z, or kc, is greater than $2\cdot4$, we may get an idea of the course of events by falling back upon the case where the refractivity $(\mu - 1)$ is very small, treated in my 1881 paper. In our present notation the light dispersed in direction θ depends upon

$$\frac{\pi c^2}{z \cos \frac{1}{2}\theta} J_1(2z \cos \tfrac{1}{2}\theta). \qquad \ldots\ldots\ldots\ldots\ldots\ldots(29)$$

When $\theta = 180°$, *i.e.* in the direction of primary propagation,

$$J_1(2z \cos \tfrac{1}{2}\theta) = z \cos \tfrac{1}{2}\theta,$$

and (29) reduces to πc^2. In this direction every element of the obstacle acts alike, and the dispersed light is a maximum[*]. In leaving this direction the dispersed light first vanishes when

$$\cos \tfrac{1}{2}\theta = 3\cdot8317/2z,$$

and afterwards when

$$2z \cos \tfrac{1}{2}\theta = 7\cdot0156,\ 10\cdot173,\ 13\cdot324,\ \text{etc.}$$

The factor (29) is applicable, whether the primary vibrations be parallel or perpendicular to the axis of the cylinder. The remaining factors may be deduced by comparison with the case of an infinitely small cylinder. Thus for vibrations parallel to the axis, we obtain from (6)

$$\psi = \left(\frac{\pi}{2ikr}\right)^{\frac{1}{2}} e^{i(nt-kr)} \times \frac{(k'c - kc) J_1(2kc \cos \tfrac{1}{2}\theta)}{\cos \tfrac{1}{2}\theta}, \qquad \ldots\ldots\ldots(30)$$

applicable however large c may be, provided $(k' - k)$ be small enough.

In like manner for vibrations perpendicular to the axis we get from (9)

$$\psi = \left(\frac{\pi}{2ikr}\right)^{\frac{1}{2}} e^{i(nt-kr)} \times \frac{(kc - k'c) \cos \theta \cdot J_1(2kc \cos \tfrac{1}{2}\theta)}{\cos \tfrac{1}{2}\theta}, \qquad \ldots\ldots(31)$$

vanishing when $\theta = 90°$, whatever may be the value of kc. It will be seen that (30) and (31) differ only by the factor $-\cos \theta$, and that this is unity in the direction of the primary light.

[* The successive maxima occur at the roots of $J_2(2z \cos \tfrac{1}{2}\theta) = 0$, viz. $2z \cos \tfrac{1}{2}\theta = 0,\ 5\cdot135,\ 8\cdot417,\ 11\cdot620$, etc. W. F. S.]

435.

THE PERCEPTION OF SOUND.

[*Nature*, Vol. CII. p. 225, 1918.]

I DO not think that Helmholtz's theory of audition, whatever difficulties there may be in it, breaks down so completely as Dr Perrett represents*. According to him, one consequence of the theory would be that "when a tuning-fork is made to vibrate, no note can be heard, but only an unimaginable din." I cannot admit this inference. It is true that Helmholtz's theory contemplates the response in greater or less degree of a rather large number of "resonators" with their associated nerves, the natural pitch of the resonators ranging over a certain interval. But there would be no *dissonance*, for in Helmholtz's view dissonance depends upon intermittent excitation of nerves, and this would not occur. So long as the vibration is maintained, every nerve would be uniformly excited. Neither is there any difficulty in attributing a simple perception to a rather complicated nervous excitation. Something of this kind is involved in the simple perception of *yellow*, resulting from a combination of excitations which would severally cause perceptions of red and green.

The fundamental question would appear to be the truth or otherwise of the theory associated with the name of J. Müller. Whatever may be the difficulty of deciding it, the issue itself is simple enough. Can more than one kind of message be conveyed by a single nerve? Does the nature of the message depend upon how the nerve is excited? In the case of sound—say from a fork of frequency 256—is there anything periodic of this frequency going on in the nerve, or nerves, which carry the message? It is rather difficult to believe it, especially when we remember that frequencies up to 10,000 per second have to be reckoned with. Even if we could accept this, what are we to think when we come to nerves conveying the sensation of light? Can we believe that there are processes in action along the nerve repeated 10^{15} times per second?

I do not touch upon the anatomical matters treated by Sir T. Wrightson and Prof. Keith, or upon the phonetic evidence brought forward with authority by Dr Perrett.

* *Nature*, Vol. CII. p. 184, 1918.

436.

ON THE LIGHT EMITTED FROM A RANDOM DISTRIBUTION
OF LUMINOUS SOURCES.

[*Philosophical Magazine*, Vol. XXXVI. pp. 429—449, 1918.]

RECENT researches have emphasized the importance of a clear comprehension of the operation under various conditions of a group of similar unit
sources, or centres, of iso-periodic vibrations, *e.g.* of sound or of light. The
sources, supposed to be concentrated in points, may be independently excited
(as probably in a soda flame), or they may be constituted of similar small
obstacles in an otherwise uniform medium, dispersing plane waves incident
upon them. We inquire into an effect, such as the intensity, at a great
distance from the cloud, either in a particular direction, or in the average of
all directions. For convenience of calculation and statement we shall consider
especially sonorous vibrations; but most of the results are equally applicable
to electric vibrations, as in light, the additional complication being merely
such as arises from the vibrations being transverse to the direction of propagation.

If the centres, supposed to be distributed at random in a region whose
three dimensions are all large, are spaced widely enough in relation to the
wave-length (λ) to act independently, the question reduces itself to one
formerly treated*, for it then becomes merely one of the composition of a
large number (n) of unit vibrations of arbitrary phases. It is known that
the "expectation" of intensity in any direction is n times that due to a single
centre, or (as we may say) is equal to n. The word "expectation" is here
used in the technical sense to represent the mean of a large number of
independent trials, or combinations, in each of which the phases are redistributed at random. It is important to remember that it is infinitely
improbable that the expectation will be confirmed in a single trial, however
large n may be. Thus in a single combination of many vibrations of arbitrary
phase there is about an even chance that the intensity will be less than $\cdot 7n$.

---

* *Phil. Mag.* Vol. x. p. 73 (1880); *Scientific Papers*, Vol. I. p. 491. For another method see
Theory of Sound, 2nd ed. § 42 *a*, and for a more complete theory K. Pearson's *Math. Contributions
to the Theory of Evolution*, xv, Dulau, London.

The general formula is that the probability of an *amplitude* between r and $r + dr$ is

$$\frac{2}{n} e^{-r^2/n} r\,dr = \frac{1}{n} e^{-I/n} dI, \qquad\qquad\dots\dots\dots\dots\dots(1)$$

if I denote the *intensity**.

As regards the "expectation" of intensity merely, the question is very simple. If $\theta, \theta', \theta'' \dots$ be the n individual phases, the expectation is

$$\int_0^{2\pi}\int_0^{2\pi}\int_0^{2\pi} \dots \frac{d\theta}{2\pi}\frac{d\theta'}{2\pi}\frac{d\theta''}{2\pi} \dots [(\cos\theta + \cos\theta' + \dots)^2 + (\sin\theta + \sin\theta' + \dots)^2].$$

Effecting the integration with respect to θ, we have

$$\int_0^{2\pi}\int_0^{2\pi} \dots \frac{d\theta'}{2\pi}\frac{d\theta''}{2\pi} \dots [1 + (\cos\theta' + \cos\theta'' + \dots)^2 + (\sin\theta' + \sin\theta'' + \dots)^2];$$

and when we continue the process over all the n phases we get finally

$$\text{Expectation of Intensity} = n.$$

The same result follows of course from (1). The "expectation" is

$$\int_0^\infty e^{-I/n} I . dI/n = n. \qquad\qquad\dots\dots\dots\dots\dots\dots(2)$$

But if we are not to expect any particular intensity when a large number of vibrations of unit amplitude and arbitrary phase are combined, what precisely is the significance to be attached to this result? As has already been suggested, we must look to what is likely to happen when we have to do with a large number m of independent trials, in each of which the n phases are redistributed at random. By (1) the chance of the separate intensities $I_1, I_2, \dots I_m$ lying between $I_1 + dI_1$, $I_2 + dI_2$, etc. is

$$n^{-m} e^{-(I_1 + I_2 + \dots)/n} dI_1 dI_2 \dots dI_m;$$

and we may inquire what is altogether the chance of the sum of intensities, represented by J, lying between J and $J + dJ$. Over the range concerned the factor $e^{-J/n}$ may be treated as constant, and so the question is reduced to finding the value of

$$\iint \dots dI_1 dI_2 \dots dI_m$$

under the condition that $I_1 + I_2 + \dots$ lies between J and $J + dJ$. This is†

$$\frac{J^{m-1}}{(m-1)!} dJ;$$

so that the chance of $I_1 + I_2 + \dots$ lying between J and $J + dJ$ is

$$\frac{e^{-J/n} J^{m-1} dJ}{n^m . (m-1)!}; \qquad\qquad\dots\dots\dots\dots\dots\dots(3)$$

---

* An interesting example of variable intensity when phases are at random is afforded by the observations of De Haas (*Amsterdam Proceedings*, Vol. xx. p. 1278 (1918)) on the granular structure of the field when a *corona* is formed from homogeneous light. The results of various combinations are exhibited to the eye simultaneously.

† See for example Todhunter's *Int. Calc.* § 272.

or, if we employ the mean value of the I's instead of the sum, the chance of the mean, viz. $(I_1 + I_2 + \ldots)/m$, lying between K and $K + dK$ is

$$\frac{e^{-mK/n} \cdot m^{m+1} K^{m-1} dK}{n^m \cdot m!} \cdot \quad \ldots\ldots\ldots\ldots\ldots\ldots(4)$$

We may compare this with the corresponding expression when $m = 1$, where we have to do with a single I, to which K then reduces. The ratio

$$R = (4):(1) = \frac{e^{-(m-1)K/n} m^{m+1} K^{m-1}}{n^{m-1} \cdot m!} \cdot \quad \ldots\ldots\ldots\ldots(5)$$

When we treat m as very large, we may take

$$m! = m^m \sqrt{(2\pi m)} \cdot e^{-m},$$

so that (5) becomes

$$\frac{e\sqrt{m}}{\sqrt{(2\pi)}} \left\{ \frac{e^{-(K/n-1)} K}{n} \right\}^{m-1} \cdot \quad \ldots\ldots\ldots\ldots\ldots(6)$$

If in (6) $K = n$ absolutely, the second factor is unity, and since the first factor increases indefinitely with m, there is a concentration of probability upon the value n, as compared with what obtains for a single combination.

In general we have to consider what becomes of

$$\sqrt{m} \cdot \{x e^{1-x}\}^{m-1}, \quad \ldots\ldots\ldots\ldots\ldots\ldots\ldots(7)$$

when $m = \infty$, and x, written for K/n, is positive. Here $x e^{1-x}$ vanishes when $x = 0$ and when $x = \infty$, and it has but one maximum when $x = 1$, $x e^{1-x} = 1$. We conclude that $x e^{1-x}$ is a positive quantity, in general less than unity. The ratio of consecutive values when m in (7) increases to $m + 1$ is

$$x e^{1-x} \sqrt{(1 + 1/m)},$$

and thus when $m = \infty$, (7) diminishes without limit, unless $x = 1$ absolutely. Ultimately there is no probability of any mean value K which is not infinitely near the value n.

Fig. 1 gives a plot of R in (5) as a function of x, or K/n, for $m = 2, 4, 6$.

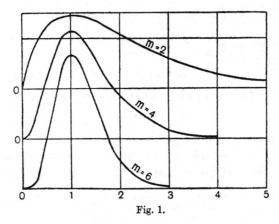

Fig. 1.

It will be observed that for $m > 2$, $dR/dx = 0$ when $x = 0$, but that for $m = 2$, $dR/dx = 4$.

The corresponding question for J may be worth a moment's notice. We have

$$R' = (3) : (1) = \frac{m J^{m-1}}{n^{m-1} . m!}; \quad \dotfill (8)$$

so that R' goes to zero as m increases, if J be comparable with n, as might have been expected.

It must not be overlooked that when the random distribution of phases is due to a random spatial distribution of centres, it fails to satisfy strictly the requirement that all the centres act independently, for some of them will lie at distances from nearest neighbours less than the number of wave-lengths necessary for approximate independence. The simple conditions just discussed are thus an ideal, approached only when the spacing is very open.

We have now to consider how the question is affected when we abandon the restriction that the spacing of the unit centres is very open. The work to be done at each centre then depends not only upon the pressure due to itself but also upon that due to not too distant neighbours. Beginning with a single source, we may take as the velocity-potential

$$\phi = - \frac{\cos k (at - r)}{4\pi r}, \quad \dotfill (9)$$

where a is the velocity of propagation, $k = 2\pi/\lambda$, and r is the distance from the centre. The rate of passage of fluid across the sphere of radius r is

$$4\pi r^2 d\phi/dr = \cos k (at - r) - kr \sin k (at - r). \dotfill (10)$$

If δp denote the variable part of the pressure at the same time and place, and ρ be the density,

$$\delta p = - \rho \frac{d\phi}{dt} = - \frac{\rho k a \sin k (at - r)}{4\pi r}. \quad \dotfill (10\,a)*$$

The rate at which work (W) has to be done is given by

$$\frac{dW}{dt} = \delta p . 4\pi r^2 \frac{d\phi}{dr} = \frac{\rho k a \sin k (at - r)}{4\pi r}$$

$$\times [kr \sin k (at - r) - \cos k (at - r)], \quad \dots (10\,b)*$$

of which the mean value depends upon the first term only. In the long run

$$W/t = \rho k^2 a/8\pi. \quad \dotfill (10\,c)*$$

It is to be observed that although the pressure is infinite at the source, the work done there is nevertheless finite on account of the pressure being in quadrature with the principal part of the rate of total flow expressed in (10).

[* In the original paper these equations were numbered (11)—(13), as well as the three following equations; to avoid confusion they have been renumbered.]

When there are two unit sources distant D from one another and in the same initial phase, the potentials may be taken to be

$$\phi = -\frac{\cos k\,(at - r)}{4\pi r}, \qquad \psi = -\frac{\cos k\,(at - r')}{4\pi r'}. \quad \ldots\ldots\ldots(11)$$

At the first source where $r = 0$

$$4\pi r^2 d\phi/dr = \cos kat - kr \sin kat,$$

$$\frac{d\phi}{dt} + \frac{d\psi}{dt} = \frac{ka \sin kat}{4\pi r} + \frac{ka}{4\pi D} \sin k\,(at - D).$$

The work done by the source at $r = 0$ is accordingly proportional to

$$1 + \frac{\sin kD}{kD}, \quad \ldots\ldots\ldots\ldots\ldots\ldots\ldots\ldots(12)$$

and an equal amount of work is done by the source at $r' = 0$. If D be infinitely great, the sources act independently, and thus the scale of measurement in (12) is such that unity represents the work done by each source when isolated. If $D = 0$, the work done by each source is doubled, and the sources become equivalent to one of doubled magnitude.

If D be equal to $\frac{1}{2}\lambda$, or to any multiple thereof, $\sin kD = 0$, and we see from (12) that the work done by each source is unaffected by the presence of the other. This conclusion may be generalized. If any number (n) of equal sources in the same phase be arranged in (say a vertical) line so that the distance between immediate neighbours is $\frac{1}{2}\lambda$, the work done by each is the same as if the others did not exist. The whole work accordingly is n, whereas the work to be done by a single source of magnitude n would be n^2. Thus if sound be wanted only in the horizontal plane where there is agreement of phase, the distribution into n parts effects an economy in the proportion of $n : 1$.

A similar calculation would apply when the initial phases differ, but we will now take up the problem in a more general form where there are any number (n) of unit sources, and by another method*. The various centres are situated at points finitely distant from the origin O. The velocity-potential of one of these at (x, y, z), estimated at any point Q, is

$$\phi = -\frac{\cos (pt + \epsilon - kR)}{4\pi R}, \ldots\ldots\ldots\ldots\ldots\ldots(13)$$

where R is the distance between Q and (x, y, z). At a great distance from the origin we may identify R in the denominator with OQ, or R_0; while under the cosine we write

$$R = R_0 - (lx + my + nz), \ldots\ldots\ldots\ldots\ldots(14)$$

* "On the Production and Distribution of Sound," *Phil. Mag.* Vol. VI. p. 289 (1903); *Scientific Papers*, Vol. V. p. 136.

l, m, n being the direction cosines of OQ. On the whole

$$-4\pi R_0 \phi = \Sigma \cos \{ pt + \epsilon - kR_0 + k \,(lx + my + nz)\}, \quad ......(15)$$

in which R_0 is a constant for all the sources, but ϵ, x, y, z vary from one source to another. The *intensity* in the direction $(l, m; n)$ is thus represented by

$$[\Sigma \cos \{\epsilon + k \,(lx + my + nz)\}]^2 + [\Sigma \sin \{\epsilon + k \,(lx + my + nz)\}]^2,$$

or by

$$n + 2\Sigma \cos \left[\epsilon_1 - \epsilon_2 + k \,\{l \,(x_1 - x_2) + m \,(y_1 - y_2) + n \,(z_1 - z_2)\}\right], \quad ...(16)$$

the summation being for all the $\tfrac{1}{2}n \,(n - 1)$ pairs of sources. In order to find the work done we have now to integrate (16) over angular space.

It will suffice if we effect the integration for the specimen term; and we shall do this most easily if we take the line through the points (x_1, y_1, z_1), (x_2, y_2, z_2) as axis of reference, the distance between them being denoted by D. If (l, m, n) make an angle with D whose cosine is μ,

$$D\mu = l \,(x_1 - x_2) + m \,(y_1 - y_2) + n \,(z_1 - z_2), \quad ....(17)^*$$

and the value of the specimen term is

$$\int_{-1}^{+1} \cos (\epsilon_1 - \epsilon_2 + kD\mu)\, d\mu,$$

that is

$$\frac{2 \sin kD \cos (\epsilon_1 - \epsilon_2)}{kD}. \quad ....(18)$$

The mean value of (16) over angular space is thus

$$n + 2\Sigma \frac{\sin kD \cos (\epsilon_1 - \epsilon_2)}{kD}, \quad ....(19)$$

where ϵ_1, ϵ_2 refer to any pair of sources and D denotes the distance between them. If all the sources are in the same initial phase, $\cos (\epsilon_1 - \epsilon_2) = 1$. If the distance between every pair of sources is a multiple of $\tfrac{1}{2}\lambda$, $\sin kD = 0$, and (19) reduces to its first term.

We fall back upon a former particular case if we suppose that there are only two sources and that they are in the same phase.

If the question of the phases of the two sources be left open, (19) gives

$$2 + 2 \cos (\epsilon_1 - \epsilon_2) \frac{\sin kD}{kD}. \quad ....(20)$$

If D be small, this reduces to

$$2 + 2 \cos (\epsilon_1 - \epsilon_2),$$

which is zero if the sources be in opposite phases, and is equal to 4 if the phases be the same.

---

* In the paper referred to, equation (19), μ was inadvertently used in two senses.

If in (20) the phases are 90° apart, the cosine vanishes. The work done is then simply the double of what would be done by either source acting alone, and this whatever the distance D may be. If this conclusion appear paradoxical, it may be illustrated by considering the case where D is very small. Then

$$-4\pi R_0 \phi = \cos(pt + \epsilon - kR_0) + \cos(pt + \epsilon \pm \tfrac{1}{2}\pi - kR_0) = \sqrt{2}.\cos(pt + \epsilon \pm \tfrac{1}{4}\pi - kR_0),$$

representing a single source of strength $\sqrt{2}$, giving intensity 2 simply.

We have seen that the effect of a number n of unit sources depends upon the initial phases and the spatial distribution, and this not merely in a specified direction, but in the mean of all directions, representing the work done. We have now to consider what happens when the initial phases are at random, or when the spatial distribution is at random within a limited region. Obviously we cannot say what the effect will be in any particular case. But we may inquire what is the expectation of intensity, that is the mean intensity in a great number of separate trials, in each of which there is an independent random distribution.

The question is simplest when the individual initial phases are at random in separate trials, and the result is then the same whether the spatial distribution be at random or prescribed. For the mean value of every single term under the sign of summation in (19) is then zero, D meanwhile being constant for a given pair of sources, while

$$\int_0^{2\pi} \cos(\epsilon_1 - \epsilon_2) \frac{d\epsilon_2}{2\pi} = 0.$$

The mean intensity, whether reckoned in all directions, or even in a specified direction (16), reduces to n simply.

If the sources are all in the same phase, or even if each individual source retains its phase, $\cos(\epsilon_1 - \epsilon_2)$ in (19) remains constant in the various trials for each pair, and we have to deal with the mean value of $\sin kD \div kD$ when the spatial distribution is at random. We may begin by supposing two sources constrained to lie upon a straight line of limited length l, where, however, l includes a very large number of wave-lengths (λ).

If the first source occupies a position sufficiently remote from the ends of the line, so that the two parts on either side (l_1 and l_2) are large multiples of λ, the mean required, represented by

$$\frac{l_1}{l}\int_0^{l_1} \frac{\sin kD}{kD}\frac{dD}{l_1} + \frac{l_2}{l}\int_0^{l_2} \frac{\sin kD}{kD}\frac{dD}{l_2}, \quad\dots\dots\dots\dots(21)$$

may be identified with π/kl, since both upper limits may be treated as infinite. Moreover, π/kl may be regarded as evanescent, kl being by supposition a large quantity.

So far positions of the first source near the ends of the line have been excluded. If the neglect of these positions can be justified, (20) reduces to 2 simply.

It is not difficult to see that the suggested simplification is admissible under the conditions contemplated. If x, x' be the distances of the two sources from one end of the line, the question is as to the value of

$$\int_0^l \frac{dx}{l} \int_0^l \frac{dx'}{l} \frac{\sin k (x' - x)}{k (x' - x)} , \qquad (22)$$

where the integration with respect to x' may be taken first. Let X denote a length large in comparison with λ, but at the same time small in comparison with l. If x lie between X and $l - X$, the integral with respect to x' may be identified with π/kl, and neglected, as we have seen. We have still to include the ranges from $x = 0$ to $x = X$, and from $x = l - X$ to $x = l$, of which it suffices to consider the former. The range for x' may be divided into two parts, from 0 to x, and from x to l. For the latter we may take

$$\int_x^l \frac{dx'}{l} \frac{\sin k (x' - x)}{k (x' - x)} = \frac{\pi}{2kl} ,$$

so that this part yields finally after integration with respect to x,

$$\frac{X}{l} \cdot \frac{\pi}{2kl} . \qquad (23)$$

As regards the former part, we observe that since $\theta^{-1} \sin \theta$ can never exceed unity,

$$\int_0^x \frac{dx'}{l} \frac{\sin k (x' - x)}{k (x' - x)} < \frac{x}{l} , \qquad (24)$$

in which again $x < X$. The result of the second integration leaves us with a quantity less than X^2/l^2. The anomalous part, both ends included, is less than

$$\frac{2X}{l} \left(\frac{X}{l} + \frac{\pi}{2kl} \right) , \qquad (25)$$

which is small in comparison with the principal part*, of the order π/kl and itself negligible. We conclude that here again the mean intensity in a great number of trials is 2 simply. It may be remarked that this would not apply to the mean intensity in a specified direction, as we may see from the case where the initial phases are the same. In a direction perpendicular to the line on which the sources lie, the phases on arrival are always in agreement, and the intensity is 4, wherever upon the line the sources may be situated. The conclusion involves the mean in *all* directions, as well as the mean of a large number of trials.

Under a certain restriction this argument may be extended to a large number n of unit sources, since it applies to every term under the summation

[* Provided that X/l is small compared with λ/X; if these ratios are of the same order, (25) is comparable with π/kl. W. F. S.]

in (19). But inasmuch as the evanescence is but approximate, we have to consider what may happen when n is exceedingly great. The number of terms is of order n^2, so that the question arises whether $n^2\pi/kl$ can be neglected in comparison with n. The ratio is of the order $n\lambda/l$, and it cannot be neglected unless the mean distance of consecutive sources is much greater than λ. It is only under this restriction that we can assert the reduction of the mean intensity to the value n when the initial phases are not at random.

The next problem proposed is the application of (19) when the n sources are distributed at random over the volume of a *sphere* of radius R. In this case the distinction between the mean in one direction and in the mean of all directions disappears. If for the moment we limit our attention to a single pair of sources, the chance of the first source lying in the element of volume dV is dV/V, and similarly of the second source lying in dV' is dV'/V. As the individual sources may be interchanged, the chance of the pair occupying the elements dV, dV' is $2dVdV'/V^2$, so that from the second part of (19) we get for a single pair the expectation of intensity

$$4 \iint \frac{\sin kr}{kr} \frac{dV}{V} \frac{dV'}{V},$$

and for the $\frac{1}{2}n(n-1)$ pairs

$$\frac{2n(n-1)}{V^2} \iint \frac{\sin kr}{kr} dV dV'. \quad \dots\dots\dots\dots\dots(26)$$

Here V is the whole volume of the sphere, viz. $\frac{4}{3}\pi R^3$, and r is written in place of D. The function of r may be regarded as a kind of potential, so that the integral in (26) represents the work required to separate thoroughly every pair of elements. As in *Theory of Sound*, § 302, we may estimate this by successive removals to infinity of outer thin shells of thickness dR. The first step is the calculation of the potential at O, a point on the surface of the sphere.

The polar element of volume at P is $r^2 \sin\theta\, d\omega d\theta dr$, where $r = OP$, $\theta =$ angle COP. The integration with respect to ω will merely introduce the factor 2π. For the integration with regard to r, we have

$$\int_0^r \frac{\sin kr}{kr} r^2 dr = \frac{\sin kr - kr \cos kr}{k^3},$$

Fig. 2.

r now standing for OQ. In terms of $\mu (= \cos\theta)$, $r = 2R\mu$, and we have next to integrate with respect to μ. We get

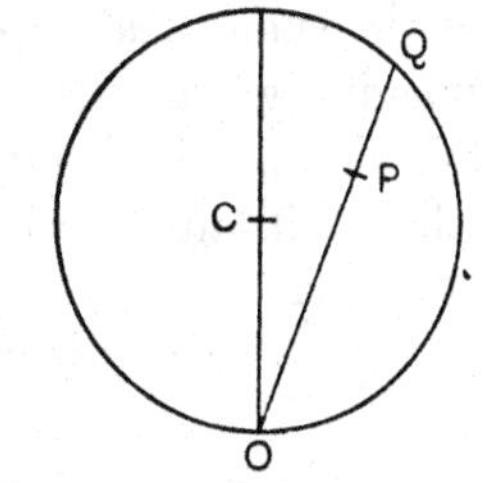

$$\int_0^1 \frac{\sin kr - kr \cos kr}{k^3} d\mu = \frac{1 - \cos 2kR - kR \sin 2kR}{k^4 R},$$

which, multiplied by 2π, now expresses the potential at O.

This potential is next to be multiplied by $4\pi R^2 dR$ and integrated from 0 to R. We find

$$\iint \frac{\sin kr}{kr} dV dV' = \frac{8\pi^2}{k^6}(\sin kR - kR \cos kR)^2. \quad\ldots\ldots\ldots\ldots(27)$$

We have now to divide by V^2, or $16\pi^2 R^6/9$; and finally we get

$$(19) = n + \frac{9n(n-1)}{k^6 R^6}(\sin kR - kR \cos kR)^2, \quad\ldots\ldots\ldots(28)*$$

where kR will now be regarded as very large. When n is moderate, or at any rate does not exceed $k^3 R^3$, the second term is relatively negligible, that is reduction occurs to n simply, provided n be not higher than of order R^3/λ^3, corresponding to one source for each cubic wave-length†. But evidently n may be so great that this reduction fails, unless otherwise justified by a random distribution of initial phases.

At the other extreme of an altogether preponderant n, the second term in (19) dominates the first, and we get in the case of constant initial phases and a very large kR,

$$(19) = \frac{9n^2 \cos^2 kR}{k^4 R^4}. \quad\ldots\ldots\ldots\ldots\ldots\ldots(29)$$

Under the suppositions hitherto made of a random spatial distribution within the sphere (R), and of uniformity of initial phases, there is no escape from the conclusion that the reduction to the simple value n fails when n is great enough. Nevertheless, there is a sense in which the reduction may take place, and the point is of importance, especially in the application to the dispersal of primary waves by a cloud of small obstacles. In order better to understand the significance of the term in n^2, let us calculate the intensity due to an absolutely uniform distribution of source of total amount n over the spherical volume. Since there is complete symmetry, it suffices to consider a single specified direction which we take as axis of z. As in (15), we have

$$-4\pi R_0 \phi = \frac{ne^{i(pt-kR_0)}}{V} \iiint e^{ikz} dx\, dy\, dz, \quad\ldots\ldots\ldots\ldots(30)$$

as the symbolical expression for the velocity-potential, from which finally the imaginary part is to be rejected. The integral over the sphere is easily evaluated, either as it stands, or with introduction of polar coordinates (r, θ, ω) which will afterwards be required. Thus with μ written for $\cos\theta$,

$$\iiint e^{ikz} dx\, dy\, dz = 2\pi \int_0^R \int_{-1}^{+1} e^{ikr\mu} r^2 dr\, d\mu$$

$$= \frac{4\pi}{k} \int_0^R \sin kr . r\, dr = \frac{4\pi}{k^3}(\sin kR - kR \cos kR). \ldots\ldots(31)$$

---

* We may confirm (28) by supposing kR very *small*, when the right-hand member reduces to n^2.

† The number of molecules per cubic wave-length in a gas under standard conditions is of the order of a million.

Accordingly

$$-4\pi R_0 \phi = \frac{3n}{k^3 R^3} (\sin kR - kR \cos kR), \quad \ldots\ldots\ldots\ldots(32)$$

reducing to n simply when kR is very small. The intensity due to the uniform distribution is thus

$$\frac{9n^2}{k^6 R^6} (\sin kR - kR \cos kR)^2, \quad \ldots\ldots\ldots\ldots\ldots\ldots(33)$$

exactly the n^2 term of (28). The distinction between (28) and (32), at least when kR is very great, has its origin in the circumstance that in the first case the n separate centres, however numerous, are discrete and scattered at random, while in the second case the distribution of the same total is uniform and continuous.

When we examine more attentively the composition of the velocity-potential ϕ in (30), we recognize that it may be regarded as originating at the *surface* of the sphere R. Along any line parallel to z, the phase varies uniformly, so that every complete cycle occupying a length λ contributes nothing. Any contribution which the entire chord may make depends upon the immediate neighbourhood of the ends, where incomplete cycles may stand over. And, since this is true of every chord parallel to z, we may infer that the total depends upon the manner in which the volume terminates, viz. upon the surface. At this rate the n^2 term in (28) must be regarded as due to the surface of the sphere, and if we limit attention to what originates in the interior this term disappears, and (kR being sufficiently large) (19) reduces to n.

When we speak of an effect being due to the *surface*, we can only mean the discontinuity of distribution which occurs there, and the best test is the consideration of what happens when the discontinuity is eased off. Let us then in the integration with respect to r in (31) extend the range beyond R to R' with introduction of a factor decreasing from unity (the value from 0 to R), as we pass outwards from R to R'. The form of the factor is largely a matter of mathematical convenience.

As an example we may take $e^{-h'(r-R)}$, or $e^{-hk(r-R)}$, which is equal to unity when $r = R$ and diminishes from R to R'. The complete integral (31) is now

$$\frac{4\pi}{k} \int_0^R \sin kr . r\, dr + \frac{4\pi}{k} \int_R^{R'} e^{-hk(r-R)} \sin kr . r\, dr. \quad \ldots\ldots(34)$$

From the second integral we may extract the constant factor e^{hkR}, and if we then treat $\sin kr$ as the imaginary part of e^{ikr}, we have to evaluate

$$\int_R^{R'} e^{(i-h)kr} r\, dr.$$

We thus obtain for (34)

$$\frac{4\pi}{k^3}(\sin kR - kR \cos kR)$$

$$- \frac{4\pi e^{-h'(R'-R)}}{k^3(1+h^2)^2}[\cos kR'\{(h^2+1)kR' + 2h\} + \sin kR'\{(h^2+1)hkR' + h^2 - 1\}]$$

$$+ \frac{4\pi}{k^3(1+h^2)^2}[\cos kR\{(h^2+1)kR + 2h\} + \sin kR\{(h^2+1)hkR + h^2 - 1\}].$$

$$\dots\dots(35)$$

When we combine the first and third parts, in which R' does not appear, we get

$$\frac{4\pi}{k^3(1+h^2)^2}[\cos kR\{2h - h^2(h^2+1)kR\} + \sin kR\{h^4 + 3h^2 + h(h^2+1)kR\}].$$

$$\dots\dots(36)$$

The first part of (35), representing the effect due to the sphere R suddenly terminated, is of order kR; and our object is to ascertain whether by suitable choice of h and R' we can secure the relative annulment of (35). As regards (36), it suffices to suppose h small enough. In the second part of (35) the principal term is of relative order $(R'/R)e^{-hk(R'-R)}$ and can be annulled by sufficiently increasing R', however small h may be.

Suppose, to take a numerical example, that $h = \frac{1}{1000}$, and that $e^{-hk(R'-R)}$ is also $\frac{1}{1000}$. Then

$$R' - R = \frac{3\lambda}{2\pi h \log_{10} e} = \frac{1\cdot 1\lambda}{h} = 1100\lambda.$$

With such a value of $R' - R$ the factor R'/R may be disregarded*.

It appears then that it is quite legitimate to regard the intensity due to the simple sphere, expressed in (33), as a surface effect; and this conclusion may be extended to the corresponding term involving n^2 in (28), relating to discrete centres scattered at random.

This extension being important, it may be well to illustrate it further. Returning to the consideration of n sources in the same initial phase distributed at random along a limited straight line, let us inquire what is to be expected at a distant point along the line produced. The first question which suggests itself is—Are the phases on arrival distributed at random? Not in all cases, but only when the limited line contains exactly an integral number of wavelengths. Then the phases on arrival are absolutely at random over the whole period, and accordingly the expectation of intensity is n precisely. If, however, there be a fractional part of a wave-length outstanding, the arrival phases are no longer absolutely at random, and the conclusion that the expectation of intensity is n simply cannot be maintained. Suppose further that n is so great that the average distance between consecutive sources is a very small

* The application to light is here especially in view.

fraction of a wave-length. The conclusion that when an exact number of wave-lengths is included the expectation is n remains undisturbed, and this although the effect due to any small part, supposed to act alone, is proportional to n^2. But the influence of any outstanding fraction of a wave-length is now of increased importance. If we do not look too minutely, the distribution of sources is approximately uniform. If it were completely so, the whole intensity would be attributable to the fractions at the ends[*], and would be proportional to n^2. In general we may expect a part proportional to n^2 due to the ends and another part proportional to n due to incomplete uniformity of distribution over the whole length. When n is small the latter part preponderates, but when n is great the situation is reversed, unless the number of wave-lengths included be very nearly integral. And it is apparent that the n^2 part has its origin in the discontinuity involved in the sharp limitation of the line, and may be got rid of by a tapering away of the terminal distribution.

Similar ideas are applicable to a random distribution in three dimensions over a volume, such as a sphere, which may be regarded as composed of chords parallel to the direction in which the effect is to be estimated. The n^2 term corresponds to what would be due to a continuous uniform distribution over the volume of the same total source, and it may be regarded as due to the discontinuity at·the surface. In addition there is a term in n, due to the lack of complete uniformity of distribution and issuing from every part of the interior.

Thus far we have been considering the operation of given unit sources, by which in the case of sound is meant centres where a given periodic introduction (and abstraction) of fluid is imposed. We now pass to the problem of equal small obstacles distributed at random and under the influence of primary plane waves. It is easy to recognize that these obstacles act as secondary sources, but it is not so obvious that the strength of each source may be treated as given, without regard to the action of neighbours. I apprehend, however, that this assumption is legitimate; in the case of aerial waves it may be justified by a calculation upon the lines of *Theory of Sound*, § 335. For this purpose we may suppose the density σ of the gas to be unchanged at the obstacles, while the compressibility is altered from m to m', so that the secondary disturbance issuing from each obstacle is symmetrical, of zero order in spherical harmonics. The expressions for the primary waves and of the disturbance inside the spherical obstacle under consideration remain as if the obstacle were isolated. But for the secondary disturbance external to the obstacle we must include also that due to neighbours. On forming the conditions to be satisfied at the surface of the sphere, expressing the equality on the two sides of pressure (or potential) and of

---

[*] It is indifferent how the fraction is divided between the two ends.

radial velocity, we find that when the radii are small enough, the obstacle acts as a source whose strength is independent of neighbours.

The operation of a cloud of similar particles may now be deduced without much difficulty from what has already been proved. We suppose that the individual particles are so small that the cloud has no sensible effect upon the progress of the primary waves. Each particle then acts as a source of given strength. But the initial phase for the various particles is not constant, being dependent upon the situation along the primary rays. This is, in fact, the only new feature of which we have to take account.

Perhaps the most important difference thence arising is that there is no longer equality of radiation in various directions, even from a spherical cloud, and that, whatever may be the shape of the cloud, the radiation in the direction of the primary rays produced is specially favoured. In this direction any retardation along the primary ray is exactly compensated by a corresponding acceleration along the secondary ray, so that on arrival at a distant point the phases due to all parts are the same. But, except in this direction and in others approximating to it, the argument that the effect may be attributed to the *surface* still applies. If in a continuous uniform distribution we take chords in the direction, for example, of either the incident or the scattered rays, we see as before that the effect of any chord depends entirely on how it terminates*. In forming an integral analogous to that of (30), in addition to the factor e^{ikz} expressive of retardation along the secondary ray, we must include another in respect of the primary ray. If the direction cosines of the latter be α, β, γ, the factor in question is $e^{ik(\alpha x + \beta y + \gamma z)}$, γ being -1 when the directions of the primary and secondary rays are the same. The complete exponent in the phase-factor is thus

$$ik\{\alpha x + \beta y + (\gamma+1)z\} = ik\sqrt{(2+2\gamma)} \cdot \frac{\alpha x + \beta y + (\gamma+1)z}{\sqrt{\{\alpha^2 + \beta^2 + (\gamma+1)^2\}}}.$$

The fraction on the right represents merely a new coordinate (ζ), measured in a direction bisecting the angle between the primary and secondary rays, so that the phase-factor may be written $e^{i\sqrt{(2+2\gamma)} \cdot k\zeta}$, γ being the cosine of the angle (χ) between the rays. In integrating for the sphere the only change required in the integrand is the substitution of $2k\cos\frac{1}{2}\chi$ for k. With this alteration equations (31), (32), (33) are still applicable. When the secondary ray is perpendicular to the primary,

$$2k\cos\tfrac{1}{2}\chi = \sqrt{2} \cdot k.$$

In order to find the mean intensity in all directions we have to integrate (33) over angular space and divide the result by 4π. It may be remarked

---

* It may be remarked that the same argument applies to the particles of a crystal forming a regular space lattice. If the wave-length be large in comparison with the molecular distance, no light can be scattered from the interior of such a body. For X rays this condition is not satisfied, and regular reflexions from the interior are possible. Comparison may be made with the behaviour of a grating referred to below.

that although $\cos^6 \frac{1}{2}\chi$ appears in the denominator of (33), it is compensated when $\cos \frac{1}{2}\chi = 0$ by a similar factor in the numerator. In the integration with respect to χ

$$\sin \chi \, d\chi = -4 \cos \tfrac{1}{2}\chi \cdot d \left(\cos \tfrac{1}{2}\chi\right).$$

If we write ψ for $2kR \cos \frac{1}{2}\chi$, the mean sought may be written

$$\frac{9n^2}{2k^2R^2} \int \frac{(\sin \psi - \psi \cos \psi)^2}{\psi^5} \, d\psi, \quad\dots\dots\dots\dots\dots(37)$$

the range for ψ being from 0 to $2kR$. The integration can be effected by "parts." We have

$$\int \frac{(\sin \psi - \psi \cos \psi)^2}{\psi^5} \, d\psi = -\frac{\sin^2 \psi - 2\psi \sin \psi \cos \psi + \psi^2}{4\psi^4}. \quad\dots(38)$$

When ψ is small, the expression on the right becomes

$$-\frac{1}{4} + \frac{\psi^2}{18},$$

so that the integral between 0 and ψ is $\psi^2/18$ simply. In general, the mean intensity is

$$\frac{9n^2}{8k^2R^2} \frac{2\psi \sin \psi \cos \psi - \sin^2 \psi - \psi^2 + \psi^4}{\psi^4}, \quad\dots\dots\dots\dots(39)$$

in which ψ stands for $2kR$.

That the intensity, whether in one direction or in the mean of all directions, should be proportional to n^2 is, of course, what was to be expected. And, since the effect is here a surface effect, it may be identified with the ordinary surface reflexion which occurs at a sudden transition between two media of slightly differing refrangibilities, and is proportional to the square of that difference. If, as in a former problem, we suppose the discontinuity of the transition to be eased off, this reflexion may be attenuated to any extent until finally there is no dispersed wave at all*.

When we pass from the continuous uniform distribution to the random distribution of n discrete and very small obstacles, the term in n^2 representing reflexion from the surface remains, and is now supplemented by the term in n, due to irregular distribution in the interior. It is the latter part only with which we are concerned in a question such as that of the blue of the sky.

It must never be forgotten that it is the "expectation" of intensity which is proved to be n. In any particular arrangement of particles the intensity may be anything from 0 to n^2. But in the application to a gas dispersing light, the motion of the particles ensures that a random redistribution of phases takes place any number of times during an interval of time less than any which the eye could appreciate, so that in ordinary observation we are concerned only with what is called the expectation.

* *Conf. Proc. Lond. Math. Soc.* Vol. xi. p. 51 (1880); *Scientific Papers*, Vol. i. p. 460.

It is hoped that the explanations and calculations here given may help to remove the difficulties which have been felt in connexion with this subject. The main point would seem to be the interpretation of the n^2 term as representing the surface reflexion when a cloud is supposed to be abruptly terminated. For myself, I have always regarded the light internally dispersed as proportional to n, even when n is very great, though it may have been rather by instinct than on sufficiently reasoned grounds. Any other view would appear to be inconsistent with the results of my son's recent laboratory experiments on dust-free air.

The reader interested in optics may be reminded of the application of similar ideas to a *grating* on which fall plane waves of homogeneous light. If the spacing be quite uniform, the light behind is limited to special directions. Seen from other directions the interior of the grating appears dark. But if the ruling be irregular, light is emitted in all directions and the interior of the grating, previously dark, becomes luminous.

In the problems considered above the space occupied by a source, whether primary or secondary, has been supposed infinitely small. Probably it would be premature to try to include sources of finite extension, but merely as an illustration of what is to be expected we may take the question of n phases distributed at random over a complete period (2π), but under the limitation that the distance between neighbours is never to be less than a fixed quantity δ. All other situations along the range are to be regarded as equally probable.

As we have seen, the expectation of intensity may be equated to

$$n + 2 \iiint \ldots \Sigma \cos(\theta_\sigma - \theta_r)\, d\theta_1 d\theta_2 \ldots d\theta_n \div \iiint \ldots d\theta_1 \ldots d\theta_n, \ldots (40)$$

and the question turns upon the limits of the integrals.

The case where there are only two phases $(n = 2)$ is simple. Taking θ_1, θ_2 as coordinates of a representative point, Fig. 3, the sides of the square $OACB$ are 2π. Along the diagonal OC, θ_1 and θ_2 are equal. If DE, FG be drawn parallel to OC, so that OD, OF are equal to δ, the prohibited region is that part of the square lying between these lines. Our integrations are to be extended over the remainder, viz. the triangles FBG, DAE, and every point, or rather every infinitely small region of given area, is to be regarded as equally probable.

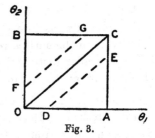

Fig. 3.

Evidently it suffices to consider one triangle, say the upper one, where $\theta_2 > \theta_1$.

For the denominator in (40) we have

$$\iint d\theta_1 d\theta_2 = \text{area of triangle } FBG = \tfrac{1}{2}(2\pi - \delta)^2.$$

In the double integral containing the cosine, let us take first the integration with respect to θ_2, for which the limits are $\theta_1 + \delta$ and 2π. We have

$$\int_{\theta_1+\delta}^{2\pi} \cos(\theta_2 - \theta_1)\,d\theta_2 = -\sin\theta_1 - \sin\delta\,;$$

and since the limits for θ_1 are 0 and $2\pi - \delta$, we get as the expectation of intensity

$$2 - 4\frac{1 - \cos\delta + (2\pi - \delta)\sin\delta}{(2\pi - \delta)^2}. \quad\ldots\ldots\ldots\ldots\ldots(41)$$

If δ^2 be neglected, this reduces to

$$2(1 - \delta/\pi). \quad\ldots\ldots\ldots\ldots\ldots\ldots\ldots(42)$$

If $\delta = \pi$, we have $2(1 - 4/\pi^2)$; and if $\delta = 2\pi$, we have 4, the only available situations being $\theta_1 = 0$, $\theta_2 = 2\pi$, equivalent to phase identity.

This treatment might perhaps be extended to a greater value, or even to the general (integral) value, of n; but I content myself with the simplifying supposition that δ is very small.

In (40) the integration with respect to θ_n supposes $\theta_1, \theta_2 \ldots \theta_{n-1}$ already fixed. If $\delta = 0$, every term such as

$$\iiint \ldots \cos(\theta_\tau - \theta_\sigma)\,d\theta_1 \ldots d\theta_n \div \iiint \ldots d\theta_1 \ldots d\theta_n$$

$$= \iint \cos(\theta_\tau - \theta_\sigma)\,d\theta_\sigma d\theta_\tau \div \iint d\theta_\sigma d\theta_\tau$$

$$= \int_0^{2\pi} d\theta_\sigma \{\sin(2\pi - \theta_\sigma) + \sin\theta_\sigma\} \div 4\pi^2 = 0,$$

and the expectation is n simply, as we have already seen. In the next approximation the correction to n will be of order δ, and we neglect δ^2.

In evaluating (40) there are $\frac{1}{2}n(n-1)$ terms under the sign of summation, but these are all equal, since there is really nothing to distinguish one pair from another. If we put $\sigma = 1$, $\tau = 2$, we have to consider

$$\iiint \ldots \cos(\theta_2 - \theta_1)\,d\theta_1 d\theta_2 \ldots d\theta_n \div \iiint \ldots d\theta_1 d\theta_2 \ldots d\theta_n. \quad\ldots\ldots(43)$$

The integration with respect to θ_n extends over the range from 0 to 2π with avoidance of the neighbourhood of $\theta_1, \theta_2, \ldots \theta_{n-1}$. For each of these there is usually a range 2δ to be omitted, but this does not apply when any of them happen to be too near the ends of the range or too near one another. This complication, however, may be neglected in the present approximation. Then

$$\int \cos(\theta_2 - \theta_1)\,d\theta_n = \cos(\theta_2 - \theta_1) \cdot \{2\pi - 2\delta(n-1)\},$$

and in like manner

$$\int d\theta_n = 2\pi - 2\delta\,(n-1),$$

so that this factor disappears. Continuing the process, we get approximately

$$\iint \cos(\theta_2 - \theta_1)\,d\theta_1\,d\theta_2 \div \iint d\theta_1\,d\theta_2,$$

as when there were only two phases to be regarded.

Accordingly, the expectation of intensity for n phases is

$$n\left\{1 - (n-1)\,\delta/\pi\right\}, \quad\ldots\ldots\ldots\ldots\ldots\ldots\ldots\ldots\ldots\ldots(44)$$

less than when $\delta = 0$, as was to be expected, since the cases excluded are specially favourable. But in order that this formula may be applicable, not merely δ, but also $n\delta$, must be small relatively to 2π.

A similar calculation is admissible when the whole range is $2m\pi$, instead of 2π, where m is an integer.

437.

THE PERCEPTION OF SOUND.

[*Nature*, Vol. CII. p. 304, 1918.]

I REGRET that I overlooked Prof. Bayliss's letter in *Nature* of October 17, in which he made an appeal for my opinion. But, if I rightly understand, the question at issue seems to be mainly one of words. Can we properly speak of the propagation of sound through an incompressible fluid? I should answer, Yes. There may be periodic motion and periodic variation of pressure; the fact that there are no variations of *density* seems immaterial. Consider plane waves, corresponding with a pure tone, travelling through air. In every thin layer of air—and thin means thin relatively to the wavelength—there are periodic motion and periodic compression, approximately uniform throughout the layer. But the compression is not essential to the travelling of the sound. The substitution of an incompressible fluid of the same density for the gas within the layer would be no hindrance. Although there is no compression, there remain a periodic pressure and a periodic motion, and these suffice to carry on the sound.

The case is even simpler if we are prepared to contemplate an incompressible fluid without mass, for then the layer need not be thin. The interposition of such a layer has absolutely no effect, the motion and pressure at the further side being the same as if the thickness of the layer were reduced to zero. To all intents and purposes the sound is propagated through the layer, though perhaps exception might be taken to the use of the word *propagation*.

As regards the ear, we have to consider the behaviour of water. From some points of view the difference between air and water is much more one of density than of compressibility. The velocities of propagation are only as 4 or 5 to 1, while the densities are as 800 to 1. Within the cavities of the ear, which are small in comparison with the wave-lengths of musical sounds, the water may certainly be treated as incompressible; but the fact does not seem to be of fundamental importance in theories of audition.

438.

ON THE OPTICAL CHARACTER OF SOME BRILLIANT ANIMAL COLOURS.

[*Philosophical Magazine*, Vol. XXXVII. pp. 98—111, 1919.]

IT is singular that the explanation of some of the most striking and beautiful of optical phenomena should be still matters of controversy. I allude to the brilliant colours displayed by many birds (*e.g.* humming-birds), butterflies, and beetles, colours which vary greatly with the incidence of the light, and so cannot well be referred to the ordinary operation of dyes. In an early paper*, being occupied at the time with the remarkable coloured reflexions from certain crystals of chlorate of potash described by Stokes, and which I attributed to a periodic twinning†, I accepted, perhaps too hastily, the view generally current among naturalists that these colours were "structure-colours," more or less like those of thin plates, as in the soap-bubble. Among the supporters of this view‡ in more recent times may be especially mentioned Poulton and Hodgkinson. In Poulton's paper§ the main purpose was to examine the history of the very remarkable connexion between the metallic colours of certain pupæ (especially *Vanessa urticæ*) and the character of the light to which the larvæ are exposed *before* pupation. In a passage describing the metallic colour itself he remarks :

"*The Nature of Effects Produced.*—The gilded appearance is one of the most metal-like appearances in any non-metallic substance. The optical explanation has never been understood. It has, however, been long known that it depends upon the cuticle, and needs the presence of moisture, and that it can be renewed when the dry cuticle is moistened. Hence it can be preserved for any time in spirit. If a piece of dry cuticle is moistened on its upper surface the colour is not renewed, but almost instantly follows the application of spirit to the lower surface. Sections of the cuticle resemble those of *Papilio machaon* described in a previous paper (*Roy. Soc. Proc.* Vol. XXXVIII. p. 279, 1885), and show an upper thin layer and a lower, much

---

* *Phil. Mag.* Vol. XXIV. p. 145 (1887) ; *Scientific Papers*, Vol. III. p. 13, see footnote.
† *Phil. Mag.* Vol. XXVI. p. 256 (1888) ; *Scientific Papers*, Vol. III. p. 204.
‡ Distinctly suggested by Hooke in his *Micrographia* (1665).
§ *Roy. Soc. Proc.* Vol. XLII. p. 94 (1887).

thicker, finely laminated layer which is also striated vertically to the surface. With Prof. Clifton's kind assistance I have been able to show that the appearances follow from interference of light, due to the presence of films of liquid between the lamellæ of the lower layer. The microscope shows brilliant red and green tints by reflected light, while in transmitted light the complementary colours are distinct, but without brilliancy. The latter colours are seen to change when pressure is applied to the surface of the cuticle, and when the process of drying is watched under the microscope, owing in both cases to the liquid films becoming thinner. In the dry cuticle the solid lamellæ probably come into contact, and prevent the admission of air, which, if present, would cause even greater brilliancy than liquid. The spectroscope shows broad interference-bands in the transmitted light, which change their position on altering the angle of incidence of the light which passes through the cuticle. Precisely similar colours, metallic on reflexion, non-metallic and with complementary tints on transmission, with the same spectroscope appearances and changes induced by the same means, are seen in the surface films which are formed on bottle-glass after prolonged exposure to earth and moisture. In the alternating layers of the pupa the chitinous lamellæ are of higher, the liquid films of lower refractive index; hence water or alcohol produces brilliant appearances, while liquids of higher refractive indices produce less effect."

I owe to Prof. Poulton the opportunity of repeating some of these observations, such as the loss of metallic appearance on drying and of recovery under alcohol. On substitution of benzol with a little bisulphide of carbon for alcohol, the surface became very dark, but regained the golden glitter on going back to alcohol.

Of a specimen of another kind Prof. Poulton writes that the bug has been in the Oxford Museum Collection for 30 or 40 years, judged by the pin. It is brown when dry, but when soaked in water becomes green like a leaf with bright iridescent green stripes on the under side. This observation also I have been able to repeat. All of which, it need hardly be said, is strongly suggestive of interference.

Dr A. Hodgkinson also has described interesting observations. In his early papers* he distinctly refers the colours to Newton's scale, which in strictness would imply a limitation to a *single* thin plate. He emphasizes the importance, for purposes of identification, of recording the colours of feathers etc. as seen by *perpendicular* reflexion, a condition best secured by illumination from a small perforated mirror, behind which the eye is placed. When daylight is used, it often suffices to examine the object with one's back to the window and at some distance from it. I shall have occasion later to refer again to Hodgkinson's work.

* *Manchester Memoirs*, 1889; 1892, p. 149.

The first, so far as I know, to challenge the "structure" theory was Dr B. Walter, whose tract* includes an elaborate discussion, accompanied by original observations, of the colours which may arise in the act of reflexion, and decides unequivocally that the colours now in question, with one or two possible exceptions, are due to surface, or quasi-metallic, reflexion as described by Haidinger, Brewster, and Stokes. The first of these writers formulated a law, named after him, which identifies the surface-colour with those rays which would be most intensely absorbed within the substance. The theory of "anomalous dispersion" since developed shows, however, that the matter does not stop there, and Walter emphasizes that much of the surface-colour may be ascribed to rays which are not themselves intensely absorbed, but being situated *near* an absorption-band, are abnormally *refracted*, and hence in accordance with Fresnel's laws are abnormally reflected. On the red side of the band the refractive index is increased and on the blue side diminished, so that when the substance is in air the surface reflexion is redder than according to Haidinger's law; but this conclusion may need to be modified when the substance is in contact with a strongly refractive solid, as when a dye spread upon a glass plate is examined from the glass side. In some cases it appears that the surface-colour is due as much, or even more, to these rays excessively refracted (and consequently reflected) as to those which would be intensely absorbed and are reflected in accordance with Haidinger's rule.

The departure from Haidinger's rule is specially important when we consider what happens at oblique incidences and with polarized light. The rays reflected in virtue of the extreme opacity of the substance to them are comparatively unaffected, and are indeed rendered more prominent by the appropriate use of a nicol. As Stokes says†: "In the case of the substances at present considered, the reflected light does not vanish, but at a considerable angle of incidence the pencil polarized perpendicularly to the plane of incidence becomes usually of a richer colour, in consequence of the removal, in a great measure, of that portion of the reflected light which is independent of the metallic properties of the medium; it commonly becomes, also, more strictly related to that light which is absorbed with such great intensity." But, as Walter appears to have been the first to explain, there is a further important change of colour with the angle of incidence, when the light-vibrations are in the plane of incidence, in virtue of the abnormal refraction with its accompanying abnormal polarizing angle. In the usual case, where the dye is in contact with air, the polarizing angle for the rays lying on the red side of the absorption-band is unusually high, so that these rays, which at moderate angles of incidence contribute largely to the resultant colour, are extinguished at incidences of from 60° to 70°. In consequence,

---

* *Die Oberflächen oder Schillerfarben*, Braunschweig, 1895.

† *Phil. Mag.* Vol. VI. Dec. 1853, p. 393; *Math. and Phys. Papers*, Vol. IV. p. 42.

the colour of the reflected light moves towards the blue with increasing obliquity.

As an example, fuchsin may be referred to, a dye specially studied by Walter, who thus (p. 52) describes the surface-colour as seen from the air side :

"(*a*) For light polarized in the plane of incidence :

"At small angles of incidence the reflexion is yellow-green, and at increasing angles becomes ever yellower and brighter.

"(*b*) For light polarized perpendicularly to the plane of incidence (that is, vibrating *in* this plane) :

"At perpendicular incidence the reflexion is the same as under (*a*), and remains approximately so up to incidences of 50°. At about 60° it becomes rapidly blue-green and at 70° an almost pure blue, attaining its greatest purity at about 72°. At still greater angles the colour passes rapidly into a bright violet, and at 85° into white.

"When ordinary unpolarized light is employed, the colour of the reflexion is intermediate between (*a*) and (*b*), but always nearer to (*a*) than to (*b*) on account of the greater intensity of reflexion under (*a*)."

It is this movement of surface-reflexions towards the blue with increasing obliquity which is regarded by Walter and Michelson* as annulling the presumption in favour of the structure theory of the animal colours, which also move in this direction ; and it must, of course, be admitted that the criterion is somewhat blurred thereby. Walter, indeed, maintains that thin plate colours change too much with angle to meet the requirements of the case. To this point I will return presently ; but what I wish to remark at the moment is that with ordinary unpolarized light the surface-colours appear to change too little. Neither in the case of fuchsin nor of diamond green G— the second dye specially discussed by Walter,—or with any other dye hitherto examined †, have I seen an adequate change of colour without the use of the nicol to eliminate vibrations in the plane perpendicular to that of incidence. In the absence of a nicol there is little sign of the blue seen with it from fuchsin at 70° incidence. Much greater changes with more saturated colour are exhibited by the wing-cases of beetles when so examined.

As to the adequacy of the surface-colours Michelson himself remarks :— "indeed, it may perhaps be objected that the (animal) colours are far more vivid than any of the reflexion hues of the aniline dyes, or of any other case

---

* *Phil. Mag.* Vol. xxi. p. 554 (1911). "On Metallic Colouring in Birds and Insects."

† Through the kindness of Sir J. Dewar I have had the opportunity of experimenting with a good many dyes from the Badische Anilin-Fabrik. Following Walter, I have used warm alcoholic solutions spread upon previously warmed glass plates. Latterly I have examined some more dyes, for which I am indebted to Prof. Green. In no case have I seen any considerable change of well-developed colour unless the light was polarized.

of 'surface-colour' hitherto observed." But perhaps this objection should not be very much insisted on in our ignorance of nature's operations and with regard to the known existence of powerful dyes, *e.g.* in feathers. It is rather the rapid loss of purity with obliquity in surface-colour which appears significant.

If a dye capable of surface-reflexion is present, there are still alternatives open. The pure or nearly pure dye may be on the outside so as to be in contact with air, or it may be overlaid by a colourless skin of horny material (chitin) in optical contact with it. The former case would be the more favourable for vivid and variable colour, but then one would expect to be able to remove the dye by solvents. So far as I am aware this has not been done, and my own trials with various solvents upon the wing-cases of beetles have not succeeded. The most satisfactory demonstration of the surface-colour theory would indeed be the extraction of the dye and its exhibition as a thin layer spread upon glass.

If, on the other hand, the dye is imprisoned within a layer of colourless chitin, the range of obliquities available in ordinary observation would be restricted and the difficulty of accounting for the variety of nearly saturated hues actually seen would be increased, more especially when we remember the dilution with white light reflected at the external surface.

There is still another view, which indeed is that actually maintained by Walter, whose argument and conclusion* it may be well to quote:

"A further striking and at the same time more instructive proof of the equivalence of the lustre of butterfly-scales and the surface-colours of strongly absorbing dyes is to be found in the changes which the colours of these organs exhibit when immersed in fluids of varied refrangibility. These experiments are instructive because they disclose the manner in which the dye is contained in animal substances.

"The experiments show that, except when it is deep blue or violet, the lustre moves one or two colour-intervals in the direction from the blue towards the red end of the spectrum with increasing refrangibility of the surrounding medium, but at the same time becomes weaker. For example, the scales of *Morpho menelaus*, L., which glitter green-blue in air, become in ether ($n = 1\cdot36$) a pure green, shining less strongly, again in chloroform ($n = 1\cdot45$) a yellowish green and now decidedly weaker than in ether. In benzol ($n = 1\cdot52$) and in bisulphide of carbon ($n = 1\cdot64$) the weak yellow-green lustre is perceptible only with direct sunshine in a dark room. In a similar manner the scales of *Urania ripheus* shining green in air, in ether, alcohol or water become golden yellow, the yellow red and the red blue, while in benzol and bisulphide of carbon scarcely a trace of glitter remains.

* *Loc. cit.* p. 96.

"Where we know that the cause of the lustre is a dye, the latter facts admit of but one interpretation—that in the case of butterfly-scales we have to do with *solutions of the dyes in chitin*, solutions whose refractivity for most of the spectrum colours is nearly equal to those of benzol and bisulphide of carbon, so that these colours, unless they are very strongly absorbed by the solution, are practically not reflected in their passage from the colourless liquids. Accordingly, the dyes which give rise to lustre in the chitin-skin of insects, and, as we shall see presently, in the horny skin in birds, are dissolved in the same fashion as cobalt oxide in blue glass or organic dyes in a layer of solid gelatine, a conception suggested in the simple observation of the scales by transmitted light and confirmed by the facts above adduced."

If Walter's argument and conclusion are accepted, the difficulty, already considerable, of explaining the richness of the animal colours is enhanced by the supposed dilution of the dyes, and one can hardly fail to observe that a simpler explanation is to reject the dye theory and refer the colours to interference. The facts recorded agree pretty closely with what happens in the case of films of old decomposed glass.

Indeed, Walter, in a later passage, very candidly admits a difficulty. He says (p. 98):

"Finally, it must not be passed over in silence that there is a circumstance which makes a difficulty for the view here propounded of the lustre colours of butterflies. This is the fact that the lustre practically disappears in benzol and bisulphide of carbon, whereas in treating the theory of surface-colours we have several times insisted that a ray strongly absorbed must under all circumstances be vigorously reflected."

Before leaving the question of the colours it may be well to consider an objection strongly urged by Walter against the interference theory, viz., that the colours of thin plates change too much with obliquity. As regards a *single* thin plate, which alone Walter seems to have contemplated, it is true, I think, that the more pronounced colours of the 2nd and 3rd order in Newton's scale change more rapidly with the retardation* than could well be harmonized with what is observed of the animal colours. But the difficulty disappears when we admit a structure several times repeated with approximate periodicity. The changes in chlorate of potash crystals with obliquity seem to agree well enough with what is required, and this form of the interference theory has the advantage of greater elasticity, *e.g.* meeting Walter's objection that the colours of a single thin plate constitute a simple series with but one independent variable. Indeed, the purity of the *reds* often to be observed from beetles' wing-cases seems to exclude an interference theory limited to a single thin plate, inasmuch as the reds from such a plate are distinctly inferior,

* See a diagram of the Colours of Thin Plates, *Ed. Trans.* Vol. xxxiii. p. 157 (1886); *Scientific Papers*, Vol. ii. p. 498.

especially when diluted with white light reflected from an outer surface not forming part of the boundary of the thin plate *.

Michelson, who with his great authority supports the surface-colour theory, mentions several tests under four headings (p. 561). To my mind these tests are as well, if not better, borne by an interference theory. But reliance seems to be chiefly placed upon "the more rigorous optical test of the measurement of the phase-difference and amplitude-ratios" when polarized light is reflected. I agree that this is a cogent argument, and unless it can be met the balance of evidence derived from simple observation would perhaps incline to the surface-colour theory. It is, I think, the fact that many beetles exhibit a less well-marked polarizing-angle than could be reconciled with the usual theory of thin plates constituted of non-absorbent material. An escape from the difficulty might perhaps be found in imagining a stratification composed of more than two materials, so that, for instance, the polarizing-angle for the first and second might differ considerably from that corresponding to the second and third. But such a structure seems rather improbable, and any combination of thin plates composed of two transparent materials only should give a definite polarizing-angle, abstraction being made from the minor deviations observed by Airy and Jamin.

At this point it may be recalled that a well-marked polarizing-angle and a sudden change of relative phase through two right angles are more closely connected than is sometimes realized. The latter without the former would involve a physical discontinuity. Michelson considers that in practice the phase-change affords the more delicate criterion†, and that in most cases it is decisive in favour of surface-colour.

A circumstance which may perhaps be regarded as telling upon the other side is afforded by the variety of colouring at different parts, but at the same angle (e.g. at perpendicular incidence) seen in certain beetles—Dr Hodgkinson mentions Chrysochroa fulminans. The "colours vary in an indescribable manner when attentively examined at different angles of incident light with the eye alone; with the mirror (viz., at perpendicular incidence) the wing-cases are seen to be coloured successively from base to tip iridescent green, yellow, orange, and red, and these tints remain unaltered by change of position of the object." I have confirmed generally this observation, and other beetles show something similar. The explanation makes large demands upon the surface-colour theory; but a moderate change of structure is all that would be required by interference.

A caution is perhaps required against regarding the two theories as mutually exclusive. Both Walter and Michelson admit exceptions, and

* Ed. Trans. loc. cit.

† Some of Michelson's diagrams are rather confusing in that they suggest a phase-difference of 180° between the two polarized components reflected perpendicularly, when evidently the distinction between the two components disappears.

certainly there is no improbability in surface-reflexion playing a part. It may be that both causes are operative in a single specimen and even at the same part of it.

The next contribution to the discussion is an important one by Mallock *, who brings to bear the instinct and experience of a naturalist as well as of a physicist. His observations were mainly on the feathers of birds and the scales of insects, and they lead him to regard interference rather than selective reflexion as the origin of the iridescent colours. "The transparency or, at any rate, the vanishing of the characteristic transmitted colour in the case of all animal tissues when immersed and permeated by a fluid of the same refractive index is strongly in favour of interference being the source of the colour, but even stronger evidence is given by the behaviour of the structures under mechanical pressure.

"If the grain or peculiarities which favour the reflexion or transmission of particular colours is of molecular size, there is no reason to suppose that pressure insufficient to cause molecular disruption would alter the action of the material on light. On the other hand, if the colours are due to interference, that is, to cavities or strata of different optical properties, compression would alter the spacing of these, and thus give rise either to different colours or, with more than a very slight compression, to the transmission and reflexion of white light.

"In every experiment of this kind which I have made either on feathers or insect scales the effect of pressure has been to destroy the colour altogether....

"With many feathers the colour returns when the pressure is taken off, but with insect scales the structure seems to be permanently injured by compression, and though when allowed to expand again the material is not colourless the brilliancy which belonged to the uninjured scale is gone, and the colour in general changed.

"The facts above mentioned seem to offer stronger reasons in favour of interference than the polarization phenomena referred to by Michelson and Walter do against it."

I have already commented on the importance of the evidence afforded by observations with polarized light; and if we have to choose between selective reflexion and thin plates of the type usually considered in theoretical writings, we may find ourselves in a position of much difficulty. The question then arises, Is there any loophole for escape? I think there may be. The polarizing angle, as given by Brewster's law, depends much upon what we may call the smoothness of the reflecting surface. A moderate curvature is of no significance in this connexion, but when the radius of curvature becomes comparable with the wave-length of the light it is another matter. Thus in the case of smooth glass the polarizing angle is about 57°—that is, light

* Proc. Roy. Soc. A, Vol. LXXXV. p. 598 (1911).

incident at this angle with the normal and vibrating in the plane of incidence is not reflected. In this observation the reflected light (if there were any) would be deviated from its original direction through an angle of $2 (90° - 57°) = 66°$, and this is the direction in which light initially unpolarized would appear completely polarized. Now replace the flat glass by a sphere of the same material, whose diameter is small in comparison with the wave-length. Light is now scattered in various directions, but the direction in which light originally unpolarized becomes completely polarized is at 90° with the original direction, instead of 66°. As the sphere grows, the polarization ceases to be complete, and the direction of best polarization moves oppositely to what would be expected—that is, still further *away* from 66°. When the circumference of the sphere is equal to twice the wave-length, the polarization, still pretty good, occurs at an angle of 135° with the original direction of the light*. In order to carry out the suggestion, we must abandon the supposition of uniform plane strata, inapplicable anyhow in its integrity to the case where one of the alternate plates is of air, and substitute a structure in which one of the alternatives takes a form such as the spherical. A layer of equal spheres, with centres disposed upon a plane, would give a specular reflexion and a polarizing angle dependent upon the diameter of the spheres and upon the intervals between them. In certain cases, *e.g.* when the circumference of the sphere (of glass) is equal to $1·75 \times$ wave-length, the polarization is very imperfect. To explain a brilliant and highly-coloured reflexion there would need to be several layers of spheres, and it might be supposed that the diameter varied in different layers. In this way it would seem possible to combine a specular and highly-coloured reflexion with a very imperfectly developed polarization, and thus to evade the difficulty which meets us when we confine ourselves to "thin plates." Spheres have been spoken of for simplicity and because some of the effects have been calculated in this case, but it is evident that similar phenomena would be produced by obstacles of other and perhaps more probable forms. The obstacles must have a different index from that of the medium in which they are embedded, and there is no need for absorption.

It may perhaps be objected that though a layer of spheres may give a specular reflexion there would be an accompaniment of light dispersed at other angles, forming in the case of a regular pattern "diffraction spectra." It is uncertain whether or not this occurs. If it does not, the explanation may be that the pattern is too fine.

The above remarks are intended merely to attenuate the difficulty arising

---

* *Phil. Mag.* Vol. xii. p. 81 (1881); *Proc. Roy. Soc.* A, Vol. lxxxiv. p. 25 (1910); *Scientific Papers*, Vol. i. p. 518; Vol. v. p. 564.

[It would appear that for "135°" we should read "120°." According to the diagram on p. 564 of Vol. v. the maximum polarization for $\eta = 2\pi R/\lambda = 2$ occurs when $\mu = +·5$, where μ is the cosine of the angle between the secondary (or scattered) ray and the backward direction of the incident ray. W. F. S.]

from the absence of a well-marked polarizing-angle, and the details need not be insisted on. No surprise is felt at the deficiency of polarization in the light reflected from unpressed and unglazed paper, of which the fibres are quite large enough to be the seat of interference effects. In illustration the transverse reflexion from glass rods and fibres may be mentioned. When we examine with a nicol the reflexion from a rod ¼ inch (6 mm.) in diameter, we can verify the extinction at a suitable angle of the light reflected from the first surface, although abundance of other light still reaches the eye. When we replace the rod by a fine fibre, this discrimination is lost, and the rotation of the nicol may make no difference, or even a difference in the wrong direction.

The greater part of the preceding discussion was written about a year and a half ago. I am now able to supplement it with further observations of my own and of others who have been kind enough to help me. Most of my experiments have been made on wing-cases of beetles found in my garden (June and July 1917). Usually attention is first attracted by the display of a vivid green coloration, but on indoor examination the variation with angle is found to be about the same as is observed with brilliant specimens from abroad. At perpendicular incidence the colour is an orange with approach to red, passing with increasing obliquity through yellow and green to a blue-green. Ordinary solvents such as water even at the boiling-point, ether, alcohol, benzol, bisulphide of carbon, acetic acid, etc., seem to be without effect, even when the precaution is taken to separate a wing-case into two parts so as to allow access to the interior of the cuticle. A treatment with hot caustic potash has more effect, in one experiment shifting the colour at perpendicular incidence from orange to a brilliant scarlet. By the action of hot somewhat diluted nitric acid the black underlying pigment may be removed without much affecting the dye, or the structure, which is the seat of the coloration.

Several experiments were made to test whether air-cavities existed. For this purpose the wing-case was exposed for some time to the action of vacuum, into which afterwards water or benzol was admitted. But no distinct evidence of the penetration of liquid could be recorded. I understand that Prof. Poulton has had a similar experience.

Again, it has been noticed by Mr H. Onslow and myself that considerable pressure fails to alter the colour of beetles and of the wings of some iridescent dragon-flies, though (Poulton, Mallock) effective in some other cases. It would seem that the hypothesis of air-cavities must be abandoned.

In the absence of air-cavities the alternating structure demanded by the interference theory would require two kinds of matter capable of resisting pressure and of sensibly different refractive indices. Probably both would be solids; and since the range of relative index is then much restricted, the brightness of the reflected light could hardly be explained without supposing more than the two or three alternations which might suffice were air in question. Mr Onslow thinks that there may then be a difficulty in finding room for the alternating structure and the protective covering.

An important question is whether the change of colour with angle is such as can plausibly be attributed to a periodic structure. As Walter points out, a good deal depends upon whether, or not, there is a limitation upon the obliquity of the rays *within* the thin plate, or plates. In the ordinary arrangement for Newton's rings there is no limitation, the direction in the air-film being parallel to that of the rays before incidence upon the first plate. The optical retardation may then vary from its maximum at perpendicular incidence to zero at 90° obliquity. According to this, it should always be possible to push the colour out of the spectrum at the blue end by sufficiently increasing the obliquity, but it must be remembered that unless special provision is made the colour effects would be overlaid by the white light reflected at these angles from the first glass surface encountered.

From what we have seen in the case of the beetle colours where we must suppose that the refractive index does not differ greatly from that (1·6) of the chitin, there is a limit to the obliquity within the thin films even when externally the incidence is grazing. If θ be the angle in the thin film and μ the refractive index, the retardation is proportional to $\cos \theta$, and in the limiting case

$$\cos \theta = \sqrt{(\mu^2 - 1)}/\mu,$$

If we take $\mu = 1\cdot5$, the minimum retardation is represented by ·746, the maximum retardation at perpendicularity being taken as unity. It may be remarked that the minimum retardation may practically be secured without pushing very far the obliquity outside. If we suppose the maximum retardation to give a coloration corresponding to the Fraunhofer line C ($\lambda = 6563$), the minimum will correspond to $\lambda = 4896$, pretty close to the line F. According to the interference theory, then, the range of coloration should be from the full red of C to the blue-green of F, and this is just about what is observed. The agreement must be admitted to be a strong argument in favour of the theory. So far as I have seen, so great a range cannot be found in the surface-colour of any dye, even with the aid of polarized light.

I have already mentioned that the opaque backing behind the seat of coloration can be attacked, and for the most part removed, with nitric acid, so as to allow the transmission colour to be observed. But a much superior effect has been obtained by Dr Eltringham, using *eau de javelle* (hypochlorite)*.

---

* Dr Eltringham's label runs :

Mimela leei. Elytron after prolonged eau-de-javelle. Only surface-film left. Transmits complementary colours to those it reflects, and reflects same colours from both sides (1917).

After removal of the backing, the wing-case was mounted with balsam in a slide, which Dr Eltringham has kindly left in my possession. Close observation of this specimen has yielded results which I think interesting and telling. Seen by transmitted light with the aid of a Coddington lens, the slide shows a pale green over the larger part of the area, which by perpendicular reflexion is a full red. The green is fairly uniform except where it appears perforated with small circular spots, which look reddish, but perhaps only by contrast. Especially to be noted is the fact that there is no colour seen by transmission at all comparable in saturation with those exhibited by reflexion. For observation of the reflected light it is advantageous, though not necessary, to renew the opaque backing, which was done by coating the under surface of the *glass* with gelatine darkened with ink. A good deal depends upon the source of light. In the first detailed examination, the source (a gas-mantle) happened to be highly localized, and I was puzzled to reconcile the highly spotty character of the reflexion, varying from red to green or green-blue according to the incidence, with the uniformity of the transmission tint. Similar appearances could of course be observed in direct sunlight. But when the slide was held very close to a large window facing a nearly uniform sky, the intervals between the spots filled up with colour, for the most part of approximately the same hue, and the reflexion was nearly uniform except for the small round holes already mentioned. Evidently the reflexion of the gas-mantle had failed to reach the eye, except from a relatively small area presenting the proper angle, thus explaining the spotty appearance observed with this illumination.

The colours reflected at moderate angles seem highly saturated. At perpendicular incidence the prism shows next to nothing beyond the uninterrupted red and red-orange, and on inclination the green region appears well isolated. The impression left upon my mind is that the phenomena cannot plausibly be explained as due to surface-colour, which in my experience is always *less* saturated than the transmission colour, and that, on the other hand, the interference theory presents no particular difficulty, unless it be that of finding sufficient room within the thickness of the cuticle. But the alternations cannot be those of plane strata, extending without interruption over the whole area of the colour.

As regards the difficulty of finding room sufficient for an optical structure of the kind contemplated, Mr Onslow estimates the available thickness at from 0.75μ* to 2μ in the case of many butterfly scales, and this is little enough. Even the larger estimate would amount to only about 9 or 10 half wave-lengths, even when allowance is made for the wave-length being less than in air, and the lower limit would apparently not suffice. But these measurements are not easy, and may perhaps be disturbed by refraction

* $\mu = \frac{1}{1000}$ mm.

effects. Mr Onslow has shown me many drawings of sections in planes perpendicular to the surface from many butterfly scales and from two or three beetle wing-cases. Most of these exhibit structures approximately periodic along the surface, but in no case a structure periodic in going inwards along the normal. But a structure of the latter kind adequate to the purpose may probably lie close upon the microscopic limit, unless, indeed, it could be made evident in a section cut very obliquely.

It must be confessed that much still remains to be effected towards a complete demonstration of the origin of these colours. Even if we admit an interference character, questions arise as to the particular manner, and there are perhaps possibilities not hitherto contemplated.

439.

ON THE POSSIBLE DISTURBANCE OF A RANGE-FINDER BY ATMOSPHERIC REFRACTION DUE TO THE MOTION OF THE SHIP WHICH CARRIES IT.

[*Transactions of the Optical Society*, Vol. xx. pp. 125—129, 1919.]

THE suggestion has been put forward (as I understand by Lt.-Col. A. C. Williams) that the action of a range-finder, adjusted for a quiescent atmosphere, may be liable to disturbance when employed upon a ship in motion, as a result of the variable densities in the air due to such motion and consequent refraction of the light. That this is *vera causa* must be admitted; but the question arises as to the direction and magnitude of the effect, and whether or not it would be negligible in practice. It is not to be supposed that any precise calculation is feasible for the actual circumstances of a ship; but I have thought that a simplified form of the problem may afford sufficient information to warrant a practical conclusion. For this purpose I take the case of an infinite cylinder moving transversely through an otherwise undisturbed atmosphere, and displacing it in the manner easily specified on the principles of ordinary hydrodynamics. When the motion of the fluid is known, the corresponding pressures and densities follow, and the refraction of the ray of light, travelling from a distance in a direction parallel to the ship's motion, may be calculated as in the case of astronomical and prismatic dispersion or of mirage. It is doubtless the fact that in the rear of the disturbing body the motion differs greatly from that assumed; but in front of it the difference is much less, and not such as to nullify the conclusions that may be drawn. The first step is accordingly to specify the motion, and to determine the square of the total velocity (q) on which depends the reduction of pressure.

The motion of the fluid, due to that of the cylinder of radius c and moving parallel to x with velocity U, is well known[*]. The problem may conveniently be reduced to one of "steady motion" by supposing the cylinder to be at rest while the fluid flows past it—a change which can make no

---

[*] See Lamb's *Hydrodynamics*, § 68.

difference to the refraction. The velocity-potential is then expressed in polar coordinates by

$$\phi = U\left(r + \frac{c^2}{r}\right)\cos\theta, \quad\text{.........................(1)}$$

where r is measured from the centre ; whence

$$\frac{d\phi}{dr} = U\left(1 - \frac{c^2}{r^2}\right)\cos\theta, \quad \frac{d\phi}{r\,d\theta} = - U\left(1 + \frac{c^2}{r^2}\right)\sin\theta. \quad\text{.........(2)}$$

Here $d\phi/dr$ vanishes when $r = c$ at the surface of the cylinder, and at a great distance the resultant velocity is U, parallel to x. In general

$$\frac{q^2}{U^2} = 1 + \frac{c^4}{r^4} + \frac{2c^2}{r^2}(\sin^2\theta - \cos^2\theta) = 1 + \frac{c^4}{r^4} + \frac{2c^2}{r^4}(y^2 - x^2). \quad\text{......(3)}$$

As will be stated more at length presently, the reduction of pressure and density is proportional to $q^2 - U^2$, so that the (abnormal) optical retardation of a ray parallel to x is proportional to $\int (q^2 - U^2)\,dx$, in which the upper limit for x may be treated as infinite. The difference of retardations for two infinitely near rays parallel to x, divided by the distance between them, gives the angle of refraction, which is thus represented by $\dfrac{d}{dy}\int q^2 dx$, or by $\int \dfrac{dq^2}{dy}\,dx$, since the limits of x are the same for both. Now

$$\frac{d}{dy}\left(\frac{q^2}{U^2}\right) = (c^4 - 2c^2 x^2)\frac{d}{dy}\left(\frac{1}{r^4}\right) + 2c^2\frac{d}{dy}\left(\frac{y^2}{r^4}\right);$$

and, since
$$dr/dy = y/r,$$

$$\frac{d}{dy}\left(\frac{1}{r^4}\right) = -\frac{4y}{r^6}, \quad \frac{d}{dy}\left(\frac{y^2}{r^4}\right) = \frac{2y(x^2 - y^2)}{r^6},$$

so that
$$\frac{d}{dy}\left(\frac{q^2}{U^2}\right) = \frac{4c^2 y}{r^6}(3x^2 - y^2 - c^2). \quad\text{.....................(4)}$$

In the integration with respect to x, y is constant, say β, and if α, β be the rectangular coordinates of the point P in the figure at which the refraction is to be estimated, the limits for x are α and ∞. Thus

$$\int_\alpha^\infty \frac{d}{dy}\left(\frac{q^2}{U^2}\right)dx = -4c^2\beta\,(\beta^2 + c^2)\int_\alpha^\infty \frac{dx}{r^6} + 12c^2\beta\int_\alpha^\infty \frac{x^2 dx}{r^6}. \quad\text{......(5)}$$

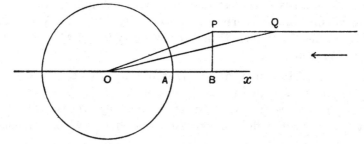

$OA = c,\ OB = a,\ PB = \beta,\ QOB\ \text{or}\ POB = \theta,\ OQ\ \text{or}\ OP = r.$

At this point it is convenient to re-introduce θ, where $x = \beta \cot \theta$, and

$$dx = -\beta \sin^{-2} \theta \, d\theta.$$

We have
$$\int \frac{dx}{r^6} = -\int \frac{\beta \sin^{-2} \theta \, d\theta}{\beta^6 (1 + \cot^2 \theta)^3} = -\frac{1}{\beta^5} \int \sin^4 \theta \, d\theta.$$

The upper limit ∞ for x corresponds to $\theta = 0$; the lower limit α corresponds to what after the integration we may still denote by θ, where $\alpha = \beta \cot \theta$. Thus

$$\int_a^\infty \frac{dx}{r^6} = \frac{1}{\beta^5} \int_0^\theta \sin^4 \theta \, d\theta = \frac{1}{4\beta^5} (\tfrac{3}{2}\theta - \sin 2\theta + \tfrac{1}{8} \sin 4\theta). \quad\ldots\ldots\ldots(6)$$

Also
$$\int \frac{x^2 dx}{r^6} = -\frac{1}{\beta^3} \int \cos^2 \theta \sin^2 \theta \, d\theta = -\frac{1}{8\beta^3} (\theta - \tfrac{1}{4} \sin 4\theta),$$

and
$$\int_a^\infty \frac{x^2 dx}{r^6} = \frac{1}{8\beta^3} (\theta - \tfrac{1}{4} \sin 4\theta), \quad\ldots\ldots\ldots\ldots\ldots(7)$$

θ now referring, as above, to the point for which $x = \alpha$. Using (6) and (7) in (5), we have

$$\int_a^\infty \frac{d}{dy}\left(\frac{q^2}{U^2}\right) dx = -\frac{c^2 (\beta^2 + c^2)}{\beta^4} (\tfrac{3}{2}\theta - \sin 2\theta + \tfrac{1}{8}\sin 4\theta) + \frac{3c^2}{2\beta^2}(\theta - \tfrac{1}{4}\sin 4\theta)$$

$$= \frac{c^2}{\beta^2}(\sin 2\theta - \tfrac{1}{2}\sin 4\theta) - \frac{c^4}{\beta^4}(\tfrac{3}{2}\theta - \sin 2\theta + \tfrac{1}{8}\sin 4\theta), \ldots\ldots(8)$$

in which, if we please, we may substitute $r \sin \theta$ for β, so that r, θ are the polar coordinates of the point of observation.

Thus

$$\int_a^\infty \frac{d}{dy}\left(\frac{q^2}{U^2}\right) dx = \frac{c^2}{r^2}\frac{\sin 2\theta - \tfrac{1}{2}\sin 4\theta}{\sin^2 \theta} - \frac{c^4}{r^4}\frac{\tfrac{3}{2}\theta - \sin 2\theta + \tfrac{1}{8}\sin 4\theta}{\sin^4 \theta}$$

$$= \frac{c^2}{r^2}f(\theta) - \frac{c^4}{r^4}F(\theta), \quad\ldots\ldots\ldots\ldots\ldots\ldots\ldots\ldots\ldots(9)$$

as we write it for brevity. It now remains to discuss (9) as a function of r and θ, under the limitation, however, that $r > c$.

When θ is small, 　　$f(\theta) = 4\theta$, 　$F(\theta) = \tfrac{4}{5}\theta$;

so that (9) becomes 　　　　$4\theta \left(\frac{c^2}{r^2} - \frac{c^4}{5r^4}\right)$, 　　　　$\ldots\ldots\ldots\ldots\ldots\ldots\ldots(10)$

vanishing when $\theta = 0$, as was to be expected, since this is a line of symmetry. Inasmuch as $r > c$, (10) takes its sign from θ.

The table gives values of f and F for certain angles of θ. In the fourth column are entered the values of (9) when $r = c$, that is on the surface of the cylinder. So far as $\theta = 50°$ these are the highest admissible. For example at $50°$

$$(9) = \cdot8173\,(2\cdot4098c^2/r^2 - c^4/r^4)$$
$$= \cdot8173\,\{(1\cdot2049)^2 - (1\cdot2049 - c^2/r^2)^2\}.$$

So long as $r > c$, the value of (9) increases as r diminishes, and the greatest admissible value occurs at the limit $r = c$. This state of things continues so long as $f > 2F$. At 60° this condition has ceased to hold, and the maximum value of (9) occurs when $r > c$. We may write generally

$$(9) = F\{(f/2F)^2 - (f/2F - c^2/r^2)^2\}.$$

θ	$f(\theta)$	$F(\theta)$	(9) when $r=c$	(9) max.
0°	0	0	—	—
10	·6840	·1404	·5436	—
20	1·2856	·2859	·9997	—
30	1·7320	·4419	1·2901	—
36	1·9021	·5436	1·3585	—
40	1·9292	·6159	1·3133	—
50	1·9696	·8173	1·1523	—
60	1·7320	1·0605	·6715	·7072
70	1·2855	1·3680	− ·0825	—

The maximum of (9) occurs when $c^2/r^2 = f/2F$, and the maximum value is $f^2/4F$. For $\theta = 60°$, the maximum is ·7072. When $\theta = 70°$, (9) itself has changed sign, the transition occurring when $F = f$.

For our present purpose we may take the highest value of (9) as 1·36.

The general dynamical equation connecting pressure (p) and density (ρ) with velocity (q) in steady motion is

$$\int \frac{dp}{\rho} = C - \tfrac{1}{2}q^2. \quad\quad\quad\quad\quad\quad (11)$$

It will suffice if we employ Boyle's law for the connexion of p and ρ, that is $p = V^2\rho$, where V is the velocity of sound*. Thus

$$V^2 \log(\rho/\rho_0) = \tfrac{1}{2}(U^2 - q^2)$$

if ρ_0 correspond with $q = U$, as at a distance from the cylinder. Or, since the variations of density here contemplated are very small,

$$\frac{\rho - \rho_0}{\rho_0} = \frac{U^2 - q^2}{2V^2}. \quad\quad\quad\quad\quad\quad (12)$$

Passing now to the optical side of the question, we have to consider the retardation experienced by a ray parallel to x, due to the variable density. In accordance with a general principle, this when small enough may be calculated along the original ray, although the actual ray now follows a somewhat different course†. Thus if $\delta\mu$ be the change of refractive index due to q, the retardation may be taken to be $\int_a^\infty \delta\mu\, dx$, and the angle χ through which the ray at P is turned is

$$\chi = \int_a^\infty \frac{d\,\delta\mu}{dy}\, dx. \quad\quad\quad\quad\quad\quad (13)$$

[* *I.e.* the "Newtonian" velocity of sound, $=280$ metres per second. W. F. S.]

† For an application to the resolving power of prisms, reference may be made to *Phil. Mag.* Vol. VIII. p. 269 (1879); *Scientific Papers*, Vol. I. p. 425.

Now for air at 0° C. and 760 mm. pressure $\mu = 1\cdot000292$, so that

$$\delta\mu = \cdot000292\,(\rho - \rho_0)/\rho_0 = -\cdot000146\,(q^2 - U^2)/V^2. \quad\dots\dots\dots(14)$$

Accordingly

$$\chi = -\frac{\cdot000146\,U^2}{V^2}\int_a^\infty \frac{d}{dy}\left(\frac{q^2}{U^2}\right)dx, \quad\dots\dots\dots\dots(15)$$

and the integral is the quantity already calculated under (9). It will be observed that the effect is independent of the absolute size of the obstacle, and is proportional to the square of the fluid velocity U. As regards the sign of the effect, we see that, so long as (9) is positive, the retardation diminishes as y increases, and thus a ray originally parallel to x is bent *inwards*. This action in a range-finder leads to an over-estimate of the distance.

We may now consider the actual magnitude of χ. If we take the highest value of (9), viz. $1\cdot36$, we have

$$\chi = -\cdot000199\,U^2/V^2. \quad\dots\dots\dots\dots\dots\dots(16)$$

To take a numerical example, let us suppose that

$$U = 25\,\frac{\text{miles}}{\text{hours}} = 1120\,\frac{\text{cm.}}{\text{sec.}},$$

and

$$V\,(0°\,\text{C.},\,760\,\text{mm.}) = 33100\,\frac{\text{cm.}}{\text{sec.}}*.$$

The angle is then

$$-\chi = 2\cdot28*\times10^{-7}.$$

This is in circular measure, and it corresponds to $\cdot0470*$ second. An error of a twentieth part of a second may be considered negligible.

The corresponding problem with a sphere in place of the cylinder could be treated, but it is rather more complicated. It does not appear necessary to enter upon it here.

The objection may present itself that the motion in two dimensions round a cylinder does not well represent the case of a ship. This must be admitted; but it seems clear that a range-finder, carried well above the deck of a ship, would be much *less* liable to be disturbed in the manner discussed.

Addendum.—In answer to a criticism the author wrote:

There may be different opinions as to how far conclusions for an infinite cylinder throw light on the case of the ship. Certainly my own impression is that the cutting away of the further parts of the cylinder would diminish the disturbance under discussion.

[* Apparently we should take $V = 28,000$ cm. per sec. (see footnote * on p. 600). We then find $-\chi = 3\cdot18 \times 10^{-7}$ in circular measure $= \cdot0656$ second, or about one-fifteenth of a second. W.F.S.]

440.

REMARKS ON MAJOR G. I. TAYLOR'S PAPERS ON THE DISTRIBUTION OF AIR PRESSURE.

T. 646, T. 1277.

[*Advisory Committee for Aeronautics*, T. 1296, 1919.]

IN response to the request for comments on this work, I may say that I have read these papers with interest.

The experiments recorded relate directly to the air pressures at various distances in the close neighbourhood of the surface of a long board parallel to the current of air in a wind channel. Three series of such pressures, and deduced air velocities, were obtained at different distances (*A*, *B*, *C*) from the leading edge. Major Taylor suggests that these experiments should be repeated and extended at the National Physical Laboratory, and in this recommendation I fully concur, as it is of importance to improve our imperfect comprehension of the character of " skin-friction."

From the air pressure measurements the author deduces by calculation the actual force exercised upon the board, and in the second paper corrects, as far as possible, some deficiencies in the first. These calculations are a little difficult to follow as approximations are introduced whose validity is difficult to estimate, at any rate, without the instinct which familiarity with the subject matter may bring with it. For example, at some points, the problem is treated as if it were two-dimensional, which the actual dimensions of the board do not seem to justify.

Without undervaluing the interest of connecting the forces experienced by the various parts of the board with the air pressures in its neighbourhood, I am inclined to prefer—or at any rate to recommend as alternative—direct measurements of the forces on the board, more as in Zahm's experiments.

The objection made in T. 646, p. 7, that in Zahm's shorter boards the influence of the end pieces is too important, could, I think, be met by suspending the end pieces separately from the rest of the board, on the principle of the " guard ring " in electrometers. In the case of short boards the two end pieces could be rigidly connected together by rods passing freely through the thickness of the board.   The suspension both of the end piece and of the board in a vertical plane would be by four wires reducing the original six degrees of freedom to two.   (Compare *Theory of Sound*, § 62 ; *Phil. Mag.* Vol. XI. p. 127 (1906), *Sci. Papers*, Vol. V. p. 283.)   Of the remaining two, one relates to the motion parallel to the wind, which is the subject of observation, and the other (rotation about the upper edge of the board) would be controlled by gravity.   When the board is long, there is less objection to the rigid attachment of the end pieces to it.

This method would doubtless require care in execution and would involve the measurement of very small displacements along the wind, but for this optical resources should be adequate.   It might also be applied to various parts of the length of the board, so as to separate the frictions there incurred.

441.

ON THE PROBLEM OF RANDOM VIBRATIONS, AND OF RANDOM FLIGHTS IN ONE, TWO, OR THREE DIMENSIONS.

[*Philosophical Magazine*, Vol. XXXVII. pp. 321—347, 1919.]

WHEN a number (n) of isoperiodic vibrations of unit amplitude are combined, the resultant depends upon the values assigned to the individual phases. When the phases are at random, the resultant amplitude is indeterminate, and all that can be said relates to the *probability* of various amplitudes (r), or more strictly to the probability that the amplitude lies within the limits r and $r + dr$. The important case where n is very great I considered a long time ago* with the conclusion that the probability in question is simply

$$\frac{2}{n} e^{-r^2/n} r\, dr. \quad \ldots\ldots\ldots\ldots\ldots\ldots\ldots\ldots\ldots\ldots(1)$$

The phase (θ) of the resultant is of course indeterminate, and all values are equally probable.

The method then followed began with the supposition that the phases of the unit components were limited to 0° and 180°, taken at random, so that the points (r, θ), representative of the vibrations, lie on the axis $\theta = 0$, and indifferently on both sides of the origin. The resultant x, being the difference between the number of positive and negative components, is found from Bernoulli's theorem to have the probability

$$\frac{1}{\sqrt{(2\pi n)}} e^{-x^2/2n} dx. \quad \ldots\ldots\ldots\ldots\ldots\ldots\ldots\ldots\ldots(2)\dagger$$

The next step was to admit also phases of 90° and 270°, the choice between these two being again at random. If we suppose $\frac{1}{2}n$ components at random along $\pm x$, and $\frac{1}{2}n$ also at random along $\pm y$, the chance of the representative point of the resultant lying within the area $dx\,dy$ is evidently

$$\frac{1}{\pi n} e^{-(x^2+y^2)/n} dx\, dy, \quad \ldots\ldots\ldots\ldots\ldots\ldots\ldots\ldots\ldots(3)$$

---

* *Phil. Mag.* Vol. X. p. 73 (1880); *Scientific Papers*, Vol. I. p. 491.
† See below.

or in terms of r, θ,

$$\frac{1}{\pi n} e^{-r^2/n} r \, dr \, d\theta. \quad\ldots\ldots\ldots\ldots\ldots\ldots\ldots\ldots\ldots(4)$$

Thus all phases are equally probable, and the chance that the resultant amplitude lies between r and $r + dr$ is

$$\frac{2}{n} e^{-r^2/n} r \, dr. \quad\ldots\ldots\ldots\ldots\ldots\ldots\ldots\ldots\ldots(1)$$

This is the same as was before stated, but at present the conditions are limited to a distribution of precisely $\frac{1}{2}n$ components along x and a like number along y. It concerns us to remove this restriction, and to show that the result is the same when the distribution is perfectly arbitrary in respect to all four directions.

For this purpose let us suppose that $\frac{1}{2}n + m$ are distributed along $\pm x$ and $\frac{1}{2}n - m$ along $\pm y$, and inquire how far the result is influenced by the value of m. The chance of the representative point lying in $r \, dr \, d\theta$ is now expressed by

$$\frac{1}{\pi \sqrt{(n^2 - 4m^2)}} e^{-nr^2/(n^2-4m^2)} e^{+2mr^2 \cos 2\theta/(n^2-4m^2)} r \, dr \, d\theta *.$$

Since r is of order $\sqrt{n}$, and m/n is small, the exponential containing θ may be expanded. Retaining the first four terms, we have on integration with respect to θ,

$$\frac{2r \, dr}{\sqrt{(n^2 - 4m^2)}} e^{-nr^2/(n^2-4m^2)} \left\{ 1 + \frac{m^2 r^4}{(n^2 - 4m^2)^2} + \ldots \right\},$$

as the chance of the amplitude lying between r and $r + dr$. Now if the distribution be entirely at random along the four directions, all the values of m of which there is a finite probability are of order not higher than $\sqrt{n}$, n being treated as infinite. But if m is of this order, the above expression becomes the same as if m were zero; and thus it makes no difference whether the number of components along $\pm x$ and along $\pm y$ are limited to be equal, or not. The previous result is accordingly applicable to a thoroughly arbitrary distribution along the four rectangular directions.

The next point to notice is that the result is symmetrical and independent of the directions of the rectangular axes, from which we may conclude that it has a still higher generality. If a total of n components, to be distributed along one set of rectangular axes, be divided into any number of large groups, it makes no difference whether we first obtain the probabilities of various resultants of the groups separately and afterwards of the final resultants, or whether we regard the whole n as one group. But the probability in each group is the same, notwithstanding a change in the system of rectangular axes; so that the probabilities of various resultants are unaltered, whether

---

[* A correction of sign here made, viz. "$+2mr^2 \cos 2\theta$" for "$-2mr^2 \cos 2\theta$," applies also to Vol. I. p. 494, lines 10 and 12. W. F. S.]

we suppose the whole number of components restricted to one set of rectangular axes or divided in any manner between any number of sets of axes. This last state of things is equivalent to no restriction at all; and we conclude that if n unit vibrations of equal pitch and of thoroughly arbitrary phases be compounded, then when n is very great the probability of various resultant amplitudes is given by (1).

If the amplitude of each component be l, instead of unity, as we have hitherto supposed for brevity, the probability of a resultant amplitude between r and $r + dr$ is

$$\frac{2}{nl^2} e^{-r^2/nl^2} r\, dr. \quad\dots\dots\dots\dots\dots\dots\dots\dots\dots\dots(5)$$

In *Theory of Sound*, 2nd edition, § 42a (1894), I indicated another method depending upon a transition from an equation in finite differences to a partial differential equation and the use of a Fourier solution. This method has the advantage of bringing out an important analogy between the present problems and those of gaseous diffusion, but the demonstration, though somewhat improved later*, was incomplete, especially in respect to the determination of a constant multiplier. At the present time it is hardly worth while to pursue it further, in view of the important improvements effected by Kluyver and Pearson. The latter was interested in the "Problem of the Random Walk," which he thus formulated :—"A man starts from a point O and walks l yards in a straight line; he then turns through any angle whatever and walks another l yards in a second straight line. He repeats this process n times. I require the probability that after these n stretches he is at a distance between r and $r + dr$ from his starting point O.

"The problem is one of considerable interest, but I have only succeeded in obtaining an integrated solution for *two* stretches. I think, however, that a solution ought to be found, if only in the form of a series in powers of $1/n$, when n is large†." In response, I pointed out that this question is mathematically identical with that of the unit vibrations with phases at random, of which I had already given the solution for the case of n infinite‡, the identity depending of course upon the vector character of the components.

In the present paper I propose to consider the question further with extension to *three* dimensions, and with a comparison of results for one, two, and three dimensions§. The last case has no application to random vibrations but only to random *flights*.

---

* *Phil. Mag.* Vol. xlvii. p. 246 (1899); *Scientific Papers*, Vol. iv. p. 370.

† *Nature*, Vol. lxxii. p. 294 (1905).

‡ *Nature*, Vol. lxxii. p. 318 (1905); *Scientific Papers*, Vol. v. p. 256.

§ It will be understood that we have nothing here to do with the direction in which the vibrations take place, or are supposed to take place. If that is variable, there must first be a resolution in fixed directions, and it is only after this operation that our present problems arise.

One Dimension.

In this case the required information for any finite n is afforded by Bernoulli's theorem. There are $n + 1$ possible resultants, and if we suppose the component amplitudes, or stretches, to be unity, they proceed by intervals of *two* from $+ n$ to $- n$, values which are the largest possible. The probabilities of the various resultants are expressed by the corresponding terms in the expansion of $(\frac{1}{2} + \frac{1}{2})^n$. For instance the probabilities of the extreme values $\pm n$ are $(1/2)^n$. And the probability of a combination of a positive and b negative components is

$$(\tfrac{1}{2})^n \frac{n!}{a!\,b!}, \quad \dots\dots\dots\dots\dots\dots\dots\dots\dots(6)$$

in which $a + b = n$, making the resultant $a - b$. The largest values of (6) occur in the middle of the series, and here a distinction arises according as n is even or odd. In the former alternative there is a unique middle term when $a = b = \frac{1}{2}n$; but in the latter a and b cannot be equated, and there are two equal middle terms corresponding to $a = \frac{1}{2}n + \frac{1}{2}$, $b = \frac{1}{2}n - \frac{1}{2}$, and to $a = \frac{1}{2}n - \frac{1}{2}$, $b = \frac{1}{2}n + \frac{1}{2}$. The values of the second fraction in (6) are the series of integers in what is known as the "arithmetical triangle."

We have now to consider the values of

$$\frac{n!}{a!\,b!} \quad \dots\dots\dots\dots\dots\dots\dots\dots\dots\dots(7)$$

to be found in the neighbourhood of the middle of the series. If n be even, the value of the term counted s onwards from the unique maximum is

$$\frac{n!}{(\frac{1}{2}n - s)!\,(\frac{1}{2}n + s)!}. \quad \dots\dots\dots\dots\dots\dots(8)$$

If n be odd, we have to choose between the two middle terms. Taking for instance, $a = \frac{1}{2}n + \frac{1}{2}$, $b = \frac{1}{2}n - \frac{1}{2}$, the sth term onwards is

$$\frac{n!}{\{\frac{1}{2}n - (s - \frac{1}{2})\}!\,\{\frac{1}{2}n + (s - \frac{1}{2})\}!}. \quad \dots\dots\dots(9)$$

The expressions (8) and (9) are brought into the same form when we replace s by the resultant amplitude x. When n is even, $x = -2s$; when n is odd, x is $-2(s - \frac{1}{2})$, so that in both cases we have on restoration of the factor $(\frac{1}{2})^n$

$$\frac{n!}{2^n.(\frac{1}{2}n - \frac{1}{2}x)!\,(\frac{1}{2}n + \frac{1}{2}x)!}. \quad \dots\dots\dots\dots\dots(10)$$

The difference is that when n is even, x has the $(n + 1)$ values

$$0, \quad \pm 2, \quad \pm 4, \quad \pm 6, \dots \pm n;$$

and when n is odd, the $(n + 1)$ values

$$\pm 1, \quad \pm 3, \quad \pm 5, \dots \pm n.$$

The expression (10) may be regarded as affording the complete solution of the problem proposed; it expresses the probability of any one of the possible resultants, but for practical purposes it requires transformation when we contemplate a very great n.

The necessary transformation can be obtained after Laplace with the aid of Stirling's theorem. The process is detailed in Todhunter's *History of the Theory of Probability*, p. 548, but the corrections to the principal term there exhibited (of the first order in x) do not appear here where the probabilities of the *plus* and *minus* alternatives are equal. On account of the symmetry, no odd powers of x can occur. I have calculated the resulting expression with retention of the terms which are of the order $1/n^2$ in comparison with the principal term. The resultant x itself may be considered to be of order not higher than $\sqrt{n}$.

By Stirling's theorem

$$n! = \sqrt{(2\pi)}\, n^{n+\frac{1}{2}} e^{-n} C_n, \quad\dots\dots\dots\dots\dots\dots(11)$$

where

$$C_n = 1 + \frac{1}{12n} + \frac{1}{288n^2} + \cdots, \quad\dots\dots\dots\dots\dots(12)$$

with similar expressions for $(\frac{1}{2}n - \frac{1}{2}x)!$ and $(\frac{1}{2}n + \frac{1}{2}x)!$ For the moment we omit the correcting factors C. Thus

$$\frac{1}{(\frac{1}{2}n - \frac{1}{2}x)!\,(\frac{1}{2}n + \frac{1}{2}x)!} = \frac{e^n}{2\pi}\left(\frac{n}{2}\right)^{-n-1}\left(1 - \frac{x^2}{n^2}\right)^{-\frac{1}{2}n-\frac{1}{2}}\left(\frac{1 - x/n}{1 + x/n}\right)^{\frac{1}{2}x}.$$

For the logarithm of the product of the last two factors, we have

$$\frac{n+1}{2}\left\{\frac{x^2}{n^2} + \frac{x^4}{2n^4} + \frac{x^6}{3n^6} + \cdots\right\} - \frac{x^2}{n} - \frac{x^4}{3n^3} - \frac{x^6}{5n^5} - \cdots$$

$$= -\frac{x^2}{2n} + \frac{x^2}{2n^2} - \frac{x^4}{4n^3}\left(\frac{1}{3} - \frac{1}{n}\right) - \frac{x^6}{6n^5}\left(\frac{1}{5} - \frac{1}{n}\right) - \cdots,$$

and for the product itself

$$e^{-x^2/2n}\left\{1 + \frac{1}{2n}\left(\frac{x^2}{n} - \frac{x^4}{6n^2}\right) + \frac{1}{8n^2}\left(\frac{3x^4}{n^2} - \frac{3x^6}{5n^3} + \frac{x^8}{36n^4}\right)\right\}. \quad\dots\dots(13)$$

The principal term in (10) is

$$\frac{\sqrt{(2\pi)}\cdot n^{n+\frac{1}{2}}e^{-n}}{2^n}\cdot\frac{e^n}{2\pi}\left(\frac{n}{2}\right)^{-n-1}e^{-x^2/2n} = \sqrt{\left(\frac{2}{n\pi}\right)}\,e^{-x^2/2n}.$$

There are still the factors C to be considered. We have

$$\frac{C_n}{C_{\frac{1}{2}(n-x)}\,C_{\frac{1}{2}(n+x)}} = \left\{1 + \frac{1}{12n} + \frac{1}{288n^2}\right\}$$

$$\left\{1 + \frac{1}{6(n-x)} + \frac{1}{72(n-x)^2}\right\}^{-1}\left\{1 + \frac{1}{6(n+x)} + \frac{1}{72(n+x)^2}\right\}^{-1}$$

$$= \left\{1 + \frac{1}{12n} + \frac{1}{288n^2}\right\}\left\{1 - \frac{1}{3n} + \frac{1}{3n^2}\left(\frac{1}{6} - \frac{x^2}{n}\right)\right\}$$

$$= 1 - \frac{1}{4n} + \frac{1}{32n^2}\left(1 - \frac{32x^2}{3n}\right). \quad\dots\dots\dots\dots\dots(14)$$

Finally we obtain

$$\sqrt{\left(\frac{2}{n\pi}\right)} e^{-x^2/2n} \left\{1 - \frac{1}{4n}\left(1 - \frac{2x^2}{n} + \frac{x^4}{3n^2}\right) + \frac{1}{32n^2}\left(1 - \frac{44x^2}{3n} + \frac{38x^4}{3n^2} - \frac{12x^6}{5n^3} + \frac{x^8}{9n^4}\right)\right\},$$

$$\ldots\ldots(15)$$

as the probability when n is large of the resultant amplitude x. It is to be remembered that x is limited to a series of discrete values with a common difference equal to 2, and that our approximation has proceeded upon the supposition that x is not of higher order than $\sqrt{n}$.

If the component amplitudes or stretches be l, in place of unity, we have merely to write x/l in place of x.

The special value of the series (15) is realized only when n is very great. But it affords a closer approximation to the true value than might be expected when n is only moderate. I have calculated the case of $n = 10$, both directly from the exact expression (10) and from the series (15) for all the admissible values of x.

TABLE I.

$n = 10.$

x	From (10)	From (15)
0......	·24609	·24608
2......	·20508	·20509
4......	·11719	·11722
6......	·04394	·04392
8......	·00977	·00975
10......	·00098	·00102

The values for $x = 0$ and twice those belonging to higher values of x should total unity. Those above from (10) give 1·00001 and those from (15) give 1·00008. It will be seen that except in the extreme case of $x = 10$, the agreement between the two formulæ is very close. But, even for much higher values of n, the actual calculation is simpler from the exact formula (10).

When l is very small, while n is very great, we may be able for some purposes to disregard the discontinuous character of the probability as a function of x, replacing the isolated points by a continuous representative curve. The difference between the abscissæ of consecutive isolated points is $2l$; so that if dx be a large multiple of l, we may take

$$\sqrt{\left(\frac{1}{2n\pi}\right)} e^{-x^2/2nl^2} dx/l \qquad\ldots\ldots\ldots\ldots\ldots(16)$$

as the approximate expression of the probability that the resultant amplitude lies between x and $x + dx$.

Two Dimensions.

If there is but one stretch of length l, the only possible value of r is of course l.

When there are two stretches of lengths l_1 and l_2, r may vary from $l_2 - l_1$ to $l_2 + l_1$, and then if θ be the angle between them

$$r^2 = l_1^2 + l_2^2 - 2l_1 l_2 \cos \theta, \quad \dots \dots \dots \dots (17)$$

and
$$\sin \theta \, d\theta = r \, dr / l_1 l_2. \quad \dots \dots \dots \dots (18)$$

Since all angles θ between 0 and π are deemed equally probable, the chance of an angle between θ and $\theta + d\theta$ is $d\theta/\pi$. Accordingly the chance that the resultant r lies between r and $r + dr$ is

$$\frac{r \, dr}{\pi l_1 l_2 \sin \theta}, \quad \dots \dots \dots \dots (19)$$

or if with Prof. Pearson* we refer the probability to unit of area in the plane of representation,

$$\phi_2(r^2) = \frac{1}{2\pi^2 l_1 l_2 \sin \theta}$$
$$= \frac{1}{\pi^2 \sqrt{\{2r^2(l_1^2 + l_2^2) - r^4 - (l_1^2 - l_2^2)^2\}}}, \quad \dots \dots (20)$$

$\phi_2(r^2) \, dA$ denoting the chance of the representative point lying in a small area dA at distance r from the origin.

If the stretches l_1 and l_2 are equal, (20) reduces to

$$\phi_2(r^2) = \frac{1}{\pi^2 r \sqrt{\{4l^2 - r^2\}}}, \quad \dots \dots \dots \dots (21)$$

Prof. Pearson's expression, applicable when $r < 2l$. When $r > 2l$, $\phi_2(r^2) = 0$.

When there are three equal stretches ($n = 3$), $\phi_3(r^2)$ is expressible by elliptic functions† with a discontinuity in form as r passes through l.

For values of n from 4 to 7 inclusive, Pearson's work is founded upon the general functional relation‡

$$\phi_{n+1}(r^2) = \frac{1}{\pi} \int_0^\pi \phi_n(r^2 + l^2 - 2rl \cos \theta) \, d\theta. \quad \dots \dots \dots (22)$$

Putting $r = 0$, he deduces the special conclusion that

$$\phi_{n+1}(0) = \phi_n(l^2), \quad \dots \dots \dots \dots (23)$$

as is indeed evident *a priori*.

---

* *Drapers' Company Research Memoirs*, Biometric Series III., London, 1906.

† Pearson (*loc. cit.*) attributes this evaluation to G. T. Bennett.

‡ Compare *Theory of Sound*, § 42 a.

From (22) the successive forms are determined graphically. For values of n higher than 7 an analytical expression proceeding by powers of $1/n$ is available, and will be further referred to later.

A remarkable advance in the theory of random vibrations and of flights in two dimensions, when the number (n) is finite, is due to J. C. Kluyver[*], who has discovered an expression for the probability of various resultants in the form of a definite integral involving Bessel's functions. His exposition is rather concise, and I think I shall be doing a service in reproducing it with some developments and slight changes of notation. It depends upon the use of a discontinuous integral evaluated by Weber, viz.

$$\int_0^\infty J_1(bx)\, J_0(ax)\, dx = u \quad \text{(say)}.$$

To examine this we substitute from

$$\pi \,.\, J_1(bx) = 2 \int_0^{\frac{1}{2}\pi} \cos\theta \sin(bx \cos\theta)\, d\theta \dagger,$$

and take first the integration with respect to x. We have[‡]

$$\int_0^\infty dx \sin(bx \cos\theta)\, J_0(ax) = 0, \qquad \text{if } a^2 > b^2 \cos^2\theta,$$

or $\qquad\qquad = (b^2 \cos^2\theta - a^2)^{-\frac{1}{2}}, \qquad \text{if } b^2 \cos^2\theta > a^2.$

Thus, if $\qquad a^2 > b^2, \quad u = 0.$ If $b^2 > a^2,$

$$u = \frac{2}{\pi} \int \frac{d\theta \cos\theta}{\sqrt{(b^2 \cos^2\theta - a^2)}} = \frac{2}{\pi b} \sin^{-1} \frac{b \sin\theta}{\sqrt{(b^2 - a^2)}}.$$

The lower limit for θ is 0, and the upper limit is given by $\cos^2\theta = a^2/b^2$. Hence $u = 1/b$, and thus

$$\left. \begin{aligned} b \int_0^\infty J_1(bx)\, J_0(ax)\, dx &= 1, \quad (b^2 > a^2) \\ \text{or} \qquad\qquad &= 0, \quad (a^2 > b^2) \end{aligned} \right\} \quad \dots\dots\dots\dots(24)$$

A second lemma required is included in Neumann's theorem, and may be very simply arrived at. In Fig. 1, G and E being fixed points, the function at F denoted by

$$J_0(g), \quad \text{or} \quad J_0 \sqrt{(e^2 + f^2 - 2ef \cos G)},$$

is a potential satisfying everywhere the equation $\nabla^2 + 1 = 0$, and accordingly may be expanded round G in the Fourier series

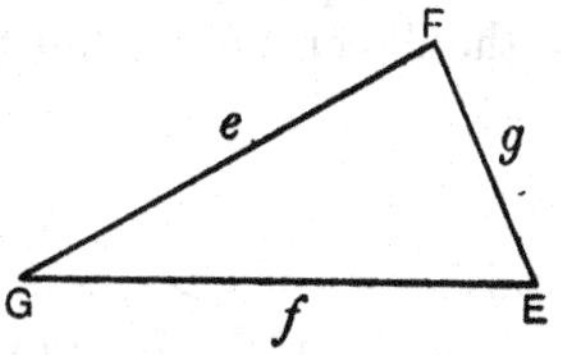

$$A_0 J_0(e) + A_1 J_1(e) \cos G + A_2 J_2(e) \cos 2G + \dots,$$

the coefficients A being independent of e and G. Thus

$$\frac{1}{2\pi} \int_0^{2\pi} J_0 \sqrt{(e^2 + f^2 - 2ef \cos G)}\, dG = A_0 J_0(e).$$

Fig. 1.

[*] Koninklijke Akademie van Wetenschappen te Amsterdam, Verslag van de gewone vergaderingen der Wis-en-Natuurkundige Afdeeling, Deel xiv, 1st Gedeelte, 30 September, 1905, pp. 325–334.

[†] Gray and Mathews, *Bessel's Functions*, p. 18, equation (46).   [‡] G. and M., p. 73.

By parity of reasoning when E and F are interchanged, the same integral is proportional to $J_0(f)$, and may therefore be equated to $A_0' J_0(e) J_0(f)$, where A_0' is now an absolute constant, whose value is at once determined to be unity by making e, or f, vanish. The lemma

$$\int_0^{2\pi} J_0 \sqrt{(e^2 + f^2 - 2ef \cos G)}\, dG = 2\pi J_0(e) J_0(f) \quad ....(25)$$

is thus established*.

We are now prepared to investigate the probability

$$P_n(r;\ l_1, l_2, \ldots l_n)$$

that after n stretches $l_1, l_2, \ldots l_n$ taken in directions at random the distance from the starting-point O (Fig. 2) shall be less than an assigned magnitude r. The direction of the first stretch l_1 is plainly a matter of indifference. On the other hand the probability that the angles θ lie within the limits θ_1 and $\theta_1 + d\theta_1$, θ_2 and $\theta_2 + d\theta_2$, ... θ_{n-1} and $\theta_{n-1} + d\theta_{n-1}$ is

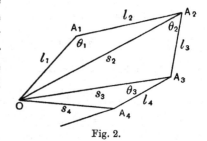

Fig. 2.

$$\frac{1}{(2\pi)^{n-1}} d\theta_1 d\theta_2 \ldots d\theta_{n-1}, \ldots(26)$$

which is now to be integrated under the condition that the nth radius vector s_n shall be less than r.

Let us commence with the case of two stretches l_1 and l_2. Then

$$P_2(r;\ l_1, l_2) = \frac{1}{2\pi} \int d\theta_1,$$

the integration being taken within such limits that $s_2 < r$, where

$$s_2^2 = l_1^2 + l_2^2 - 2l_1 l_2 \cos \theta_1.$$

The required condition as to the limits can be secured by the introduction of the discontinuous function afforded by Weber's integral. For

$$r \int_0^\infty J_1(rx) J_0(s_2 x)\, dx$$

vanishes when $s_2 > r$, and is equal to unity when $s_2 < r$. After the introduction of this factor, the integration with respect to θ_1 may be taken over the complete range from 0 to 2π. Thus

$$P_2(r;\ l_1, l_2) = \frac{r}{2\pi} \int_0^{2\pi} d\theta_1 \int_0^\infty dx\, J_1(rx) J_0(s_2 x).$$

---

* Similar reasoning shows that if $D_0(g)$ represent a symmetrical purely divergent wave,

$$\int_0^{2\pi} D_0 \sqrt{(e^2 + f^2 - 2ef \cos G)}\, dG = 2\pi J_0(e) D_0(f),$$

provided that $f > e$.

Taking first the integration with respect to θ_1, we have by (25)

$$\frac{1}{2\pi}\int_0^{2\pi} d\theta_1 J_0(s_2 x) = J_0(l_1 x) J_0(l_2 x),$$

and thus

$$P_2(r; l_1, l_2) = r\int_0^\infty dx J_1(rx) J_0(l_1 x) J_0(l_2 x). \quad\ldots\ldots\ldots(27)$$

The method can be extended to any number (n) of stretches. Beginning with the integration with respect to θ_{n-1} in (26), we have as before

$$\frac{1}{2\pi}\int d\theta_{n-1} = \frac{r}{2\pi}\int_0^{2\pi} d\theta_{n-1}\int_0^\infty dx J_1(rx) J_0(s_n x)$$

$$= r\int_0^\infty dx J_1(rx) J_0(l_n x) J_0(s_{n-1} x).$$

The next integration gives

$$\frac{1}{(2\pi)^2}\iint d\theta_{n-2} d\theta_{n-1} = r\int_0^\infty J_1(rx) J_0(l_n x) J_0(l_{n-1} x) J_0(s_{n-2} x)\, dx,$$

and so on.   Finally

$$P_n(r; l_1, l_2, \ldots l_n) = \frac{1}{(2\pi)^{n-1}}\iint\ldots d\theta_1 d\theta_2 \ldots d\theta_{n-1}$$

$$= r\int_0^\infty J_1(rx) J_0(l_1 x) J_0(l_2 x)\ldots J_0(l_n x)\, dx, \quad\ldots\ldots(28)$$

—the expression for P_n discovered by Kluyver.

It will be observed that (28) is symmetrical with respect to the l's; the order in which they are taken is immaterial.

When all the l's are equal,

$$P_n(r; l) = r\int_0^\infty J_1(rx)\{J_0(lx)\}^n dx. \quad\ldots\ldots\ldots\ldots(29)$$

If in (29) we suppose $r = l$,

$$P_n(l; l) = -\int_0^\infty \{J_0(lx)\}^n dJ_0(lx)$$

$$= -\frac{\{J_0(lx)\}^{n+1}}{n+1}\Big|_0^\infty = \frac{1}{n+1}; \quad\ldots\ldots\ldots\ldots(30)$$

so that after n equal components have been combined the chance that the resultant shall be less than one of the components is $1/(n+1)$, an interesting result due to Kluyver. The same author notices some of the discontinuities which present themselves, but it will be more convenient to consider this in a modified form of the problem.

The modification consists in dealing, not with the chance of a resultant less than r, but with the chance that it lies between r and $r + dr$. It may seem easy to pass from the one to the other, as it involves merely a differentiation with respect to r.   We have

$$\frac{d}{dr}\{rJ_1(rx)\} = -\frac{d}{dr}\{rJ_0'(rx)\}$$

$$= -J_0'(rx) - rxJ_0''(rx) = rxJ_0(rx),$$

in virtue of the differential equation satisfied by J_0.   Thus, if the differentiation under the integral sign is legitimate,

$$\frac{dP_n}{dr} = 2\pi r\phi_n(r^2) = r\int_0^\infty x\,dx\,J_0(rx)\,J_0(l_1 x)\,J_0(l_2 x)\ldots J_0(l_n x)\ldots, \quad\ldots(31)$$

and, if all the l's are equal,

$$\phi_n(r^2) = \frac{1}{2\pi}\int_0^\infty x\,dx\,J_0(rx)\,\{J_0(lx)\}^n, \quad\ldots\ldots\ldots\ldots\ldots(32)$$

the form employed by Pearson, whose investigation is by a different method. If we put $n = 1$ in (32),

$$\phi_1(r^2) = \frac{1}{2\pi}\int_0^\infty x\,dx\,J_0(rx)\,J_0(lx), \quad\ldots\ldots\ldots\ldots\ldots(33)$$

and this is in fact the equation from which Pearson starts.   But it should be remarked that the integral (33), as it stands, is *not convergent*.   For when z is very great,

$$J_0(z) = \sqrt{\left(\frac{2}{\pi z}\right)}\cos\left(\frac{1}{4}\pi - z\right), \quad\ldots\ldots\ldots\ldots\ldots(34)$$

so that $(r \neq 0)$

$$\frac{1}{2\pi}\int_0^x x\,dx\,J_0(rx)\,J_0(lx) = \frac{1}{2\pi^2\sqrt{(rl)}}\int_0^x dx\,\{\sin(r+l)\,x + \cos(r-l)\,x\},$$

and this is not convergent when $x = \infty$.

The criticism does not apply to (29) itself when $n = 1$, but it leads back to the question of differentiation under the sign of integration.   It appears at any rate that any number of such operations can be justified, provided that the integrals, resulting from these and the next following operation, are finite for the values of r in question.   But this condition is not satisfied in the differentiation under the integral sign of (29) when $n = 1$.   For the next operation upon (32) then yields

$$\int_0^\infty x^2\,dx\,J_1(rx)\,J_0(lx).$$

When we substitute for $J_0(lx)$ from (34) and for $J_1(rx)$ from

$$J_1(z) = \sqrt{\left(\frac{2}{\pi z}\right)}\cos\left(\frac{3\pi}{4} - z\right),$$

we get $$\int_0^x x\,dx\,\cos\left(\frac{3\pi}{4} - rx\right)\cos\left(\frac{\pi}{4} - lx\right),$$

which becomes infinite with x, even for general values of r and l.

So much by way of explanation; but of course we do not really need to discuss the cases $n=1$, $n=2$, or even $n=3$, for which exact solutions can be expressed in terms of functions which may be regarded as known.

For higher values of n it would be of interest to know how many differentiations with respect to r may be made under the sign of integration. It may be remarked that since all J's and their derivatives to any order are less than unity, the integral can become infinite only in virtue of that part of the range where x is very great, and that there we may introduce the asymptotic values.

We have thus to consider

$$\frac{d^p}{dr^p}\,\phi_n\,(r^2) = \frac{1}{2\pi}\int_0^\infty dx\, x^{p+1} J_0^p\,(rx)\,\{J_0\,(lx)\}^n. \quad\ldots\ldots\ldots\ldots(35)$$

For the leading term when z is very great, we have

$$J_0^p\,(z) = \frac{d^p}{dz^p}\left\{\sqrt{\left(\frac{.2}{\pi z}\right)}\cos\left(\frac{1}{4}\pi - z\right)\right\}$$

$$= \sqrt{\left(\frac{2}{\pi z}\right)}\cos\left(\frac{1}{4}\pi - z - \frac{1}{2}\,p\pi\right), \quad\ldots\ldots\ldots\ldots(36)$$

$$\{J_0\,(z)\}^n = \left(\frac{2}{\pi z}\right)^{\frac12 n}\cos^n\left(\frac{1}{4}\pi - z\right), \quad\ldots\ldots\ldots\ldots\ldots\ldots(37)$$

so that with omission of constant factors our integral becomes

$$\int^\infty dx\, x^{p+\frac12-\frac12 n}\cos\left(\frac{1}{4}\pi - rx - \frac{1}{2}\,p\pi\right)\cos^n\left(\frac{1}{4}\pi - lx\right).\ldots\ldots\ldots(38)$$

In this $\cos^n\left(\frac14\pi - lx\right)$ can be expanded in a series of cosines of multiples of $(\frac14\pi - lx)$, commencing with $\cos n\,(\frac14\pi - lx)$ and ending when n is odd with $\cos\left(\frac14\pi - lx\right)$, and when n is even with a constant term. The various products of cosines are then to be replaced by cosines of sums and differences. The most unfavourable case occurs when this operation leaves a constant term, which can happen only for values of r which are multiples of l. We are then left with

$$\int^\infty dx\, x^{p+\frac12-\frac12 n} = \left.\frac{x^{p+\frac32-\frac12 n}}{p+\frac32-\frac12 n}\right|^\infty .$$

The integral is thus finite or infinite according as

$$p< \text{ or } >\tfrac12\,(n-3).$$

If, however, there arise no constant term, we have to consider

$$\int^\infty dx\, x^s \cos mx = \left.\frac{x^s}{m}\sin mx\right|^x - \frac{s}{m}\int^\infty dx\, x^{s-1}\sin mx,$$

where m is finite; and this is finite if s, that is $p+\frac12-\frac12 n$, be negative. The differentiations are then valid, if

$$p<\tfrac12\,(n-1).$$

We may now consider more especially the cases $n=4$, etc. When $n=4$, $s = p+\frac12-\frac12 n = p-\frac32$.

If $p = 1$, $s = -\frac{1}{2}$, and the cosine factors in (38) become

$$\cos\left(\tfrac{1}{4}\pi + rx\right)\cos^4\left(\tfrac{1}{4}\pi - lx\right),$$

yielding finally

$$\cos\left(\frac{5\pi}{4} + rx - 4lx\right), \quad \cos\left(\frac{3\pi}{4} - rx - 4lx\right),$$

$$\cos\left(\frac{3\pi}{4} + rx - 2lx\right), \quad \cos\left(\frac{\pi}{4} - rx - 2lx\right), \quad \cos\left(\frac{\pi}{4} + rx\right),$$

so that there is no constant term unless $r = 4l$, or $2l$. With these exceptions, the original differentiation under the integral sign is justified.

We fall back upon ϕ_4 itself by putting $p = 0$, making $s = -\frac{3}{2}$. The integral is then finite in all cases ($r \neq 0$), in agreement with Pearson's curve.

Next for $n = 5$, $s = p - 2$.

When $p = 1$, $s = -1$, and we find that the cosine factors yield a constant term only when $r = 3l$. Pearson's curve does not suggest anything special at $r = 3l$; it may be remarked that the integral with $p = 1$ is there only logarithmically infinite.

If $n = 5$, $p = 0$, $s = -2$; and the integral for ϕ_5 is finite for all values of r.

When $n = 6$, $s = p - 2\frac{1}{2}$. In this case, whether $p = 1$, or 0, no question can arise. The integrals are finite for all values of r.

A fortiori is this so, when $n > 6$.

If we suppose $p = 2$, $s = \frac{1}{2}(5 - n)$. Thus $n = 7$ makes $s = -1$, and infinities might occur for special values of r. But if $n > 7$, $s < -\frac{3}{2}$, and infinities are excluded whatever may be the value of r.

Similarly if $p = 3$, infinities are excluded if $n > 9$, and so on.

Our discussion has not yet yielded all that could be wished; the subject may be commended to those better versed in pure mathematics. Probably what is required is a better criterion as to the differentiation under the integral sign*.

We may now pass on to consider what becomes of Kluyver's integral when n is made infinite. As already remarked, Pearson has developed for it a series proceeding by powers of $1/n$, and it may be convenient to give a version of his derivation, without, however, carrying the process so far.

[* The criterion enunciated on p. 614 appears to have been devised to meet the case when $s = 0$ and the integral, though finite, does not converge to a definite value when $x = \infty$. If, however, $s < -1$, or < 0, respectively, according as the cosine factors in (38) do or do not produce a constant term, the integral (38) has been shown to be finite; it is also convergent; and the integrals obtained by omitting before each successive differentiation the factor to be differentiated, viz.: $\cos\left(\tfrac{1}{4}\pi - rx - \tfrac{1}{2}q\pi\right)$ where $q < p$, are also finite (cf. Todhunter's *Integral Calculus*, 1889, Arts. 214, 284). In these circumstances it would appear that (38) is itself valid, and that it is unnecessary to consider the integral obtained by "the next following operation" ($s = p + \tfrac{3}{2} - \tfrac{1}{2}n$). It would seem then that the above considerations are sufficient to justify the differentiation by which ϕ_4 is obtained ($p = 0$, $s = -\tfrac{3}{2}$), and *a fortiori* that for ϕ_5 ($p = 0$, $s = -2$), etc. W. F. S.]

The evaluation of the principal term depends upon a formula due, I think, to Weber*, viz.

$$u = \int_0^\infty J_0(rx) e^{-p^2 x^2} x \, dx = \frac{1}{2p^2} e^{-r^2/4p^2}, \quad \dots\dots\dots\dots(39)$$

making†

$$\frac{du}{dr} = \int_0^\infty J_0'(rx) e^{-p^2 x^2} x^2 \, dx = -\frac{1}{2p^2} \int_0^\infty J_0'(rx) \, x \, de^{-p^2 x^2}$$

$$= \frac{1}{2p^2} \int_0^\infty e^{-p^2 x^2} \{J_0'(rx) + rx J_0''(rx)\} \, dx$$

$$= -\frac{r}{2p^2} \int_0^\infty J_0(rx) e^{-p^2 x^2} x \, dx = -\frac{r}{2p^2} u$$

Hence

$$u = Ce^{-r^2/4p^2}.$$

To determine C we have merely to make $r = 0$. Thus

$$C = u_{r=0} = \int_0^\infty e^{-p^2 x^2} x \, dx = \frac{1}{2p^2},$$

by which (39) is established.

Unless lx is small, the factor $\{J_0(lx)\}^n$ in (32) diminishes rapidly as n increases, inasmuch as $J_0(lx)$ is less than unity for any finite lx. Thus when n is very great, the important part of the range of integration corresponds to a small lx.

Writing s for $\frac{1}{2}nl^2$, we have

$$\log J_0(lx) = \log\left(1 - \frac{sx^2}{2n} + \frac{s^2 x^4}{16n^2} - \frac{s^3 x^6}{288n^3} + \dots\right)$$

$$= -\frac{sx^2}{2n} - \frac{s^2 x^4}{16n^2} - \frac{s^3 x^6}{72n^3} + \dots;$$

so that

$$\{J_0(lx)\}^n = e^{-\frac{1}{2}sx^2}\left(1 - \frac{s^2 x^4}{16n} - \frac{s^3 x^6}{72n^2} + \frac{s^4 x^8}{512n^2}\right),$$

making

$$2\pi\phi_n(r^2) = \int_0^\infty x \, dx \, J_0(rx) \, e^{-\frac{1}{2}sx^2}\left(1 - \frac{s^2 x^4}{16n} - \frac{s^3 x^6}{72n^2} + \frac{s^4 x^8}{512n^2}\right). \quad \dots(40)$$

Calling the four integrals on the right I_1, I_2, I_3, and I_4, we have by (39)

$$I_1 = \int_0^\infty x \, dx \, J_0(rx) \, e^{-\frac{1}{2}sx^2} = \frac{1}{s} e^{-r^2/2s}, \quad \dots\dots\dots\dots(41)$$

$$-I_2 = \frac{s^2}{4n} \frac{d^2 I_1}{ds^2} = \frac{s^2}{4n} \frac{d^2}{ds^2}\left(\frac{1}{s} e^{-r^2/2s}\right), \quad \dots\dots\dots\dots(42)$$

$$I_3 = \frac{s^3}{9n^2} \frac{d^3 I_1}{ds^3} = \frac{s^3}{9n^2} \frac{d^3}{ds^3}\left(\frac{1}{s} e^{-r^2/2s}\right), \quad \dots\dots\dots\dots(43)$$

$$I_4 = \frac{s^4}{32n^2} \frac{d^4 I_1}{ds^4} = \frac{s^4}{32n^2} \frac{d^4}{ds^4}\left(\frac{1}{s} e^{-r^2/2s}\right). \quad \dots\dots\dots\dots(44)$$

---

* Gray and Mathews, loc. cit. p. 77.

† I apprehend that there can be no difficulty here as to the differentiation, the situation being dominated by the exponential factor.

Thus

$$2\pi\,\phi_n\,(r^a) = \frac{e^{-r^2/2s}}{s}\left\{1 - \frac{1}{4n}\left(2 - \frac{2r^2}{s} + \frac{r^4}{4s^2}\right) - \frac{1}{9n^2}\left(6 - \frac{9r^2}{s} + \frac{9r^4}{4s^2} - \frac{r^6}{8s^3}\right)\right.$$

$$\left. + \frac{1}{32n^2}\left(24 - \frac{48r^2}{s} + \frac{18r^4}{s^2} - \frac{2r^6}{s^3} + \frac{r^8}{16s^4}\right)\right\}$$

$$= \frac{e^{-r^2/2s}}{s}\left\{1 - \frac{1}{4n}\left(2 - \frac{2r^2}{s} + \frac{r^4}{4s^2}\right) + \frac{1}{12n^2}\left(1 - \frac{6r^2}{s} + \frac{15r^4}{4s^2} - \frac{7r^6}{12s^3} + \frac{3r^8}{128s^4}\right)\right\},$$

$$\dots\dots(45)$$

in agreement (so far as it goes) with Pearson, whose σ^2 is equal to our s. The leading term is that given in 1880.

Three Dimensions.

We may now pass on to the corresponding problem when flights take place in three dimensions, where we shall find, as might have been expected, that the mathematics are simpler. And first for *two* flights of length l_1 and l_2. If μ be the cosine of the angle between l_1 and l_2 and r the resultant,

$$r^2 = l_1^2 + l_2^2 - 2l_1 l_2 \mu,$$

giving
$$r\,dr = -l_1 l_2 d\mu. \dots\dots\dots\dots\dots(46)$$

The chance of r lying between r and $r + dr$ is the same as the chance of μ lying between μ and $\mu + d\mu$, that is $-\frac{1}{2}d\mu$, since all directions in space are to be treated as equally probable. Accordingly the chance of a resultant between r and $r + dr$ is

$$\frac{r\,dr}{2l_1 l_2}. \dots\dots\dots\dots\dots(47)$$

The corresponding volume is $4\pi r^2 dr$, so that in the former notation

$$\phi_2\,(r\,;\ l) = \frac{1}{8\pi l^2 r}, \dots\dots\dots\dots\dots(48)$$

l_1 and l_2 being supposed equal. It will be seen that this is simpler than (21). It applies, of course, only when $r < 2l$. When $r > 2l$, $\phi_2 = 0$.

In like manner when l_1 and l_2 differ, the chance of a resultant less than r is zero, when r falls short of the difference between l_2 and l_1, say $l_2 - l_1$. Between $l_2 - l_1$ and $l_2 + l_1$ the chance is

$$\int_{l_2-l_1}^{r} \frac{r\,dr}{2l_1 l_2} = \frac{r^2 - (l_2 - l_1)^2}{4l_1 l_2}. \dots\dots\dots\dots(49)$$

When r has its greatest value $(l_2 + l_1)$, (49) becomes

$$\frac{(l_2 + l_1)^2 - (l_2 - l_1)^2}{4l_1 l_2} = 1. \dots\dots\dots\dots(50)$$

The " chance " is then a certainty, as also when $r > l_1 + l_2$.

In proceeding to the general value of n, we may conveniently follow the analogy of the two-dimensional investigation of Kluyver, for which purpose we require a function that shall be unity when $s < r$, and zero when $s > r$. Such a function is

$$\frac{2}{\pi} \int_0^\infty dx \frac{\sin sx}{sx} \frac{\sin rx - rx \cos rx}{x}; \qquad\qquad (51)$$

for it may be written

$$-\frac{2r}{\pi s} \int_0^\infty \sin sx \, d\left(\frac{\sin rx}{rx}\right) = \frac{2}{\pi} \int_0^\infty \frac{\sin rx}{x} \cos sx \, dx$$

$$= \frac{1}{\pi} \int_0^\infty \frac{\sin(s+r)x - \sin(s-r)x}{x} \, dx = 1 \text{ or } 0,$$

according as s is less or greater than r.

In like manner for a second lemma, corresponding with (25), we may reason again from the triangle GFE (Fig. 1). $J_0(g)$ is replaced by $\sin g/g$, a potential function symmetrical in three dimensions about E and satisfying *everywhere* $\nabla^2 + 1 = 0$. It may be expanded about G in Legendre's series[*]

$$A_0 \frac{\sin e}{e} + A_1 \mu \left(\frac{\sin e}{e^2} - \frac{\cos e}{e}\right) + \dots,$$

μ being written for $\cos G$, and accordingly

$$\frac{1}{2} \int_{-1}^{+1} d\mu \frac{\sin \sqrt{(e^2 + f^2 - 2ef\mu)}}{\sqrt{(e^2 + f^2 - 2ef\mu)}} = A_0 \frac{\sin e}{e}.$$

When E and F are interchanged, the same integral is seen to be proportional to $\sin f/f$, and may therefore be equated to

$$A_0' \frac{\sin e}{e} \frac{\sin f}{f},$$

where A_0' is now an absolute constant, whose value is determined to be unity by putting e, or f, equal to zero. We may therefore write

$$\frac{1}{2} \int_{-1}^{+1} d\mu \frac{\sin \sqrt{(e^2 + f^2 - 2ef\mu)}}{\sqrt{(e^2 + f^2 - 2ef\mu)}} = \frac{\sin e}{e} \frac{\sin f}{f}. \qquad\qquad (52)$$

As in the case of two dimensions, similar reasoning shows that

$$\frac{1}{2} \int_{-1}^{+1} d\mu \frac{\cos \sqrt{(e^2 + f^2 - 2ef\mu)}}{\sqrt{(e^2 + f^2 - 2ef\mu)}} = \frac{\sin e}{e} \frac{\cos f}{f}, \qquad\qquad (53)$$

provided $e < f$.

With appropriate changes, we may now follow Kluyver's argument for two dimensions. The same diagram (Fig. 2) will serve, only the successive triangles are no longer limited to lie in one plane. Instead of the angles θ, we have now to deal with their cosines, of which all values are to be regarded

[*] *Theory of Sound*, § 330.

as equally probable. The probability that these cosines shall lie within the interval μ_1 and $\mu_1 + d\mu_1$, μ_2 and $\mu_2 + d\mu_2$, ... μ_{n-1} and $\mu_{n-1} + d\mu_{n-1}$ is

$$\frac{1}{2^{n-1}} d\mu_1 d\mu_2 \dots d\mu_{n-1}, \dots\dots\dots\dots\dots(54)$$

which is now to be integrated under the condition that the nth radius s_n shall be less than r.

We begin with two stretches l_1 and l_2. Then, in the same notation as before, we have

$$P_2(r; l_1, l_2) = \frac{1}{2} \int d\mu,$$

the integration being within such limits as make $s_2 < r$, where

$$s_2{}^2 = l_1{}^2 + l_2{}^2 - 2l_1 l_2 \mu.$$

Hence, by introduction of the discontinuous function (51),

$$P_2(r; l_1, l_2) = \frac{1}{\pi} \int_{-1}^{+1} d\mu \int_0^\infty dx \frac{\sin s_2 x}{s_2 x} \frac{\sin rx - rx \cos rx}{x}.$$

But by (52)

$$\frac{1}{2} \int_{-1}^{+1} d\mu \frac{\sin s_2 x}{s_2 x} = \frac{\sin l_1 x}{l_1 x} \frac{\sin l_2 x}{l_2 x},$$

and thus $\quad P_2(r; l_1, l_2) = \dfrac{2}{\pi} \displaystyle\int_0^\infty dx \dfrac{\sin rx - rx \cos rx}{x} \dfrac{\sin l_1 x}{l_1 x} \dfrac{\sin l_2 x}{l_2 x}. \quad \dots\dots(55)$

A simpler form is available for dP_2/dr, since

$$\frac{d}{dr}(\sin rx - rx \cos rx) = rx^2 \sin rx.$$

Thus $\qquad \dfrac{dP_2}{dr} = \dfrac{2r}{\pi l_1 l_2} \displaystyle\int_0^\infty \dfrac{dx}{x} \sin rx \sin l_1 x \sin l_2 x, \quad \dots\dots\dots\dots(56)$

in which we replace the product of sines by means of

$$4 \sin rx \sin l_1 x \sin l_2 x = \sin(r + l_2 - l_1)x$$
$$+ \sin(r - l_2 + l_1)x - \sin(r + l_2 + l_1)x - \sin(r - l_2 - l_1)x.$$

If r, l_2, l_1 are sides of a real triangle, any two of them together are in general greater than the third, and thus when the integration is effected by the formula

$$\int_0^\infty \frac{\sin u}{u} du = \tfrac{1}{2}\pi,$$

we obtain three positive and one negative term. Finally

$$\frac{dP_2}{dr} = \frac{r}{2l_1 l_2},$$

in agreement with (47). The expression is applicable only when the triangle is possible. In the contrary case we find dP/dr equal to zero when r is less than the difference and greater than the sum of l_1 and l_2.

This argument must appear very roundabout, if the object were merely to obtain the result for $n = 2$. The advantage is that it admits of easy extension to the general value of n. To this end we take the last stretch l_n and the immediately preceding radius s_{n-1} in place of l_2 and l_1 respectively, and then repeat the operation with l_{n-1}, s_{n-2}, and so on, until we reach l_2 and $s_1 (= l_1)$. The result is evidently

$$P_n(r; l_1, l_2, \ldots l_n) = \frac{2}{\pi} \int_0^\infty dx \frac{\sin rx - rx \cos rx}{x} \frac{\sin l_1 x}{l_1 x} \frac{\sin l_2 x}{l_2 x} \cdots \frac{\sin l_n x}{l_n x},$$

$$\ldots \ldots (57)$$

or if we suppose, as for the future we shall do, that the l's are all equal,

$$P_n(r; l) = \frac{2}{\pi} \int_0^\infty dx \frac{\sin rx - rx \cos rx}{x} \left(\frac{\sin lx}{lx}\right)^n. \qquad \ldots \ldots (58)$$

This is the chance that the resultant is less than r. For the chance that the resultant lies between r and $r + dr$, we have, as the coefficient of dr,

$$\frac{dP_n}{dr} = \frac{2r}{\pi l^n} \int_0^\infty \frac{dx}{x^{n-1}} \sin rx \sin^n lx. \qquad \ldots \ldots (59)$$

Let us now consider the particular case of $n = 3$, when

$$\frac{dP_3}{dr} = \frac{2r}{\pi l^3} \int_0^\infty \frac{dx}{x^2} \sin rx \sin^3 lx. \qquad \ldots \ldots (60)$$

In this we have

$$\sin rx \sin^3 lx = \tfrac{1}{8}\{3 \cos (r - l) x - 3 \cos (r + l) x - \cos (r - 3l) x + \cos (r + 3l) x\}.$$

And

$$\int_0^\infty \frac{dx}{x^2} \{\cos (r - l) x - \cos (r + l) x\}$$

$$= 2 \int_0^\infty \frac{dx}{x^2} \left\{\sin^2 \frac{(r + l) x}{2} - \sin^2 \frac{(r - l) x}{2}\right\}$$

$$= \tfrac{1}{2}\pi \{r + l - |r - l|\};$$

and in like manner for the second pair of cosines.

Thus

$$\frac{dP_3}{dr} = \frac{r}{8l^3} \{2r - 3|r - l| + |r - 3l|\} \qquad \ldots \ldots (61)$$

expresses the complete solution. When

$$r < l, \quad dP_3/dr = r^2/2l^3,$$
$$3l > r > l, \quad dP_3/dr = (3lr - r^2)/4l^3,$$
$$r > 3l, \quad dP_3/dr = 0.$$

It will be observed that dP_3/dr is itself continuous; but the next derivative changes suddenly at $r = l$ and $r = 3l$ from one finite value to another.

Next take $n = 4$. From (59)

$$\frac{dP_4}{dr} = \frac{2r}{\pi l^4} \int_0^\infty \frac{dx}{x^3} \sin rx \sin^4 lx,$$

and
$$-\frac{d^2}{dr^2}\left(\frac{1}{r}\frac{dP_4}{dr}\right) = \frac{2}{\pi l^4}\int_0^\infty \frac{dx}{x}\sin rx \sin^4 lx$$

$$= \frac{1}{8\pi l^4}\int_0^\infty \frac{dx}{x}\{\sin(r+4l)x + \sin(r-4l)x$$

$$- 4\sin(r+2l)x - 4\sin(r-2l)x + 6\sin rx\}$$

$$= \frac{1}{16l^4}\{1 \pm 1 - 4 \mp 4 + 6\} = \frac{1}{16l^4}\{3 \pm 1 \mp 4\},$$

the alternatives depending upon the signs of $r - 4l$ and $r - 2l$.

When
$$r < 2l, \quad -16l^4\frac{d^2}{dr^2}\left(\frac{1}{r}\frac{dP_4}{dr}\right) = 6,$$

$$4l > r > 2l, \quad -16l^4\frac{d^2}{dr^2}\left(\frac{1}{r}\frac{dP_4}{dr}\right) = -2,$$

and when $r > 4l$, the value is zero. In no case can the value be infinite, from which we may infer that

$$\frac{d}{dr}\left(\frac{1}{r}\frac{dP_4}{dr}\right) \quad \text{and} \quad \frac{1}{r}\frac{dP_4}{dr}$$

must be continuous throughout.

From these data we can determine the form of dP_4/dr, working backwards from the large value of r, where all derivatives vanish.

$$(4l > r > 2l) \quad -16l^4\frac{d}{dr}\left(\frac{1}{r}\frac{dP_4}{dr}\right) = -2(r-4l),$$

$$(2l > r) \quad -16l^4\frac{d}{dr}\left(\frac{1}{r}\frac{dP_4}{dr}\right) = 6(r-2l) + 4l = 6r - 8l,$$

giving continuity at $r = 4l$ and $r = 2l$. Again

$$(4l > r > 2l) \quad -16l^4\frac{1}{r}\frac{dP_4}{dr} = -(r^2 - 16l^2) + 8l(r-4l)$$

$$= -(r-4l)^2,$$

$$(2l > r) \quad -16l^4\frac{1}{r}\frac{dP_4}{dr} = 3(r^2 - 4l^2) - 8l(r-2l) - 4l^2$$

$$= 3r^2 - 8rl.$$

Finally
$$\frac{dP_4}{dr} = 4\pi r^2 \phi_4(r;\ l) = \frac{r^2(8l-3r)}{16l^4} \quad (r < 2l)$$

$$\text{or} \quad\quad = \frac{r(4l-r)^2}{16l^4} \quad (4l > r > 2l) \quad\quad\quad \Bigg\}, \dots\dots(62)$$

and vanishes, of course, when $r > 4l$.

From (61), (62) we may verify Pearson's relation [*], $\phi_4(o;\ l) = \phi_3(l;\ l)$.

[* This implies that Pearson's relation (p. 610) holds for three dimensions. We have in fact, for flights in three dimensions,

$$\phi_{n+1}(r;\ l) = \tfrac{1}{2}\int_0^\pi \phi_n(\sqrt{r^2 + l^2 - 2rl\cos\theta};\ l)\sin\theta\,d\theta,$$

whence $\phi_{n+1}(o;\ l) = \phi_n(l;\ l)$. W. F. S.]

From these examples the procedure will be understood. When n is even, we differentiate (59) $(n-2)$ times, thus obtaining

$$-\frac{d^{n-2}}{dr^{n-2}}\left(\frac{1}{r}\frac{dP_n}{dr}\right) = (-1)^{\frac{n}{2}}\frac{2}{\pi l^n}\int_0^\infty \frac{dx}{x}\sin rx \sin^n lx, \quad\ldots\ldots\ldots(63)$$

in which $\sin^n lx$ is replaced by the series containing $\cos nlx$, $\cos(n-2)lx$, ... and ending with a constant term. When this is multiplied by $\sin rx$, we get sines of $(r \pm nl)x$, $\{r \pm (n-2)l\}x$, ... $\sin rx$, and the integration can be effected. Over the various ranges of $2l$ the values are constant, but they change discontinuously when r is an *even* multiple of l. The actual forms for dP_n/dr can then be found, as already exemplified, by working backwards from $r > nl$, where all derivatives vanish, and so determining the constants of integration as to maintain continuity throughout. These forms are in all cases algebraic.

When n is odd, we differentiate $(n-3)$ times, thus obtaining a form similar to (60) where $n = 3$. A similar procedure then shows that the result assumes constant values over finite ranges with discontinuities* when r is an *odd* multiple of l. On integration the forms for dP_n/dr are again algebraic.

I have carried out the detailed calculation for $n = 6$. It will suffice to record the principal results. For the values of

$$-2^6 l^6 \frac{d^4}{dr^4}\left(\frac{1}{r}\frac{dP_6}{dr}\right)$$

we find for the various ranges:

$$(r < 2l),\ -20\ ;\quad (2l < r < 4l),\ +10\ ;$$
$$(4l < r < 6l),\ -2\ ;\quad (6l < r),\ 0.$$

And on integration for

$$-2^6 l^6 \left(\frac{1}{r}\frac{dP_6}{dr}\right),\qquad\ldots\ldots\ldots\ldots\ldots\ldots\ldots(64)$$

$$(0-2l)\qquad -\frac{5r^4}{6} + 4lr^3 - 16l^3 r,$$

$$(2l-4l)\qquad +\frac{5r^4}{12} - 6lr^3 + 30l^2 r^2 - 56l^3 r + 20l^4,$$

$$(4l-6l)\dagger\ -\frac{r^4}{12} + 2lr^3 - 18l^2 r^2 + 72l^3 r - 108l^4,$$

$$(r > 6l)\qquad 0.$$

[* There are, however, no discontinuities in the value of $\frac{d^{n-3}}{dr^{n-3}}\left(\frac{1}{r}\frac{dP_n}{dr}\right)$, since the integrals $\int_0^\infty \frac{dx}{x^2}\sin^2\frac{(r-nl)x}{2}$, $\int_0^\infty \frac{dx}{x^2}\sin^2\frac{\{r-(n-2)l\}x}{2}$, etc., which appear in the result when n is odd, are continuous for all values of r (cf. the solution for $\frac{dP_3}{dr}$ on p. 621).

† The result for $(4l-6l)$ may be written $-\frac{1}{12}(6l-r)^4$. And in general, when $(n-2)l < r < nl$, we find that $2^n l^n \left(\frac{1}{r}\frac{dP_n}{dr}\right) = \frac{2}{(n-2)!}(nl-r)^{n-2}$, whether n be even or odd. W. F. S.]

We may now seek the form approximated to when n is very great. Setting for brevity $l = 1$ in (59), we have

$$\log \left(\frac{\sin x}{x}\right)^n = n \left\{ -\frac{x^2}{6} + h_4 x^4 + h_6 x^6 + \ldots \right\},$$

where

$$h_4 = -\frac{1}{180}, \quad h_6 = -\frac{1}{35.81}, \quad \ldots\ldots\ldots\ldots\ldots(65)$$

and

$$\left(\frac{\sin x}{x}\right)^n = e^{-nx^2/6} \left\{ 1 + n h_4 x^4 + n h_6 x^6 + \tfrac{1}{2} n^2 h_4^2 x^8 + \ldots \right\},$$

so that

$$\frac{1}{r}\frac{dP_n}{dr} = \frac{2}{\pi} \int_0^\infty x\, dx \sin rx\, e^{-nx^2/6} \left\{ 1 + n h_4 x^4 + n h_6 x^6 + \tfrac{1}{2} n^2 h_4^2 x^8 + \ldots \right\}. \ldots(66)$$

The expression for the principal term is a known definite integral, and we obtain for it

$$\frac{dP_n}{dr} = \frac{3\sqrt{6}\,.\,r^2}{\sqrt{\pi}\,.\,n^{\frac{3}{2}}} e^{-3r^2/2n}, \quad \ldots\ldots\ldots\ldots\ldots(67)$$

which may be regarded as the approximate value when n is very large. To restore l, we have merely to write r/l for r throughout.

In pursuing the approximation we have to consider the relative order of the various terms. Taking nx^2 as standard, so that x^2 is regarded as of the order $1/n$, nx^3 is of order n^{-3} and is omitted. But n^2x^8 is of order n^{-2} and is retained. The terms written down in (66) thus suffice for an approximation to the order n^{-2} inclusive.

The evaluation of the auxiliary terms in (66) can be effected by differentiating the principal term with respect to n. Each such differentiation brings in $-x^2/6$ as a factor, and thus four operations suffice for the inclusion of the term containing x^8. We get

$$\frac{dP_n}{dr} = \frac{3\sqrt{6}\,.\,r^2}{\sqrt{\pi}\,.\,l^3} \left[N + n h_4 \,.\, 6^2 \frac{d^2 N}{dn^2} - n h_6 \,.\, 6^3 \frac{d^3 N}{dn^3} + \tfrac{1}{2} n^2 h_4^2 \,.\, 6^4 \frac{d^4 N}{dn^4} \right], \quad \ldots(68)$$

where

$$N = n^{-\frac{3}{2}} e^{-3r^2/2nl^2}. \ldots\ldots\ldots\ldots\ldots\ldots(69)$$

Finally

$$\frac{dP_n}{dr} = \frac{3\sqrt{6}\,.\,r^2 e^{-3r^2/2nl^2}}{\sqrt{\pi}\,.\,l^3 \,.\, n^{\frac{3}{2}}} \left\{ 1 - \frac{3}{20n}\left(5 - \frac{10r^2}{nl^2} + \frac{3r^4}{n^2 l^4} \right) \right.$$

$$\left. + \frac{1}{40 n^2}\left(\frac{29}{4} - \frac{69 r^2}{nl^2} + \frac{981 r^4}{10 n^2 l^4} - \frac{1341 r^6}{35 n^3 l^6} + \frac{81 r^8}{20 n^4 l^8} \right) \right\}. \ldots(70)$$

Here $dP_n/dr\,.\,dr$ is the chance that the resultant of a large number n of flights shall lie between r and $r + dr$. In Pearson's notation,

$$4\pi r^2 \phi_n = dP_n/dr.$$

The maximum value of the principal term (67) occurs when $r/l = \sqrt{(2n/3)}$.

It is some check upon the formulæ to compare the exact results for $n = 6$ in (64) with those derived for the case of n great in (70), although with such a moderate value of n no precise agreement could be expected. The following table gives the numerical results for $l\,dP_6/dr$ in the two cases:

TABLE II.

$n = 6$.

r/l	From (64)	From (70)
0......	$\cdot2500\,r^2/l^2$	$\cdot2483\,r^2/l^2$
$\cdot5$...	$\cdot05900$	$\cdot05886$
1......	$\cdot2005$	$\cdot2007$
2......	$\cdot4167$	$\cdot4169$
3......	$\cdot2930$	$\cdot2922$
4......	$\cdot0833$	$\cdot1055$
5......	$\cdot00652$	$\cdot00716$
6......	$\cdot00000$	......

So far as the principal term in (70) is concerned, the maximum value occurs when $r/l = 2$.

It will be seen that the agreement of the two formulæ is in fact very good, so long as r/l does not much exceed $\sqrt{n}$. As the maximum value of r/l for which the true result differs from zero, is approached, the agreement necessarily falls off. Beyond $r/l = n$, when the true value is zero, (70) yields finite, though small, values.

P.S. March 3rd.—In (45) we have the expression for the probability of a resultant (r) when a large number (n) of isoperiodic vibrations are combined, whose representative points are distributed at random along the circumference of a circle of radius l, so that the component amplitudes are all equal. It is of interest to extend the investigation to cover the case of a number of groups in which the amplitudes are different, say a group of p_1 components of amplitude l_1, a group containing p_2 of amplitude l_2, and so on to any number of groups, but always under the restriction that every p is very large. The total number (Σp) may still be denoted by n. The result will be applied to a case where the number of groups is infinite, the representative points of the components being distributed at random over the *area* of a circle of radius L. We start from (31), now taking the form

$$2\pi\phi_n(r^2) = \int_0^\infty x\,dx\,J_0(rx)\,\{J_0(l_1x)\}^{p_1}\{J_0(l_2x)\}^{p_2}\dots \quad\dots\dots\dots(71)$$

The derivation of the limiting form proceeds as before, where only one l was considered. Writing $s_1 = \tfrac{1}{2}p_1l_1^2$, $s_2 = \tfrac{1}{2}p_2l_2^2$, etc., we have

$$\log\left[\{J_0(l_1x)\}^{p_1}\{J_0(l_2x)\}^{p_2}\dots\right] = -\frac{x^2}{2}\Sigma(s) - \frac{x^4}{16}\Sigma\left(\frac{s^2}{p}\right) - \frac{x^6}{72}\Sigma\left(\frac{s^3}{p^2}\right),$$

and thus

$$2\pi\phi_n(r^2) = \int_0^\infty x\,dx\,J_0(rx)\,e^{-\frac{1}{2}x^2\Sigma(s)}\left[1 - \frac{x^4}{16}\,\Sigma\left(\frac{s^2}{p}\right)\right.$$
$$\left. - \frac{x^6}{72}\,\Sigma\left(\frac{s^3}{p^2}\right) + \frac{x^8}{512}\left\{\Sigma\left(\frac{s^2}{p}\right)\right\}^2\right]. \quad\ldots(72)$$

As before, the leading term on the right is

$$I_1 = \frac{1}{\Sigma(s)}\,e^{-\frac{1}{2}r^2/\Sigma(s)}, \quad\ldots\ldots\ldots\ldots\ldots\ldots\ldots(73)$$

and the other integrals can be derived from it by differentiations with respect to $\Sigma(s)$.   So far as the first two terms inclusive, we find

$$2\pi\phi_n(r^2) = \frac{e^{-\frac{1}{2}r^2/\Sigma(s)}}{\Sigma(s)}\left\{1 - \frac{\Sigma(s^2/p)}{4}\left(\frac{2}{\{\Sigma(s)\}^2} - \frac{2r^2}{\{\Sigma(s)\}^3} + \frac{r^4}{4\{\Sigma(s)\}^4}\right)\right\}, \quad\ldots(74)$$

from which we may fall back upon (45) by dropping the Σ and making $p = n$. In general $\Sigma(p) = n$.   The approximation could be pursued.

Let us now suppose that the representative points are distributed over the area of a circle of radius L, all infinitesimal equal areas being equally probable.   Of the total n the number (p) which fall between l and $l + dl$ should be $n\,.\,(2l\,dl/L^2)$, and thus

$$\Sigma(s) = \tfrac{1}{2}\Sigma(pl^2) = \frac{n}{L^2}\int_0^L l^3\,dl = \frac{nL^2}{4}, \quad\ldots\ldots\ldots\ldots(75)$$

$$\Sigma(s^2/p) = \tfrac{1}{4}\Sigma(pl^4) = \frac{n}{2L^2}\int_0^L l^5\,dl = \frac{nL^4}{12}. \quad\ldots\ldots\ldots\ldots(76)$$

Introducing these values in (74), we get

$$2\pi\phi_n(r^2) = \frac{4e^{-2r^2/nL^2}}{nL^2}\left\{1 - \frac{2}{3n}\left(1 - \frac{4r^2}{nL^2} + \frac{2r^4}{n^2L^4}\right)\right\}. \quad\ldots\ldots\ldots(77)^*$$

A similar extension may be made in the problem where the component vectors are drawn in three dimensions.

---

* The applicability of the second term (in $1/n$) to the case of an entirely random distribution over the area of the circle L is not over secure.

442.

ON THE RESULTANT OF A NUMBER OF UNIT VIBRATIONS, WHOSE PHASES ARE AT RANDOM OVER A RANGE NOT LIMITED TO AN INTEGRAL NUMBER OF PERIODS.

[*Philosophical Magazine*, Vol. XXXVII. pp. 498—515, 1919.]

A NUMBER (n) of points is distributed at random on a straight line of length a. When n is very great, the centre of gravity of the points tends to coincidence with the middle point of the line, which is taken as origin of coordinates. What is the probability that the error of position, that is its deviation from the origin, lies between x and $x + dx$?

Divide the length a into a large odd number $(2s + 1)$ of parts, each equal to b. The number of points to be expected on each b is nb/a. This expectation would be fulfilled in the mean of a large number of independent trials, but in a single trial it is subject to error. If the actual number be $nb/a + \xi$, the chance that ξ lies between ξ and $\xi + d\xi$ is by Bernoulli's theorem

$$\frac{d\xi}{\sqrt{(2\pi nb/a)}} e^{-a\xi^2/2nb}, \qquad \dots\dots\dots\dots\dots\dots(1)$$

in which it is assumed that while b/a is very small, nb/a is nevertheless very great*. In the language of the Theory of Errors, the modulus, proportional to "probable error," is $\sqrt{(2nb/a)}$.

The points which fall on any small part b may be treated as acting at the middle of the part. For instance, those which fall on the part which includes the origin are supposed to act at the origin and so make no contribution to the sum of the moments; while on other parts the moment is proportional to the distance between the middle of the part and the origin. Thus if

$$\xi_{-s}, \ \xi_{-s+1}, \ \xi_{-s+2}, \ \dots \ \xi_{-1}, \ \xi_0, \ \xi_1, \ \dots \ \xi_s$$

be the values of the various ξ's, the coordinate x of the centre of gravity is given by

$$x = \frac{b(\xi_1 - \xi_{-1}) + 2b(\xi_2 - \xi_{-2}) + \dots + sb(\xi_s - \xi_{-s})}{n + \xi_{-s} + \xi_{-s+1} + \dots + \xi_s} . \qquad \dots\dots\dots(2)$$

* Compare *Phil. Mag.* Vol. XLVII. p. 246 (1899); *Scientific Papers*, Vol. IV. p. 370.

If the whole number of the points be n exactly, the sum of the ξ's in the denominator of (2) must vanish exactly; but if we assume this beforehand, the various ξ's are not independent, as is required by the rules of the Theory of Errors. We may evade the difficulty by supposing the value of ξ on any part to be the result of an *independent* distribution of n points over the whole length. The total of the ξ's is then not necessarily zero, but if we select those cases in which the total is zero, or nearly enough zero, the original requirement is fulfilled. In point of fact no selection is required, inasmuch as the probable error of the sum of ξ's is $\sqrt{(2s+1)}$ times the probable error of each and therefore proportional to $\sqrt{(2s+1)} \cdot \sqrt{(2nb/a)}$, or $\sqrt{(2n)}$, so that no error of which there is a finite probability is comparable with n. We may accordingly take (2) in the simplified form

$$x = \frac{b}{n} \{\xi_1 - \xi_{-1} + 2(\xi_2 - \xi_{-2}) + \ldots + s(\xi_s - \xi_{-s})\}; \quad \ldots\ldots\ldots\ldots(3)$$

and the (modulus)² for the composite error x is given by

$$\frac{\text{Mod}^2 x}{\text{Mod}^2 \xi} = \frac{2b^2}{n^2}(1^2 + 2^2 + 3^2 + \ldots + s^2).$$

For our purpose the sum of the series may be identified with $\int_0^s s^2 ds$, or $s^3/3$, or if we prefer it, $(2s+1)^3/24$, that is $a^3/24b^3$, and thus

$$\text{Mod}^2 \text{ for } x = a^2/6n, \quad \ldots\ldots\ldots\ldots\ldots\ldots(4)$$

s, as well as n, being regarded as infinitely great.

The probability of an error between x and $x + dx$ in the position of the centre of gravity of the n points is accordingly

$$\sqrt{\left(\frac{6n}{\pi}\right)} e^{-6nx^2/a^2} dx/a, \quad \ldots\ldots\ldots\ldots\ldots(5)$$

showing in what manner the probability of a finite x becomes infinitely small as n increases without limit.

The method hitherto employed requires that the total number (n) of points be very great. It is of interest also to inquire what are the various probabilities when n is small or moderate. In dealing with this problem it seems more convenient to reckon the distances from one end of the line a, and to calculate in the first instance the chances for the *sum* (σ) of the distances. We take $\phi_n(\sigma) \, d\sigma/a$ to represent the chance that for n points this sum lies between σ and $\sigma + d\sigma$, and we commence with a sequence formula connecting ϕ_{n+1} with ϕ_n. If for the moment we suppose ϕ_n known and consider the inclusion of an additional point, we see that

$$\phi_{n+1}(\sigma) = \int_{\sigma-a}^{\sigma} \phi_n(\sigma) \, d\sigma/a. \quad \ldots\ldots\ldots\ldots\ldots(6)$$

By means of (6) the various functions may be built up in order.

We start from $\phi_1(\sigma)$. This is zero, unless $0 < \sigma < a$, and then is unity. Hence between 0 and a

$$\phi_2(\sigma) = \int_0^\sigma \phi_1(\sigma)\, d\sigma/a = \sigma/a.$$

If σ lies between a and $2a$,

$$\phi_2(\sigma) = \int_{\sigma-a}^a \phi_1(\sigma)\, d\sigma/a = \frac{2a - \sigma}{a}.$$

Thus

$$\left. \begin{aligned} &\phi_2(\sigma) = 0, \quad (\sigma < 0); \quad \phi_2(\sigma) = \sigma/a, \quad (0 < \sigma < a); \\ &\phi_2(\sigma) = (2a - \sigma)/a, \quad (a < \sigma < 2a); \quad \phi_2(\sigma) = 0, \quad (2a < \sigma) \end{aligned} \right\} , \quad \ldots(7)$$

by which ϕ_2 is completely determined; and it will be seen that there is no breach of continuity in the values of ϕ_2 itself at the critical places. These values are symmetrical on the two sides of $\sigma = a$, and can be represented on a diagram by two straight lines passing through $\sigma = 0$ and $\sigma = 2a$, and meeting at $\sigma = a$. (See Fig. 1.)

In like manner we can deduce ϕ_3 from ϕ_2. If $\sigma < 0$, $\phi_3 = 0$, and indeed generally $\phi_n = 0$. If $0 < \sigma < a$,

$$\phi_3(\sigma) = \int_0^\sigma \frac{\sigma\, d\sigma}{a^2} = \frac{\sigma^2}{2a^2}.$$

If $a < \sigma < 2a$,

$$\phi_3(\sigma) = \int_{\sigma-a}^a \frac{\sigma\, d\sigma}{a^2} + \int_a^\sigma \frac{2a - \sigma}{a^2}\, d\sigma = (-\tfrac{3}{2}a^2 + 3a\sigma - \sigma^2)/a^2.$$

From the symmetry it follows that when $2a < \sigma < 3a$,

$$\phi_3(\sigma) = (3a - \sigma)^2/2a^2.$$

When $\qquad\qquad \sigma > 3a, \quad \phi_3(\sigma) = 0.$

It may be remarked that in this case not only is ϕ_3 continuous, but also the first derivative ϕ_3'. The representative curves for all three portions are parabolic. The maximum of ϕ_3, occurring at $\sigma = 3a/2$, is $\tfrac{3}{4}$.

These problems might also be attacked in another and perhaps more direct manner by expressing the probabilities as multiple definite integrals. Thus in the case of two points the chance of distances x and y from the chosen end is $dx\,dy/a^2$, and what we require is the integral of this taken between the proper limits. If we treat x and y as rectangular coordinates of a point lying within the square whose side is a, the probability we seek is represented by the length of the line within the square which is drawn perpendicular to the diagonal through the origin, σ itself corresponding to the position of the line as measured along the diagonal*.

For three points we have to consider a *cube* of side a, when the chance is represented in like manner by the *area* within the cube of a plane drawn perpendicularly to the diagonal through the origin. At first, that is near the

[* $\sigma = \sqrt{2} \times$ shortest distance of the line from the origin. For the cube (next paragraph of text) $\sigma = \sqrt{3} \times$ shortest distance of the area from the origin. W. F. S.]

origin, the area is triangular and increases as σ^2; afterwards it becomes hexagonal, and after passing through the form of a regular hexagon, when its area is a maximum, returns backwards through the same phases.

The calculations by the sequence formula present no difficulty of principle. When $n = 4$, I find

$$(0 < \sigma < a), \quad \phi_4(\sigma) = \sigma^3/6a^3;$$
$$(a < \sigma < 2a), \quad \phi_4(\sigma) = \{\sigma^3 - 4(\sigma - a)^3\}/6a^3;$$

when $2a < \sigma < 4a$, the above values are repeated symmetrically. In this case there is no discontinuity either in ϕ_4, or ϕ_4', or ϕ_4''. When $\sigma = 2a$, that is in the middle of the range,

$$\phi_4 = \tfrac{2}{3}, \quad \phi_4' = 0.$$

The calculations might be pursued to higher values of n without much trouble. In all cases there is symmetry with respect to the middle of the range. The functions ϕ_n are algebraic and rise in degree by a unit at each step. At the beginning of the range $\phi_{n+1}(\sigma) = (\sigma/a)^n/n!$, so that the contact at both ends of the representative curves with the line of abscissæ becomes of high order.

Again, since σ must lie somewhere between 0 and na, we must have

$$\int_0^{na} \phi_n(\sigma) \, d\sigma/a = 1; \quad \ldots\ldots\ldots\ldots\ldots\ldots(8)$$

from the above expressions we may test this in the cases of $n = 2, 3, 4$.

A plot of the curves for these cases is given in Fig. 1. The ordinate represents $\phi(\sigma)$ and the abscissa represents σ itself with a taken as unity, so that the area of each curve is unity.

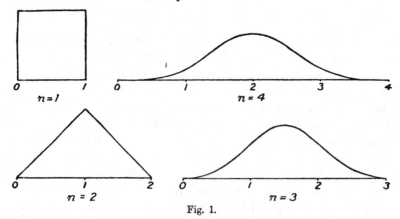

Fig. 1.

In order to pass from these curves in which σ is the *sum* of the distances from one end to the representative curves for the *mean* distance, which must lie between 0 and a, we have merely to reduce the scale of the abscissæ in the ratio $n : 1$, and to increase the scale of the ordinates in the same ratio, so that the area is preserved. For instance, when $n = 4$, the middle ordinate will be increased from $\tfrac{2}{3}$ to $\tfrac{8}{3}$.

The sequence formula (6) serves well enough for the derivation of the facility curves appropriate to moderate values of n, but it does not lend itself readily to examination of the passage towards the final form when n is great. This purpose is better attained by an adaptation of a remarkable method due to Laplace[*], and employed by him and by Airy[†] for the derivation of the usual exponential formula for the facility of error. Here again it will be the *sum* of the distances of the points, now reckoned from the middle of the line, that we consider in the first instance.

The distances, instead of being continuously distributed, are supposed to be limited to definite values, all equally probable,

$$- sb, \; (- s + 1)\, b, \; (- s + 2)\, b, \; \ldots - b, \; 0, \; b, \; 2b, \; \ldots sb,$$

where $2sb = a$, and ultimately s will be made infinite. The question is— What is the chance that the sum of the distances of n points shall be equal to lb, where l is a positive or negative integer? On examination it appears that the combination follows the same laws "as the addition of indices in the successive multiplications of the polynomial

$$e^{-is\theta} + e^{-i(s-1)\theta} + e^{-i(s-2)\theta} + \ldots + e^{i(s-2)\theta} + e^{i(s-1)\theta} + e^{is\theta}$$

by itself, supposing the operation repeated $n - 1$ times. And therefore the number of combinations required will be the coefficient of $e^{il\theta}$ (which is also the same as the coefficient of $e^{-il\theta}$) in the expansion of

$$\{e^{-is\theta} + e^{-i(s-1)\theta} + \ldots + e^{i(s-1)\theta} + e^{is\theta}\}^n."$$

"The number of combinations required is therefore the same as the term independent of θ in the expansion of

$$\tfrac{1}{2}\left(e^{il\theta} + e^{-il\theta}\right)\{e^{-is\theta} + e^{-i(s-1)\theta} + \ldots + e^{i(s-1)\theta} + e^{is\theta}\}^n,$$

or the same as the term independent of θ," when

$$\cos l\theta \left\{1 + 2\cos\theta + 2\cos 2\theta + \ldots + 2\cos s\theta\right\}^n$$

is expanded and arranged according to cosines of multiples of θ. By summing the series and application of Fourier's theorem this term is found to be

$$\frac{1}{\pi}\int_0^\pi \cos l\theta \left\{\frac{\sin \tfrac{1}{2}(2s+1)\,\theta}{\sin \tfrac{1}{2}\theta}\right\}^n d\theta. \quad\ldots\ldots\ldots\ldots\ldots(9)$$

This is the number of combinations which gives rise to a sum equal to l, and in order to obtain the probability of l it must be divided by the whole number of combinations equally probable, that is $(2s + 1)^n$. What we have to consider is accordingly the value of

$$\frac{1}{\pi(2s+1)^n}\int_0^\pi \cos l\theta \left\{\frac{\sin \tfrac{1}{2}(2s+1)\,\theta}{\sin \tfrac{1}{2}\theta}\right\}^n d\theta. \quad\ldots\ldots\ldots(10)$$

---

[*] See Todhunter's *History of the Theory of Probability*, p. 521.

[†] *Theory of Errors of Observations*, Macmillan, 1861, p. 8. In a comparison of the present notation with that of Laplace and Airy, the symbols n and s will be seen to be interchanged.

In their discussion, Laplace and Airy regard both n and s as infinite. Here it is proposed to make s infinite, so as to attain a continuous distribution of the points, but without limitation upon the value of n, which may be any integer. If, as before, σ denote the sum of the distances,

$$\sigma = lb = la/2s.$$

When s is very great, $\sin s\theta$ alternates with great rapidity, so that the integral comes to depend upon that part of the range where θ is very small. We may then replace $\sin \frac{1}{2}\theta$ by $\frac{1}{2}\theta$, and taking $\psi = \theta s$, we find

$$\frac{1}{s\pi} \int_0^\infty \cos \frac{2\sigma\psi}{a} \frac{\sin^n \psi}{\psi^n} d\psi \quad \dots\dots\dots(11)$$

as the equivalent of (10) when s becomes infinite. This is the probability which attaches to a single integral value of l, or to a change $d\sigma$, where $d\sigma = a/2s$. Thus the probability that σ lies between σ and $\sigma + d\sigma$ may be written

$$\frac{2d\sigma}{a\pi} \int_0^\infty \cos \frac{2\sigma\psi}{a} \frac{\sin^n \psi}{\psi^n} d\psi, \quad \dots\dots\dots(12)$$

which is the required result for a continuous distribution and is applicable to any value of n. In our former notation,

$$\phi_n(\sigma) = \frac{2}{\pi} \int_0^\infty \cos \frac{2\sigma\psi}{a} \frac{\sin^n \psi}{\psi^n} d\psi, \quad \dots\dots\dots(13)$$

in which, however, σ now represents the sum of the distances from the *centre* of the line, instead of from one *end* of it.

If $n = 1$, (13) reduces to

$$\phi_1(\sigma) = \frac{1}{\pi} \int_0^\infty \frac{\sin(1 + 2\sigma/a)\psi + \sin(1 - 2\sigma/a)\psi}{\psi} d\psi,$$

which is unity when σ lies between $\pm \frac{1}{2}a$, but otherwise vanishes.

Again, if $n = 2$, we find that $\phi_2'(\sigma) = \pm 1/a$, if σ lies between $\pm a$, and otherwise vanishes, and so on.

More generally, the sequence formula may be deduced from (13), but to obtain it in the original form (6), where the distances are measured from the end of the line, we must write $\sigma - \frac{1}{2}na$ for σ in (13). Then we have

$$\frac{2}{\pi} \int_0^\infty \int_{\sigma-a}^\sigma \cos \frac{2\psi}{a} (\sigma - \frac{1}{2}na) \cdot \frac{\sin^n \psi}{\psi^n} d\psi \, d\sigma/a,$$

in which

$$\int_{\sigma-a}^\sigma \cos \frac{2\psi}{a} (\sigma - \frac{1}{2}na) \, d\sigma = a \cos \frac{2\psi}{a} \left(\sigma - \frac{n+1}{2}a \right) \cdot \frac{\sin \psi}{\psi},$$

so that (6) is verified.

We may now examine the form assumed by ϕ_n in (13), when n is very large. The process is almost the same as that followed in a recent paper*. By taking logarithms we find

$$\frac{\sin^n \psi}{\psi^n} = e^{-n\psi^2/6} \{1 + nh_4 \psi^4 + nh_6 \psi^6 + \tfrac{1}{2} n^2 h_4^2 \psi^8\}, \quad \ldots\ldots\ldots(14)$$

where
$$h_4 = -\frac{1}{180}, \quad h_6 = -\frac{1}{35.81}. \quad \ldots\ldots\ldots\ldots\ldots(15)$$

Retaining for the moment only the leading term, we get

$$\phi_n(\sigma)\, d\sigma/a = \frac{d\sigma}{a} \frac{2}{\pi} \int_0^\infty \cos(2\sigma\psi/a)\, e^{-n\psi^2/6} d\psi$$

$$= \sqrt{(6/n\pi)}\, e^{-6\sigma^2/na^2} d\sigma/a. \quad \ldots\ldots\ldots\ldots(16)$$

In comparing this with (5), we must observe that there x denotes the mean of the distances of which σ is the sum, so that $\sigma = nx$, and thus the two results are in agreement.

If we denote the leading term in ϕ_n by Φ, we obtain from (13) and (14)

$$\phi_n = \Phi + 6^2 n h_4 \frac{d^2\Phi}{dn^2} - 6^3 n h_6 \frac{d^3\Phi}{dn^3} + \tfrac{1}{2} 6^4 n^2 h_4^2 \frac{d^4\Phi}{dn^4}, \quad \ldots\ldots\ldots(17)$$

by means of which the approximation in powers of $1/n$ can be pursued. The terms written would suffice for a result correct to $1/n^2$ inclusive, but we may content ourselves with the term which is of the order $1/n$ in comparison with the leading term. We have

$$\Phi = \sqrt{\left(\frac{6}{n\pi}\right)} e^{-6\sigma^2/na^2},$$

$$\frac{d^2\Phi}{dn^2} = \sqrt{\left(\frac{6}{n^5\pi}\right)} e^{-6\sigma^2/na^2} \left\{\frac{3}{4} - \frac{18\sigma^2}{na^2} + \frac{36\sigma^4}{n^2a^4}\right\},$$

and accordingly

$$\phi_n(\sigma) = \sqrt{\left(\frac{6}{n\pi}\right)} e^{-6\sigma^2/na^2} \left\{1 - \frac{3}{5n}\left(\frac{1}{4} - \frac{6\sigma^2}{na^2} + \frac{12\sigma^4}{n^2a^4}\right)\right\}. \quad \ldots\ldots(18)$$

Here $\phi_n(\sigma)\, d\sigma/a$ expresses the probability that the sum of the distances, measured from the centre of the line, shall lie between σ and $\sigma + d\sigma$.

In terms of the mean (x) of the distances, we should have

$$\sqrt{\left(\frac{6n}{\pi}\right)} e^{-6nx^2/a^2} \left\{1 - \frac{3}{5n}\left(\frac{1}{4} - \frac{6nx^2}{a^2} + \frac{12n^2x^4}{a^4}\right)\right\} dx/a \quad \ldots\ldots(19)$$

as the probability that x shall lie between x and $x + dx$. It should be observed that in virtue of the exponential factor only moderate values of nx^2/a^2 need consideration.

---

* *Phil. Mag.* Vol. xxxvii. p. 344 (1919), equations (65), (66), etc. [This Volume, p. 624.]

As a check upon (19) we may verify that it becomes unity when integrated with respect to x between $-\infty$ and $+\infty$. Starting from

$$\sqrt{\left(\frac{u}{\pi}\right)} \int_{-\infty}^{+\infty} e^{-ux^2} dx = 1,$$

and differentiating with respect to u, we get

$$\frac{2u^{\frac{3}{2}}}{\pi^{\frac{1}{2}}} \int_{-\infty}^{+\infty} x^2 e^{-ux^2} dx = 1,$$

and differentiating again $\dfrac{4u^{\frac{5}{2}}}{3\pi^{\frac{1}{2}}} \displaystyle\int_{-\infty}^{+\infty} x^4 e^{-ux^2} dx = 1.$

Using these integrals in (19) with $a = 1$, $u = 6n$, the required verification follows.

The above verification suggests a remark which may have a somewhat wide application. In many cases we can foresee that a facility function will have a form such as $A e^{-ux^2} dx$, and then, since

$$A \int_{-\infty}^{+\infty} e^{-ux^2} dx = 1,$$

it follows that $A = \sqrt{(u/\pi)}$. According to this law, the expectation of x is zero, but the expectation of x^2 is finite. If we know this latter expectation, we may use the knowledge to determine u. For

$$\text{Expectation of } x^2 = 2\sqrt{(u/\pi)} \int_0^\infty x^2 e^{-ux^2} dx = 1/2u.$$

We may take an example from the problem, just considered, of the position of the centre of gravity of points distributed along a line. If $x_1, x_2, \ldots x_n$ be the coordinates of these points reckoned from the middle and x that of the centre of gravity,

$$\text{Mean } x^2 = \iiint \ldots \frac{dx_1 dx_2 \ldots dx_n}{a^n} \frac{(x_1 + x_2 + \ldots + x_n)^2}{n^2},$$

the integrations being in each case from $-\frac{1}{2}a$ to $+\frac{1}{2}a$. Taking first the integration with respect to x_n, we find that

$$\text{Mean } x^2 = \frac{a^2}{12n^2} + \text{the corresponding expression with } x_n \text{ omitted,}$$

so that $\text{Mean } x^2 = a^2/12n.$

Accordingly $u = 6n/a^2$, as in (19).

A similar argument might be employed for the law of facility of various resultants (r) of n unit vibrations with phases entirely arbitrary, starting with $A e^{-ur^2} r dr$, and assuming that the mean value of r^2 is n.

My principal aim in attacking the above problem was an introduction to the question of random vibrations when the phases of the unit components are distributed along a circular arc not constituting an entire circle. When the circle is complete the solution has already been given *, and the same solution obviously applies when the circular arc covers any number of complete revolutions. All phases of the resultant are then equally probable, and the only question relates to the probability of various amplitudes, or intensities. But if the arc over which the representative points are distributed is not a multiple of 2π, all values of the resultant phase are not equally probable and the question is in many respects more complicated.

There is an obvious relation between the question of the resultant of random vibrations and that of the position of the centre of gravity of the representative points of the components. For if θ denote the phase of a unit component, the intensity of the resultant is given by

$$R^2 = (\Sigma \cos \theta)^2 + (\Sigma \sin \theta)^2.$$

If we suppose unit masses placed at angles θ round the circular arc of radius unity, the rectangular coordinates of the centre of gravity are

$$\bar{x} = (\Sigma \cos \theta)/n, \qquad \bar{y} = (\Sigma \sin \theta)/n;$$

and r, the distance of the centre of gravity from the centre of the circle, is related to R according to $r = R/n$. And in like manner the phase of the resultant corresponds with the angular position of the centre of gravity.

The analogy suggests that a mechanical arrangement might be employed to effect vector addition. A disk, supported after the manner of a compass-card, would carry the loads, and the resulting deflexion from the horizontal would be determined by mirror reading. Perhaps there would be a difficulty in securing adequate delicacy.

To return to the theoretical question, if we suppose the circular arc to be very small, we see that the probability of various phases of the resultant, within the narrow limits imposed, follows the laws determined for the centre of gravity of points distributed at random along a straight line. In this case the amplitude of the resultant is n to a high degree of approximation, n being the number of unit components.

But when the circular arc (α) is so large that $\sin \alpha$ deviates appreciably from α, the question is materially altered. We may, however, frame an argument on the lines followed in equations (1) and (2). Thus with α replacing a and β replacing b, we have for the resultant whose amplitude is R and phase (reckoned from the middle) Θ,

$$R \sin \Theta = \sin \beta . (\xi_1 - \xi_{-1}) + \sin 2\beta . (\xi_2 - \xi_{-2}) + \ldots + \sin s\beta . (\xi_s - \xi_{-s}). \quad \ldots (20)$$

$$R \cos \Theta = \cos \beta . \left(\frac{2n\beta}{\alpha} + \xi_1 + \xi_{-1}\right) + \ldots + \cos s\beta . \left(\frac{2n\beta}{\alpha} + \xi_s + \xi_{-s}\right). \quad \ldots \ldots (21)$$

* *Phil. Mag.* Vol. x. p. 73 (1880); *Scientific Papers*, Vol. i. p. 491. See also *Phil. Mag.* Vol. xxxvii. p. 321 (1919). [This Volume, p. 604.]

Here Θ is a small angle, whose probability is under consideration, but R is in general large and may then be reckoned as if the distribution were uniform. Thus

$$R = \frac{2n}{\alpha} \int_0^{\frac{1}{2}a} \cos \beta \, d\beta = (2n/\alpha) \sin \tfrac{1}{2}\alpha, \quad \ldots\ldots\ldots\ldots(22)$$

and

$$\Theta = \frac{\alpha}{2n \sin \frac{1}{2}\alpha} \{\sin \beta \, (\xi_1 - \xi_{-1}) + \ldots + \sin s\beta \, (\xi_s - \xi_{-s})\}. \quad \ldots\ldots(23)$$

By the rules of the Theory of Errors, we have

$$\frac{\text{Mod}^2 \, \Theta}{\text{Mod}^2 \, \xi} = \frac{\alpha^2}{2n^2 \sin^2 \frac{1}{2}\alpha} \{\sin^2 \beta + \sin^2 2\beta + \ldots + \sin^2 s\beta\}. \quad \ldots\ldots(24)$$

In (24) $\text{Mod}^2 \, \xi = 2n\beta/\alpha$, as before, and the series of $(\sin)^2$ may be replaced by

$$\frac{1}{\beta} \int_0^{\frac{1}{2}a} \sin^2 \beta \, d\beta = \frac{\alpha - \sin \alpha}{4\beta}.$$

Thus

$$\text{Mod}^2 \, \Theta = \frac{\alpha \, (\alpha - \sin \alpha)}{4n \sin^2 \frac{1}{2}\alpha}. \quad \ldots\ldots\ldots\ldots\ldots\ldots(25)$$

If α is small, this reduces to $\alpha^2/6n$, as in (4). If $\alpha = \pi$, that is if the distribution be over a semicircle, we get $\pi^2/4n$. If we make $\alpha = 2\pi$ in (25), the result is indeterminate, since, although $\sin \frac{1}{2}\alpha = 0$, n is infinite. There is a like indeterminateness when α is any multiple of 2π, and this was to be expected. When the arc of distribution consists of entire revolutions, the phase of the resultant is arbitrary. But if the arc differs, even a little, from an integral number of revolutions, there is a definite phase favoured for the resultant, and $\text{Mod}^2 \, \Theta$ diminishes as n increases.

The case where the arc consists of entire revolutions is exceptional also as regards the amplitude, or intensity, of the resultant. As we know, in that case no definite value is approached, however great n may be, and the *expectation* of intensity is n. But if there be a fractional part of a revolution outstanding, the intensity does tend to a definite value, that namely which corresponds to a uniform distribution over the arc, and this value is proportional to the *square* of n.

We may go further and calculate what exactly is the expectation of intensity. We have to evaluate

$$\iiint \ldots \frac{d\theta \, d\theta' \, d\theta''}{\alpha \ \alpha \ \alpha} \ldots [(\cos \theta + \cos \theta' + \cos \theta'' + \ldots)^2 + (\sin \theta + \sin \theta' + \sin \theta'' + \ldots)^2]$$

$$= \frac{1}{\alpha^n} \iiint \ldots d\theta \, d\theta' \, d\theta'' \ldots [n + 2 \cos (\theta - \theta')$$

$$+ 2 \cos (\theta - \theta'') + \ldots + 2 \cos (\theta' - \theta'') + \ldots], \quad \ldots(26)$$

the integration being in each case from $-\frac{1}{2}\alpha$ to $+\frac{1}{2}\alpha$. Taking first the integration with respect to θ, we have

$$\frac{1}{\alpha} \int_{-\frac{1}{2}a}^{+\frac{1}{2}a} d\theta \, [n + 2 \cos (\theta - \theta') + 2 \cos (\theta - \theta'') + \ldots + 2 \cos (\theta' - \theta'') + \ldots]$$

$$= 4\alpha^{-1} \sin \tfrac{1}{2}\alpha \{\cos \theta' + \cos \theta'' + \ldots\} + n + 2 \cos (\theta' - \theta'') + \ldots.$$

On continuing the integration the first part yields finally

$$8 (n-1) \alpha^{-2} \sin^2 \tfrac{1}{2} \alpha;$$

while the remaining parts give the original terms over again with omission of those containing θ. Thus

$$\text{Expectation of intensity} = n + 8\alpha^{-2} \sin^2 \tfrac{1}{2}\alpha \{n-1+n-2+n-3+ \dots +1\}$$

$$= n + 4n (n-1) \alpha^{-2} \sin^2 \tfrac{1}{2} \alpha. \quad\quad\quad\quad\quad\quad (27)$$

If $\alpha = 0$, this becomes n^2, as was to be expected. If $\alpha = 2\pi$, or any multiple of 2π, the expectation is n, as we knew. In general, when α becomes great, so as to include many complete revolutions, the importance of the n^2 part decreases. In (27) n may have any integral value.

In the case of $n = 2$, we may go further and find the expression for the probability of a given amplitude (r) taken always positive, and phase (θ). The amplitude of the components is unity, and the phases, measured from the centre of the arc, θ_1 and θ_2. The probability that these phases shall lie between θ_1 and $\theta_1 + d\theta_1$, θ_2 and $\theta_2 + d\theta_2$ is $\alpha^{-2} d\theta_1 d\theta_2$. We have now to replace the two variables θ_1, θ_2 by r, θ, where

$$r = 2 \cos \tfrac{1}{2} (\theta_1 - \theta_2), \quad \theta = \tfrac{1}{2} (\theta_1 + \theta_2),$$

or

$$\theta_1 = \theta \pm \cos^{-1} (\tfrac{1}{2} r), \quad \theta_2 = \theta \mp \cos^{-1} (\tfrac{1}{2} r),$$

making

$$\frac{d\theta_1}{d\theta} = 1, \quad \frac{d\theta_2}{dr} = \frac{\pm 1}{\sqrt{(4-r^2)}}, \quad \frac{d\theta_1}{dr} = \frac{\mp 1}{\sqrt{(4-r^2)}}, \quad \frac{d\theta_2}{d\theta} = 1.$$

Accordingly

$$\frac{d\theta_1 d\theta_2}{\alpha^2} = \frac{\pm 2 d\theta\, dr}{\alpha^2 \sqrt{(4-r^2)}}. \quad\quad\quad\quad\quad\quad (28)$$

The interchange of θ_1 and θ_2 makes no difference to r and θ, so that we may take

$$\frac{4 d\theta\, dr}{\alpha^2 \sqrt{(4-r^2)}} \quad\quad\quad\quad\quad\quad\quad\quad (29)$$

as the chance that the amplitude of the resultant shall lie between r and $r + dr$ and the phase between θ and $\theta + d\theta$. In (29) α is supposed not to exceed 2π.

As a check, we may revert to the case where $\alpha = 2\pi$. The limits for θ are then independent of the value of r, and are taken to be $-\pi$ and $+\pi$. And

$$\frac{4 dr}{\sqrt{(4-r^2)}} \int_{-\pi}^{+\pi} \frac{d\theta}{\alpha^2} = \frac{2}{\pi} \frac{dr}{\sqrt{(4-r^2)}} = \frac{2\pi r\, dr}{\pi^2 r \sqrt{(4-r^2)}} \quad \dots\dots (30)$$

represents the chance that r shall lie between r and $r + dr$ independently of what θ may be, in agreement with Pearson's expression*. Integrating again with respect to r, we find

$$\int_0^2 \frac{2 dr}{\pi \sqrt{(4-r^2)}} = 1,$$

as should be, all cases being now covered.

* Compare *Phil. Mag.* Vol. xxxvii. p. 328 (1919), equation (21). [This Volume, p. 610.]

In the general case the limits for r and θ are interdependent. The possible range for θ is from $-\frac{1}{2}\alpha$ to $+\frac{1}{2}\alpha$ $(\alpha < \pi)$, but we require the range when r is prescribed. In virtue of the symmetry it suffices to consider a positive θ, and we begin by supposing α less than π, so that the extreme values of r are $2\cos\frac{1}{2}\alpha$ and 2. We proceed to consider the relations by which the limiting values of r and θ are connected.

For a given (positive) θ less than $\frac{1}{2}\alpha$ the upper limit of r is 2 and the lower limit is $2\cos(\frac{1}{2}\alpha - \theta)$. When $\theta > \frac{1}{2}\alpha$, there are no corresponding values of r. In Fig. 2, where α is taken to be $\frac{1}{2}\pi$, the shaded area gives the possible values of r corresponding to any θ, or conversely the values of θ corresponding to a prescribed r.

In order to find the chances of a given θ, we integrate with respect to r in (29). We find

$$\frac{4d\theta}{\alpha^2} \int_{2\cos(\frac{1}{2}\alpha-\theta)}^{2} \frac{dr}{\sqrt{(4-r^2)}} = \frac{4d\theta}{\alpha^2}(\tfrac{1}{2}\alpha - \theta), \quad \ldots\ldots\ldots\ldots(31)$$

as the chance that θ, if positive, lies between θ and $\theta + d\theta$. If we integrate (31) again with respect to θ between 0 and $\frac{1}{2}\alpha$, we get $\frac{1}{2}$, the correct value, as there is an equal chance of θ being negative.

Again, in order to find the chance of a prescribed r, when θ is free to vary, we have to integrate (29) first with respect to θ. Referring to Fig. 2, we see that when $r < 2\cos(\frac{1}{2}\alpha)$, there are no corresponding values of θ, and that when r lies between $2\cos(\frac{1}{2}\alpha)$ and 2, the limits for θ are 0 and $\frac{1}{2}\alpha - \cos^{-1}(\frac{1}{2}r)$. In the first case there is no possibility of r lying between r and $r + dr$; in the second case the probability is

$$\frac{4dr}{\alpha^2\sqrt{(4-r^2)}}\{\tfrac{1}{2}\alpha - \cos^{-1}(\tfrac{1}{2}r)\}, \quad \ldots\ldots\ldots\ldots\ldots(32)$$

which must be doubled when we admit, as we must, negative values of θ. If we integrate (32) as it stands again, with respect to r, we find the correct value, since

$$\int_{2\cos\frac{1}{2}\alpha}^{2} \frac{4dr}{\alpha^2\sqrt{(4-r^2)}}\{\tfrac{1}{2}\alpha - \cos^{-1}(\tfrac{1}{2}r)\} = \tfrac{1}{2}.$$

We may regard (31) and (32) as the solution of the problem in the case where $\alpha < \pi$.

When $\alpha > \pi$, θ may lie outside the limits $\pm\frac{1}{2}\alpha$ applicable to θ_1 and θ_2, and the question becomes more complicated. It appears that we must distinguish two cases under this head, (i) where $\pi < \alpha < 3\pi/2$, and (ii) where $3\pi/2 < \alpha < 2\pi$.

First for $\pi < \alpha < 3\pi/2$, Fig. 4, where α is supposed to be $5\pi/4$.

From $\theta = 0$ to $\theta = \frac{1}{2}(\alpha - \pi)$, r ranges from 0 to 2. From $\theta = \frac{1}{2}(\alpha - \pi)$ to $\theta = \frac{1}{2}\alpha$, r ranges from $2\cos(\frac{1}{2}\alpha - \theta)$ to 2. At $\theta = \frac{1}{2}\alpha$ the lower and upper

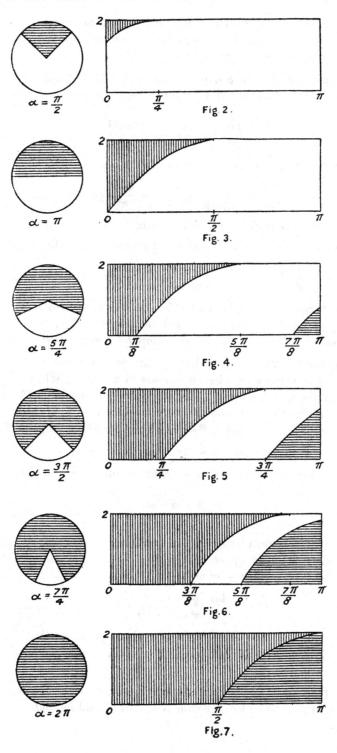

$\alpha = \dfrac{\pi}{2}$

Fig. 2.

$\alpha = \pi$

Fig. 3.

$\alpha = \dfrac{5\pi}{4}$

Fig. 4.

$\alpha = \dfrac{3\pi}{2}$

Fig. 5

$\alpha = \dfrac{7\pi}{4}$

Fig. 6.

$\alpha = 2\pi$

Fig. 7.

limits coincide. From $\theta = \frac{1}{2}\alpha$ to $\theta = 3\pi/2 - \frac{1}{2}\alpha$, there are no corresponding values of r. At the latter limit a zero value of r enters, and from $\theta = 3\pi/2 - \frac{1}{2}\alpha$ to $\theta = \pi$, r ranges from 0 to $2\cos(2\pi - \frac{1}{2}\alpha - \theta)$.

The whole range from $\theta = 0$ to $\theta = \pi$ thus divides itself into four parts. In the first part from $\theta = 0$ to $\theta = \frac{1}{2}(\alpha - \pi)$, we get as the chance of θ from (29)

$$\frac{4d\theta}{\alpha^2}\int_0^2 \frac{dr}{\sqrt{(4 - r^2)}} = \frac{2\pi d\theta}{\alpha^2}. \quad \dots\dots\dots\dots(33)$$

In the second part from $\theta = \frac{1}{2}(\alpha - \pi)$ to $\theta = \frac{1}{2}\alpha$, the chance is

$$\frac{4d\theta}{\alpha^2}\int_{2\cos(\frac{1}{2}\alpha - \theta)}^2 \frac{dr}{\sqrt{(4 - r^2)}} = \frac{4d\theta}{\alpha^2}\left(\frac{1}{2}\alpha - \theta\right). \quad \dots\dots\dots(34)$$

For the third part, from $\theta = \frac{1}{2}\alpha$ to $\theta = 3\pi/2 - \frac{1}{2}\alpha$, there is no possibility.

For the fourth part, from $\theta = 3\pi/2 - \frac{1}{2}\alpha$ to $\theta = \pi$, the chance for θ is

$$\frac{4d\theta}{\alpha^2}\int_0^{2\cos(2\pi - \frac{1}{2}\alpha - \theta)} \frac{dr}{\sqrt{(4 - r^2)}} = \frac{4d\theta}{\alpha^2}\left(\frac{1}{2}\alpha + \theta - \frac{3\pi}{2}\right). \quad \dots\dots\dots(35)$$

If we integrate (33), (34), and (35) over the (positive) ranges to which they apply and add the results, we get the correct value, viz. $\frac{1}{2}$. This part of the question might be treated more simply without introducing r at all.

We have next to consider what in this case, viz. $\pi < \alpha < 3\pi/2$, are the probabilities of various r's when θ is allowed to vary. When r is less than its value at $\theta = \pi$, viz. $2\cos(\pi - \frac{1}{2}\alpha)$, the corresponding range for θ is made up of two parts, the first from $\theta = 0$ to $\theta = \frac{1}{2}\alpha - \cos^{-1}(\frac{1}{2}r)$, and the second from $\theta = 2\pi - \frac{1}{2}\alpha - \cos^{-1}(\frac{1}{2}r)$ to $\theta = \pi$, so that the whole range of θ is

$$\frac{1}{2}\alpha - \cos^{-1}(\frac{1}{2}r) + \pi - \{2\pi - \frac{1}{2}\alpha - \cos^{-1}(\frac{1}{2}r)\} = \alpha - \pi.$$

Thus from $r = 0$ to $r = 2\cos(\pi - \frac{1}{2}\alpha)$ the chance of r lying between r and $r + dr$ is

$$\frac{4dr(\alpha - \pi)}{\alpha^2\sqrt{(4 - r^2)}}. \quad \dots\dots\dots\dots\dots(36)$$

When r lies between $2\cos(\pi - \frac{1}{2}\alpha)$ and 2, the second part disappears and we have only the one range of θ, equal to $\frac{1}{2}\alpha - \cos^{-1}(\frac{1}{2}r)$, so that the chance of r lying between r and $r + dr$ is

$$\frac{4dr\{\frac{1}{2}\alpha - \cos^{-1}(\frac{1}{2}r)\}}{\alpha^2\sqrt{(4 - r^2)}}. \quad \dots\dots\dots\dots(37)$$

Expressions (36) and (37), obtained on the supposition that θ is positive, are to be doubled when we allow for the equally admissible negative values of θ.

When (36), (37), as they stand, are integrated over the ranges of r to which they apply and added, the sum is $\frac{1}{2}$, as it should be under the suppositions made.

It still remains to consider the case where $3\pi/2 < \alpha < 2\pi$. From $\theta = 0$ to $\theta = \frac{1}{2}(\alpha - \pi)$, r (as before) ranges from 0 to 2. From $\theta = \frac{1}{2}(\alpha - \pi)$ to $\theta = \frac{1}{2}(3\pi - \alpha)$, r ranges from $2\cos(\frac{1}{2}\alpha - \theta)$ to 2. At this point (Fig. 6) a second range enters for r. From $\theta = \frac{1}{2}(3\pi - \alpha)$ to $\theta = \frac{1}{2}\alpha$, the first range is, as before, from $2\cos(\frac{1}{2}\alpha - \theta)$ to 2, and the second range is from 0 to $2\cos(2\pi - \frac{1}{2}\alpha - \theta)$. Lastly, from $\theta = \frac{1}{2}\alpha$ to $\theta = \pi$, the first range of r disappears, while the second continues to be from 0 to $2\cos(2\pi - \frac{1}{2}\alpha - \theta)$.

The probabilities of various θ's being positive and lying within specified ranges can be obtained as before. For the range from $\theta = 0$ to $\theta = \frac{1}{2}(\alpha - \pi)$ we get the expression (33), and from $\theta = \frac{1}{2}(\alpha - \pi)$ to $\theta = \frac{1}{2}(3\pi - \alpha)$ we get (34). For the third range from $\theta = \frac{1}{2}(3\pi - \alpha)$ to $\theta = \frac{1}{2}\alpha$, we get

$$\frac{4d\theta}{\alpha^2}\left[\int_{2\cos(\frac{1}{2}\alpha - \theta)}^{2} + \int_{0}^{2\cos(2\pi - \frac{1}{2}\alpha - \theta)}\right]\frac{dr}{\sqrt{(4 - r^2)}} = \frac{4d\theta}{\alpha^2}\left(\alpha - \frac{3\pi}{2}\right); \quad \ldots(38)$$

and from $\theta = \frac{1}{2}\alpha$ to $\theta = \pi$,

$$\frac{4d\theta}{\alpha^2}\int_{0}^{2\cos(2\pi - \frac{1}{2}\alpha - \theta)}\frac{dr}{\sqrt{(4 - r^2)}} = \frac{4d\theta}{\alpha^2}\left(\frac{1}{2}\alpha + \theta - \frac{3\pi}{2}\right). \quad \ldots\ldots\ldots(39)$$

If the integrations with respect to θ are effected over the appropriate ranges and the results added, we get $\frac{1}{2}$, as was to be expected.

Finally, for the probabilities of various r's when θ is left open, we get for r between 0 and $2\cos(\pi - \frac{1}{2}\alpha)$ two ranges for θ, viz. from 0 to $\frac{1}{2}\alpha - \cos^{-1}(\frac{1}{2}r)$, and again, from $\theta = 2\pi - \frac{1}{2}\alpha - \cos^{-1}(\frac{1}{2}r)$ to $\theta = \pi$, making altogether $(\alpha - \pi)$. Thus for these values of r the probability is that expressed in (36).

When r lies between $2\cos(\pi - \frac{1}{2}\alpha)$ and 2, we recover in like manner (37). And as before we may verify the results by showing that when the second integrations are carried out over the appropriate ranges and the integrals added, we recover $\frac{1}{2}$.

It may be remarked that the latter results may be applied to the complete circle by making $\alpha = 2\pi$ (Fig. 7). The second range for r then disappears, and for the whole range now extending for all values of θ from $r = 0$ to $r = 2$ we get

$$\frac{dr}{\pi\sqrt{(4 - r^2)}}, \quad \ldots\ldots\ldots\ldots\ldots\ldots\ldots\ldots\ldots(40)$$

which needs to be doubled in order to take account of negative values of θ.

This completes the investigation for an arbitrary α (less than 2π), when $n = 2$. Since even for the complete circle ($\alpha = 2\pi$) the case $n = 3$ leads to elliptic integrals, there is no encouragement to try an extension to other values of α.

443.

PRESIDENTIAL ADDRESS.

[Proceedings of the Society for Psychical Research, Vol. XXX. pp. 275—290, 1919.]

BEFORE entering upon the matters that I had intended to lay before you, it is fitting that I should refer to the loss we have sustained within the last few days in the death of Sir William Crookes, a former President of the Society during several years from 1896—1899, and a man of world-wide scientific reputation. During his long and active life he made many discoveries in Physics and Chemistry of the first importance. In quite early days his attention was attracted by an unknown and brilliant green line in the spectrum, which he succeeded in tracing to a new element named Thallium, after its appearance. Later he was able so to improve vacua as to open up fresh lines of inquiry with remarkable results in more than one direction. The radiometer, a little instrument in which light, even candle-light, or ordinary day-light, causes the rotation of delicately suspended vanes, presents problems even yet only partially solved. And his discoveries relating to electric discharge in high vacua lie near the foundation of the modern theories of electricity as due to minute charged particles called electrons, capable of separation from ordinary chemical atoms, and of moving with speeds of the order of the speed of light. One is struck not only by the technical skill displayed in experiments more difficult at the time they were made than the younger generation of workers can easily understand, but also by the extraordinary instinct which directed Crookes' choice of subjects. In several cases their importance was hardly realized at the time, and only later became apparent.

I shall have occasion presently to notice in some little detail his early "Notes on Phenomena called Spiritual." It was these that attracted my own attention to the subject. In 1889 he published further "Notes of Séances with D. D. Home" in Vol. VI. of our *Proceedings*. I fancy that he was disappointed with the reception that his views met with, having been sanguine enough to expect that he would obtain the same credence when he wrote on psychical matters as when he was dealing with Physics or Chemistry.

In later years I understand he did not often introduce the subject, but when questioned was firm that he had nothing to retract. One would give much to know whether this attitude is still maintained.

Any hesitation that I may have felt in undertaking the honourable office to which you have called me was largely due to the fact that I have no definite conclusions to announce, and that such experiences as I have had were long ago, and can hardly now carry weight as evidence to anyone but myself. But I have always taken an interest in questions such as those considered by the Society, and I may perhaps as well give a short account of what I have seen, for it will at any rate help to explain my attitude and serve as a foundation for comment.

I may begin with what is now called hypnotism. This is an old story; but many have forgotten, or never realized, the disbelief which was general in the fifties of the last century both on the part of the public and of medical men. As to the former, reference may be made to *Punch**, and as to the latter I suppose there can be no doubt, although of course there were distinguished exceptions. At the present day orthodox medical opinion has so far shifted its ground as to claim for the profession control of what was formerly dismissed as impossible and absurd—certainly a less unreasonable position.

It was some ten or eleven years from the date of *Punch's* cartoon that I witnessed in a friend's rooms at Cambridge an exhibition of the powers of Madame Card. I think eight or ten of us were tried, including myself. We were made to gaze for a time at a "magnetic" disk; afterwards she made passes over our closed eyes, and finally defied us to open them. I and some others experienced no difficulty; and naturally she discarded us and developed her powers over those—about half the sitters—who had failed or found difficulty. Among the latter were personal friends of my own and two well-known University athletes. One was told that he could not give his name, another that he would have to cross the room towards her when she beckoned, and so on. In spite of obvious efforts to resist her influence they had to obey. In conversation afterwards they assured me that they could not help it; and indeed they made such fools of themselves that I had no difficulty in believing them. From that evening I have never felt any doubt as to the possibility of influencing unwilling minds by suggestion; and I have often wished that on other occasions, where dubious phenomena were in question, some of which I shall presently refer to, conviction one way or the other had followed this precedent. I ought to add that, although stories were afloat to that effect, I

---

* Vol. XXIV. p. 120 (1853).—*Lecturer on Electro-Biology.* "Now, Sir! You can't jump over that Stick! Ahem!" *Subject.* "Jump? Eh! Ugh! Lor bless me, Jump? No, I know I can't—never could jump—Ugh!"

[*Thunders of Applause from the Gentlemen in the cane-bottom chairs—(i.e. believers).*

never saw the influence of Madame Card conveyed otherwise than by word or gesture.

After this experience I was not disinclined to believe that what was, or at any rate had recently been, orthodox opinion might be quite wrong, and accordingly became interested in what I heard from friends of the doings of Home and other so-called mediums. Some of the stories could, as it seemed, be explained away only on the supposition of barefaced lying, or more charitably as the result of hallucination, whether self-induced, or due to the suggestion and influence of others. The possibility of the latter view cannot be left out of account, but I have never seen anything to show that it has the remotest application to my own experience or that of the friends with whom I have co-operated.

The interest that I felt was greatly stimulated by the appearance of Sir W. Crookes' "Notes of an Enquiry into the Phenomena called Spiritual during the years 1870—73*." I was acquainted with some of the author's scientific work, and knew that he was a skilful experimenter and likely to be alive to the precautions required in order to guard against sense illusions. Presumably also he would feel the difficulty of accepting conclusions so much out of harmony with ordinary and laboratory experience. If heavy tables in a dining-room can leave the floor, how is it that in the laboratory our balances can be trusted to deal with a tenth of a milligram?

I have lately read over again Sir W. Crookes' article, and I do not wonder at the impression it produced upon me. I am tempted to quote one or two passages against which I find my old pencil marks. Under the heading— The Appearance of Hands, either Self-luminous or Visible by Ordinary Light, he writes, "I have retained one of these hands in my own, firmly resolved not to let it escape. There was no struggle or effort made to get loose, but it gradually seemed to resolve itself into vapour, and faded in that manner from my grasp." I believe that the rationalistic explanation is that the hand was an inflated glove, like a rubber balloon, from which the air gradually leaked away, but I gave Sir W. Crookes credit for being able to retain the rubber.

Another incident of an entirely different character is thus described. "A lady was writing automatically by means of the planchette. I was trying to devise a means of proving that what she wrote was not due to 'unconscious cerebration.' The planchette, as it always does, insisted that, although it was moved by the hand and arm of the lady, the *intelligence* was that of an invisible being who was playing on her brain as on a musical instrument, and thus moving her muscles. I therefore said to this intelligence, 'Can you see the contents of this room?' 'Yes,' wrote the planchette. 'Can you see to read this newspaper?' said I, putting my finger on a copy of the

* *Quarterly Journal of Science*, Jan. 1874.

Times, which was on the table behind me, but without looking at it. 'Yes,' was the reply of the planchette. 'Well,' I said, 'if you can see that, write the word which is now covered by my finger, and I will believe you.' The planchette commenced to move. Slowly and with great difficulty, the word 'however' was written. I turned round, and saw the word 'however' was covered by the tip of my finger."

"I had purposely avoided looking at the newspaper when I tried this experiment, and it was impossible for the lady, had she tried, to have seen any of the printed words, for she was sitting at one table, and the paper was on another table behind, my body intervening."

The two mediums whose names are mentioned in the article, and with whom most of the observations were made, are Home and Miss Fox, afterwards Mrs Jencken. A highly desirable characteristic of Home's mediumship was the unusual opportunity allowed to the sense of sight. Home always objected to darkness at his séances. "Indeed," says Sir William Crookes, "except on two occasions...everything that I have witnessed with him has taken place in the light."

I found (and indeed still find) it difficult to accept what one may call the "knave and fool theory" of these occurrences; but failing that, it would seem to follow that one must admit the possibility of much that contrasts strongly with ordinary experience, and I was naturally anxious to obtain first hand information on which I could form an independent judgment. Home was no longer available, but I was able to obtain the co-operation of Mrs Jencken, who stayed in my country house as guest during two or three visits extending altogether, I suppose, over fourteen days or so. She was accompanied by a nurse and baby, and for a small part of the time by Mr Jencken, who seemed curiously slow to understand that we had to regard him as well as his wife with suspicion, when I explained that we could not attach importance to séances when both were present. It may be well to add that they received nothing beyond the usual courtesy and entertainment due to guests.

The results were upon the whole disappointing, and certainly far short of those described by Sir W. Crookes. Nevertheless, there was a good deal not easy to explain away. Very little of importance occurred in a good light. It is true that at any hour of the day Mrs Jencken was able to get raps upon a door by merely placing her fingers upon it. The listener, hearing them for the first time, felt sure there was someone on the other side, but it was not so. The closest scrutiny revealed no movement of her fingers, but there seemed nothing to exclude the possibility of bone-cracking with the door acting as sounding-board. However, on one or two occasions loud thumps were heard, such as one would hardly like to make with one's knee. With the exception of her fingers Mrs Jencken seemed always to stand quite clear, and the light was good.

On the other hand, during séances the light was usually bad—gas turned very low. But in some other respects the conditions may be considered good. Before commencing, the room was searched and the doors locked. Besides Mrs Jencken, the sitters were usually only Lady Rayleigh and myself. Sometimes a brother or a friend came. We sat close together at a small, but rather heavy, pedestal table; and when anything appeared to be doing we held Mrs Jencken's hands, with a good attempt to control her feet also with ours; but it was impracticable to maintain this full control during all the long time occupied by the séances. In contrast to some other mediums, Mrs Jencken was not observed to fidget or to try to release her limbs.

As I have said, the results were disappointing; but I do not mean that very little happened or that what did happen was always easy to explain. But most of the happenings were trifling, and not such as to preclude the idea of trickery. One's coat-tails would be pulled, paper cutters, etc., would fly about, knocks would shake our chairs, and so on. I do not count messages, usually of no interest, which were spelt out alphabetically by raps that seemed to come from the neighbourhood of the medium's feet. Perhaps what struck us most were lights which on one or two occasions floated about. They were real enough, but rather difficult to locate, though I do not think they were ever more than six or eight feet away from us. Like some of those described by Sir W. Crookes, they might be imitated by phosphorus enclosed in cotton wool; but how Mrs Jencken could manipulate them with her hands and feet held, and it would seem with only her mouth at liberty, is a difficulty.

Another incident hard to explain occurred at the close of a séance after we had all stood up. The table at which we had been sitting gradually tipped over until the circular top nearly touched the floor, and then slowly rose again into the normal position. Mrs Jencken, as well as ourselves, was apparently standing quite clear of it. I have often tried since to make the table perform a similar evolution. Holding the top with both hands, I can make some, though a bad, approximation; but it was impossible that Mrs Jencken could have worked it thus. Possibly something better could be done with the aid of an apparatus of hooks and wires; but Mrs Jencken was a small woman, without much apparent muscular development, and the table for its size is heavy. It must be admitted that the light was poor, but our eyes were then young, and we had been for a long time in the semi-darkness.

In common, I suppose, with most witnesses of such things, I repudiate altogether the idea of hallucination as an explanation. The incidents were almost always unexpected, and our impressions of them agreed. They were either tricks of the nature of conjuring tricks, or else happenings of a kind very remote from ordinary experience.

A discouraging feature was that attempts to improve the conditions usually led to nothing. As an example, I may mention that after writing, supposed to be spirit writing, had appeared, I arranged pencils and paper inside a large glass retort, of which the neck was then hermetically sealed. For safety this was placed in a wooden box, and stood under the table during several séances. The intention was to give opportunity for evidence that would be independent of close watching during the semi-darkness. It is perhaps unnecessary to say that though scribbling appeared on the box, there was nothing inside the retort. Possibly this was too much to expect. I may add that on recently inspecting the retort I find that the opportunity has remained neglected for forty-five years.

During all this time I have been in doubt what interpretation to put upon these experiences. In my judgment the incidents were not good enough, or under good enough conditions, to establish occult influences; but yet I have always felt difficulty in accepting the only alternative explanation. Some circumstances, if of secondary importance, are also worthy of mention. Unlike some other mediums that I have known, Mrs Jencken never tried to divert one's attention, nor did she herself seem to be observant or watching for opportunities. I have often said that on the unfavourable hypothesis her acting was as wonderful as her conjuring. Seldom, or never, during the long hours we were together at meals or séances did she make an intelligent remark. Her interests seemed to be limited to the spirits and her baby.

Mr Jencken is another difficulty. He, an intelligent man, was a spiritualist, and, I have no reason to doubt, an honest one, before he married his wife. Could she have continued to deceive him? It seems almost impossible. He bore eye-witness to the baby—at the age of three months I think it was—taking a pencil and writing a spirit message, of which we saw what purported to be a photograph. If, on the other hand, he had found her out, would he have permitted her to continue her deceptions?

After the death of Home and Mrs Jencken, so-called physical manifestations of a well attested kind seem rather to have fallen into abeyance, except in the case of Eusapia Palladino. Although I attended one or two of her séances at Cambridge and saw a few curious things, other members of the Society have had so much better opportunities that I pass them by. There is no doubt that she practised deception, but that is not the last word.

One of the difficulties which beset our inquiry is the provoking attitude of many people who might render assistance. Some see nothing out of the way in the most marvellous occurrences, and accordingly take no pains over the details of evidence on which everything depends. Others attribute all these things to the devil, and refuse to have anything to say to them. I have sometimes pointed out that if during the long hours of séances we could keep the devil occupied in so comparatively harmless a manner we deserved well of our neighbours.

A real obstacle to a decision arises from the sporadic character of the phenomena, which cannot be reproduced at pleasure and submitted to systematic experimental control. The difficulty is not limited to questions where occult influences may be involved. This is a point which is often misunderstood, and it may be worth while to illustrate it by examples taken from the history of science.

An interesting case is that of meteorites, discussed by Sir L. Fletcher, formerly Keeper of Minerals in the British Museum, from whose official pamphlet (published in 1896) some extracts may be quoted:—" 1. Till the beginning of the present [i.e. 19th] century, the fall of stones from the sky was an event, the actuality of which neither men of science nor the mass of the people could be brought to believe in. Yet such falls have been recorded from the earliest times, and the records have occasionally been received as authentic by a whole nation. In general, however, the witnesses of such an event have been treated with the disrespect usually shown to reporters of the extraordinary, and have been laughed at for their supposed delusions: this is less to be wondered at when we remember that the witnesses of a fall have usually been few in number, unaccustomed to exact observation, frightened by what they both saw and heard, and have had a common tendency towards exaggeration and superstition."

After mention of some early stones, he continues:

" 3. These falls from the sky, when credited at all, have been deemed prodigies or miracles, and the stones have been regarded as objects for reverence and worship. It has even been conjectured that the worship of such stones was the earliest form of idolatry....The Diana of the Ephesians, ' which fell down from Jupiter,' and the image of Venus at Cyprus appear to have been, not statues, but conical or pyramidal stones."

" 5. Three French Academicians, one of whom was the afterwards renowned chemist Lavoisier, presented to the Academy in 1772 a report on the analysis of a stone said to have been seen to fall at Lucé on September 13, 1768. As the identity of lightning with the electric spark had been recently established by Franklin, they were in advance convinced that 'thunder-stones' existed only in the imagination; and never dreaming of the existence of a ' sky-stone' which had no relation to a ' thunder-stone,' they somewhat easily assured both themselves and the Academy that there was nothing unusual in the mineralogical characters of the Lucé specimen, their verdict being that the stone was an ordinary one which had been struck by lightning."

" 6. In 1794 the German philosopher Chladni, famed for his researches into the laws of sound, brought together numerous accounts of the fall of bodies from the sky, and called the attention of the scientific world to the fact that several masses of iron, of which he specially considers two, had in all probability come from outer space to this planet."

In 1802 Edward Howard read a paper before the Royal Society of London giving an account of the comparative results of a chemical and mineralogical investigation of four stones which had fallen in different places. He found from the similarity of their component parts " very strong evidence in favour of the assertion that they had fallen on our globe. They have been found at places very remote from each other, and at periods also sufficiently distant. The mineralogists who have examined them agree that they have no resemblance to mineral substances properly so called, nor have they been described by mineralogical authors." After this quotation from Howard, Fletcher continues:

" 13. This paper aroused much interest in the scientific world, and, though Chladni's theory that such stones come from outer space was still not accepted in France, it was there deemed more worthy of consideration after Poisson (following Laplace) had shown that a body shot from the moon in the direction of the earth, with an initial velocity of 7592 feet a second, would not fall back upon the moon, but would actually, after a journey of sixty-four hours, reach the earth, upon which, neglecting the resistance of the air, it would fall with a velocity of about 31,508 feet a second."

" 14. Whilst the minds of the scientific men of France were in this unsettled condition, there came a report that another shower of stones had fallen, this time...within easy reach of Paris. To settle the matter finally, if possible, the physicist Biot was directed by the Minister of the Interior to inquire into the event on the spot. After a careful examination...Biot was convinced that on Tuesday, April 26, 1803, about 1 p.m., there was a violent *explosion* in the neighbourhood of l'Aigle...that some moments before...a *fire ball* in quick motion was seen...that on the same day many stones fell in the neighbourhood of l'Aigle. Biot estimated the number of the stones at two or three thousand....With the exception of a few little clouds of ordinary character, the sky was quite clear. The exhaustive report of Biot, and the conclusive nature of his proofs, compelled the whole of the scientific world to recognise the fall of stones on the earth from outer space as an undoubted fact."

I commend this history to the notice of those scientific men who are so sure that they understand the character of Nature's operations as to feel justified in rejecting without examination reports of occurrences which seem to conflict with ordinary experience. Every tiro now knows that the stones to be seen in most museums had an origin thought impossible by some of the leading and most instructed men of about a century ago.

Other cases of strange occurrences, the nature or reality of which is, I suppose, still in doubt, are "Globe lightning" and "Will of the wisp." The evidence for globe lightning is fairly substantial, but in the judgment of many scientific men is outweighed by the absence of support in laboratory experience. At one time I was more disposed to believe in it than I am now, in view of the great extension of electrical experimenting during the last thirty years. Kelvin thought it might be explained as an ocular illusion. By a lightning flash the retina is powerfully impressed, it may be excentrically, with the formation of a prolonged positive "spectrum" or image which, as the eye tries to follow it, appears to sail slowly along. Some seconds later, the arrival of the sound of thunder causes a shock, under which the luminous globe disappears and is thought to have burst explosively. I think this explanation, which would save the good faith and to some extent the good sense of the observers, deserves attention.

Then again the Will of the wisp, for which I take it there used to be plenty of evidence. I have been told by the Duke of Argyll—the friend and colleague of Gladstone—that in his youth it was common at Inveraray, but had been less seen latterly, owing, he thought, to drainage operations. Chemists will not readily believe in the spontaneous inflammation of "marsh gas," but I have heard the suggestion made of phosphoric gases arising from the remains of a dead sheep that had got entangled.

The truth is that we are ill equipped for the investigation of phenomena which cannot be reproduced at pleasure under good conditions. And, a clue is often necessary before much progress can be made. Men had every motive for trying to understand malaria. Exposure at night on low ground was known to be bad; and it had even been suggested that mosquito nets served as a protection; but before Pasteur, and indeed for some years after, it seems never to have occurred to any one that the mosquito itself was the vehicle. Sir A. Geikie has remarked that until recent times the study of the lower forms of life was regarded with something like contempt. Verily, the microbes have had their revenge.

But when all this has been said we must not forget that the situation is much worse when it is complicated by the attempts of our neighbours to mislead us, as indeed occasionally happens in other matters of scientific interest where money is involved. Here also the questions before this Society differ from most of those dealt with by scientific men, and may often need a different kind of criticism.

Such criticism it has been the constant aim of the Society to exercise, as must be admitted by all who have studied carefully our published matter. If my words could reach them, I would appeal to serious inquirers to give more attention to the work of this Society, conducted by experienced men and women, including several of a sceptical turn of mind, and not to indulge in hasty conclusions on the basis of reports in the less responsible newspaper press or on the careless gossip of ill-informed acquaintances. Many of our members are quite as much alive to *a priori* difficulties as any outsider can be.

Of late years the published work of the Society has dealt rather with questions of another sort, involving telepathy, whether from living or other intelligences, and some of the most experienced and cautious investigators are of opinion that a case has been made out. Certainly some of the cross-correspondences established are very remarkable. Their evaluation, however, requires close attention and sometimes a background of information, classical and other, not at the disposal of all of us. In this department I often find my estimate of probabilities differing from that of my friends. I have more difficulty than they feel over telepathy between the living, but if I had no doubts there I should feel less difficulty than many do in going further. I think emphasis should be laid upon the fact that the majority of scientific men do not believe in telepathy, or even that it is possible. We are very largely the creatures of our sense-organs. Only those physicists and physiologists who have studied the subject realize what wonderful instruments these are. The eye, the ear, and the nose—even the human nose—are hard to beat, and within their proper range are more sensitive than anything we can make in the laboratory. It is true that with long exposures we can photograph objects in the heavens that the eye cannot detect; but the fairer comparison is between what we can see and what can be photographed in say $\frac{1}{10}$th second—all that the eye requires. These sense-organs, shared with the higher animals, must have taken a long time to build up, and one would suppose that much development in other directions must have been sacrificed or postponed in that interest. Why was not telepathy developed until there could be no question about it? Think of an antelope in danger from a lion about to spring upon him, and gloating over the anticipation of his dinner. The antelope is largely protected by the acuteness of his senses and his high speed when alarmed. But would it not have been simpler if he could know something telepathically of the lion's intention, even if it were no more than vague apprehension warning him to be on the move?

By telepathy is to be understood something more than is implied in the derivation of the word, the conveying of feeling or information otherwise than by use of the senses, or at any rate the known senses. Distance comes into the question mainly because it may exclude their ordinary operation. Some appear to think that all difficulty is obviated by the supposition of an unknown physical agency capable of propagating effects from one brain to another, acting like the transmitter and receiver in wireless telegraphy or telephony. On a physical theory of this kind one must expect a rapid attenuation with distance, not suggested by the records. If distance is an important consideration, one might expect husbands and wives with their heads within two or three feet of one another to share their dreams habitually. But there is a more fundamental objection. Specific information is, and can only be, conveyed in this manner by means of a *code*. People seem to forget that all speaking and writing depend upon a code, and that even the voluntary or involuntary indications of feeling by facial expression or gestures involve something of the same nature. It will hardly be argued that telepathy acts by means of the usual code of common language, as written or spoken.

The conclusion that I draw is that no pains should be spared to establish the reality of telepathy on such sure ground that it must be generally admitted by all serious inquirers. It is quite natural that those who have already reached this position should be more interested in the question of communications from the dead. To my mind telepathy with the dead would present comparatively little difficulty when it is admitted as regards the living. If the apparatus of the senses is not used in one case, why should it be needed in the other?

I do not underrate the difficulties of the investigation. Very special conditions must be satisfied if we are to be independent of the good faith of the persons primarily concerned. The performance of the Zanzigs may be recalled. When there could be no question of confederates, answers respecting objects suddenly exhibited were given with such amazing rapidity that secret codes seemed almost excluded. But when a party, in which I was included, attempted to get a repetition under stricter conditions, there was an almost entire failure. Our requirement was simply that the husband should not speak *after* he had seen the object that was to be described by the wife. But I must add the inevitable qualification. Towards the end of the evening cards were correctly told several times, when we were unable to detect anything that could serve as audible signals.

I have dwelt upon the difficulties besetting the acceptance of telepathy, but I fully recognize that a strong case has been made out for it. I hope that more members of the Society will experiment in this direction. It is work that can be done at home, at odd times, and without the help of mediums, professional or other. Some very interesting experiences of this kind have been recorded by a former President, Prof. Gilbert Murray. With perhaps an excess of caution, he abstained from formulating conclusions that must have seemed to most readers to follow from the facts detailed. I trust we may hear still more from him.

It is hardly necessary to emphasize that in evaluating evidence it is quality rather than quantity with which we are concerned. No one can doubt the existence of apparently trustworthy reports of many occult phenomena. For this there must be a reason, and our object is to find it. But whatever it may be, whether reality of the phenomena, or the stupidity or carelessness or worse of the narrators, a larger sweep is sure to add to the material. However, we may hope that such additions will occasionally afford clues, or at least suggestions for further inquiry. And if the phenomena, or any of them, are really due to supernormal causes, further solid evidence of this will emerge. I feel that I ought to apologize for giving utterance to what must seem platitudes to the more experienced working members of the Society.

Some of the narratives that I have read suggest the possibility of prophecy. This is very difficult ground. But we live in times which are revolutionary in science as well as in politics. Perhaps some of those who accept extreme " relativity" views reducing time to merely one of the dimensions of a four-dimensional manifold, may regard the future as differing from the past no more than north differs from south. But here I am nearly out of my depth, and had better stop.

I fear that my attitude, or want of attitude, will be disappointing to some members of the Society who have out-stripped me on the road to conviction, but this I cannot help. Scientific men should not rush to conclusions, but keep their minds open for such time as may be necessary. And what was at first a policy may become a habit. After forty-five years of hesitation it may require some personal experience of a compelling kind to break the crust. Some of those who know me best think that I ought to be more convinced than I am. Perhaps they are right.

However this may be, I have never felt any doubt as to the importance of the work carried on by the Society over many years, and I speak as one who has examined not a few of the interesting and careful papers that have been published in the *Proceedings*. Several of the founders of the Society were personal friends, and since they have gone the same spirit has guided us. Our goal is the truth, whatever it may turn out to be, and our efforts to attain it should have the sympathy of all, and I would add especially of scientific men.

444.

THE TRAVELLING CYCLONE.

[*Philosophical Magazine*, Vol. XXXVIII. pp. 420—424, 1919.]

[*Note.*—The concluding paragraphs of this paper were dictated by my father only five days before his death. The proofs therefore were not revised by him. The figure was unfortunately lost in the post, and I have redrawn it from the indications given in the text.—RAYLEIGH.]

ONE of the most important questions in meteorology is the constitution of the travelling cyclone, for cyclones usually travel. Sir N. Shaw* says that "a velocity of 20 metres/second [44 miles per hour] for the centre of a cyclonic depression is large but not unknown, a velocity of less than 10 metres/second may be regarded as smaller than the average. A tropical revolving storm usually travels at about 4 metres/second." He treats in detail the comparatively simple case where the motion (relative to the ground) is that of a solid body, whether a simple rotation, or such a rotation combined with a uniform translation; and he draws important conclusions which must find approximate application to travelling cyclones in general. One objection to regarding this case as typical is that, unless the rotating area is infinite, a discontinuity is involved at the distance from the centre where it terminates. A more general treatment is desirable, which shall allow us to suppose a gradual falling off of rotation as the distance from the centre increases; and I propose to take up the general problem in two dimensions, starting from the usual Eulerian equations as referred to uniformly rotating axes†. The density (ρ) is supposed to be constant, and gravity can be disregarded. In the usual notation we have

$$\frac{1}{\rho}\frac{dp}{dx} = \omega^2 x + 2\omega v - \frac{Du}{Dt}, \quad \dots\dots\dots\dots\dots\dots(1)$$

$$\frac{1}{\rho}\frac{dp}{dy} = \omega^2 y - 2\omega u - \frac{Dv}{Dt}, \quad \dots\dots\dots\dots\dots\dots(2)$$

where $\qquad D/Dt = d/dt + u\,d/dx + v\,d/dy. \quad \dots\dots\dots\dots(3)$

---

* *Manual of Meteorology*, Part IV. p. 121, Cambridge, 1919.
† Lamb's *Hydrodynamics*, § 207, 1916.

Here x, y are the coordinates of a point, referred to axes revolving uniformly in the plane xy with angular velocity ω*, u and v are the components of relative velocity of the fluid in the directions of the revolving axes; that is the components of *wind*. We have now to define the motion for which we wish to determine the balancing pressures.

We contemplate a motion (relatively to the ground) of rotation about a centre C, Fig. 1, situated on the axis of x, the successive rings P at distance

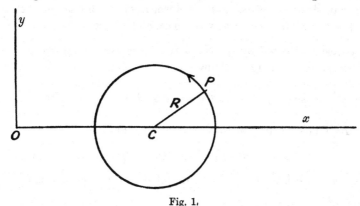

Fig. 1.

R from C revolving with an angular velocity ζ, which may be a function of R. And upon this is to be superposed a uniform velocity of translation U, parallel to x and carrying everything forward. If initially C be at O, the fixed origin, its distance from O along Ox at time t will be Ut. Thus

$$u = U - \zeta y, \quad v = \zeta(x - Ut), \quad\quad\quad\quad\dots\dots\dots(4)$$

ζ being a known function of R, where

$$R^2 = y^2 + (x - Ut)^2 = y^2 + X^2. \quad\quad\quad\dots\dots\dots(5)$$

These equations give u and v in terms of the coordinates and of the time, and the values are to be introduced into (1) and (2). From the manner in which x and t enter (representing a uniform translation of the entire system) it is evident that $d/dt = - U d/dx$. We have

$$\frac{du}{dx} = - \frac{\zeta' X y}{R}, \quad \frac{du}{dy} = - \zeta - \frac{\zeta' y^2}{R},$$

$$\frac{dv}{dx} = \zeta + \frac{\zeta' X^2}{R}, \quad \frac{dv}{dy} = \frac{\zeta' X y}{R},$$

ζ' being written for $d\zeta/dR$; and

$$\frac{Du}{Dt} = \frac{\zeta' U X y}{R} - u \frac{\zeta' X y}{R} - v\left(\zeta + \frac{\zeta' y^2}{R}\right) = - \zeta^2 X,$$

$$\frac{Dv}{Dt} = - U\left(\zeta + \frac{\zeta' X^2}{R}\right) + u\left(\zeta + \frac{\zeta' X^2}{R}\right) + v \frac{\zeta' X y}{R} = - \zeta^2 y.$$

* In the application to a part of the earth's atmosphere, ω is the earth's angular velocity multiplied by the sine of the *latitude*.

Hence

$$\frac{1}{\rho}\frac{dp}{dx} = \omega^2 x + 2\omega\zeta X + \zeta^2 X, \quad \frac{1}{\rho}\frac{dp}{dy} = \omega^2 y - 2\omega(U - \zeta y) + \zeta^2 y, \quad \dots(6)$$

and on integration

$$\frac{p}{\rho} = \tfrac{1}{2}\omega^2(x^2 + y^2) - 2\omega U y + \int(2\omega\zeta + \zeta^2)\,R\,dR. \quad \dots\dots\dots(7)$$

As might have been expected, the last term in (7) is the same function of R as when $U = 0$, but R itself is now a function of U and t.

In the case considered by Sir N. Shaw, ζ is constant and may be removed from under the integral sign. Thus

$$\frac{p}{\rho} = \tfrac{1}{2}\omega^2(x^2 + y^2) - 2\omega U y + (\omega\zeta + \tfrac{1}{2}\zeta^2)\{y^2 + (x - Ut)^2\}. \quad \dots\dots(8)$$

If $U = 0$, R^2 identifies itself with $x^2 + y^2$, and we get

$$p/\rho = \tfrac{1}{2}(\omega + \zeta)^2(x^2 + y^2). \dots\dots\dots\dots\dots\dots(9)$$

A constant as regards x and y, which might be a function of t, may be added in (8) and (9).

We see that if $\omega + \zeta = 0$, that is if the original terrestrial rotation is annulled by the superposed rotation, p is constant, the whole fluid mass being in fact at rest. It was for the purpose of this verification that the terms in ω^2 were retained. We may now omit them as representing a pressure independent of the motion under consideration. In the strictly two-dimensional problem there is a pressure increasing outwards due to "centrifugal force." In the application to the earth's atmosphere, this pressure is balanced by a component of gravity connected with the earth's ellipticity. Thus in Shaw's case we have

$$\frac{p}{\rho} = \text{const.} + (\omega\zeta + \tfrac{1}{2}\zeta^2)\left\{\left(y - \frac{\omega U}{\omega\zeta + \tfrac{1}{2}\zeta^2}\right)^2 + (x - Ut)^2\right\}, \quad \dots\dots(10)$$

showing that the field of pressure, though still circular, is no longer centred at O as when $U = 0$, or even at C, where $x = Ut$, $y = 0$, but is displaced sideways to the point where $x = Ut$, $y = \omega U/(\omega\zeta + \tfrac{1}{2}\zeta^2)$. Shaw calls this the *dynamic centre*; it is the point which is conspicuous on the weather map as the centre of the system of circular isobars.

As a case where the circular motion diminishes to nothing as we go outwards, let us now suppose that $\zeta = Ze^{-R^2/a^2}$, falling off slowly at first but afterwards with great rapidity. We have

$$\int_0^R \zeta R\,dR = \tfrac{1}{2}Za^2(1 - e^{-R^2/a^2}), \quad \int_0^R \zeta^2 R\,dR = \tfrac{1}{4}Z^2 a^2(1 - e^{-2R^2/a^2});$$

and thus from (7)

$$\frac{p}{\rho} = \text{const.} - 2\omega Uy - \tfrac{1}{4}a^2\,(Ze^{-R^2/a^2} + 2\omega)^2, \quad \dots\dots\dots\dots(11)$$

where, as usual, $R^2 = y^2 + (x - Ut)^2$.

May 17th.

The completion of this paper was interrupted by illness.

The two-dimensional solution requires a ceiling, as well as a floor, to take the pressure. In the absence of a ceiling we must introduce gravity, and since in the supposed motion no part of the fluid is vertically accelerated, the third equation of motion gives simply

$$\frac{p}{\rho} = \text{const.} - gz.$$

Thus (10) is altered merely by the addition of the term $-gz$.

I had supposed too that the solution would remain substantially unaltered even though ρ were variable as a function of p. But these conclusions seem to be at variance with those put forward by Dr Jeffreys in the January No. of the *Philosophical Magazine*. I am not able to pursue the comparison at present.

June 25th, 1919.

[The following note was contributed by Sir Joseph Larmor, and was appended to the paper as originally published.

This paper was left incomplete on Lord Rayleigh's decease on June 30. It may therefore be permissible to direct attention to its main conclusion from another aspect, by way of paraphrase. Two questions are involved. If a vortical system can persist at rest, in an atmosphere rotating with the Earth, can it also persist, slightly modified, with a translatory velocity U? And if so, how will the distribution of pressure in it be modified? The equations of fluid motion relative to the ground are (1) and (2); in them the last terms Du/Dt and Dv/Dt express the components of relative acceleration, and these are clearly the centrifugal accelerations $-\zeta^2 X$, $-\zeta^2 y$ in the relative orbits assumed to be circular, as found analytically lower down. On substituting these values, the equations give for δp an exact differential form which is integrated in (7); therefore a modified motion is possible, and the first question is answered in the affirmative, in agreement so far with fact*. The displacement of the pressure-system due to the progressive motion is

---

* The conditions of stability for flow of liquid with varying vorticity had been considered in a series of papers, for which reference may be made to the section Hydrodynamics of the catalogue appended to this volume.

then examined for two special cases by the formulæ (10) and (11), showing also general agreement with fact as regards displacement of the centre of the vortex.   But the value of U is not determined by these considerations, which refer to frictionless fluid.   When viscosity in the fluid is taken into account, the general argument seems to remain applicable; for the velocity of convection U, being uniform, will not modify the viscous stresses.   But, in any case, internal viscosity is negligible in meteorological problems.   It is the friction against land or ocean, introducing turbulence which spreads upward, that disturbs and ultimately destroys the cyclonic system; and the high degree of permanence of the type of motion seems to permit that also to be left out of account.   As remarked in the postscript, the changes of pressure arising from convection involve changes of density, which will modify the motion, but perhaps slightly.   There does not seem to be definite discordance with Dr Jeffreys' detailed discussion.]

445.

PERIODIC PRECIPITATES.

[*Philosophical Magazine*, Vol. XXXVIII. pp. 738—740, 1919.]

[*Note.*—This paper was found in the author's writing-table drawer after his death. It is not dated, but was probably written in 1917. It was no doubt withheld in the hope of making additions.]

I owe my knowledge of this subject, as well as beautiful specimens, to Prof. S. Leduc of Nantes. His work on the Mechanism of Life* gives an account of the history of the discovery and a fairly detailed description of the *modus operandi*. "According to Prof. Quincke of Heidelberg, the first mention of the periodic formation of chemical precipitates must be attributed to Runge in 1885†. Since that time these precipitates have been studied by a number of authors, and particularly by R. Liesegang of Düsseldorf, who in 1907 published a work on the subject, entitled *On Stratification by Diffusion.*" In 1901 and again in 1907 Leduc exhibited preparations showing concentric rings, alternately transparent and opaque, obtained by diffusion of various solutions in a layer of gelatine.

"The following is the best method of demonstrating the phenomenon. A glass lantern slide is carefully cleaned and placed absolutely level. We then take 5 c.c. of a 10 per cent. solution of gelatine and add to it one drop of a concentrated solution of sodium arsenate. This is poured over the glass plate whilst hot, and as soon as it is quite set, but before it can dry, we allow a drop of silver nitrate solution containing a trace of nitric acid to fall on it from a pipette. The drop slowly spreads in the gelatine, and we thus obtain magnificent rings of periodic precipitates of arsenate of silver....The distance between the rings depends on the concentration of the diffusing solution. The greater the fall‡ of concentration, the less is the interval between the rings."

* Translated by W. Deane Butcher, Rebman Limited, Shaftesbury Avenue, London.

[† The year "1885" agrees with the source of the quotation (l. c. p. 67), but it appears that the correct date is "1855." For in the original chronologically arranged historical passage in Prof. Quincke's paper "Über unsichtbare Flüssigkeitsschichten u.s.w." in *Annalen der Physik*, Vierte Folge, Band 7 (1902), pp. 643—647, the first mention of the subject is attributed to F. F. Runge in 1855; in agreement with the translation of this passage given at length in subsequent pages of *The Mechanism of Life* (*see* p. 118).

‡ The words "fall of" had been omitted from the quotation as printed in the original publication in the *Philosophical Magazine*. With this omission, the statement appears still to be valid, and to express more simply the fundamental property of the phenomenon. W. F. S.]

In considering an explanation, the first question which presents itself is why should the precipitate be intermittent at all? I suppose the answer is to be found in the difficulty of precipitation without a nucleus. At a place where the second material (silver nitrate) has only just penetrated, there may be indeed a chemical interchange, but the resultant (silver arsenate) still remains in a kind of solution. Only when further concentration has ensued, can a precipitate in the usual sense be formed, and a visible line of silver arsenate constituted. But this line will not thicken itself far outwards, since the silver arsenate forming a little beyond, as the diffusion progresses, will prefer to diffuse back and deposit itself upon the nucleus already in existence. In this way the space just outside the nucleus becomes denuded of the weaker ingredient (sodium arsenate). This process goes on for a time, but ultimately when the stronger solution has penetrated to a place where a sufficiency of the weaker still remains, a condition of things arises where a new precipitation becomes possible. But between these lines of precipitation there is a clear space. The process then recurs and, as it appears, with much regularity. This view harmonizes with the observed diminution of the linear period as the concentration increases.

We may perhaps carry the matter a little further, considering for simplicity the case where the original boundary is a straight line, the strong solution occupying the whole of the region on one side where x (say) is negative. For each line of precipitation x is constant, and the linear period may be called dx. According to the view taken, the data of the problem involve three concentrations—the two concentrations of the original solutions and that of arsenate of silver at which precipitation occurs without a nucleus. The three concentrations may be reckoned chemically. There are also three corresponding coefficients of diffusion. Let us inquire how the period dx may be expected to depend on these quantities and on the distance x from the boundary at which it occurs. Now dx, being a purely linear quantity, can involve the concentrations only as ratios; otherwise the element of mass would enter into the result uncompensated. In like manner the diffusibilities can be involved only as ratios, or the element of time would enter. And since these ratios are all pure numbers, dx must be proportional to x. In words, the linear period at any place is proportional, *cœteris paribus*, to the distance from the original boundary. In this argument the thickness of the film— another linear quantity—is omitted, as is probably for the most part legitimate. In imagination we may suppose the film to be infinitely thin or, if it be of finite thickness, that the diffusion takes place strictly in one dimension.

The specimens that I have prepared, though inferior to M. Leduc's, show the leading features sufficiently well. I have used the arsenate of silver procedure, and the broadening of the intervals in passing outwards is very evident when the plate is viewed through a Coddington lens.

Hookham's Crystals.

Another remarkable example of fine periodic structure was brought to my notice by Mr George Hookham. In this case double refraction plays an important part and a careful study of the crystals requires the use of a polarizing microscope. I have had the advantage not only of receiving interesting specimens and a sample of one of the solutions employed, but also of witnessing for myself Mr Hookham's procedure.

The active ingredient is copper sulphate; but, as it is desired to obtain a film which is initially amorphous, other ingredients must be added. In the solution given me there is both salycine and sugar. Mr Hookham describes it as consisting of a solution saturated (in the cold) with copper sulphate and salycine, to which is added 3 per cent. of strong syrup. A few drops are placed upon a strip of glass, such as are ordinarily used for microscopic slides, and are spread with the finger. The slide is then warmed over a spirit lamp, when any excess of liquid may be thrown off. By a further application of heat the whole is then dried somewhat rapidly. There is usually immediate formation of crystals at the edges, but throughout a space in the interior the film should be amorphous and nearly invisible. At this stage the amorphous film shows nothing in the polariscope, but in a short time after cooling developments set in and proceed with rapidity. There is much here to excite admiration and perplexity, as in other similar phenomena of crystallization, but the feature in which I am specially interested, viz. the formation of a structure periodic several thousand times in the inch, does not appear to present itself unless the plate is kept warm until crystallization has set in. Mr Hookham mentions a temperature about 30° F. above that of the room. I have usually placed the slides over hot water pipes or on the mantelpiece.

446.

ON RESONANT REFLEXION OF SOUND FROM A PERFORATED WALL.

[Philosophical Magazine, Vol. XXXIX. pp. 225—233, 1920.]

[*Note.*—This paper, written in 1919, was left by the Author ready for press except that the first two pages were missing. The preliminary sentences, taken from a separate rough sheet, were perhaps meant to be expanded.

Prof. Wood* had observed highly coloured effects in the reflexion from a granular film of sodium or potassium, which he attributed to resonance from the cavities of a serrated structure of rod-like crystals.]

THIS investigation was intended to illustrate some points discussed with Prof. R. W. Wood. But it does not seem to have much application to the transverse vibrations of light. Electric resonators could be got from thin conducting rods $\frac{1}{2}\lambda$ long; but it would seem that these must be disposed with their lengths perpendicular to the direction of propagation, not apparently leading to any probable structure.

The case of sound might perhaps be dealt with experimentally with bird-call and sensitive flame. A sort of wire brush would be used.

The investigation follows the same lines as in *Theory of Sound*, 2nd ed. § 351 (1896), where the effect of porosity of walls on the reflecting power for sound is considered. In the complete absence of dissipative influences, what is not transmitted must be reflected, whatever may be the irregularities in the structure of the wall. In the paragraph referred to, the dissipation regarded is that due to gaseous viscosity and heat conduction, both of which causes act with exaggerated power in narrow channels. For the present purpose it seems sufficient to employ a simpler law of dissipation.

Let us conceive an otherwise continuous wall, presenting a flat face at $x=0$, to be perforated by a great number of similar narrow channels, uniformly

---

* [See *Phil. Mag.* July 1919, pp. 98—112, especially p. 111, where a verbal opinion of Lord Rayleigh is quoted that in certain cases the grooves of gratings might possibly act as resonators. The explanation of the absorption of sound by porous bodies such as curtains, given in *Theory of Sound*, second edition, §§ 348—351, dates back to 1883: see *Scientific Papers*, Vol. II. No. 103, pp. 220—5, "On porous bodies in relation to Sound."]

distributed, and bounded by surfaces everywhere perpendicular to the face of the wall. If the channels be sufficiently numerous relatively to the wave-length of vibration, the transition, when sound impinges, from simple plane waves on the outside to the waves of simple form in the interior of the channels occupies a space which is small relatively to the wave-length, and then the connexion between the condition of things outside and inside admits of simple expression.

On the outside, where the dissipation is neglected, the velocity potential (ϕ) of the plane waves, incident and reflected in the plane of xy, at angle θ, is subject to

$$d^2\phi/dt^2 = a^2 (d^2\phi/dx^2 + d^2\phi/dy^2), \quad\dots\dots\dots\dots\dots(1)$$

or if $\phi \propto e^{int}$, where n is real,

$$d^2\phi/dx^2 + d^2\phi/dy^2 + k^2\phi = 0, \quad\dots\dots\dots\dots\dots(2)$$

k being equal to n/a. The solution of (1) appropriate to our purpose is

$$\phi = e^{i(nt + ky\sin\theta)} \{A e^{ikx\cos\theta} + B e^{-ikx\cos\theta}\}, \quad\dots\dots\dots\dots(3)$$

the first term representing the incident wave travelling towards $-x$, and the second the reflected wave. From (3) we obtain for the velocity u parallel to x, and the condensation s, when $x = 0$,

$$u = \frac{d\phi}{dx} = e^{i(nt + ky\sin\theta)} ik \cos\theta (A - B), \quad\dots\dots\dots\dots(4)$$

$$as = -\frac{1}{a}\frac{d\phi}{dt} = -\frac{in}{a} e^{i(nt + ky\sin\theta)} (A + B), \quad\dots\dots\dots(5)$$

so that

$$\frac{u}{as} = \cos\theta \, \frac{B - A}{B + A}. \quad\dots\dots\dots\dots\dots\dots(6)$$

For the motion inside a channel we introduce in (1) on the left a term $hd\phi/dt$, h being positive, to represent the dissipation. Thus, if ϕ be still proportional to e^{int}, we have in place of (2)

$$d^2\phi/dx^2 + d^2\phi/dy^2 + d^2\phi/dz^2 + k'^2\phi = 0, \quad\dots\dots\dots(7)$$

where k'^2 is now complex, being given by

$$k'^2 = k^2 - inh/a^2. \quad\dots\dots\dots\dots\dots\dots(8)$$

If we write $k' = k_1 - ik_2$, where k_1, k_2 are real and positive, we have

$$k_1^2 - k_2^2 = k^2, \quad k_1 k_2 = \tfrac{1}{2} nh/a^2. \quad\dots\dots\dots\dots\dots(9)$$

At a very short distance from the mouth of the channel $d^2\phi/dy^2$, $d^2\phi/dz^2$ in (7) may be neglected, and thus

$$\phi = e^{int} \{A' \cos k'x + B' \sin k'x\}. \quad\dots\dots\dots\dots(10)$$

If the channel be closed at $x = -l$,

$$A' \sin k'l + B' \cos k'l = 0,$$

and we may take
$$\phi = A'' \cos k' (x + l) e^{int}. \qquad (11)$$

From (11) when x is very small,
$$u = d\phi/dx = - k'A'' \sin k'l . e^{int}, \qquad (12)$$
$$as = - a^{-1} d\phi/dt = - ikA'' \cos k'l . e^{int}, \qquad (13)$$

so that
$$\frac{u}{as} = \frac{k'}{ik} \tan k'l. \qquad (14)$$

Now, under the conditions supposed, where the transition from the state of things outside to that inside, at a distance from the mouth large compared with the diameter of a channel, occupies a space which is small compared with the wave-length, we may assume that s is the same in (6) and (14), and that

$$(\sigma + \sigma')u \text{ in } (6) = \sigma u \text{ in } (14),$$

where σ represents the perforated area and σ' the unperforated. Accordingly, if we put $A = 1$, as we may do without loss of generality, the condition to determine B is

$$\frac{B-1}{B+1} = \frac{\sigma}{(\sigma + \sigma') \cos \theta} \frac{k' \tan k'l}{ik}. \qquad (15).$$

If there be no dissipation in the channels, $h = 0$, and $k' = k$. In this case
$$B = \frac{(\sigma + \sigma') \cos \theta \cos kl - i\sigma \sin kl}{(\sigma + \sigma') \cos \theta \cos kl + i\sigma \sin kl}. \qquad (16)$$

Here Mod $B = 1$, or the reflexion is total, as of course it should be. If in (16) $\sigma = 0$, $B = 1$, the wall being unperforated. On the other hand, if $\sigma' = 0$, the partitions between the channels being infinitely thin,

$$B = \frac{\cos \theta \cos kl - i \sin kl}{\cos \theta \cos kl + i \sin kl}. \qquad (17)$$

In the case of perpendicular incidence $\theta = 0$, and
$$B = e^{-2ikl}, \qquad (18)$$
the wall being in effect transferred from $x = 0$ to $x = - l$.

We have now to consider the form assumed when k' is complex. In (15)
$$\left. \begin{aligned} \cos k'l &= \cos k_1 l \cos ik_2 l + \sin k_1 l \sin ik_2 l, \\ \sin k'l &= \sin k_1 l \cos ik_2 l - \cos k_1 l \sin ik_2 l. \end{aligned} \right\} \qquad (19)$$

Before proceeding further it may be worth while to deal with the case where h, and consequently k_2, is very small, but $k_2 l$ so large that vibrations in the channels are sensibly extinguished before the stopped end is reached. In this case
$$\cos ik_2 l = \tfrac{1}{2} e^{k_2 l}, \quad \sin ik_2 l = \tfrac{1}{2} i e^{k_2 l},$$
so that in (19), $\tan k'l = - i$. Also by (9), $k'/k = 1$, and (15) becomes

$$\frac{B-1}{B+1} = - \frac{\sigma}{(\sigma + \sigma') \cos \theta}, \qquad (20)$$

making $B = 0$ when, for example, $\sigma' = 0$, $\cos \theta = 1$. The reflexion may also vanish when the obliquity of incidence is such as to compensate for a finite σ'.

In examining the formula for the general case we shall write for brevity

$$\cos \theta \, (\sigma + \sigma')/\sigma = S, \quad\quad\quad\quad\quad\quad\quad\quad\dots\dots\dots\dots(21)$$

and drop l, so that k_1, k_2, k stand respectively for $k_1 l$, $k_2 l$, kl. This makes no difference to the first of equations (9), while the second becomes

$$k_1 k_2 = \tfrac{1}{2} n h l^2 / a^2. \quad\quad\quad\quad\quad\quad\quad\dots\dots\dots\dots(9 \text{ bis})$$

Thus
$$B = \frac{kS \cos k' - ik' \sin k'}{kS \cos k' + ik' \sin k'}. \quad\quad\quad\dots\dots\dots\dots(22)$$

Separating real and imaginary parts, we find for the numerator of B in (22)

$$\cos k_1 \cos ik_2 \left[kS - \frac{k_1 \tan ik_2}{i} - k_2 \tan k_1 \right.$$

$$\left. + i \left\{ kS \tan k_1 \frac{\tan ik_2}{i} - k_1 \tan k_1 + \frac{k_2 \tan ik_2}{i} \right\} \right]. \quad\dots(23)$$

The denominator of (22) is obtained (with altered sign) by writing $-S$ for S in (23).

In what follows we are concerned with the modulus of B. Leaving out factors common to the numerator and denominator, we may take

$$\text{Mod}^2 \text{ Numerator} = \left\{ kS - \frac{k_1 \tan ik_2}{i} - k_2 \tan k_1 \right\}^2$$

$$+ \left\{ \left(kS \frac{\tan ik_2}{i} - k_1 \right) \tan k_1 + \frac{k_2 \tan ik_2}{i} \right\}^2. \quad\dots(24)$$

The evanescence of B requires that of both the squares in (24), or that

$$kS = \frac{k_1 \tan ik_2}{i} + k_2 \tan k_1 = ik_1 \cot ik_2 - k_2 \cot k_1, \quad\dots\dots\dots(25)$$

or again with elimination of S,

$$ik_1 \,(\tan ik_2 + \cot ik_2) = k_2 \,(\tan k_1 + \cot k_1),$$

whence
$$k_1 \sin 2k_1 + ik_2 \sin 2ik_2 = 0, \quad\quad\quad\dots\dots\dots\dots(26)$$

or in the notation of the hyperbolic sine

$$k_1 \sin 2k_1 = k_2 \sinh 2k_2. \quad\quad\quad\quad\dots\dots\dots\dots(27)$$

If this equation, independent of σ, σ', and $\cos \theta$, can be satisfied, it allows us to find k_1 from an assumed k_2, or conversely, and thence k by means of (9).

The next step is to calculate S by means of one of equations (25). If S, so found, $> \cos \theta$, we may choose σ'/σ so that B shall vanish; but if $S < \cos \theta$, no ratio σ'/σ will serve to annul the reflexion. If the incidence be perpendicular, S must exceed unity. If S were negative, the reflexion would be finite, whatever may be the angle of incidence and the ratio σ'/σ.

It is natural to expect an evanescence of reflexion when the damping is small and the tuning such as to give good resonance. In this case we may suppose k_2 and $\pi - 2k_1{}^*$ to be small, and then (27) gives approximately

$$k_1 = \frac{\pi}{2}\left(1 - \frac{4k_2^2}{\pi^2}\right), \quad k = \surd(k_1^2 - k_2^2) = \frac{\pi}{2}\left(1 - \frac{6k_2^2}{\pi^2}\right).\ldots\ldots(28)$$

By (25)
$$\begin{aligned} kS &= k_1 \tanh k_2 + k_2 \tan k_1 \\ &= k_1 \tanh k_2 + k_2/\tan(2k_2^2/\pi) \\ &= \frac{\pi}{2k_2}(1 + k_2^2 + \ldots), \end{aligned}$$

so that
$$S = k_2^{-1}\left\{1 + \left(1 + \frac{6}{\pi^2}\right)k_2^2 + \ldots\right\}. \ldots\ldots\ldots\ldots(29)$$

Since S is large and positive, the condition for no reflexion can be satisfied by making the perforated area σ small enough.

For a more general discussion we may trace the curves (B, A, Fig. 1)

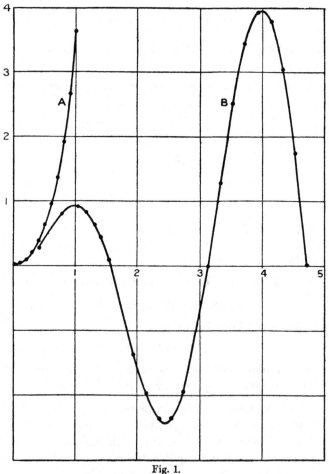

Fig. 1.

* So that wave-length is 4 times l.

representing the two members of (27), regarding k_1 and k_2 as abscissæ and taking as ordinates

$$y = k_1 \sin 2k_1, \quad y' = k_2 \sinh 2k_2. \quad \dots\dots\dots\dots\dots(30)$$

If k_1 and k_2 be both small,

$$y = 2k_1^2 (1 - \tfrac{2}{3}k_1^2), \quad y' = 2k_2^2 (1 + \tfrac{2}{3}k_2^2), \quad \dots\dots\dots(31)$$

so that at the origin both curves touch the line of abscissæ and start with the same curvature. Subsequently $y' > y$ and increases with great rapidity. On the other hand, y vanishes whenever k_1 is a multiple of $\tfrac{1}{2}\pi$, although the successive loops increase in amplitude in virtue of the factor k_1. The solutions of (27) correspond, of course, to the equality of the ordinates y and y'. It is evident that there are no solutions when y is negative. The most important occur when k_2 is small and $2k_1$ just short of π. But to the same small values of k_2 correspond also values of $2k_1$ which fall just short of $3\pi, 5\pi$, etc., or which just exceed $2\pi, 4\pi$, etc. More approximately these are

$$2k_1 = m\pi + \frac{4 \cos m\pi . k_2^2}{m\pi}, \quad \dots\dots\dots\dots\dots(32)$$

where $m = 1, 2, 3$, etc.

In order to examine whether these solutions are really available, we must calculate S. By (25)

$$kS = k_2 \left(1 - \frac{1}{3} k_2^2\right) \left(\frac{m\pi}{2} + \frac{2 \cos m\pi . k_2^2}{m\pi}\right) + k_2 \tan \left(\frac{m\pi}{2} + \frac{2 \cos m\pi . k_2^2}{m\pi}\right).$$

If m is odd, we have approximately

$$kS = \frac{m\pi}{2k_2} (1 + k_2^2); \quad \dots\dots\dots\dots\dots(33)$$

and if m is even,

$$kS = \frac{m\pi k_2}{2} \left\{1 + k_2^2 \left(\frac{8}{m^2\pi^2} - \frac{1}{3}\right)\right\}. \quad \dots\dots\dots(34)$$

Since k is approximately $\tfrac{1}{2}m\pi$, we see that when m is odd, S is large, and the condition of no reflexion can be satisfied, as when $m = 1$. On the other hand, when m is even, S is small, and here also the condition of no reflexion can be satisfied, at any rate at high angles of incidence.

It should be remarked that high values of m, leading to high values of k, correspond with overtones of the resonating channels.

A glance at Fig. 1 shows that there is no limitation upon the values of the positive quantities k_1 and k_2. And since k_1 is always greater than k_2, k, as derived from k_1 and k_2, is always real and positive.

So far we have supposed that the values of k_1, corresponding with small values of k_2, are finite, as when $m = 1, 2, 3$, etc. But the figure shows that solutions of (27) may exist when k_1, as well as k_2, is small. In this case we obtain from (31)

$$k_1{}^2 = k_2{}^2 (1 + \tfrac{4}{3} k_2{}^2), \quad\dots\dots\dots\dots\dots\dots(35)$$

making

$$k^2 = k_1{}^2 - k_2{}^2 = \tfrac{4}{3} k_2{}^4. \quad\dots\dots\dots\dots\dots\dots(36)$$

Hence by (25)

$$kS = k_1 \tanh k_2 + k_2 \tan k_1 = 2k_2{}^2 (1 + \tfrac{2}{3} k_2{}^2), \quad\dots\dots\dots(37)$$

and

$$S = \sqrt{3} \cdot (1 + \tfrac{2}{3} k_2{}^2). \quad\dots\dots\dots\dots\dots\dots(38)$$

Here again the condition of no reflexion can be satisfied, whatever the angle (θ) of incidence, by a suitable choice of σ'/σ. But the damping is no longer small, in spite of the smallness of k_2, since k_2 is not now small *in comparison with* k_1 and k. On the contrary, k_1 and k_2 are nearly equal, and B is small in comparison with k_2, so that this case stands apart.

Not only is it always possible to find a series of values of k_1 satisfying (27) with any assumed value of k_2, but the values so obtained make S positive. For in (25) k_1, k_2, $\tanh k_2$ are positive, and so also is $\tan k_1$, since

$$\tan k_1 = \sin 2k_1 / 2 \cos^2 k_1,$$

and $\sin 2k_1$ is positive.

It is a question of some importance to consider whether when σ, σ', and θ, determining S, are given, the reflexion can always be annulled by a suitable choice of k_1 and k_2. It appears that the answer is in the affirmative. Let us consider the various loops of Fig. 1 which give possible values of k_2. The ranges for $2k_1$ are from 0 to π, from 2π to 3π, from 4π to 5π, and so on. As we have seen, the intermediate ranges are excluded. In the first range between 0 and π we found that S may be made as great as we please by a sufficiently close approach to π. At the other end where $k_1 = 0$, the value of S was $\sqrt{3}$, or 1·7321. This is the smallest value which occurs. When $2k_1 = \tfrac{1}{2}\pi$, it appears that $k_2 = $ ·5656, $k = $ ·5449, and $S = 1·776$*. And again, when $2k_1 = \tfrac{3}{4}\pi$, $k_2 = $ ·5797, $S = 1·964$. We conclude that within this range some value of k_1 with its accompanying k_2 can be found which shall annul the reflexion, provided S exceed 1·7321, but not otherwise.

In each of the other admissible ranges, S takes all positive values from 0 to ∞. At the beginning of a range when $2k_1$ slightly exceeds 2π, 4π, etc., S starts from 0, as appears from (34); and at the end of a range, as 3π, 5π, etc. are approached, S is very great (33). Within each of these ranges it is possible to annul the reflexion by a suitable choice of k_1, k_2, whatever σ, σ', and θ may be.

If the actual value of S differs from that calculated, the reflexion is finite, and we may ask what it then becomes. If we denote the value of S, as calculated from k_1, k_2, by S_0, (24) gives

$$\text{Mod}^2 \text{ Numerator} = k^2 (S - S_0)^2 \{1 + \tan^2 k_1 \tanh^2 k_2\},$$

---

* [This result (1·776) is a correction of the value (1·947) given in the original. W. F. S.]

and in like manner (by changing the sign of S),

$$\text{Mod}^2 \text{ Denominator} = k^2 (S + S_0)^2 \{1 + \tan^2 k_1 \tanh^2 k_2\};$$

and hence
$$\text{Mod}^2 B = \left(\frac{S - S_0}{S + S_0}\right)^2, \quad \dots\dots\dots\dots\dots\dots\dots\dots\dots(39)$$

where
$$S = \cos\theta (\sigma + \sigma')/\sigma. \quad \dots\dots\dots\dots\dots\dots(21)$$

If σ, the perforated area, is relatively great, it makes little difference what its actual value may be, but if σ is relatively small, as in the case of strong resonance, it is otherwise.

It would be preferable to suppose S fixed at S_0 and to calculate the effect of a variation of k with h given. The resulting expressions are, however, rather complicated, and it is evident without calculation that the reflexion will be very sensitive to changes of wave-length when there is high resonance as a consequence of small dissipation and accurate tuning. The spectrum of the reflected light [in the corresponding optical circumstances] would then show a *narrow* black band.

CONTENTS OF VOLUMES I–VI

CLASSIFIED ACCORDING TO SUBJECT

(*Note.*—So much of the classification as relates to Vols. I.—IV. is almost identical with the corresponding classification on pp. 569—597 of Vol. IV. In the part which relates to Vols. V. and VI. the Articles have, on the average, been distributed amongst a somewhat larger number of Subjects than in the earlier classification.)

I. · MATHEMATICS

II. GENERAL MECHANICS

* [1917. It would be more correct to say $P_n(\cos\theta)$, where $\cos\theta$ lies between ± 1.]

II. GENERAL MECHANICS—*continued.*

IV. CAPILLARITY

IV. CAPILLARITY—*continued.*

V. HYDRODYNAMICS

V. HYDRODYNAMICS—*continued.*

V. HYDRODYNAMICS—*continued.*

V. HYDRODYNAMICS—*continued.*

VI. SOUND

VI.   SOUND—*continued*.

VI. SOUND—*continued*.

VI. SOUND—*continued.*

VI. SOUND—*continued.*

VI. SOUND—*continued.*

VII. THERMODYNAMICS

VIII. DYNAMICAL THEORY OF GASES

VIII. DYNAMICAL THEORY OF GASES—*continued.*

IX. PROPERTIES OF GASES

IX. PROPERTIES OF GASES—*continued.*

IX. PROPERTIES OF GASES—*continued.*

X. ELECTRICITY AND MAGNETISM—*continued.*

X. ELECTRICITY AND MAGNETISM—*continued.*

X. ELECTRICITY AND MAGNETISM—*continued.*

X. ELECTRICITY AND MAGNETISM—*continued.*

X. ELECTRICITY AND MAGNETISM—*continued.*

* [1914. It would have been in better accordance with usage to have said "of Relative Index differing little from Unity."]

XI. OPTICS—*continued.*

XI. OPTICS—*continued.*

XI. OPTICS—*continued.*

XI. OPTICS—*continued*.

XI. OPTICS—*continued*.

* [1914. It would have been in better accordance with usage to have said "of Relative Index differing little from Unity."]

XII. MISCELLANEOUS

XII. MISCELLANEOUS—*continued*.

XII. MISCELLANEOUS—*continued.*

INDEX OF NAMES

(NOTE.—So much of this Index as relates to Vols. I—IV is almost identical with the corresponding Index on pp. 599—604 of Vol. IV. The part which relates to Vols. V and VI has been prepared on a fuller scale. In the earlier Index a considerable number of references of relatively minor importance were not included; and it appears that as a rule, when a name occurs on several pages of the same Article, only the first or a few of those pages were cited.)

* *See* Errata p. xiv.

CAMBRIDGE : PRINTED BY J. B. PEACE, M.A., AT THE UNIVERSITY PRESS

Printed in the United States
By Bookmasters